AF892910

TA
462
.K32
2003

Helmut Kaesche

# Corrosion of Metals

**Physicochemical Principles and Current Problems**

With 338 Figures

PROPERTY OF
The Shell Global Library at
Shell Technology Center - Houston
3333 Highway Six South
Houston, Texas 77082

Springer

Prof. em. Dr. rer. nat. Helmut Kaesche
University Erlangen-Nürnberg
Institute for Materials Sciences
Chair for Surface Science and Corrosion
91058 Erlangen
Germany

Library of Congress Cataloging-in-Publication Data
Kaesche, Helmut.
Corrosion of metals: physicochemicl principles and current problems / Helmut Kaesche.
p.cm.---(Engineering materials and processes)
Includes bibliographical references and index.
ISBN 3-540-00626-5 (harc cover: alk. paper)
1. Corrosion and anti-corrosives. 2. Metals. I. Title. II. Series
TA462.K32 2003
620.1'623--dc21 2003054778

ISBN 3-540-00626-5 Springer-Verlag Berlin Heidelberg New York

This work is subject to copyright. All rights are reserved, whether the whole or part of the material is concerned, specifically the rights of translation, reprinting, reuse of illustrations, recitation, broadcasting, reproduction on microfilm or in other ways, and storage in data banks. Duplication of this publication or parts thereof is permitted only under the provisions of the German Copyright Law of September 9, 1965, in its current version, and permission for use must always be obtained from Springer-Verlag. Violations are liable for prosecution under German Copyright Law.

Springer-Verlag Berlin Heidelberg New York
a member of BertelsmannSpringer Science + Business Media GmbH

http://www.springer.de

© Springer-Verlag Berlin Heidelberg 2003
Printed in Germany

The use of general descriptive names, registered names, trademarks, etc. in this publication does not imply, even in the absence of a specific statement, that such names are exempt from the relevant protective laws and regulations and therefore free for general use.

Typesetting: Digital data supplied by author
Final processing by PTP-Berlin Protago-TeX-Production GmbH, Berlin
Illustrations: Fotosatz-Service-Köhler GmbH, Würzburg
Cover-Design: Erich Kirchner, Heidelberg
Printed on acid-free paper 62/3020 Rw 5 4 3 2 1 0

# Preface

In preparing this treatise of aqueous metallic corrosion I have followed the path set by three editions of the German predecessor 'Die Korrosion der Metalle - Physikalisch-chemische Prinzipien und aktuelle Probleme'. In the preface to the first, 1966, edition, the aim to present a fairly comprehensive introduction to corrosion science, excepting high-temperature dry oxidation, was already described, and this intention has remained unchanged ever since. The main purpose was to order theoretical and practical experiences concerning the subject, and to do so proceeding from physicochemical principles to practical applications. Also, the monograph was and is hoped to be useful for engineers, metallurgists, physicists and chemists with the usual academic background.

Almost four decades since the appearance of the first German edition, corrosion science has now developed into a highly interdisciplinary branch of the broader field of surface science and technology. Nevertheless, an attempt to once more cover the full subject by one single author appeared suggested, and the present English version is the result.

The first chapters briefly outline the principles of electrode kinetics and electrochemical thermodynamics, as relevant for aqueous corrosion processes. They introduce what, in fact, are basic essentials, indispensable for the understanding of corrosion kinetics in general, and of the origins of corrosion resistance as well as of corrosion failures in particular. These introductory chapters should be sufficiently self-consistent in order to adequately inform the reader, without requiring recourse to technical journals or conference proceedings. With respect to applications, that is with respect to the real problems of corrosion engineering, such as, mainly, breakdown of passivity by pitting, stress corrosion cracking, corrosion fatigue, and the like, the situation is different. In the first place, the chapters pertinent to these topics, indeed the greater part of the text, still aim at describing the main principles and aspects of what are difficult matters. Naturally, these are, however, the fields where research is in continuous progress. Therefore, the reader is now frequently referred to the sources of more detailed information by extensive quotations of textbooks, conference proceedings, and publications in journals.

During the years, the task of critically reviewing at least the main and most important lines of development certainly has become more demanding. The increase in volume of published contributions from many research groups is, in fact, almost intimidating. Necessarily, the book has, therefore, also once more increased in volume, though by far not in direct proportion to the growth of the relevant literature. During the years, I myself have gained experience by thirteen years of work at the Federal Institute for Materials Research and Testing, Berlin, followed by twenty-five years of continued research, combined with teaching corrosion science to a considerable number of students at the Engineering Sciences' Faculty of the University of Erlangen-Nürnberg. The work of the Erlangen corrosion research group, diploma and doctoral students as well as permanent staff, which I have much enjoyed to encourage and to direct, certainly has influenced my preferences. But then this same work always was concerned with improving the understanding the essentials of practical corrosion problems, rather than concentrating on academic aspects of the matter.

Compared to the second and third of the German editions, which already had been not only updated, but also supplemented by chapters on hydrogen embrittlement, corrosion fatigue, and impedance spectroscopy, the structure of the present English text is once more the same as before. In some chapters, sections on aqueous high-temperature kinetics and thermodynamics in general and on cracking of steels in high-temperature, high-pressure feed water in particular, as well as on noise spectroscopy, have been added. I have not compensated the increase of space devoted to special applications by abbreviating the introductory chapters. Sound knowledge of the basic essentials of corrosion science is, in fact, important for good judgment of risks, and lack of competence in this direction tends to be detrimental. This is especially true when it comes to important questions of safe planning, constructing, and maintenance of inherently sensitive equipment, such as power plants, or prestressed concrete structures, or airplanes.

The text has been completely revised and rewritten in what hopefully is tolerable, if probably continental, English. It now comes in the format imposed by a computer text program which I have learned to use, reluctantly and not to perfection, regretting previous times when publishers were ready to accept manuscripts, and when typesetters were masters of an admirable art. Nevertheless, many thanks are certainly again due to Springer Publishing Company, which as usual has been very helpful, patiently watching the slow progress of the project. Also, I am very grateful to Friedrich-Alexander University of Erlangen-Nürnberg in general, and to my colleagues and consecutive successors on the Chair for Corrosion and Surface Technology, Martin Stratmann and Patrik Schmuki, in particular, who all have for years generously allowed me to continue, as professor emeritus, using equipment of the Institute for Materials Sciences.

Erlangen, January 2003 H. Kaesche

# Contents

# 1 Introduction

*Corrosion of metals* is understood to be degradation of metals by chemical surface reactions with aggressive components of the environment. The metals may be structural materials such as steel reinforcments in concrete structures, or steel cables of suspension bridges; or they may be functional materials such as dental alloys, or copper leads for printed circuits. A typical example of corrosion is rusting of iron: Rust, a mixture of oxides and hydroxides of iron, is the product of iron surface atoms reacting with oxygen and water, both present in the surrounding moist atmosphere, or else present in a surrounding aqueous solution. Corrosion thus differs from wear, which is degradation caused by mechanical friction, as for instance abrasion. The primary products of wear are metallic particles, whereas products of corrosion always are non-metallic chemical species either solid or dissolved.

Often, attack by corrosion and by friction are simultaneous: The metal fragments produced by wear may become corroded subsequently and thus sometimes turn from a relatively soft metallic into a brittle and possibly very abrasive ceramic powder; or a protective oxide layer may fail by abrasion, with subsequent rapid corrosion. These complicated events of *corrosive wear* and *erosion corrosion* will not further be discussed in this book. It is another matter with the combined impact of corrosion and mechanical stresses either static or dynamic, i.e. with *stress corrosion cracking* and *corrosion fatigue*. These corrosive degradation processes are of paramount importance, if not for the basic theory then certainly for applied corrosion science, especially as they show that the loss of strength of a metallic structure, caused by very little overall surface reaction, can be extremely severe, when the result is cracking, with corrosion propelling the crack tip. We also shall have to consider the again rather intriguing topic of *hydrogen embrittlement* in metals, as the detrimental result of again sometimes very slight corrosion involving hydrogen uptake from the surface, and again possibly severe damage by cracking.

Only the very noble metals, such as gold or platinum, are immune against corrosion under normal environmental conditions in the sense that the reaction with atmospheric oxygen is impossible for reasons of thermodynamic principles. For the majority of metals, such as iron and steels, or aluminum and its alloys, or titanium and its alloys, etc., the chemical reaction with atmospheric oxygen is spontaneous, i.e. involving a net gain in entropy of the metal/environment system together with its surroundings. For any useful engineering material the prerequisite of course is that under normal service conditions oxydation is slow at least in dry air. For the non-noble metals this normally is the result of a microscopically thin oxide film formed by slight initial oxydation blocking further corrosion, except in moist air.

For dry oxidation rapidly to proceed, solid state diffusion of either metal, or oxygen, or both, through the air-formed oxide film must be rapid, and this normally requires temperatures far beyond the range of every-day use of materials. Hence the term *high-temperature corrosion* of e.g. gas turbine blades, and a special field of corrosion science which, however, is beyond the scope of this book.

When ambient-temperature corrosion does proceed, it obviously must involve, in the first place, the destruction of the air-formed primary oxide film, and this is what happens at the surfaces of the metals in contact with liquids such as sea-water, or rain, or contaminated moisture condensing from the atmosphere, or in contact with some other aqueous phase. Historically, the investigation of this role of the air-formed oxide film has been one of the roots of corrosion science as established by *U.R. Evans*, while the equally fundamental contributions by *A.Frumkin* and *C. Wagner* were concerned with the physical chemistry of acid corrosion in the absence of surface films. For the details of this historic development the reader is referred to literature quoted at the end of this chapter, and to Chap.4.

In principle, the liquid which attacks a metal may be other than aqueous: Methanol, or ethanol, or other organic solvents will do quite as well, if contaminated by aggressive solutes, but we will not consider these variants in detail. The main theme of this book, therefore, is what informally goes under the heading of *wet corrosion* of metals, in contact with aqueous solutions. *Corrosion protection*, then, includes all means of preventing wet corrosion by suitable choice of alloys, or by suitable treatment of the aqueous phase, or by application of protective coatings, or by other suitable methods of surface treatment including, especially, *passivation* again by oxide films. Dealing with wet corrosion, the range of temperatures to be considered is limited to the range from the freezing point to the boiling point of aqueous solutions. This is no undue narrowing of the subject: We are, in fact, considering the vast majority of corrosion processes, especially as we shall stretch the consideration to include high-pressure, high-temperature waters acting on power plant steels.

With respect to corrosion of metals in contact with aqueous solutions the first important point is that the liquid phase will normally contain dissolved salts, and/or acids, and/or bases, and therefore will be electrically conducting due to the dissociation of these electrolytes. We are, therefore, concerned with *electrolytic solutions*, and we shall see that this has very important consequences for the properties of the metal/solution interfaces and, by consequence, for the chemical kinetics of corrosion processes. A more proper denomination of wet corrosion therefore is *electrolytic corrosion*, or else, as sometimes preferred, *electrochemical corrosion.* Corrosion of metals in contact with organic electrolytic solutions or with molten salts can be expected often to be governed by essentially similar kinetics, whereas corrosion in non-electrolytic organic solvents, important for the production of organo-metallic compounds, as well as corrosion by liquid metals, especially by liquid sodium, which would have been the coolant in the fast nuclear breeders, has different principles. For aqueous solutions there remains, by the way, in principle the question of possible non-electrolytic contributions to corrosion.

Again looking at iron rusting in moist air, which will be shown really to be rusting under a thin moisture film, the chemical surface reaction is obvious: Iron is oxidized to a mixture of oxides and oxyhydroxides containing iron cations of valence two and three, while molecular oxygen, dissolved from the atmosphere into the moisture film, is reduced to hydroxyl ions. This is an example of the general principle that the chemical corrosion reaction always is an oxidation of the metal accompanied by the reduction of an oxidizing agent in the surrounding solution. It is synonymous to state that corrosion reactions always involve the loss of electrons by metal atoms and the uptake of electrons by the oxidant. The chemical corrosion products often are oxides or hydroxides, but they may be sulfates, or chlorides, or carbonates, etc., or more complicated compounds and mixtures of compounds as well, or else the metal cations together with anions may remain dissolved in the solution. Also, there are many other possible oxidizing agents besides dissolved oxygen. A very typical example is the corrosion of iron in acid solutions such as hydrochloric acid or sulfuric acid, where metal dissolution into the acid is accompanied by reduction of dissolved hydrogen ions to molecular hydrogen eventually escaping in gas bubbles. In this case, hydrogen ions are acting as oxidizing agent. Both hydrogen ions and molecular oxygen as oxidizing agents will be discussed in much more detail in subsequent chapters.

The aim of corrosion science is the quantitative theory of corrosion kinetics in general, and of corrosion rates in particular. Thus, the focus primarily is on the rates of the chemical reaction kinetics. This is not always straightforward procedure: Very often, local values of the corrosion rate may be quite different from values averaged over the entire wetted metallic surface, and both values may change with time. Also very often, local rates of metal oxidation and local rates of reduction of the oxidizing agents may substantially differ locally due to short-circuited galvanic cells in action on the metallic surface. Uniform corrosion is, in fact, the exception rather than the rule; at the same time uniform corrosion must be understood before the fundamentals of the various types of non-uniform corrosion can be attacked successfully.

A fundamental aspect of the matter is that all spontaneous chemical reactions, including corrosion, tend to come to an end by reaching chemical equilibrium, where the forward corrosion reaction is exactly balanced by its reversal, the reduction of the corrosion products to the metal. We shall, therefore, have to outline the principles of chemical corrosion reactions in terms of chemical equilibrium thermodynamics. In this context, the important physical quantity is the *free reaction enthalpy* or *Gibbs reaction energy,* as the thermodynamic driving force of reactions proceeding at approximately constant temperature and pressure. We shall also describe how certain galvanic cells represent the equilibrium of typical corrosion reactions, with an electrical cell potential equivalent to the free reaction enthalpy of the respective reaction, as quantitatively calculated by the *Nernst equation.* We shall introduce the concept of the *electrode potential* of metal/solution systems, measured against a reference electrode, particularly against the standard hydrogen electrode. If the metal is in equilibrium with the solution, the electrode potential once more obeys the Nernst equation, and this leads to such popular concepts as the potential-pH diagrams summarizing stability conditions of possibly protective metal compounds. If, instead, the metal is undergoing corrosion, then the electrode potential is a *corrosion potential,* essentially governed by two *Butler-Volmer equations,* together with equations of convection and diffusion. This leads to the exact description of uniform corrosion. From this comparatively safe starting point we shall, then, set out to explore the much more complex corrosion processes which are of primary interest in practice.

## References

In this book, corrosion science is treated from the fundamentals without aiming at completeness, and also without aiming to serve as a compendium of references for solving corrosion problems in the various industrial, public, or domestic applications. For more details in every particular case, the reader will find many other sources of useful information. Two treatises covering the complete field are [1] and [2]. Concerning the background of what is now the established field of corrosion science, special reference is made to two classical treatises of the subject [3,4].

1. Metals Handbook. 9th ed., Vol. 13: Corrosion. ASM International, Metals Park, Ohio, 1987
2. Corrosion. Shreir, L.L., Jarman, R.A., and Burstein, G.T., eds., Vol. 1: Metal/Environment Reactions, Vol. 2: Corrosion Control. 3rd ed. Butterworth-Heinemann Ltd. Oxford etc., 1994, reprinted 1995
3. Evans, U.R.: The Corrosion and Oxidation of Metals (1960, following previous editions), with First Supplementary Volume (1968) and Second Supplementary Volume (1976). E. Arnold, Publ., London
4. Wagner, C. in: Handbuch der Metallphysik (G. Masing, ed.), Vol. 1, Part 2. Akademische Verlagsgesellschaft Leipzig, 1940

# 2 Corrosion Reactions and Corrosion Products

To outline corrosion reaction chemistry we turn in the first place to *acid corrosion*, considering iron in dilute hydrochloric acid solution as a typical example. The reaction equation is

$$(\mathrm{Fe})_m + 2(\mathrm{HCl})_l \rightarrow (\mathrm{FeCl_2})_l + (\mathrm{H_2})_g \,, \tag{2.1}$$

where the subscripts $m$, $l$, and $g$ indicate the phases solid metal, liquid aqueous solution, and gas, containing the species solid pure iron, dissolved hydrochloric acid, dissolved ferrous chloride, and gaseous hydrogen, respectively. When components of different phases participate in a reaction, this indexing is useful, but we shall drop it nevertheless whenever possible without danger of confusion, in order to simplify notation. The arrow '$\rightarrow$' indicates that the reaction spontaneously proceeds as written from left to right: metallic iron and dissolved hydrochloric acid are the *reactants*, dissolved ferrous chloride and gaseous hydrogen the *products* of the overall reaction.

For correct thermodynamic reasoning it is important to understand that in Eq. (2.1) the symbols $(\mathrm{HCl})_l$ and $(\mathrm{FeCl_2})_l$ summarily denote dissolved hydrochloric acid and ferrous chloride in their actual physicochemical state. In this particular case the situation is simple: Hydrogen chloride and ferrous chloride are 'strong' electrolytes, virtually completely dissociated into dissolved ions, $H^+$, $Cl^-$, and $Fe^{2+}$. Substituting $(\mathrm{FeCl_2})_l$ by $(\mathrm{Fe^{2+}})_l + 2(\mathrm{Cl^-})_l$ and $(\mathrm{HCl})_l$ by $(\mathrm{H^+})_l + (\mathrm{Cl^-})_l$, the terms $(\mathrm{Cl^-})_l$ cancel, and the reaction equation simplifies to

$$(\mathrm{Fe})_m + 2(\mathrm{H^+})_l \rightarrow (\mathrm{Fe^{2+}})_l + (\mathrm{H_2})_g \,. \tag{2.2}$$

Instead of hydrochloric acid, sulfuric acid can be considered just as well, and the reaction equation then is

$$(\mathrm{Fe})_m + (\mathrm{H_2SO_4})_l \rightarrow (\mathrm{FeSO_4})_l + (\mathrm{H_2})_g \,. \tag{2.3}$$

Arbitrarily assuming complete dissociation of $H_2SO_4$ into $H^+$ and $SO_4^{2-}$, this equation again changes into Eq. (2.2). For other acids the result always is the same, and this suggests to write, for acid corrosion of any metal, Me, regardless of the nature of the acid, the reaction equation

$$(\mathrm{Me})_m + z\,(\mathrm{H^+})_l \rightarrow (\mathrm{Me^{z+}})_l + \frac{z}{2}\,(\mathrm{H_2})_g \,. \tag{2.4}$$

There are, however, complications, which should be recognized. In the case of sulfuric acid solution the assumption of complete dissociation neglects that, while the first dissociation, $H_2SO_4 \rightarrow H^+ + HSO_4^-$, is almost complete, the second, $HSO_4^- \rightarrow H^+ + SO_4^{2-}$, is incomplete to a degree depending on the composition of the solution. Also, ions in aqueous solution are more or less hydrated, depending on their size and charge, and a more realistic symbol for $Me^{z+}$ ions would be $Me^{z+}(H_2O)_x$, where $x$ is the hydration number indicating the number of water molecules carried by the moving ion. In aqueous solutions the very small protons $H^+$ really are hydrated hydronium ions $H_3O^+(H_2O)_y$, with again a hydration number, $y$. When the electrolyte is very 'weak' the solution predominantly contains undissociated molecules. An example is hydrogen sulfide, $H_2S$, which only weakly dissociates into $H^+$, $HS^-$, and $S^{2-}$. Of the many more details of solution chemistry, at least two should be noted: The first is that with decreasing acidity dissolved cations such as $Fe^{2+}$, but many others as well, tend to hydrolyse by the formation of hydroxy compounds, e.g. $FeOH^+$; the second is the formation of complex anions such as ferrocyanide anions $[Fe(CN)_6]^{4-}$, or complex hydroxides such as $Fe(OH)_3^-$. Equation (2.4), therefore, is straightforward for high-acidity solutions only. For consequences concerning low-acidity solutions, see Chap. 4.

In gases the particles present are molecules. In metals, the situation is more complex, because there are no uncharged atoms but metal cations on lattice sites embedded in the metallic electron gas. Still, the situation can be abbreviated using the overall symbol $(\mathrm{Me})_m$ as representing metal ions together with electrons in a combination of zero total electric charge.

With decreasing acidity and increasing concentration of dissolved metal ions, the electrolytic solution eventually will become saturated with respect to a solid product of the corrosion reaction. The deposition of solid compounds merits particular attention, as it is important for both corrosion thermodynamics and kinetics. A straightforward example is acid corrosion of lead in dilute sulfuric acid, where lead sulfate is scarcely soluble and very soon solid lead sulfate is deposited forming a protective layer blocking further corrosion. This role of the corrosion product layer is, however, incidental, as in principle the sulfate might be deposited anywhere else in the solution. The overall reaction now is

$$(\mathrm{Pb})_m + (\mathrm{H_2SO_4})_l \rightarrow (\mathrm{PbSO_4})_s + (\mathrm{H_2})_g \,, \tag{2.5}$$

where the subscript $s$ indicates a non-metallic solid. An obviously likely supposition is that this overall reaction proceeds in two consecutive steps:

$$a)\ (\mathrm{Pb})_m + 2(\mathrm{H^+})_l \rightarrow (\mathrm{Pb^{2+}})_l + (\mathrm{H_2})_g \,, \tag{2.6}$$

$$b)\ (\mathrm{Pb^{2+}})_l + (\mathrm{SO_4^{2-}})_l \rightarrow (\mathrm{PbSO_4})_s \,.$$

We shall prove later that the first of these step, as every reaction of type Eq. (2.2), proceeds by the superposition of separate *half-reactions*, or *partial reactions*: metal dissolution as an *anodic* half-reaction, and hydrogen deposition as a *cathodic* half-reaction.

Turning to the second typical group of corrosion reactions, i.e., to *oxygen corrosion*, and particularly to rusting of iron, we recall that the overall reaction is the oxydation of iron by atmospheric oxygen, via oxygen dissolved in an aqueous solution, as for instance in an aqueous film on the surface of the metal, to what summarily is described as Fe-III-oxyhydroxide:

$$4(\mathrm{Fe})_m + 3(\mathrm{O_2})_g + 2(\mathrm{H_2O})_l \rightarrow 4(\mathrm{FeOOH})_s \,. \tag{2.7}$$

In the present context the real, normally much more complicated composition of rust is not yet the important point. Instead, we note a number of consecutive steps, into which the overall reaction very probably separates: First, the aqueous phase must absorb oxygen from the atmosphere, and in the solution the dissolved oxygen must arrive at the metal surface by either diffusion or convection. It is important to recognize that these physical, i.e. non-chemical processes in fact are discernible steps in the overall chemical reaction. The next stage is a *phase boundary reaction,* or *interfacial reaction*, summarily written as

$$4(\mathrm{Fe})_m + 3(\mathrm{O_2})_l + 2(\mathrm{H_2O})_l \rightarrow 4(\mathrm{FeOOH})_s \,. \tag{2.8}$$

This again is a reaction by far too complicated to proceed without further separation into many more steps, as, especially, anodic metal dissolution and cathodic oxygen reduction. Also, one may expect that the primary reaction product is dissolved $Fe^{2+}$, followed by the oxidation to $Fe^{3+}$, followed by the deposition of rust. This should cause a sequence of steps approximately as follows:

$$\begin{aligned}
&a)\ 4(\mathrm{Fe})_m + 2(\mathrm{O_2})_l + 4(\mathrm{H_2O})_l \rightarrow 4(\mathrm{Fe^{2+}})_l + 8(\mathrm{OH^-})_l \,,\\
&b)\ 4(\mathrm{Fe^{2+}})_m + (\mathrm{O_2})_l + 2(\mathrm{H_2O})_l \rightarrow 4(\mathrm{Fe^{3+}})_l + 4(\mathrm{OH^-})_l \,,\\
&c)\ 4(\mathrm{Fe^{3+}})_l + 12(\mathrm{OH^-})_l \rightarrow 4(\mathrm{FeOOH})_s + 4(\mathrm{H_2O})_l \,.
\end{aligned} \tag{2.9}$$

In reality, the deposition of solid rust is even more complicated, with such sub-steps as the formation of hydroxide via intermediate stepwise hydrolysis, followed by formation of polybases first dissolved, then colloidal, then solid, followed by partial dehydration of the hydroxide. Also, while the orthorhombic goethite, $\alpha$-FeOOH, usually is a main constituent of rust, there also are varying amounts of orthorombic lepidokrokite, $\gamma$-FeOOH, rhomboedric hematite, $\alpha$-$Fe_2O_3$, and tetragonal maghemite, $\gamma$-$Fe_2O_3$. In rust, iron predominantly is trivalent, but from thermodynamics a lower valence state is expected at the rust/iron interface, that is, cubic magnetite, $Fe_3O_4$. Also, a considerable part of the rust normally is amorphous; also, depending on environmental pollution, rust may contain iron sulfates, chlorides, and other contaminants. A rather curious intermediate constituent is 'green rust', which is mixed iron-II,III-hydroxide containing either chloride or sulfate.

The complicated reaction path for rusting of iron in moist air explains why, in contrast to dry oxidation, the deposit of corrosion products on the metal surface normally is porous. Therefore, it also normally is non-protective, in the sense that it takes comparatively thick layers of rust to block further corrosion, or in the sense that rusting of otherwise unprotected iron will normally decay only very slowly after prolonged exposure times, and only after already severe corrosive attack. This, of course, is the typical disadvantage of iron and plain steels as structural materials. The so-called *weathering steels* with some tenths of percent of copper and chromium are exceptional, as they tend, though not always and only in the course of years, to develop dense rust layers protective without additional coatings. The situation is systematically better in the case of zinc: In contact with moist air, this metal almost always tends to develop thin, dense layers of basic carbonates which are well protective. Therefore, zinc coatings are a very useful for long-time protection of plain steels, and this is the reason for the massive use of galvanizing. A further example is the greenish patina of copper, a layer of basic carbonates, sulfates and chlorides, developing from primary layers of dark brown oxides and sulfides, but again only in the course of many years or even decades.

Generally, the chemistry of corrosion products tends to be complicated. Except for iron, and except for the chemistry of passivating oxide layers, we shall not, however, deal much with this part of inorganic chemistry, and the reader interested in more information is referred to the literature [1]. Earlier critical reviews of outdoor atmospheric corrosion products of zinc, cadmium, lead, copper, aluminum, and iron, are found in [2,3], more recent information for instance in [4], and, especially concerning zinc and copper, in [5]. Indoor atmospheric corrosion products, not normally important as detrimental to the strength of materials, but nowadays a critical factor for the safe operation of miniaturized semiconductor circuits, have become the object of much concern [6]. A quite different field is the normally only slight corrosion in the water-carrying loops of both boiling and pressurized light-water nuclear reactors, because the transport of traces of radioactive reaction products through the loops is undesirable. Hence extensive investigations into the behaviour of high-alloy steels and nickel-base alloys in contact with hot water of high purity, to be found e.g. in [7]. In contrast, hot high-pressure concentrated salt solution are encountered as very aggressive geological brines, of considerable interest for safe underground nuclear waste storage.

In the next chapter, we shall need a generalized formulation of chemical reactions, as follows: Let A, B, etc., be the reactants, and M, N, etc. the products of a chemical reaction. Dropping phase subscripts, we write

$$(-\nu_A)A + (-\nu_B)B + \ldots \rightarrow \nu_M M + \nu_N N + \ldots . \tag{2.10}$$

with *stoichiometric coefficients* $\nu_A$, $\nu_B$, etc.. As an example, we note that in Eq. (2.1), the stoichiometric coefficients of Fe, HCl, $FeCl_2$, and $H_2$, are $-1, -2, 1$, and 1, respectively. We also note that the numerical values of stoichiometric coefficients indicate the relative numbers of molecules, atoms, or ions of the respective species consumed or produced by the reaction. As the common factor of the relation may have any value, including *Avogadro's constant*, $N_A = 6.02214 \times 10^{23}$, which is the number of molecules, or atoms, or ions of any chemical species per mole, the equation also indicates the relative numbers of moles of each species consumed or produced. If for two species, X and Y, the numbers $n_X$ and $n_Y$ of moles are consumed or produced by the reaction, then $n_X/n_Y = \nu_X/\nu_Y$. We define the *extent* of a reaction by introducing a *conversion number,* $\lambda_r$, which counts the number of unit conversions or 'formula conversions' of reactants to products as specified by the reaction equation. When the reaction advances by $d\lambda_r$, the change in moles of the $i$-th substance is $dn_i = \nu_i d\lambda_r$.

When the overall reaction is stationary, advancing in what is termed *steady state*, the rate, $v_n$, of every consecutive step is equivalent to the rate of the overall reaction, $v$. But each step is proceeding forwards and backwards, with rates $\overrightarrow{v_n}$ and $\overleftarrow{v_n}$, and the overall rate of any step is $v_n = \overrightarrow{v_n} - \overleftarrow{v_n}$, with directions indicated by half arrows. Often, but not necessarily, the rate of a reverse step reaction will be small and sometimes negligible, especially when the reaction is far from equilibrium. Also often, however, the situation is such that the rate $v_n$ of a particular step is small compared with both $\overrightarrow{v_n}$ and $\overleftarrow{v_n}$. In this case, although $\overrightarrow{v_n} - \overleftarrow{v_n} = v_n$, the important point is that $\overrightarrow{v_n}, \overleftarrow{v_n} \gg v_n$, and $\overrightarrow{v_n} \simeq \overleftarrow{v_n}$. Such a step is in quasi-equilibrium, behaving approximately as in true equilibrium, where $\overrightarrow{v_n} = \overleftarrow{v_n}$, and $v = 0$. In fact, the usual assumption is that in a sequence of steps of an overall reaction all except one are in quasi-equilibrium. The calculation of the overall reaction rate then reduces to the calculation of the rate of this one non-equilibrium, *rate-determining* step. When there are parallel half-reactions such as metal dissolution and hydrogen deposition, two rate-determining steps have to be considered. For quasi-equilibria as well as for true equilibria, arrows, '→,' usually are replaced by 'equals' signs, '=' As a further rule, equilibria normally can be expected to be established especially for reactions between ions in solution, or for cation hydrolysis in solution, or for the formation of complex ions in solution, etc. Hence, equations such as $(Fe^{2+})_l + (H_2O)_l = (FeOH^+)_l + (H^+)_l$, or $(Fe^{2+})_l + 6(CN^-)_l = ([Fe(CN)_6]^{4-})_l$, will often be used in the following.

From these considerations it follows that in overall equilibrium, where the total reaction rate of any step and the overall corrosion rate is zero because every reaction step is exactly balanced, the rates of forward and reverse reactions of different steps are not necessarily the same. The absolute value of the rates of forward and reverse steps in equilibrium is the respective *equilibrium exchange rate,* and an important quantity. Obviously, the concept of equilibrium exchange rates is one of chemical kinetics, and yet it is important for applied thermodynamics: When the exchange rate is too low we cannot expect the respective equilibrium really to become established. Reaction rates are functions of, among other quantities such as temperature and activation energy, the numbers of particles participating in the reaction as reactants and products in a given volume; i.e. they are proportional to powers of the *concentrations* of reactants and products, as will be discussed more fully further below. Here we only note that below some small, though ill defined value of concentrations of reaction partners the calculation of thermodynamic equilibrium quantities becomes futile, because the reaction actually does no longer regularly proceed.

Techniques of corrosion protection are not normally aiming at chemical equilibrium of the corrosion reaction. Instead, the aim usually is to leave overall-reaction thermodynamics unchanged, and to reduce the reaction rate to a tolerable low value by blocking reaction kinetics. Formally, this is the case that for at least one step, and thus for the overall reaction as well, $v_n = v \simeq 0$ because $\overrightarrow{v}, \overleftarrow{v} \simeq 0$. A comparatively trivial but very practical method of vast economic importance is to separate metal and aggressive surroundings by a suitable coating. Many other methods will become obvious in later chapters.

## References

See the textbooks and handbooks of inorganic chemistry.

1. Meyer, H.J.; Werkstoffe u. Korr. *15* (1964) 653
2. a) Feitknecht, W.; Corr. Sci. **7** (1967) 629; b) Grauer, W.; Werkstoffe u. Korr. *31* (1980) 837; **32** (1981) 113
3. Kucera, V., and Mattson, E., in: Corrosion Mechanisms.(Mansfeld, F., ed.). Marcel Dekker, New York, Basel, Hong Kong, 1987
4. Leygraf, C., in: Corrosion Mechanisms in Theory and Practice.(Marcus, P.and Oudar, J., eds). Marcel Dekker, New York, Basel, Hongkong, 1998
5. Proc. 1st Int. Symp. on Corrosion of Electronic Materials and Devices, Sinclair, J.D., ed., The Electrochemical Society, Inc., Pennington, N.J., 1991
6. Proc. Int. Conf. on Water Chemistry in Nuclear Power Plants, Kashiwazaki, Niigate, 1998, Japan Atomic Industrial Forum, Inc., 1998

# 3 Chemical Thermodynamics of Corrosion

## 3.1 Outline of Fundamentals

In any corrosion process, the solid metal, the contacting aqueous solution and, normally, an adjacent gas phase are parts of a *system* in which the reaction causes physical and chemical changes. A description of the system in terms of thermodynamics then requires the specification of the boundaries separating the system and its *surroundings* [1–6]: These may be permeable a) to both matter and energy, where energy is work and/or heat, or b) permeable to energy alone, or c) impermeable to both matter and energy. The system is then said to be a) *open*, or b) *closed*, or c) *isolated*, respectively.

We consider closed systems. The Second Law of thermodynamics, in the form of the *Clausius inequality*, then states that a reaction changing the properties of the system is spontaneous only if it increases the total *entropy*, $S_{total}$, of the system together with its surroundings. The condition for any spontaneous infinitesimal reaction in the system therefore is that the infinitesimal change in entropy is positive: $dS_{total} \geq 0$. The tendency is to increase the entropy to a maximum, where the '=' sign applies, and where equilibrium is established. The entropy principle thus permits to predict the direction of spontaneous reactions. Without non-thermodynamic additional suppositions it does not predict rates of reactions.

We postpone the consideration of the influence of temperature fluctuations typical at least for long-time every-day corrosion, assuming, instead, constant temperature, $T$, throughout the system and its surroundings. It then follows from the entropy principle together with the First Law of thermodynamics, which is the energy conservation principle, that any spontaneous reaction in the system must decrease, in the case of constant volume, $V$, the *free energy* or *Helmholtz energy*, $F$, or, in the case of constant pressure, $p$, the *free enthalpy* or *Gibbs energy*, $G$. All properties $p$, $V$, $T$, $F$, $G$ are now and everywhere in the following referred to the system alone. The condition for a spontaneous infinitesimal change in the sytem thus is, at constant $V$, that $dF \leq 0$, or, at constant $p$, that $dG \leq 0$. Again the '=' sign applies when the reaction is in equilibrium. Free energy, $F$, and free enthalpy, $G$, are defined as $F = W - TS$, and $G = H - TS$, where $W$ is the *intrinsic energy*, $S$ the *entropy*, and $H = W + pV$ the *enthalpy* of the system. $T$ is the *absolute temperature* on the Kelvin scale. The relation between Kelvin and Celsius temperatures is $T/\mathrm{K} = \theta/^{\circ}\mathrm{C} + 273.15$.

Normally, corrosion systems are at ambient pressure, and constant-pressure thermodynamics at least approximately apply. Therefore, the Gibbs energy, $G$, is the fundamental property, and we need to calculate changes in $G$ caused by the corrosion reaction. For elevated but still constant pressures, appropriate recalculations are easily available, as described further below.

The first important point is that intrinsic energy, enthalpy, entropy, free energy, and free enthalpy all are *state functions*, determined by pressure, temperature and system composition alone, regardless of the reversible or irreversible path which brought the system to this state. A change in free enthalpy caused by a change in composition of the system due to an irreversible corrosion reaction therefore is the same as that by the same change caused by a reversible reaction. For the reversible path, the change in Gibbs energy usually can be evaluated, and this, then, is the object of this chapter.

Let $dG$ be the infinitesimal change in Gibbs energy caused by an infinitesimal advance of a reaction, labeled $r$, at constant temperature and pressure. Since $dG \leq 0$, we also have $(\partial G/\partial \lambda_r)_{p,T} \leq 0$, where $\lambda_r$ is the reaction extent defined previously. For convenience, we abbreviate this differential quotient, i.e. the *molar Gibbs reaction energy*, to:

$$(\partial G/\partial \lambda_r)_{p,T} = \Delta_r G \, . \tag{3.1}$$

The overall reaction in the direction indicated by '→' is spontaneous when $\Delta_r G < 0$; it is in equilibrium when $\Delta_r G = 0$; and it is spontaneous in the reverse direction when $\Delta_r G > 0$. Also, $\Delta_r G$ is a very remarkable quantity: It is the maximal useful work, especially the maximal electrical work, which at constant pressure and temperature can be obtained from an infinitesimal change in reaction extent, i.e., for an infinitesimal change in composition of the system, referred to the unit reaction conversion. Similarly, we define the *reaction enthalpy*, $\Delta_r H$, the *reaction entropy*, $\Delta_r S$, and the *reaction volume*, $\Delta_r V$, as

$$\Delta_r H = (\partial H/\partial \lambda_r)_{p,T} \, , \quad \Delta_r S = (\partial S/\partial \lambda_r)_{p,T} \, , \tag{3.2}$$

$$\Delta_r V = (\partial V/\partial \lambda_r)_{p,T} \, ,$$

respectively. Tabular enthalpies usually are in kilojoules per mole, (kJ $mol^{-1}$), entropies in joules per mole, (J $mol^{-1}$), reaction volumes in liter, (L). We note that $-p\Delta_r V$ is the expansion work done by the system against the external pressure. Usually, $p$ is measured in bar, i.e. in units of $10^5$ Pa. The maximal non-expansion, useful work then is

$$\Delta_r G = \Delta_r H - T\Delta_r S \, . \tag{3.3}$$

We note, for later use, two derivatives of the Gibbs reaction enthalpy:

$$a)\ (\partial \Delta_r G/\partial p)_T = \Delta_r V \, , \qquad b)\ (\partial \Delta_r G/\partial T)_p = -\Delta_r S \, , \tag{3.4}$$

As an example, we consider acid corrosion of iron according to Eq. (2.1), and a well controlled experiment as schematically shown in Fig. 3.1: A piece of pure iron, Fe, is fully immersed in a solution of hydrogen chloride, HCl, and ferrous chloride, $FeCl_2$, in water, $H_2O$. Water does not explicitly participate in the reaction as stated in Eq. (2.1). The solution is saturated with respect to dissolved hydrogen in equilibrium with gaseous hydrogen, $H_2$, which passes the

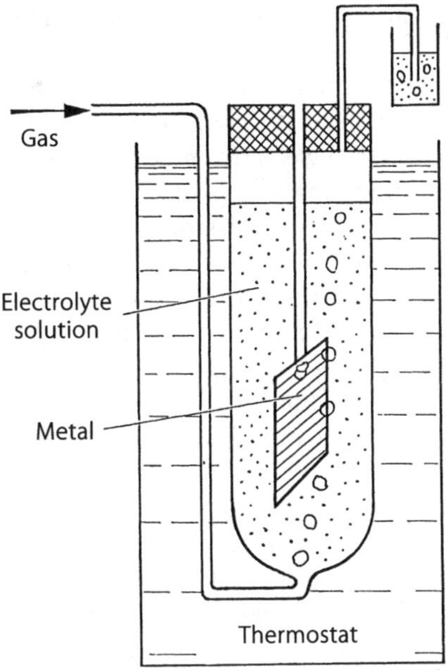


**Fig. 3.1.** Simple corrosion experiment at constant temperature and pressure with a fully immersed metal specimen

solution as a stream of bubbles and leaves the solution vessel through a bubble valve preventing the entrance of air. The system is this vessel placed in a thermostat in the role of the surroundings, essentially acting as a heat sink of large capacity, ensuring constant temperature. We neglect, as a second-order effect among others not explicitly noted here, that some water and hydrogen chloride will evaporate into the hydrogen bubbles, i.e. we consider the gaseous phase to be essentially pure hydrogen at atmospheric pressure, $p$. As the continuous gas flow does not noticeably change the composition of the system, the behaviour is the same as that of a closed system. In fact, the gas flow might be dispensed with when acid corrosion is accompanied by evolution of gaseous hydrogen against the pressure of the atmosphere. We note that slow acid corrosion may proceed without local gas evolution by a reaction path involving dissolution of molecular hydrogen into the solution, diffusion to the surface of the solution, and desorption into the gas phase. Normally, however, acid corrosion is understood to be accompanied by hydrogen bubble evolution. In any case, the reaction is completely irreversible, except for the useless work of expanding hydrogen bubbles against the external pressure.

As an extensive property, the Gibbs energy of the system is the sum of contributions from the three phases present, and evaluated as follows:

a) The gas is practically pure hydrogen, with $n_{H_2}$ moles of hydrogen present at any moment in the gas phase contained by the reaction vessel. If $G_{H_2}$ denotes the molar Gibbs energy of pure hydrogen, the Gibbs energy of the gas phase is $G_g = n_{H_2} G_{H_2}$.

b) Concerning solid iron and any other metal as well, we recall that the symbol Fe, or generally Me, indicates the actual state of metal atoms, which is cations on lattice sites and electrons in the electron gas of the metal. We do not, however, at present separately consider contributions to the Gibbs energy from ions and electrons. Instead, it is legitimate to replace pairs of these separate contributions from species with opposite electric charge by their zero-charge sum which per mole, or per 'gramatom', is the quantity $G_{Me}$ in general, or $G_{Fe}$ in the present case. If $n_{Fe}$ is the number of moles of metallic iron, the resulting contribution to the Gibbs energy therefore is $n_{Fe} G_{Fe}$.

c) The electrolytic solution is well stirred by the gas flow and thus homogeneous like the metal and the gas, but it is a mixed phase. In terms of thermodynamics it can formally be treated either as containing dissolved $FeCl_2$ and HCl, or as containing, in stoichiometric numbers of zero net charge, $H^+$, $Fe^{2+}$, and $Cl^-$.

More generally, consider a dissolved electrolyte labeled $i$ of composition $(Me^{z+}_{\nu_{i+}})(X^{z-}_{\nu_{i-}})$ which dissociates into $\nu_{i+}Me^{z+}$ cations and $\nu_{i-}X^{z-}$ anions. The differential contribution of the dissolved compound to the Gibbs energy of the solution, $G_l$, and to the Gibbs energy of the system, G, as a whole is

$$\left(\frac{\partial G}{\partial n_i}\right)_{p,T,n_{j\neq i}} = \nu_{i+}\left(\frac{\partial G}{\partial n_{Me^{z+}}}\right)_{p,T,n_{j\neq i}} + \nu_{i-}\left(\frac{\partial G}{\partial n_{X^{z-}}}\right)_{p,T,n_{j\neq i}} . \tag{3.5}$$

The expressions in brackets are *partial molar Gibbs energies.* At this point we introduce, as a key property in electrochemical reaction thermodynamics, the *electrochemical potential,* $(\tilde{\mu}_i)_x$, of any compound $i$ in a phase $x$ by setting:

$$\left(\frac{\partial G_x}{\partial n_i}\right)_{p,T} = (\tilde{\mu}_i)_x . \tag{3.6}$$

Anticipating an explanation which will be supplied shortly, we note that the electrochemical potential of a species $i$, carrying, per particle, an electric charge $z_i e_\circ$ is

$$(\tilde{\mu}_i)_x = (\mu_i)_x + z_i F \phi_x . \tag{3.7}$$

The *Faraday constant,* $F = 9.64853$ C mol$^{-1}$, is the product, $e_\circ N_A$, of the elementary charge, $1.602 \times 10^{-19}$C and the Avogadro constant. $\phi_x$ is the *macroscopic electrostatic potential* in the phase $x$, also designated as the *Galvani potential* of the phase. In this equation, the term $(\mu_i)_x$ is the *chemical potential* of the species $i$ in the phase $x$. For uncharged species, chemical and electrochemical potentials are equal. For zero-charged combinations of anions and cations, the sum of the electrochemical potentials is equal to the sum of the chemical potentials.

Substituting chemical and electrochemical potentials in Eq. (3.5) and abbreviating species indices, we obtain

$$\begin{aligned}(\tilde{\mu}_i)_x, &= \nu_{i+}(\tilde{\mu}_{i+})_x + \nu_{i-}(\tilde{\mu}_{i-})_x \\ &= \nu_{i+}((\mu_{i+})_x + z_+F\phi_x) + \nu_{i-}((\mu_{i-})_x - z_-F\phi_x) \\ &= \nu_{i+}(\mu_{i+})_x + \nu_{i-}(\mu_{i-})_x = (\mu_i)_x \,. \end{aligned} \tag{3.8}$$

This easily adapts to pure substances $i$: In this case, the partial molar Gibbs energy equals the molar Gibbs energy: $\mu_i = G_i$, and this also holds for pure metals, if applied to the zero-charge combination, Me, of metal ions, $Me^{x+}$, on lattice sites and $x$ electrons, $xe^-$, in the electron gas of the metal.

Instead of the total Gibbs energy of the system, we now consider the molar Gibbs reaction energy, $\Delta_r G$, as defined in Eq. (3.1). Since the simultaneous changes in numbers of moles of participants, $d\nu_i$, are related by $d\nu_i = \nu_i d\lambda_r$, the infinitesimal molar change is $dG = \sum \nu_i \tilde{\mu}_i d\lambda_r$, and therefore:

$$\Delta_r G = \left(\frac{\partial G}{\partial \lambda_r}\right) = \sum \nu_i \tilde{\mu}_i, \,. \tag{3.9}$$

At this point, we recall from Chap. 2 that *overall corrosion reactions* such as acid iron corrosion can formally be written as involving uncharged species, or zero-charge combinations of anions and cations in the same solution, or metal ions and electrons in the metal. We therefore see that in this case Eq. (3.9) simplifies to

$$\Delta_r G = \left(\frac{\partial G}{\partial \lambda_r}\right) = \sum \nu_i \mu_i, \textit{ for overall reactions.} \tag{3.10}$$

For the corrosion of iron in hydrochloric acid solution, the set of chemical potentials of reaction participants is

$$(\mu_{\mathrm{Fe}})_m, \quad (\mu_{\mathrm{H_2}})_g,$$

$$(\mu_{\mathrm{FeCl_2}})_l = (\mu_{\mathrm{Fe^{2+}}})_l + 2(\mu_{\mathrm{Cl^-}})_l, \quad (\mu_{\mathrm{HCl}})_l = (\mu_{\mathrm{H^+}})_l + (\mu_{\mathrm{Cl^-}})_l$$

To calculate the Gibbs reaction energy, two alternative equivalent equations are

$$a)\ \Delta_{\mathrm{Fe/Fe^{2+}}} G = (\mu_{\mathrm{FeCl_2}})_l + (\mu_{\mathrm{H_2}})_g - (\mu_{\mathrm{Fe}})_m - 2(\mu_{\mathrm{HCl}})_l \,, \tag{3.11}$$

$$b)\ \Delta_{\mathrm{Fe/Fe^{2+}}} G = (\mu_{\mathrm{Fe^{2+}}})_l + (\mu_{\mathrm{H_2}})_g - (\mu_{\mathrm{Fe}})_m - 2(\mu_{\mathrm{H^+}})_l \,.$$

Hence,

$$(\mu_{\mathrm{Fe^{2+}}})_l - 2(\mu_{\mathrm{H^+}})_l = (\mu_{\mathrm{FeCl_2}})_l - 2(\mu_{\mathrm{HCl}})_l \,. \tag{3.12}$$

Passing, for the moment, complications arising from incomplete dissociation of either acids, such as $H_2SO_4$, or corrosion products, such as $ZnCl_2$, we generalize Eq. (3.11) by writing

$$\Delta_r G = (\mu_{\mathrm{Me^{z+}}})_l + \frac{z}{2}(\mu_{\mathrm{H_2}})_g - (\mu_{\mathrm{Me}})_m - z(\mu_{\mathrm{H^+}})_l \,. \tag{3.13}$$

As shown in the following section, Eq. (3.13) can be further generalized to apply to other corrosion overall reactions, such as oxygen corrosion. Before, we digress and attend to *electrode reactions* such as

$$(\mathrm{Me})_m \rightarrow (\mathrm{Me}^{z+})_l + z(\mathrm{e}^-)_m \,, \tag{3.14}$$

and

$$\frac{z}{2}(\mathrm{H}_2)_g \rightarrow z(\mathrm{H}^+)_l + z(\mathrm{e}^-)_m + \,, \tag{3.15}$$

in which electrons explicitly appear as reactants. Formally, Gibbs reaction energies are easily formulated:

$$\begin{aligned} \Delta_r G &= (\tilde{\mu}_{\mathrm{Me}^{z+}})_m + z(\tilde{\mu}_{\mathrm{e}^-})_m - (\mu_{\mathrm{Fe}})_m \\ &= (\mu_{\mathrm{Me}^{z+}})_l + z(\mu_{\mathrm{e}^-})_m - (\mu_{\mathrm{Me}})_m - zF(\phi_m - \phi_l) \,, \end{aligned} \tag{3.16}$$

and

$$\begin{aligned} \Delta_r G &= z(\tilde{\mu}_{\mathrm{H}^+})_l + z(\tilde{\mu}_{\mathrm{e}^-})_m - (z/2)(\mu_{\mathrm{H}_2})_g \\ &= z(\mu_{\mathrm{e}^-})_m + z(\mu_{\mathrm{H}^+})_l - (z/2)(\mu_{\mathrm{H}_2})_g - zF(\phi_m - \phi_l) \,. \end{aligned} \tag{3.17}$$

We see that for electrode reactions terms containing Galvani potential differences between metal and solution do not cancel. Defining

$$\phi_{x,y} = \phi_x - \phi_y \,, \tag{3.18}$$

we obtain the general expression

$$\Delta_r G = \sum \nu_i \tilde{\mu}_i = \sum \nu_i \mu_i - z_r F \phi_{m,l} \,, \textit{for electrode reactions.} \tag{3.19}$$

Here, $z_r$ is the *reaction valence*, i.e. the number of electrons participating in the unit conversion of the reaction.

Since we are now explicitly considering Galvani potentials and electrochemical potentials, both properties merit closer description: The electrochemical potential $(\tilde{\mu}_i)_x$ is the change in Gibbs energy, referred to molar turnover, caused by the removal, at constant $p$ and $T$, of an infinitesimal amount of a charged species $i$ from an electrically conducting phase $x$ at the Galvani potential $\phi_x$. In the course of electrode reactions, charged species are transferred across the metal/solution interface, but we can alternatively consider the reversible energy change involved in transferring the species from the metal across the metal/vacuum interface to infinity and thence into the solution across the vacuum/solution interface.

To simplify matters we choose, for the moment, to reduce the problem to the calculation of the electrochemical potential of electrons in the metal phase only, i.e. to the calculation of the reversible work to transfer electrons from the interior of the metal across a metal/vacuum interface to infinity. The situation is schematically desribed in Fig. 3.2.

In the metal, $\phi_x$ is constant. We refer it to the charge-free infinity, setting $\phi_\infty = 0$, In the metal, two effects contribute to $\phi_x$: The first is an excess, $Q = n_Q\, e_\circ$, of electric charges in the phase over the number of lattice cations compensated by the number of electrons in the electron gas of the metal. The excess may be positive or negative owing to excess electrons, or electron holes, or ions. Also, the excess is accidental, as the charging may have been caused e.g. by ill defined mechanical friction. The second effect is due to the presence of

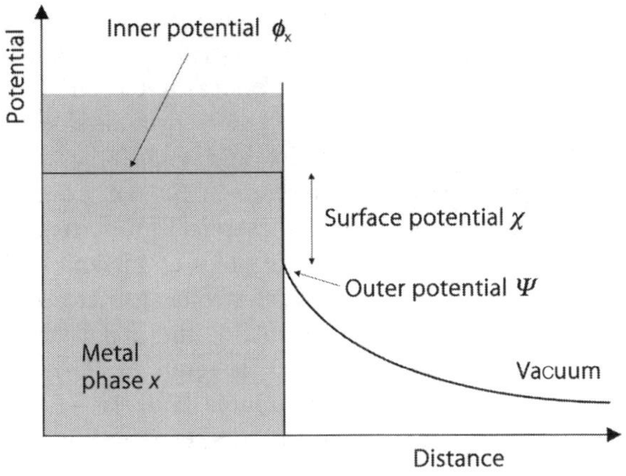


**Fig. 3.2.** Illustrating the distribution of the electrostatic Galvani, $\phi$, potential from the interior of a metal through a metal/vacuum interface across an electrical double layer into the vacuum and to infinite distance, also showing the 'inner potential', $\phi_x$, the 'outer potential', $\psi$, and the 'surface potential', $\chi$. The assumed sign of changes of the potential is arbitrary

an *electric double layer* at the metal/vacuum interface caused by an asymmetry of the electron gas distribution close to the metal surface, and possibly caused by an additional adsorbed dipole layer on the surface, such as oriented water molecules. The thickness of the double layer is sub-microscopic.

The overall charge of the double layer is zero, but the layer adds a difference, $\Delta\phi_{\mathrm{dl}}$, in Galvani potential, designated as *surface potential*, $\chi$, to $\phi_x$ as established by the excess charge. At a point just outside the metal surface, the Galvani potential in the vacuum phase is $\phi_x + \chi$, and this is the *outer potential*, $\psi$. More precisely, $\psi$ is the potential outside the metal surface just beyond the range of image forces, i.e. beyond ca. 1 µm.

Consider a metallic sphere of radius $r_\circ = 1 \times 10^{-2}$ m. In the first place, we disregard any double-layer potential difference. In this case, the external potential, $\psi$, just outside the metal is $\psi = \phi_x$. It decays with distance towards the charge-free infinity as $\chi = Q/4\pi\epsilon_\circ r$, with the *vacuum permittivity* $\epsilon_\circ = 8.854 \times 10^{-12}\ \mathrm{C^2 J^{-1}\ m^{-1}}$. From this we calculate, for the metal/vacuum interface with zero contributions from the electrical double layer, an outer potential $\psi = 1.0\,\mathrm{V}$, if $n_Q = 6.9 \times 10^7$, corresponding to $1.1 \times 10^{-17}$ mol of elementary charges. In terms of sizable chemical reaction extents, this certainly is a negligible quantity. Calculating, further below, Gibbs electrode reaction energies, we shall nevertheless always tacitly assume that infinitesimal reaction extents leave $\phi_x$ and, thus, $n_q$, virtually unchanged.

Normally, the potentials $\phi_x$ and $\psi$ are different owing to the presence of the interfacial double layer. As a difference in potential in a homogeneous phase, in this case the vacuum, $\psi$ is, in principle, a measurable quantity; $\chi$, however, is unknown. For a layer of oriented water molecules the estimate for $\chi$ would be some 0.1 V, but the contribution from the asymmetry of the electron gas in the surface is unknown and may be much larger. Thus, the Galvani potential, $\phi_x$, is also unknown. We note that the *Volta potential* is the measurable difference of two surface potentials, $(\psi)_x - (\psi)_y$, between two points just outside two conducting and contacting phases $x$ and $y$. When $x$ and $y$ are chemically identical, but at different values of $\phi$, then the difference in Galvani potential is equal to the Volta potential, since $\psi$ depends on the chemical properties of a metal only. This also applies to the chemically identical poles of galvanic cells.

The reversible work to transfer, in vacuo, 1 mol of electrons against macroscopic electrical forces from infinity to a point just outside the metal surface is $-F\psi$, and so far no other than this Coulomb work is required. The electric part of the work required to cross the surface double layer is $-F\chi$. In the metal no further Coulomb work is required, so the total Coulomb work is $-F(\psi + \chi) = -F\phi_x$.

At this point, we do not attempt to evaluate the work of chemical interaction of electrons with the bulk metal. Instead, we simply assign a name to this rest of the electrochemical potential, $\mu_{e^-}$, and call it the chemical potential. The total reversible work then is expressed as $\tilde{\mu}_{e^-} = \mu_{e^-} - F\phi_x$, which is Eq. (3.7) applied to metallic electrons. This perhaps doubtful procedure does, in fact, illustrate that in terms of rigorous thermodynamics the electrochemical potential does lack precise of definition.

With regard to absolute values of the electrochemical potential of electrons in metals, we rearrange Eq. (3.7) into $\tilde{\mu}_{e^-} = \alpha_{e^-} - F\chi$, where $\alpha_{e^-} = \mu_{e^-} - F\chi$ is the *real potential* of electrons, i.e., the molar reversible work to transport an infinitesimal number of electrons from inside the metal to a position just outside the metallic surface. Neglecting an entropy term, this is the negative of the *electronic work function*, $A_{e^-}$, a measurable quantity varying between ca 1 and 5 electron volts, eV, per electron, depending on the nature of the metal, also depending on surface conditions, and reproducible to within some 0.01 eV. Since 1 eV = 96.485 kJ mol$^{-1}$, the real potential is between ca. 100 and 500 kJ mol$^{-1}$, which gives an impression of the magnitude of effects involved. We note that this real potential of electrons in metals is the work to take electrons from the *Fermi* level of the conduction band of the metal to the outside of the surface of the metal. We also note that, once the real potential for electrons in a metal is known, values for metal ions in metals, but also for ions in electrolyte solutions can be estimated, though only with poor accuracy. We do not, however, continue this line of thought, as we are not really concerned with absolute values of potentials.

In principle, the absolute electric charge in an isolated phase is accidental, and this also applies to of a set of contacting phases such as metal/solution systems.[1] The real potential certainly is a more reliable quantity, as the magnitude of adsorption effects on the potential is comparatively small. It would, then, be interesting to split the real potentials into separately measured values of $\mu_i$-

[1] A practical method to fix conditions is charges is to 'ground' the metal phase, connecting it to the earth as arbitrary zero reference level.

and $z_iF\chi$ contributions, but this is strictly impossible. We thus have no means to know chemical potentials of electrons and ions, although they certainly are reliable quantities, affected only by the composition of the respective phase and thus by macroscopic, easily controlled properties.

In contrast to overall reactions, single electrode reactions cannot spontaneously and continuously advance, because the accompanying change in the relative electric charge of the metal with respect to the solution builds up an increasing difference in Galvani potential, until the preexisting difference in chemical potential is balanced by an opposing difference in $z_rF\phi_{m,l}$. In this state, the electrode reaction is in equilibrium. Since, as described above, very small changes in electrical charge cause large shifts in electrical potential, electrode reaction equilibria are established by negligible reaction extents, i.e. at virtually constant chemical composition of the system under consideration. In equilibrium $\Delta_rG = 0$, and therefore

$$\sum \nu_i\mu_i = z_rF(\phi_{m,l})_{eq} \text{ , } \textit{for electrode reactions in equilibrium.} \tag{3.20}$$

Thus, the Galvani potential difference, $(\phi_{m,l})_{eq}$, is entirely determined by the stoichiometric sum of chemical potentials of participants in the electrode reaction, and these contributions, although not measurable separately, are reliable quantities, unaffected by accidental changes in the total electrical charge of the system, or by accidental changes in surface dipole adsorption. Therefore, differences in Galvani potential across interfaces between conducting phases, established by electrode reaction equilibria, are reliable quantities as well, although also not measurable. As to the unknown numerical magnitude of interfacial Galvani potentials, an estimate of not more than about 1 V would seem reasonable, judging from the voltages of galvanic cells, which, as described below, are sums of a small number of single Galvani potential differences, not much exceeding, as e.g. for the lead battery, ca. 2 V.

## 3.2
## Calculating Gibbs Energies for Overall Reactions

With the aim of calculating Gibbs energies of overall chemical reactions we inspect, in the first place, the dependence of the chemical potential of a species $i$ in a phase $x$ on the composition of $x$. The strategy to do this is to calculate the molar infinitesimal work necessary to transfer the species from a hypothetical phase, where with regard to pressure, temperature and composition $i$ is in a *standard state,* into the phase of interest at the same pressure and temperature, but at different composition. The composition is measured in terms of ratios of numbers of moles of species $i$, $j$, etc., either per volume or per mass of phase $x$; and these ratios are called *concentrations.* Let $\mathscr{C}_i$ be the concentration of a species $i$ in the phase at the desired composition, $\mathscr{C}_i^\circ$ the standard state concentration. The appropriate expression for the chemical potential $\mu_i$ as function of $\mathscr{C}_i$ then is:

$$\mu_i = \mu_i^\circ + RT \ln \frac{\mathcal{C}_i}{\mathcal{C}_i^\circ} f_i \, . \tag{3.21}$$

$R$ is the universal *gas constant*, 8.3145 kJ $K^{-1}$ $mol^{-1}$, $f_i$ is a dimensionless factor, the *activity coefficient*, and the product $\mathcal{C}_i f_i$ is an effective concentration, the *thermodynamic activity* $\mathcal{A}_i$. The term $RT \ln (\mathcal{C}_i/\mathcal{C}_i^\circ)$ is the infinitesimal work, referred to 1-molar reaction extent, required to take species $i$ from the standard state to a hypothetical state where the concentration is $\mathcal{C}_i$, if there is no chemical or physical interaction between particles of species $i$, and also no interaction between such particles and particles of other species present in the phase. This behaviour is equivalent to that of an ideal gas. Accordingly, $RT \ln f_i$ is the additional work caused by chemical interaction between particles in the real phase.

The useful rule is to choose standard states such that the numerical value of $\mathcal{C}_i^\circ$ in the respective unit is 1. Not quite correctly, Eq. (3.21) equation then is usually rewritten into

$$\mu_i = \mu_i^\circ + RT \ln \mathcal{C}_i f_i \, , \tag{3.22}$$

with the tacit understanding that *here and everywhere in the following*, $\mathcal{C}_i$ always is the dimensionless quantity $\mathcal{C}_i/\mathcal{C}_i^\circ$. From Eqs. (3.9) and (3.22) the Gibbs reaction energy is expressed as a function of the concentrations of reaction participants as follows:

$$\begin{aligned} \Delta_r G &= \sum \nu_i \mu_i^\circ + RT \sum \nu_i \ln \mathcal{A}_i \\ &= \sum \nu_i \mu_i^\circ + RT \sum \nu_i \ln \mathcal{C}_i + RT \sum \nu_i \ln f_i \, . \end{aligned} \tag{3.23}$$

Usually, the first term on the right hand of this equation is the most important. We define, therefore, again as a key property in reaction thermodynamics, the *standard Gibbs reaction energy*:

$$\Delta_r G^\circ = \sum \nu_i \mu_i^\circ \tag{3.24}$$

When the reaction is in equilibrium, that is, when $\Delta_r G = 0$, the stoichiometric product has the equilibrium value $\left(\sum \nu_i \ln \mathcal{A}_i\right)_{eq}$, and the expression

$$\Delta_r G^\circ = -RT \left(\sum \nu_i \ln \mathcal{A}_i\right)_{eq} = -RT \, ln \, K \tag{3.25}$$

relates the standard Gibbs reaction energy, the equilibrium value of the stoichiometric product of reaction participants, and the constant-pressure *equilibrium constant*, $K$, of the reaction. When dealing with constants $K$ further below, we shall drop 'equilibrium' subscripts.

Attending to the stoichiometric sum $\sum \nu_i \mu_i^\circ$, we need to prove that this sum can be replaced by the stoichiometric sum of measurable *standard Gibbs energies of formation*, $\Delta_f G_i^\circ$. These are defined as molar Gibbs energies of formation from the parent elements, with all participants in their respective standard states. By non-physical, arbitrary convention, the standard Gibbs energies of the pure elements in their respective standard state are zero. The reaction of formation of dissolved hydrogen chloride from elementary gaseous hydrogen and gaseous chloride is $\frac{1}{2}H_2 + \frac{1}{2}Cl_2 \rightarrow HCl$; the reaction of formation of dissolved ferrous chloride is $Cl_2$+Fe $\rightarrow$ $FeCl_2$. Therefore, the standard Gibbs energies of formation of $FeCl_2$ and HCl are related to the chemical potentials by the expressions

$$a)\ \Delta_f G^\circ_{FeCl_2} = \mu^\circ_{FeCl_2} - \mu^\circ_{Fe} - \mu^\circ_{Cl_2}\,, \tag{3.26}$$

$$b)\ \Delta_f G^\circ_{HCl} = \mu^\circ_{HCl} - (1/2)\mu^\circ_{H_2} - (1/2)\mu^\circ_{Cl_2}\,.$$

Substituting these expressions into into Eq. (3.24), applied to acid corrosion of iron according to Eq. (2.1), the result is

$$\Delta_r G^\circ = \mu^\circ_{FeCl_2} + \mu^\circ_{H_2} - 2\mu^\circ_{HCl} - \mu^\circ_{Fe} = \Delta_f G^\circ_{FeCl_2} - 2\Delta_f G^\circ_{HCl_2}\,, \tag{3.27}$$

because $\Delta_f G^\circ_{Fe}$ and$\Delta_f$ $G^\circ_{H_2}$ are both zero by definition. The general form of this expression is, indeed,

$$\Delta_r G^\circ = \sum \nu_i \mu^\circ_i = \sum \nu_i \Delta_f G^\circ_i\,. \tag{3.28}$$

By analogy, we can also express standard molar reaction enthalpies and standard molar reaction entropies :

$$a)\ \Delta_r H^\circ = \sum \nu_i \Delta_f H^\circ_i\,, \quad b)\ \Delta_r S^\circ = \sum \nu_i \Delta_f S^\circ_i\,. \tag{3.29}$$

Again for acid corrosion of iron, but now alternatively described by Eq. (2.2), Eq. (3.28) changes into

$$\Delta_r G^\circ = \Delta_f G^\circ_{FeCl_2} - 2\Delta_f G^\circ_{HCl_2} = \Delta_f G^\circ_{Fe^{2+}} - 2\Delta_f G^\circ_{H^+}\,. \tag{3.30}$$

Simply for brevity of expressions, a further non-physical, arbitrary convention is to set $\Delta_f G^\circ_{H^+}$ equal to zero. Thus referring $\Delta_f G^\circ_{Fe^{2+}}$ to the unknown value of $\Delta_f G^\circ_{H^+}$, Eq. (3.30) reduces to $\Delta_r G^\circ = \Delta_f G^\circ_{Fe^{2+}}$. Using the label 'Me/$Me^{z+}$' for the overall acid corrosion reaction, the general form of this is

$$\Delta_{Me/Me^{z+}} G^\circ = \Delta_f G^\circ_{Me^{z+}} \tag{3.31}$$

Usually, this quantity is denoted $\Delta G^\circ_{Me^{z+}}$, and it is often called *molar standard Gibbs energy of formation of* $Me^{z+}$, though it is an overall-reaction property, and especially is not $\mu^\circ_{Me^{z+}}$. In the same way, $\Delta G^\circ_i$ values are obtained for other cations, and anions as well. For example, $\Delta G^\circ_{Cl^-}$ is the standard Gibbs energy of the reaction $\frac{1}{2}H_2 + \frac{1}{2}Cl_2 \rightarrow Cl^- + H^+$ , and $\Delta G^\circ_{OH^-}$ is the standard Gibbs energy of the reaction $\frac{1}{2}H_2 + \frac{1}{2}O_2 \rightarrow OH^- + H^+$ ; etc.. In tables, the 'f'-subscript is usually dropped here, and as well in the expressions for standard enthalpies and standard entropies of formation.

For thermodynamic calculations concerning overall reactions only, no more theoretical considerations are required, and the necessary data are readily available. Table 3.1 gives values of $\Delta G^\circ_{Me^{z+}}$, and of $\Delta G^\circ_{X^{z-}}$, and of some more species as will be used in following sections. The scatter in some data on ionic iron species is remarkable, as they are key quantities in corrosion thermodynamics. Here, as elsewhere in such compilations, it should be noted that the inaccuracy is due to scatter of data from different sources rather than the usual scatter in

**Table 3.1.** Standard enthalpies, Gibbs energies of formation, and standard entropies of pure solids of ions and of electrolytes in aqueous solution at 25°C. The standard state for ions and electrolytes is the hypothetical 1-molar ideal-state solution. Data from [9]. Note scatter of tabulated values

| formula | state | $\Delta H_i^\circ$/kJ mol$^{-1}$ | $\Delta G_i^\circ$/kJ mol$^{-1}$ | $S_i^\circ$/Jmol$^{-1}$K$^{-1}$) |
|---|---|---|---|---|
| $H^+$ | l | 0.0... | 0.0... | 0.0... |
| $H_2$ | g | 0.0... | 0.0... | 130.684 |
| $Zn^{2+}$ | l | −152.84 | −147.16 | 107.53 |
| Zn | m | 0.0... | 0.0... | 41.6 |
| $Cu^{2+}$ | l | 65.78 | 65.7 | −97.2 |
| Cu | m | 0.0... | 0.0... | 33.2 |
| $Pb^{2+}$ | l | 1.63 | −24.31 | 21.3 |
| Pb | m | 0.0... | 0.0... | 64.89 |
| $PbO_2$ | s | -276.65 | −218.99 | 76.6 |
| $PbSO_4$ | s | | -813.83 | 140.2 |
| $Cl^-$ | l | −166.8533 | −131.0663 | 56.701 |
| HCl | l | " | " | " |
| $SO_4^{2-}$ | l | −909.27 | −744.63 | 20.1 |
| $H_2SO_4$ | l | " | " | " |
| $Cl_2$ | g | 0.0... | 0.0... | 222.957 |
| $OH^-$ | l | −229.994 | −157.293 | −10.75 |
| $H_2O$ | l | −285.830 | −237.178 | 69.91 |
| $O_2$ | g | 0.0... | 0.0... | 205. 028 |
| Fe | m | 0.0... | 0.0... | 27. 2 |
| $Fe^{2+}$ | l | −89.1,... − 92.4 | −78.8,..., − 91.5 | −137.7,..., − 106. |
| $FeOH^+$ | l | −324. | −277.4,..., − 268.0 | −29.,..., − 94.6 |
| $Fe^{3+}$ | l | −48.5, − 50.2 | −4.6,..., − 16.6 | −315.9 − 380.3 |
| $FeOH^{2+}$ | l | −290.8 | −229.41 | −142. |
| $Fe(OH)_2$ | s | −569.0,574. | −486.6, − 492.017 | 87.9,29.1 |
| $Fe_3O_4$ | s | −1118.4 | −1015.5 | −146.4 |
| $\alpha$-$Fe_2O_3$ | s | −824.2 ..., − 825.5 | −742.2, − 743.6 | 87.4 |

results of repeated experiments, and it is, therefore, not quite correct to to use average values.

Table 3.1 also lists *standard enthalpies of formation*, $\Delta H_i^\circ$, and *standard entropies*, $S_i^\circ$. Analogous to $\Delta G_i^\circ$, $\Delta H_i^\circ$ is calculated from the contributions of participants to the formation of the respective compound from the parent elements, with zero standard enthalpies $H_i^\circ$ of the elements themselves and of dissolved $H^+$. Concerning standard entropies it is, instead, necessary to account for the Third Law of thermodynamics which states that the entropy of any perfectly ordered solid is zero at $T = 0$. Therefore, absolute values of entropies, also called *Third Law entropies*, are available, except for dissolved ions. For a reaction of the type $Me + 2H^+ \rightarrow Me^{z+} + \frac{z}{2}H_2$, the standard reaction entropy would have to be calculated by the equation

$$\Delta_{Me/Me^{z+}} S^\circ = S^{\circ *}_{Me^{z+}} - z S^{\circ *}_{H^+} + \frac{z}{2} S^\circ_{H_2} - S^\circ_{Me} \,. \tag{3.32}$$

where $S^{\circ *}_{Me^{z+}}$ and $S^{\circ *}_{H^+}$ are absolute entropies of ions, which, however, are unknown. Since the other terms in the equation are measurable quantities, the stoichiometric difference, $S^{\circ *}_{Me^{z+}} - z S^{\circ *}_{H^+}$, is also measurable. For brevity, it is renamed *conventional standard molar entropy* $S^\circ_{Me^{z+}}$. Generally, the conventional standard entropy $S^\circ_{X^z}$ of any cation or anion carrying the charge z is defined as

$$S^\circ_{X^z} = S^{\circ *}_{X^z} - z S^{\circ *}_{H^+} \,, \tag{3.33}$$

where z is to be taken with its sign. We note that, from theoretical considerations concerning the structure of aqueous solutions and the change caused by hydrogen ions, the absolute value of $S^{\circ *}_{H^+}$ is estimated at $-20.9$ kJ mol$^{-1}$ K$^{-1}$. This is a point of some interest when calculating high-temperature values of thermodynamic data in Sect. 3.6, but of no concern in the present context. Here, we are interested in the interrelation of thermodynamic properties, for instance of acid corrosion of iron described by Eq. (2.3), taking place between participants in their respective standard states. The result is

$$\begin{aligned} a)\ \Delta G^\circ_{Fe^{2+}} &= \Delta H^\circ_{Fe^{2+}} - T(S^\circ_{Fe^{2+}} + S^\circ_{H_2} - S^\circ_{Fe}) \,, \\ b)\ \Delta G^\circ_{Cl^-} &= \Delta H^\circ_{Cl^-} - T(S^\circ_{Cl^-} - \frac{1}{2} S^\circ_{H_2} - \frac{1}{2} S^\circ_{Cl_2}) \,, \end{aligned} \tag{3.34}$$

and these relations are readily verified by substituting data from Table 3.1 into the equations.

Using these data, an estimate of the influence of minor changes in temperature on the Gibbs reaction energy is easy. For acid corrosion of iron, $\Delta_r S^\circ$ is $-19\,(\pm 12)$ J mol$^{-1}$ K$^{-1}$. By a small change in temperature, such as $\pm 25$ K, the change $\Delta_r S^\circ \Delta T$ is $\pm 0.5\,(\pm 0.3)$ kJ mol$^{-1}$, and can be neglected in most cases of practical interest. Therefore, the absence of a thermostat in the experimental setup described by Fig. 3.1 would not seriously affect the thermodynamics of the corrosion reaction at ambient temperature.

Turning attention to standard states we note, in the first place, that the standard pressure is 1 atm = 1.013 bar. The 1%-difference between bars and atmospheres will be neglected in the following. By definition, the derivative $\partial \mu_i^\circ / \partial p$ does not exist. There is no such restraint with respect to temperature, and the standard state temperature may have any value; where not noted otherwise, it is, however, always set at $T^\circ = 25°C = 298.15$ K.

For gaseous substances, the measure of concentration is the pressure $p_i$. The standard state of a gaseous substance is the pure gas at pressure $p_i^\circ = p^\circ = 1$ bar, and temperature $T^\circ = 298.15$ K, with the properties of an ideal gas, so that the state function $p_i V = n_i RT$ exactly applies. In practice, the state of real pure gases, especially of pure hydrogen, but not much less of pure oxygen, and with sufficient accuracy for other gases as well, is quite close to ideal behaviour not only at 1 bar, but also at much higher pressures. At 1 bar, real pure gases therefore can be considered to be in there standard state. Hence, at other values of $p_i$ of the pure gas or in mixtures where the partial pressure is $p_i$, we obtain the relation

$$\mu_i = \mu_i^\circ + RT \ln p_i \text{ , } \textit{for gases, except at excessive pressures.} \tag{3.35}$$

Only for excessive pressures, such as perhaps sometimes encountered in hydrogen embrittlement, $p_i$ must be replaced by the *fugacity* $p_i f_i$, where $f_i$ is the *fugacity coefficient.*

The standard state of a solid substance is the pure solid substance. If existing in several crystalline modifications, the modification stable at 298.15 K is chosen. As an example, we note that at 25°C ferritic pure iron represents the standard state of iron.

In homogeneous alloys, i.e., in metallic solid solutions, chemical component potentials differ from there standard values. In this case, concentrations are given as *mole fractions,* or *gramatom fractions,* $\gamma_i$. When a homogeneous alloy contains $n_j$,$n_k$,$n_l$,..., moles of metals $j$,$k$,$l$,..., respectively, the mole fraction is $\gamma_j = n_j/\sum_i n_i$. Multiplication of $\gamma_i$ with the factor 100 gives alloy concentrations in 'atom percent'. On first sight, it might appear easy to calculate alloy component chemical potentials using the relation

$$\mu_i = \mu_i^\circ + RT\, ln\, \gamma_i f_i \text{ , } \textit{for alloys.} \tag{3.36}$$

The equation is correct; there is, however, a serious difficulty concerning the thermodynamics of alloy corrosion reactions: These directly involve metal atoms in the surface. But for alloys, we must, then, expect that metal dissolution from the surface is selective, initially different in rate for different alloy components. Since, however, solid state diffusion of alloy component atoms from the bulk of the metal to the surface normally will be much slower than initial selective metal dissolution from the surface, there normally is no longer thermodynamic equilibrium between the bulk and the surface of the metal, and the chemical potentials of alloy components in the surface are not known. The present chapter therefore really deals with corrosion of pure or approximately pure metals only. Eq. (3.36) is nevertheless noted, because it is applicable to corrosion of liquid amalgams, i.e. liquid homogeneous alloys of mercury with zinc or sodium, and these, while not normally interesting in practice, have played an important role in the development of corrosion science, as discussed in the following chapter. Here, suffice it to note that at low values of gramatom fractures of zinc or sodium, liquid amalgams are almost ideal metallic liquids, and the activity coefficients $f_i \simeq 1$.

The standard state of water is pure liquid water at 298.14 K and 1 bar total pressure. The pressure again is nearly irrelevant. Except for highly concentrated aqueous solutions, the chemical potential of liquid water approximately is equal to the standard potential: $\mu_{H_2O} \simeq \mu_{H_2O}^\circ$. When at high values of solute concentrations the approximation must be corrected, there is no difficulty to exactly determine the activity coefficient, $f_{H_2O}$, in the equation relating the mole fraction $\gamma_{H_2O}$ and the chemical potential $\mu_{H_2O}$,

$$\mu_{H_2O} = \mu^{\circ}_{H_2O} + RT ln\, \gamma_{H_2O} f_{H_2O} \,, \tag{3.37}$$

but this will become necessary only rarely.

In aqueous solutions, the molar concentrations of a solute $i$ usually is given as the *molarity*, $c_i$, which is number of moles of $i$ per liter of solution, mol $L^{-1}$. The concentration 1 mol $L^{-1}$ usually is abbreviated to '1 M'. The 'normality' of acids is the molarity in terms of moles of $H^+$ obtained by dissociation. Thus, a 0.5+M $H_2SO_4$ solution is '1 normal', or 1 N An alternative measure is the *molality*, which is number of moles of solute per kilogram of solvent, mol $kg^{-1}$, and independent of temperature. At low concentrations, molalities and molarities are practically equal, and the difference will be neglected in the following. The small dependence of concentrations on pressure will be neglected as well. The chemical potential of a species $i$ in aqueous solution therefore is

$$\mu_i = \mu_i^{\circ} + RT\, ln\, c_i\, f_i \,, \text{ for solutes in aqueous solution.} \tag{3.38}$$

Products, $c_i f_i$, of concentrations and activity coefficients denote thermodynamic activities $a_i$. For uncharged particles, such as dissolved molecular oxygen, the activity coefficient is $f_i \simeq 1$ with sufficient accuracy. It is only for dissolved ions that activity coefficients seriously differ from 1 even at very low concentrations as a result of electrostatic interaction between ions of opposite sign. We shall return to this problem presently.

From Eqs. (3.10), (3.11), and (3.23), together with the expressions (3.33–3.38) describing the dependence of chemical potentials on composition, two alternative expressions are obtained for the quantity $\Delta_r G$ as a function of the activities of ferrous chloride, hydrochloric acid, and hydrogen. Recalling that in pure iron $a_{Fe} = 1$, these two expressions are

$$\begin{aligned} \Delta_{Fe/Fe^{2+}} G &= \Delta G^{\circ}_{Fe^{2+}} + RT \ln \frac{a_{FeCl_2}}{(a_{HCl})^2}\, p_{H_2} \\ &= \Delta G^{\circ}_{Fe^{2+}} + RT \ln \frac{c_{FeCl_2} f_{FeCl_2}}{(c_{HCl})^2 (f_{HCl})^2}\, p_{H_2} \,, \end{aligned} \tag{3.39}$$

and

$$\begin{aligned} \Delta_{Fe/Fe^{2+}} G &= \Delta G^{\circ}_{Fe^{2+}} + RT \ln \frac{a_{Fe^{2+}}}{(a_{H^+})^2}\, p_{H_2} \\ &= \Delta G^{\circ}_{Fe^{2+}} + RT \ln \frac{c_{Fe^{2+}} f_{Fe^{2+}}}{(c_{H^+})^2 (f_{H^+})^2}\, p_{H_2} \,. \end{aligned} \tag{3.40}$$

Phase subscripts have been ignored, but it is important to recognize that there is only one homogeneous solution. Therefore, both expressions are correct in the sense of strict thermodynamics as long as no attempt is made to separately evaluate $a_{Fe^{2+}}$, $a_{H^+}$, $f_{Fe^{2+}}$, and $f_{H^+}$, which are ill-defined single-ion activities.

In equilibrium, $\Delta_r G = 0$. Since Table 3.1 lists $\Delta G^{\circ}_{Fe^{2+}} = -85 \pm 5$ kJ $mol^{-1}$, and since at 298.15 K, $RT = 2.478$ kJ $mol^{-1}$, it follows from Eqs. (3.25) that for acid dissolution of iron the equilibrium constant $K$ is between $10^{14}$ and $10^{16}$. Whenever the actual values of $a_{FeCl_2}$, $a_{HCl}$, and $p_{H_2}$ are such that $a_{FeCl_2} p_{H_2} / a^2_{HCl} \geq K$, the corrosion reaction therefore is in equilibrium or beyond equilibrium, and corrosion is impossible. For $a_{FeCl_2} = a_{HCl} = 1$ mol $L^{-1}$ an irrealistic increase in $p_{H_2}$ to between $10^{14}$ and $10^{16}$ bar would be required to achieve this thermodynamic immunity. For $a_{FeCl_2} = 1$ mol L$^{-1}$ and $p_{H_2} = 1$ bar, however, the neutralization of the acid to $a_{HCl} \leq 10^{-8}$ mol L$^{-1}$ would be quite a realistic possibility, and neutralization of acid waste solutions is, of course, standard practice for corrosion protection of iron and steel pipes.

Equation (3.40) is readily re-formulated for the general case of Me dissolving to $Me^{z+}$ under standard hydrogen electrode conditions:

$$\begin{aligned}\Delta_{Me/Me^{z+}} G &= \Delta G^{\circ}_{Me^{z+}} + RT \ln \frac{a_{Me^{z+}}}{(a_{H^+})^z} (p_{H_2})^{\frac{z}{2}} \\ &= \Delta G^{\circ}_{Me^{z+}} + RT \ln \frac{c_{Me^{z+}} f_{Me^{z+}}}{(c_{H^+})^z (f_{H^+})^z} (p_{H_2})^{\frac{z}{2}} \,. \end{aligned} \quad (3.41)$$

At this point, we can show that the dependence of the Gibbs reaction energy on pressure, described by Eq. (3.4), is low. Since $\Delta_r G^{\circ}$ is independent of pressure, and since the pressure dependence of the specific volumina of condensed phases is negligible, we only have to consider

$$d(\Delta_{Me/Me^{z+}} G) = 2.303(z/2)RT d \log p_{H_2} \,. \quad (3.42)$$

Thus, when $z = 2$, an increase in pressure of the gas participating in the reaction by 100 bar causes a change in Gibbs reaction energy of only 5.7 kJ mol$^{-1}$, which often can be neglected. This would be the small effect if in the experiment described by Fig. 2.1 the hydrogen pressure, and also the pressure in the surroundings, were increased to 100 bar. One might consider the alternative effect of flushing the vessel with a hydrogen/nitrogen mixture at 100 bar total pressure, and 1 bar partial hydrogen pressure. In this case, the pressure effect on the Gibbs reaction energy would be almost nil. In the following there will, therefore, be no more need to consider pressure effects on Gibbs energies.

A curious formal difficulty arises, when in the corrosion test described in Fig. 3.1 the flushing of the solution with gaseous hydrogen is replaced by flushing with pure nitrogen, so that $p_{H_2} = 0$, or when the solution, thought to have a very large volume, is prepared without ferrous chloride, so that $a_{Me^{z+}} = 0$. In any case, $\Delta_r G = -\infty$. This, of course, does not make sense, and the reason is that we cannot assign thermodynamic properties to chemical species which at zero concentration really are non-existent. The only real consequence, therefore, is that as long as the concentration of reaction products is kept zero, equilibrium of the reaction cannot be attained, and this is trivial. For practical purposes, the assumption $p_{H_2} = 1$ bar usually is retained as a consequence of the introduction of the standard reference hydrogen electrode, as will be seen presently. Concerning $a_{FeCl_2}$, the usual practice is to set this activity, or generally the activity $a_i$ of any dissolved salt of any corroded metal, whenever it nominally is zero, arbitrarily at $10^{-6}$ mol L$^{-1}$, and to do so without explanation. Since every decade in $a_i$ adds another 5.7 kJ mol$^{-1}$ to $\Delta_r G$, the convention does need justification, as shown in Chap. 5.

Since in this section the aim only is to arrive at reasonably accurate numerical calculations, the obstacles of rigorous thermodynamics can be disregarded by accepting approximate calculations of single-ion activity coefficients

**Table 3.2.** Single-ion activity coefficients in solutions of varying ionic strength in mol kg$^{-1}$. Data from [10]

| Ion | $2 \times I^* = 0.001$ | 0.002 | 0.005 | 0.010 | 0.020 | 0.050 | 0.100 | 0.200 |
|---|---|---|---|---|---|---|---|---|
| $H^+$ | 0.975 | 0.967 | 0.950 | 0.933 | 0.914 | 0.88 | 0.86 | 0.83 |
| $OH^-$ | 0.975 | 0.964 | 0.946 | 0.926 | 0.900 | 0.86 | 0.81 | 0.76 |
| $Cl^-$ | 0.975 | 0.964 | 0.945 | 0.924 | 0.898 | 0.85 | 0.80 | 0.75 |
| $SO_4^{2-}$ | 0.903 | 0.867 | 0.803 | 0.740 | 0.660 | 0.55 | 0.45 | 0.35 |
| $HCO_3^-$ | 0.975 | 0.964 | 0.947 | 0.928 | 0,902 | 0.86 | 0.82 | 0.78 |
| $Pb^{2+}$ | 0.903 | 0.868 | 0.805 | 0.742 | 0.665 | 0.55 | 0.45 | 0.37 |
| $Cd^{2+}$ | 0.903 | 0.868 | 0.805 | 0.744 | 0.67 | 0.56 | 0.47 | 0.41 |
| $Fe^{2+}$ | 0.905 | 0.870 | 0.809 | 0.749 | 0.675 | 0.57 | 0.49 | 0.41 |
| $Fe^{3+}$ | 0.802 | 0.738 | 0.632 | 0.54 | 0.46 | 0.33 | 0.26 | 0.18 |

and thus, because single-ion concentrations are readily measured by chemical analysis, of single-ion activities as well. For dilute electrolyte solutions, there are, in fact, excellent calculations of single-ion activity coefficients available by the *Debye-Hueckel* theory of anion/cation Coulomb interaction. In an extended form using a number of well-founded additional corrections, the theory is applicable up to a limit of the *ionic strength*, $I^*$, of the solution, defined as $I^* = \frac{1}{2}\sum_i c_i z_i^2$. For 1-1 electrolytes such as NaCl, $I^* = c$; for 2-1 electrolytes such as $FeCl_2$ it is $\frac{1}{2}c(1 \times 2^2 + 2 \times 1) = 3\,c$, etc. From the extended Debye-Hueckel theory single-ion activity coefficients are well available for values $I^* \leq 0.1$. Since domestic tap water is a solution mostly of calcium bicarbonate at $c_i \leq 1$ mmol L$^{-1}$; $I^* \leq 3$ mmol L$^{-1}$, tap-water single-ion activities are readily calculated with quite sufficient accuracy, and so are single-ion activities for boiler feed water. The composition of seawater is mostly sodium chloride at about $c = I^* = 0.5$ mol L$^{-1}$, and thus beyond the range of applicability of the theory. The same applies to geological brines extremely rich in chlorides.

Where Debye-Hueckel calculations apply, the factor $RT \ln f_{\mathrm{Me}^{z+}}/(f_{\mathrm{H}^+})^z$ in Eq. (3.39) is readily calculated using data in Table 3.2. For $Fe^{2+}$ and $2I^* = 0.002$, 0.02, and 0.2, the result is $-0.2$, $-0.5$, $-1.3$ kJ mol$^{-1}$, respectively. These numbers are examples of errors introduced when activity coefficients are ignored and activities considered to be approximately equal to concentrations. Normally, though not always, the consideration of corrosion processes will not be much affected by this type of simplification. Especially when, as previously noted, in the case of $c_{\mathrm{Me}^{z+}} \simeq 0$ an arbitrary convention is necessary, it does not matter whether $a_{\mathrm{Me}^{z+}}$, or else $c_{\mathrm{Me}^{z+}}$ is chosen to be $10^{-6}$.

At higher values of the ionic strength, recourse to equations of the type Eq. (3.39) is always possible. In this case the question is how to define the concentration and activity of dissolved metal salt, when the solution contains other compounds with the same anions. An example is ferrous chloride dissolved in hydrochloric acid, or sodium chloride solution. The answer is the definition of *mean concentrations*, $c_\pm$, *mean activities*, $a_\pm$, and *mean activity coefficients*, $f_\pm$.

**Table 3.3.** Mean activity coefficient $f_\pm = \sqrt{f_{H^+} f_{Cl^-}}$ of pure hydrochloric acid aqueous solutions of varying ionic strength at 25°C. Data from [11]

| I* | 0.002 | 0.010 | 0.100 | 1.000 | 2.000 | 4.000 | 8.000 |
|---|---|---|---|---|---|---|---|
| $f_\pm$ | 0.95 | 0.91 | 0.78 | 0.81 | 1.01 | 1.76 | 5.90 |

Passing general formulations we note, as examples, that for dissolved ferrous chloride

$$a_{Fe^{2+}}(a_{Cl^-})^2 = (a_\pm)^3, \quad c_{Fe^{2+}}(c_{Cl^-})^2 = (c_\pm)^3, \quad f_{Fe^{2+}}(f_{Cl^-})^2 = (f_\pm)^3 .$$

The hidden complication is that, except for symmetrical 1,1-, 2,2-, etc. electrolytes, the standard state now is set at $c_\pm = 1$, instead of $c = 1$. Otherwise, mean thermodynamic quantities are well defined, and there are many tabular mean activity coefficients for concentrated solutions available [11]. As an example, Table 3.3 gives values for pure concentrated hydrochloric acid solutions. The tendency for the mean activity coefficient to go through a minimum and then to values $> 1$ is typical for many electrolytes as the result of repulsive crowding of electrolyte anions and cations at high concentrations, but then it is not the general rule.

As noted previously, a case of practical interest in corrosion often is that $c_{Me^{z+}} \simeq 0$. The convention to account for this condition by arbitrarily assuming $10^{-6}$ instead of zero then might be applied to the mean activity of the metal salt just as well as to its activity, or else its concentration, mean or otherwise.

In previous sections, the answer to the question of how to account for incomplete dissociation of electrolytes and other complications of electrolyte chemistry has been postponed. We can now present the usual answer: It is to disregard these complications, if only with caution, since incomplete dissociation, or hydrolysis of cations, etc., may cause severe changes of activity coefficients, too large to be neglected. When unusually small values of activity coefficients really are encountered, and respected, appropriately corrected calculations are always possible. Some of these, such as complex anion formation, will be considered in Sect. 3.5.

## 3.3 Equilibrium Galvanic Cells

As noted previously and proved in the following chapter, corrosion processes proceed by the superposition of anodic and cathodic electrode reactions at one and the same metal surface. Such half-reactions also proceed, but separately at different electrodes, in galvanic cells; and to understand the functioning of such cells is useful for understanding corrosion processes.

A thermodynamically well defined cell is the familiar lead acid battery. The current-providing overall reaction or *cell reaction* is the simultaneous oxidation

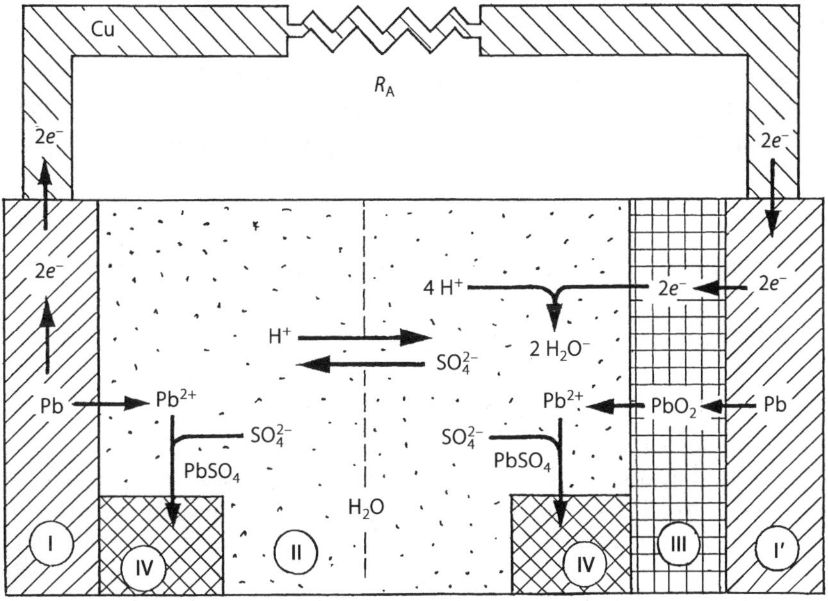


**Fig. 3.3.** Phase scheme and electrode reactions of the lead acid battery

of lead and reduction of lead oxide, with solid lead sulfate as the product of both half-reactions:

$$Pb + PbO_2 + 2H_2SO_4 \rightarrow 2PbSO_4 + 2H_2O \tag{3.43}$$

Fortunately, a side reaction, which is acid corrosion of lead according to Eq. (2.3), is negligibly slow due to an effect which will be explained as 'high hydrogen overpotential' in the following chapter. Using data listed in Table 2.1, the Gibbs reaction energy is readily calculated as

$$\begin{aligned} \Delta_r G &= \Delta_r G^\circ + 2.303\ RT\ \log \frac{(a_{H_2O})^2}{(a_{H_2SO_4})^2}\,, \qquad (3.44) \\ \Delta_r G^\circ &= 2\mu^\circ_{PbSO_4} + 2\mu^\circ_{H_2O} - \mu^\circ_{PbO2} - 2\mu^\circ_{H_2SO_4}\,, \\ &= 2G^\circ_{PbSO_4} + 2G^\circ_{H_2O} - G^\circ_{PbO_2} - 2G^\circ_{H_2SO_4}\,, \\ &= -393.77\ \text{kJmol}^{-1}\,. \end{aligned}$$

Figure 3.3 schematically shows the phases participating in the reaction: The system is composed of a lead plate *I* in contact with an aqueous solution *II* of sulfuric acid, and a lead plate *I'*, covered with lead dioxide *III* in the same solution. The composition of the solution is assumed to be the same everywhere, and the presence of some dissolved and dissociated lead sulfate is neglected. The layer of lead dioxide is assumed to have electronic conductivity and to cover, without pores, the plate *I'*. Solid lead sulfate *IV*, resp. *IV'*, is present at both plates. As shown in the figure, the battery is being discharged through a copper wire between the poles and the electrical resistance of some working device such as an electric engine.

The half-system composed by the phases *I*, *II*, and *IV* is the 'Pb/$PbSO_4$-electrode' of the battery, the half-system composed by *I'*,*II*,*III* and *IV'* is the 'Pb/$PbO_2$,$PbSO_4$-electrode'. During discharge, the reaction at the Pb/$PbSO_4$-electrode is

$$(Pb)_I + (SO_4^{2-})_{II} \rightarrow (PbSO_4)_{IV} + 2(e^-)_I , \tag{3.45}$$

leaving electrons in the lead plate *I*. The reaction at the Pb/$PbO_2$,$PbSO_4$-electrode is

$$(PbO_2)_{III} + 4(H^+)_{II} + (SO_4^{2-})_{II} + 2(e^-)_{I'} \\ \rightarrow (PbSO_4^{2-})_{IV'} + 2(H_2O)_{II} , \tag{3.46}$$

taking electrons out of the lead plate *I'*.

In the lead acid battery, the Pb/$PbO_2$,$PbSO_4$-electrode is the *plus pole* and the *cathode*, the Pb/$PbSO_4$- electrode is the *minus pole* and the *anode*. More generally, in any galvanic cell the plus pole and cathode is the electrode where during discharge of the cell a reduction is taking place which, therefore, is also called *cathodic half-reaction*, and the minus pole and anode is the electrode, where an oxidation or *anodic half-reaction* is taking place. The sum of the half-reactions Eqs. (3.45) and (3.46) is not yet the complete cell reaction, which requires the transfer of electrons through the outer circuit, that is the additional step

$$2(e^-)_I \rightarrow 2(e^-)_{I'} . \tag{3.47}$$

When the battery is working, electric current is flowing throughout the cell, including the acid solution. While ohmic resistance potential drops in metallic conductors at low currents can be neglected, there is, due to the comparatively much poorer conductivity of even concentrated electrolyte solutions, normally an appreciable ohmic voltage (usually designated more briefly as 'ohmic drop') in the solution already at low current flow. The conductivity of, e.g., a 30% $H_2SO_4$ solution is $\kappa \simeq 1\ \Omega^{-1}\ cm^{-1}$, while for iron it is $\simeq 10^5\ \Omega^{-1}\ cm^{-1}$. In the solution, current flow is by movement of all anions and cations present, with contributions from different ionic species depending on the respective concentration, charge, and mobility.

The steady-state rate of *electrolytic migration* of ions is governed by the balance of acceleration by the electric field and retardation by frictional drag. The characteristic quantity is the *mobility*, $u_i$, of ions of species *i*, which is the rate of migration, usually measured in cm $s^{-1}$, in a field of unit strength, 1 V $cm^{-1}$. The fraction $t_i$ of the current transported by ions *i* is the *transference number* of the respective ion. For binary electrolytes, the numbers $t_{anion}$ and $t_{cation}$, while not equal to 0.5, are rather close to this value except when the cation is $H^+$, because then $t_{H^+} \simeq 0.8$, or when the anion is $OH^-$, and $t_{OH^-} \simeq 0.8$, in both cases due to an exceptionally high mobility of the respective ion. Overall ionic current flow is electroneutral, but it can cause local shifts in the composition of solutions, and this will become an issue of major concern when later considering corrosion in holes, crevices and cracks. At present, the interesting quantity rather is the voltage existing between the poles of the battery, i.e., the *cell potential.*

When the external metallic circuit is interrupted, electron transport according to Eq. (3.47) is blocked. The *open-circuit potential* of the battery then might be measured using an electrostatic voltmeter; and this would be a measurement of the Volta potential, $\psi_{I,I'}$, between the chemically identical poles of the battery. When producing current, the voltage of any battery decreases, tending, however, to the *open circuit voltage* as the current tends to zero. Therefore, the use of a voltmeter with an input resistance $> 10^6\Omega$ will normally measure the open circuit voltage with any reasonably required accuracy. The quantity thus measured directly is the difference of Galvani potentials, $\phi_{I,I'}$, between two copper wires leading to the voltmeter, which is the same as the difference of Galvani potentials between the lead poles of the battery. In the present context, the theoretically much more interesting experimental procedure is the *Poggendorff* compensation method, where a voltage opposing the cell voltage is inserted into the external circuit and varied until the current can be changed from discharging to charging the cell by slight, in principle infinitesimal, changes in this voltage compensation. At this point, the system as a whole then is operating at zero current, i.e., at zero rate of the cell reaction, in a state of reversibility by infinitesimal changes of conditions, and, therefore, in equilibrium.

Since the cell potential obviously remains the same when the Poggendorff compensation setup is removed, the conclusion is that the open cell voltage of the lead acid battery is an example, of a *reversible cell potential,* $U_{\mathrm{rev}}$. Therefore, the work, calculated as product of electrical charge taken from the battery and $U_{rev}$, must equal the maximum useful work available from the cell reaction. Calculating this work for the unit reaction conversion, the result is $|\Delta_r G| = z_r F|U_{\mathrm{rev}}|$. This leaves the sign open to an arbitrary convention. As in this text chemical reactions always are written so that corrosion advances from left to right[2], the choice necessary to conform with usual standards is

$$\Delta_r G = z_r F U_{\mathrm{rev}} \; . \tag{3.48}$$

In abbreviated notation, the cell is described as

$Pb,PbSO_4|H_2SO_4,H_2O|PbSO_4,PbO_2,Pb,$

also to be read from left to right. We note, that the battery may be seen as a system in which metallic lead is corroded by lead dioxide in the role of the oxidant, which in turn is reduced.

Accepting that the open-circuit potential and the compensated voltage of the battery are the same, it follows that in the open cell the cell reaction is in equilibrium except for the transfer of electrons between the poles. Therefore, in

[2] Elsewhere, e.g. in [9], the convention is to write reaction equations in reverse, and to reverse the sign in relating Gibbs energies and cell potentials.

the open cell both the anodic half-reaction, Eq. (3.45), and the cathodic half-reaction Eq. (3.46) are in equilibrium regardless of the chemical composition of the phases, as long as at open circuit the overall cell reaction cannot proceed. The relative electric charge of the poles $I$ and $I'$ is such that the chemical driving force $\Delta_r G$ is exactly balanced by the opposing electric driving force $z_r F \phi_{I,I'}$, as previously described for single electrode equilibria.We therefore have the relation

$$U_{\text{rev}} = \phi_I - \phi_{I'} = \phi_{I,I'} \tag{3.49}$$

It is intuitive that the open-cell potential of any reversible or non-reversible galvanic cell must be the sum of single Galvani potential differences, $\phi_{x.y}$, existing in series between the poles. This measurable sum of separately unmeasurable contributions will be more closely inspected in Sect. 3.4. For the moment, we generalize the exemplary case of the lead acid battery and conclude that for any electrochemical overall reaction a galvanic cell, hypothetical or really available, representing the open-circuit equilibrium of the reaction, has the cell potential

$$U_{\text{rev}} = \frac{\Delta_r G^\circ}{z_r F} + \frac{RT}{z_r F} \sum \ln \mathscr{A}_i^{\nu_i} . \tag{3.50}$$

This is known as the *Nernst equation,* and the term $\Delta_r G^\circ / z_r F$ as the *standard cell potential,* $U_r^\circ$. At 25°C, $RT/z_r F$ is $25.693/z_r$ (mV). Often it is convenient to replace natural by common logarithms, and, since $\ln x = \ln 10 \times \log x$, we then use $2.303\ RT/z_r F = 59{,}160/z_r$ (mV).

## 3.4 Galvanic Cells with Transference

Both zinc in contact with a zinc sulfate solution, and copper in contact with a copper sulfate solution are half-cells with readily established equilibrium of the respective electrode reaction. Therefore, the *Daniell* cell combining the two half-cells is a good current source. The phase system is schematically shown in Fig. 3.4a. Zinc poles of the cell are arbitrarily assumed rather than the usual copper poles. The overall cell reaction equation is

$$(\text{Zn})_{\text{I}} + (\text{Cu}^{2+})_{\text{III}} \rightarrow (\text{Zn}^{2+})_{\text{II}} + (\text{Cu})_{\text{IV}} , \tag{3.51}$$

and the abbreviated description of the cell is

$Zn|ZnSO_4,H_2O||CuSO_4,H_2O|Cu|Zn$.

The cell potential is unaffected by this choice of the metal forming the poles. It is also unaffected by adding still another metallic phase, such as copper wires, on both sides of the cell. Assuming thermodynamic reversibility, the open-cell potential would be expected to be

$$U_{\text{rev}} = U_{\text{rev}}^\circ + \frac{RT}{2F} \ln \frac{(a_{\text{Zn}^{2+}})_{II}}{(a_{\text{Cu}^{2+}})_{III}} , \tag{3.52}$$

where $U_{rev}^\circ = \Delta_r G^\circ / z_r F$, and $\Delta_r G^\circ = \Delta G_{\text{Zn}^{2+}}^\circ - \Delta G_{\text{Cu}^{2+}}^\circ$. Using data listed in Table 3.1, $U_{\text{rev}}^\circ = -1.102\ V$, and for equal concentrations of copper and zinc sulfate

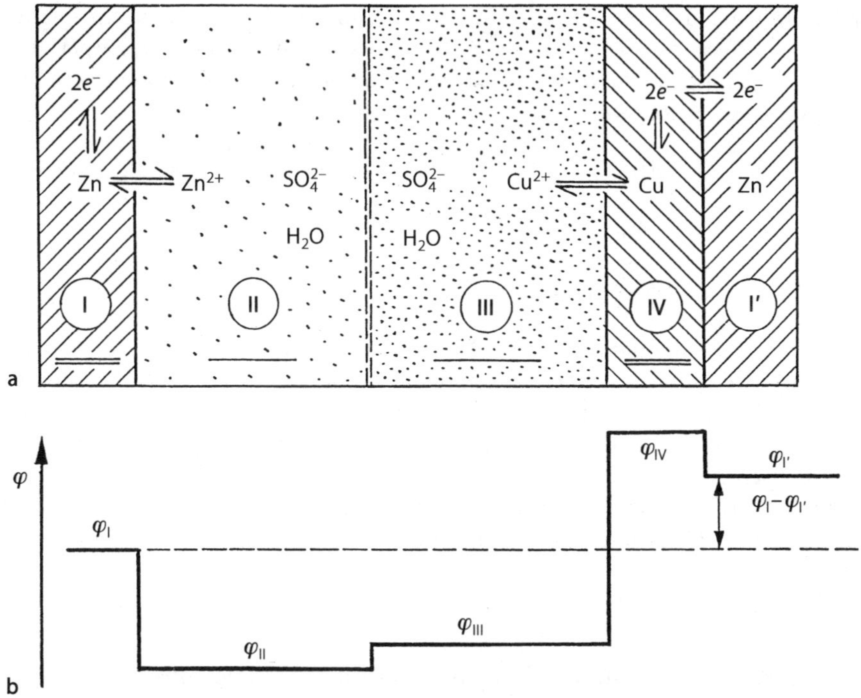


**Fig. 3.4.** **a** Phase scheme and electrode reactions of Daniell cell, **b** a possible distribution of the Galvani potential throughout the cell

in the cell compartments the observed open-cell potential is, indeed, close to this value. However, $U_{rev}$ is not exactly the cell potential $\phi_{I.I'}$, because in terms of thermodynamics the Daniell cell is flawed: There is a phase boundary *II/III* between two aqueous solutions of different composition, indicated by the symbol '||'. This *liquid junction* is permeable for migrating current-carrying ions; but it also is permeable for diffusing ions tending to equalize the composition of the two phases. Therefore, this interface is not in equilibrium. At the same time, the Galvani potentials $\phi_{II}$ and $\phi_{III}$ will be different in the open cell due to the existence of an electric double layer across the liquid junction. It is, then, impossible to exactly evaluate $\phi_{II,III}$ in terms of equilibrium thermodynamics.

Still, it is reasonable to assume, for the open Daniell cell, equilibrium of the electrode reactions

$$(\mathrm{Zn})_{\mathrm{I}} = (\mathrm{Zn}^{2+})_{\mathrm{II}} + 2(\mathrm{e}^-)_{\mathrm{I}}, \tag{3.53}$$

$$(\mathrm{Cu}^{2+})_{\mathrm{III}} + 2(\mathrm{e}^-)_{\mathrm{IV}} = (\mathrm{Cu})_{\mathrm{I}}\mathrm{V}, \tag{3.54}$$

and of the electron transfer reaction

$$(\mathrm{e}^-)_{\mathrm{IV}} = (\mathrm{e}^-)_{\mathrm{I'}}\ . \tag{3.55}$$

Relating what in terms of strict thermodynamics really are unknown quantities, the differences in Galvani potential across the interfaces $I/II$, $III/IV$, and $IV/I'$, as a function of single-ion standard potentials and single-ion activities are evaluated as follows:

$$a)\quad \phi_{I,II} = \frac{1}{2F}\{\mu^{\circ}_{Zn^{2+}} + 2(\mu_{e^-})_I - \mu^{\circ}_{Zn} + 2.303RT\log(a_{Zn^{2+}})_{II}\} \tag{3.56}$$

$$b)\quad \phi_{II,IV} = \frac{1}{2F}\{\mu^{\circ}_{Cu} - 2(\mu_{e^-})_{IV} - \mu^{\circ}_{Cu^{2+}} - 2.303RT\log(a_{Cu^{2+}})_{III}\}$$

$$c)\quad \phi_{IV,I'} = \frac{1}{F}\{(\mu_e^-)_{IV} - (\mu_e^-)_{I'}\}$$

The chemical potential of electrons has not been split into a standard potential and an activity term; hence, the phase subscript was retained. The additional single Galvani potential difference $\phi_{II,III}$, across interface between the two solutions, $II$ and $III$, is the *liquid junction potential* or *diffusion potential,* $\varepsilon_{diff}$.

At open circuit no current is flowing through the cell, and there is no ohmic potential drop in either phase. The cell potential then must be the sum of contributions from double layers in series throughout the cell. Therefore,

$$\phi_{I,I'} = \sum \phi_{x,y} = U_{rev} + \varepsilon_{diff}\ , \tag{3.57}$$

where $U_{rev}$ is the same as in Eq. (3.52).

Figure 3.4b schematically shows the open-circuit cell potential, $\phi_{I,I'}$, as such a sum. We note that this is but one arbitrarily assumed distribution of the potential through the cell. We especially note that nothing is directly known about the sign of the charge of zinc, which is the anode of the cell, or about the sign of the charge of copper, which is the cathode of the cell, with respect to the adjacent solution. What we do know is that the charge in the phase $I'$ is positive with respect to the charge in the chemically identical phase $I$, because we know that on short-circuiting the cell electrons will flow in the external circuit from pole $I$ to pole $I'$.

Because of the existence of the liquid junction the Daniell cell is a *cell with transference,* while the lead acid battery is a cell without transference and thus better defined in terms of thermodynamics. As to the diffusion potential, $\varepsilon_{diff}$, the first note is that in practice there must be some diaphragm separating the two solution phases, preventing rapid mixing but allowing ionic migration. The thickness of the interdiffusion zone in the diaphragm is macroscopic, while the thickness of metal/solution electric double layers is of the order of diameters of hydrated ions (see Chap. 5). In the interdiffusion zone the different ions tend to diffuse to lower concentration, but they will do so at initially different rates due to different ionic mobilities. By anions and cations moving at different rates, a gradient of the Galvani potential in the interdiffusion zone will accelerate anion diffusion and decelerate cation diffusion, or vice versa, until all ions move at the same rate. This then is the steady state, and the steady state potential difference is the diffusion potential. We imply, not always quite safely, that a steady diffusion state does exist.

Fortunately, $\varepsilon_{\text{diff}}$ is a small quantity at least in two examples where the calculation is straightforward, though, of course, only approximate: The first is a junction between solutions of one and the same electrolyte dissociating into ions of equal valency, $\pm z$. As an example, we consider the junction between a HCl-solution A with $c_A = 10^{-3}$ M, and a HCl-solution B with $c_B = 10^{-2}$ M. In this case we have.

$$\phi_{A,B} = \varepsilon_{\text{diff}} = \frac{u_{\text{H}^+} - u_{\text{Cl}^-}}{u_{\text{H}^+} + u_{\text{Cl}^-}} \frac{RT}{zF} \ln \frac{c_B}{c_A} . \tag{3.58}$$

Neglecting the dependence of $u_i$ on the concentration of the solutions we use tabular values, $u_{\text{H}^+} \simeq 4 \times 10^{-3}$, $u_{\text{Cl}^-} \simeq 8 \times 10^{-4}$ cm$^2$ s$^{-1}$ V$^{-1}$, and $\varepsilon_{\text{diff}} = 39$ mV. This is small indeed but, due to the high mobility of $H^+$, not negligible.

For more precise measurements, there is a practical way to minimize the diffusion potential, provided that the solutions *II* and *III* are dilute: it is to add between solutions *II* and *III* a 'salt bridge', which is a U-tube with frits at both ends, containing concentrated dissolved KCl or, alternatively, $NH_4NO_3$ solution. By this procedure, two new phase boundaries are introduced in series between the cell compartments, and thus also two diffusion potentials dominated by KCl diffusion, or else $NH_4NO_3$ diffusion, each very small because $u_+ \simeq u_-$, and of opposite sign, thus cancelling to within about 1 mV. A second method is to add, to dilute solutions in the two cell compartments, an excess amount of a 'neutral salt' or 'supporting electrolyte' with anions and cations not participating in the cell reaction, such as NaCl or $Na_2SO_4$. Dominated by the total salt content practically equal in the two compartments, the diffusion potential in the diaphragma between the compartments again is negligible.

The second straightforward case is that of two dilute solutions at equal concentration of different binary electrolytes dissociating again into ions of equal valence $\pm z$. This is the case of the Daniell cell and the expression for the diffusion potential now is

$$\varepsilon_{\text{diff}} = \frac{RT}{2F} \ln \frac{u_{\text{Cu}^{2+}} + u_{\text{Cl}^-}}{u_{\text{Zn}^{2+}} + u_{\text{Cl}^-}} . \tag{3.59}$$

Since $u_{\text{Cu}^{2+}} \simeq u_{\text{Zn}^{2+}}$, the diffusion potential does not exceed some few mV.

The situation is not always as favorable as that: For solutions high in concentration of reacting ions, a presupposition of the relations quoted above, which is concentration-independence of transference numbers, fails. Marked errors also are to be expected when an aqueous solution contacts a nonaqueous solution, which, however, is beyond the scope of the present text. On the whole, the usual practice is to neglect $\varepsilon_{\text{diff}}$, expecting that the error introduced in cell potential measurements does not normally exceed some 10 or at most 20 mV, and often is smaller, and this supposition will be adopted almost everywhere in the following.

## 3.5 More on Equilibrium Electrode Potentials

In the open lead acid battery, and in the open Daniell cell, both half-cell electrodes are in equilibrium. This is, however, by far not the general rule for $Me/Me^{z+}$ electrodes. As an example, the $Fe/Fe^{2+}$ electrode equilibrium is difficult to achieve, and this is the reason for the inaccurate $\Delta G^{\circ}_{Fe^{2+}}$ value in Table 2.1. As noted in the previous chapter, reversibility essentially requires a high equilibrium exchange rate of the the reaction, and this quantity is rather more often low than high, so that reversible electrode reactions, and thus reversible cells like the lead acid battery and the Daniell cell are rare. A further example of reasonably good reversibility is the *hydrogen electrode* with the half-cell reaction

$$(H_2)_g = 2(H^+)_l + 2(e^-)_m \ . \tag{3.60}$$

For a hydrogen electrode, platinum is used as the metal in contact with an acid solution continuously flushed with pure gaseous hydrogen. In this system platinum is thermodynamically immune with respect to corrosion and is acting as a source and sink for electrons only. To improve reversibility, the surface of the metal is greatly increased by spongy platinizing, but this is incidental. With other electrode metals acting – in the chosen acid – as electron source/sink only, reversibility is poorer, but the thermodynamics of the system remain unchanged when platinum is replaced by, e.g., gold.

A highly irreversible electrode is the *oxygen electrode.* The reaction equation is

$$2(H_2O)_l = (O_2)_g + 4(H^+)_l + 4(e^-)_m \ . \tag{3.61}$$

The construction would be the same as for hydrogen electrodes, replacing flushing with hydrogen by flushing with oxygen, but to achieve reversibility requires very considerable experimental efforts. Therefore, the cell

$$\mathrm{Pt,H_2|acid,H_2O|O_2,Pt}\ ,$$

which would be an ideal *fuel cell* representing the overall reaction

$$2(H_2O)_l = 2(H_2)_g + (O_2)_l \ , \tag{3.62}$$

does not readily establish the reversible cell potential. The thermodynamic data necessary to calculate $U_{rev}$ are, however, available from other than electrochemical, viz. thermochemical measurements, and already listed in Table 3.1. The reversible cell potential is calculated as

$$U_{rev} = U^{\circ}_{rev} + \frac{RT}{4F} \ln \frac{(p_{H_2})^2\, p_{O_2}}{(a_{H_2O})^2} \ , \quad U^{\circ}_{rev} = \frac{2\Delta G^{\circ}_{H_2O}}{4F} = 1.229\,V \ , \tag{3.63}$$

where $a_{H_2O} \simeq 1$. It follows, that an equilibrium oxygen electrode is much more positive than a hydrogen electrode in the same solution and $O_2$ thus a much stronger oxidizing agent. When at 25°C, $p_{H_2} = 1$ bar, and, as in air, $p_{O_2} = 0.2$ bar, we have $U_{rev} = 1.21$ *V*, regardless of the acidity of the solution.

We now consider a cell with copper poles, composed of a $Me/Me^{z+}$ electrode and a hydrogen electrode, which we briefly describe as

$$Cu\,|Me\,|Me^{z+}\;\|H^{+}\,|H_2,Pt\,|Cu,$$

where '‖' again indicates a liquid junction and a neglected diffusion potential, and where $Me^{z+}$ and $H^+$ indicate solutions of an Me-salt, and of an acid, respectively. The reversible cell potential, to be either measured, when the reversible cell actually is available, or else calculated from thermochemical data, is

$$U_{rev} = U_{rev}^{\circ} + \frac{RT}{zF}\ln\frac{a_{Me^{z+}}}{(a_{H^+})^z}\,(p_{H_2})^{\frac{z}{2}}, \quad U_{rev}^{\circ} = \frac{\Delta G_{Me^{z+}}^{\circ}}{zF}\,, \tag{3.64}$$

because in this case $z_r = z$.

At this point, it is usual but arbitrary convention, without reference to real separate contributions to the cell potential, to rewrite this into

$$\begin{aligned} U_{rev} &= E_{Me/Me^{z+}} - E_{H_2/H^+} \\ &= E_{Me/Me^{z+}}^{\circ} + \frac{RT}{zF}\ln a_{Me^{z+}} - \frac{RT}{F}\ln\frac{a_{H+}}{\sqrt{p_{H_2}}}\,. \end{aligned} \tag{3.65}$$

In words: The reversible cell potential, $U_{rev}$, is formulated as the difference: $E_{Me/Me^{z+}} - E_{H_2/H^+}$, of the *equilibrium electrode potentials* of the $Me/Me^{z+}$ electrode and of the hydrogen electrode, respectively. The so-called 'contribution from the $Me/Me^{z+}$ electrode' is the sum

$$E_{Me/Me^{z+}} = E_{Me/Me^{z+}}^{\circ} + \frac{RT}{zF}\ln a_{Me^{z+}}\,, \tag{3.66}$$

where $E_{Me/Me^{z+}}^{\circ}$ is the *standard electrode potential* of the $Me/Me^{z+}$-electrode.

Concerning the activity, $a_{Me^{z+}}$, of ions of the corroded metal in a solution which contains these ions only as a result of the corrosion process, we remember that the obvious assumption $a_{Me^{z+}} = c_{Me^{z+}} \simeq 0$ has been ruled out as impracticable previously, alternatively suggesting, by convention, $1 \times 10^{-6}$ instead of 0 mol $L^{-1}$ .

The so-called 'contribution from the hydrogen electrode' is

$$E_{H_2/H^+} = \frac{RT}{F}\ln\frac{a_{H^+}}{\sqrt{p_{H_2}}}\,, \tag{3.67}$$

implying, again by arbitrary convention, that the standard electrode potential of the hydrogen electrode $E_{H_2/H^+}^{\circ}$ is

$$E_{H_2/H^+}^{\circ} = 0\,, \quad \textit{for all temperatures,} \tag{3.68}$$

corresponding to the previous convention to assume $\Delta G_{H^+}^{\circ}$ to be zero regardless of temperature. In other words: $E_{Me/Me^{z+}}$ is the equilibrium electrode potential of a $Me/Me^{z+}$ electrode referred to the *standard hydrogen reference electrode* at $a_{H^+} = 1$, $p_{H_2} = 1$; activities measured in mol $L^{-1}$, pressures in bar.

Generally, the equilibrium potential $E_r$ of an electrode reaction labeled $r$ is the potential of a cell consisting of the real or hypothetical equilibrium electrode and the standard hydrogen electrode. To calculate $E_r^{\circ}$ it is convenient to write the

**Table 3.4.** Standard electrode potential series of aqueous $Me/Me^{z+}$ electrodes at 25°C. Data from [9]

| $Me/Me^{z+}$ | $E_r^\circ/V$ | $Me/Me^{z+}$ | $E_r^\circ/V$ | $Me/Me^{z+}$ | $E_r^\circ/V$ |
|---|---|---|---|---|---|
| $Li/Li^+$ | −3.045 | $Cr/Cr^{3+}$ | −0.90 | $H_2/H^+$ | ±0.0... |
| $Na/Na^+$ | −2.714 | $Fe/Fe^{2+}$ | −0.44 | $Cu/Cu^{2+}$ | +0.340 |
| $Mg/Mg^{2+}$ | −2.356 | $Cd/Cd^{2+}$ | −0.403 | $Cu/Cu^+$ | +0.520 |
| $Be/Be^+$ | −1.97 | $In/In^{3+}$ | −0.338 | $Hg/Hg_2^{2+}$ | +0.796 |
| $Al/Al^{3+}$ | −1.67 | $Co/Co^{2+}$ | −0.277 | $Ag/Ag^+$ | +0.799 |
| $Ti/Ti^{2+}$ | −1.63 | $Ni/Ni^{2+}$ | −0.257 | $Pd/Pd^{2+}$ | +0.915 |
| $Zr/Zr^{4+}$ | −1.55 | $Sn/Sn^{2+}$ | −0.136 | $Pt/Pt^{2+}$ | +1.19 |
| $Mn/Mn^{2+}$ | −1.18 | $Pb/Pb^{2+}$ | −0.125 | $Au/Au^{3+}$ | +1.52 |
| $Zn/Zn^{2+}$ | −0.763 | $Fe/Fe^{3+}$ | −0.037 | $Au/Au^+$ | +1.83 |

respective electrode reaction $\nu_A A + \ldots = \nu_M M + \ldots + z_r e^-$, and to calculate the sum $(1/z_r F) \sum \Delta_f G_i^\circ$ ignoring electrons. The mathematical routine will become more obvious from some examples of such calculations presently.

In Eq. (3.67), the single-ion activity, $a_{H^+}$, needs closer attention than $a_{Me^{z+}}$ in Eq. (3.66), because numerical values of $a_{H^+}$ used in this equation must be compatible to values introduced in many other thermodynamic calculations, such as solubility products of oxides and hydroxides. Therefore, we have to consider the concept of 'pH', defined by the expression

$$pH \simeq -\log a_{H^+} \ . \tag{3.69}$$

Measurements of pH are calibrated by comparison with specified standard buffer solutions. These are solutions standardized with the aim of minimizing inaccuracies in thermodynamic calculations caused by approximating $\log a_{H^+}$ values. Buffer solutions are concentrated electrolytic solutions such as carbonate/bicarbonate solutions, where $a_{H^+}$ is stable, governed by, in this case, the $CO_3^{2-}/HCO_3^-$ equilibrium. pH values are then determined by measurements with special electrode systems changing their cell potential by 59 mV $(pH)^{-1}$ at 25°C. Nowadays, these are commercial *glass electrodes,* which really are cells with half-cells separated by a glass membrane reproducibly responding to $a_{H^+}$ in the solution to be tested, in comparison to buffer solutions in the same place. Carefully measured, pH is reproducible to within 0.01 units in the range $0 \leq pH \leq 12$, at least at room temperature. Less accurate routine measurements probably should rather be regarded as being significant to within about 0.1 units only.

Table 3.3 lists an ordered set of standard potentials, $E_{Me/Me^{z+}}$. Negative values indicate that the metals, in contact with a solution at $a_{Me^{z+}} = 1$, and at $a_{H^+} = 1$, tend to spontaneous acid corrosion accompanied by evolution of gaseous hydrogen at atmospheric pressure. Such metals (more precisely, such $Me/Me^{z+}$ electrodes) are called 'less noble than hydrogen'. When $E_{Me/Me^{z+}}$ is positive, the metal is said to be 'more noble than hydrogen'. Any further conclusion as to the rate of the respective reactions would be erroneous. Also, while galvanic cells with dissimilar metal electrodes will later be seen to play an important role for many corrosion processes, the supposition that dissimilar metal contact is increasingly dangerous with increasing difference of standard potentials is distinctly misconcepted.

When $a_{Me^{z+}}$ is $10^{-6}$ instead of 1 mol $L^{-1}$, the calculated equilibrium electrode potential of the metal, $E_{Me/Me^{z+}}$, always is more negative by $(2.303RT/zF)\log 10^{-6} = -0.354/z$ (V) than the standard potential.

In aqueous solutions, the dissociation of water

$$H_2O = H^+ + OH^- \tag{3.70}$$

is in equilibrium. The equilibrium constant is calculated from the standard Gibbs energies of formation listed in Table 3.1:

$$\Delta_w G^\circ = \Delta G^\circ_{OH^-} - \Delta G^\circ_{H_2O} = -RT \ln K_w \, . \tag{3.71}$$

At 25°C the constant is $K_w = a_{H^+} a_{OH^-} = 0.98 \times 10^{-14}$.

When the solution is neutral, $a_{H^+} = a_{OH^-}$, and $pH = -\sqrt{K_w} = 7$, whereas $pH = 0$ for $a_{H^+} = 1$. Changing from pH 0 to pH 7, the calculated equilibrium electrode potential of the cell representing corrosion with accompanying hydrogen evolution changes to a value less negative by $0.413/z_r$ V. This, then, is no longer acid corrosion, and it would seem intuitive that corrosion accompanied by hydrogen evolution ceases, because the concentration of hydrogen ions becomes negligible, causing the rate of hydrogen evolution to tend to zero for reasons of kinetics rather than thermodynamics. In fact, this is correct reasoning in most cases of practical interest. There are, however, exceptions, because water is also available as a source of hydrogen by the reaction

$$2H_2O + 2e^- \rightarrow H_2 + 2OH^- \, . \tag{3.72}$$

$H_2O$ being a very stable molecule, this reaction normally is slow. In terms of thermodynamics, the reaction is, however, identical with hydrogen evolution by reduction of $H^+$. To see this, we simply have to consider that $OH^-$ always is in equilibrium with $H^+$ and $H_2O$. Adding $2OH^- + 2H^+ = 2H_2O$ to Eq. (3.72) we are back to the equation $2H^+ + 2e^- \rightarrow H_2$. Hence, $E_{H_2/H_2O} = E_{H_2/H^+}$, and in terms of thermodynamics water is the same oxidizing species as hydrogen ions. Also, the fact is by no means academic only: It is by this direct reaction that in contact with water sodium almost explodes by catastrophic corrosion. Even in contact with moist air, the metal is highly unstable.

As remarked previously, cathodic hydrogen deposition can proceed without formation of gas bubbles, via dissolved molecular hydrogen desorbing into the atmosphere, but we ignore this as a minor effect. Instead, we turn to oxygen corrosion with the overall reaction

$$Me + \frac{z}{2}O_2 + zH^+ \rightarrow Me^{z+} + \frac{z}{2}H_2O \, . \tag{3.73}$$

A galvanic cell representing this reaction would have an $Me/Me^{z+}$-anode, and an oxygen cathode with the electrode reaction already noted in Eq. (3.61). The shorthand description of this cell is

$$\mathrm{Cu|Me|Me^{z+}||H^+|O_2,Pt|Cu}\,,$$

and the hypothetical equilibrium cell potential is

$$U_{rev} = E_{Me/Me^{z+}} - E_{O_2} \tag{3.74}$$

The oxygen equilibrium electrode potential is

$$E_{O_2} = E^\circ_{O_2} + \frac{RT}{4F}\ln\frac{(a_{H^+})^4}{(a_{H_2O})^2}\,p_{O_2}\,, \tag{3.75}$$

and we know from Eq. (3.63) together with (3.68) that $E^\circ_{O_2} = 1.229$ V. For dilute solutions in contact with air, where $p_{O_2} = 0.2$ bar, the equation reduces to

$$E_{O_2} = 1.21 + \frac{RT}{F}\ln a_{H^+} = 1.21 - 0.059\ \mathrm{pH}\ (\mathrm{V}) \tag{3.76}$$

The equilibrium potential of the cell representing oxygen corrosion of a metal Me is thus always 1.21 V more negative than that of the corresponding cell representing acid corrosion of Me, indicating a large increase in the thermodynamic force driving the reaction. From this it is seen that only few metals at the high end of the standard potential series are immune with respect to corrosion in the sense that the spontaneous corrosion reaction is strictly impossible.

All electrodes so far considered in this section are classified as being *electrodes of the first kind* with either metal/metal ion or ion/gas electrode reactions. A different class are the *electrodes of the second kind* such as the $Pb/PbSO_4$, $H^+$ and the $Pb/PbO_2,PbSO_4$ electrodes, where the ions $Me^{z+}$ react with anions present in the solution to deposits of salts, or oxides, or hydroxides, etc., with the consequence that the activity, $a_{Me^{z+}}$, rather than being open to an arbitrary convention, now is governed by the *thermodynamic solubility* of the solid chemical compound in question. The electrode potential of these electrodes then responds to the activity of the anion participating in the reaction. Some electrodes of the second kind are very useful because the equilibrium of the electrode reaction is easily and reproducibly established even without particularly careful handling. One of the best of these is the *silver/silver chloride electrode*, where the electrode reaction $Ag = Ag^+ + e^-$ and the deposition reaction $Ag^+ + Cl^- = AgCl$ readily attain equilibrium. When the solution is saturated with respect to solid silver chloride, the silver ion activity is governed by the *solubility product*

$$K_{AgCl} = a_{Ag^+}a_{Cl^-}\,, \tag{3.77}$$

where $a_{Ag^+}$ and $a_{Cl^-}$ are equilibrium activities. At room temperature, $K_{AgCl} = 1.77 \times 10^{-10}$. Therefore, when $a_{Cl^-} = 1$, $a_{Ag^+} = 1.77 \times 10^{-10}$ mol L$^{-1}$, and we explicitly note that, while it is useless to consider species at exactly zero activity, numerical values of activities, and of corresponding concentrations as well, very close to zero are perfectly well defined in terms of chemical thermodynamics. In terms of reaction kinetics the situation is different: As shown in Chap. 4, rates of elementary reaction steps are proportional to the concentration or sometimes to powers of the concentration of reacting species. This also applies to the rate-determining step of the equilibrium reaction of an electrode, i.e. to the equilibrium exchange rate introduced in Chap. 2, which must be reasonably high. Since the equilibrium of the Ag/AgCl-electrode is readily established, the rate-determining step cannot be $Ag = Ag^+ + e^-$, because, at negligibly low $c_{Ag^+}$, the rate of the reverse reaction would be negligibly slow. Instead, the rate-determining step probably involves $Cl^-$-ions, as their concentration is high. It may be of the type $Ag + Cl^- = AgCl + e^-$, but this aspect leads to solid-state electrochemical kinetics beyond what we can discuss at present.

The equilibrium potential of the Ag/AgCl-electrode is

$$\begin{aligned} E_{Ag/AgCl} &= E^\circ_{Ag/Ag^+} + \frac{RT}{F}\ln\frac{K_{AgCl}}{a_{Cl^-}} = E^\circ_{Ag/AgCl} - \frac{RT}{F}\ln a_{Cl^-} \\ &= 0.222 - 0.059\log a_{Cl^-}\ (V)\ . \end{aligned} \tag{3.78}$$

We might, in principle, continue to calculate the $Ag/Ag^+$-potential from the very small $a_{Ag^+}$-values calculated from the solubility product. We might even ignore this thermodynamic relation and numerically pack its effect into the activity coefficient $f_{Ag^+}$, which then would be found to be of the order $10^{-10}$ and to depend on the chloride concentration. In this particular case, the method does not make much sense; often, when reactions reducing $a_{Me^{z+}}$ by, e.g. hydrolysis, or by complex anion formation, etc., are less drastic, it is quite convenient.

Another very useful reference electrode is the *calomel electrode*, i.e., the mercury-mercurous chloride electrode. The equilibrium electrode reaction is

$$2Hg + 2Cl^- = Hg_2Cl_2 + 2e^-\ , \tag{3.79}$$

and the electrode potential is

$$\begin{aligned} E_{Hg/Hg_2Cl_2} &= E^\circ_{Hg/Hg_2Cl_2} - \frac{RT}{F}\ln a_{Cl^-} \\ &= 0.268 - 0.059\log a_{Cl^-}\ (V)\ . \end{aligned} \tag{3.80}$$

The solubility product used for the calculation of this potential is $1.43 \times 10^{-18}$. The calomel electrode is schematically depicted in Fig. 3.4 showing the experimental setup for the measurement of the equilibrium potential of a copper-copper sulfate electrode referred to a calomel electrode saturated with respect to both mercurous chloride potassium chloride as well. The potential, $E_{Hg/Hg_2Cl_2}$, of this 'saturated calomel electrode', a very common tool in experimental electrochemistry, is 0.245/V on the SHE scale. From the potential $U_{rev}$ of the cell, the potential $E_{Cu/Cu^{2+}}$ is obtained by subtracting 0.245/V. Generally, the equilibrium potential, $E_r$, of an electrode reaction labeled $r$, is calculated from the potential $U_{rev}$ of a cell with one the half-cell representing the reaction $r$, and a second reference half-cell which is at $E_{ref}$ with respect to the standard hydrogen electrode (SHE). Then, $E_r = U_{rev} - E_{ref}$.

In the experiment described by Fig. 3.5, we can replace the $Cu/Cu^{2+}$ electrode, or any other equilibrium half-cell, by an arbitrary metal/electrolyte solution-system, usually called *working electrode*, such as iron in hydrochloric acid, or iron in sea-water, etc.. The cell, now composed of an irreversible half-cell and a reversible reference cell, still has a potential, $U$, which again can

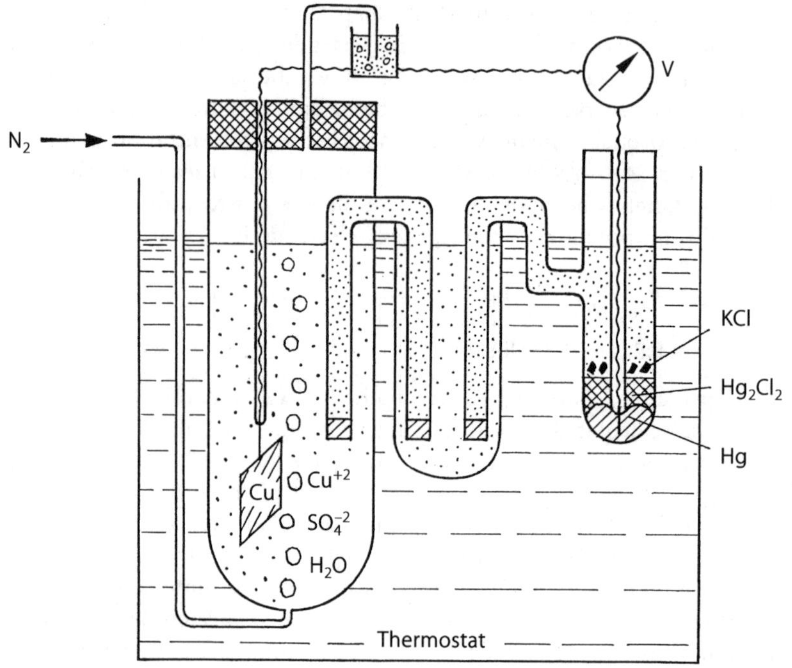


**Fig. 3.5.** Experimental setup for measuring the electrode potential of a $Cu/CuSO_4$-electrode referred to a calomel electrode

be recalculated referred to SHE. This is called the *electrode potential* $\varepsilon$ of the irreversible electrode; hence $\varepsilon = U - E_{\text{ref}}$[3].

When $\varepsilon$ is governed by a corrosion overall reaction, which is the case of interest in our context, we designate it as a *corrosion potential*, $\varepsilon_{\text{corr}}$, and this quantity will play a major role in all of the following. It will be seen in Chap. 4 that for an overall reaction with equilibrium potentials, $E_{\text{Me/X}}$, of the anodic half-reaction turning Me into some chemical compound, X, and $E_{\text{ox/red}}$ of the cathodic half-reaction reducing an oxidant, $\varepsilon_{corr}$ obeys the inequality

$$E_{\text{Me/X}} \leq \varepsilon_{\text{corr}} \leq E_{\text{ox/red}} \tag{3.81}$$

The important subject of Chap. 4 will be to relate corrosion rates and corrosion potentials. In the present chapter, we shall use Eq. (3.81) only to predict directions of reactions, and to predict corrosion products. Such predictions are of considerable interest. As an example we recall from Chap. 2 that in sulfuric acid lead may become fully protected against corrosion by a layer of lead sulfate

[3] Thus, galvanic cell potentials are noted as follows: Generally, $U$ is the potential of a galvanic cell; $U_{rev}$ is the potential of a reversible cell; $E_r$ is the potential of a reversible electrode reaction, labeled $r$, referred to a standard hydrogen electrode (SHE); $\varepsilon$ is the potential of an arbitrary working electrode, referred to SHE. Occasionally, $\varepsilon'$ is used when the reference electrode is not SHE.

produced by an initial reaction with the solution. The $Pb/PbSO_4$ system is an electrode of the second kind, and we can readily calculate conditions of equilibrium between solid lead sulfate and sulfuric acid. If from experience we know that solid sulfate tends to form a protective layer, we can, then, predict conditions of corrosion protection from equilibrium thermodynamics. Of course, the calculations are the same for any metal/metal compound electrode, be the compound protective or non-protective. An important case is iron in the presence of rust, which, as already described in Chap. 2, usually is non-protective. Quite generally, metal/metal oxide and metal/metal hydroxide systems, with protective or, else non-protective oxydic and hydroxydic corrosion products merit special attention, as follows:

For any $Me/Me(OH)_z$ electrode the reaction is

$$Me + zOH^- = Me(OH)_z + ze^- \ . \tag{3.82}$$

We add the equilibrium $H_2O = H^+ + OH^-$ to obtain the equivalent formulation

$$Me + zH_2O = Me(OH)_z + zH^+ + ze^- \ , \tag{3.83}$$

and we see that the equilibrium potential is

$$E_{Me/Me(OH)_z} = E^\circ_{Me/Me(OH)_z} + \frac{RT}{F} \ln a_{H^+} \ , \tag{3.84}$$

$$E^\circ_{Me/Me(OH)_z} = \frac{1}{zF} (\Delta G^\circ_{Me(OH)_z} - z\Delta G^\circ_{H_2O}) \ .$$

For oxides $Me_xO_y$, with valence $z+$ of the cations , and $2y = xz$, the reaction is

$$xMe + yH_2O = Me_xO_y + 2yH^+ + 2ye^- \ , \tag{3.85}$$

and

$$E_{Me/Me_xO_y} = E^\circ_{Me/Me_xO_y} + \frac{RT}{F} \ln a_{H^+} \ ,$$

$$E^\circ_{Me/Me_xO_y} = \frac{1}{2yF} (\Delta G^\circ_{Me_xO_y} - y\Delta G^\circ_{H_2O}) \ . \tag{3.86}$$

Some oxides dissolve in alkaline solutions, forming complex anions. This is observed for *amphoteric metals* such as zinc or aluminum, which in acid solutions dissolve to $Zn^{2+}$ or $Al^{3+}$, and in alkaline solutions to zincate, $Zn(OH)_4^{2-}$], and aluminate, $Al(OH)_4^-$, respectively. In this case, complex anion formation is the result of *hydrolysis* according to: $Me(OH)_z + OH^- = Me(OH)_{z+1}^-$, or

$$Me(OH)_z + H_2O = Me(OH)_{(z+1)}^- + H^+ \ . \tag{3.87}$$

The equilibrium constant is

$$K_{hdrl} = a_{Me(OH)_{(z+1)}^-} a_{H^+} \ . \tag{3.88}$$

For the resulting electrode reaction

$$Me + (z+1)H_2O = Me(OH)_{z+1}^- + (z+1)H^+ + ze^- \ , \tag{3.89}$$

the equilibrium potential is:

$$E_{\mathrm{Me/Me(OH)}_{z+1}^-} = E^\circ_{\mathrm{Me/Me(OH)}_{z+1}^-} + \frac{RT}{zF} \ln a_{\mathrm{Me(OH)}_{z+1}^-} (a_{\mathrm{H}^+})^{(z+1)} , \tag{3.90}$$

$$E^\circ_{\mathrm{Me/Me(OH)}_{z+1}^-} = \frac{1}{zF} \left( \Delta G^\circ_{\mathrm{Me(OH)}_{z+1}^-} - (z+1)\Delta G^\circ_{\mathrm{H_2O}} \right) .$$

Solid oxide deposition can be seen as a hydrolytic reaction step as well, and we also note preceding hydrolytic steps such as $Me^{z+} + H_2O = Me(OH)^{z-1} + H^+$, which will become a point of interest further below.

For pure metals and oxides or hydroxides as reaction products, it is instructive to consider, following Pourbaix et al. [13], the stability of specified compounds as function of solution pH and metal electrode potential, $\varepsilon$, where the potential is the corrosion potential, $\varepsilon_{corr}$, between the equilibrium potentials of anodic metal dissolution and of cathodic reduction of the oxidant. We demonstrate the procedure of constructing *potential-pH diagrams* or *Pourbaix diagrams* for the case of iron in Fig. 3.6. It should be noted that this and other such diagrams should mainly be used as summary overviews rather than as sources of precise and precisely predictive information.

The plot is $\varepsilon$ vs. pH. The equilibrium potentials of the hydrogen and the oxygen electrode as function of pH have been included, since the differences between these and the equilibrium potentials for metal dissolution or oxide/hydroxide formation measure the thermodynamic driving force for acid and for oxygen corrosion. Equilibrium lines for the electrode reactions of iron and iron ions have been calculated using values listed in Table 3.1, taken from [9], but the diagram is essentially the same as the previous version [13], especially as it is presented for $0 \leq \mathrm{pH} \leq 14$ only, ignoring iron dissolution to complex ferrite and/or ferrate ions in solutions of higher pH. For a recent version based on a consistent set of revised thermodynamic data, also considering hydrolysis of dissolved $Fe^{2+}$ to $FeOH^+$ and to dissolved undissociated $Fe(OH)_2$, see [14]. For the extension of the diagram to higher pH values see Chap. 15.

From the possible oxide phases we select, as is standard practice in the corrosion literature, magnetite, $Fe_3O_4$, and hematite, $\alpha$-$Fe_2O_3$, although, as noted in Chap. 2, other oxides and hydroxides both crystalline and amorphous are found in rust in various amounts depending on conditions of exposure of iron to aggressive environments. The choice is arbitrary, especially as it would be erroneous to assume that the oxide with the lowest Gibbs energy of formation always should be expected to be present. Oxide deposition is solid phase formation, requiring nucleation, which may be easier for metastable than for stable phases, especially when favored by crystallografic epitaxy of metastable phase and metal substrate. Also, the solution may tend to supersaturation again as a result of difficult solid-phase nucleation, and then the more soluble metastable phase again would be preferred. This type of sequence of events is, of course, in accord with the Second Law of thermodynamics; it is only the reverse change of a stable into a metastable phase which is forbidden.

Concerning reactions involving dissolved iron species, the calculation usually is made for activities $a_{\mathrm{Fe}}$ varying between 1 and $10^{-6}$ mol $L^{-1}$. The steps in the construction of the diagram then are the following:

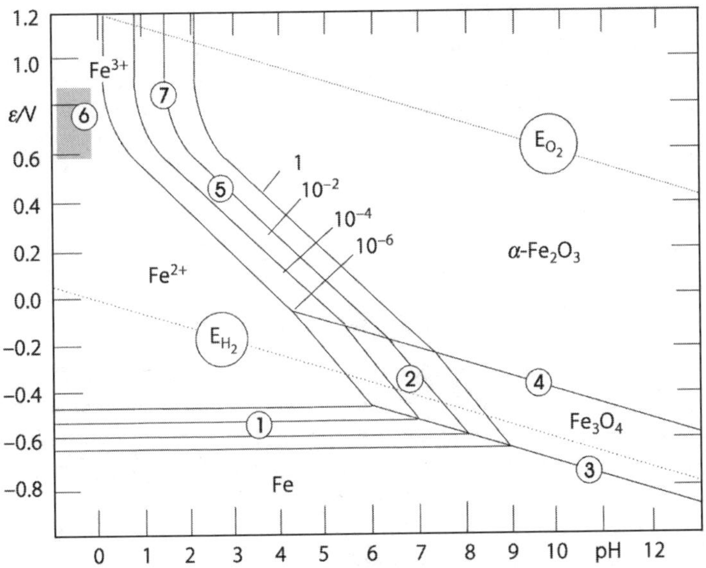


**Fig. 3.6.** Potential-pH-diagram for iron in contact with aqueous solutions at 25°C. Chemical species considered: Fe,$Fe_3O_4$,α-$Fe_2O_3$,$Fe^{2+}$,$Fe^{3+}$,$H^+$,$OH^-$,$H_2O$. $FeOH^+$ is disregarded. Equilibria, numbered as in the text, have been calculated using data from [9]. $E_{H_2}$,$E_{O_2}$: Hydrogen and oxygen electrode equilibrium potentials. Concentrations in mol $L^{-1}$ of dissolved $Fe^{2+}$ or $Fe^{3+}$ ions

1) In acid solutions, the electrode reaction is

$$\mathrm{Fe} = \mathrm{Fe}^{2+} + 2\mathrm{e}^- \,, \tag{3.91}$$

and the equilibrium potential

$$\begin{aligned} E_{\mathrm{Fe/Fe^{2+}}} &= E^\circ_{\mathrm{Fe/Fe^{2+}}} + \frac{RT}{2F} \ln a_{\mathrm{Fe^{2+}}} \\ &= -0.44\,(\pm 0.04) - 0.029 \log a_{\mathrm{Fe^{2+}}}\ (\mathrm{V}) \,. \end{aligned} \tag{3.92}$$

In this equation, and elsewhere in this section, scatter indicated for $E^\circ_r$ does not reflect the limits of a regular error distribution. Instead, it indicates, without weighting confidence levels, the range of values contributed from different authors, as noted previously.

For any value of $a_{\mathrm{Fe^{2+}}}$, the calculated equilibrium potential indicates the border of a domain, $\varepsilon \leq E_{\mathrm{Fe/Fe^{2+}}}$, of thermodynamic stability, i.e., immunity with respect to corrosion. For $\varepsilon \geq E_{\mathrm{Fe/Fe^{2+}}}$ iron is expected to dissolve, and we are in a domain which we may label 'corrosion', although nothing is known about the corrosion rate. When the solution is supposed to contain virtually no $Fe^{2+}$, then $E_{\mathrm{Fe/Fe^{2+}}}$ as calculated for $10^{-6}$ mol $L^{-1}$ is a reasonable estimate of potential below which practically negligible corrosion rates can be expected. Thus, measuring the corrosion potential, $\varepsilon_{\mathrm{corr}}$, alone should be helpful to make predictions, so far regardless of pH.

2) Consider, in the domain labeled 'corrosion', an increase in pH at constant $a_{Fe^{2+}}$. In this direction, the solubility of oxides decreases until solid oxide is deposited from the saturated solution. In the case of Fe-II,III-oxide $Fe_3O_4$, the oxide is the product of an electrode reaction:

$$3Fe^{2+} + 4H_2O = Fe_3O_4 + 8H^+ + 2e^- \ , \tag{3.93}$$

and the equilibrium potential is

$$\begin{aligned} E_{Fe^{2+}/Fe_3O_4} &= E^\circ_{Fe^{2+}/Fe_3O_4} + \frac{RT}{2F} \ln \frac{(a_{H^+})^8}{(a_{Fe^{2+}})^3} \\ &= 0.97\,(\pm 0.12) - 0.089 \log a_{Fe^{2+}} - 0.237\,\mathrm{pH}\,(\mathrm{V})\ . \end{aligned} \tag{3.94}$$

Scatter of the standard potential is mainly caused by the scatter of $\Delta G^\circ_{Fe^{2+}}$-values. The reaction does not involve metallic iron, the metal serving as a sink and source of electrons only. To study the reaction, platinum might be used just as well in contact with solid $Fe_3O_4$ and with the solution containing dissolved $Fe^{2+}$.

Depending on the choice of values for $\Delta G^\circ_{Fe^{2+}}$- and $\Delta G^\circ_{FeOH^+}$, which both are not very well known, $FeOH^+$ may or may not be an important species beyond about pH 7. The effect would be to lower $a_{Fe^{2+}}$ in favor of $a_{FeOH^+}$. Assuming total constant activity $a_{Fe} = a_{Fe^{2+}} + a_{FeOH^+}$ and $a_{Fe} = 10^{-6}$ mol $L^{-1}$, oxide deposition would be expected to shift to a pH-value higher by ca. 0.5 or 1.0 units. This complication is neglected in the following; it will, however, reappear when elevated-temperature diagrams are considered in the following section.

3) When, in the domain of $Fe_3O_4$- stability, $\varepsilon$ is lowered at constant pH, the equilibrium of the reaction

$$3Fe + 4H_2O = Fe_3O_4 + 8H^+ + 8e^- \tag{3.95}$$

is reached at the equilibrium potential

$$E_{Fe/Fe_3O_4} = E^\circ_{Fe/Fe_3O_4} + \frac{RT}{F} \ln a_{H^+} = -0.086 - 0.059\ \mathrm{pH}\ (\mathrm{V})\ . \tag{3.96}$$

Below this potential, metallic iron once more is the stable phase, i.e., thermodynamically immune with respect to corrosion.

4) When, instead, $\varepsilon$ is increased in the domain of $Fe_3O_4$-stability, the equilibrium of oxidation of magnetite to hematite is reached, with equations of the reaction and of the equilibrium potential as follows:

$$2Fe_3O_4 + H_2O = 3\alpha\text{-}Fe_2O_3 + 2H^+ + 2e^- \ , \tag{3.97}$$

$$\begin{aligned} E_{Fe_3O_4/\alpha\text{-}Fe_2O_3} &= E^\circ_{Fe_3O_4/\alpha\text{-}Fe_2O_3} + \frac{RT}{F} \ln a_{H^+} \\ &= 0.203 - 0.059\ \mathrm{pH}\ (\mathrm{V})\ . \end{aligned} \tag{3.98}$$

5) Dissolved $Fe^{2+}$ can be oxidized to hematite by the electrode reaction

$$2Fe^{2+} + 3H_2O = \alpha\text{-}Fe_2O_3 + 6H^+ + 2e^- \;, \tag{3.99}$$

if the potential is lowered to values below

$$E_{Fe^{2+}/\alpha\text{-}Fe_2O_3} = E^\circ_{Fe^{2+}/\alpha\text{-}Fe_2O_3} + \frac{RT}{2F} \ln \frac{(a_{H^+})^6}{(a_{Fe^{2+}})^2}$$

$$= 0.72(\pm\, 0.04) - 0.059 \log a_{Fe^{2+}} - 0.178\,\text{pH (V)} \;. \tag{3.100}$$

Steps 4 and 5 again do not involve metallic iron; therefore again, the half-cell representing the respective electrode potential is platinum in a solution in equilibrium with both solid magnetite and hematite, or with hematite alone, respectively.

Two more steps not involving metallic iron remain to be considered:

6) The reaction

$$Fe^{2+} = Fe^{3+} + e^- \tag{3.101}$$

is readily established e.g. at a platinum electrode in a solution containing both $Fe^{2+}$ and $Fe^{3+}$, and

$$\begin{aligned} E_{Fe^{2+}/Fe^{3+}} &= E^\circ_{Fe^{2+}/Fe^{3+}} + \frac{RT}{F} \ln \frac{a_{Fe^{3+}}}{a_{Fe^{2+}}} \\ &= 0.771 + 0.059 \log \frac{a_{Fe^{3+}}}{a_{Fe^{2+}}} \text{ (V)} \;. \end{aligned} \tag{3.102}$$

In equilibrium, the ratio $a_{Fe^{3+}}/a_{Fe^{2+}}$ is 1 for $\varepsilon = 0.771$(V), indicating that $Fe^{3+}$ is a strong oxidizing agent. The $Fe^{2+}/Fe^{3+}$ couple is a typical *redox couple*, representing an electrode reaction involving a change in the valence of an ion together with a rearrangement of the ion's hydration shell only, and has a high equilibrium exchange rate. The figure indicates the levels of $\varepsilon$, where the activity ratio varies between 0.01 and 100.

7) When, finally, beyond $E_{Fe^{2+}/Fe^{3+}}$, the activity $a_{Fe^{2+}}$ is negligible, $\alpha$-$Fe_2O_3$ is precipitated by the reaction

$$2\,Fe^{3+} + 3\,H_2O = \alpha\text{-}Fe_2O_3 + 6\,H^+ \;. \tag{3.103}$$

For this reaction, the equilibrium lines calculated and plotted in the figure for varying values of $a_{Fe^{3+}}$ once more are approximate only, due to the scatter of values reported for $\Delta G^\circ_{Fe^{3+}}$. At the same time, hydrolysis of $Fe^{3+}$ to $FeOH^{2+}$ apparently does not, at room temperature, interfere with oxide deposition.

The original choice of magnetite and hematite as oxide phases to be considered was mainly based on the results of immersion tests with iron coupons mostly in harmless solutions such as, among others, tap water, or even sodium bicarbonate solutions as model tap water. The observation was that at pH values between 8 and 12 the test specimen remained unattacked, and this was explained as being caused by protective magnetite and/or hematite layers. This then was labeled *passivity*, and it is mainly from these observations that it has become standard practice to use the same label for other pure metals in $\varepsilon$-pH domains indicating oxide stability. Passivity of metals, including passivity of alloys, will become a topic of major concern in later chapters. The effect is always due to microscopically thin layers of oxides of the respective metal, and sometimes these passivating oxide films are, in fact, in equilibrium with the solution. A prominent case is steel in moist concrete, where pH is 12.6, and where passivity is due to thermodynamically stable magnetite. A further important case is power-plant heat-exchange tubing, where steel in contact with neutral or weakly alkaline high-temperature water is protected again by thermodynamically stable magnetite. Examples of applicability of calculations used for constructing $\varepsilon$-pH diagrams thus are certainly available. At the same time, iron in 0.5 M $H_2SO_4$-solution is an intensively studied model system showing passivity of the metal, and yet the passivating oxide, which is $\gamma$-$Fe_2O_3$, is soluble in the solution (see Chap. 10). On the other hand, iron rust is the prominent example of a thermodynamically metastable mixture of oxides which is non-protective for reasons already explained in Chap. 2. Domains of so-called passivity in $\varepsilon$-pH diagrams should, therefore, only tentatively be labeled as domains of possible corrosion protection by thin layers of oxidic corrosion products, including, sometimes, passivation.

Concerning the success of immersion tests, such as tests in tap water, we recall remarks in Chap. 2 on the role of air-formed oxide films actually formed by dry oxidation. If the film remains unattacked, the metal will indeed appear passive. Starting, instead, from an oxide-free surface the important point probably is that passivation requires oxide film growth by a mechanism similar to that of dry oxidation rather than by the metal-dissolution/oxide-precipitation mechanisms producing porous rust. Such a process of initial oxidation should become more probable with decreasing conductivity of the solution, because then the electrolytic mechanism of corrosion becomes increasingly difficult. Detailed information so far is lacking but may become available from in-situ tunnel microscopy. A further suggestion would be to regard passivation as the limiting case of the electrolytic redeposition mechanism, when the distance of redeposition from the metal surface tends to zero. As described in Chap. 12, this quantitatively describes conditions of passivation of lead by dichromate solutions [15]. All this is kinetics rather than thermodynamics, and left to description further below.

Whenever passivating films are pure oxides in equilibrium with the adjacent solution, a measurement of $\varepsilon$ and pH should suffice to confirm safe corrosion protection. If instead, for reasons of kinetics, passivation fails owing to oxide porosity, or by such processes as pitting, crevice corrosion, and corrosion cracking, then, of course, kinetics still cannot contradict thermodynamics. A study of local values of $\varepsilon$ and pH then necessarily must reveal that nothing is happening against the Second Law, as was shown for a number of examples already in the early literature [13].

With respect to the $Fe\text{-}H_2O$ diagram, we add a note concerning the electrode reaction producing solid $Fe(OH)_2$, and the corresponding equilibrium potential:

$$Fe + 2H_2O = Fe(OH)_2 + 2H^+ + 2e^- \ , \tag{3.104}$$

$$\begin{aligned} E_{Fe/Fe(OH)_2} &= E^\circ_{Fe/Fe(OH)_2} + \frac{RT}{F} \ln a_{H^+} \\ &= -0.08\,(\pm 0.01) - 0.059 \text{ pH (V)} \ . \end{aligned} \tag{3.105}$$

Comparison with Eq. (3.96) shows that, possibly, $Fe(OH)_2$ is slightly more stable than $Fe_2O_3$. This has been claimed to be significant [14], with the added supposition that $Fe(OH)_2$ is less protective than magnetite. Here, as generally, the rating of merits of corrosion product as being more or less protective, even if perhaps based on experimental evidence, certainly is non-thermodynamic. Also, the original supposition that, instead, $Fe(OH)_2$ is an intermediate metastable reaction product preceding magnetite formation still does seem quite as attractive [16].

As products of aqueous corrosion, pure oxides are the exception rather than the rule. More complicated compounds such as basic carbonates or basic chlorides produced on such metals as copper and on zinc by atmospheric corrosion require calculations in more dimensions than potential and pH only. At the same time, equilibrium conditions for reactions such as malachite formation on copper,

$$2Cu^{2+} + HCO_3^- + 2H_2O = CuCO_3 \cdot Cu(OH)_2 + 3H^+ \ , \tag{3.106}$$

can be readily evaluated, and they can be represented, if desired, in $\varepsilon$-pH diagrams at various constant values of bicarbonate activity [13]. We do not follow this line of argument and only note that such calculations, previously cumbersome, are now rapidly executed with the aid of computerized banks of thermodynamic data using algorithms for the search for the Gibbs energy minimum, where $\Delta_r G = 0$, of multi-component, multi-reaction systems.

As a last item in this section we consider conditions of equilibrium between a homogeneous binary alloy and an adjacent electrolyte solution. As noticed further above, this does not normally concern solid alloys because selective corrosion of alloy components will change the surface composition and destroy bulk/surface equilibria. The situation is different for liquid alloys of mercury with such metals as sodium, or zinc, or indium, which can be mechanically stirred, ensuring the desired equilibria. Such liquid *amalgams* have been important for the development of fundamental corrosion science, as described in Chap. 4. An example is zinc amalgam with only a few atom percent of zinc, that is, at low values of the gramatom fraction $\gamma_{Zn}$ and corresponding high values of $\gamma_{Hg} = 1 - \gamma_{Zn}$. These are very nearly ideal alloys with activity coefficients $f_{Zn} \simeq f_{Zn} \simeq 1$. Equilibrium in contact with a solution containing dissolved $Hg_2^{2+}$ as well as $Zn^{2+}$ requires that the electrode is at equilibrium both as a $Hg/Hg_2^{2+}$ and as a $Zn/Zn^{2+}$ electrode; hence

$$E_r = E^\circ_{Hg/Hg_2^{2+}} + \frac{RT}{2F} \ln \frac{a_{Hg_2^{2+}}}{1 - \gamma_{Zn}} = E^\circ_{Zn/Zn^{2+}} + \frac{RT}{2F} \ln \frac{a_{Zn^{2+}}}{\gamma_{Zn}} \ . \tag{3.107}$$

Substituting values from Table 3.3, the equilibrium concentration of dissolved mercury is found to be negligible for any realistic concentration $\gamma_{Zn}$:

$$a_{Hg_2^{2+}} = 3.0 \times 10^{-54} \frac{1-\gamma_{Zn}}{\gamma_{Zn}} a_{Zn^{2+}} \,. \tag{3.108}$$

It follows that for an initial value $a^\circ_{Hg_2^{2+}} = 0$ there is almost no reaction, $2Hg + Zn^{2+} \rightarrow Hg_2^{2+} + Zn$, required to establish equilibrium. Therefore, the equilibrium potential essentially is $E_{Zn/Zn^{2+}}$ as calculated for the initial values of composition of amalgam and solution. Since the equilibrium exchange rate of the zinc electrode reaction is high, the equilibrium is, in fact, readily established.

## 3.6 Calculating Elevated-Temperature Gibbs Energies

The calculation of elevated-temperature standard Gibbs reaction energies, $\Delta_r G^\circ$, and, from these, equilibrium constants and equilibrium electrode potentials, based on room-temperature data and on heat capacity data is routine in physical chemistry. Reviews of elevated-temperature thermodynamics of corrosion products are also available [13c,17–19]. In the present context, the problem is simplified without unduly adding to inaccuracy, if an approximation concerning the influence of pressure is accepted: At elevated temperatures, such as e.g. 600 K in power-plant feed-water loops, the water vapor pressure is, of course, far beyond 1 bar; it is, e.g., 88 bar, and we may take this value as the pressure of dilute solutions at 600 K as well. Still, as long as only solid or liquid phases participate in reactions, the partial reaction volume $\Delta_r V$ in Eq. (3.2) is a small quantity because the compressibility of the phases is low. We therefore disregard, for condensed phases, the influence of total pressure on $\Delta_r G$, and consider pure condensed phases as always being approximately in their standard state although the pressure exceeds 1 bar. Pressurized liquid water, instead of water vapor at 1 bar, is also considered to be in its elevated-temperature standard state, and solutions are assumed to be dilute, so that the solute water also does not much depart from standard state conditions. For concentrated geological brines, this assumption probably has to be checked for a possibly higher level of inaccuracy.

As before, we refer electrode potentials to the standard hydrogen electrode at pH = 0, at partial pressure $p_{H_2} = 1$ bar in the high-pressure water vapor, and at the desired temperature, $T$. The reference potential, $E^\circ_{H_2/H^+}$, is zero by definition, regardless of temperature. Should we require that the reference electrode reaction involves evolution of gaseous hydrogen against the high vapor pressure, $E_{H_2/H^+}$ would be more negative than $E^\circ_{H_2/H^+}$ by $(RT/F)\ln\sqrt{p_{H_2O}}$ (V). We note that at 600 K, $RT/F$ is 0.052 V, and $2.303 \times RT/F = 0.119$ V.

As before, the standard Gibbs reaction energy is given by Eq. (3.28), where the standard Gibbs energies of formation are now to be taken at the desired elevated temperature. Normally, this requires a calculation of the change in Gibbs energies over the range of temperatures between $T_\circ = 298.15$ K and $T$. In the following, we describe a basic step-by step procedure, disregarding available and useful faster algorithms.

From Eq. (3.4b) it follows that we have to evaluate, for every $\Delta G_i^\circ$, the integral

$$(\Delta G_i^\circ)_T = (\Delta G_i^\circ)_{T_\circ} - \int_{T_\circ}^{T} \Delta S_i^\circ \, dT \,. \tag{3.109}$$

$\Delta S_i^\circ$ is the temperature-dependent stoichiometric sum, $\sum \nu_j S_j^\circ$, of contributions from the participants $j$ in the formation of species $i$ from its parent elements. As entropy essentially is heat divided by temperature, it is probably obvious that, excluding phase transformations such as recrystallisation or melting, the expression for the standard-state entropy is

$$(S_j^\circ)_T = (S_j^\circ)_{T_\circ} + \int_{T^\circ}^{T} \frac{C_p(j)}{T} \, dT \,, \tag{3.110}$$

where $C_p(j)$ is the constant-pressure specific heat capacity of the substance $j$ in $\mathrm{J\,mol^{-1}}$. Phase transformations would add terms $\Delta H_{trs}/T_{trs}$, with transformation enthalpies $\Delta H_{trs}$, and transformation temperatures $T_{trs}$, but will not be considered in the following. Heat capacities, $C_p(j)$, depend on temperature; but for differences in temperature of only some 300 K the change is not very important, and the integral $\int C_p(j) dT/T$ can be replaced by the expression $\overline{C}_p(j) \int dt/T$, where $\overline{C}_p(j)$ is the heat capacity averaged over the range of temperatures. Therefore,

$$(S_j^\circ)_T = (S_j^\circ)_{T_\circ} + \overline{C}_p(j) \ln \frac{T}{T_\circ} \,. \tag{3.111}$$

Introducing the stoichiometric sum $\Delta C_p(i) = \sum \nu_j C_p(j)$, we eventually arrive at:

$$\begin{aligned}(\Delta G^\circ)_T = {} & (\Delta G_i^\circ)_{T_\circ} - (\Delta S_i^\circ)_{T_\circ} (T - T_\circ) \\ & + \overline{\Delta C}_p(i) (T - T_\circ) - T \overline{\Delta C}_p(i) \ln \frac{T}{T_\circ} \,. \end{aligned} \tag{3.112}$$

When specific heat data are available over the range of temperatures considered, the evaluation is straightforward. As an example, we calculate for pressurized liquid water, but disregarding pressure, $\Delta G_{H_2O}^\circ$ at 600 K, using data from [17]: $(\Delta G_{H_2O}^\circ)_{298.15} = -237.1$ kJ mol$^{-1}$, $(\Delta S_{H_2O}^\circ)_{298.15} = -163.3$ J K$^{-1}$ mol$^{-1}$. Neglecting $\Delta C_p$ terms we obtain $(\Delta G_{H_2O}^\circ)_{600} = -187.90$ kJ mol$^{-1}$. Considering $\Delta C_p$ terms we use $\overline{C}_p(H_2) = 29.1$, and $\overline{C}_p(O_2) = 30.7$ J mol$^{-1}$. Extrapolating from lower-temperature data we assume $\overline{C}_p(H_2O) = 76.4$ kJ mol$^{-1}$, so that $\overline{\Delta C}_p(j) = 31.9$ kJ mol$^{-1}$. From this a sufficiently accurate value $(\Delta G_{H_2O}^\circ)_{600} = -191.7$ kJ mol$^{-1}$ is calculated.

We also note that, at 400, 500, and 600 K, the equilibrium constant of water dissociation, Eq. (3.71), is $5.0 \times 10^{-13}$, $5.0 \times 10^{-12}$, and $5.0 \times 10^{-12}$, respectively.

For most of the interesting pure substances, elevated-temperature Gibbs energies, obtained using Eq. (3.112), are tabulated [17]. As an example, we once more turn to the Fe/$H_2O$-system, and we find, for 600 K, $\Delta G^\circ_{Fe_3O_4} = -914{,}4$ and $\Delta G^\circ_{\alpha\text{-}Fe_2O_3} = -661.4$ kJ mol$^{-1}$. Directly proceeding to the calculation of equilibrium electrode potentials we calculate from these data that

$$(E_{Fe/Fe_3O_4})_{600} = -0.191 - 0.119 \text{ pH (V)} , \tag{3.113}$$

$$(E_{Fe_3O_4/\alpha\text{-}Fe_2O_3})_{600} = 0.187 - 0.119 \text{ pH (V)} . \tag{3.114}$$

Comparing these expressions with Eqs. (3.96) and (3.100) we see that the change in standard electrode potentials caused by a 300 K-increase in temperature is not very important.

For an estimate of the 600 K-value of $\Delta G^\circ_{Fe/Fe^{2+}}$ we again apply the first step suggested by Eq. (3.112), using, as in Sect. 3.2, $\Delta_{Fe/Fe^{2+}} S^\circ = -19\,(\pm 12)$ J K$^{-1}$ mol$^{-1}$, and the result is

$$(E_{Fe/Fe^{2+}})_{600} = -0.41\,(\pm\,0.05)) + 0.060 \log a_{Fe^{2+}} . \tag{3.115}$$

In this case, as always when ions participate in a reaction, the second step of using Eq. (3.114), which is considering $\Delta C_p$-contributions, is not directly applicable, since partial molar heat capacities of ions are not measurable. Several propositions have been made to overcome this obstacle with non-thermodynamic assumptions. One of these, Criss and Cobble's 'correspondence principle' implies the use of absolute Third Law entropies of ions, estimated using Eq. (3.33) and assuming, as noted in Sect. 3.2, $S^{\circ\star}_{H^+} = -20.9$ J K$^{-1}$ mol$^{-1}$ [20]. The next step is to assume that for ions

$$(S^\circ_i)_T = a_T + b_T\,(S^{\circ\star}_i)_{298.15} , \tag{3.116}$$

where $a$ and $b$ depend on $T$ and on the type of ions considered [20]. For the Fe/$Fe^{2+}$-couple, the calculation gives $E^\circ_{Fe/Fe^{2+}} = -0.39$ (V) [19] and, thus, at 600 K, $\Delta G^\circ_{Fe^{2+}} = -75.3$ kJ mol$^{-1}$. Since the second approximation is not much different from the first, we may pass more elaborated propositions for the calculation of ionic entropies. References can be found in [14].

With this value for $\Delta G^\circ_{Fe^{2+}}$, the equilibrium potentials of the $Fe^{2+}/Fe_3O_4$ couple and of the $Fe^{2+}/\alpha\text{-}Fe_2O_3$ couple are calculated as

$$(E_{Fe^{2+}/Fe_3O_4})_{600} = 0.41\,(\pm\,0.12) - 0.180 \log a_{Fe^{2+}} - 0.476 \text{ pH} , \tag{3.117}$$

$$(E_{Fe^{2+}/\alpha\text{-}Fe_2O_3})_{600} = 0.33\,(\pm\,0.08) - 0.119 \log a_{Fe^{3+}} - 0.357 \text{ pH} . \tag{3.118}$$

Finally, we use the first-approximation estimate that at 600 K $\Delta G^\circ_{Fe^{3+}} \simeq$ 72.2 kJ mol$^{-1}$, and arrive at an approximate expression for the equilibrium constant of $\alpha$-$Fe_2O_3$-precipitation from $Fe^{3+}$-solutions:

$$(K_{\alpha\text{-}Fe_2O_3})_{600} = \frac{(a_{H^+})^6}{(a_{Fe^{3+}})^2} \simeq 1.6 \times 10^{10} , \tag{3.119}$$

which completes the recalculation of data used in Sect. 3.5 for the Fe/$H_2O$-system. The task of reconstructing the diagram is, however, not yet accomplished: For 600 K we can no longer assume that the total activity of dissolved iron species is approximately equal to either $a_{Fe^{2+}}$ or to $a_{Fe^{3+}}$. Instead, we now must recognize that hydrolysis reactions, starting with $Fe^{2+} + OH^- = FeOH^+$, followed by further steps of reactions with water, precede oxide precipitation. The effect is due to an increase of the dielectric constant of water with temperature causing an increase in Coulomb interaction between ions, favoring ion pairing including hydrolysis. Therefore, the activities $a_{Fe^{2+}}$ and $a_{Fe^{3+}}$ are, already at low pH, distinctly lower than the total content of dissolved iron species, and the onset of oxide deposition is shifted to higher values of pH. The effect is most pronounced at low values of of total dissolved iron content, such as $10^{-6}$ mol $L^{-1}$. The standard entropies of hydrolized ions once more show considerable scatter (see Table 3.1), and the construction of the diagram tends to become meaningless if accumulated scatter from various sources is respected. In order to obtain a more useful compilation of data it needs the choice of a consistent set of data, though the choice may involve arbitrary assumptions as to the oxide phases precipitated, and arbitrary fit parameters for estimating entropies, and thus always remains open to discussion. Also, accounting for consecutive steps of hydrolysis rapidly increases the amount of calculations necessary, especially if redissolution of oxides in alkaline solutions is also considered. All this, then, is beyond the scope of an outline of fundamentals and will not be pursued presently. Instead, reference is made to recent work where the full apparatus of computerized data handling has been used to reinspect the Fe-$H_2O$ system at temperatures between 298.15 and 600 K, taking into account all steps of hydrolysis up to $Fe(OH)_4^-$, and $Fe(OH)_4^{2-}$ for the full range of pH between 0 and 12, with some interesting changes of data as compared with previous simpler calculations [14]. The interpretation in terms of effects of high-pressure, high-temperature aqueous solutions on the corrosion of unalloyed steel is, however, as usual difficult due to lack of experimental evidence. In view of rather demanding safety considerations involved, more such evidence would be desirable.

The electrode potential, $\varepsilon$, of the working electrode, which is, e.g., unalloyed steel in contact with high-pressure high-temperature water, usually will not be measured using a reference hydrogen electrode, but using some suitable second-kind electrode, such as the Ag/AgCl-electrode, allowing the calculation of $\varepsilon$ from its own known potential referred to SHE at temperature $T$. Reference electrodes used at working conditions are called 'internal'. For experimental details, see the literature [21]. An interesting point arises when, instead, 'external' reference electrodes are used at ambient temperature $T_\circ$, and the measuring circuit thus is non-isothermal. We consider the complications only for the simple case of two standard hydrogen/platinum electrodes at different temperatures. At zero current, four additional contributions to the cell potential are possible. These are: 1) The metallic thermocouple potential between two Cu/Pt junctions at different temperatures, 2) the Thomson potential caused by a gradient in temperature in homogeneous metallic conductors, 3) the Soret potential or thermal junction potential, caused by temperature-gradient induced thermodiffusion in electrolyte solutions, and 4) the true change in Galvani potential at the platinum/solution interfaces with temperature. The first two effects are small. As reviewed in [19], the true change in $E_{H_2/H^+}$ with temperature is probably of the order 1 mV $K^{-1}$ [22]. Similar to the diffusion potential, the thermal junction potential depends on structure and composition of the junction. Apparently, it is negligible when KCL-salt solution bridges are used; it is between 0.01 and 0.05 mV $K^{-1}$ for other electrolytes, and may be up to 0.5 mV $K^{-1}$ for more concentrated acids or bases [22].

pH is either calculated using the product $c_{H^+} f_{H^+}$ of measurable concentration and activity coefficient, with the latter obtained for dilute solutions from the extended Debye-Hückel theory (see Sect. 3.2) [23]; or else it is measured using pH-sensitive electrodes. At elevated temperatures, glass electrodes are no longer useful; instead, recourse is made to the more classical method of using the pH-dependence of metal/metal oxide electrodes, with oxides, especially zirconium dioxide containing yttrium oxide [24,25]. Very carefully measured, data on pH are accurate within some 0.01 units; more often, the accuracy may be considerably lower.

## References


1. Jost, W. and Troe, J.: Kurzes Lehrbuch der physikalischen Chemie. 18. Aufl., Verlag Steinkopf, Darmstadt 1973
2. Atkins, P. W.: Physical Chemistry. 6th ed., Oxford University Press, Oxford 1998
3. Wedler, G.: Lehrbuch der physikalischen Chemie, 3. Aufl., VCH Verlagsges., Weinheim, 1987
4. Lange, E. and Göhr, F.: Thermodynamische Elektrochemie. Hüthig Verl., Heidelberg 1962
5. Kortüm, G.: Lehrbuch der Elektrochemie. 5. Aufl., Verlag Chemie, Weinheim 1972
6. Comprehensive Treatise of Electrochemistry. (Bockris, J.O'M., Conway, B.E., Yeager, E., and R.E. White, eds.). Plenum Press, New York and London, 1981
7. Bockris, J. O'M. and Khan, S.U. M.: Surface Electrochemistry, a Moleculear Approach. Plenum Press, New York, London 1993
8. Bagotzki, V.S.: Fundamentals of Electrochemistry. Plenum Press, New York 1993 (Khimiya Press, Moscow 1988)
9. Bard, A.J., Parsons, R., and Jordan, J.: Standard Potentials in Aqueous Solutions. Int. Union of Pure and Applied Chemistry, Marcel Decker, Inc., New York, Basel 1985
10. Kielland, J.: Amer. Chem. Soc. *59* (1937) 1675
11. Zemaitis, J.F., Clark, D.M., Rafal, M., and Scriverner, N.C.: Handbook of Aqueous Electrolyte Thermodynamics. American Inst. of Chemical Engineers, Inc., New York 1986
12. Bates, R.C.: Determination of pH. Theory and Practice. J. Wiley & Sons, New York etc., 1964, 1973

13. a) Pourbaix, M. et al.: Atlas d'Équilibres Électrochimiques a 25°C. Gauthier-Villars, Paris, 1963 (English translation: NACE, Houston, CEBELCOR, Bruxelles, 1974); b) Pourbaix, M.: Leçons en Corrosion Électrochimiques, 2. édn., CEBELCOR,Bruxelles, 1975; c) v. Muylder, J.: Thermodynamics of Corrosion; in: loc. cit. [6], Vol. 4 (Electrochemical Materials Science)
14. Beverskog, B. and Puigdomenech, I.: Corr. Sci. *38* (1996) 2121
15. Rickert, H. and Holzaepfel, G.: a) Ber. Bunsenges. Phys. Chemie *70* (1966) 171); b) Werkstoffe u. Korr. *17* (966) 376
16. Schikorr, G.: Z. Elektrochem. *25* (1929) 65
17. Barin, J.: Thermochemical Properties of Pure Substances. VCH-Verlags-Ges., Weinheim, 1989
18. Townsend, H.E.: in: Proc. 4th Int. Congr. Metallic Corrosion, Amsterdam. NACE, Houston, 1972. p. 477
19. Macdonald, D.D.: The Electrochemistry of Metals in Aqueous Systems at Elevated Temperatures; in: Modern Aspects of Electrochemistry. (Conway, B.E. and Bockris, J.O'M., eds.). No. 11. Plenum Press, New York and London, 1975, p. 141
20. Criss, C.M. and Cobble, J.W.: J. Amer. Chem. Soc. *86* (1964) 5385, 5390
21. a) Macdonald, D.D.: J. Electrochem. Soc.*126* (1979) 1618; b) MacDonald, D.D., Scott, A.C., and Wentrcek, P.: Corr. NACE *34* (1978) 75
22. de Bethune, A. Licht, S., and Swendeman, N.: J. Electrochem. Soc. *106* (1959) 616
23. Staples, B.R., Holcomb, G.R., and Cramer, S.D.: Corrosion NACE, *48* 1992) 35
24. Makela, K., Hettiarachchi, S., and MacDonald, D.D.: in: Proc. Int. Conf. Interaction of Iron Based Materials with Water and Steam, Heidelberg, 1992, APRI Palo Alto, VGB, Essen, 1993, Nr. 11
25. Danielson, M.J., Koski, O.H., and Myers, J.: ibid. *132* (1985) 1866

# 4 The Electrolytic Mechanism of Corrosion

## 4.1
## Overview

In this chapter, we shall verify the supposition that corrosion proceeds by the superposition of kinetically independent electrode half-reactions: anodic metal dissolution and cathodic reduction of an oxidizing species, such as hydrogen ions, or oxygen molecules, or others.

Historically, the concept was based on observations of corrosion by obvious action of short-circuited galvanic cells, with distinct anodes and cathodes as visible places of of separate metal dissolution and oxidant reduction. A good example is acid corrosion of pure zinc: Immersed in cold, deaerated sulfuric acid solution the metal is not attacked, and there is no visible hydrogen evolution. When, however, zinc is contacted with a platinum wire in the same solution, hydrogen evolution is seen at the platinum surface, and this must be accompanied by zinc dissolution. This is a case of *contact corrosion* (see Chap. 11), and schematically shown in Fig. 4.1. In this model arrangement, the short-circuit electric current can be measured using a null-resistance ammeter switched into the external zinc/platinum connection. Owing to the *condition of electroneutrality*, a current of equal strength, but opposite sign, is passing through the cell, carried across the zinc/acid interface by zinc ions, across the acid/platinum interface by electrons, and by all ions present in the electrolytic

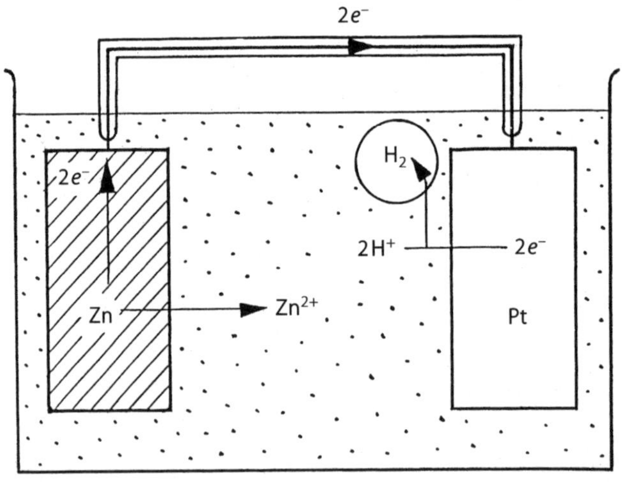


**Fig. 4.1.** Reaction scheme of a short-circuited zinc/acid solution/platinum galvanic cell

solution. The current measured in the external circuit must be equivalent to the rate of anodic zinc dissolution, as determined by chemical analysis of the dissolved amount of zinc in a given time interval, and to the rate of hydrogen evolution available by measuring hydrogen evolution with a manometer. The calculations use *Faraday's law*, as explained in Sect. 4.2.

We note that impure zinc would be seen to undergo acid corrosion already without contact corrosion. Even pure zinc rapidly dissolves in hot concentrated hydrochloric acid solution. In both cases, the dissolution kinetics come close to what is *uniform corrosion*. True uniform corrosion will be discussed later; before, we continue sketching history. In this connection, early work by de La Rive, and especially by Palmaer [1] has been important, although it was based on the erroneous assumption that galvanic cell action is the principle of aqueous corrosion. Cases of obvious uniform corrosion without separation of macroscopic anodes and cathodes then where explained by the action of what was, and still is, termed *local cells*. By definition, local cells are galvanic cell with tiny, microscopic anodes and cathodes present on the metal surface. There are, in fact, many cases of local cell action between, e.g., the matrix of a metal and heterogeneous inclusions, such as cementite inclusions in steel, or graphite inclusion in cast iron. Quite often, local cell action is indeed important, and may even quantitatively account the rate of corrosion. It still is erroneous to extend the local-cell concept to true uniform corrosion, where the metal is anode and cathode simultaneously. In the limiting cases of uniform corrosion case local electrodes would be single surface atoms, but this is a misconception, because parts of galvanic cells must be phases in the sense of thermodynamics [2].

In neutral electrolyte solutions the predominant type of corrosive attack of metals is oxygen corrosion. In this case again, galvanic cell action often governs the mechanism of corrosion, and in this connection early extensive investigations by Evans et al. have been fundamental for the development of corrosion science [3]. The metal anode usually was plain steel in contact with nickel, copper, and lead acting as cathodes. The latter metals, if immersed alone in an aerated neutral salt solution, would themselves be subject to oxygen corrosion; in contact with steel they are *cathodically protected*, which is an electrolytic effect needing an explanation of its own given in Chap. 5. By this effect, these metals act, coupled to steel, as inert sources of electrons for oxygen reduction only. Oxygen is reduced at the steel surface as well as on the cathodes. The mechanism of such galvanic cells is depicted schematically in Fig. 4.2, simplifying a rather more complex sequence of events occurring during oxygen corrosion of iron already described in Chap. 2.

The breakthrough of physicochemical kinetics applied to electrolytic corrosion came by the investigation of acid corrosion of liquid amalgams originally by Frumkin [4] and especially by Wagner and Traud [5]. Liquid amalgams have absolutely uniform surfaces, and there is no question of a separation of the surface into anodic and cathodic areas. Still, as shown presently, the mechanism of corrosion is electrolytic in the sense that anodic metal dissolution and cathodic hydrogen deposition are independent processes, with elementary events of both

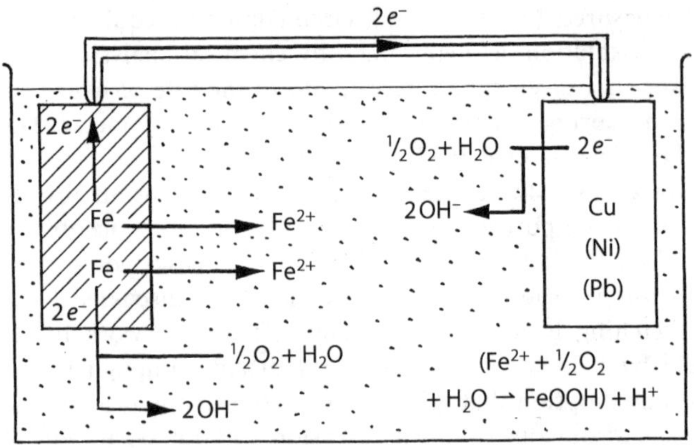


**Fig. 4.2.** Reaction scheme of a short-circuited galvanic iron/copper(nickel,lead) cell in a neutral aerated salt solution

half-reactions occurring in random and continuously changing distribution over the surface.

In any case the important notion is to see electrode reactions as electric currents, $j_r$, carrying charges across metal/solution interfaces, i.e., across electric double layers, and thus passing Galvani potential differences, $\phi_{m,l}$, (see Chap. 3). Therefore, it is intuitive that the current density, $i_r$, referred to the metal surface area, of electrode-reaction currents and the Galvani potential differences across the interfaces are interrelated. Unfortunately, these interrelations are seriously flawed because, as explained previously, single differences, $\phi_{m,l}$, of Galvani potentials cannot be determined experimentally. We shall see that we can, however, measure changes $di_r/d\phi_{m,l}$ since experiments can be controlled so that these differentials are equal to $di_r/d\varepsilon$, where $\varepsilon$ is the measurable electrode potential. Hence the fundamental importance of functions relating electrode reaction current densities and electrode potentials. In the following, we shall usually abbreviate the correct designation of these functions, which is *current density-electrode potential functions*, to *current-potential functions* or *current-potential curves*. The quantitative expressions of these functions, $i(\varepsilon)$, or $\varepsilon(i)$, will be seen to dominate corrosion kinetics.

To illustrate the role of current-potential curves for corrosion processes we once more return to the functioning of a galvanic cell representing a corrosion reaction with metal dissolution and oxidant reduction occurring at separate electrodes. A well-suited and well-investigated example is a cell modeling oxygen corrosion by a mechanism of considerable practical importance, which is coupling, as the anode, iron in a neutral deaerated solution, with, as the cathode, platinum in an aerated solution of otherwise the same composition, as described in the following section.

## 4.2
## Reactions, Currents, and Potentials in Galvanic Cells

The experimental setup used to study a short-circuited cell modeling *corrosion by differential aeration* is shown in Fig. 4.3 [6a]. In this work, one half-cell, the anode, was iron or plain steel in a 0.1 M NaCl solution continuously flushed with nitrogen and thus kept free of dissolved oxygen. Out of contact with the cathode, the corrosion rate of the iron anode would have been close to zero, since in neutral solutions hydrogen deposition by reduction of water is negligible. The second half-cell, the cathode, was platinum, also in 0.1 M NaCl solution, but flushed with air and thus saturated with respect to oxygen at 0.2 bar. Mixing of the solutions was prevented by a 'cellophane' diaphragm sufficiently slowing interdiffusion without too much blocking ionic conductivity. The external metallic connection between the half-cells contained an ammeter with an input resistance, $R_a = 0.5\Omega$, which was small compared with the internal resistance, $R_{\text{int}} \simeq 100\ \Omega$, of the solution between the metal electrodes. The temperature was kept at 18°C; single experiments lasted 10 hours. During each run, the current $j$

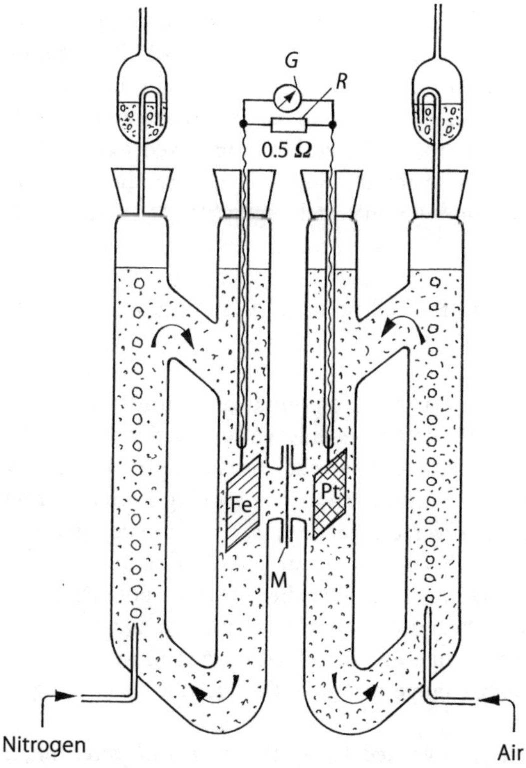


**Fig. 4.3.** Experimental differential aeration cell. (From Grubitsch and Sneck)

was recorded for the duration, $\tau$, of the experiments. From these data, the total current flow, $Q = \int j dt$, was evaluated with an error not exceeding 3%. This total flow of electricity was compared with the total effect of corrosion determined from the weight loss, $\Delta M$, of the steel specimen within $\pm 0{,}2$ mg. From this, the number of moles of dissolved iron was evaluated as $n_{\mathrm{Fe}} = \Delta M / M_{\mathrm{Fe}}$, where $M_{\mathrm{Fe}} = 55.85$ g mol$^{-1}$ is the molar or gramatom mass of iron.

Since without contact with the cathode the corrosion of the iron anode is negligible, $\Delta M$ should be exclusively due to anodic electrolysis with the total amount $Q$ of electric charge. *Faraday's law* then states that

$$Q = n_{\mathrm{Me}} z F \,, \tag{4.1}$$

where $z = 2$ is the valence of iron cations dissolved according to: $Fe \rightarrow Fe^{2+} + 2e^-$. A typical set of results was $Q = 62.7$ C, $\Delta M = 19.6$ mg. Since Faraday's law predicts 18.1 mg, the observed weight loss was slightly higher than expected, probably due to some residual dissolved oxygen in the anode compartment of the cell. Still, the result can be taken as confirming the supposition.

An interesting point is that in the cathode compartment solution, or in the 'catholyte', pH always increased during the experiment from, e.g., initially 6.2 to finally 8.6. This effect is due to the cathodic reaction producing $OH^-$ ions: $O_2 + 2H_2O + 4e^- \rightarrow 4OH^-$. As discussed in Chap. 3, hydrolysis of $Fe^{2+}$ is not important at room temperature, so the reaction: $Fe^{2+} + OH^- \rightarrow FeOH^+$ should be to weak to acidify the solution in the anode compartment, or in the 'anolyte'. A sometimes observed decrease from, e.g, pH 6.4 to 5.1 was probably caused by hydrolysis of some $Fe^{3+}$ present due to, again, residual dissolved oxygen [6b]. The current, $j$, produced by the cell was about 1.6 mA. Therefore, the resistance voltage, $jR_a$, in the ammeter was about 0.5 mV and negligible. At currents even much higher than those produced by this cell, ohmic potentials in metallic leads and also in the metals of electrodes are generally negligible. Therefore, the only important resistance contribution to the cell potential is a potential drop

$$\phi_x - \phi_y = \phi_{x,y} \tag{4.2}$$

in in the electrolytic solution, where $x$ and $y$ are points adjacent to the surfaces of iron and platinum, respectively. Using *Ohm's law,* we summarily express $\phi_{x,y}$ as the product $jR_{\mathrm{int}}$ of current strength and internal resistance of the cell. We note that, since in the homogeneous solution cations are moving from the anode to the cathode, and anions in the reverse direction, $\phi_x$ is more positive than $\phi_y$ and, thus, $\phi_{x,y}$ is positive. For more on current distribution in solutions, see Chap. 11. We also note that, at least initially, the diffusion potential, $\varepsilon_{\mathrm{diff}}$, across the diaphragm is negligible, because anolyte and catholyte have the same composition except for the presence of some dissolved oxygen in the catholyte. In the following we ignore a slight change in $\varepsilon_{\mathrm{diff}}$ with time due to a change in pH of the catholyte.

The current flow of the cell is governed by a) the *open cell potential, U,* measured on opening the short circuit, b) by $R_{\mathrm{int}}$ and $R_a$, and c) by the current-potential curves. $U$, incidentally, is much smaller than the equilibrium potential,

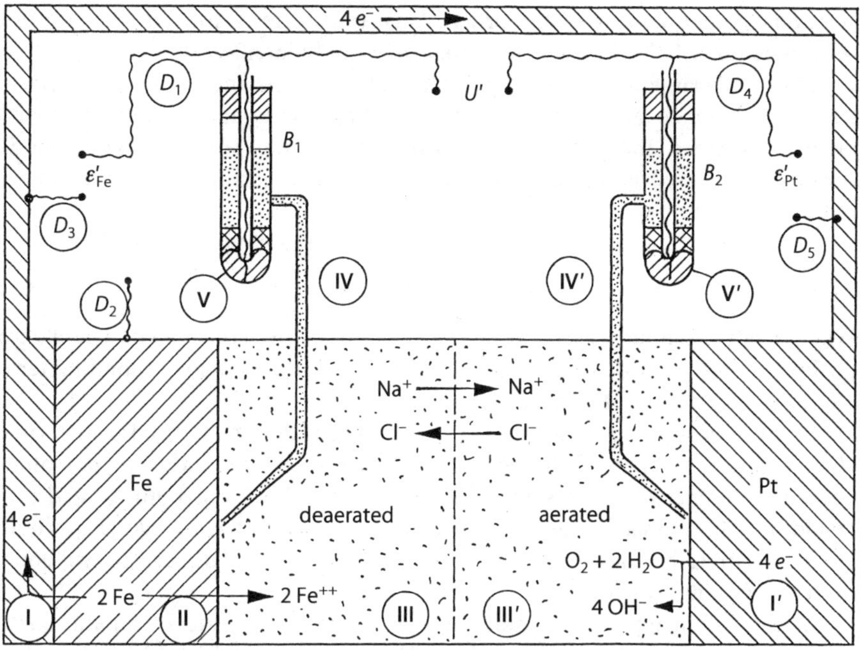


**Fig. 4.4.** Phase scheme of a model differential aeration cell equipped with identical reference electrodes for the measurement of the electrode potentials and of the resistance potential drop in the solution

$U_{rev}$, of the cell reaction: $Fe + (1/2)O_2 + H_2O \rightarrow Fe^{2+} + 2OH^-$, because for open circuit neither the anode nor the cathode are equilibrium electrodes.

For a better notion of the conditions in the working cell, Fig. 4.4 shows the phase scheme of the short-circuited cell which now has been equipped with reference electrodes for the measurement of the electrode potentials $\varepsilon_{an}$ and $\varepsilon_{cath}$. To simplify the argument, the input resistance, $R_i$, of the ammeter has been ignored and the short-circuit is assumed to be platinum. As indicated, the reference electrodes usually would not be hydrogen electrodes but electrodes of the second kind, such as calomel or Ag/AgCl electrodes. Measured potentials, then, are of type $\varepsilon'$, to be recalculated into type $\varepsilon$ using the known potential of the reference electrode with respect to SHE.

In this arrangement, there is no current flow through the reference electrodes, provided that the input resistance of the voltmeters used is high. Another important point is that ohmic voltages caused by current flow in the solutions must be excluded from the measurement of the electrode potentials. To achieve this, the reference electrodes are equipped with capillary probes ending in narrow tips directly adjacent to the metal surfaces without, however, shielding the surfaces. In a proper arrangement, with average-concentration solutions, the residual resistance of the solution between the tips of these *Haber-Luggin-capillaries* is of the order of 1 Ω, so that the error in the measured values of electrode potentials $\varepsilon$ is of the order 1 mV mA$^{-1}$ cm$^2$. More on the point of resistance potentials will have to be discussed only later.

In the experiments described here, the internal resistance almost exclusively was the resistance of the diaphragm separating anode and cathode compartments. Therefore, the potential $\phi_{\text{an}}$ in the anolyte was constant in the anode compartment, $\phi_{\text{cath}}$ was constant in the cathode compartment; and we can replace $\phi_{x,y}$, by $\phi_{\text{III, III}'}$.

In the complete cell circuit, we have, then, in the notation of Fig. 4.4,

$$\phi_{\text{I,II}} + \phi_{\text{II,III}} + \phi_{\text{III,III}'} + \phi_{\text{III}',\text{I}} = \varepsilon_{\text{an}} + jR_{\text{int}} - \varepsilon_{\text{cath}} = 0 \,. \tag{4.3}$$

When a non-negligible external resistance $R_a$ must be considered, a non-zero positive ohmic term $\phi_{I',I}$ has to be added to the left, and a corresponding term $jR_a$ to the right side of this equation.

In these experiments, Ag/AgCl reference electrodes where used in the same NaCl solution as that of the cell so that there were no diffusion potentials between reference electrodes and the cell solution. Otherwise both $\varepsilon_{\text{an}}$ and $\varepsilon_{\text{cath}}$ would be slightly inaccurate due to such diffusion potentials, while $jR_{\text{int}}$ would still be accurate because in the difference of electrode potentials the two equal diffusion potentials cancel.

In the following, we are more interested in in the rate of anodic metal dissolution than in the total weight loss by corrosion. Using the conversion number $\lambda_r$ introduced in Chap. 2 we note that $d\lambda_r/dt$ or, briefly, $\dot{\lambda}_r$ (mol s$^{-1}$) conveniently measures rates of a reactions labeled $r$. It is also convenient to use, for electrode reactions, *equivalent electrode reaction rates* defined by $\dot{\tilde{\lambda}}_r = z_r\dot{\lambda}_r$ where $z_r$ is the reaction valence introduced in Chap. 3. In the case of the anode in the cell discussed presently, $z_r = z = +2$ because there is only metal dissolution. Also, with the number $n_{\text{Fe}}$ of gram atoms of iron involved, we have $\dot{n} = \dot{\lambda}_r$.

The current $j$ flows across the Fe/solution interface by the metal dissolution reaction, and across the solution/platinum interface by the oxygen reduction reaction. At this point, we introduce the convention to take anodic, oxidizing current as positive and cathodic, reducing currents as negative. The convention is arbitrary; especially, it is not determined by the sign of either $\phi_{\text{II/III}}$ or $\phi_{\text{III}'/\text{I}'}$, which both are unknown quantities. We thus have, for any anodic current $j_{\text{an}}$ such as the iron dissolution current $j_{\text{Fe}}$, and for any cathodic current $j_{\text{cath}}$ such as the oxygen reduction current $j_{O_2}$,

$$j_{\text{an}} = +|j_{\text{an}}| \,, \quad j_{\text{cath}} = -|j_{\text{cath}}| \,. \tag{4.4}$$

By Faraday's law we also have

$$j_r = F\dot{\tilde{\lambda}}_r = z_r F\dot{\lambda}_r \,, \tag{4.5}$$

where $z_r$ is taken with its sign. In the present case, $j_r = j_{\text{Fe}} = -j_{O_2}$.

$j_{\text{Fe}}$ is a measure of the rate of iron dissolution. Always in the present case, dissolution, and thus corrosion, of the iron anode is practically uniform, with the same rate everywhere on the iron surface. Let $\dot{n}_{\text{Fe}}$ be the number of moles

**Table 4.1.** Mass loss rate, $\Delta\dot{M}$, and thickness loss rate, $\dot{D}$ equivalent to the corrosion current density $i_{corr} = 1$ A m$^{-2}$ = 0.1 mA cm$^{-2}$ = 100 $\mu$A cm$^{-2}$

| reaction | $\Delta\dot{M}$ g m$^{-2}$ d$^{-1}$ | $\dot{D}$ mm a$^{-1}$ |
|---|---|---|
| $Cu \rightarrow Cu^{2+}$ | 28.48 | 1.16 |
| $Fe \rightarrow Fe^{2+}$ | 25.02 | 1.16 |
| $Zn \rightarrow Zn^{2+}$ | 29.29 | 1.50 |
| $Al \rightarrow Al^{3+}$ | 8.05 | 1.09 |

of iron dissolved per second, and $A_{Fe}$ the anode area. Then, the local rate of iron dissolution,

$$v_{Fe} = \dot{n}_{Fe}/A_{Fe}\ (\mathrm{mol\, s^{-1} m^{-2}})\ , \tag{4.6}$$

and the local *current density* of iron dissolution,

$$i_{Fe} = j_{Fe}/A_{Fe}\ (\mathrm{A\, m^{-2}})\ , \tag{4.7}$$

are constant everywhere on the surface $A_{Fe}$. By analogy, we likewise define the current density, $i_r$, of any electrode reaction labeled $r$ at an electrode surface of area $A_r$ as $i_r/A_r$. For the moment, we disregard complications in the case of non-uniform distribution of rates on the surface $A_r$.

In the cell presently under consideration, the anode dissolution current density also is the anode corrosion rate. Generally, when $j_{Me}$ and $A_{Me}$ are metal dissolution strength and metal surface, respectively, and when metal dissolution is uniform, or else approximately uniform, it makes sense to define the *corrosion current density* as

$$i_{corr} = j_{Me}/A_{Me}\ (\mathrm{A\, m^{-2}})\ . \tag{4.8}$$

In practice, corrosion is usually judged from the loss in thickness of metal parts or from the loss of metallic mass. Therefore, rates of corrosion, $v_{corr}$, are usually expressed as *thickness loss rate* $\dot{D}$ (mm a$^{-1}$), or *mass loss rate* $\Delta\dot{M}$ (g m$^{-2}$ d$^{-1}$), where 'a' and 'd' indicate 'year' and ' day', respectively. With the molar mass $M_{Me}$ (g mol$^{-1}$), the specific mass $\rho_{Me}$ (g cm$^{-3}$) of the respective metal, and the valence $z$ of the respective dissolved metallic cation, current densities $i_{corr}$ are converted as follows:

$$\begin{aligned} 1\,\mathrm{A\,m^{-2}} &= 8.64 \times 10^4 \frac{M_{Me}}{z\,F}\ \mathrm{g\,m^{-2}\,d^{-1}} \\ &= 3.154 \times 10^4 \frac{M_{Me}}{z\,F\,\rho_{Me}}\ \mathrm{mm\ a^{-1}}\ . \end{aligned} \tag{4.9}$$

Table 4.1 lists some examples of corrosion rates $\Delta\dot{M}$ and $\dot{D}$ equivalent to the corrosion current density 1 A m$^{-2}$. From these numbers it is seen that, concerning the strength of structural metals, corrosion is negligible whenever $i_{corr}$ does not exceed about 0.01 A m$^{-2}$ = 1 $\mu$A cm$^{-2}$.

In a further series of the experiments as described above, the platinum electrode was replaced by an iron electrode. This, then, represents corrosion by

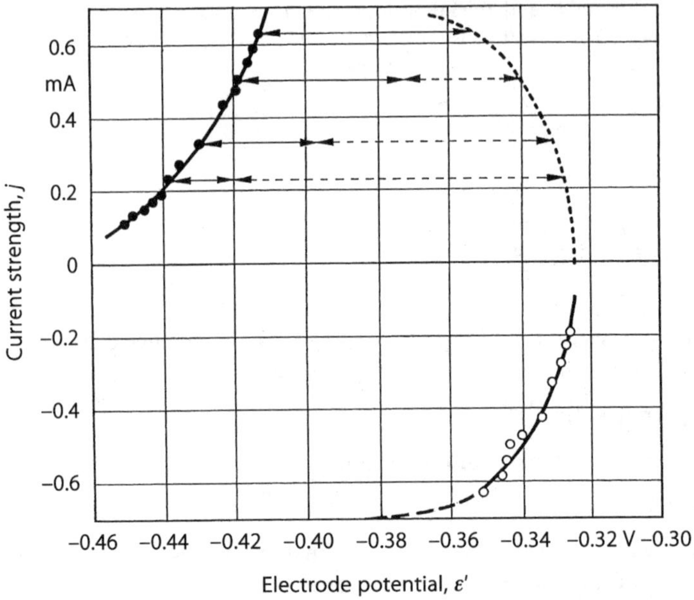


**Fig. 4.5.** Investigating a model differential aeration cell: Anodic current-potential curve for iron in deaerated solution (*solid circles*) and cathodic current-potential curve for iron in aerated solution (*open circles*). The solution was 0.1 M NaCl, 0.05 M $CH_3COONa$. *Solid arrows*: Internal cell resistance voltage $jR_{int}$; broken arrows: external resistance voltage $jR_a$. Electrode potentials are referred to Ag/AgCl-electrodes in the same solution. (Data from Grubitsch and Sneck)

differential aeration in its pure form. Again, iron in deaerated solution is the anode of the cell; iron in aerated solution, the cathode. On open circuit, this cathode would be subject to oxygen corrosion; short-circuited with the anode, *cathodic protection* again suppresses corrosion, but once more we postpone the explanation of this effect. In these experiments, the current produced by the cell was decreased in steps by inserting calibrated resistances $R_a$ up to 500 Ω into the outer circuit adding terms $\phi_{I',I}$ and $JR_a$ to Eq. (4.3). Some observations obtained with $R_a > 500\,\Omega$ have been ignored as they seem irregular. Figure 4.5 shows a typical result: The cell current, $j$, decreased with increasing $R_a$, casing an decrease in $jR_{\text{int}}$, an increase in $jR_a$. At the same time, we see an increase in the anode electrode potential, $\varepsilon_{\text{an}}$, and a decrease in the cathode electrode potential, $\varepsilon_{\text{cath}}$, and these changes are examples of current-potential curves of electrodes.

Current-potential curves will be derived from theoretical principles in later chapters. For the moment, we very summarily state that we may formally express any such function as the sum of an open-circuit potential, $\varepsilon(j = 0)$, and a current-dependent deviation, $\pi(j)$, which we introduce as the *electrode polarization*:

$$\varepsilon = \varepsilon(j = 0) + \pi(j) \, . \tag{4.10}$$

In our case, the polarizing current is the short-circuit cell current; the anode polarization is positive, and the cathode polarization, negative:

$$a) \; \varepsilon_{\mathrm{an}} = \varepsilon_{\mathrm{an}}(j = 0) + |\pi_{\mathrm{an}}(j)| \, , \quad b) \; \varepsilon_{\mathrm{cath}} = \varepsilon_{\mathrm{cath}}(j = 0) - |\pi_{\mathrm{cath}}(j)| \, . \tag{4.11}$$

Polarization, $\pi$, must not be confused with *overpotential,* $\eta_r$, which for any electrode reaction, $r$, is defined as the difference of actual electrode potential and the equilibrium potential, $E_r$, of the $r$-reaction:

$$\eta_r(j) = \varepsilon - E_r \, . \tag{4.12}$$

As noted above, the open-circuit differential aeration cell is not in equilibrium. By extrapolating the curves in Fig. 4.5, it is indeed seen that $\varepsilon_{\mathrm{an}}(0) = -0.47$ V, referred to Ag/AgCl reference electrodes in the same solution. Substituting $\varepsilon_{\mathrm{Ag/AgCl}} = +0.28$, we have $\varepsilon_{an} = -0.19$ V, referred to SHE. Since $E^{\circ}_{\mathrm{Fe/Fe^{2+}}} \simeq -0.44$ V, and $a_{\mathrm{Fe^{2+}}} \ll 1$ mol L$^{-1}$, the open-circuit potential of the anode is much more positive than the equilibrium potential of the anodic electrode reaction. Likewise, the open-circuit cathode potential is much more negative than the oxygen equilibrium potential. Both $\varepsilon_{\mathrm{an}}(0)$ and $\varepsilon_{\mathrm{cath}}(0)$ really are corrosion potentials, as will be explained further below.

## 4.3 Measuring Current-Potential Curves

Since current-potential curves essentially govern 'wet' corrosion as a branch of electrode reaction kinetics, we need to briefly describe current-potential measurements. In the first place, interest is focused on steady-state curves, as they govern steady-state corrosion. We note that, as an experimental tool very characteristic for the entire field of electrochemical kinetics, nonsteady-state measurements, including measurements of the response of electrode behaviour to pulsed changes in either potential or current, or various other nonsteady-state methods play an important role in the elucidation of the mechanism of corrosion processes. More details will become apparent later in their proper place. Reference is also made to the extensive literature [7–9].

The typical experimental arrangement is described by Figs. 4.6 and 4.7: The electrode to be studied is proceeding at the *working electrode* Me. The electrode potential is measured between points B and C of the circuit, referred to a *reference electrode.* Unless otherwise noted, electrode potentials measured against a reference electrode suitable for the respective purpose are always recalculated as values $\varepsilon$ referred to SHE. We recall the condition of null-current in the measuring circuit and also the role of diffusion potentials possibly interfering with the $\varepsilon$-measurement, if not minimized by a salt bridge as indicated in the figure. Polarizing currents can be made to flow from a *counter electrode* to the working electrode from a current source introduced between points A and B of the circuit. Sintered porous glass frits usually are used to prevent mixing of working-electrode and reference-electrode solutions across the salt bridge. The

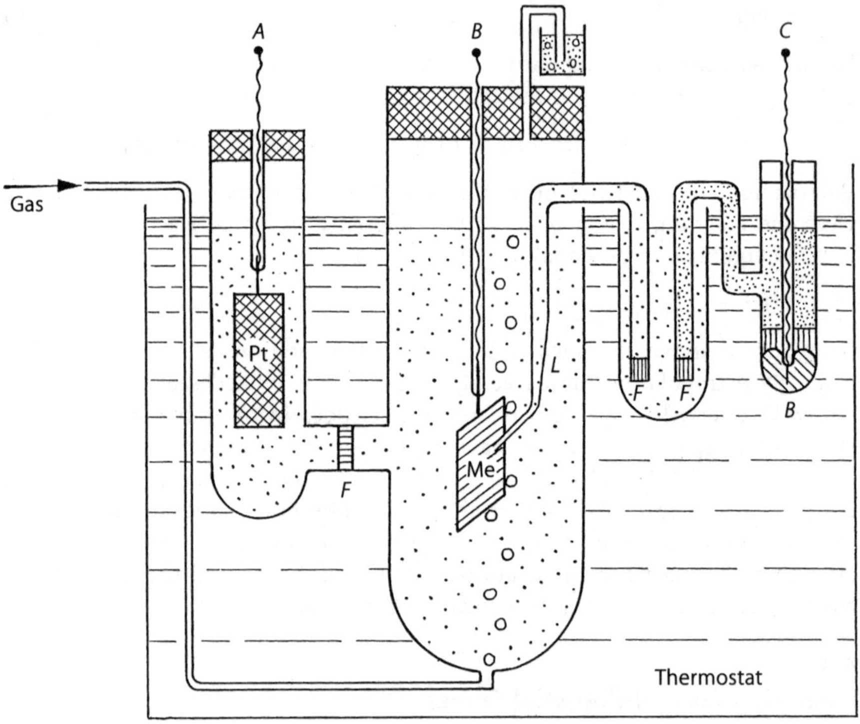


**Fig. 4.6.** Measuring current-potential curves: The typical three-electrodes setup with working electrode Me, counter electrode Pt, reference electrode, Haber-Luggin capillary L, glass frits F. For the external current source see Fig. 4.7

reference electrode is equipped with a Haber-Luggin probe eliminating most of the resistance voltage caused by the flow of the polarizing current in the solution between the working electrode and a counter electrode. Usually, but not necessarily, the counter electrode is an inert noble metal such as platinum. A further sintered porous glass plug usually separates the counter electrode solution from the working-electrode solution as a measure to prevent mixing solutions. Current flow in the polarizing circuit requires an electrode reaction occurring at the counter-electrode/solution interface, such as cathodic hydrogen evolution or anodic oxygen evolution, etc., and products of this reaction must not interfere with the electrode reactions occurring at the working electrode/solution interface.

Two types of devices producing the desired polarizing current in the working-electrode/counter-electrode circuit are important: As indicated in Fig. 4.7a, a relatively large d.c. voltage, e.g. 100 V, may be used in series with an ammeter and an adjustable resistance large compared to the resistance of the cell. In this case, the polarizing current is governed by the the external resistance and practically independent of changes of the cell resistance. This,

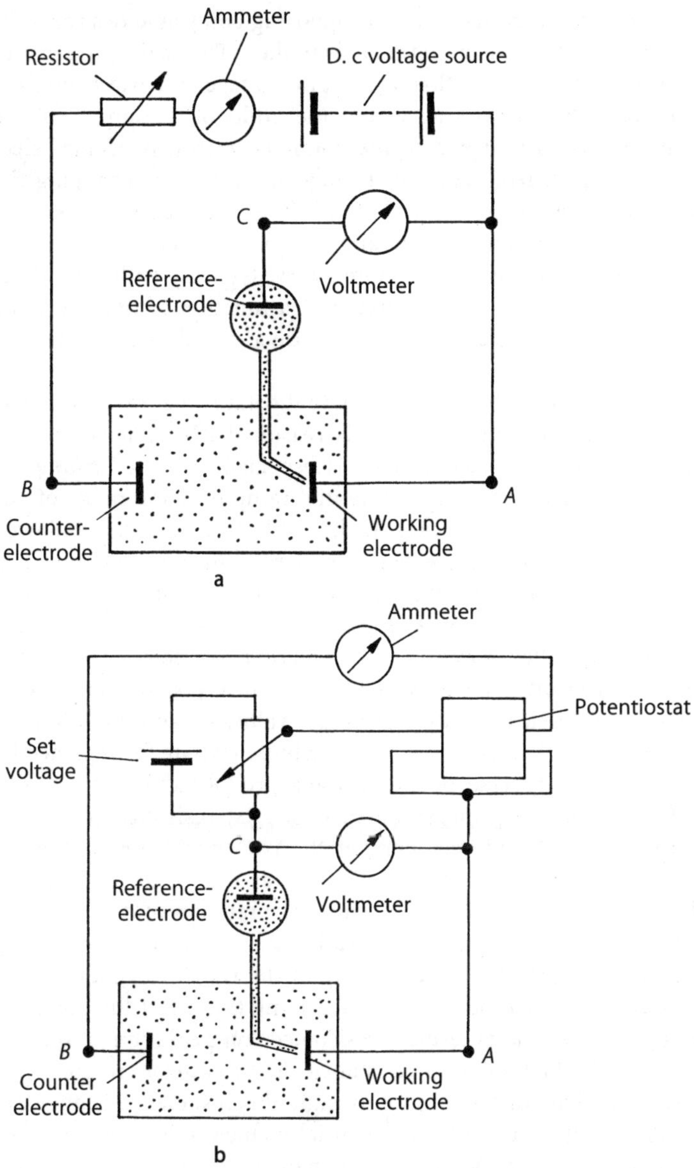


**Fig. 4.7a,b.** Basic circuits used for the **a** galvanostatic and **b** potentiostatic measurement of current-potential curves

therefore, is a *galvanostat* which measures potentials as function of currents. It is quite convenient for many purposes, but has largely fallen in disuse owing to the oncoming of the electronic *potentiostat* which measures currents as function of potentials. As schematically shown in Fig. 4.7b, in this case the difference of the electrode potential $\varepsilon$ and an opposing adjustable desired voltage $U_{ex}$ is fed to the input of an operational amplifier. The main property of this device is a very high gain so that the input voltage is always virtually zero whenever the amplifier is working within the limits of its output voltage limits. The output voltage is returned to the counter-electrode/working- electrode circuit, forcing a polarizing current through the cell, thus changing the electrode potential of the working electrode according to the current-potential curve, until $\varepsilon = U_{ex}$. The response of the amplifier is also very rapid so that the control of the potential is essentially momentary, typically within milliseconds. This also holds when $U_{ex}$ is programmed as a function of time by some suitable function generator replacing the voltage divider indicated in the figure.

The particular advantage of the electronic potentiostat is that measurements are easily automated, and it is mainly for this reason that they have become standard practice in corrosion science. This is especially true for *triangular potential sweep measurements* applying a potential ramp over the range of potentials interesting in the respective case, with $d\varepsilon/dt = \dot{\varepsilon} = const$ followed by a reversal with $\dot{\varepsilon} = -const$. This experimental procedure usually is called *potentiokinetic* or *potentiodynamic*, although at any moment it is expected to be potentiostatic. Often, the potential sweep is reversed periodically, and the resulting set of curves is called a *cyclic voltammogram*. For low values of the scan rate $\dot{\varepsilon}$ the result will often be close to the true steady state curve, i.e. it may be considered as quasi-steady state. For higher values it is essential to have some knowledge of the deviations from steady-state behaviour for instance by checking that forward and reverse potential scans produce the same results. This checking of potential sweep measurements for hysteresis effects is, however, often ignored, much to the detriment of the respective work. A typical error of this type is overrunning pitting potentials, as described in Chap. 12. Without knowledge of at least the scan rate, the interpretation of data thus obtained tends to become difficult. Another possible source of error is the ohmic resistance voltage in the solution between the tip of the Haber-Luggin capillary and the working electrode. As a measure counteracting this effect the potentiostat often is equipped with what is termed 'resistance compensation', requiring the knowledge of the residual solution resistance in the control circuit of the setup. When the conductivity of the solution is high, this residual resistance can be determined from oscillographs of the potential change following galvanostatic current steps: Because ohmic voltages then change virtually instantly, while electrode reaction potentials follow only within milliseconds, the separation of ohmic effects often is successful. Ohmic effects tend to become a very serious problem when it comes to solutions of low conductivity, such as boiler feed water.

Depending on the performance of the potentiostat, the scan rate can be made quite high, allowing nonsteady-state measurements particularly suitable for the analysis of electrode reaction kinetics. One such type of investigations is the a.c. resistance or impedance measurement described in the Appendix.

A potentiostat can be used as a galvanostat. In principle, this means removing the d.c. voltage source in Fig. 4.7a and changing from controlling the potential to controlling the voltage across a series resistance across the series resistance $R_a$ by a potentiostat.

Above, we have already introduced the electrode polarization, $\pi$, as the change in potential, $\varepsilon$, caused by a polarizing current, $j$, referred to the electrode area, $A_{metal}$. As a measure of the degree of polarizibility we now introduce the *polarization resistance* $R_\pi$, distinguishing between the *differential polarization resistance,*

$$(R_\pi)_{\text{diff}} = d\varepsilon/di\,, \tag{4.13}$$

and the *integral polarization resistance*

$$(R_\pi)_{\text{integral}} = \frac{\varepsilon_{i=i_1} - \varepsilon_{i=0}}{i_1}\,. \tag{4.14}$$

$R_\pi$ usually is expressed in $\Omega$ cm$^2$. Polarization resistances normally are non-ohmic, changing with $i$. They are, by definition, direct-current quantities without capacitive or inductive components. While $(R_\pi)_{\text{integral}}$ always is positive, $(R_\pi)_{\text{diff}}$ sometimes is negative, as in the case of active/passive transitions (see Chap. 10). A special case interesting in corrosion monitoring is $(R_\pi)_{\text{diff}}$ at $i = 0$, because it is sometimes directly related to the steady state corrosion rate (see Chap. 6).

## 4.4 Uniform Corrosion: The Work of Wagner and Traud

As described in Chap. 3, acid corrosion of a dilute amalgams, such as solutions of sodium, indium or zinc in liquid mercury is the model example of uniform corrosion, where the metal surface is anode and cathode at the same time. In this context, Wagner and Traud have investigated the corrosion of dilute, stirred zinc amalgam in hydrochloric acid solution [5]. The aim was to unequivocally show that the mechanism is electrolytic, proceeding by simultaneous anodic zinc dissolution and cathodic hydrogen deposition as kinetically independent partial reactions coupled by the condition of electoneutrality.

The overall corrosion reaction is: $Zn + 2H^+ \rightarrow Zn^{2+} + H_2$. Within the classical concepts of formal chemical reaction kinetics, the obvious supposition would have been that the course of this reaction requires the formation of intermediate compounds between reactants, for instance the formation of a complex compound $ZnH_2^{2+}$, changing, by shifting electrons within the complex, to $H_2Zn^{2+}$, and eventually dissociating into the reaction products Zn and $H_2$. The method to disprove this mechanism was to separately check the kinetics of zinc dissolution, $Zn \rightarrow Zn^{2+}$, and hydrogen deposition, $2H^+ + 2e^- \rightarrow H_2$, and to show that during corrosion these half-reactions proceed simultaneously at equal equivalent rates, $\dot{\lambda}_{Zn^{2+}}$ and $\dot{\lambda}_{H_2}$ or, which is the same, with an anodic, positive current $j_{\text{an}}$ of zinc dissolution passing the metal/ solution interface simultaneously with a cathodic, negative current $j_{\text{cath}}$ of hydrogen deposition.

By the condition of electroneutrality, $j_{an} + j_{cath} = 0$. Also, since the amalgam surface is uniform, the second condition is that the electrode potential of the amalgam is the same everywhere on the surface. If all this is true it makes physicochemical sense to talk, as has been done already in preceding chapters, of quantities $j_{an}$ and $j_{cath}$ as measurable electric *partial currents* and, after division by the metal surface area, of quantities $i_{an}$ and $i_{cath}$ as measurable partial current densities.

Similarly, it then also makes sense to describe the kinetics of electrolytic corrosion as being governed by the superposition of anodic and cathodic *partial current-potential curves*. We note that electrodes, where several electrode reactions occur simultaneously, are also called *mixed electrodes*. Later, we shall quantitatively inspect current-potential functions of electrode reactions and their superposition. For the moment we only need to consider that the potential of a mixed electrode polarized with an external current must be governed by the superposition of the partial current-potential curves to the *total current-potential curve*.

The experimental proof of the theorem of independent superposition of partial current-potential curves in the case of acid amalgam corrosion was obtained using two suppositions which, somewhat luckily, eventually turned out to be legitimate in the particular case investigated. The first was that at an amalgam surface, where most surface metal atoms are Hg, the elementary, atomic events of hydrogen deposition mostly occur at Hg surface sites. Therefore, there is hope that the mechanism of hydrogen deposition is the same on pure mercury as on the amalgam. Cathodic hydrogen deposition then was studied by measuring the cathodic current-potential curve, $j_H(\varepsilon)$, of hydrogen deposition from acid solutions on mercury in a arrangement essentially as described in Fig. 4.7, with a stirred mercury pool as the working electrode. The solution was HCl, and the reference electrode was an Ag/AgCl electrode in the same solution. For reasons which will become more obvious in Chap. 5, KCl was added, so that the total ionic concentration was always the same. The current-potential curves measured at different concentrations of HCl are plotted in Fig. 4.8 as solid lines. As shown in Fig. 4.9, replotting these curves in a log $j$-$\varepsilon$ diagram reveals that a relation

$$j_H = -A\,exp\,(-B\,\varepsilon) \tag{4.15}$$

is obeyed in every case. Also, $A \simeq const \times c_{H^+}$. This type of current-potential relation will become a point of major interest in the following chapters. We note that the slope $d\varepsilon/d\log|j_H|$ of the lines always is 0.12 V.

The second suppositions was that the mechanism of zinc dissolution from the amalgam can be studied separately in the absence of hydrogen deposition by lowering the concentration of $H^+$ ions to about $10^{-7}$ M, i.e., by changing from the acid to a neutral solution, at the same time varying, in a series of experiments, both the zinc content, $\gamma_{Zn}$, of the amalgam and the zinc concentration, $c_{Zn^{2+}}$, of the solution. The sum of concentrations of dissolved $ZnCl_2$ and $CaCl_2$ was kept constant in an attempt to keep the activity coefficient $f_{Zn^{2+}}$ constant. Again the current-potential curves of what now was a $Zn(Hg)/Zn^{2+}$-electrode

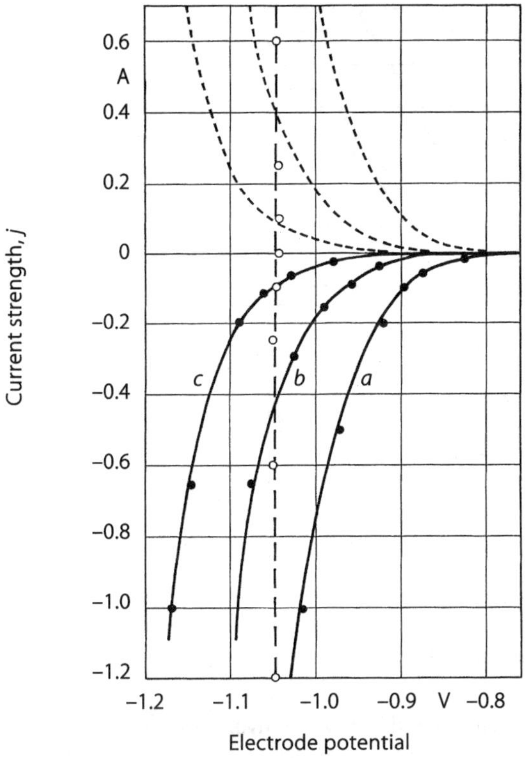


**Fig. 4.8.** Current-potential diagram of the corrosion of zinc amalgam in hydrochloric acid solution. $\varepsilon'$ referred to an Ag/AgCl-electrode in the same solution. *Solid curves:* Partial current-potential curves of hydrogen deposition on pure mercury. Solution composition: a: 1.0 M $CaCl_2$, 1.0 M HCl; b: 1.0 M $CaCl_2$, 0.1 M HCl, 0.9 M KCl; c: 1.0 M $CaCl_2$, 0.01 M HCl, 0.99 M KCl. *Broken curve:* Total current-potential curve of zinc amalgam with 1.6 wt.% zinc, practically identical with partial current-potential curve of zinc dissolution and deposition. Solution composition: 0.9 M $CaCl_2$, 1.0 M $ZnCl_2$, 0.1 M HCl, 0.9 M KCl. (Data from Wagner and Traud)

where measured. On first sight surprisingly, the equation relating current and potential now is reduced, for any given set of concentrations, to $\varepsilon = const$, and this holds for positive and negative currents as well, i.e., the potential is independent of the rate of either anodic zinc dissolution or cathodic zinc deposition. Also, and what is much more, it turns out that $const \simeq E_{Zn/Zn^{2+}}$ : The potential of the zinc amalgam is always practically equal to the zinc equilibrium potential, as described by Eq. (3.107). Therefore, the kinetic mechanism of either zinc dissolution or zinc deposition actually does not matter for the establishment of the electrode potential. The reason is that at the amalgam/solution interface the electrode reaction of zinc dissolution and zinc deposition always is in quasi-equilibrium due to a high equilibrium exchange rate of the reaction. As a matter of fact, this was crucial for the success of the desired proof,

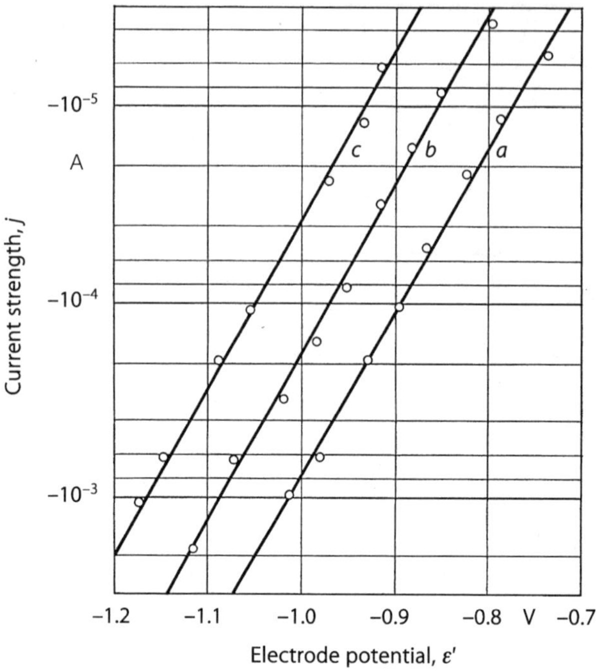


**Fig. 4.9.** Semi-logarithmic plot of current-potential curves of hydrogen deposition. (From Wagner and Traud)

because assuming pH-independent kinetics of $Me/Me^{z+}$-electrodes is quite unsafe, as shown in Chap. 5. An electrode which does not change its potential with polarizing currents is called *non-polarizable*. Since the quasi-equilibrium is not much disturbed by slow overall corrosion, the current-potential curve of the amalgam remains the same when acid corrosion occurs. Therefore the total current-potential curve of the amalgam in acid solutions virtually is the same as the partial current-potential curve of zinc dissolution and redeposition. The broken line in Fig. 4.8 actually is one of the total current-potential curves measured in this work.

Since now for various sets of values for pH, $c_{Zn^{2+}}$, and $\gamma_{Zn}$, the partial current-potential curves, $j_{Zn}$ and $j_{H_2}$ are known, it remains to check the conclusion that the potential of the corroded amalgam, i.e., the corrosion potential $\varepsilon_{corr}$, must obey the condition $j_{an}(\varepsilon_{corr}) = -j_{cath}(\varepsilon_{corr})$, and that the corrosion rate, expressed as corrosion current strength, $j_{corr}$, must equal the partial currents passing the amalgam/solution interface at the corrosion potential. A simple method of doing this evaluation is indicated in Fig. 4.8 for some of the measurements: As shown there, absolute values of cathodic hydrogen deposition currents are plotted, and values of both the corrosion current and of the corrosion potential are obtained from the intersection of anodic and cathodic current-potential

**Table 4.2.** Comparison of experimental $(\dot{\lambda}_{H_2})_{exp}$ and calculated $(\dot{\lambda}_{H_2})_{calc}$ values of the rate of hydrogen deposition (and metal dissolution) during acid corrosion of zinc amalgam. Data from Wagner and Traud

| $\gamma_{Zn}$ wt.% | $c_{H^+}$ mol $L^{-1}$ | $c_{Zn^{2+}}$ mol $L^{-1}$ | $10^5 \times \dot{\lambda}_{exp}$ mol $h^{-1}$ | $10^5 \times \dot{\lambda}_{calc}$ mol $h^{-1}$ |
|---|---|---|---|---|
| 0.4 | 0.10 | 0.10 | 1.23 | 1.19 |
| 0.4 | 0.01 | 0.10 | 0.25 | 0.23 |
| 0.4 | 0.10 | 0,01 | 2.40 | 2.24 |
| 0.4 | 0.10 | 0.10 | 1.15 | 1.19 |
| 1.6 | 0.10 | 0.10 | 1.81 | 1.83 |
| 1.6 | 0.01 | 0.10 | 0.36 | 0.32 |
| 1.6 | 0.10 | 0.01 | 3.00 | 3.24 |
| 1.6 | 1.00 | 0.10 | 5.67 | 5.36 |
| 1.6 | 0.10 | 0.10 | 1.81 | 1.83 |

curves. Since the metal as an anode is non-polarizable, corrosion is, in this case, governed by the kinetics of cathodic hydrogen deposition. It is, thus, entirely under *cathodic control.* Cases of *anodic* and of *mixed control* will be described later.

In a final step, the calculated corrosion rate of the amalgam had to be compared with experimental values obtained by manometric measurement of hydrogen evolution in the course of corrosion. These numbers are easily obtained, and they can be compared with values calculated from the superposition of pairs of current-potential curves described above. As shown in Table 4.2, the two sets of results compare well, verifying the electrolytic mechanism of corrosion in aqueous solutions of sufficient conductivity.

More proofs will be reported in Chap. 5. Generally, the principle of aqueous uniform corrosion proceeding by simultaneous anodic and cathodic half-reactions controlled by electroneutrality is very well established. There are, however, some observations indicating that the principle may fail in the case of corrosion of iron, chromium and chromium-nickel steels in hot concentrated acids [10]. Apparently, non-electrolytic overall corrosion reactions can sometimes contribute to the kinetic mechanism. More information to this point would be very desirable.

## References

1. Palmer, W.: The Corrosion of Metals. Ingeniörs Vetenkaps Akademien, Handlingar Nr. *93 108*. Svenska Bokhandelszentralen A.-B., 1929, 1931
2. Kaesche, H.: Z. Metallkunde *61*(1970)94
3. Evans, U.R.: loc.cit. Chap. 2 [3]
4. Frumkin, A.Z.: Z.phys. Chemie Abt. A *160* (1932) 116
5. Wagner, C. and Traud, W.: Z. Elektrochemie *44* (1938) 391; Wagner, C.: loc. cit. Chap. 2 [4]

6. a) Grubitsch, H. and Sneck T.: Monatshefte Chemie *86* (1955) 752; b) Bohnenkamp, K., private communication 1980
7. Bard, A.J. and Faulkner, L.R.: Electrochemical Methods. Fundamentals and Applications. J. Wiley & Sons; New York etc., 1980
8. Reef, R., Peat, R. Peter, L.M., Pletcher, D., and Robinson, J.: Instrumental Methods in Electrochemistry. J. Wiley & Sons; New York etc., 1985
9. Yeager, E.; Bockris, J.O'M., Conway, B.E., and Sarangapani, S., eds.: Electrodics: Experimental Techniques. Comprehensive Treatise of Electrochemistry, Vol. 9. Plenum Press; New York and London, 1984
10. Florianovitch, G.M. and Kolotyrkin, Y.M.: Dokl. Akad. Nauk SSSR *157* (1964) 422; Mansfeld, F. and Kendall, J.V.: Corr. Sci. *16* (1967) 653

# 5 The Kinetics of Electrode Reactions

## 5.1 Hydrogen Deposition and Hydrogen Ionisation

In this section, the discussion of the mechanism of hydrogen evolution at metallic surfaces will be used to outline the principles of electrode reaction kinetics. This follows the historic development, since originally hydrogen evolution was considered to be the model example of a simple electrode reaction. The view is not really up-to-date: The theory of electrode kinetics is much more advanced concerning *redox reactions* such as: $Fe^{2+} = Fe^{3+} + e^-$, involving only a tunneling exchange of electrons with an ion undergoing a change in valence and a rearrangement of the hydration shell only. It is for this type of reactions that an advanced theory is available which, however will be disregarded in the following. Hydrogen evolution certainly is an important half-reaction of corrosion processes, and the principles of kinetics can be demonstrated for this example quite well, if only at the expense of a quantitative quantum-mechanical treatment. For a deeper understanding of electrode kinetics the reader is referred to the literature [1–5].

The complete electrode reaction

$$2H^+ + 2e^- \rightarrow H_2 \tag{5.1}$$

involves a number of discernible steps. The first is the transport of $H^+$ from the bulk of the solution to the metal surface by electrolytic migration, diffusion, and convection. We indicate this step by writing

$$H^+ \rightarrow (H^+)_* \,. \tag{5.2}$$

Sometimes the migrating or diffusing species is not $H^+$, but undissociated acid, dissociating only upon arrival at the metal surface. This applies to both dissolved carbon dioxide, $CO_2$, and hydrogen sulfide, $H_2S$, which both are very weak acids, as explained in Chap. 2. Due to weak dissociation, the concentration of acid available for the corrosion process is much higher than indicated by the pH of the solution measuring $H^+$ activity. This has important consequences for instance for hydrogen embrittlement (see Chap. 14), but will be ignored presently.

The '$*$'-subscript indicates the position of closest approach of hydronium ions to the metal surface, which essentially is the radius of several Å units, i.e. several 0.1 nm, of the hydration shell carried by the ion. From this position, the ions are discharged to atomic hydrogen adsorbed at the surface in the *charge-transfer reaction,*

$$(H^+)_* + e^- \rightarrow H_{ad} \,, \tag{5.3}$$

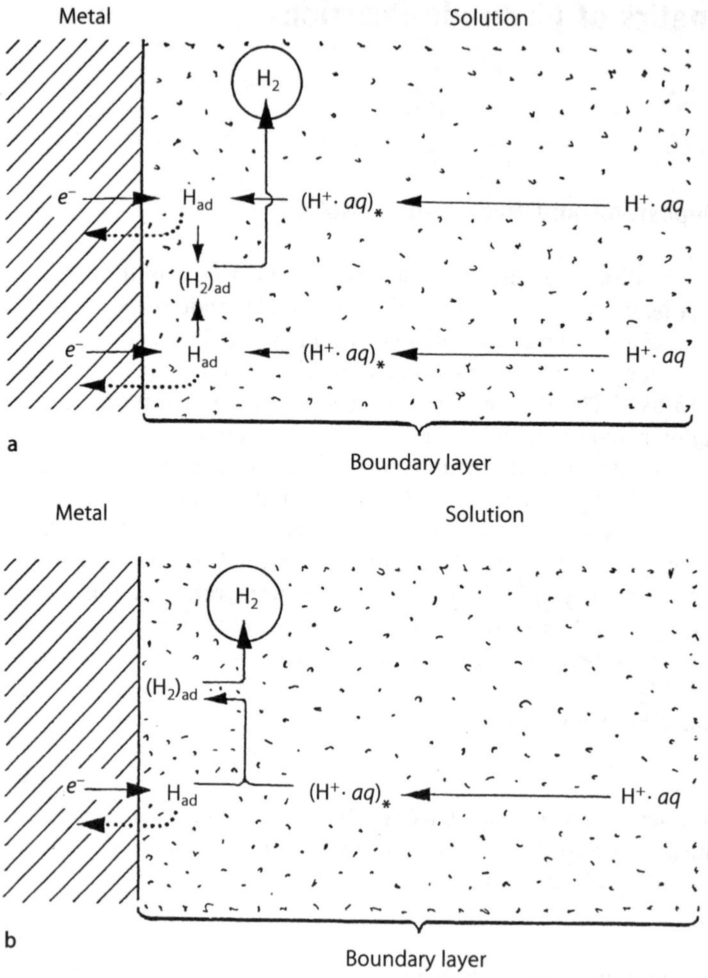


**Fig. 5.1.** Two mechanisms of cathodic hydrogen deposition: **a** Volmer-Tafel, **b** Volmer-Heyrovsky mechanism

also called the *Volmer reaction*. Adsorbed atomic hydrogen recombines to molecular hydrogen by the *Tafel reaction*,

$$H_{ad} + H_{ad} \rightarrow H_2 . \tag{5.4}$$

Molecular hydrogen, $H_2$, escapes in bubbles. The sequence of steps according to Eqs. (5.3) and (5.4) is called *Volmer-Tafel* mechanism, and schematically shown in Fig. 5.1a. The second step may involve sub-steps such as intermediate formation of adsorbed molecular hydrogen and desorption of molecular hydrogen into the solution. Especially when the rate of $H^+$- reduction is low, dissolved molecular hydrogen may diffuse to the surface of the solution and escape into the surrounding atmosphere without preceding formation of bubbles. Since, apparently, such post-deposition effects do not normally determine the hydrogen evolution overpotential developed below, we do not, in Eq. (5.2) and elsewhere, specify whether $H_2$, is adsorbed, or dissolved, or gaseous. The more important species is adsorbed atomic hydrogen and the degree, $\theta_H$, of surface coverage, with $0 \leq \theta_H \leq 1$. It is probably intuitive that, when $\theta_H$ is high, the probability for $H^+$ arriving at the surface to meet $H_{ad}$ instead of metal surface atoms is also high. In this case, an *electrochemical recombination* by the *Heyrovsky reaction*

$$(H^+)_* + H_{ad} + e^- \rightarrow H_2 \tag{5.5}$$

follows the Volmer reaction in a sequence of events called *Volmer-Heyrovsky mechanism* as shown in Fig. 5.1b.

For more quantitative insight we first inspect the preceding step of transport of $H^+$ to the metal surface. For simplicity we assume an acid solution containing an excess of a strong electrolyte so that the transference number $t_{H^+}$ is close to zero. Transport of $H^+$ then is by diffusion and/or convection only. To describe mass transfer we use the *Nernst boundary-layer model*. The suggestion is to approximate boundary-layer theory by considering a solution well stirred everywhere except in a thin layer of definite thickness, $\delta_{Nernst}$, adjacent to the metal surface. As in the *Prandtl boundary layer* there is, within the Nernst layer, no motion of the solution. Beyond the Nernst layer the concentration $c_{H^+}$ of dissolved $H^+$ is assumed constant, $(c_{H^+})_o$. Through the layer, the concentration decreases linearly to $(c_{H^+})_*$ at the distance $x = 0$ from the metal surface. In average-concentration solutions, the thickness of this diffusion boundary layer $\delta_{Nernst}$, exceeds that of the the electrical double layer thickness $\delta_{dl}$, by orders of magnitude, so referring to $x = 0$ for the '$*$'-position is permitted.

We note that a more refined view is to distinguish the Nernst layer, within which the concentration of a reacting chemical species changes towards a limiting surface value, and the Prandtl layer within which the flow velocity drops to zero at the metal surface due to frictional drag. The diffusion layer actually is a sublayer of the hydrodynamic Prandtl layer, but we need not consider this as long as we take $\delta_{Nernst}$ as an empirical rather than a calculated quantity. For more details, see the Appendix.

Assuming a flat and large metal surface with the normal vector pointing into the solution along an axis, $x$, we use Fick's first law to describe the rate, $\dot{n}_{H^+}$ mol s$^{-1}$ of diffusion of $H^+$ to the metal surface. We have

$$\dot{n}_{H^+} = -A_{metal} D_{H^+} \frac{d\,c_{H^+}}{d\,x}\,, \tag{5.6}$$

where $A_{metal}$ is the surface area, and $D_{H^+}$ the diffusion coefficient of $H^+$ ions. As long as $H^+$ transport through the boundary layer is by diffusion only, and for steady-state conditions, when $(c_{H^+})_*$ does not change with time, the current density, $i_H$, of hydrogen deposition is equal to $F \times \dot{n}_{H^+}/A_{metal}$. Therefore,

$$i_H = -F D_{H^+} \left( \frac{d\,c_{H^+}}{d\,x} \right)_{x=0}\,. \tag{5.7}$$

Also in steady state, the gradient $d\,c_{\mathrm{H^+}}/d\,x$ is constant throughout the diffusion boundary layer, equal to $((c_{\mathrm{H^+}})_\circ - (c_{\mathrm{H^+}})_*)/\delta_{\mathrm{Nernst}}$, and

$$i_{\mathrm{H}} = -F D_{\mathrm{H^+}} \frac{(c_{\mathrm{H^+}})_\circ - (c_{\mathrm{H^+}})_*}{\delta_{\mathrm{Nernst}}} . \tag{5.8}$$

Usually, $\delta_{\mathrm{Nernst}}$ is calculated in cm, $D_{\mathrm{H^+}}$ in $\mathrm{cm^2\,s^{-1}}$. Then, concentrations $c_{\mathrm{H^+}}$ must be inserted in mol $\mathrm{cm^{-3}}$, instead of the usual mol $\mathrm{L^{-1}}$.

Corresponding to the concepts of forward and reverse reactions, we distinguish the (negative) current density $\vec{i}_{\mathrm{H}}$ of hydrogen deposition and the (positive) current density $\overleftarrow{i}_{\mathrm{H}}$ of hydrogen ionisation. At present, however, our case more precisely is $i_{\mathrm{H}} = \vec{i}_{\mathrm{H}}$, requiring that the potential is considerably more negative than the hydrogen equilibrium potential, so that the reverse current is negligible.

Due to a special mechanism of transport over hydrogen bridges between water molecules, diffusion of $H^+$ is unusually rapid with $D_{\mathrm{H^+}} \simeq 10^{-4}\ \mathrm{cm^2\,s^{-1}}$. As noted above, the Nernst diffusion layer thickness is not readily calculated. Experience shows that a value of about $10^{-3}$ cm is a good guess for rapidly stirred solutions. In stagnant solutions, especially those filling pores, cracks, or some other recess, it can be larger by orders of magnitude. The important point is that $\delta_{\mathrm{Nernst}}$ enters an estimate of the maximum rate of acid corrosion in solutions of low acidity, which is arrived at by considering that the limiting case of Eq. (5.8) is $(c_{\mathrm{H^+}})_* = 0$. Therefore, the current density of hydrogen deposition by reduction of $H^+$ ions cannot exceed the *diffusion-limited current density*

$$i_{\mathrm{H,D}} = -F D_{\mathrm{H^+}} \frac{(c_{\mathrm{H^+}})_\circ}{\delta_{\mathrm{Nernst}}} . \tag{5.9}$$

From Eqs. (5.8) and (5.9) follows that

$$\frac{(c_{\mathrm{H^+}})_*}{(c_{\mathrm{H^+}})_\circ} = 1 - \frac{i_{\mathrm{H}}}{i_{\mathrm{H,D}}} . \tag{5.10}$$

When $|i_{\mathrm{H}}|$ is small, the concentration $(c_{\mathrm{H^+}})_*$ is practically the same as the bulk concentration , and there is equilibrium of the diffusion transport. More precisely, it is quasi-equilibrium, but we shall from now on drop the 'quasi'-prefix .

When, instead, at higher rates of hydrogen deposition, $(c_{\mathrm{H^+}})_* < (c_{\mathrm{H^+}})_\circ$ we have to consider that the activity $(a_{\mathrm{H^+}})_*$ is *potential-determining* in the sense that it enters the Nernst equation, which is $E_{\mathrm{H_2/H^+}} = (RT/zF)\ln(a_{\mathrm{H^+}})_*$ if we assume hydrogen bubble formation at $p_{\mathrm{H_2}} = 1\,\mathrm{bar}$. Here, we recall Eq. (4.12), defining the overpotential of an electrode reaction as the difference between actual potential and equilibrium potential. Also, we tentatively assume equilibrium for all reaction steps of hydrogen deposition, except the preceding transport of $H^+$ to the metal surface. For this case, we calculate $\eta_{\mathrm{diff}}$, of hydrogen deposition as the difference of equilibrium potentials, $E(a_{\mathrm{H^+}})_* - E(a_{\mathrm{H^+}})_\circ$. The result is

$$\eta_{\mathrm{H}} = \frac{RT}{F} \ln \frac{(c_{\mathrm{H^+}})_*}{(c_{\mathrm{H^+}})_\circ} , \quad \textit{for rate-determining diffusion,} \tag{5.11}$$

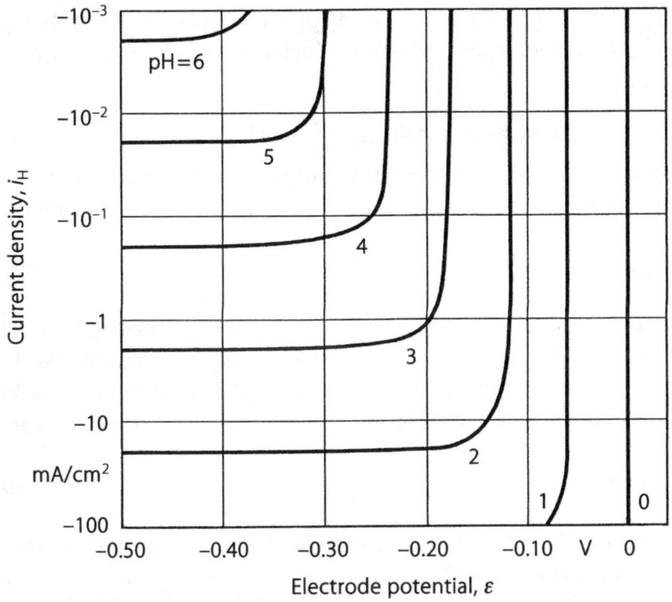


**Fig. 5.2.** Cathodic current-potential curves for hydrogen deposition governed by rate-determining diffusion of $H^+$ from a well stirred solution to the metal surface. $D_{H^+} = 10^{-4}\ cm^2\ s^{-1}$, $\delta_{Nernst} = 5 \times 10^{-3}$ cm; parameter: pH

because the activity coefficients approximately cancel. Substituting Eq. (5.10) we also find that

$$\eta_H = \frac{RT}{F} \ln\left(1 - \frac{i_H}{i_{H,D}}\right) , \quad \textit{for rate-determining diffusion.} \tag{5.12}$$

This equation for *diffusion overpotential* or *concentration polarization* and should be obeyed whenever in a well-stirred solution diffusion is the rate-determining step in hydrogen deposition. A set of such curves is presented in Fig. 5.2. But experience shows that, while in low-pH solutions the diffusion-limited current $i_{H,D}$ readily shows up, Eq. (5.12) does not describe $\varepsilon(i_H)$ as observed in acid solutions, proving that some other rate-determining step predominates. Actually, this behaviour is typical for most other electrode reactions as well, except for, e.g., redox couples of the type $Fe^{2+}/Fe^{3+}$.

Continuing to consider cathodic hydrogen deposition, we now assume equilibrium of the charge-transfer reaction across the metal/solution interface and of the transport reaction, and tentatively inspect the consequences of non-equilibrium of the Tafel reaction Eq. (5.4). If, as assumed, the preceding Volmer reaction is in equilibrium we have, in terms of electrochemical potentials,

$$\tilde{\mu}_{H^+} + \tilde{\mu}_{e^-} = \mu_{H^+} + F\phi_l + \mu_{e^-} - F\phi_m = (\mu_H)_{ad} . \tag{5.13}$$

In this equation, $\mu_{H^+} = (\mu_{H^+})^\circ + RT \ln a_{H^+}$, and $(\mu_H)_{ad} = (\mu_H)^\circ_{ad} + RT \ln(a_H)_{ad}$. We simplify conditions by neglecting activity coefficients, assuming $a_{H^+} = c_{H^+}$, and $(a_H)_{ad} = \theta_H$. In this case,

$$F\phi_{m,l} = (\mu_{H^+})^\circ + RT \ln c_{H^+} + \mu_{e^-} - (\mu_H)^\circ_{ad} - RT \ln \theta_H \,, \tag{5.14}$$

Contracting constant terms and constant contributions from the working ele-crode/reference electrode circuit into a constant, we have a Nernst equation:

$$\varepsilon = E_{H_{ad}/H^+} = const + \frac{RT}{F} \ln \frac{c_{H^+}}{\theta_H} \,. \tag{5.15}$$

When the reference electrode is SHE, *const* is the (unknown) standard potential $E^\circ_{H_{ad}/H^+}$. In a next step, Tafel's original and quite rational supposition was to assume that the rate of the recombination reaction should be proportional to $\theta^2$. Setting $i_H \sim \theta^2$ and once more contracting constants the result is:

$$i_H = const' \, c^2_{H^+} \exp\left(-\frac{2F}{RT}\,\varepsilon\right) \,. \tag{5.16}$$

In a semi-logarithmic $\log|i|$-$\varepsilon$ plot, these are straight lines. While they do have the general form of the experimental curves Eq. (4.15), they predict a slope $d\varepsilon/d\log|i_H| = 2.303RT/2F = 0.029\,\mathrm{V}$, which is far off the experimental value 0.12 V observed for hydrogen deposition on mercury. Also, the predicted dependence of the reaction rate on $c_{H^+}$ disagrees with experience. Therefore, the Tafel reaction, too, was not rate-determining for hydrogen deposition on mercury, and the observed overpotential was not what is termed *reaction overpotential.* The same is generally true for hydrogen deposition on solid metals. Therefore, we now consider the last possibility, which is rate-determining charge-transfer reaction Eq. (5.3) or *activation overpotential.*

As usual in formal reaction kinetics, we assume that the rate of the forward reaction, $H^+ + e^- \rightharpoonup H_{ad}$, is proportional to the concentration of $H^+$ ions in positions suitable for discharge, while the reverse reaction, $H_{ad} \rightharpoonup (H^+)_* + e^-$, should be proportional to the coverage of adsorbed atomic hydrogen, $\theta_H$. Expressing rates as current densities, we thus assume that

$$v_H = \overset{\rightharpoonup}{v}_H - \overset{\leftharpoonup}{v}_H = \overset{\rightharpoonup}{k}\,(c_{H^+})_* - \overset{\leftharpoonup}{k}\,\theta_H \,, \tag{5.17}$$

where the rate constants $\overset{\rightharpoonup}{k}$ and $\overset{\rightharpoonup}{k}$ are positive quantities. More precisely, we should consider that $\overset{\rightharpoonup}{k} = \overset{\rightharpoonup}{k}'\,(1-\theta_H)$, since Volmer charge transfer occurs at the bare metal surface only. Ignoring this, the following reasoning applies to small values of $\theta_H$ only.

To describe the influence of potential on the Volmer reaction we have to consider that the reactants must pass an energy barrier separating the energy states of reactants and products. In this context, energy is free enthalpy or Gibbs energy, and the difference of stable energy states of reactants and products is free reaction enthalpy, i.e., Gibbs reaction energy. Describing formal reaction kinetics, we shall not, however, in this chapter consider entropy effects. Instead we summarily note that molecular, or atomic, or ionic reactants must be 'activated', i.e., that they must, for instance by collisions with other particles, have at acquired least the *activation energy*, A, before they can pass over the barrier. According to *Boltzmann* statistics, the fraction of activated reactants is proportional to $\exp(-A/RT)$, and this leads to the *Arrhenius equation:*

$$k = k_\circ \exp(-A/RT) \,, \tag{5.18}$$

relating constants of type $k$ in Eq. (5.17) and temperature. Typically, activation energies have numerical values up to about 150 kJ mol$^{-1}$. We postpone the discussion of the nature of what is termed the 'activated complex' in a reaction and directly proceed to introduce activation energies of the forward and of the reverse reaction into Eq. (5.17). Also, we proceed to expressing rates as current densities by considering the conventions concerning the sign of currents, and by multiplying the constants with the Faraday constant: $Fk = K'$. The result is

$$i_\mathrm{H} = \vec{i}_\mathrm{H} + \overleftarrow{i}_\mathrm{H} = -(c_{\mathrm{H}^+})_* \, \vec{K}' \exp\left(-\frac{\vec{A}}{RT}\right) + \theta_\mathrm{H} \, \overleftarrow{K}' \exp\left(-\frac{\overleftarrow{A}}{RT}\right) \tag{5.19}$$

We now consider that in the course of the Volmer reaction hydrated hydronium ions are discharged to adsorbed hydrogen atoms which do not carry a hydration shell. Very small, with a radius of about 0.05 nm, they are much closer to the surface than the hydrated ions. In other words: During the Volmer reaction the center of hydrogen ions moves closer to the surface along a *reaction coordinate*, $\xi$, parallel to the axis $x$. In the course of this move, the energy of the species changes from a stable minimum for the hydrated hydronium ion ion at closest distance to the metal surface at $\xi = \xi_*$ through the activation peak to a stable minimum for the adsorbed hydrogen atom at $\xi = 0$. Schematically, this course of energy is indicated by the solid curve in Fig. 5.3. It is qualitative only, arbitrary in any quantitative detail. Since the species is a univalent cation, the energy contains, at distance $\xi$ from the metal surface, elecrostatic energy, $F\phi_\xi$.

For further progress, we simplify the situation by assuming a linear decrease of $\phi_\xi$ from $\phi_m$ to $\phi_l$ between $\xi = 0$ to $\xi = \xi_*$. Thus, the double layer is thought to be a capacitor with the metal surface as one plate carrying a - positive or negative - charge due to an asymmetry in metal electron distribution, as discussed in Chap. 3. The opposite plate is the solution surface adjacent to the metal at the distance of closest approach of ions including hydration shells, carrying an equal but opposite – negative or positive – charge caused by excess anions or cations. We must, incidentally, not assume that hydrogen ions, though present, are necessarily important for the charge of solution side of the capacitor. Instead, the charge may be that of an excess of other ions present in the solution.

In terms of electrical engineering, the capacitor 'leaks' due to the passage of charges carried by the transfer reaction. Therefore, an *equivalent circuit* would be a capacitor switched parallel to a resistor representing the polarization resistance, $R_\pi$. Since as a rule steady-state current-potential curves are non-linear, $R_\pi$ usually is non-ohmic, i.e current dependent. According to the model, the thickness of the double layer, $\delta_\mathrm{dl}$, is of the order of several Å. Therefore, the

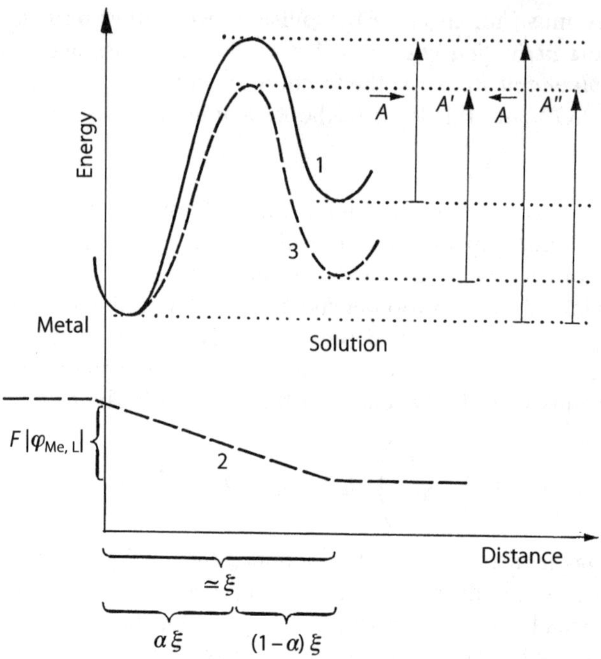


**Fig. 5.3.** Energy change during forward and reverse Volmer reaction as function of the distance across the electrical double layer at the metal/solution interface. (See text)

double layer is expected to have a high capacity. Experimental values are in the range of 20 microfarad (1 $\mu$F = 1 $\mu$C $V^{-1}$) and thus very large, indeed.

Also according to this model, the double layer coincides with the distance through which the Volmer reaction proceeds: $\xi_* = \delta_{dl}$. From this it follows that the electrostatic contribution, $F\phi_\xi$, to the energy of the ionic species going through the Volmer reaction changes linearly from $F\phi_m$ to $F\phi_l$, as also shown in Fig. 5.3. The graph of $F\phi$ vs. $\xi$ again is arbitrary, and we remember that Galvani potentials are quantities evading measurement. Therefore, the broken curve 3 representing non-electrostatic, chemical energy also is arbitrary. Still, the schematic figure clearly shows that the activation energies for the reaction of $H^+$ ions differ from what would apply to an uncharged species. We are unable to tell more about these changes than that one activation energy is increased by some fraction $\alpha F \phi_{m,l}$, while the other is lowered by some fraction $\beta F \phi_{m,l}$. The numbers, $0 \leq \alpha,\beta \leq 1$, are the *transfer coefficients* of the electrode reaction. As seen from the figure, we expect that $\beta = 1 - \alpha$, but we shall continue to write $\beta$. By arbitrary choice, which does not matter because $\phi_{m,l}$ may be positive as well as negative, we further assume that

$$a)\ \overrightarrow{A} = A' + \beta F \phi_{m,l}\ , \quad b)\ \overleftarrow{A} = A'' - \alpha F \phi_{m,l}\ . \qquad (5.20)$$

Substituting these relations into Eq. (5.18) we obtain an expression for the current densities of hydrogen deposition and hydrogen ionisation relating current densities and potential:

$$i_{\mathrm{H}} = -\overrightarrow{K}' \exp\left(-\frac{A'}{RT}\right) (c_{\mathrm{H}^+})_* \exp\left(-\frac{\beta F}{RT}\phi_{m,l}\right) + \\ + \overleftarrow{K}' \left(-\frac{A''}{RT}\right) \theta_{\mathrm{H}} \exp\left(\frac{\alpha F}{RT}\phi_{m,l}\right) . \tag{5.21}$$

At constant temperature, terms $\exp(-A/RT)$ are constants. Also, while differences, $\phi_{m,l}$, of Galvani potentials are not measurable, we know that, by virtue of the three-electrode experimental arrangement for measuring current-potential curves described in Fig. 4.6, changes $\Delta\phi_{m,l}$ are equal to changes $\Delta\varepsilon$ of the electrode potential, because $\phi_{m,l} = \varepsilon + const$, where *const* is the sum of contributions to $\varepsilon$ from the null-current reference-electrode/working-electrode circuit. Accounting for these contributions by once more contracting constants and abbreviating, temporarily, $F/RT = \Phi'$ we arrive at:

$$i_{\mathrm{H}} = -\overrightarrow{K}\,(c_{\mathrm{H}^+})_* \exp(-\beta\,\Phi'\,\varepsilon) + \overleftarrow{K}\,\theta_{\mathrm{H}} \exp(\alpha\,\Phi'\,\varepsilon) . \tag{5.22}$$

This is the fundamental *Butler-Volmer* (or, more correct: *Erdey Grusz-Butler-Volmer*) equation, as applied to the special case of the hydrogen electrode.

We recall that the hydrogen overpotential is $\eta_{\mathrm{H}} = \varepsilon - E_{\mathrm{H_2/H^+}}$, by definition, and that hydrogen deposition and hydrogen ionisation proceed at equal rates for $\varepsilon = E_{\mathrm{H_2/H^+}}$, or $\eta_{\mathrm{H}} = 0.$, Corresponding to the concept of equilibrium exchange rates introduced in Chap. 2, we define the *equilibrium exchange current density* of the hydrogen electrode as

$$i_{\mathrm{H}}^{\circ} = |\,\overrightarrow{i}_{\mathrm{H}}\,(E_{\mathrm{H_2/H^+}})| = \overleftarrow{i}_{\mathrm{H}}\,(E_{\mathrm{H_2/H^+}}) . \tag{5.23}$$

At equilibrium $(c_{\mathrm{H}^+})_* = (c_{\mathrm{H}^+})_\circ$, and $\theta_{\mathrm{H}}$ is at some equilibrium value $(\theta_{\mathrm{H}})_{\mathrm{eq}}$. Therefore,

$$i_{\mathrm{H}}^{\circ} = \overrightarrow{K}\,(c_{\mathrm{H}^+})_\circ \exp\left(-\beta\Phi'\,E_{\mathrm{H_2/H^+}}\right) = \overleftarrow{K}\,(\theta_{\mathrm{H}})_{\mathrm{eq}} \exp(\alpha\Phi'\,E_{\mathrm{H_2/H^+}}) . \tag{5.24}$$

Combining Eqs. (5.24) and (5.22) we finally have

$$i_{\mathrm{H}} = -i_{\mathrm{H}}^{\circ}\left\{\frac{(c_{\mathrm{H}^+})_*}{(c_{\mathrm{H}^+})_\circ} \exp\left(-\beta\,\Phi'\,\eta_{\mathrm{H}}\right) - \frac{\theta_{\mathrm{H}}}{(\theta_{\mathrm{H}})_{\mathrm{eq}}} \exp\left(\alpha\,\Phi'\,\eta_{\mathrm{H}}\right)\right\} . \tag{5.25}$$

The equation allows for non-equilibrium conditions for both the transport of $\mathrm{H}^+$ to the metal surface and the Tafel recombination. However, when the charge transferring Volmer reaction alone is rate-determining, both pre-exponential ratios in the equation are 1 and the equation conveniently reduces to:

$$i_{\mathrm{H}} = -i_{\mathrm{H}}^{\circ}\left\{\exp(-\beta\,\Phi'\,\eta_{\mathrm{H}}) - \exp(\alpha\,\Phi'\,\eta_{\mathrm{H}})\right\} . \tag{5.26}$$

The charge transfer coefficients $\alpha$,$\beta$, depend on the $\phi$-$\xi$ function through the double layer. A likely supposition is that the maximum of the energy barrier is at $\xi = \delta_{\mathrm{dl}}/2$, and that $\alpha = \beta = 0.5$. The superposition of the now symmetrical

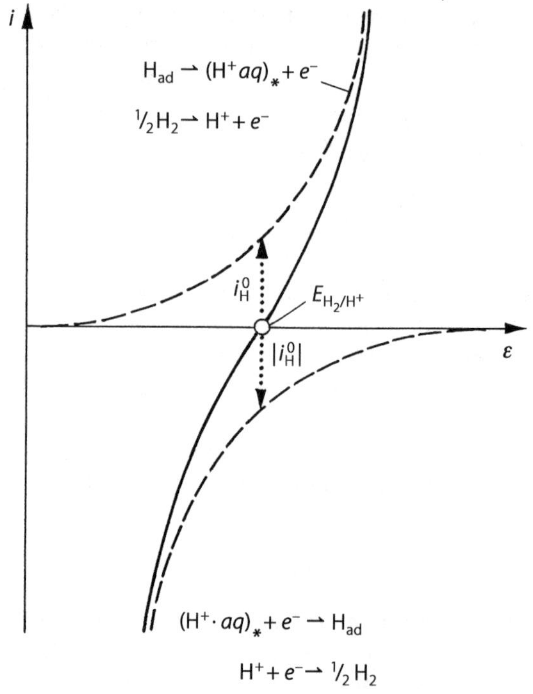


**Fig. 5.4.** *Broken curves:* Partial current-potential curves for activation-controlled cathodic hydrogen deposition and anodic hydrogen ionisation, assuming equilibrium for both $H^+$-transport to the surface, and recombination of atomic to molecular hydrogen. Transfer coefficients $\alpha = \beta = 0.5$. *Solid curve:* $i_H = \vec{i}_H + \overleftarrow{i}_H$

current-potential curves $\vec{i}_H\ (\varepsilon)$ and $\overleftarrow{i}_H\ (\varepsilon)$ to $i_H(\varepsilon)$ is schematically shown in Fig. 5.4. Whether $i_H$ should be expressed as a function of $\varepsilon$, as in Eq. (5.22), or as a function of $\eta_H$, as in Eq. (5.26), depends on circumstances. In the following chapter it will be seen that for the discussion of corrosion processes electrode potentials rather than overpotentials are the suitable independent variable.

For sufficiently negative values of the overpotential, $\eta_H$, the second term of the right hand side of Eqs. (5.22–5.26) is negligible regardless of the magnitude of $i_H^\circ$, and $i_H = \vec{i}_H$ . With $\beta = 0.5$ this already applies for $\eta_H \leq -0.12\,\text{V}$, and in this range of potentials the current-potential curve Eq. (5.22) reduces to to

$$i_H = -\vec{K}\,(c_{H^+})_*\exp(-\beta\,\Phi'\,\varepsilon)\,. \tag{5.27}$$

Furthermore, when the Volmer charge transfer reaction alone is rate-determining, $(c_{H^+})_* = (c_{H^+})_\circ$, and

$$i_H = -\vec{K}\,(c_{H^+})_\circ\exp(-\beta\,\Phi'\,\varepsilon) = -i_H^\circ\exp(-\beta\,\Phi'\,\eta_H)\,. \tag{5.28}$$

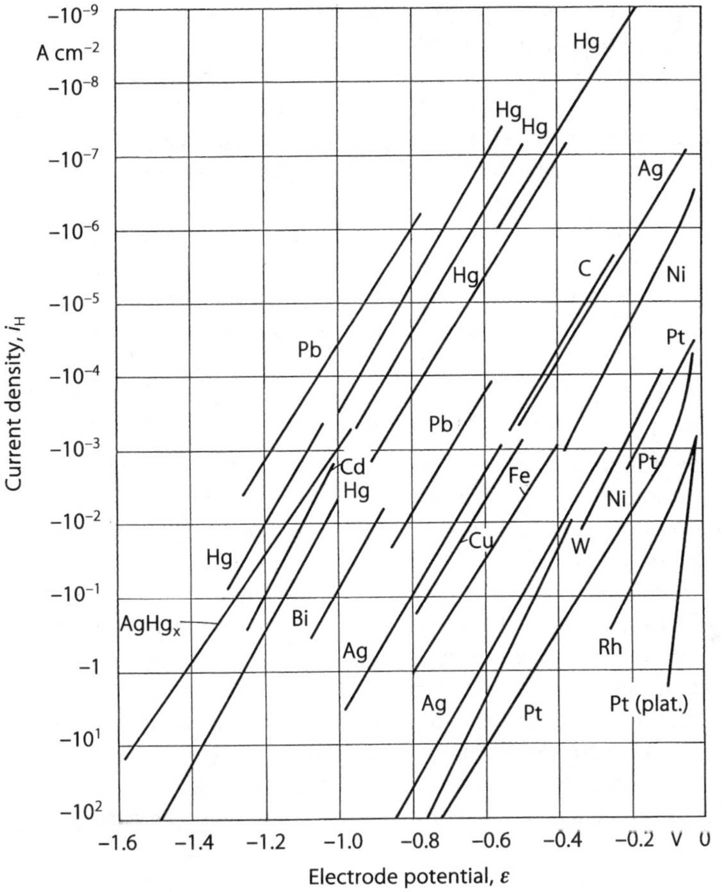


**Fig. 5.5.** Current-potential curves for cathodic hydrogen deposition observed on various metals from various solutions. (Compilation from Vetter)

Plotting log $|i|$ vs. $\varepsilon$, the graphs of these functions again are straight lines, with slopes

$$\frac{d\varepsilon}{d\log|i|} = \frac{2.303}{0.5\,\Phi'} = 0.118\text{V}\ . \tag{5.29}$$

Both the predicted dependence on $c_{H^+}$ and the slope confirm the experimental observations for cathodic hydrogen deposition on mercury described in Chap. 4, and there is little doubt that in this system the rate determining step was the charge transfer reaction. Also, as documented by Fig. 5.5, this appears to apply to many other metals as well. We note that, rather confusingly, the straight lines in semi-logarithmic plots of current-potential curves and their slope often are called 'Tafel lines' and 'Tafel slopes', respectively.

At this point, a warning appears appropriate. It is that not only for the hydrogen electrode, but quite generally for electrode reactions at bare metal surfaces, spurious effects caused by impurities, if not controlled by very careful preparation of both metal and electrolytic solutions, tend to seriously influence current-potential measurements, impairing reproducibility. The determination of reproducible, useful values of Tafel slopes therefore is no easy task, as current-potential curves obtained in experiments carried out without sufficient precaution tend to be distorted. The first useful, but not always followed rule to avoid errors is to check measurements for hysteresis effects caused by slow adsorption and/or desorption of impurities. We shall, incidentally, see shortly that such impurity effects sometimes are important, when the result is corrosion inhibition.

As shown by Fig. 5.5, values of cathodic hydrogen overpotential differ, at any given potential, by many orders of magnitude for different metals. This reflects an equally large range of exchange current densities, $i^\circ_{\mathrm{H}}$, or else of $\vec{K}$ values, which by far outweighs changes caused by variations of the solution composition. As an example, we quote that in 1 M HCl a value of about $10^{-3}$ A cm$^2$ was observed for platinum, but only about $2 \times 10^{-13}$ A cm$^2$ for lead [5]. We note that the latter number easily explains why the lead acid battery is not seriously impaired by acid corrosion of lead. At the same time, the high value for platinum explains why this metal is a preferred choice for hydrogen/oxygen fuel cells as well as for the construction of the standard hydrogen electrode. We add that the range of exchange current densities as well as the sensitivity to impurities is typical for many electrode reactions other than hydrogen deposition, and a source of much disagreement in work on electrode kinetics.

The variation of exchange current densities in one and the same solution depending on the nature of the metal reflects similar variations in constants $K$ appearing in relations of type Eq. (5.22), reappearing, mutatis mutandis, in Butler-Volmer equations applying to charge transfer kinetics of oxygen reduction or metal dissolution, and other electrode reactions. As yet, theoretical calculations of these quantities from principles of metal physics, obviously highly desirable, are hardly available. For technical metals and alloys interesting in actual corrosion service, even if nominally pure, accidental surface properties produced by varying methods of surface preparation, by grain boundaries, by emerging screw dislocations, by precipitates, and other factors, no such possibility really exists, and the magnitude of exchange current densities, or constants $K$, must be taken as they are found empirically in every case.

When an exchange current density of some reaction, labeled $r$, is very low, e.g. $\leq 10^{-10}$ A cm$^{-2}$, so that the reaction is highly irreversible, the current densities of both forward and reverse reaction are negligible at least in the vicinity of the equilibrium potential $E_r$. In the case of the hydrogen electrode, a further point will interest further below: When the reaction is highly irreversible, anodic hydrogen ionisation will be negligible because any supply of adsorbed atomic hydrogen will immediately be exhausted. Overrunning a calculated equilibrium potential $E_{\mathrm{H_2/H^+}}$ in the anodic direction, quite normal for studies of passivity, (see Chap. 10), does not, then, imply a noticeable current of hydrogen ionisation.

Further above, the methods described as useful to analyze the rate-determining mechanism of hydrogen deposition were to study a) the influence of the concentration of the reactant $H^+$, i.e. of the *reaction order* with respect to this species, and b) the slope of the logarithmic current-voltage curves. Both quantities are steady-state properties of the system under investigation, and we especially point out that steady-state corrosion is, of course, governed by steady-state current-voltage curves. At the same time, non steady-state properties such as the response of electrode reactions to pulsed polarization, or else to alternating-current polarization are important or may even have more power for the analysis of electrode reaction kinetics.

We continue to follow classical procedures by attending to the Heyrovsky electrochemical recombination reaction Eq. (5.5). Considering the cathodic current, $\vec{i}_H$, only, assuming equilibrium of transport of $H^+$ to the metal surface, and assuming $\beta = 0.5$, the rate law is expected to be

$$i_H = \vec{i}_H = -k\,\theta_H\,(c_{H^+})_*\,\exp(-0.5\,\Phi'\,\varepsilon)\;. \tag{5.30}$$

The coverage, $\theta_H$, of the surface with atomic hydrogen probably depends on $i_H$, but usually the relation is not explicitly known. When $\theta_H$ is small the Heyrovsky reaction is negligible; when, instead, $\theta_H = 1$, the rate law has the same form as for the Volmer reaction. At intermediate values of the coverage, we are at a loss to quantify the relation except for an interesting special case: It is that, even if probably valid for small values of $\theta_H$ only, the Volmer reaction preceding the Heyrovsky reaction may be in equilibrium, so that Eq. (5.15) applies, with the equilibrium value $\theta_H = (\theta_H)_{eq}$. Solving Eq. (5.15) for $(\theta_H)_{eq}$ and contracting constants, the equation for the current-voltage curve in the case of rate-determining Heyrovsky reaction then is:

$$i_H = const\; c_{H^+}^2\,\exp(-\Phi'\varepsilon)\exp(-0.5\,\Phi'\,\varepsilon)\;, \tag{5.31}$$

predicting a slope $d\varepsilon/d\log|i| = 1.5/2.303\Phi = 0.041\,\mathrm{V}$. As an apparently rare example, Fig. 5.6 shows a current-voltage curve for cathodic hydrogen deposition on pure iron from a acid solution containing a corrosion inhibitor changing from an initial 41 mV-slope to a 116 mV-slope [6]. This may be explained by assuming that initially Eq. (5.31) applies, but that eventually the iron surface becomes saturated with respect to adsorbed atomic hydrogen. The assumption is arbitrary, and the original supposition that, instead, $\theta_H$ is small throughout, is also legitimate. In this case, the change in slope of the current-potential curve would indicate a change to rate-determination from the Heyrovsky to the Volmer reaction.

Increasing the rate of cathodic hydrogen deposition, $(c_{H^+})_*$ eventually becomes smaller than $(c_{H^+})_\circ$ due to slow diffusion of $H^+$. Using Eq. (5.11), once more assuming a rate-determining Volmer reaction, and ignoring anodic hydrogen ionisation, Eq. (5.25) changes to:

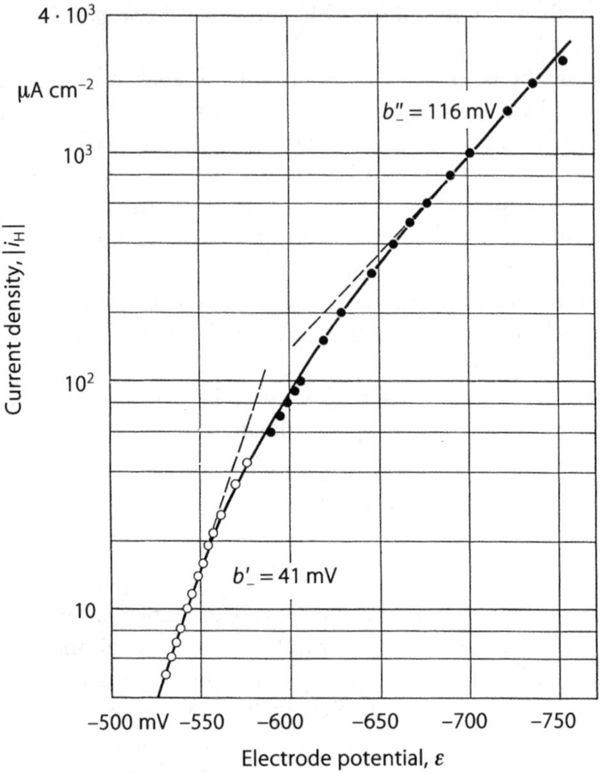


**Fig. 5.6.** Cathodic current-potential curve of hydrogen deposition on carbonyl iron from a $HClO_4/NaClO_4$ solution, pH 2.25, 25°C, with $1.8 \times 10^{-4}$ mol $L^{-1}$ phenylthiourea. (From Kaesche)

$$i_H = \vec{i}_H = -i_H^\circ \left(1 - \frac{i_H}{i_{H,D}}\right) \exp(-\beta\,\Phi'\,\eta_H)\,, \tag{5.32}$$

or

$$i_H = -i_H^\circ \exp(-\beta\,\Phi'\,\eta_H) \left\{1 - \frac{i_H^\circ}{i_{H,D}} \exp(-\beta\,\Phi'\,\eta_H)\right\}^{-1}\,. \tag{5.33}$$

A set of such curves, calculated for varying values of $i_H^\circ$, and constant pH and $i_{H,D}$, is shown in Fig. 5.7. The curves are governed by the ratio $i_H^\circ/i_{H,D}$, changing, as $i_H^\circ$ decreases, from pure concentration overpotential to pure charge transfer overpotential.

As mentioned previously, the reproducibility of $i_H^\circ$ values is poor. With respect to $i_{H,D,}$ he situation is better, because diffusion towards a surface is independent of incidental changes of surface properties. According to Eqs. (5.32) or (5.33), the diffusion overpotential of hydrogen deposition should become predominant for low values of $i_H^\circ$ and weakly acid solutions. As an example, Fig. 5.8 shows a set of cathodic curves for hydrogen deposition on iron in

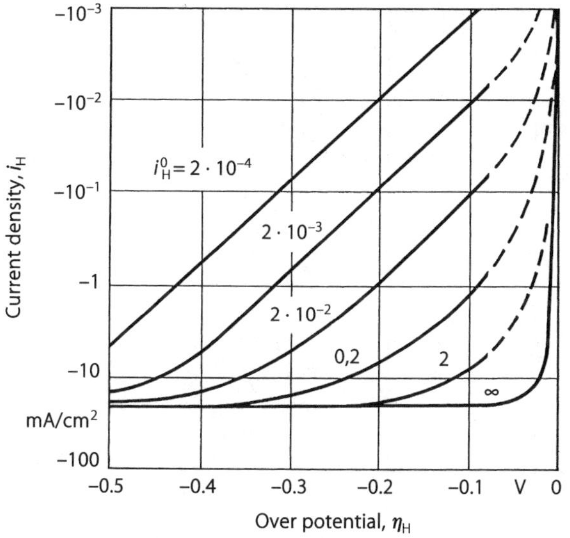


**Fig. 5.7.** Current-potential curves for cathodic hydrogen deposition from an acid solution at pH 2. *Solid curves:* $\vec{i}_H$ calculated from Eq. (5.33) for $i_{H,D} = 20 \text{mA cm}^{-2}$ and varying values of $i^\circ_H$. *Broken curves:* Estimated sum $\vec{i}_H + \overleftarrow{i}_H$ close to $E_{H_2/H^+}$

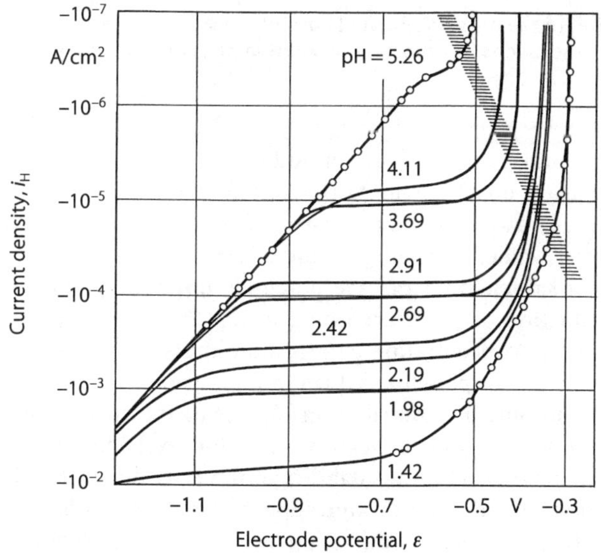


**Fig. 5.8.** Cathodic total current-potential curves for iron in stirred, deaerated HCl/4%NaCl solutions. Below the hatched band, the curves coincide with the cathodic current-potential curve for hydrogen deposition. Above the band, anodic iron dissolution is superposed. For $i \rightarrow 0$, the potential tends to the corrosion potential, $\varepsilon_{corr}$. (From Stern)

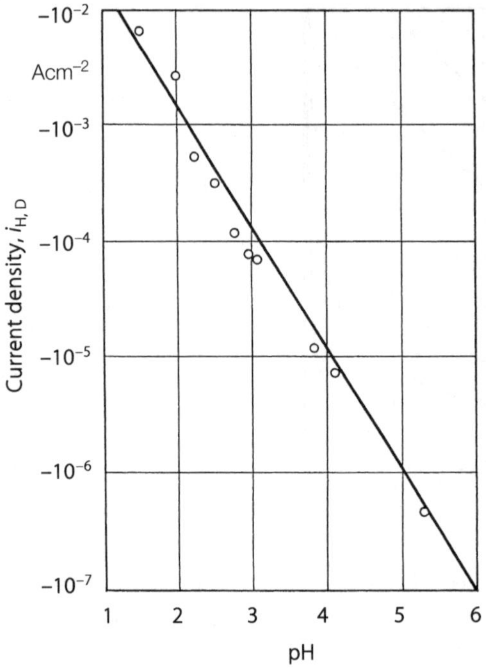


**Fig. 5.9.** Plotting the logarithm of the diffusion-limited current density of hydrogen deposition on iron from HCl/NaCl solutions as a function of solution pH, confirming Eq. (5.9). (From Stern)

HCl/NaCl solutions, confirming expectations. In addition, Fig. 5.9 proves the linear dependence of $\log |i_{H,D}|$, on pH required by Eq. (5.9).

Surprisingly, Fig. 5.8 also shows that currents of hydrogen deposition can surpass the diffusion-limited value, if only the potential is sufficiently decreased. The reason is that, given a sufficiently large overpotential, $\eta_H$, hydrogen deposition can proceed by reduction of water according to Eq. (3.72). As the concentration of water is high, no concentration polarization limits the rate of this reaction. Closer to the open-circuit potential, which in this case is the corrosion potential of iron in the solutions, the contribution $(i_H)_{H_2O}$ to the current density of hydrogen deposition is negligible, indicating a very low exchange current density of this electrode reaction due to the stability of the $H_2O$ molecule. The reverse current of oxidation of hydrogen to water does not play any role.

While hydrogen deposition by water reduction generally does not influence corrosion of iron, the situation is different for metals such as sodium reacting with water at a very negative $\varepsilon_{corr}$ (see Chap. 6); also, it limits the possibilities of practicing *cathodic corrosion protection* (see Appendix).

Here, we insert consequence of the existence of diffusion-limited current densities of hydrogen deposition. The calculation of $i_{H,D}$ certainly applies to the case of spontaneous acid corrosion with superposed anodic metal dissolution and cathodic hydrogen deposition proceeding at equivalent rates. It follows that, as long as hydrogen deposition occurs by $H^+$ reduction only, i.e. as long as $i_H$ cannot exceed $i_{H,D}$, the anodic current density, $i_{Fe}$, also cannot exceed $|i_{H,D}|$. We thus have an estimate of maximal rates of corrosion. For acid corrosion, the point is of minor importance, but the analogous reasoning applied to oxygen corrosion will further below be seen to be very useful.

As apparent from Fig. 5.8, the current-potential curve $(i_H)_{H_2O}(\varepsilon)$ is a Tafel line, suggesting a rate-determining charge transfer reaction. The slope $d \log | i |/d\varepsilon \simeq 0.19$ V is rather high, and the details of the reaction mechanism are open to discussion. In the case of hydrogen deposition from NaCl solutions on sodium and on indium amalgam it has been claimed that the reaction path involves initial deposition of sodium followed by a non-electrochemical reaction of metallic sodium with water [7]. This reaction path would, however, hardly seem probable for water reduction on iron. Disregarding these and other details, we account for the additional source of hydrogen by summarily writing, for negligible rate of anodic hydrogen ionisation:

$$(i_H)_{H_2O} = -(i_H^\circ)_{H_2O} \exp(-\beta \, n_e \, \Phi' \, \eta_H) \,, \tag{5.34}$$

and, wherever necessary, calculating the total current density of cathodic hydrogen deposition as the sum of contributions from $H^+$-reduction according to Eq. (5.28) and from $H_2O$-reduction according to Eq. (5.34).

In Eq. (5.34), a number $n_e$ has been tacitly introduced. This is the number of charges involved in the charge transfer reaction, designated as *charge transfer reaction valence*, not necessarily equal to the reaction valence, $z_r$ . In the simple case of hydrogen deposition, $n_e = z_r = z = 1$. For other reactions, the choice is not directly obvious, as will be seen for anodic iron dissolution. Apparently there is reason to conclude from quantum mechanics that always $n_e = 1$ , but there also is controversial discussion concerning this point.

## 5.2
## Oxygen Reduction and Oxygen Evolution

In this section, we mainly consider approximately neutral electrolytic solutions open to air, and saturated with respect to dissolved oxygen corresponding to the atmospheric oxygen pressure of about 0.2 bar. The corrosion process is metal dissolution, or else metal oxidation, accompanied by oxygen reduction. For the case of iron and mild steel, the sequence of elementary events in the cause of this type of corrosion has already been outlined in Chap. 2. We now inspect more closely the cathodic partial reaction described by Eq. (3.61).

The first reaction step, which is dissolution of gaseous oxygen into the solution, will be assumed to be in equilibrium. As for the discussion of hydrogen deposition, we also assume that the solution is well stirred and at constant concentration of all components, including dissolved oxygen, everywhere, except in the Nernst diffusion boundary layer. The rate of transport of dissolved oxygen across this layer to the distance of closest approach to the metal surface, symbolized by the equation

$$O_2 \rightarrow (O_2)_* \tag{5.35}$$

is again calculated using Fick's first law of steady-state diffusion. Since in this case $z_r = 4$, the rate of diffusion, expressed in terms of the cathodic, negative current density $i_{O_2} = \vec{i}_{O_2}$, of oxygen reduction is

$$i_{O_2} = -4FD_{O_2} \frac{dc_{O_2}}{dx} = -4FD_{O_2} \frac{(c_{O_2})_\circ - (c_{O_2})_*}{\delta_{\text{Nernst}}} . \tag{5.36}$$

As before, subscripts '∘' and '∗' denote bulk and surface concentrations, respectively. From this follows the existence of an upper limit for $|i_{O_2}|$, the diffusion-limited current density of oxygen reduction:

$$i_{O_2,\text{D}} = -4FD_{O_2} \frac{(c_{O_2})_\circ}{\delta_{\text{Nernst}}} . \tag{5.37}$$

At this point, we once more digress to corrosion, in this case to oxygen corrosion by superposition of anodic metal dissolution, especially iron dissolution, and cathodic oxygen reduction, proceeding at equivalent rates. In analogy to what has been explained for acid corrosion, it should be obvious that

$$i_{\text{corr}} \leq |i_{O_2,\text{D}}|; \; \textit{for uniform oxygen corrosion} , \tag{5.38}$$

indicates the upper limit of the corrosion rate. $(c_{O_2})_\circ$ is a small quantity, because the solubility of oxygen in aqueous solutions is low. At constant temperature, the solubility obeys *Henry's law*, i.e., $(c_{O_2})_\circ = k\, p_{O_2}$ [9]. From Fig. 5.10 we take, as a typical value, a solubility at 25°C of 25 cm$^3$ $O_2$/kg $H_2O$, or $1 \times 10^{-3}$ mol L$^{-1}$, at $p_{O_2} = 1$ bar. At 0.2 bar, the number is $0.2 \times 10^{-3}$ mol L$^{-1}$, or $2 \times 10^{-7}$ mol cm$^{-3}$. For the diffusion coefficient of oxygen in water we substitute ca. $1 \times 10^{-5}$ cm$^2$ s$^{-1}$, for $\delta_{\text{Nernst}}$, assuming a moderately well stirred solution, ca. $5 \times 10^{-3}$ cm. With these numbers the diffusion-limited current density of oxygen reduction, and also the upper limit of the current density of uniform oxygen corrosion is estimated as ca. 1.5 A m$^{-2}$. For iron, this limits the thickness loss rate of uniform oxygen corrosion to ca. 1 mm a$^{-1}$. Vigorous stirring may increase the estimate by a factor of about 5, but normal service conditions, with more or less stagnant solutions, can decrease it by a factor of 10 or more. We shall see below that in the case of iron and mild steel all these estimates are quite realistic.

To obtain an expression for the diffusion overpotential of oxygen reduction we only need to consider that in the relation Eq. (3.75) for the oxygen equilibrium potential the pressure $p_{O_2}$ can be replaced, owing to Henry's law, by the potential-determining concentration $c_{O_2}$, but we do not need to do this explicitly. In equilibrium, $c_{O_2} = (c_{O_2})_\circ$; in non-equilibrium, $(c_{O_2})_*$ corresponds to a fictitious lowered oxygen pressure; and the difference of potentials thus calculated is the diffusion overpotential:

$$\eta_{O_2} = \frac{RT}{4F} \ln \frac{(c_{O_2})_*}{(c_{O_2})_\circ} , \quad \textit{for rate-determining oxygen diffusion.} \tag{5.39}$$

As pointed out in Chap. 3, the oxygen electrode is notoriously irreversible, apparently due to the high stability of the oxygen molecule and a high activation energy of the rate-determining charge-transfer reaction. Since the oxygen

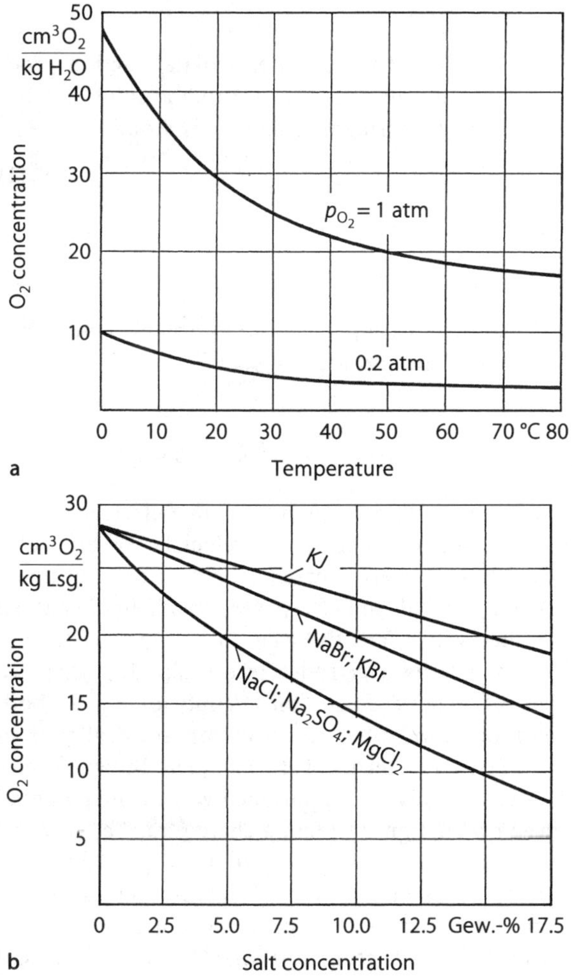


**Fig. 5.10a,b.** Solubility of oxygen **a** in water at oxygen pressures 1 and 0.2 bar between 0 and 80°C, **b** in salt solutions at oxygen pressure 1 bar, and 25°C. Data from Landolt-Börnstein

electrode is the cathode of the hydrogen/oxygen fuel cell, the mechanism of cathodic oxygen reduction has been the object of many investigations, which we shall not, however, consider in much detail [10]. In neutral solutions, where the electrode reaction is: $O_2 + 2H_2O \rightarrow 4OH^-$, a possible sequence of one-electron discharge steps is:

$$O_2 + e^- \rightarrow O_2^- , \tag{5.40}$$

$$O_2^- + H_2O \rightarrow HO_2 + OH^- ,$$

$$HO_2 + H_2O + e^- \rightarrow 2OH + OH^- ,$$

$$OH + e^- \rightarrow OH^- \ .$$

Concerning the unstable radicals $HO_2$ and OH as intermediate reaction products, there is spectroscopic evidence not directly for electrolytic solutions, but for gas phase reactions. An alternative intermediate species is atomic oxygen:

$$O_2^- + H_2O \rightarrow OH + O + OH^- \ , \qquad (5.41)$$

$$O + H_2O + e^- \rightarrow OH + OH^- \ ,$$

$$OH + e^- \rightarrow OH^- \ .$$

Also, hydrogen peroxide, $H_2O_2$, sometimes appears as a more stable and, therefore, more easily recognized intermediate product:

$$O_2 + 2H_2O + 2e^- \rightarrow H_2O_2 \qquad (5.42)$$

$$H_2O_2 + 2e^- \rightarrow 2OH^- \ ,$$

every two-electron step probably proceeding through more one-electrode steps.

For pH $\simeq 7$ the oxygen equilibrium potential, $E_{O_2/H^+}$, calculated by Eq. (3.75) is 0.80 V. Exchange current densities, $i^\circ_{O_2}$, certainly dependent on pH, but much more dependent on the nature of the metal, and always very small, do not exceed about $10^{-10}$ A cm$^{-2}$, and smaller values are the rule. For oxygen reduction at semiconducting oxide films of passive electrodes (see Chap. 10), they are usually still smaller by orders of magnitude. In the vicinity of $E_{O_2/H^+}$ both cathodic oxygen reduction and and anodic oxygen evolution are negligible in any case, and a quantitative evaluation in terms of the complete Butler-Volmer equation is of no practical interest. Only an important decrease in potential causes noticeable rates of oxygen reduction. If, then, a charge-transfer reaction is rate-determining we expect a relation $i \simeq \exp(-\beta\, n_e\, \Phi'\, \varepsilon)$, probably with $n_e = 1$, and probably with $\beta \simeq 0.5$. As before, $\Phi' = F/RT$. As long as we do not know the rate determining charge transfer step, we do not know how to specify '$\simeq$'; still we assume, probably accepting considerable inaccuracy, and for negligible rate of anodic oxygen evolution only, that

$$i_{O_2} = -\vec{k}\,(c_{O_2})_*\, \exp(-0.5\,\Phi'\,\varepsilon) \ . \qquad (5.43)$$

For small rates of cathodic oxygen reduction, $(c_{O_2})_* = (c_{O_2})_\circ$ and, by analogy to Eq. (5.28),

$$i_{O_2} = -i^\circ_{O_2}\, \exp(-0.5\,\Phi'\,\eta_{O_2}) \ ; \qquad (5.44)$$

For an exchange current density of, e.g., $10^{-10}$ A cm$^{-2}$, calculated current densities, $|i_{O_2}|$, of oxygen reduction exceed 1 $\mu$ cm$^{-2}$ only for $|\eta_{O_2}| > 0.5$ V.

When the rate of oxygen reduction increases, the surface concentration $(c_{O_2})_*$ drops to values lower than $(c_{O_2})_\circ$. In analogy to Eq. (5.32), the equation of the current-potential curve then turns into

$$i_{O_2} = \vec{i}_{O_2} = -i^\circ_{O_2}\left(1 - \frac{i_{O_2}}{i_{O_2,D}}\right)\exp(-\beta\,\Phi'\,\eta_{O_2}) \ , \qquad (5.45)$$

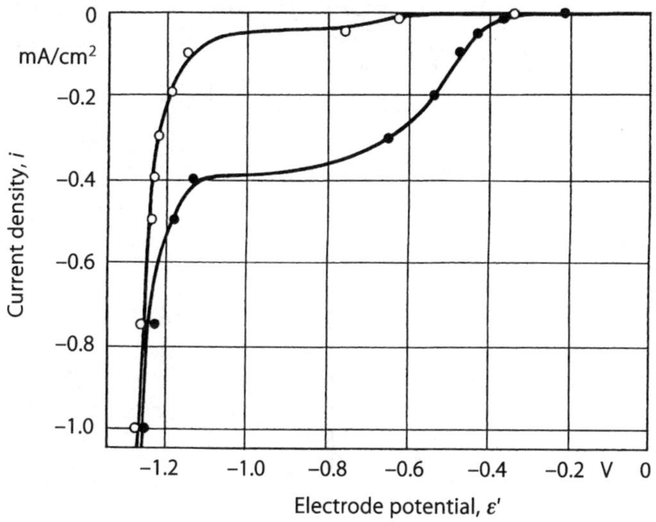


**Fig. 5.11.** Current-potential curve of cathodic oxygen reduction and cathodic hydrogen deposition by water reduction on stainless steel in stirred NaOH/0.5 M NaCl-solution, pH 11, 25°C. *Solid circles:* aerated solution; *open circles:* nearly deaerated solution. ε' referred to saturated calomel electrode. (From Kaesche)

or, after solving for $i_{O_2}$,

$$i_{O_2} = -i^{\circ}_{O_2} \exp\left(-\beta\,\Phi\,\eta_{H}\right)\left\{1 - \frac{i^{\circ}_{O_2}}{i_{O_2,D}} \exp\left(-\beta\,\Phi'\,\eta_{O_2}\right)\right\}^{-1} . \tag{5.46}$$

As an example of experimental evidence, Fig. 5.11 shows the cathodic current-potential curve of stainless steel (18 wt.% Cr, 8 wt.% Ni) in an aerated alkaline solution, where hydrogen deposition by reduction of $H^+$ is negligible [11]. In such solution, stainless steel initially carries a passivating oxide film (see Chap. 10), and anodic metal dissolution is negligible. By cathodic polarization the oxide film is reduced, but the effect does not show up in the steady-state current-potential curves. The observed current is the sum, $i_{\mathrm{cath}} = i_{O_2}+(i_{H}))_{H_2O}$, of cathodic reduction of oxygen and of cathodic reduction of water, corresponding to what is expected from Eqs. (5.45) or (5.46) and (5.34). The figure also shows the effect of almost complete deaeration of the solution, and there is little doubt that the cathodic plateau almost suppressed by this procedure is the cathodic diffusion-limited current density $i_{O_2,D}$.

In weakly acid solutions, hydrogen deposition by reduction of $H^+$ adds to the total cathodic current, which then is $i_{\mathrm{cath}} = i_{O_2} + (i_H)_{H^+} + (i_H)_{H_2O}$. We do not explicitly give the complete equation evaluated from Eq. (5.33), (5.34), and (5.46), but Fig. 5.12 shows that observed current-potential curves again correspond to expectations: At low pH, activation-controlled cathodic hydrogen deposition predominates; at intermediate values the diffusion-limited currents for hydrogen deposition by $H^+$ reduction and oxygen reduction are superposed,

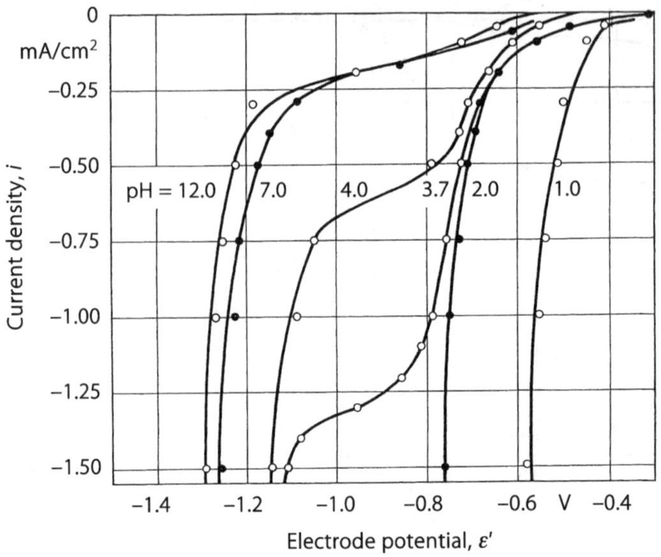


**Fig. 5.12.** Current-potential curves for cathodic hydrogen deposition by $H^+$ and $H_2O$ reduction and by cathodic oxygen reduction on stainless steel in stirred, aerated $H_2SO_4/NaOH/0.5$ M $Na_2SO_4$ solutions for varying values of pH, 25°C. ε' referred to saturated calomel electrode. (From Kaesche)

with the contribution from hydrogen deposition decreasing with increasing pH. Cathodic hydrogen deposition by water reduction eventually predominates at sufficiently negative potentials in low-acidity solutions.

Often, equations of type Eq. (5.46) oversimplify real oxygen reduction kinetics. An example is the double-wave of the current-potential curve shown in Fig. 5.13. In this case, the metal was a dropping mercury electrode, as used in polarography in order to avoid working with a metal surface possibly disturbed by slow contamination by solution impurities. Apparently, the double wave represents the diffusion-limited currents of the consecutive steps in Eq. (5.42) for oxygen reduction via intermediate metastable hydrogen peroxide [12]. If hydrogen peroxide reduction is retarded, the reaction valence, $z_r$, is 2 instead of 4. Also, a sluggish sequence of reaction steps can obscure the plateau of the diffusion-limited current-potential curve. All this, of course, affects the applicability of Eq. (5.38), but then the error made in estimating maximal corrosion rates will not be wrong by more than a tolerable factor 2.

So far, the diffusion-layer thickness $\delta_{\text{Nernst}}$ was introduced as an empirical quantity, estimated for the boundary layer at a metal surface in a solution stirred by conventional means. An experimental arrangement allowing, instead, the quantitative evaluation is the *rotating disk electrode*, i.e. a circular disk of radius $r$ touching the surface of a solution of kinematic viscosity $\nu$ and rotating at a speed $\omega$. The rotation induces a flow of the solution calculable from hydrodynamics (see Appendix). The flow is laminar up to a critical range of val-

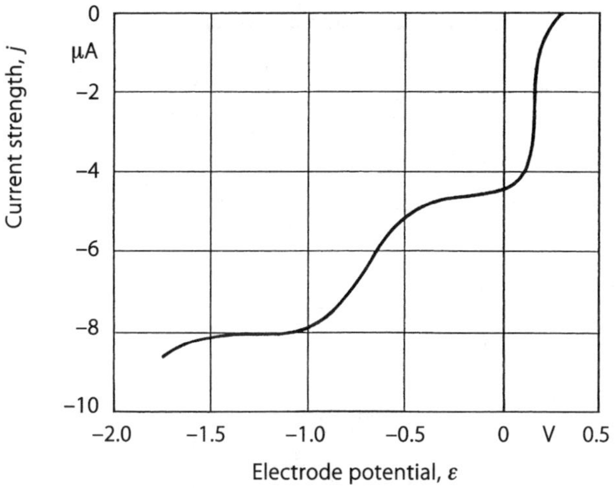


**Fig. 5.13.** Current-potential curve of oxygen reduction from 0.1 M KCl-solution at a dropping mercury electrode. (From Kolthoff and Miller)

ues of the *Reynolds* number, *Re*, defined as the ratio $\omega\, r^2/\nu$. For laminar flow, $\delta_{\text{Nernst}} \sim \omega^{-1/2}$, and $|i_{O_2,D}| \sim \omega^{1/2}$. This method has been applied to nickel electrodes rotating in the surface of otherwise stagnant, aerated 0.1 M NaOH solutions. In these, nickel initially is passive at a corrosion potential of −0.08 V [13]. Figure 5.14 summarizes the results of the experiments: During slow potentials scans in the cathodic direction, the current density of oxygen reduction at first is very low due to an almost negligible value of the exchange current density at the oxide film, which is supposed to be $Ni(OH)_2$. On passing the nikel/oxide equilibrium potential the passivating film is reduced, and cathodic hydrogen deposition by water reduction sets on. Anodic metal dissolution always is unimportant. Reversing the potential scan at −1.05 V, the cathodic diffusion-limited current of oxygen reduction now is prominent, because the nickel surface is bare, and the oxygen exchange current density relatively high. In this region, current densities are diffusion-limited and depend on *Re*, as expected. On re-passing the potential of oxide formation in the anodic direction, the electrodes reoxidise, and the cycle can be repeated.

At sufficiently high values of the rotational speed, $\omega$, oxygen reduction is under pure charge-transfer control, and Eq. (5.44) applies. We designate this as a 'kinetic current density', and substitute this abbreviation in Eq. (5.45), which we now rewrite into

$$\frac{1}{i_{O_2}} = \frac{1}{i_{O_2,\text{kin}}} + \frac{1}{i_{O_2,D}}\,, \quad i_{O_2,\text{kin}} = i^{\circ}_{O_2}\, \exp(-\beta\,\Phi'\,\eta_{O_2})\,. \tag{5.47}$$

Using this relation, charge-transfer data can be extracted from current-potential measurements of oxygen-reduction kinetics under mixed diffusion and charge transfer control, as shown in Fig. 5.15 [14]. In this graph, the $1/\sqrt{\omega}$-dependence

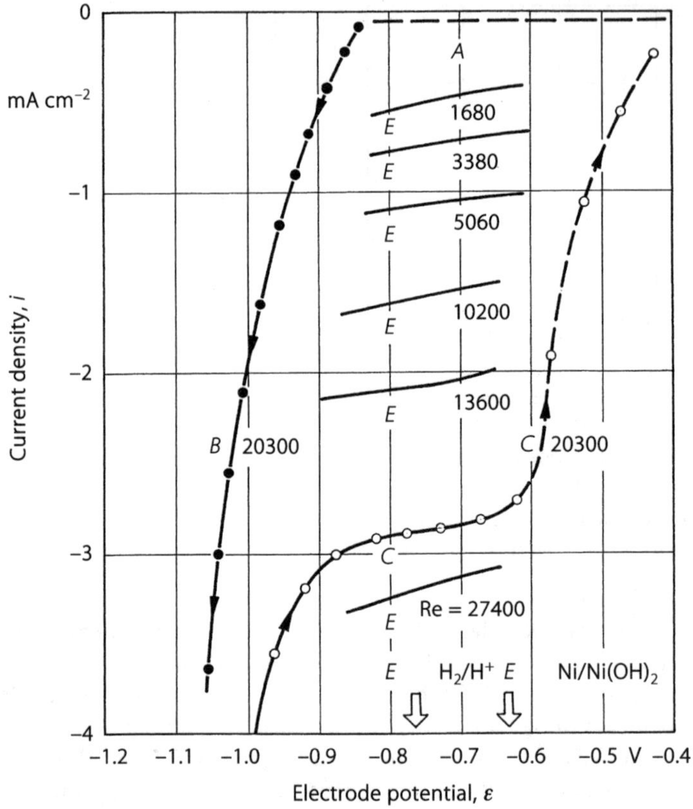


**Fig. 5.14.** Quasi-stationary current-potential curves for cathodic oxygen reduction and hydrogen deposition by water reduction on rotating nickel disk electrodes in oxygen-saturated 0.1 M NaOH solutions. *A*: Oxygen-reduction on passive metal. *B*: Hydrogen deposition by water reduction. *C*: Hydrogen deposition, oxygen reduction and repassivation. *E*: Diffusion-limited currents of oxygen reduction. Parameter: *Reynolds* number. (Data from Postlethwaite and Sephton)

of the reciprocal current is clearly seen. $1/i_{O_2,\mathrm{kin}}$ is the intercept of the straight lines with the ordinate .

## 5.3 Metal Dissolution and Metal Deposition

Inspecting the kinetics of metal dissolution and redeposition electrode reactions, we recall that dissolved metal ions, $Me^{z+}$, carry a hydration shell. We presently disregard complications such the formation of complex copper-ammonia ions, $Cu(NH_3)_4^{2+}$ or cyanide complexes such as $Cd(CN)_4^{2-}$. Doing so, we continue to concentrate on main principles, disregarding, as previously for hydrogen deposition and oxygen reduction, many more subtle details. We also disregard for

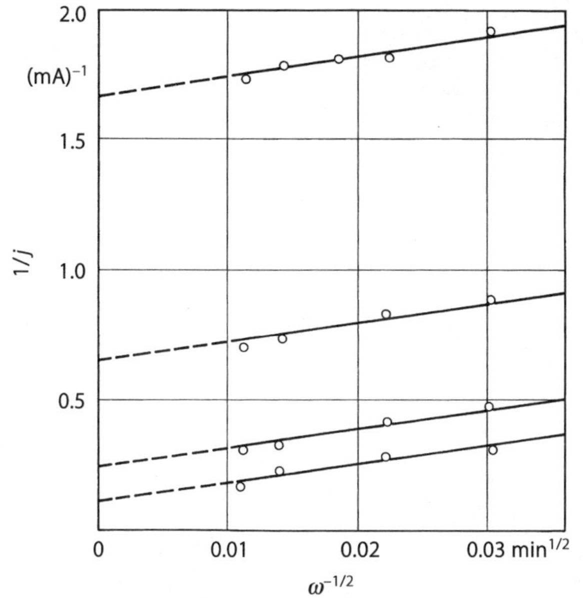


**Fig. 5.15.** The reciprocal current density of reduction of oxygen on iron rotating disk electrodes as function of the reciprocal rotational speed. Neutral boric acid/borate buffer solution, 1 bar oxygen pressure, 0.5 cm disk radius. Parameter: Electrode potential, from bottom upwards: −0.68, −0.58, −0.48, −0.43 V. (From Jovancicevic and Bockris)

the moment the influence of surface impurities, grain boundaries, dislocations, etc. We continue using a simplified approach, recalling that quantities such as exchange current densities evade calculation from metal-physical surface principles at least in the case of technical metals and alloys anyway.

The model of a metal surface used to describe $Me/Me^{z+}$ kinetics is schematically shown in Fig. 5.16: We are looking at a monocrystalline area of a metal surface, i.e. at the face of a single grain of an otherwise polycrystalline metal. The surface has ledges of atomic height, and the ledges have kinks. Kink-site atoms are said to occupy 'half-crystal positions' because the crystal binding force is one-half of that in the bulk crystal. Also, kink sites are, at least at low overpotential, energetically preferred sites of both anodic metal dissolution and cathodic metal deposition. During these elementary processes, kink sites always are recreated until a ledge is completely dissolved or completed. Since many kinked ledges are available at every moment, anodic metal dissolution does not easily exhaust kink sites. Other sites of elementary $Me/Me^{z+}$ half-reactions, viz. terrace sites, are available at higher expense of energy, but this will become a point of major interest only in Chap. 8. Also, atomic surface ledges often will belong to screw dislocations emerging in the surface, but this again will not be explicitly considered in the following. Terefore, the model used presently is exceedingly simple: As in the preceding sections, the metal/solution interface

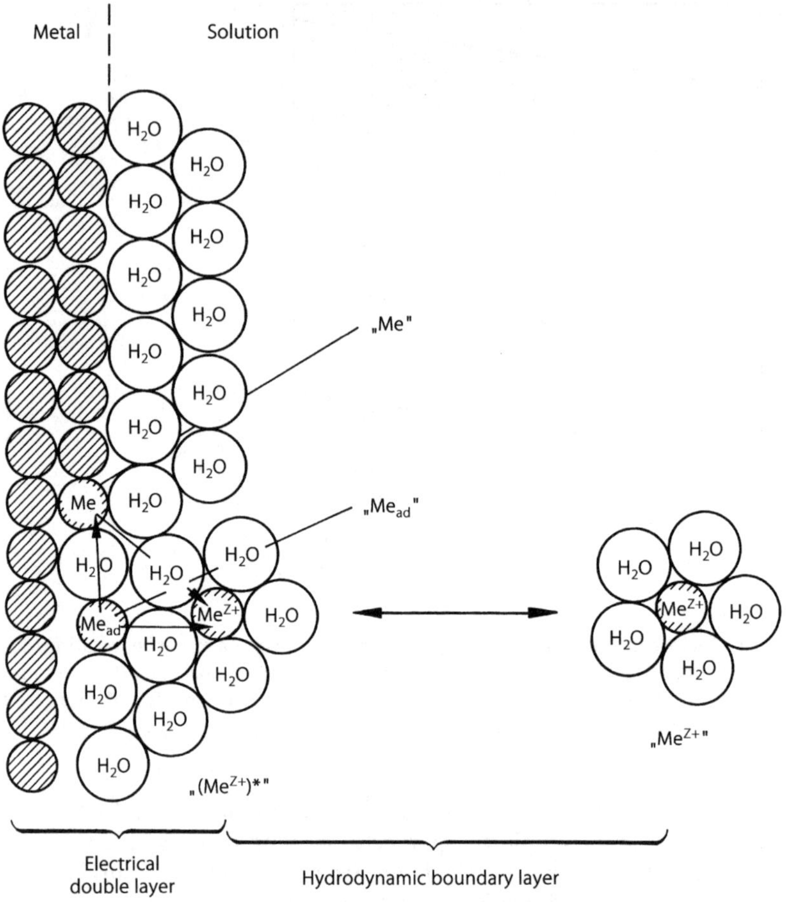


**Fig. 5.16.** A simple picture of the transfer of metal cations of a monocrystalline surface across the electric double layer into the solution either directly from kink sites, or via an intermediate stage of adsorption on the surface

is thought to be a simple leaking capacitor. In the overall charge-transfer half-reaction of metal dissolution the reactants are atoms in kink sites, $Me_{kink}$, in the metal side of the electric double layer, and the products are hydrated ions, $(Me^{z+})_*$, in the solution side of the layer at closest distance, as determined by the radius of the hydration shell. Leaving kink sites, atoms have two possibilities: Remaining in the interior, metal side of the interfacial double layer, they can move from their surface crystallografic site into an adsorption layer, where they freely move laterally as still uncharged 'ad-atoms', $Me_{ad}$, before they cross the double layer in the charge-transfer reaction step. Or else kink-site atoms may shortcut the two consecutive steps and directly go through the charge-transfer step across the double layer. In any case, ions $(Me^{z+})_*$ then move into the bulk of the solution to the $Me^{z+}$-position. Hence, the two possible reaction paths for anodic metal dissolution are:

$$Me_{kink} \rightarrow Me_{ad} \rightarrow (Me^{z+})_* \rightarrow Me^{z+} , \tag{5.48}$$

and

$$Me_{kink} \rightarrow (Me^{z+})_* \rightarrow Me^{z+} . \tag{5.49}$$

When the lifetime of ad-atoms is short, and/or the distances they can travel away from the kink sites is small, the two paths blend into one. At the same time, the charge-transfer step may be more complicated and separate into more sub-steps, as described below for the case of iron dissolution. Also at the same time, cathodic metal deposition is expected to occur through the same but reversed reaction steps.

Assuming metal dissolution via ad-atoms, an interesting case is that of equilibrium of both charge-transfer across the double layer and transport of dissolved cations into the bulk of the solution, but non-equilibrium of exchange of atoms between kink sites and the adsorption layer. Using arguments analogous to those described in Sect. 5.1 for the case of non-equilibrium between adsorbed atomic hydrogen and the gas phase, we then expect a deviation of the thermodynamic activity of ad-atoms from the equilibrium value. Limiting considerations to low degrees of coverage, $\theta_{Me,ad}$, we may set the activity of ad-atoms equal to the coverage, to obtain, by analogy to Eq. (5.15), an expression describing equilibrium between ad-atoms and cations dissolved in the bulk of the solution:

$$\varepsilon = const + \frac{RT}{zF} \ln \frac{a_{Me^{z+}}}{\theta_{Me,ad}} . \tag{5.50}$$

When the exchange between kink sites and adsorbed layer is not in equilibrium, the coverage of ad-atoms will depend on the current density of metal dissolution. This leads to an expression relating overpotential, equilibrium coverage and non-equilibrium coverage with the current density, which is the current-potential curve for metal dissolution and metal deposition for the case of rate-determining *crystallization overpotential.* It needs an explicit evaluation of the function $\theta_{Me,ad}(i)$. The many attempts in this direction are well refereed especially in [1]. We do not follow this line of research, as it is of only minor concern for corrosion kinetics, though obviously important for the kinetics of cathodic metal deposition and, therefore, for the techniques of metal plating.

When the sequence of reaction steps Eq. (5.48) blends into the sequence Eq. (5.49) we may, in a first approximation, pass the distinction between kink-site atoms and ad-atoms, and account for the influence of both species, $Me_{kink}$ and $Me_{ad}$, by a quantity $\theta_{Me}$ designating atoms generally able to cross the double layer. $\theta_{Me}$ may or may not be a function of the current density; if it is, then crystallisation overpotential contributes to charge-transfer overpotential. We then still expect, by analogy to the kinetics of the hydrogen electrode, an expression for the sum of the – positive, anodic – current density of metal dissolution and the – negative, cathodic – current density of metal deposition which reads:

$$\begin{aligned} i_{\mathrm{Me}} &= \overrightarrow{i}_{\mathrm{Me}} + \overleftarrow{i}_{\mathrm{Me}} \\ &= \overrightarrow{K}_{\mathrm{Me}}\ \theta_{\mathrm{Me}}\ \exp(\alpha\, n_e\, \Phi'\ \varepsilon) - \overleftarrow{K}_{\mathrm{Me}}\ (c_{\mathrm{Me}^{z+}})_*\ \exp(-\beta\, n_e\, \Phi'\ \varepsilon)\,. \end{aligned} \tag{5.51}$$

Constants $K$ are positive quantities. As before, $(c_{\mathrm{Me}^{z+}})_*$ is the concentration of dissolved $\mathrm{Me}^{z+}$ cations in the position of closest approach to the metal surface, the transfer coefficients $\alpha$,$\beta$ probably are close to 0.5, the charge transfer valence $n_e$ normally is 1, but sometimes perhaps 2.

Once more proceeding as previously for the hydrogen and the oxygen electrode, we define exchange current densities of $\mathrm{Me}/\mathrm{Me}^{z+}$ electrodes as

$$\begin{aligned} i^{\circ}_{\mathrm{Me}} &= \overrightarrow{K}_{\mathrm{Me}}\ (\theta_{\mathrm{Me}})_{\mathrm{eq}}\ \exp(\alpha\, n_e\, \Phi'\, E_{\mathrm{Me}/\mathrm{Me}^{z+}}) \\ &= \overleftarrow{K}_{\mathrm{Me}}\ (C_{\mathrm{Me}^{z+}})_*\ (c_{\mathrm{Me}^{z+}})_{\circ}\ \exp(-\beta\, n_e\, \Phi'\, E_{\mathrm{Me}/\mathrm{Me}^{z+}})\,. \end{aligned} \tag{5.52}$$

Exchange current densities depend on solution composition and surface properties, but again much more on the nature of the metal. A typical value for metals such as iron or nickel or chromium would be about $10^{-5}\mathrm{Acm}^{-2}$. Much smaller values are encountered as well, except for reversibel electrodes such as the $\mathrm{Zn}/\mathrm{Zn}^{2+}$ or the $\mathrm{Cu}/\mathrm{Cu}^{2+}$-electrode with exchange current densities higher by several orders of magnitude.

Recalling that, by definition, $\eta_{\mathrm{Me}} = \varepsilon - E_{\mathrm{Me}/\mathrm{Me}^{z+}}$, we finally find that

$$i_{\mathrm{Me}} = i^{\circ}_{\mathrm{Me}} \left\{ \frac{\theta_{\mathrm{Me}}}{(\theta_{\mathrm{Me}})_{\mathrm{eq}}} \exp(\alpha n_e\, \Phi'\, \eta_{\mathrm{Me}}) - \frac{(c_{\mathrm{Me}^{z+}})_*}{(c_{\mathrm{Me}^{z+}})_{\circ}} \exp(-\beta n_e\, \Phi'\, \eta_{\mathrm{Me}}) \right\}. \tag{5.53}$$

For large negative current densities of metal deposition, $(c_{\mathrm{Me}^{z+}})_* < (c_{\mathrm{Me}^{z+}})_{\circ}$. Then, the rate of anodic metal dissolution is negligible, the current, $i_{\mathrm{Me}}$, is equal to the negative current of metal deposition, and

$$\frac{(c_{\mathrm{Me}^{z+}})_*}{(c_{\mathrm{Me}^{z+}})_{\circ}} = 1 - \frac{|i_{\mathrm{Me}}|}{i_{\mathrm{Me,D}}}; \quad i_{\mathrm{Me,D}} = -zFD_{\mathrm{Me}} \frac{(c_{\mathrm{Me}^{z+}})_{\circ}}{\delta_{\mathrm{Nernst}}}\,. \tag{5.54}$$

Here, $i_{\mathrm{Me,D}}$ is the diffusion-limited current density of cathodic metal deposition. The derivation implies that cations $\mathrm{Me}^{z+}$ move through the diffusion boundary layer by diffusion only, electrolytic transference being suppressed by excess neutral salt additions. The diffusion coefficient, $D_{\mathrm{Me}}$, for monovalent cations is ca. $10^{-5}$; for divalent cations it is smaller, between $10^{-5}$ and $10^{-6}$ $\mathrm{cm}^2\ \mathrm{s}^{-1}$.

Anodic metal dissolution involves transport of $\mathrm{Me}^{z+}$ ions through he diffusion boundary layer into the bulk of the solution. An increase in $(c_{\mathrm{Me}^{z+}})_*$ is not limited by diffusion, but eventually the saturation concentration of some compound $\mathrm{Me_pX_q}$ will be reached, where X is an anion present in the solution. The deposition of the solid compound on the surface will normally block a further increase in the rate of anodic metal dissolution. This effect is disregarded presently, but will prominently reappear in the context of passivity.

For low current densities, we assume $\theta_{\mathrm{Me}}/(\theta_{\mathrm{Me}})_{\mathrm{eq}} = 1$. This turns Eq. (5.55) into

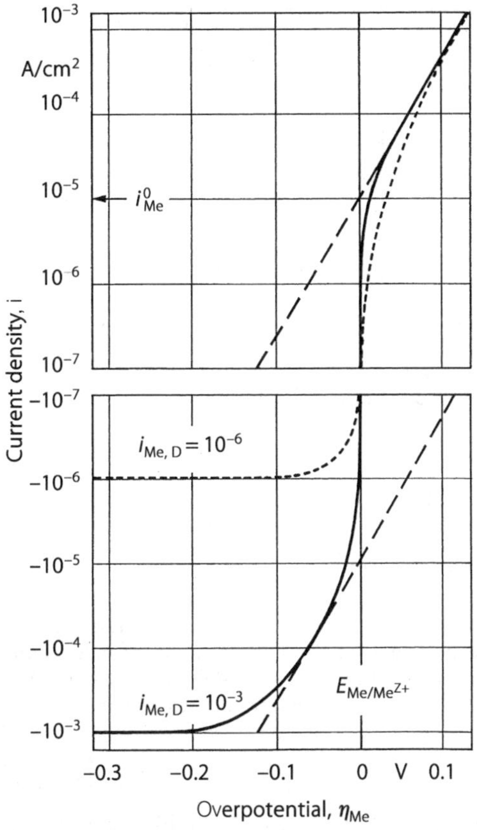


**Fig. 5.17.** Current-potential diagram for a $Me/Me^{z+}$ electrode according to Eq. (5.56), calculated with $n_e = 2, \alpha = \beta = 0.5$, $i^{\circ}_{Me} = 10^{-5}$ A cm$^{-2}$. $i_{Me,D} = 10^{-3}$ (*solid curve*) resp. $10^{-6}$ A cm$^{-2}$ (*dotted curve*). *Broken lines*: Tafel lines for forward and reverse partial reactions and pure charge-transfer control

$$i_{Me} = i^{\circ}_{Me} \frac{\exp(\alpha n_e \, \Phi \, \eta_{Me}) - \exp(-\beta n_e \, \Phi' \, \eta_{Me})}{1 - \frac{i^{\circ}_{Me}}{i_{Me,D}} \exp(\alpha n_e \, \Phi' \, \eta_{Me})} \,. \tag{5.55}$$

Figure 5.17 presents curves computed with Eq. (5.56) for arbitrary values of the constants as indicated in the legend. The graph shows two characteristic cases, viz. $i^{\circ}_{Me}/i_{Me,D} \ll 1$, and $\gg 1$, respectively. The abscissa is overpotential, $\eta_{Me}$, not potential, $\varepsilon$. In these coordinates, changes in $i_{Me,D}$ may be caused either by changes of the bulk concentration $(c_{Me^{z+}})_{\circ}$, or by changes of $\delta_{Nernst}$. The difference would directly show up if the abscissa were potential rather than overpotential, because changing $\delta_{Nernst}$ leaves the equilibrium potential $E_{Me/Me^{z+}}$ unaffected.

As discussed in Chap. 3, the bulk concentration of dissolved cations of the corroded metal usually is virtually zero. Substituting 0 for $(c_{Me^{z+}})_{\circ}$ in Eq. (5.55),

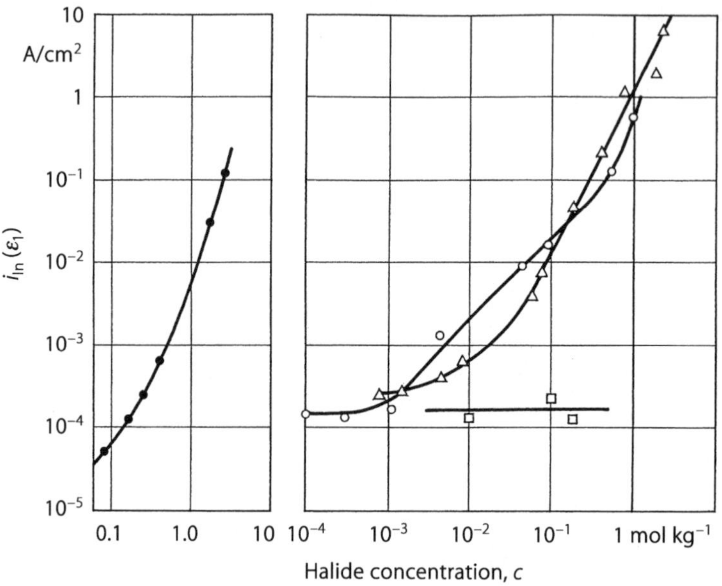


**Fig. 5.18.** Dependence of the anodic anodic current density for indium dissolution from indium amalgam in $HClO_4$ at constant potential on the concentration of halogenides $Cl^-$, $Br^-$, $J^-$, and $F^-$, forming complex anions with $In^{3+}$ ions. ●● NaCl at −0.330 V. △△ NaBr, ○○ NaJ, □□ NaF, all at −0.276 V. (From Lossew and Molodow)

we have $i_{Me,D} = 0$, and, thus, formally an increase in diffusion overvoltage to $-\infty$. At the same time the equilibrium potential $E_{Me/Me^{z+}}$ formally shifts to $-\infty$. While this does not directly affect the term in the equation describing the anodic partial current-potential curve for metal dissolution, it leaves the exchange current density also formally tending to zero. The practical consequence of all this only is that for negligible concentrations $c_{Me^{z+}}$ the partial current-voltage curve for metal deposition can be altogether neglected.

If, instead of hydrated cations, $Me^{z+}(H_2O)_x$, i.e. instead of 'aquo-complexes', cations with other ligands are formed, an influence of the concentration of the respective species is expected. As an example, Fig. 5.18 illustrates the influence of additions of sodium chloride, bromide, iodide, and fluoride on the anodic current density of dissolution of indium from indium amalgam at constant potential in perchloric acid solutions [15]. All halides form complex ions with dissolved $In^{3+}$ ions. $Cl^-$, $Br^-$, and $I^-$ also cause an important increase in dissolution on current density observed at constant potential, indicating the participation of these halides in the rate-determining step of anodic metal dissolution, apparently by complexing the indium species going through the charge transfer reaction. $F^-$ has no effect, although it also does complex dissolved $In^{3+}$. Apparently, $F^-$ does not enter the charge transfer reaction, and the indium species going through this reaction step is the hydrated rather than the fluorided ion.

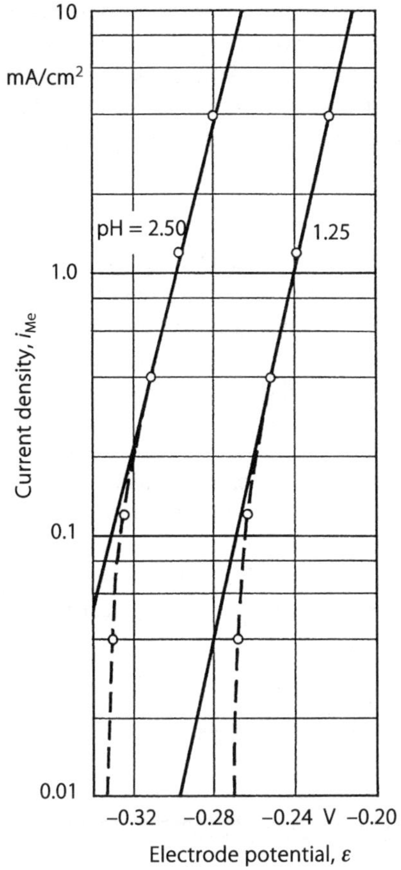


**Fig. 5.19.** Steady-state anodic current-potential curves for carbonyl iron in deaerated $HClO_4/NaClO_4$ solutions at 20°C. *Solid curves:* Partial current densities for anodic metal dissolution. *Broken curves:* Total current-potential curve for anodic metal dissolution and cathodic hydrogen deposition. (From Bonhoeffer and Heusler)

The case of indium dissolution is but one example of observations complicating the actual course of events as compared with the simple picture developed before deriving Eq. (5.52) and the following equations. Still more intricate are cases where $Me^{z+}$ eventually does leave the electrode reaction as hydrated cation, but where intermediate reactions with other species present in the solution enter the rate equation. By definition, these are examples of *reaction catalysis* when the effect is an increase, while they are examples of *reaction inhibition* when the effect is a decrease in reaction rate. A case of catalysis which over the years has experienced great interest is pure iron in acid solutions [16]. The main observation, shown in Fig. 5.19, is that at constant potential the partial current density of anodic metal dissolution increases with increasing pH [16a].

Not the only possible, but the generally accepted conclusion is that anodic metal dissolution is accelerated by $OH^-$ ions by the intermediate formation of a hydroxylated iron compound. In the first place this does seem unlikely, because in acid solutions $OH^-$ obviously is not available in any important amount. The difficulty can be evaded by postulating that on the iron surface adsorbed $OH^-$ are present by a reaction

$$H_2O \rightarrow OH^-_{ad} + (H^+)_* \,, \tag{5.56}$$

which may even be assumed to be in equilibrium, in which case '→' can be replaced by '='. Subscripts in this and the following equations have their usual meaning; $OH^-_{ad}$, although adsorbed, is on the solution side of the interfacial double layer. Next, $OH^-_{ad}$ adjacent to kink sites is assumed to react, again in an equilibrium reaction, forming an intermediate compound:

$$Fe_{kink} + OH^-_{ad} = FeOH_{ad} + e^- \,. \tag{5.57}$$

$FeOH_{ad}$ is on the metal side of the double layer. The charge transfer reaction then is the two-electron step of charging $Fe_{kink}$ to $Fe^{2+}$ which simultaneously reacts with $OH^-_{ad}$. The reaction product is $FeOH^+$ on the solution side of the double layer. This transfer reaction is catalysed by $FeOH_{ad}$ always present by the preceding reaction Eq. (5.57). The catalyst re-emerges unchanged from the reaction :

$$Fe_{kink} + OH^-_{ad} + FeOH_{ad} \rightarrow (FeOH^+)_* + FeOH_{ad} + 2e^- \,. \tag{5.58}$$

In acid solutions, $(FeOH^+)_*$ is unstable and de-hydroxylates:

$$(FeOH^+)_* + (H^+)_* \rightarrow (Fe^{2+})_* + H_2O \,. \tag{5.59}$$

The sequence of reaction steps ends with the transport of iron cations into the bulk of the solution:

$$(Fe^{2+})_* \rightarrow Fe^{2+} \,. \tag{5.60}$$

If step Eq. (5.58) is rate-determining, we expect an expression for the anodic partial current density of iron dissolution which is

$$\vec{i}_{Fe} = \vec{K}'_{Fe} \, \theta_{Fe} \, \theta_{FeOH} \, \theta_{OH^-} \exp(2\alpha \, \Phi' \, \varepsilon) \,, \tag{5.61}$$

indicating coverages of reaction participants as $\theta_i$. We may tentatively assume that the coverage of the surface with adsorbed $OH^-_{ad}$ does not, or does not much depend on the potential, so that $\theta_{OH^-}$ is constant. We may also assume that $\theta_{Me}$ is constant. Analogous to the derivation of Eq. (5.15) we then may write a Nernst equation for the equilibrium of reaction Eq. (5.57):

$$\varepsilon = E_{Fe,OH^-/FeOH} = \mathit{const} + \frac{RT}{F} \ln \frac{\theta_{FeOH}}{\theta_{OH^-}} \,, \tag{5.62}$$

where *const* is the unknown but constant standard reaction potential. Solving this equation for $\theta_{OH^-}$, substituting the resulting expression into Eq. (5.61), and contracting constants, the expression for the current-potential curve in the case of rate-determining step Eq. (5.58) turns into:

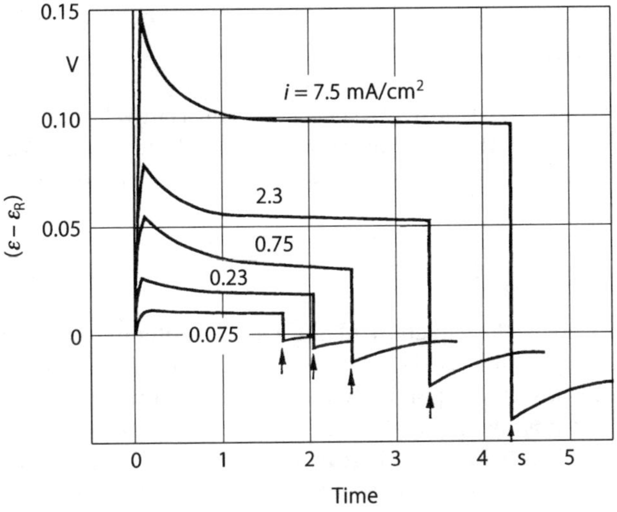


**Fig. 5.20.** Change of potential of carbonyl iron in $HClO_4/NaClO_4$ solutions, accompanying galvanostatic anodic pulses switched on at t = 0, switched off at ↑. (From Heusler)

$$\vec{i}_{Fe}=\vec{K}_{Fe}\ \theta_{Fe}\theta^2_{OH^-} exp(\Phi'\varepsilon)\exp(2\alpha\Phi'\varepsilon)\ . \tag{5.63}$$

This is a Tafel line which should represent the solid lines in Fig. 5.19, since the reverse currents of both metal deposition and hydrogen deposition are negligible. Expecting $\alpha = 0.5$, the predicted slope is $d\varepsilon/d\log i = 2.303RT/2F = 0.029$ V, which compares surprisingly well with the experimental 30 mV. If, particularly for small values of the coverage, $\theta_{OH^-} \sim c_{OH^-}$, the order of the reaction with respect to the bulk-concentration of $OH^-$ should be the even number 2. Observed values are between 1.5 and 2 and thus not very convincing. However, more evidence in favor of what has come to be known as the 'Heusler' mechanism comes from pulsed galvanostatic polarization experiments shown in Fig. 5.20: On pulsing the anodic current, the potential rapidly goes to a maximum and then decays to a steady-state value. The interpretation is that the equilibrium of step Eq. (5.57) is established only comparatively slowly, and that the overpotential at first follows Eq. (5.61) at constant $\theta_{FeOH}$. The expected initial slope then is $d\varepsilon/d\log i = 2.303RT/F = 0.059$ V, again very well comparing with experimental values of 60 mV. The steady-state potentials once more confirm Eq. (5.63).

The mechanism of anodic iron dissolution is, however, controversial, or else it may change depending on circumstances such as surface preparation. A variant is to assume a sequence of one-electron steps with FeOH not as a catalyst but as an intermediate reaction product [17]. The supposed sequence of steps is:

$a)\ Fe + OH^- \rightarrow FeOH + e^-$ ,

$$b)\ \mathrm{FeOH} \rightarrow \mathrm{FeOH}^+ + e^- , \tag{5.64}$$

$$c)\ \mathrm{FeOH}^+ \rightarrow \mathrm{Fe}^{2+} + \mathrm{OH}^- .$$

This is known as the 'Bockris, Drazic, Despic'-, or 'consecutive', or 'non-catalytic' mechanism, though $OH^-$ does act as a catalyst. If step Eq. (5.64a) is in equilibrium, and step Eq. (5.64b) rate-determining, the equation for anodic partial current density is

$$\vec{i}_{\mathrm{Fe}} = \vec{K}'_{\mathrm{Fe}}\ \theta_{\mathrm{OH}^-} \exp(\Phi\,\varepsilon) \exp(\alpha\,\Phi\,\varepsilon) . \tag{5.65}$$

When $i_{\mathrm{Fe}} = \vec{i}_{\mathrm{Fe}}$, this also is the equation of the current-potential curve. If $\alpha = 0.5$, the calculated slope then is $d\varepsilon/d\log i = 2.303RT/1.5\ F = 0.040$ V. This value has also been observed, apparently depending on surface preparations producing more or less abundant dislocations [19]. A further point of interest is the influence of halide ions on the reaction. As described above, $Cl^-$, $Br^-$, and $I^-$, catalyse the anodic indium dissolution; but the same ions inhibit iron dissolution apparently because they are adsorbed on the iron surface displacing adsorbed $OH^-$ [20]. Still more refined information revealing still more details of the mechanism has been gained from impedance spectroscopy [18,21, Appendix].

Apparently, the energetics of the different variants of the mechanism differ only relatively slightly. Such fine details may appear to have secondary importance only, but species such as adsorbed $OH^-$, $FeOH^+$, or FeOH really are precursors of passivating oxide films, more fully described in Chap. 10, and their reactions are of considerable practical interest. If passivation fails and saturation of the solution with respect to some iron oxide is reached by sufficient iron dissolution, the result is, instead, rusting as described in Chap. 2, and as further discussed in Chap. 9, again pointing to the importance of precursor reactions.

Not many direct observations of atomic-scale morphological changes on iron surfaces accompanying anodic dissolution are, so far, available. As an interesting example, Fig. 5.21 shows the appearance of a crystal plane of iron, vicinal to (211), during anodic dissolution in perchloric acid [22]. During dissolution, the plane has coarsened by the formation of three-sided pyramids with (112) faces. Further electron-microscopic studies using replica techniques revealed a situation as sketched in Fig. 5.22: The faces of the pyramids have steps of probably one-atomic height. The dissolution occurring from these steps requires a step-orientation parallel to [113]-directions, but the steps are slightly disoriented due to kinks of always the same sign, and this misorientation increases with increasing anodic overpotential. The observations convincingly show that in this case dissolution occurs directly from kink sites rather than via ad-atoms, and that the kink density, i.e. the property $\theta_{\mathrm{kink}}$ Eq. (5.61), increases with both overpotential and solution-pH.

So far it would seem that anodic iron dissolution is governed by charge-transfer control until the solution becomes saturated with respect to some iron

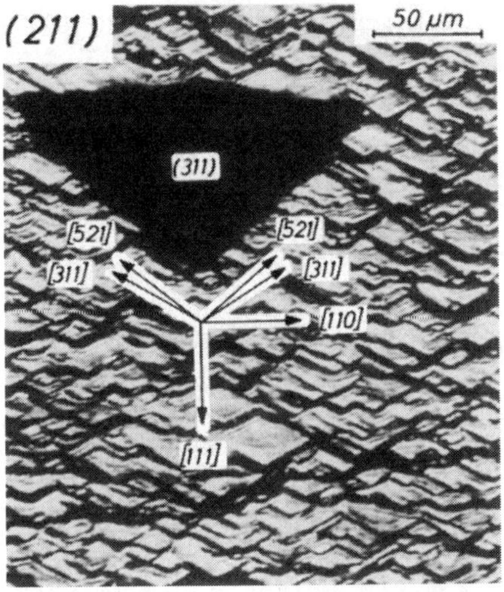


**Fig. 5.21.** Optical micrograph showing the morphology of an iron crystal plane originally vicinal to (211), misorientation ca. 1°, after anodic iron dissolution in 1M $HClO_4$, pH 0.85, $\varepsilon = -0.465$ V, $i = 4.38$ mA $cm^{-2}$. (From Allgeier and Heusler)

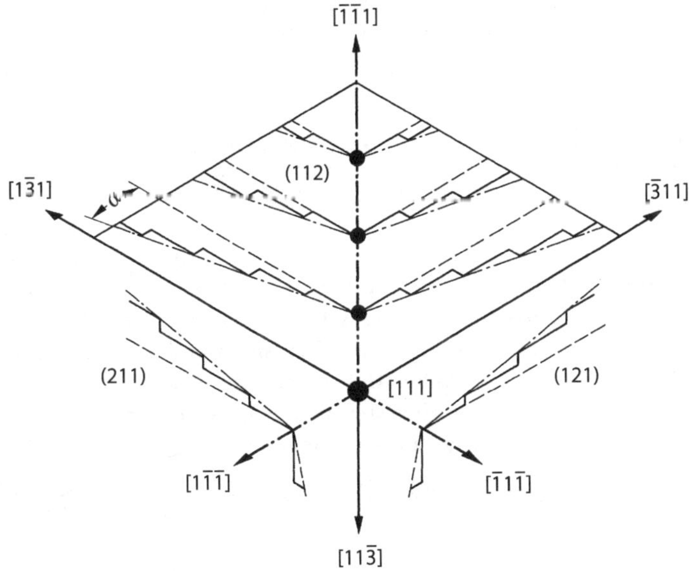


**Fig. 5.22.** Schematic view in a [111]-direction onto a dissolution pyramid on iron in perchloric acid, with its three (112) faces projected onto (111). Faces are stepped by edges which should be parallel to [311], but are slightly disoriented due to kinks of always the same sign. (From Allgeier and Heusler)

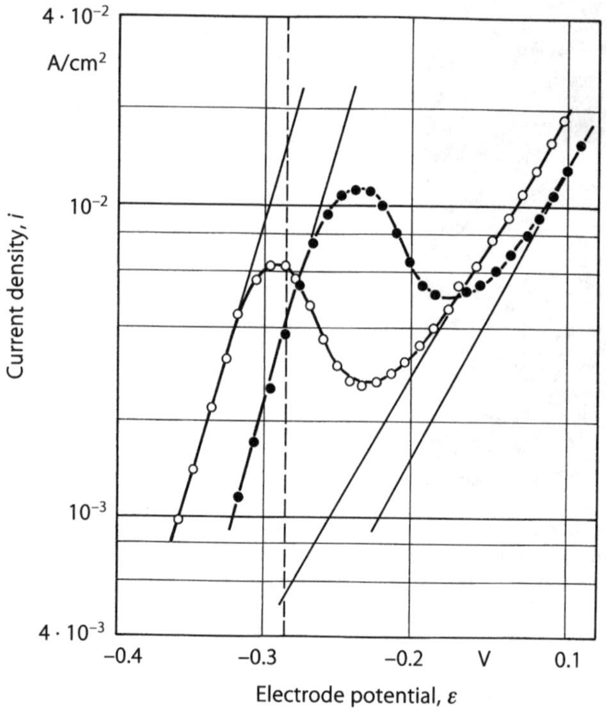


**Fig. 5.23.** Current-potential curve of anodic iron dissolution in acetic acid/1 M $NaClO_4$. *Solid circles:* pH 3.25, *open circles:* pH 3.85. (From Nord and Bech-Nielsen)

oxide or hydroxide. Circumstances once more are, however, found to be quite more intricate: As shown in Fig. 5.23, presenting data typical for weakly acid solutions, the rate of anodic dissolution passes through a primary maximum and then changes to a Tafel line exhibiting a lower slope and also a weaker dependence on pH [23]. A possible explanation is that the mechanism of iron dissolution changes from rate-determining step Eq. (5.58) to rate-determining formation of $Fe(OH)_2$ as another metastable intermediate reaction product,

$$Fe + (Fe(OH)_2)_{ad} \rightarrow (Fe(OH)^+)_* + (Fe(OH)_{ad} + e^- , \quad (5.66)$$

adding still more versatility to the mechanism.

From their position in the periodic table of chemical elements, chromium, cobalt, and nickel would be expected to follow kinetics similar to that for iron dissolution. As reviewed, e.g., in [16c], hydroxylated intermediates in fact do play an important role, though the information available at present is by far not as abundant as for iron.

To illustrate more possible variants of dissolution kinetics, two more examples merit attention: The first is zinc in perchloric acid [23]. As shown in Fig. 5.24, the partial current-potential curve of anodic zinc is bent. In the case

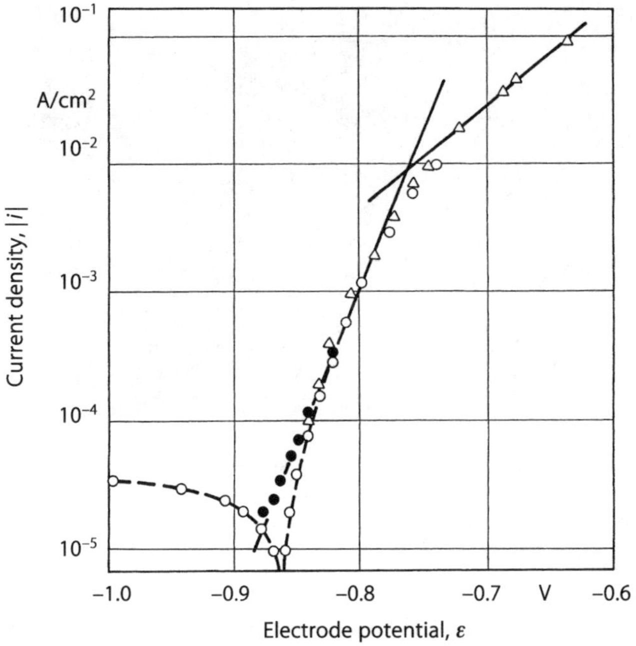


**Fig. 5.24.** Current-potential curve for zinc in $HClO_4$/1 M $Na_2ClO_4$, pH 3.25, 25°C. *Broken curve:* Total current-potential curve for zinc dissolution and hydrogen deposition, showing a cathodic diffusion-limited current of hydrogen deposition. *Solid lines:* Partial current-potential curves for anodic zinc dissolution following two different rate-determining steps. (From Gaiser and Heusler)

of hydrogen deposition described in Sect. 5.1, a similar bend was explained by assuming a change in the rate-determining step. Similarly, the tentative explanation now is that zinc dissolution proceeds through two consecutive steps, $Zn \rightarrow Zn^{+} + e^{-}$, and $Zn^{+} \rightarrow Zn^{2+} + e^{-}$. When the first step is rate-determining, and if $\alpha = 0.5$, the slope of the Tafel line is $2.303RT/0.5\,F = 0.118$ V, but when the first step is in equilibrium and the second rate determining, the slope is $2.303\,RT/1.5\,F = 0.040$ V.

A similar case is copper, which also reaches the stable state of divalent dissolved cations via two consecutive one-electron steps: $Cu \rightarrow Cu^{+} + e^{-}$, and $Cu^{+} \rightarrow Cu^{2+} + e^{-}$ [25]. In acid $CuSO_4$ solutions the anodic partial current-potential curve has the 0.040 V-slope, as expected for equilibrium of the first reaction step. Cathodic copper deposition begins with the step $Cu^{2+} + e^{-} \rightarrow Cu^{+}$. The rate of this step is independent of the $Cu^{+}$ concentration; hence the partial current-potential curve of cathodic copper deposition should have and does have a 0.118 V-slope. As a point of interest for Chap. 6 we note that for anodic copper dissolution the electrode reaction step $Cu^{+} \rightarrow Cu^{2+}$ disappears when the acid solution contains dissolved oxygen: In this case a

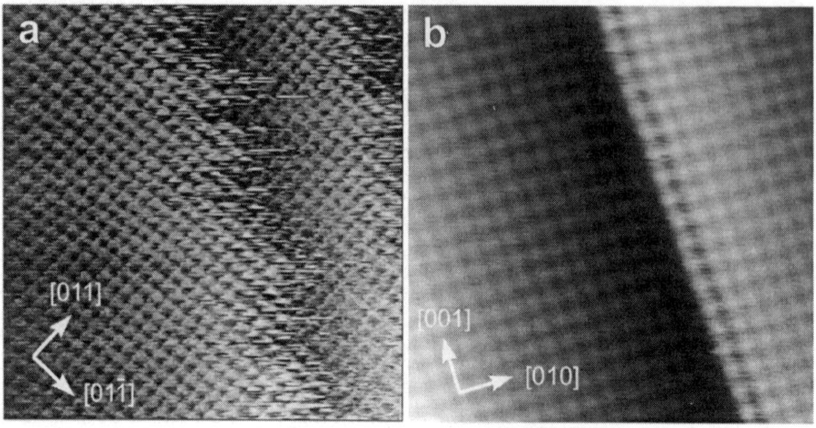


**Fig. 5.25a,b.** High-resolution STM images showing the atomic surface structure of a Cu(100) surface. The imaged area is 7 × 7 nm. Solutions and potentials: **a** 10 mM $H_2SO_4$, −0.34 V; **b** 1 mM HCl, −02 V. (From Magnussen, Vogt, Scherer, Lachenwitzer, and Behm)

reaction $2Cu^+ + \frac{1}{2}O_2 + 2H^+ \rightarrow 2Cu^{2+} + H_2O$ takes place in the solution adjacent to the metal surface, which does not require an exchange of electrons with metallic copper. By consequence, the dissolution rate increases by the effect of a non-electrolytic reaction following the electrode reaction $Cu \rightarrow Cu^+ + e^-$, and this, then, is the rate-determining step.

In recent years, in-situ scanning tunnel microscopy (STM) has become a powerful tool to image atomic events of copper dissolution and redeposition on copper surfaces in dilute acidic solutions [26–28]. We quote work on monocrystalline copper in dilute hydrochloric acid [26,27], where metal dissolution is slowed by chloride adsorption. In fact, the primary aim of this work was to study the pattern of the monomolecular adsorption layers on (100) surfaces. In [29], the crystallografy of copper dissolution was compared in dilute sulfuric and dilute hydrochloric acid. As an example, Fig. 5.25 shows STM images of an atomic ledge on (100) surfaces. Observed in sulfuric acid, the ledge appears blurred, owing to rapid movement of atoms along the ledge during the comparatively slow scanning procedure. This probably reflects a high exchange current density for copper dissolution and redeposition. The right-hand figure illustrates the effect of chloride ions: These are seen to occupy the shallow sites between copper surface atoms, thus forming a quadratic array in a c(2 × 2) structure, rotated through 45° with respect to the parent copper surface lattice. At the same time, the observed surface ledge now is well defined, apparently due to a decrease of the exchange current density caused by chloride adsorption. The effect is an example of inhibition of electrode reactions by adsorbed species, more fully described in Chap. 6. Refining imaging techniques, more details have become available: As reported in [25d], Cu(100) dissolution and redeposition in dilute hydrochloric acid proceeds by the removal and deposition of the primitive surface unit cell containing two Cu and an adsorbed Cl atom, at kinks in the {001}-oriented steps.

## 5.4
## Closer Inspection of the Electrical Double Layer

So far, the model used to describe the electric double layer at the metal/solution interface was the simple capacitor. This implies that the charge transfer reaction is subject to the full Galvani potential difference, $\phi_{m,l}$, across the metal/solution interface. The success of this treatment of the matter is surprising, as the model seriously oversimplifies the real situation. It is true that in the metal the charges are placed in the surface which thus can be considered as similar to a capacitor plate. In the much less conductive electrolytic solution, however, the distribution of excess ions representing the opposite charge of the capacitor is 'smeared out' due to the distributive action of diffusion, and the result is a space charge. This leads to the more realistic model of the double layer shown in Fig. 5.25a. On the solution side the double layer is two-layered with a 'rigid' inner part, the *Helmholtz layer*, which as before is a capacitor of fixed distance, $\xi$, of its plates, and an outer 'diffuse' part, the *Gouy layer* or, more precisely, the *Gouy-Chapman-Stern layer* within which the charge decays along the distance, $x$, beyond the Helmholtz plane. In the absence of specific adsorption of other species, the potential drop across the Helmholtz layer still is linear between the potential in the metal surface, $\phi_m$, and the potential in the Helmholtz, solution-side plane, $\phi_\xi$. The latter quantity usually is called *zeta potential*, $\zeta$, so that $\phi_{m,\xi} = \phi_m - \zeta$. Accordingly, the potential difference across the Gouy layer, between the Helmholtz plane and the bulk of the solution is $\phi_{\xi,l} = \zeta - \phi_l$. But the distance, $\xi$, still is the distance of closest approach of ions entering charge transfer reactions. Therefore we must, in principle, replace, in all Volmer-Butler type expressions, the total potential drop, $\phi_{m,l}$, by the corrected value, $\phi_{m,\xi} = \phi_m - \zeta$. Terms $\exp(\pm\,\alpha n_e F\varepsilon/RT)$ then always change into terms $\exp(\mp\,\alpha n_e F\zeta/RT)\exp(\pm\alpha n_e F\varepsilon/RT)$. Rate constants in current-potential functions therefore contain terms $\exp(\mp\,\alpha n_e F/RT\zeta)$, which in principle depend on potential and solution composition.

Zeta potentials can be calculated applying the concepts of the Debye-Hückel theory, which were developed to describe the electrostatic field around solvated ions, applied to plane metal surfaces in solutions. The calculations describe an exponential decay of the Galvani potential across the Gouy layer. The characteristic property is the *Debye length*, $\kappa_{\text{Debye}}$, which is the distance of decay to $(1/e)\zeta$. Typical $\kappa_{\text{Debye}}$-values, calculated for 1,1-electrolytes such as NaCl in 1 M, 0.1 M, 0.01 M, and 0.001 M solution are 0.3, 1.3, and 10 nm, respectively. The ratio $\zeta/\phi_{m,l}$ is related to the distance, $x$, from the metal surface by the expression

$$\frac{\zeta}{\phi_{m},l} = \frac{1}{1 + x/\kappa_{\text{Debye}}}\,. \tag{5.67}$$

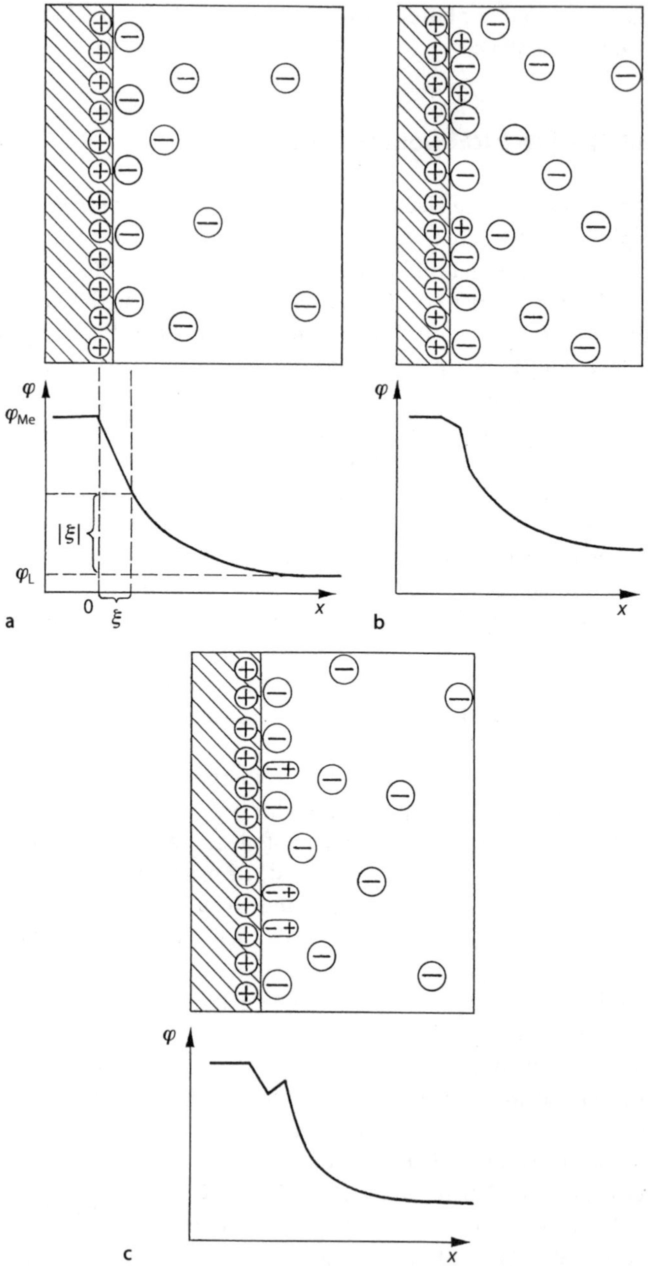


**Fig. 5.26a–c.** Schematic structure of the electrical double layer, and the change in Galvani potential across the electric double layer at metal/solution interfaces, assuming a positive surface charge in the metal, a negative surface charge in the rigid Helmholtz layer adjacent to the metal, and a negative space charge in the diffuse adjoining Gouy layer, also showing the zeta-potential. **a** Without additional adsorption; **b** with specific cation adsorption; **c** with specific dipole adsorption. (According to Vetter)

In dilute solutions, where the Debye length is large, most of the potential drop is in the Gouy layer, and an explicit correction of Volmer-Butler type equations is necessary, though usually disregarded. In concentrated solutions, Eq. (5.67) holds only approximately; it still is true that most of the potential drop now is in the Helmholtz layer, and this more or less justifies the treatment given to the matter in all of the preceding derivations concerning charge transfer controlled electrode reactions.

Above, bulk concentrations of dissolved substances were labeled $c_\circ$, concentrations in what we now know to be the Helmholtz plane of the double layer, $c_*$. The equilibrium condition was $c_* = c_\circ$. More precisely this equation holds for ions carrying a charge $\pm z$ only if the $\zeta$-potential is negligible. Otherwise the electric field causes an equilibrium distribution of the respective ions in the Gouy layer resulting in the relation

$$(c_{\text{ion}})_* = (c_{\text{ion}})_\circ \exp(\mp zF\zeta/RT) \,, \qquad (5.68)$$

and this adds a second effect to the influence of the $\zeta$-potential on the rate of electrode reactions described above. As an example we note that considering both effects we have to rewrite Eq. (5.28) describing the current-potential curve of cathodic hydrogen deposition when anodic hydrogen ionisation is negligible into

$$i_{\text{H}} = \vec{i}_{\text{H}} = -\vec{K}\,(c_{\text{H}^+})_\circ \exp(-F\zeta/RT)\exp\{\beta F(\varepsilon - \zeta)/RT\} \,. \qquad (5.69)$$

In solutions where the concentration of $H^+$ is small compared with the total ionic content due to an excess of dissolved supporting electrolyte, the influence of the $\zeta$-potential is constant, and the overall effect is a change in the value of the rate constant, which is an empirical quantity anyway. For acid solutions, $\zeta$ depends on pH. Under certain conditions as to potential and solution composition, $\exp(F\zeta)RT$ is proportional to $(c_{\text{H}^+})_\circ$. When this approximation holds, Eq. (5.69) re-simplifies to Eq. (5.28) except for a change in the dependence of the current on pH: The term $(c_{\text{H}^+})_\circ$ changes into $\sqrt{(c_{\text{H}^+})_\circ}$. However, all this is highly variable matter especially when applied to other electrode reactions, and mainly shows the limits of strict applicability of simple Butler-Volmer equations except for rather narrow ranges of potential.

In principle, the $\zeta$-potential is a measurable quantity. From its sign the sign of the charge in the respective metal surface can be determined, though this is not the usual way to arrive at this information (see below). The magnitude of $\phi_{m,\xi}$ and, thus, of $\phi_{m,l}$ as well remains unknown, and insofar Fig. 5.25a is arbitrary. Furthermore, Fig. 5.25a describes the simplest case only: As indicated in Fig. 5.25b and 5.25c, specific adsorption of, e.g., cations and of dipoles complicate the change in potential across the double layer. Further above, we have also noted that a part of the potential drop is expected to occur in a dipole layer in the metal surface. We do not expect this effect to be important for charge-transfer reactions, and it is ignored in Fig. 5.25. On the solution side, oriented adsorption of water molecules as dipoles is almost certainly always to be expected, but in view of the high concentration of water we may expect saturation adsorption, and, therefore, a constant contribution to rate constants.

In electrodics, the concept of the *null-electrode*, where $\phi_{m,l} = 0$, has been a point of considerable theoretical interest in the search for a scale of electrode potentials less arbitrary than the standard hydrogen scale. Unfortunately, we can only control conditions such that $\zeta = 0$, indicating zero excess charge in the metal surface. The electrode potential $\varepsilon_{\zeta=0}$, referred to SHE, is the *potential of zero charge*, $\varepsilon_{pzc}$, of the respective metal/solution system. Owing to possibly remaining dipole layers both in the metal and in the solution side of the double layer, $\varepsilon_{pzc}$ is not, however, generally the null-potential.

The zero charge potential, $\varepsilon_{pzc}$, will be seen in Chap. 7 to be an important point considering the action of adsorption inhibition in acid solutions. The determination is straightforward in the case of mercury in contact with inert solutions, i.e, in the absence of electrode reactions, because it coincides with the potential of maximal surface tension observed at 0.19 V. For solid metals, various methods such as measuring the double layer capacity, or the contact angle of adhering gas bubbles, or the $\zeta$-potential, and others have been used. From various sources reviewed in [1,3,29] we quote as token examples −0.37 V for iron, +0.3 V for gold. −0.6 V for zinc. However, $\varepsilon_{pzc}$ is very sensitive to surface conditions, and changes with solution composition due to specific adsorption of cations or anions, and of dipoles other than water. Also, the nature of crystal faces is important: For silver in NaF solution the values are −0.46, −0.61, −0.77 for (111), (100), and (110) faces, respectively [30].

At this point a remark concerning transfer coefficients, $\alpha$ and $\beta = 1 - \alpha$, as they appear in Volmer-Butler equations, is appropriate. As described above, the choice $\alpha = \beta = 0.5$ is often well suited to fit experimental values of Tafel slopes. However, other values between 0 and 1 are allowed, depending on the structure of the electrical double layer. Therefore, changes in values of transfer coefficients do not necessarily indicate changes in rate-determining steps. They may, instead, simply reflect deficiencies in the simple plate capacitor model for the electric double layer.

## References

1. Vetter, K.J.: Elektrochemische Kinetik. Springer Verlag, Berlin etc., 1961 (Electrochemical Kinetics. Academic Press New York and London, 1967)
2. Bockris, J.O'M. and Khan, S.U.M.: loc. cit. Chap. 3 [7]
3. a) Schmickler. W.: Electrode Kinetics. VCH, New York etc., 1993; b) Interfacial Electrochemistry. Oxford University Press. New York, Oxford 1998
4. Marcus, P. and Oudar, J., eds.: Corrosion Mechanisms. Theory and Practice. Marcel Dekker, New York etc., 1996
5. Uhlig, H.H.: Corrosion and Corrosion Control. J. Wiley, New York and London, 1963
6. Kaesche, H.: Z. Elektrochemie, Ber. Bunsenges. Phys. Chem. *63* (1959) 492
7. Stern, M.J.: Electrochem. Soc. *102* (1955) 609
8. Frumkin, A.N., Koshunov, V., and Bagozkaya, I.: Electrochim. acta *15* (1970) 289

9. Landolt-Börnstein: Zahlenwerte und Funktionen; 6th ed., 2nd Part, b: Lösungsgleichgewichte I. Springer, Berlin, Göttingen, etc., 1961
10. a) Vielstich, W.: Brennstoffelemente. VCh, Weinheim, 1965; b) Bockris, J.O'M. and Srinisan, S.: Fuel Cells. McGraw Hill, 1969; b) Hoare, J.P.: The Electrochemistry of Oxygen. Interscience Publ., J. Wiley & Sons, 1968
11. Kaesche, H.: unpublished work, 1961
12. Kolthoff, J.M., and Miller, C.S.: J. Am. Chem. Soc. *63* (1941) 3110
13. Postlethwaite, J. and Sephton, I.; Corr. sci. *10* (1970) 775
14. Jovancicevic, V. and Bockris, J.O'M.: J. Electrochem. Soc. *113* (1986) 1797
15. Lossew, W.W. and Molodov, A.J.: Dokl. Akad. Nauk. SSSR *130* (1960) 11, *198* (1963) 1114; Kolotyrkin, Y.M.: J. Electrochem. Soc. *108* (1961) 209
16. a) Bonhoeffer, K.F. and Heusler, K.E.: Z. phys. Chemie N. F. *8* (1956) 390; b) Heusler, K.E.: Iron. In: Bard, A.J., ed.: Encyclopedia of Electrochemistry of the Elements, Vol. IX, Part A, Marcel Dekker, New York, Basel, 1982; c) Lorenz, W. and Heusler, K.E.: Anodic Dissolution of Iron Group Elements. In: Corrosion Mechanisms.(Mansfeld, F., ed.). Marcel Dekker, New York, Basel, 1987
17. Bockris, J.O'M., Drazic, D., and Despic, A.R.: Electrochim. acta *4* (1961) 325; see also [16c,18]
18. Keddam, M.: Anodic Dissolution. In: loc.cit [4]
19. Eichkorn G., Lorenz, W.J., Albert,L., and Fischer, H.: Electrochim. acta *13* (1968) 183
20. Geana, D. and Lorenz, W.J.: Corr. Sci. *13* (1973) 50, *14* (1974) 657; Electrochim.acta *20* (1975) 273
21. Schweickert, W.J., Lorenz, W.J., and Friedburg, H.: J. Electrochem. Soc. *127* (1980) 1693; Keddam, M., Mattos, O.R., and Takenouti, H.: ibid. *128* (1981) 257, 266
22. Allgeier, W. and Heusler, K.E.: Z.phys. Chemie N. F. *98* (1975) 161; J. Appl. Electrochem. *9* (1979) 155; Folleher, B. and Heusler, K.E.: J. Electroanal. Chem. *180* (1984) 77
23. Nord, H. and Bech-Nielsen, G.: Electrochim. acta *16* (1970) 161
24. Gaiser, L. and Heusler, K.E.: Electrochim. acta *15* (1970) 161
25. Bockris, J.O'M. and Enyo, M.: Trans. Farad. Soc. *58* (1962) 1187; see also: loc. cit. Chap. 3 [4], Vol. 4: Electrochemical Materials Sciences
26. Suggs, D.W., Bard, A.J.: a) J. Amer. Chem. Soc. *116* (1994) 10725; b) J. phys. Chem. *99* (1995) 8349
27. Moffat, T.P.: J. Phys. Chem. B *102* (1998) 10020
28. a) Vogt, M.R., Möller, F., Schilz, C.M., Magnussen, O.M., and Behm, R.J.: Surf. Sci. *367* (1996) L33; b) Magnussen, O.M., Vogt, M.R., Scherer, J., Lachenwitzer, A., and Behm, R.J.: Materials and Corrosion *49* (1998) 169; c) Vogt, M.R., Lachenwitzer, A., Magnussen, O.M., and Behm, R.J.: Surface sci. *399* (1998) 49; d) Magnussen, O.M., Zitzler, L., Gleich, B., Vogt, M.R., and Behm, R.J.: Electrochim. acta *46* (2001) 3725
29. Frumkin, A., Damaskin, B., Grigoriev, N. and Bagotzkaya, I.: Electrochim. acta *19* (1974) 69

30. Trasattis, S.: in: Gerischer, H. and Tobias, W., eds.: Advances in Electrochemistry and Electrochemical Engineering, Vol. 10. Wiley Interscience, New York 1977
31. Frumkin, A., Damaskin, B., Grigoriev, N, and Bagotzkaja, I.: Electrochim. acta *19* (1974) 69

# 6 Uniform Electrolytic Corrosion

## 6.1
## Acid and Neutral Solutions

Continuing to verify the principles of electrolytic corrosion, we turn from liquid amalgams to solid metals and to solutions with pH up to ca. 7, which is the overall range of 'acid' corrosion by superposition of anodic metal dissolution and cathodic hydrogen deposition by reduction of hydrogen ions rather than water. To exclude superposed oxygen reduction, the solution is assumed to be deaerated. The temperature is at 25°C or at some similar ambient value. The solution volume is supposed to be large, so that initial concentrations remain virtually constant during the course of the corrosion reaction, which is: $Me + zH^+ \rightarrow Me^{z+} + (z/2)H_2$. The metal surface is thought to be bare, i.e. free of films such as air-formed oxide films, and free of layers of solid corrosion products such as oxides or hydroxides.

The metal undergoing acid corrosion is a *multiple electrode*, since it is, at the same time, both a $Me/Me^{z+}$ and a $H_2/H^+$ electrode. Without external polarization electroneutrality requires that the (positive) current, $j_{Me,}$ of metal dissolution is equal to the value, $|j_H|$, of the (negative) current of hydrogen deposition. Both $j_{Me}$ and $|j_H|$ are equivalent to the corrosion rate, i.e. equal to the corrosion current, $j_{corr}$. Considering that the current $j_r$ of an electrode reaction labeled $r$ always is the sum $\vec{j}_r + \overleftarrow{j}_r$ of the (positive) current of the forward and (negative) current of the reverse reaction, we therefore have, without external polarization:

$$\vec{j}_{Me} + \overleftarrow{j}_{Me} = j_{Me} = -(\vec{j}_H + \overleftarrow{j}_H) = -j_H = j_{corr} \tag{6.1}$$

As to nomenclature, we note that electrode-reaction currents, $j_r$, are 'partial' currents of overall corrosion reactions. In this connection we can, but we will not normally denote quantities $\vec{j}_r$ and $\overleftarrow{j}_r$, i.e. partial currents of electrode reactions, as 'sub-partial' currents of corrosion reactions. Similarly, we shall not classify 'sub-partial' current-potential functions, $\vec{j}_r(\varepsilon)$ and $\overleftarrow{j}_r(\varepsilon)$, versus 'partial' current-potential functions, $j_r(\varepsilon)$. We further recall that current densities, $i$, are defined as ratios, $j/A_{metal}$, of current strength and metal surface area. With respect to kinetics the important properties are current densities, $i$, rather than currents strengths, $j$. For brevity, we shall usually refer to $i(\varepsilon)$-functions as 'current-potential curves'.

The study of electrolytic corrosion is governed by the study of these current-potential curves, and we need to consider the appropriate analysis by external polarization with 'total' currents, $j_s$, forced across the metal/solution interface in the three-electrode arrangement introduced in Chap. 4. Electroneutrality then requires that $j_s = \Sigma j_r$.

Concerning the distribution of the rates of partial reactions on the metal surface, we turn to the special case of uniform corrosion, and assume that both the current density of anodic metal dissolution and the current density of cathodic hydrogen deposition have, at any moment, the same value everywhere on the metal surface. Therefore,, the electrode potential, $\varepsilon$, also is the same everywhere; and we also have $i_s = \Sigma i_r$. For acid corrosion, and in terms of current-potential curves, we thus have, for the total current-potential curve, the expression:

$$i_s(\varepsilon) = i_{\mathrm{Me}}(\varepsilon) + i_{\mathrm{H}}(\varepsilon) = \overrightarrow{i}_{\mathrm{Me}}(\varepsilon) + \overleftarrow{i}_{\mathrm{Me}}(\varepsilon) + \overrightarrow{i}_{\mathrm{H}}(\varepsilon) + \overleftarrow{i}_{\mathrm{H}}(\varepsilon) \,. \tag{6.2}$$

For liquid amalgams, corrosion is uniform down to atomic entities of the surface. For solid pure metals, the true degree of uniformity depends on circumstances: Usually, the metal will be polycrystalline, exposing to the solution an array of differing grain surfaces of, normally, microscopic size having different crystallografic orientations, and we expect slightly different kinetics for differently oriented grain surfaces. Also, an initially polished surface will roughen in the course of corrosion. But if we disregard minor local discrepancies in corrosion rate by averaging current densities over surface areas comprising many grains, the description as a macroscopically uniform process is quite successful, as shown presently.

For pure or nearly pure metals in acid solutions and for sufficiently small current densities, we expect charge-transfer reactions to be rate-determining for both hydrogen deposition and metal dissolution. Therefore, we can attempt to summarize the detailed expressions describing electrode kinetics, developed in Chap. 5, by writing:

$$i_{\mathrm{H}} = B_{\mathrm{H}} \exp\left(\varepsilon / b'_{\mathrm{H}}\right) - A_{\mathrm{H}} \exp\left(-\varepsilon / a'_{\mathrm{H}}\right) , \tag{6.3}$$

$$i_{\mathrm{Me}} = B_{\mathrm{Me}} \exp\left(\varepsilon / b'_{\mathrm{Me}}\right) - A_{\mathrm{Me}} \exp\left(-\varepsilon / a'_{\mathrm{Me}}\right) . \tag{6.4}$$

For a given set of conditions, and in steady state, quantities of type $A$,$B$,$a'$, and $b'$ can be expected to be constant, or approximately constant. From Chap. 5 we recall that the pre-exponential factors escape theoretical calculation, and essentially are empirical parameters. Slope factors were described as better predictable, for instance as being equal to $\alpha F/RT$, but variations must be expected nevertheless, so again they are essentially empirical.

Substituting Eq. (6.3) and Eq. (6.4) into Eq. (6.2) we arrive at the equation of the total current-potential curve for acid corrosion as the sum of four exponential terms for the four partial current densities $\overrightarrow{i}_{\mathrm{Me}}$, $\overleftarrow{i}_{\mathrm{Me}}$, $\overrightarrow{i}_{\mathrm{H}}$, and $\overleftarrow{i}_{\mathrm{H}}$, i.e., as the sum of four Tafel lines. The situation is schematically shown in Fig. 6.1. When $\overrightarrow{i}_{Me} = -\overleftarrow{i}_{Me}$, the electrode potential, $\varepsilon$, is equal to the equilibrium potential $E_{\mathrm{Me/Me^{z+}}}$; when $\overrightarrow{i}_{H} = -\overleftarrow{i}_{H}$, it is equal to $E_{\mathrm{H_2/H^+}}$.

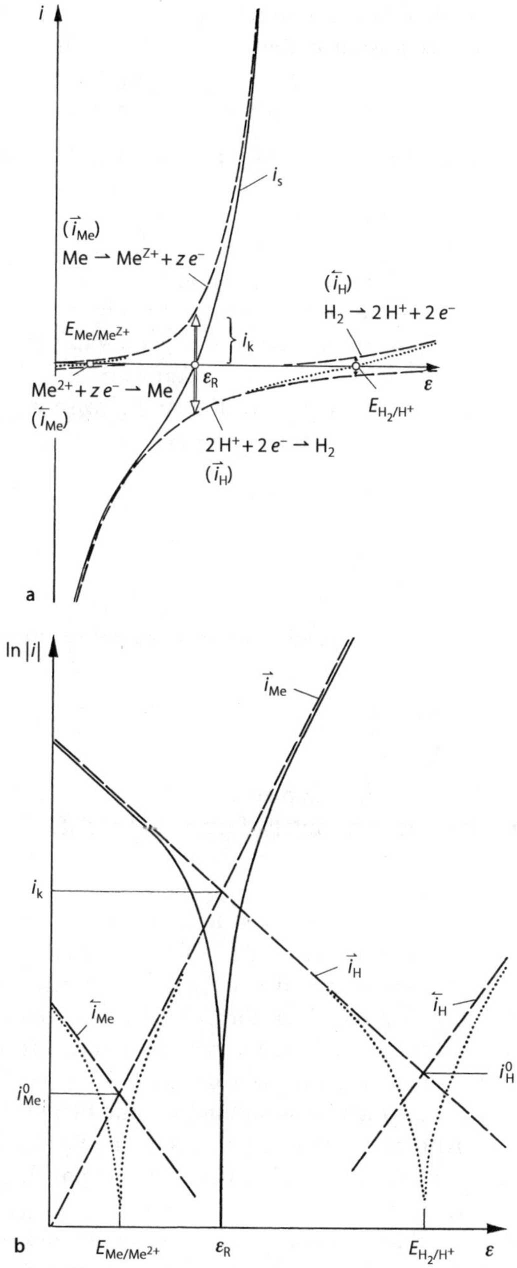


**Fig. 6.1a,b.** Complete current density-potential diagram ('current-potential diagram') for uniform acid corrosion, plotting electrode potential $\varepsilon$ versus **a** current densities $i$, **b** logarithms ln $|i|$. Charge-transfer reaction control is assumed for both metal dissolution and hydrogen deposition. Cathodic metal deposition and anodic hydrogen ionisation are assumed negligible in the vicinity of the corrosion potential, as typical for acid corrosion of iron-group elements

For the corrosion potential, we note the condition $E_{Me/Me^{z+}} \leq \varepsilon_{corr} \leq E_{Me/Me^{z+}}$, satisfying Eq. (3.81). At the corrosion potential, $\varepsilon_{corr}$, the total current density is zero, $i_s = 0$. Since this is the density of the current flowing in the the polarizing circuit, $\varepsilon_{corr}$, also is the 'open circuit potential'. On open circuit, both $i_{Me} = \vec{i}_{Me} + \overleftarrow{i}_{Me}$ and $-i_H = -(\vec{i}_H + \overleftarrow{i}_H)$ are equal to the corrosion current density, $i_{corr}$.

Figure 6.1 is drawn for the case that in the vicinity of the corrosion potential both the cathodic reaction of metal deposition and the anodic reaction of hydrogen ionisation are negligible. As shown below, this is, in fact, typical for iron group metals. It is not, however, a general principle, as already demonstrated previously for the case of acid corrosion of zinc amalgam, where the corrosion rate was found to be small compared with the exchange rate of the metal dissolution/metal deposition electrode reaction. In the special case now considered, the equation for the total current-potential curve reduces to

$$i_s = \vec{i}_{Me} + \vec{i}_H = B_{Me} \exp(\varepsilon / b'_{Me}) - A_H \exp(\varepsilon / - b'_H) . \tag{6.5}$$

Also considering that

$$i_{corr} = B_{Me} \exp(\varepsilon_{corr} / b'_{Me}) = A_H \exp(-\varepsilon_{corr} / a'_H) , \tag{6.6}$$

we can rewrite Eq. (6.5) by a simple shift of the origin of the coordinate system to the corrosion potential [6]:

$$i_s = i_{corr} \left[ \exp\left(\frac{\varepsilon - \varepsilon_{corr}}{b'_{Me}}\right) - \exp\left(\frac{\varepsilon - \varepsilon_{corr}}{a'_H}\right) \right] . \tag{6.7}$$

Since the corrosion potential is the potential adopted by the metal spontaneously, without external polarization, it is useful to further abbreviate this expression by substituting the electrode polarization $\pi$, defined by Eq. (4.10): $(\varepsilon - \varepsilon_{corr}) = \pi$.

When Eq. (6.7) is satisfied, we therefore do not need to know the overpotentials, $\eta_r = \varepsilon - E_r$,. This is convenient, because as a rule in actual service, and as discussed as a problem concerning thermodynamics in Chap. 3, the activity $a_{Me^{z+}}$ of dissolved cations often is virtually zero and, formally, $E_{Me/Me^{z+}}$ tends to $-\infty$. At the same time, Eq. (6.7) suggests a convenient method of measuring corrosion current densities: It is to suppress cathodic hydrogen evolution by anodic polarization, or to suppress anodic metal dissolution by cathodic polarization, and to determine from a logarithmic plot the Tafel lines of anodic metal dissolution resp. of cathodic hydrogen deposition. The corrosion current density $i_{corr}$ then is obtained by extrapolating the respective Tafel line back to $\varepsilon_{corr}$. The procedure was first used by Kolotyrkin and Frumkin to determine rates of acid corrosion of nickel, checking the method by simultaneous determination of $i_{corr}$ from the cathodic Tafel line and from manometric measurements of hydrogen evolution [1]. More measurements showing that anodic and the cathodic Tafel lines intersect rather exactly at the corrosion potential have been reported for $HClO_4/NaClO_4$- and HCl/NaCl solutions [2,3]. Another typical example of what can be observed in well purified solutions by careful measurements is

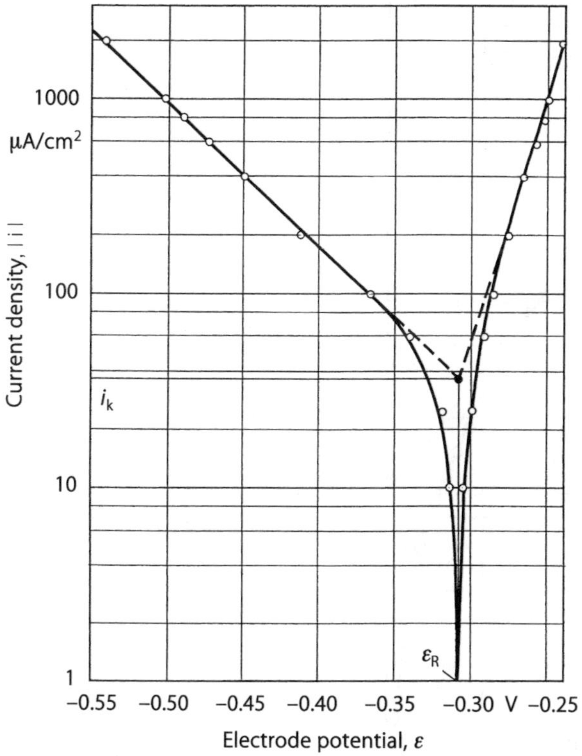


**Fig. 6.2.** Steady-state total current-potential curve of carbonyl iron in $H_2SO_4$/ 0.45 *M* $Na_2SO_4$ solution, pH 1.7, deaerated, 10°C. (From Kaesche)

shown in Fig. 6.2 [4]. It might, therefore, seem that Eq. (6.7) is particularly well founded, and it is, in fact, sometimes overestimated as basic for corrosion kinetics. Therefore, the previous warning (see Chap. 5) concerning the limitations of applicability of simple exponential functions for partial current-potential curves, i.e. 'Tafel behaviour', due to, among other effects, hysteresis effects by impurity adsorption is explicitly repeated.

The validity of the principles of electrolytic corrosion are generally accepted as evident without further proof. As a matter of fact, checks by independent measurements have been reported only rarely. Figure 6.3 shows that Tafel-line extrapolation does, indeed, produce correct results, since in this case, the procedure was checked by colorimetric chemical analysis of of the solution composition, and the results of the two independent experiments compare very well [5]. More such measurements have been reported for iron in acid solutions containing additions of the pickling inhibitor phenylthiourea (see Chap. 7), and the data Table 3.1 prove the quite satisfying correspondence of analytical corrosion rates and of values obtained by extrapolating total current-potential curves [6].

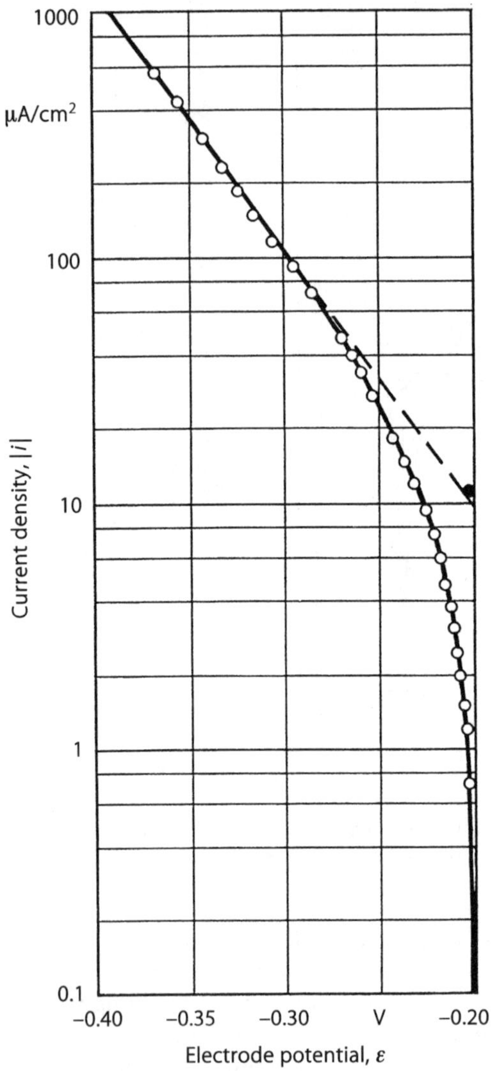


**Fig. 6.3.** Cathodic branch of the total current-potential curve for iron in deaerated HCl/4%NaCl-solution, pH 2. Solid circle: Corrosion rate-determined through colorimetric analysis of dissolved iron. (From Stern)

In principle, the method of determining corrosion rates from polarization curves does not require Tafel kinetics of partial reactions. Generally, it suffices a) to polarize the electrode into a region of potential where the current is equal to one of the partial currents, b) to evaluate the mathematical relation describing this curve, and c) to extrapolate this relation back to the corrosion potential. The method has not, however, been used for other than Tafel-line behaviour of

**Table 6.1.** Corrosion rate of carbonyl iron in $HClO_4/0.5MNaClO_4$-solutions, pH 2.0, 25°C with added phenylthiourea (PTU). **a** Values from colorimeric analysis of solution iron content; **b** values from evaluation of total current-potential curves. (From Kaesche)

| PTU concentration $c/\text{molL}^{-1}$ | corrosion rate $i_{\text{corr}}\ \mu\text{A cm}^{-2}$ a) | b) |
|---|---|---|
| 0 | $190 \pm 10$ | $210 \pm 10$ |
| $1 \times 10^{-7}$ | $210 \pm 10$ | $220 \pm 10$ |
| $1 \times 10^{-6}$ | $70 \pm 5$ | $70 \pm 5$ |
| $1 \times 10^{-4}$ | $14 \pm 3$ | $8 \pm 1$ |

partial reactions, except in the case of cathodic oxygen reduction in the region of rate determining oxygen diffusion (see below).

By the principle of superposition of partial current-potential curves, the total current, $i_s$, generally is the sum of two terms $i_{\text{corr}} \times f(\pi)$ and $-i_{\text{corr}} \times g(-\pi)$, with $f(0) = |g(0)| = 1$, where $f(\pi)$ and $g(\pi)$ are functions of the polarization, $\pi$. As long as both terms increase monotonously with $\pi$, there always is a monotonous function relating $i_{\text{corr}}$ to the tangent $(\partial\varepsilon/\partial i_s)_{\pi=0}$ indicating an increase of the rate of corrosion with an increase in the slope of the total current-potential curve. We recall from Eq. (4.13) that the tangent is the reciprocal of the differential polarization resistance, $(R_\pi)_{\text{diff}}$ When this function can be evaluated, it therefore offers means of determining corrosion rates by differential polarization measurements with only minute disturbance of the surface state of the corroded electrode.

In our case, where Eq. (6.7) is expected to hold, the evaluation is simple, because for small values of x the approximation $\exp x \simeq 1 + x$ can be used. Considering that $d\pi = d(\varepsilon - \varepsilon_{\text{corr}}) = d\varepsilon$, this leads to

$$\left(\frac{\partial\varepsilon}{\partial i_s}\right)_{\varepsilon_{\text{corr}}} = i_{\text{corr}} \left(\frac{1}{b'_{\text{Me}}} + \frac{1}{a'_{\text{H}}}\right) . \tag{6.8}$$

We remember that $b'_{\text{Me}} = d\varepsilon/d\ln \vec{i}_{\text{Me}}$, and $a'_{\text{H}} = d\varepsilon/d\ln \vec{i}_{\text{H}}$. More conveniently, we use $b_{\text{Me}} = d\varepsilon/d\log \vec{i}_{\text{Me}}$, and $a_{\text{H}} = d\varepsilon/d\log \vec{i}_{\text{H}}$. Dropping the 'diff'-subscript of $R_\pi$, we arrive at

$$i_{\text{corr}} = 0.434 \frac{a_{\text{H}}\, b_{\text{Me}}}{a_{\text{H}} + b_{\text{Me}}} \frac{1}{R_\pi} , \tag{6.9}$$

where $R_\pi$ is the differential polarization resistance of the corroded electrode at the open circuit corrosion potential. At the corrosion potential, the total current-potential curve is approximately linear over a range of potentials of several mV, and the determination of the tangent is easy. Hence the usual description of the method of evaluation as 'linear polarization' method. As one of the apparently rare examples of close inspection of the method, Table 6.2 shows a comparison of corrosion rate values obtained from Eq. (6.7) and Eq. (6.9), i.e., from Tafel-line extrapolations and from polarization restistance measurements [6]. The

**Table 6.2.** Evaluation of the **a** partial current-potential curves and **b** polarization resistance of iron in $HClO_4$/0.5MNa$ClO_4$-solutions, pH 2, with additions of phenylthiourea (PTU). Temperature 25°C, resp. 15°C (*). (From Kaesche)

| $c_{PTU}$ (mol L$^{-1}$) | $b_{Me}$ (mV) | $a_H$ (mV) | $R_\pi$ ($\mu$A mV$^{-1}$cm$^{-2}$) | $i_{corr}$ ($\mu$A cm$^{-2}$) Eq.(6.7) | Eq.(6.9) |
|---|---|---|---|---|---|
| $1 \times 10^{-6}$ | $71 \pm 5$ | $92 \pm 5$ | $3.8 \pm 0.1$ | $50 \pm 5$ | $55 \pm 5$ |
| $7 \times 10^{-6}$ | $80 \pm 5$ | $85 \pm 5$ | $2.0 \pm 0.1$ | $37 \pm 5$ | $39 \pm 5$ |
| $1 \times 10^{-6*}$) | $64 \pm 5$ | $83 \pm 5$ | $0.47 \pm 0.03$ | $7.6 \pm 1$ | $8 \pm 2$ |
| $1 \times 10^{-5*}$) | $92 \pm 5$ | $82 \pm 5$ | $0.25 \pm 0.02$ | $4.7 \pm 0.2$ | $5 \pm 1$ |
| $1 \times 10^{-4*}$) | $95 \pm 5$ | $78 \pm 5$ | $0.14 \pm 0.02$ | $2.6 \pm 0.2$ | $3 \pm 1$ |

good correspondence of the two sets of values is, of course, trivial and a check of correct experimenting only, since the compared methods are not mutually independent.

Equation (6.9) is due to Wagner and Traud [7]. As a tool suitable for rapid corrosion monitoring it later has, for some years, received much attention [8]. The suggestion is to substitute estimated or empirical values for the Tafel slopes, $a_H$ and $b_{Me}$, into Eq. (6.9), and experimental values for the polarization resistance, $R_\pi$. Nowadays, the direct-current polarization resistance, $R_\pi$, is sometimes obtained by extrapolation of the alternating-current resistance to zero voltage frequency (see Appendix). The limitations of the method should, however, be reognized: Even when Eq. (6.7) and, therefore, Eq. (6.9) really are valid, which is not at all generally to be expected, the slope constants may vary, depending on details of transfer-reaction electrode kinetics, between about 20 and 120 mV, or perhaps more. By consequence, rates calculated from Eq. (6.9) may vary by up to a factor 3 or more. This may be tolerable when the aim of monitoring only is to ascertain that corrosion is either negligible, or else not negligible and, thus, dangerous. In this case, even crude estimates based on the simple assumption that very often, though by far not always, a decrease in polarization resistance qualitatively indicates an increase in corrosion rate, may sufficiently answer the purpose. As a counter-example, the case of zinc, or indium, or sodium dissolution from dilute amalgams in acids containing dissolved metal cations in well defined concentration is instructive: As explained previously in Sect. 4.4, overall metal dissolution now is a secondary side-effect, small in comparison to the high equilibrium exchange current densities of the respective $Me/Me^{z+}$- reactions. Therefore, the polarization resistance is close to zero and the corrosion potential close to the equilibrium potential $E_{Me/Me^{z+}}$. There then is no useful relation between $R_\pi$ and $i_{corr}$, although the corrosion rate is well defined by the superposition of partial electrode reactions.

In the case of iron-group metals, Me, in solutions at zero or near-zero concentration of dissolved $Me^{z+}$ cations and at low pH the position of the corrosion potential between $E_{Me/Me^{z+}}$ and $E_{H_2/H^+}$ is usually such that the partial currents of both anodic hydrogen ionisation and cathodic metal deposition are

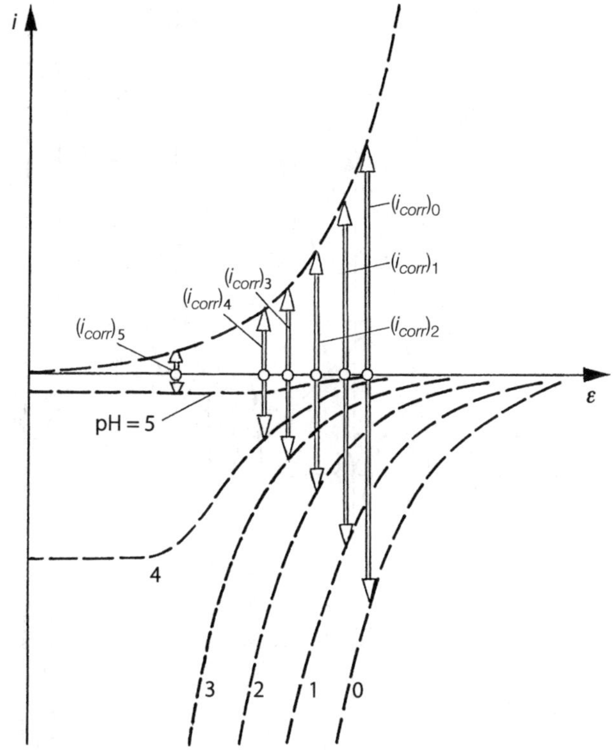


**Fig. 6.4.** Schematic current-potential diagram showing the change of corrosion rate and corrosion potential in deaerated acid solutions in the case of negligible rates of both anodic hydrogen ionisation and cathodic metal dissolution, assuming pH-independent kinetics of the rate of anodic metal disssolution, and assuming hydrogen deposition fro $H^+$-reduction only

negligible indeed, and Eq. (6.7) applies. If, at the same time, the partial current of anodic metal dissolution does not depend on pH, as apparently true for nickel [1], the dependence of $i_{corr}$ on pH is as schematically shown in Fig. 6.4, indicating a decrease of $i_{corr}$ with increasing pH, which is, of course, intuitive. More precisely, the decrease is due to a cathodic partial current-potential curve obeying Eq. (5.32), provided that an excess neutral salt content of the solution minimizes an influence of the $\zeta$ potential.

The situation depicted in Fig. 6.4 is corroborated by experimental evidence obtained for nickel in acid solutions nickel [1]. A typically different case is iron in acid solutions in the absence of dissolved halides, due to the dependence of rate of anodic metal dissolution on pH described in Sect. 5.3. As a well-documented example, Fig. 6.5 shows the shift of both the anodic partial current-potential curve for metal dissolution and of the cathodic partial current voltage curve for hydrogen deposition to more negative potentials accompanying an

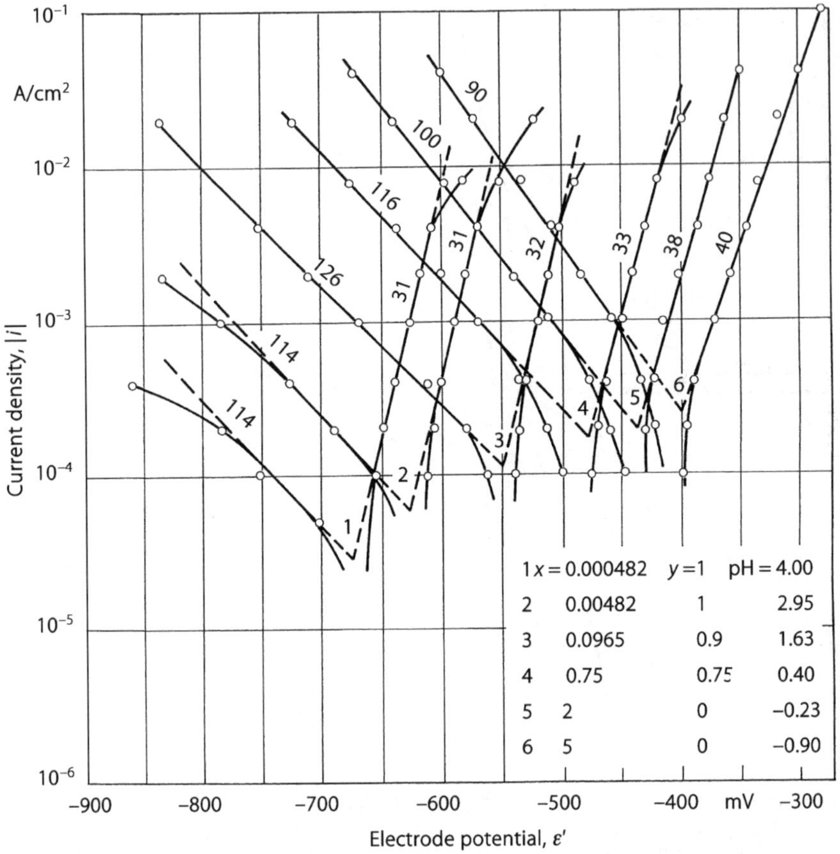


**Fig. 6.5.** Total current-potential curves, and anodic and cathodic partial current-potential curves for iron in x-molar $H_2SO_4$/ y-molar$Na_2SO_4$-solutions showing weak dependence of corrosion rate on pH due to $OH^-$-influence on anodic metal dissolution. Also indicated: Tafel slopes of partial current-potential curves. Potential referred to saturated calomel electrode. (From Voigt)

increase in pH [9]. The result is a much weaker decrease in corrosion rate as compared to the case of anodic metal dissolution without acceleration by hydroxo-compounds as intermediates of the anodic electrode reaction.

## 6.2 Neutral and Alkaline Solutions

We continue to inspect corrosion rates as function of pH under the same conditions as in the preceding section, but allowing a) for dissolved oxygen in the solution and b) for an increase in pH up to values beyond 7 into the region of alkaline solutions. Under these conditions, we have to consider, besides the currents of metal dissolution, $i_{Me}$, and of hydrogen deposition by $H^+$-reduction, $i_H$, the currents of oxygen reduction, $i_{O_2}$, and of hydrogen deposition by $H_2O$-reduction, $(i_H)_{H_2O}$. For the kinetics of these additional electrode reactions, see Sects. 5.1 and 5.2.

A particularly interesting case is corrosion of iron and unalloyed steel in solutions of pH between about 6 and 8, i.e., in approximately neutral solutions, because these are the aqueous phases in everyday corrosion by atmospheric moisture, or rain, or domestic water, or sea water, etc. In these solutions, hydrogen deposition by $H^+$ reduction is negligible regardless of the potential, due to the low diffusion-limited current estimated from Eq. (5.9). Hydrogen deposition by reduction of water is also negligible, because at the corrosion potential of typical structural materials (see Chap. 11) the cathodic overpotential, $\eta_H = \varepsilon_{corr} - E_{H_2/H^+}$, is not negative enough to overcome the high energy barrier for water reduction, provided that the temperature does not by far exceed room temperature. Anodic oxygen evolution is ruled out, because at the corrosion potential the cathodic overpotential for oxygen reduction, $\eta_{O_2} = \varepsilon_{corr} - E_{O_2,H_2O/H^+}$, is very negative, viz. $\eta_{O_2} = \eta_H - 1{,}21(\mathrm{V})$. Regardless of potential, anodic hydrogen ionisation is ruled out because without cathodic hydrogen deposition there is no gaseous hydrogen, unless the solution is intentionally flushed with the gas. On the whole, we therefore have to consider, for iron and unalloyed steel in the vicinity of $\varepsilon_{corr}$, the partial currents of anodic metal dissolution and cathodic oxygen reduction only. However, because current-potential measurements usually will be carried out, as in Figs. 5.11 and 5.12, into the potential range of hydrogen deposition by reduction of water we also consider, if only for experimental convenience, the partial current of hydrogen deposition by water reduction.

In abbreviated notation the equation for the total current-potential curve then reads

$$\begin{aligned} i_s &= i_{Me} + i_{O_2} + (i_H)_{H_2O} = \vec{i}_{Me} + \vec{i}_{O_2} + (\vec{i}_H)_{H_2O} \\ &= B_{Me} \exp\left(\frac{\varepsilon}{b'_{Me}}\right) - A_{O_2}\left(1 - \frac{i_{O_2}}{i_{O_2,D}}\right)\exp\left(\frac{\varepsilon}{-a'_{O_2}}\right) - \\ &\qquad -A_{H_2O}\exp\left(\frac{\varepsilon}{-a'_{H_2O}}\right) . \end{aligned} \tag{6.10}$$

The solid curve in Fig. 6.6 is the total current-potential curve for oxygen corrosion in the range of the cathodic diffusion-limited current density. By a slight shift of curve A in the anodic direction, the superposition of curves C and, now, $A_4$ will shift the corrosion potential into the range of mixed transfer-reaction/diffusion control of cathodic oxygen reduction. In any of these cases, Eq. (5.38), in the form $i_{corr} \lesssim |i_{O_2,D}|$, applies as an estimate of the corrosion rate,

In Sect. 3.5 the catastrophic reaction of sodium with water was briefly mentioned as an example of corrosion by reduction of water. Though certainly not really obeying simple Tafel kinetics as described by Eq. (6.7), the case may

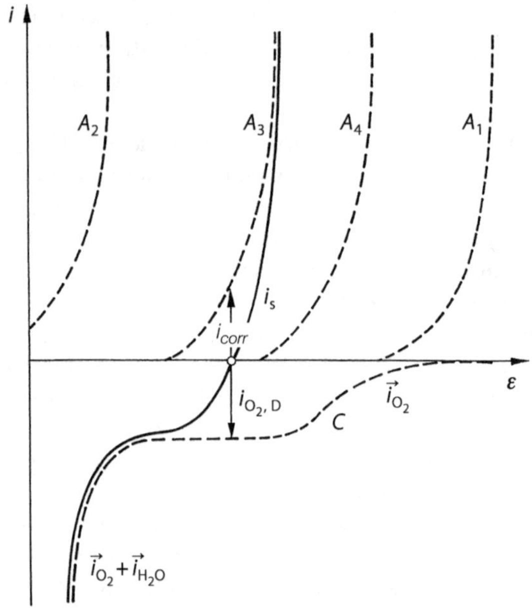


**Fig. 6.6.** Schematic current-potential curves for corrosion of metals in approximately neutral solutions by superposition of partial currents of anodic metal dissolution (A) and cathodic oxygen reduction and water reduction (C), assuming negligible reverse reaction for all electrode reactions. $A_{1...4}$ represent typical positions of the anodic curves depending on the magnitude of $B_{Me}$ in Eq. (6.10), establishing the corrosion potential in the range of charge-transfer controlled oxygen reduction ($A_1$), charge-transfer controlled water reduction, ($A_2$), diffusion controlled oxygen reduction ($A_3$), an mixed charge-transfer/diffusion controlled oxygen reduction ($A_4$), respectively. Cases $A_1$ and $A_2$ reflect behaviour of copper and sodium, respectively, both $A_3$ and $A_4$ that of iron and mild steel

roughly be considered as an example of superposition of curves $A_2$ and C in Fig. 6.6. Resistance to this type of corrosion by reduction of water is, of course, a trivial prerequisite for any structural material. Still, hydrogen is seen slowly evolving from magnesium at least in hot water. It also has been shown that iron shavings develop some hydrogen in contact with water, though only very slowly, during months [10]. Slow hydrogen evolution is also sometimes observed in the circulating water of closed central heating systems. A more detrimental corrosive action of this type is hydrogen evolution from zinc in hot water, causing the formation of bubbles underneath hot galvanized layers on steel line pipes [11,12].

As an example of superposition of curves $A_1$ and C, shifting the corrosion potential into the range of charge-transfer reaction control we have already presented the case of acid corrosion of iron in the preceding section. In neutral solutions, the case is represented by copper. The long-time corrosion resistance of this metal is, however, probably much more due to protective layers of corrosion products than to the overpotential of anodic metal dissolution.

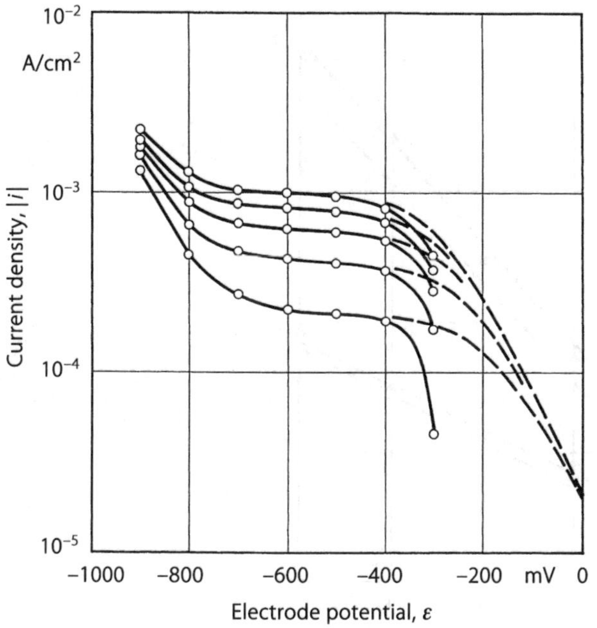


**Fig. 6.7.** Cathodic branches of the total current-potential curve for rotating disks of mild steel in 0.5 M NaCl solution flushed with a 20%$O_2$/80%$N_2$ gas mixture. Root of disk rotational number, from top to bottom: $\sqrt{\omega} = 5,4,3,2,1\ s^{-1}$. Broken curves: Estimated partial current-potential curves for oxygen reduction. (From Bohnenkamp)

In the present context, the more important experience is that the superposition of curves C and either $A_3$ or $A_4$ essentially describes corrosion of iron and mild steel in approximately neutral solutions. As an example illustrating this observation, Figure 6.7 presents a logarithmic plot of the cathodic branches of total current-potential curves for mild steel in an aerated neutral NaCl-solution, observed with rotating disks [13]. The curves show diffusion-limited currents in the range of potentials between −400 and −700 mV, increasing with increasing rotational speed. At more negative potentials, the current increases beyond the limiting plateau, indicating cathodic hydrogen evolution by water reduction. At more positive potentials, the current decreases towards zero at the corrosion potential. In this region and beyond, the broken curves indicate the partial current-potential curves for oxygen reduction with mixed diffusion/charge-transfer reaction control and, eventually, charge-transfer control only. Apparently, the corrosion potential was, though only nearly so, in the range of diffusion control of cathodic oxygen reduction. Then, Eq. (5.38) should apply in the form $i_{corr} = |i_{O_2,D}|$, According to Eq. (5.37), we expect $i_{corr} \sim c_{O_2}/\delta_{Nernst}$. In Sect. 5.2 we have already noted that for rotating disk cathodes the Nernst diffusion layer thickness is $\delta_{Nernst} \sim 1/\sqrt{\omega}$. The relation with $1/\sqrt{\omega}$ has been verified using rotating iron disks, in this particular case

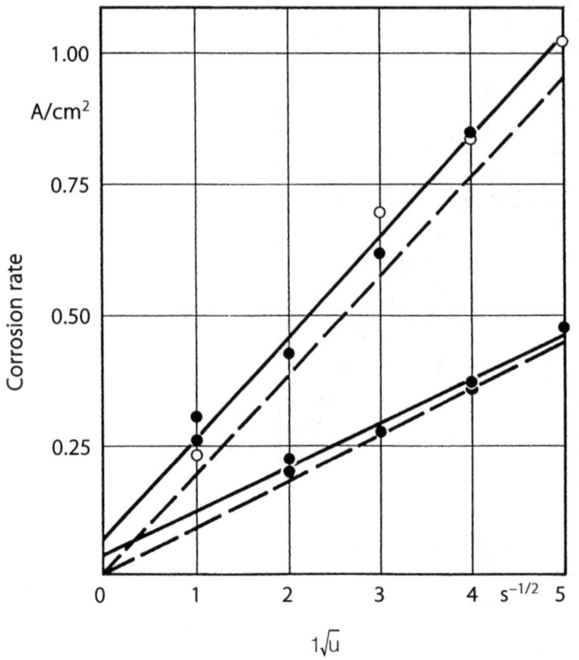


**Fig. 6.8.** Corrosion rate of rotating disks of mild steel in $H_2SO_4$/0.4 M $Na_2SO_4$ solution, pH 2.7, 20°C, as function of the square root of the disk rotational number, determined from the iron content of the solutions after 6 h (*solid circles*), resp. 2 h (*open circles*). Solutions were flushed with air (*upper line*), resp. 10%$O_2$/90%$N_2$ mixtures (*lower lines*). Broken lines calculated from Eq. (5.38). (From Bohnenkamp)

in sulfuric acid solutions, at two values, 0.2 and 0.1 atm, of the oxygen partial pressure, thus also verifying the dependence of the kinetics on $c_{O_2}$. The slight acidity of the solutions did not yet interfere with reaction control by oxygen diffusion; it was chosen to avoid the deposition of rust which would have slowed oxygen diffusion to the metal surface. More measurements confirming Eq. (5.37) for mild steel corrosion in neutral aerated solutions are shown in Fig. 6.9 [14]. In this case, $CaCl_2$ solutions were chosen, aiming at modeling corrosion in domestic tap water. Again the correspondence of theory and experiment is quite convincing. Still, it must be recognized that slight changes in conditions may shift the corrosion potential of mild steel in approximately neutral solutions into the range of mixed diffusion/charge-transfer reaction control.

When the corrosion of iron and mild steels in approximately neutral solutions is precisely, or else at least approximately governed by oxygen diffusion, the equation for the total current voltage curve in the vicinity of the corrosion potential is

$$i_s(\varepsilon) = B_{\mathrm{Fe}} \exp(\varepsilon/b'_{\mathrm{Fe}}) + i_{\mathrm{O_2,D}} = i_{\mathrm{corr}} \left\{\exp(\pi)/b'_{\mathrm{Fe}}) - 1\right\} , \tag{6.11}$$

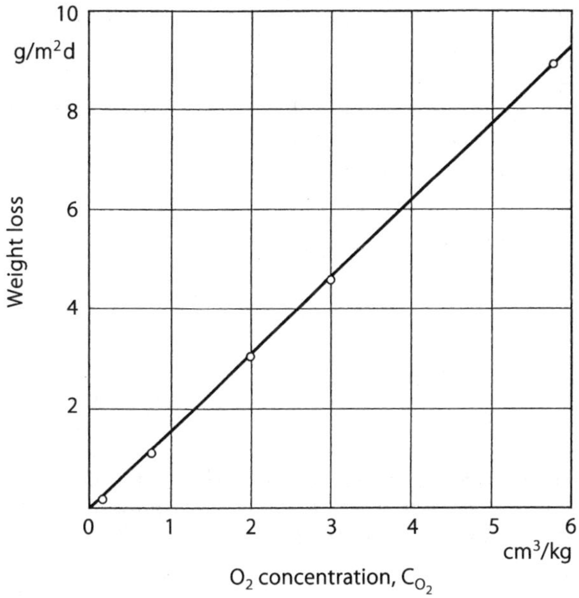


**Fig. 6.9.** Weight loss of mild steel during 48 h in a stirred solution of 165 mg kg$^{-1}$ $CaCL_2$ in water as a function of the oxygen content of the solution. (From Uhlig, Triadis, and Stern)

where, as before, $\pi = \varepsilon - \varepsilon_{corr}$. From this it follows that the corrosion rate is related to the differential polarization resistance observed at $\varepsilon_{corr}$, i.e. to $R_\pi$, and to the slope constant, $b_{Fe} = 2.303b'_{Fe}$, as derived in [8a]:

$$i_{corr} = 0.434\, b_{Fe}\, (1/R_\pi)\ . \tag{6.12}$$

Accuracy is improved, if $b_{Fe}$ is explicitly determined in the vicinity of $\varepsilon_{corr}$, by measuring the curvature of the total current-potential curve. To do so, it suffices to record two changes $\Delta i_s$ caused by two imposed changes $\pm\Delta\pi$ [15]. The method has been used to monitor corrosion of mild steel in sea-water, checking the procedure by chemical analysis of the amount of iron dissolved in the solution. The latter, averaging method gave a value of 0.162 mA cm$^{-2}$. As shown in Fig. 6.10 it compares well with the momentary values calculated from polarization data. The decrease in corrosion rate observed during the experiment is probably due to a thin layer of rust. Apparently, this did not impede the applicability of Eq. (6.11), but only decreased $i_{O_2,D}$. From this it appears that, within limits, the influence of a thin rust layer can be taken care of by empirically correcting the ratio $D_{O_2}/\delta_{Nernst}$ in Eq. (5.37) to lower values. Rust, then, simply acts as an inert porous layer, and this, indeed, is a model frequently used to describe its properties as far as kinetics are concerned. In Chap. 9, we shall, however, see that conditions often are quite more complicated.

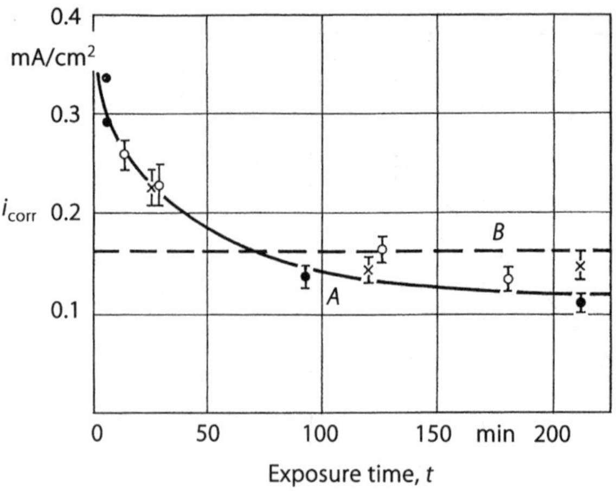


**Fig. 6.10.** The corrosion current density of Siemens Martin steel in artificial, aerated sea-water. Solid curve: Values determined from polarization measurements. Broken line: average corrosion current density calculated from iron content of the solution. (From Engell)

Whenever corrosion is truly diffusion-controlled, measuring $i_{O_2,D}$ is certainly quite as useful for determining corrosion rates as measuring the polarization resistance. Corrosion monitoring by $i_{O_2,D}$-recording, once popular, has, however, come into disuse. At the same time, the qualitative control of safe corrosion protection by continuous monitoring of sufficiently low residual oxygen contents in deaerated solutions such has power plant feed water is, of course, standard practice.

At this stage, we may attempt to summarize predictions as to the dependence of the rate of corrosion of iron on pH of aerated, stirred electrolytic solutions assuming, as appropriate for iron-group metals, conditions such that in the vicinity of the corrosion potential reverse reactions of metal deposition, hydrogen ionisation, and oxygen evolution are all negligible. Also assuming, again as appropriate for these metals, anodic rate control by charge-transfer reactions, and cahodic rate control by diffusion and/or charge-transfer, we arrive at the schematic current-potential diagram in Fig. 6.11. The prediction then is that at low pH the rate of corrosion is high due to rapid cathodic hxdrogen deposition, decreasing with increasing pH to lower values, and eventually tending to a plateau value, where oxygen diffusion is rate-determining for the overall reaction. If the mechanism of anodic metal dissolution is independent of pH, no further change of the rate of corrosion with pH is expected. If, instead, some $OH^-$-catalyzed mechanism is operative, one possibility would be that the corrosion rate eventually increases at high pH due to hydrogen deposition by water reduction. For comparison, Fig. 6.12 shows the typical behaviour of iron, mild steels and low-alloy steels in aqueous electrolytic solutions: As predicted, the rate of corrosion decreases, from low values of pH, to a pH-independent

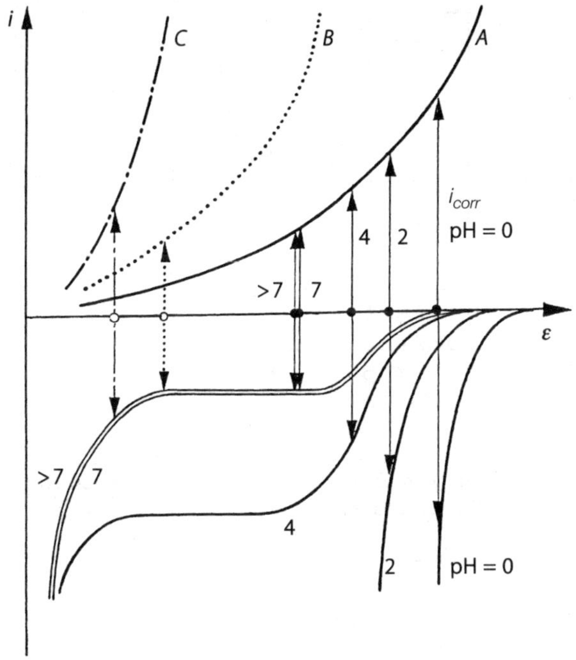


**Fig. 6.11.** Estimated dependence of the corrosion rate of an iron-group metal in aerated, stirred electrolytic solutions at ambient temperatures, assuming kinetics governed by anodic transfer-reaction kinetics of metal dissolution, and mixed diffusion/transfer-reaction kinetics of cathodic hydrogen deposition and oxygen reduction. **a** Metal dissolution independent of pH; **b** metal dissolution catalyzed by pH-dependent intermediate reaction products, but cathodic water reduction still negligible; **c** similar to be, but with noticeable cathodic water reduction

value in the range of near-neutral solutions. Beyond pH 10, however, it drops to virtually zero [16]. The reason for this effect, unexpected from what has so far been explained, is *passivation* by non-porous ultra-thin oxide films, fully described in Chap. 10.

The typical change in corrosion potential of iron and low-alloy steels in the range of non-passivating solutions up to about pH 11 is shown in Fig. 6.13 [4]. The decrease of the corrosion potential up to about pH 5 corresponds to expectations. For an $OH^-$-catalyzed mechanism of iron dissolution, the decrease should continue to higher values of pH. Since, instead, it tends to increase, although passivation sets in only beyond pH 10, the influence of pH on active anodic iron dissolution in neutral solutions may appear doubtful. It is, however, necessary to consider that in the range of the cathodic diffusion-limited current of oxygen reduction the pH of the solution adjacent to the metal surface differs from bulk pH, because the cathodic partial reaction produces $OH^-$ and thus increases pH at the metal surface, which is where iron atoms react to hydroxylated compounds as reaction intermediates. When surface-pH is high, the metal

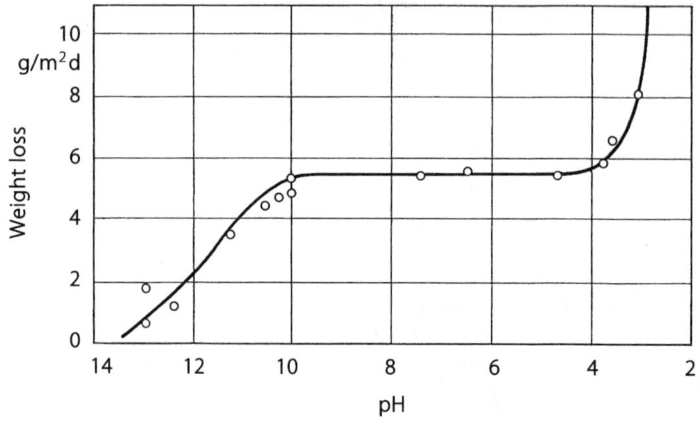


**Fig. 6.12.** Experimental dependence of the corrosion rate of mild steel in aerated NaOH and HCl solutions. (From Whitmann, Russel, and Altieri)

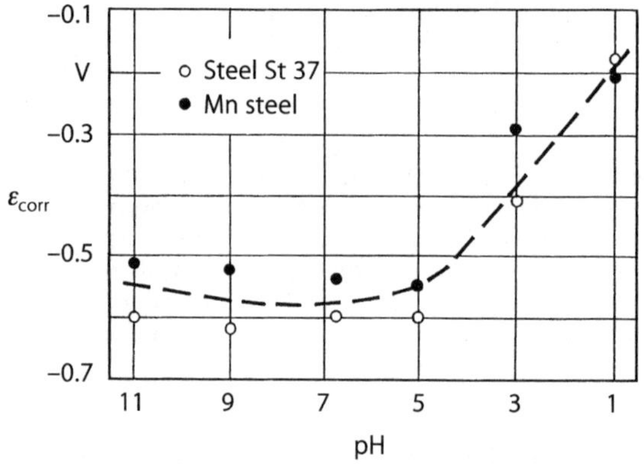


**Fig. 6.13.** Steady-state corrosion potential of two low-alloy steels in aerated, stirred $H_2SO_4$/NaOH/0.5 M $Na_2SO_4$l solutions. (From Kaesche)

surface may well become saturated with respect to adsorbed $OH^-$, preventing any further dependence of the anodic partial current on pH.

More typical evidence concerning the dependence of corrosion rates on pH is shown in Fig. 6.14 for aluminum [17], and in Fig. 6.15 for zinc [18]. For aluminum, the effect of passivation once more is prominent, in this case in near-neutral solutions. The passivating oxide is, however, amphoteric, and it dissolves in alkaline solutions to complex $(AlOH)_4^-$ ions, causing a steep increase in corrosion rates accompanied by hydrogen deposition through water reduction.

For zinc, the rate of corrosion in near-neutral solutions again is small as compared to the the rate of corrosion of iron, i.e., by far not equivalent to

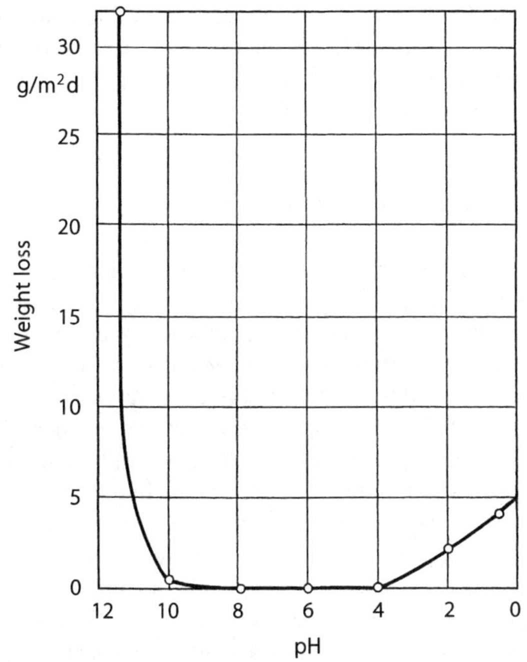


**Fig. 6.14.** Dependence of the corrosion rate of pure aluminum on solution pH in aerated $H_2SO_4$/NaOH/1 M NaCl solutions. (From Pryor and Keir)

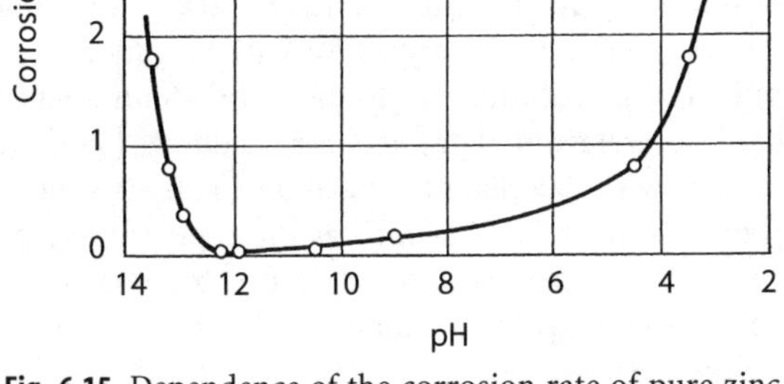


**Fig. 6.15.** Dependence of the corrosion rate of pure zinc on solution pH in aerated NaOH and HCl solutions. (From Roetheli, Cox, and Littreal)

the cathodic diffusion-limited current density of oxygen reduction. Passivation is not as pronounced as with aluminum, while again the protective, though insufficiently passivating, layers of corrosion products are amphoteric, so that in alkaline solutions they dissolve to complex $(ZnOH)_4^{2-}$-ions.

## 6.3 The Dependence of Corrosion Rates on Temperature

For a discussion of the influence of temperature on the rate of corrosion reactions we return to the Arrhenius equation Eq. (5.18). As noted previously, the underlying concept is a mechanism of chemical reactions involving the passage of reaction species through an intermediate 'activated complex' subsequently dissociating into reaction products. The activated complex is at an energy level exceeding the energy of the reactants by the 'activation energy', $A$ (see Fig. 5.3). As noted previously, this really is an activation Gibbs energy, $\Delta G$, with an entropy term $T\Delta S$ and an enthalpy term $\Delta H$ (see Chap. 3). When the valency of the rate-determining step is zero, i.e. $n_e = 0$, the evaluation of the activation energy is straightforward, because differentiating Eq. (5.18) yields $d \ln k/d(1/T) = -A/R$. When, instead, $n_e$ is positive as for anodic partial reactions, or negative, as for cathodic partial reactions, electrostatic energy enters the activation energy, as described by Eq. (5.20) for the special (and most probable ) case $n_e = \mp 1$. The difference $\phi_{m,l}$ of Galvani potentials is an unknown quantity, but there is no reason to assume that it does not depend on temperature. Differentiation then produces expressions such as

$$\left(\frac{\partial \ln \vec{i}_r}{\partial(1/T)}\right)_\varepsilon = -\frac{A'}{R} + \frac{\gamma n_e F}{R}\left\{\phi_{m,l} + \frac{1}{T}\left(\frac{\partial \phi_{m,l}}{\partial(1/T)}\right)_\varepsilon\right\}, \tag{6.13}$$

with $0 \leq \gamma \leq 1$. Provided that the simple capacitor model for the electric double layer is valid, $\gamma$ is the the same as $\alpha$, rsp. $\beta$, as introduced in Chap. 5, and $\gamma = 0.5$ is a useful estimate. The two terms on the right side of this equation cannot be determined separately, but the sum is available from current-potential measurements as an apparent activation energy. Figure 6.16 shows current-potential curves for acid corrosion of iron at different temperature. Plotting values of $d \ln i/d(1/T)$, the apparent activation energies for cathodic hydrogen deposition on iron and for anodic iron dissolution are found to be about 25 and 100 kJ mol$^{-1}$, respectively. For other relevant data, see, for instance, [19a]. It should be recalled that the capacitor with a constant capacitance is the simplest possible model of the electrical double layer. Details such as the $\zeta$-potential (see Sect. 5.4), which certainly is expected to change with temperature, are neglected. In the present context aiming at an overview rather than a quantitative theory, this does not, however, matter very seriously.

More insight into the nature of activation energies is obtained from reaction rate theory in the form of the *activated complex theory* or *transition state theory* [19]. As it does not usually supply quantitative predictions except for reactions in the gas phase, a brief comment may suffice. Two main suppositions are

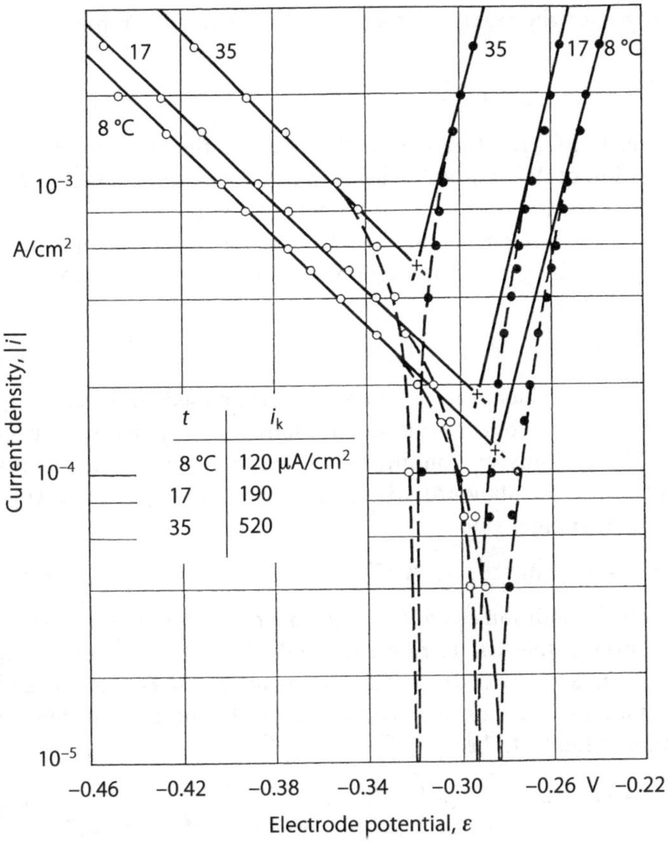


**Fig. 6.16.** Total current-potential curves for iron in deaerated $H_2SO_4/Na_2SO_4$ solution, pH 2, at different temperatures. (From Kaesche)

essential. The first is that the activated complex, usually indexed as $M^{\ddagger}$, passes through a transition state and decomposes into reaction products at a rate equal to the product of a transition coefficient, $\kappa \leq 1$ (and probably $\kappa \simeq 1$), the decomposition frequency, $f$, and the concentration of the activated complex, $c_{\ddagger}$. We note that in the case of electrode reactions, $M^{\ddagger}$, is a charged species: $(M^{n_e})^{\ddagger}$. The current density, $\vec{i}_r$, of any forward partial reaction then is

$$\vec{i}_r = n_e F \kappa f c_{\ddagger} . \tag{6.14}$$

For reverse partial reactions an analogous relation holds for $\overleftarrow{i}_r$.

The second proposition is that the decomposition of the activated complex is slow compared to the exchange rate of the reaction producing the activated complex, so that $M^{\ddagger}$ is in equilibrium, or else quasi-equilibrium, with the reactants A, B, etc., and with electrons in the electrode. In this case, equilibrium thermodynamics apply for the reaction producing the activated complex. As in Eq. (2.10), we write

$$(-\nu_A)A + (-\nu_B)B + ... \rightarrow M^{\ddagger} + n_e e^- \tag{6.15}$$

During the formation of $M^{\ddagger}$, reacting entities move along the reaction coordinate, $\xi$, and the position of $M^{\ddagger}$ is $\gamma \times \xi$. Therefore, the reaction producing $M^{\ddagger}$ passes only through a part, $\gamma \times \phi_{m,l}$, of the interfacial potential difference, while the complete electrode reaction passes through $\phi_{m,l}$, always with the supposition of negligible influence of the $\zeta$-potential. By analogy to Eq. (3.19) we find that the Gibbs reaction energy is

$$\Delta_r G = \Sigma \nu_i \tilde{\mu}_i = \Sigma \nu_i \mu_i - \gamma n_e F \phi_{m,l}\,, \quad 0 \leq \gamma \leq 1\,. \tag{6.16}$$

In equilibrium, $\Delta_r G = 0$. Chemical potentials are expressed as $\mu = \mu^\circ + RT \ln cf$, where $c$ and $f$ are concentrations and activitiy coefficients, respectively, as in Eq. (3.21). Simplifying conditions, we assume activity coefficients to be 1, and we introduce the standard Gibbs reaction energy $\Delta_{\ddagger} G^\circ = \Sigma \nu_i \mu_i{}^\circ$. Equation (6.16) then rearranges into

$$RT \ln c_{\ddagger} = RT\Sigma(-\nu_{i \neq M^{\ddagger}}) \ln c_{i \neq M^{\ddagger}} - \Delta_{\ddagger} G^\circ + \gamma n_e F \phi_{m,l}\,. \tag{6.17}$$

Supposing experimental conditions such that the concentration of all reactants remains virtually constant, the first term on the right side of the relation is a constant quantity. Also, $\Delta_{\ddagger} G^\circ = \Delta_{\ddagger} H^\circ - T\Delta_{\ddagger} S^\circ$. Using this in Eq. (6.17), and substituting the resulting expression for $c_{\ddagger}$ into Eq. (6.14), the current density of a forward reaction is found to be

$$\vec{i_r} = const \times zF\kappa f \exp\left(\frac{\Delta_{\ddagger} S^\circ}{R}\right) \exp\left(-\frac{\Delta_{\ddagger} H^\circ - \gamma n_e F \phi_{m,l}}{RT}\right)\,. \tag{6.18}$$

As a result of statistical thermodynamics, which we mention without explanation, $f = kT/h$, where $k$ is *Boltzmann's constant*, $1.38 \times 10^{-23}$ J K$^{-1}$ and $h$ is *Planck's constant*, $6.23 \times 10^{-34}$ J s$^{-1}$. In our context, this has no direct application, and the practical access to the temperature dependence of electrode reaction rates is to use experimental apparent activation energies.

When, as in the case of acid corrosion of iron, the mechanism of corrosion at the corrosion potential is governed by charge-transfer control of both the anodic and the cathodic partial reaction, the knowledge of two such activation energies is needed. When, instead, as often for iron in neutral aerated solutions, only the anodic partial reaction is charge-transfer controlled, while the rate of the cathodic partial reaction is practically equal to the diffusion-limited current density, $i_{O_2,D}$, and Eq. (5.38) applies, we need to know only the temperature dependence of diffusion of dissolved oxygen, which is readily available.

Generally, diffusion in solutions can, at least within the range of temperatures between room temperature and ca. 100°C, be described as approximately falling into the class of thermally activated reactions, and the Arrhenius equation is approximately applicable within these limits. In the case of oxygen dissolved in aqueous solutions at atmospheric pressure, a complication arises because, by Eq. (5.37), $i_{O_2,D}$ depends on the concentration, $c_{O_2}$, as well as on the diffusion

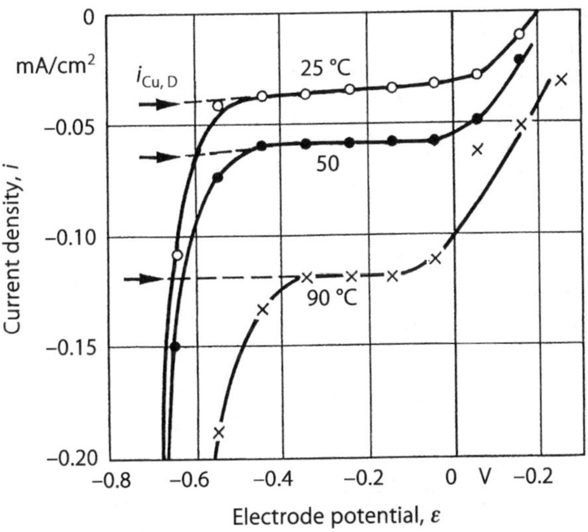


**Fig. 6.17.** Current-potential curves for cathodic deposition of copper on platinum, with superposed cathodic hydrogen deposition by water reduction at potentials below −0.4 V, in deaerated 0.8 M $CuSO_4$/0.1 $Na_2SO_4$ solution at different temperatures. (From Kaesche)

coefficient, $D_{O_2}$, and $c_{O_2}$ may change with temperature. To more clearly demonstrate the influence of temperature on the diffusion coefficient alone, Fig. 6.17 shows cathodic current-potential curves for copper deposition on platinum from dilute $Cu^{2+}$ solutions at constant concentration. The diffusion-limited current is well defined; at sufficiently negative potentials cathodic water decomposition masks copper deposition. Applying the Arrhenius equation, Eq. (5.18), a very low activation energy of ca. 15 kJ $mol^{-1}$, is estimated. Using this equation implies, that the mechanism of diffusion in solutions involves the passage of diffusing molecules over energy barriers between adjacent positions in the array of water molecules. Typically, the barrier height in this short-range order of solution molecules is low, much lower than in the long-range order of crystalline solids. The picture is complicated by the thermal expansion of water with increasing temperature lowering viscosity, and the calculation is approximate only.

Since the change in $D_{O_2}$ with temperature is comparatively low, it should, in the case of oxygen corrosion in air-saturated solutions, be largely compensated and even over-compensated by the decrease in $c_{O_2}$. Experiments with mild steel disks rotating in NaCl solutions at temperatures between 20 and 80°C confirm this prediction, as seen in Fig. 6.18 [13].

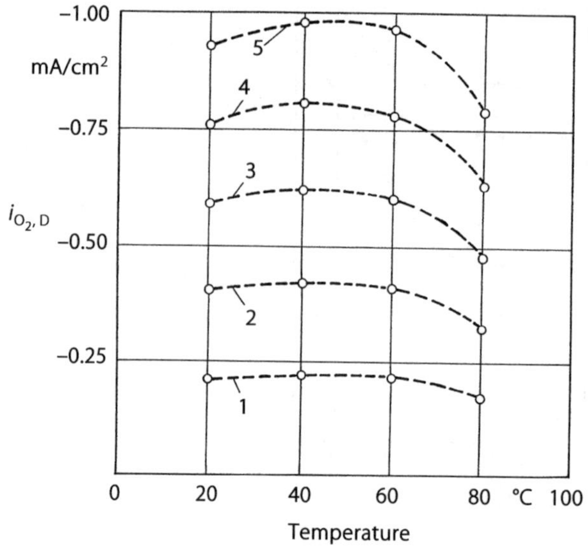


**Fig. 6.18.** Diffusion-limited current density for oxygen reduction at rotating steel disks from 0.5 M NaCl solution at different temperatures. Solutions flushed with $20\%O_2/80\%N_2$ mixtures. Parameter: Square root, $\sqrt{n}/s^{-\frac{1}{2}}$, of disk rotations per second.(From Bohnenkamp)

## References


1. Kolotyrkin, Ya.M. and Frumkin, A.: Dokl. Akad. Nauk SSSR *33* (1941) 446
2. Bonhoeffer, K.F. and Heusler, K.E.: Z. phys. Chem. N. F. *8* (1960) 390
3. Fischer, H. and Yamaoka, H.: Chem.Ber. *94* (1961) 1477
4. Kaesche, H.: ERP Report 1577, Bundesanstalt f. Materialprüfung, Berlin 1961
5. Stern, M. and Geary, A.L.: J. Electrochem. Soc. *102* (1955) 609
6. Kaesche, H.: Z. Elektrochemie Ber. Bunsenges. phys. Chem. *63* (1959) 495
7. Wagner, C. and Traud, W.: loc. cit. Chap. 4 [5]
8. a) Stern, M. and Eisert, E.D.: Proc. Amer. Soc. Testing Materials 59 (1959) 1280 29 (1973) 397, 403; b) Mansfeld, F.; Corr. NACE 29 (1973) 397, 403; 32 (1973) 143; c) Prazak, M.: Werkstoffe u. Korr. 25 (1974) 104; d) Palombarini, G., Felloni, L., and Cammarota, G.P.; Corr. NACE 32 (1973) 245; e) Leroy, R.L.: ibid. 29 (1973) 272; f) Bandy, R. and Jones, D.A.: ibid.32 (1976) 126; g) Heitz, E. and Schwenk, W.: Werkstoffe u. Korr. 27 (1976) 241; h) Mansfeld, F.: in: Advances in Corrosion Science and Technology. Fontana, G. and Staehle, W., eds., Plenum Publ. Comp., New York 1976, p. 163
9. Voigt, Ch.: Doctoral thesis, Dresden (1968), quoted from Schwabe, K., ed.: Korrosionsschutzprobleme. VEB Deutscher Verlag f. Grundstoffin-dustrie, Leipzig, 1969
10. Schikorr, G.: Z. Elektrochemie *35* (1929) 62, 65
11. Friehe, G.: Sanitär- u. Heizungstechnik *3* (1963) 69

12. Kaesche, H.: Werkstoffe u. Korr. *26* (1975) 175
13. Bohnenkamp, K.: Arch. Eisenhüttenwesen *26* (1955) 393
14. Uhlig, H.H., Triadis, D, and. Stern, M.: J. Electrochem. Soc. *102* (1955) 59
15. Engell, H.J.: Archiv Eisenhüttenwesen *29* (1958) 553
16. Whitmann, W., Russel, R., and Altieri, V.: Ind. Engng. Chem *16* (1924) 665
17. Pryor, M.J. and Keir, D.S.: J. Electrochem. Soc. *105* (1957) 629
18. Roetheli, B., Cox, G., and Littreal, W.: Metals and Alloys *3* (1932) 73
19. See for instance: a) Glasstone, S., Laidler, K.J., and Eyring, H.: Theory of Rate Processes. McGraw-Hill Book Co., New York 1940; b) Atkins, P.W.: loc. cit. Chap. 3 [2]; c) Conway, B.E.: Theory and Practice of Electrode Processes. The Ronald Press Comp., New York,1965; Krishtalik, L.I.: loc. cit. Chap. 3 [6], Vol. 7, Kinetics and Mechanism of Electrode Processes; e) Heusler, K.E.: in: Int. Conf. High Temperature High Pressure Electrochemistry in Aqueous Solutions. The University of Surrey, 1973. NACE, Houston, 1975

# 7 Adsorption Inhibitors of Acid Iron Corrosion

By definition, *corrosion inhibitors* are soluble substances which, added to the aggressive solution, reduce the corrosion rate without participating in the overall reaction. As they are not reactants in the overall corrosion reaction they are negative catalysts. In this chapter we shall not deal with inhibitors such as sodium carbonate, or other substances increasing the pH of the solution and favoring spontaneous passivation, or hydrazine as a reagent reducing dissolved oxygen and thus removing the aggressive species in many neutral-solution corrosion processes. Also, we shall disregard inhibitors as components of protective paint systems, such as the classical read lead oxide, $Pb_3O_4$, an intensely oxidizing species again aiming at passivating iron and steel surfaces. Instead, the object will presently be the mechanism of corrosion protection by inhibitors of acid corrosion acting through adsorption on the metal surface from a solution containing the inhibitive substance in millimolar or even smaller concentration only. The typical industrial application of adsorption inhibitors is protection of hot-rolled steel against heavy metal loss by corrosion during acid pickling. Another important application is protection of underground gas and oil well piping against corrosion and accompanying embrittlement by attack by underground acid hydrogen sulfide solutions.

Discussing adsorption inhibitors of acid corrosion, we still deal with bare metal surfaces, while the usual case of everyday corrosion of non-noble metals is that of metals carrying an oxide film or other films of corrosion products. Filmed metals will be the object of later chapters, especially concerning passivity, or, as it is usually put, metals in the *passive state*. Consequently, corrosion of non-oxidized metal surfaces often is also called that of metals in the *active state*. It is confusing that, at the same time, the term 'passive' corrosion protection also is used to embrace all methods of protection which leave the composition of the aggressive solution unchanged, achieving protection by such 'passive' means as painting the surfaces. In this sense, adding inhibitors to the solution would belong to the 'active' methods of protection. The reader interested in more comprehensive treatises on the subject is referred to the literature [1–3]. Special reference should be made to the work by H. Fischer [4]. This author also has pointed out the need to distinguish 'primary' inhibition directly by the substances as they have been added to the solution, and 'secondary' inhibition by products of chemical reactions of these substances, as for instance by cathodic reduction. The principles of adsorption at electrodes are covered in much detail, including the influence of adsorption on the structure of the electrical double layer, in treatises on electrode kinetics [5]. Of these details, only few will be discussed in the following, and the simple model of the electrical double layer as a plate capacitor will be retained, although many inhibitors, and especially industrial inhibitors, are large molecules or ions probably seriously interfering with the distribution of the Galvani potential across the metal/solution interface.

Even for the narrow field of adsorption inhibition of acid corrosion the number of suitable substances is huge, as they may be heterocycles containing oxygen or sulfur or nitrogen, or high-molecular alcohols, aldehydes, amines and amides, sulfonic or fatty acids and their derivatives, thiourea and its derivatives, thiazoles and thioureazoles, quaternary nitrogen compounds and phosphonium compounds, unsaturated ring and chain systems, thioamides and thiosemicarbazides, high-molecular nitriles, mercaptanes, sulfides, sulfoxides, thiazine, etc. [6].

Typically, adsorption inhibitors are organic molecules with a carbon chain or ring system, but still with some solubility in water, and containing functional groups tending to interact with metal surfaces. Therefore, the action essentially is that of screening the metal surface against the solution, blocking the passage of ions and/or electrons between metal and solution. The typical action of an organic adsorption inhibitor of acid corrosion was already documented by the data presented in Tables 6.1 and 6.2 showing the decrease of the corrosion rate of iron caused by very low concentrations of phenylthiourea. The *inhibitor efficiency*, $W_I$, defined by

$$W_I = \frac{(i_{corr})_0 - (i_{corr})_I}{(i_{corr})_0} 100\% \tag{7.1}$$

from the corrosion rates $(i_{corr})_0$ and $(i_{corr})_I$ observed in the absence and in the presence of the inhibitor, was 95% already for a solution concentration of the inhibitor of only $10^{-4}$ mol $L^{-1}$.

From the supposition that inhibitors of acid corrosion act by adsorption and thus by blocking electrode reactions it follows that the efficiency should increase with increasing size of inhibitor molecules. As seen in Fig. 7.1, this is verified by early experiments with mild steel in sulfuric acid solutions [7].

Acid corrosion of iron and unalloyed steel is a case of active uniform, or near-uniform corrosion. As shown in the preceding chapter, it also is a case where the principle of independent superposition of partial anodic and cathodic current-potential curves, based on the assumption that the metal surface is anode and cathode at the same time everywhere, is applicable. At the same time, we have seen in Sect. 5.3, that in the important case of iron in acid solutions we have good reason to recognize that on the atomic scale the elementary steps of metal dissolution preferentially occur at kinks in mono-atomic ledges at the surface. We have not discussed the nature of surface lattice places preferred for the step of cathodic hydrogen ion reduction, but we would not expect that these are the same as for anodic elementary steps. Just as well, we may expect that, at least at low coverage of the surface, inhibitor adsorption is favored on certain lattice positions. Therefore, inhibition is not necessarily symmetrical with respect to anodic and cathodic partial reactions. We must also allow for possible changes in mechanism of partial reactions caused by inhibitor adsorption. Therefore, when corrosion is of the type as described by Eq. (6.5), we may,

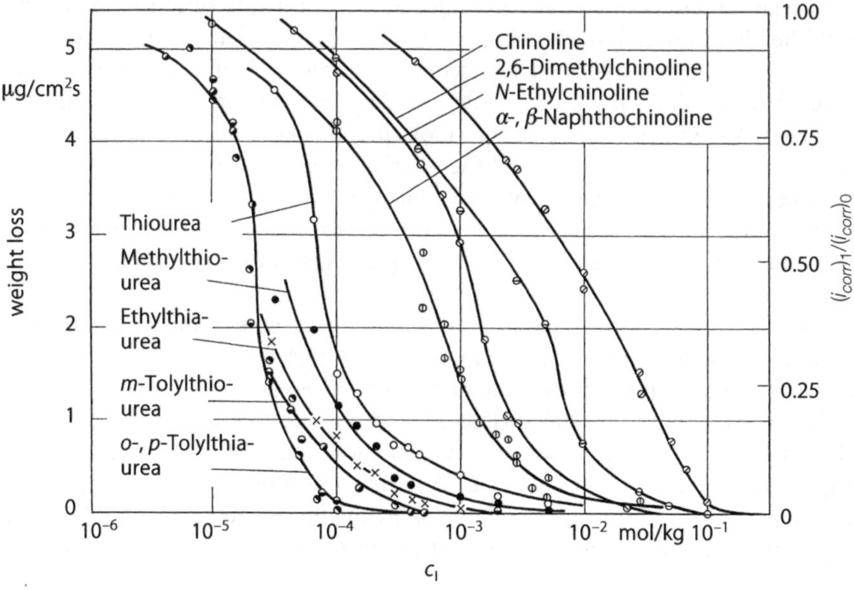


**Fig. 7.1.** Weight loss of 0.1% carbon steel in 5% $H_2SO_4$ solution at 40°C showing inhibition by thioureas and chinolines. (From Hoar and Holiday)

in principle, expect that all constants, $B_{Me}$,$A_H$,$b'_{Me}$, and $a'_H$ are functions of the coverage, $\theta_I$, of the metal surface by adsorbed inhibitor particles.

Changes of Tafel slope constants caused by inhibitors have, in fact, already been documented in Table 6.2. More such data are reported elsewhere [8], and also shown in some figures in the present section.

In this situation, the primary point of interest is the *adsorption isotherm* of the inhibitor, that is the constant-temperature relation between the coverage, $\theta_I$, of the metallic surface with adsorbed inhibitor particles and the concentration of the inhibitor in the solution, $c_I$.

This relation is governed by the interaction of inhibitor particles with the metal surface and by their mutual interaction. One possibility is that mutual interaction is negligible, and interaction with the metal substrate independent of coverage. In this case, the *Langmuir isotherm* applies, which is

$$\theta_I = kc_I/(1 + kc_I) \quad \text{resp.} \quad \theta_I/(1 - \theta_I) = kc_I \, . \tag{7.2}$$

This relation is easily derived assuming that the rates of adsorption and desorption are $v_{ad} = k_1c_I(1 - \theta_I)$ and $v_{des} = k_2\theta_I$, and that in steady state $v_{ad} = v_{des}$. We also note the *Frumkin isotherm*,

$$\frac{\theta_I}{1 - \theta_I} = kc_I \exp(-\kappa\theta_I) \, , \tag{7.3}$$

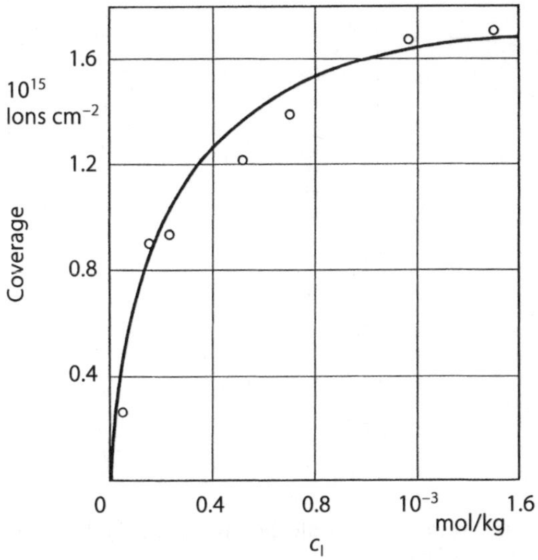


**Fig. 7.2.** Adsorption isotherm for iodine ions on iron from deaerated 0.5 M $H_2SO_4$ solution containing radioactive KJ. (From Heusler and Cartledge)

were $\kappa$ is an adjustable parameter accounting for interaction between adsorbed particles, indicating repulsion or attraction by values $\kappa > 0$, or $\kappa < 0$, respectively.

Frequently, the the *Temkin isotherm*

$$\theta_I = k_1 + k_2 \ln(k_2 c_I) , \tag{7.4}$$

is useful, which reflects a decrease in adsorption Gibbs energy with increasing coverage, indicating preferential adsorption at the energetically most favorable sites. The isotherm often applies for intermediate values of the coverage, but it coincides, in the same range, with the Frumkin isotherm for the case of repulsion between adsorbed particles.

As one of few examples of a directly measured isotherms, Fig. 7.2 shows the coverage of iron by adsorption of iron by radioactive marked iodine ions from deaerated 0.25 M $H_2SO_4$ solutions [9]. The curve is a Langmuir isotherm. Figure 7.3 shows the coverage of pure mercury with hexylalcohol adsorbed from 1 M NaCl solutions [10a]. In this case, $\theta_I$ was determined from the double layer capacitance of the mercury/solution interface, and the method was unambiguous because, due to lack of reactions occurring across the interface, the electrode was ideally polarizable. In a second step, the equilibrium exchange current density, $i^\circ_{Cd}$, of the reaction $Cd = Cd^{2+} + 2e^-$ was determined after adding cadmium to the mercury, using current-potential measurements. In this figure, the efficiency, W is defined as $W = 1 - (i^\circ_{Cd})_I/(i^\circ_{Cd})_\circ$, calculated from the exchange current densities $i_{Cd}$ in the presence and in the absence of inhibiting iodine. The direct correlation of $\theta_I$ and W is obvious. Also, the mechanism of inhibition appears

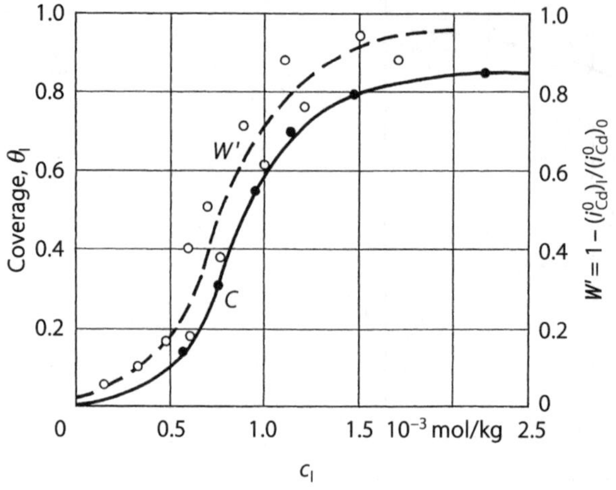


**Fig. 7.3.** Solid curve: Coverage, $\theta_I$, for adsorption of hexylalcohol from 1 M NaCl solution on mercury, determined from the double layer capacitance of pure mercury. *Broken curve:* Accompanying change of of the equilibrium exchange current density, $i^\circ_{Cd}$, for cadmium dissolution and redeposition at a cadmium amalgam electrode,expressed as coefficient $W = 1 - (i^\circ_{Cd})_I/(i^\circ_{Cd})_I$. (From Müller and Lorenz)

to be quite simple: The data are well understood assuming that the anodic current density of cadmium dissolution is a Tafel relation: $i = B_{Me} \exp(\varepsilon/b'_{Me})$, that iodine adsorption does not change the slope constant in the exponential function, and that the pre-exponential factor is proportional to the non-covered part of the surface:

$$i_{Me} = (1 - \theta_I)\, B^\circ_{Me} \exp\left\{\varepsilon/b'_{Me}\right\} . \tag{7.5}$$

While the determination of the coverage thus was straightforward, the dependence on the iodide concentration did not correspond to any of the above quoted isotherms. Instead, a more elaborate relation considering association of adsorbed particles is suggested [10b].

In a study of inhibition of acid iron corrosion by phenylthiourea and by $\beta$-naphtochinoline, the coverage, $\theta_I$, was evaluated from the shift of the cathodic Tafel line of hydrogen deposition using, at constant potential, the relation

$$i_H = -(1 - \theta_I)\, A^\circ_H \exp\left\{\varepsilon/a'_H\right\} . \tag{7.6}$$

In the log-log plot in Fig. 7.4, the data for $\beta$-naphtochinoline fall on a line with slope 1, as required by the Langmuir isotherm [8b]. For thiourea, the slope is 2, possibly indicating Langmuir-type adsorption of iihibitor particles dissociated into two half-particles.

For acid corrosion obeying Eq. (6.5), and inhibition symmetrical with respect to both anodic metal dissolution and cathodic hydrogen deposition, Eqs. (7.5) and (7.6) apply simultaneously. I this case, inhibition leaves the corrosion potential unchanged, and the consequence is that

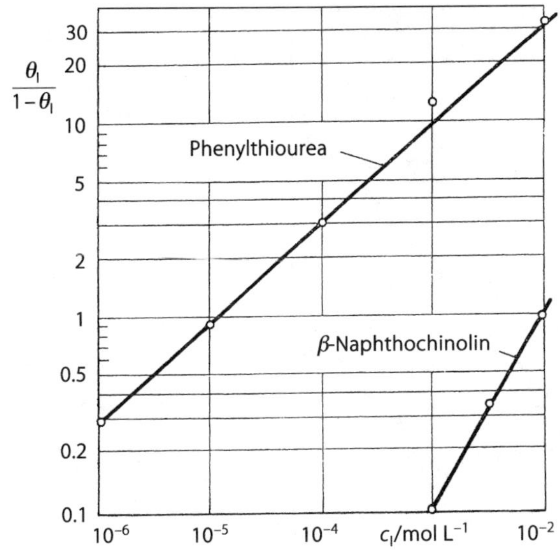


**Fig. 7.4.** Langmuir plot of cathodic overpotential data taken at constant potential for iron in $H_2SO_4$/0.5 M $Na_2SO_4$ solution, pH 2, 25°C, in the presence of phenylthiourea resp. $\beta$-naphtochinoline. (From Kaesche)

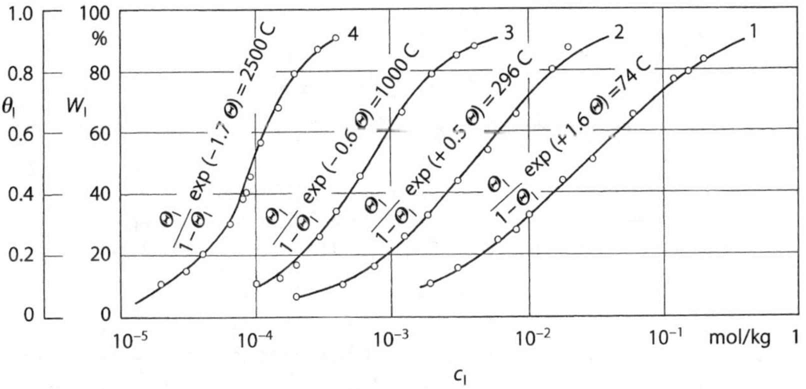


**Fig. 7.5.** Inhibitor efficiency, W, resp. coverage, $\theta_I$, of aliphatic amines adsorbed on unalloyed steel from 0.5 M $H_2SO_4$ solution. 1: n-hexylamin, 2: n-octylamine, 3: n-decylamine, 4: n-dodeylamine. (From Szklarska-Smialowska and Wieczorek)

$$(i_{corr})_I = (1 - \theta_I)\,(i_{corr})_\circ \,. \tag{7.7}$$

This prediction has been verified in a series of measurements of the corrosion rate of unalloyed steel in 0.5 M $H_2SO_4$ solutions with additions of aliphatic amines of varying carbon chain length. The result is shown in Fig. 7.5 [11]. The curves relating inhibitor concentration and coverage, resp. inhibitor efficiency,

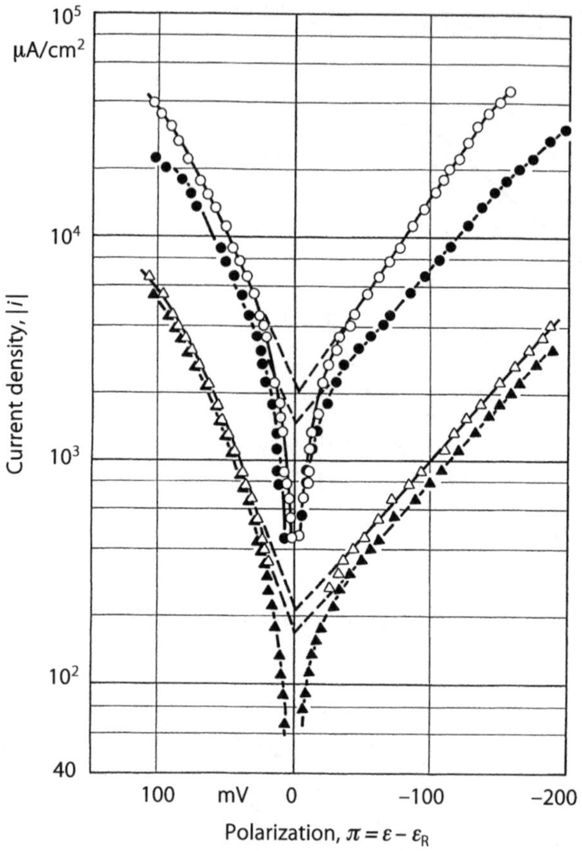


**Fig. 7.6.** Total current-potential curves for iron in 6 M HCl solution, 25°C without inhibitor and with added homopiperazine. *Open circles:* No inhibitor. *Solid circles:* 0.011, *open triangles:* 0.023, *solid triangles:* 0.048 mol $L^{-1}$. (From Hackerman, Justice, and McCafferty)

have been calculated as Frumkin isotherms. Obviously, the fit is quite convincing; also it is interesting that, as the chain length of the amines increases, the factor $\kappa$ goes from negative to positive values, indicating a change from intermolecular attraction to repulsion in the adsorbate. In other, and apparently in most cases, conditions are more complicated. An example is iron in hydrochloric acid with additions of the diamine homopiperazine. Again, the corrosion potential is virtually unaffected by inhibitor adsorption, indicating symmetrical inhibition [12]. As shown in Fig. 7.6 there is, however, a surge in inhibitor efficiency when $c_I$ is increased from 0.01 to 0.02 M. A possible explanation is a change of the orientation of adsorbed particles on the surface at a critical coverage, resulting in a denser packing.

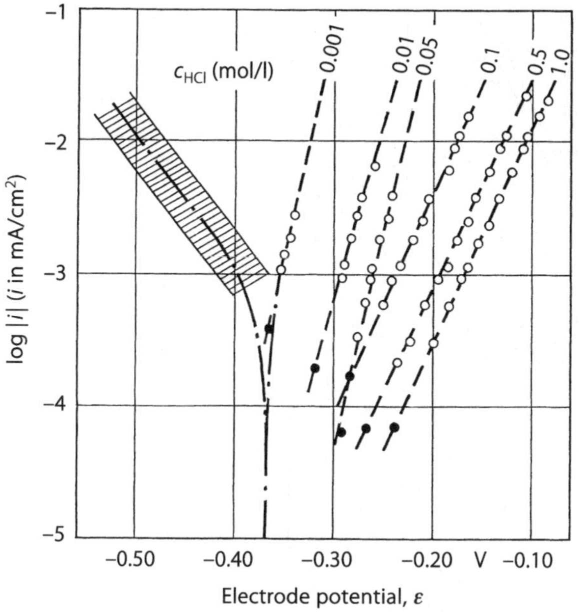


**Fig. 7.7.** Current-potential curves for pure iron in deaerated HCl solutions. *Open circles:* Anodic partial current-potential curves for iron dissolution at various values of $c_{HCl}$. *Dash-dot curve:* Total current-potential curve for iron in 0.001 M HCl. *Cross hatched:* Scatter of the position of cathodic partial current -potential curves for cathodic hydrogen deposition for all values of $c_{HCl}$. *Solid circles:* Corrosion rates. (From Lorenz, Yamaoka, and Fischer)

An interesting case is iron in hydrochloric acid [13]. As seen in Fig. 7.7 the increase in acid concentration, $c_{HCl}$, causes an increase in anodic overpotential, probably by replacing adsorbed $OH^-$ ions, and thus interfering with the catalytic mechanism described in Sect. 5.3. At the same time, there also is an increase in the anodic Tafel slope, possibly indicating a change in the dissolution mechanism. With increasing acid concentration, an obvious expectation would have been a decrease in cathodic overpotential for hydrogen deposition. Instead, this effect apparently is compensated by an increase in overpotential by inhibitive chloride adsorption. On the whole, inhibitor action by chloride is distinctly asymmetric. Apparently, this is the rule rather than the exception. As an extreme example, Fig. 7.8 shows that in the case of iron in sulfuric acid inhibited by dimethylchinoline there is only anodic inhibition [7]. This suggests inhibitor adsorption at surface kink sites only, leaving crystallografic planes in the surface free for cathodic hydrogen deposition.

Phenylthiourea or, more precisely, the protonated cation, $(C_6H_5)$-NH-CS-$NH_3^+$, is a very effective inhibitor in sulfuric or perchloric acid, although it enhances corrosion, when added in very low concentration, perhaps by decomposition of the molecule producing free sulfide [14]. In the molecule, CS is

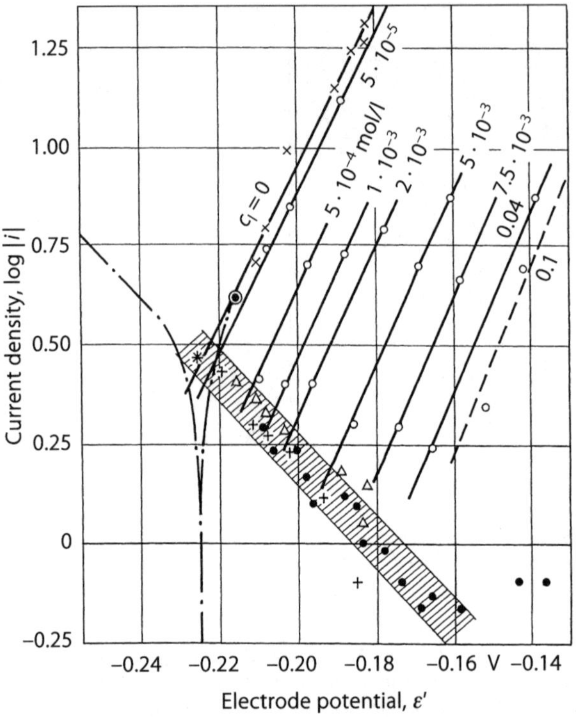


**Fig. 7.8.** Current-potential diagram for corrosion of mild steel in 5% $H_2SO_4$ solution, 40°C, with additions of 2,6-dimethylchinoline. *Crosses, open circles:* Anodic partial current-potential curves for iron dissolution in the absence and in the presence of the inhibitor. *Solid circles:* Corrosion rates. Cross-hatched: Scatter of position of cathodic current-potential curve for hydrogen deposition for all inhibitor concentrations. *Dash-dot curve:* Total current-potential curve in the absence of the inhibitor. (From Hoar and Holliday)

the active group causing adsorption by electronic interaction with the metal surface, ($C_6H_5$) causes inhibition by screening the surface, and $NH_3^+$ ensures sufficient though small solubility. A further interesting effect is that inhibitor adsorption enhances diffusion of atomic hydrogen into the metal, pointing to an increase in surface activity of adsorbed atomic hydrogen [15]. As indicated by the Tafel slope of the cathodic partial current-potential curve (see Fig. 7.9) the inhibitor apparently causes a change in the reaction mechanism. Similar to iodide and carbon monoxide [9], inhibition by phenylthioruea is destroyed by anodic polarization [8b,d,f]. The effect is demonstrated in Fig. 7.9 By these obervations, a threshold potential, $\varepsilon_u$, of a change in kinetics from inhibited to unhinhibited dissolution is well defined. It is improbable that this reflects a discontinuity in the adsorption isotherm. Instead, the effect probably is one of kinetics rather than thermodynamics. A very similar case is the inhibition of acid corrosion of iron by iodide, described in [9]. Here the explanation uses the concepts of the catalytic mechanism for iron dissolution described previously,

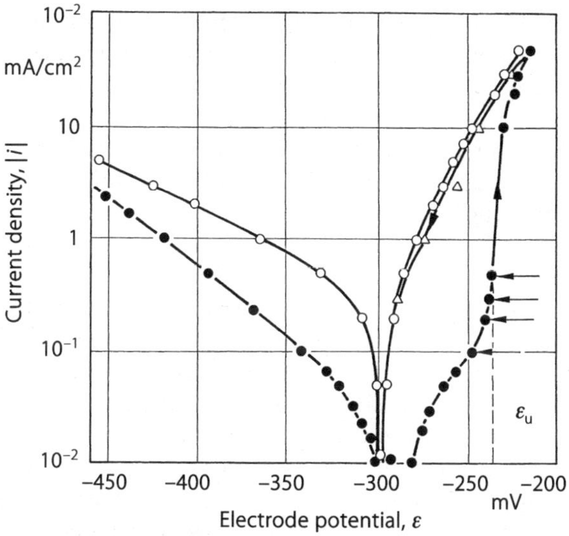


**Fig. 7.9.** Total current-potential curve for iron in $H_2SO_4$/0.5 $Na_2SO_4$ solution. *Open circles:* Stationary curve in solution without inhibitor. *Solid circles:* Effect of addition of $2 \times 10^{-5}$ mol $L^{-1}$ phenylthiourea on the stationary current-potential curve, indicating inhibitor desorption beyond ca. 250 mV. *Open triangles:* Non-stationary current-potential curve rapidly measured after inhibitor desorption. *Arrows* indicate direction of steps of galvanostatic polarization. Note also the change in Tafel slope of cathodic current-potential curve caused by inhibitor adsorption. (From Heidemeyer and Kaesche)

and especially the rate-determining role of reaction according to Eq. (5.58). The essential assumption is that anodic iron dissolution can also proceed by the reaction

$$\mathrm{Fe}_{\mathrm{kink}} + \mathrm{J}^-_{\mathrm{ad}} + \mathrm{OH}^-_{\mathrm{ad}} \rightarrow \mathrm{Fe}^{2+} + \mathrm{J}^- + \mathrm{OH}^-_{\mathrm{ad}} + 2\mathrm{e}^- , \tag{7.8}$$

introducing a catalytic dissolution step. When the coverage of the surface with adsorbed iodide is $\theta_I$, the remaining surface available for reaction Eq. (5.58) is reduced to the fraction $(1 - \theta_I)$, and this factor enters Eq. (5.63). At the same time, the rate of reaction Eq. (7.8) should be proportional to $\theta_I(1 - \theta_I)$. For negligible reverse reaction of iron deposition, the equation describing the partial anodic current of iron dissolution thus becomes

$$i_{\mathrm{Fe}} = k_1(1 - \theta_I)\exp((1 + 2\alpha)F/RT\varepsilon) + k_2\theta_I(1 - \theta_I)\exp(2\alpha F/RT\varepsilon) . \tag{7.9}$$

To complete the model, Langmuir adsorption is assumed considering, however, that iron dissolution by Eq. (7.9) removes iodide ions from adsorption sites. Therefore, while the adsorption rate remains $k_{\mathrm{ad}}(1 - \theta_I)$, the desorption rate increases from $k_{\mathrm{des}}\theta_I$ to a higher value, and the adsorption equilibrium now is described by

$$k_{\mathrm{ad}}(1 - \theta_I) = k_{\mathrm{des}}\theta_I + k_2\theta_I(1 - \theta_I)\exp(2\alpha F/RT\varepsilon) . \tag{7.10}$$

From this, $\theta_I$ can be evaluated as a function of $\varepsilon$, provided that the constants $k_{ad}$ and $k_{des}$ are independent of the potential over the comparatively narrow range between $\varepsilon_{corr}$ and $\varepsilon_u$. Here, and in other cases, conditions may change for high values of $\theta_I$ to a state where the rate determining step of metal dissolution is determined by the energy required to produce pores in an otherwise dense and perhaps multimolecular layer of adsorbed inhibitor [16]. This indicates the limits of treating inhibition in terms of simple electrode kinetics.

Adsorption of inhibitors such as thiourea derivatives, but also carbon monoxide, and even iodide ions probably involves covalent chemical bonding to the metal substrate. In other cases of charged-particle adsorption, the bonding is by Coulomb attraction. Then, the excess charge of the metal, as described in Chap. 3, should become important: When beyond the potential of zero charge, $\varepsilon_{pzc}$, (see Sect. 5.4), it is positive, adsorption of anions should be preferred, when below $\varepsilon_{pzc}$ it is negative, adsorption of cations. Hence, one reason of interest in $\varepsilon_{pzc}$. In this context, the ideal system to study is mercury as an ideally polarizable electrode. This applies to the metal in contact with solutions allowing no electrode reactions, at least in the range of potentials between anodic oxygen evolution and cathodic hydrogen deposition. Changing the potential at zero polarizing current changes the excess charge in the metal surface. This, then, lowers surface tension, $\sigma$, for both negative or positive excess charges, since their own Coulomb interaction is repulsion in any case. Measuring $\sigma$ with a *Lippman capillary electrometer*, $\varepsilon_{pzc}$ then is found as the maximum of a normally parabolic curve. The curve is distorted by Coulomb adsorption of ions because the adsorbed charges compensate the charges in the metal surface. Figure 7.10 shows the shift of $\varepsilon_{pzc}$ from −0.19 V in the 'capillary inactive' sodium sulfate solution, to −0.35 V by iodide adsorption, and to −0.05 V by tetraethylammonia adsorption [17]. For solid metals, measuring surface tension is difficult and will not be discussed here. Instead, Fig. 7.11 shows radiotracer data for the adsorption of iodide ions from sulfuric acid solutions [18]. As expected, adsorption of the anions decreases with decreasing potential, tending, however, to zero only at potentials where the charge in the metal surface already is considerably negative.

As shown by Fig. 7.12 the electrocapillary curve of mercury is depressed in the presence of neutral adsorbates such as hexylalcohol in the vicinity of $\varepsilon_{pzc}$ only [19]. Apparently, adsorbed neutral molecules are displaced from the interface by both high positive and negative charges in the metal surface.

Iron in acid solutions is a particularly special case [20]: As shown in Fig. 7.13 the high corrosion rate observed in 3 M $H_2SO_4$ solution is almost unaffected by the addition of tetrabutylammonium (TBA) sulfate, because apparently the TBA cation is not adsorbed at the metal surface. This is explained by considering that the corrosion potential, $\varepsilon_{corr} = -0.22$ $(V)$, is more positive than $\varepsilon_{pzc} = -0.37$ $(V)$, leaving the surface negatively charged with respect to the solution [21]. When, instead, potassium iodide is added to the acid at zero TBA concentration, strong inhibition, and also anodic inhibitor desorption, is observed as already described above. If, now, TBA is addded together with KJ,

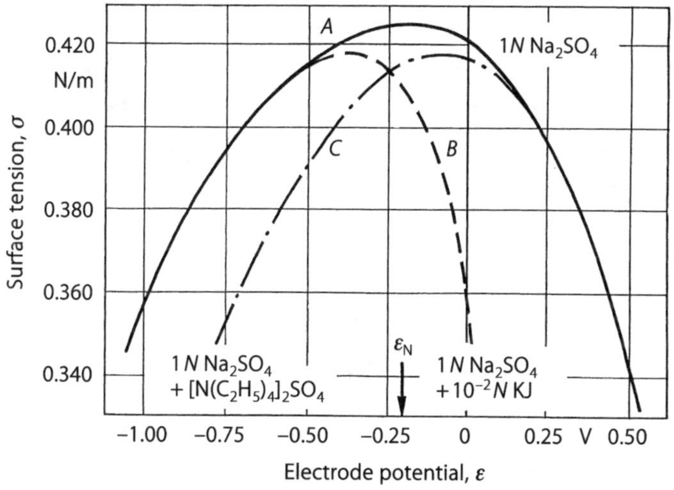


**Fig. 7.10.** Electrocapillary curve for mercury in $Na_2SO_4$ solution without specific adsorption (*A*), and in the same solution with adsorption of iodide anions (*B*), resp. tetraethylammonia cations (*C*), showing shifts in the potential of zero charge. (From Gouy)

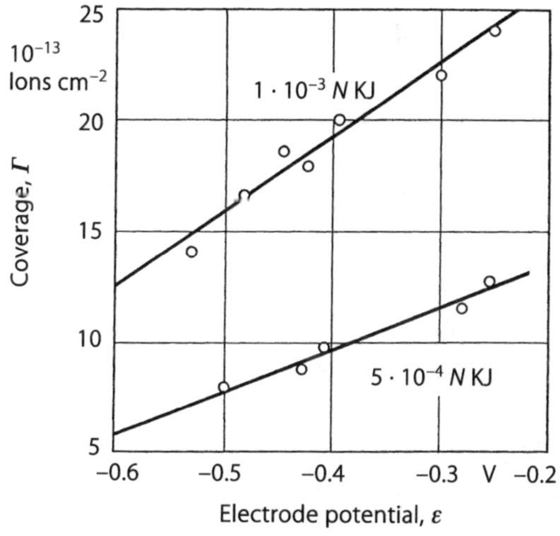


**Fig. 7.11.** Adsorption of iodide on lead from 0.5 M $H_2SO_4$ solution containing KJ, determined by radiotracer measurements. (From Kolotyrkin)

inhibition is considerably increased, i.e., the effects of KJ and TBA are synergistic, and this does throw interesting light on theoretical concepts useful for the development of commercial inhibitors. As to physical chemistry, the effect is explained by assuming a shift of $\varepsilon_{pzc}$ to a value more positive than −0.2 (V). This, however, is unexpected if Coulomb attraction of iodide is responsible for

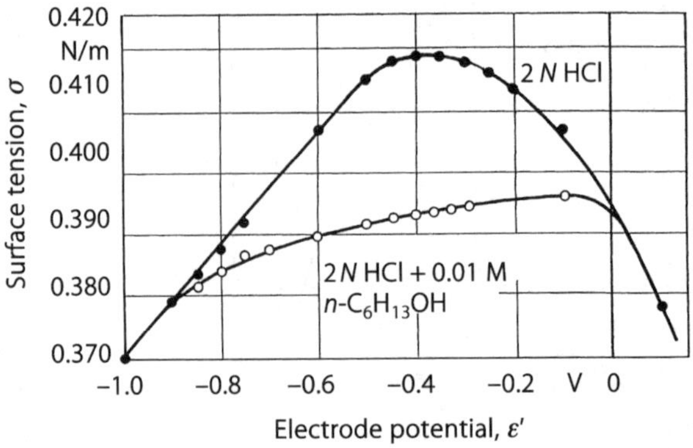


**Fig. 7.12.** Change of the elecrocapillary curve for mercury in 2 N HCl solution (solid circles) by addition of $10^{-2}$ mol $kg^{-1}$ of hexylalcohol as a typical uncharged adsorbate. (From Nikolajewa, Frumkin, and Jofa)

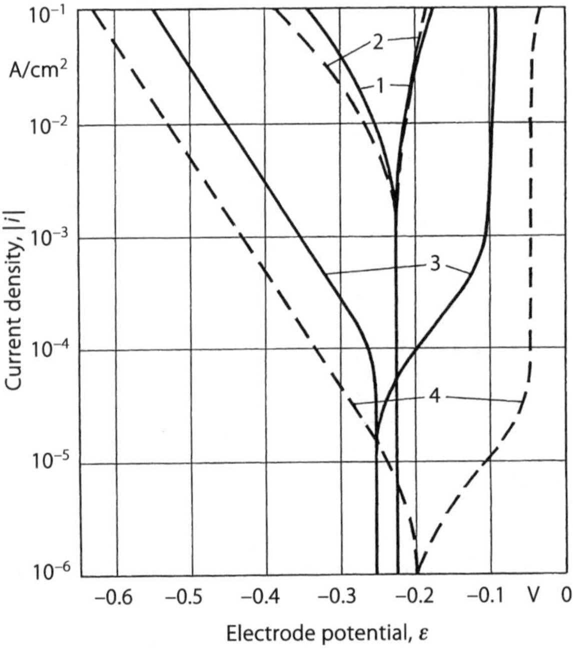


**Fig. 7.13.** Total current-potential curves of iron in 3 M $H_2SO_4$ solutions. 1: pure acid, 2: $10^{-3}$ mol $kg^{-1}$ tetrabutylammonium sulfate (TBA) 3:+ $10^{-3}$ mol $kg^{-1}$ potassium iodide (KJ), 4: $10^{-3}$ mol $kg^{-1}$ TBA + $10^{-3}$ mol $kg^{-1}$ KJ. Note synergistic effect of TBA and KJ; note also inhibitor desorption by anodic polarization. (From Jofa and Roshdestwenskaja)

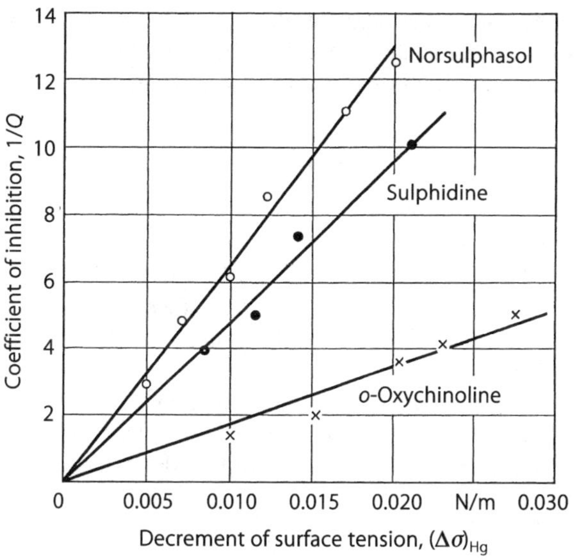


**Fig. 7.14.** Correlation of corrosion data for iron and surface tension data for mercury in 0.5 M $H_2SO_4$ solution with and without inhibitors as indicated in the figure.The correlation is linear if the ratio $(i_{\text{corr}})_\circ/(i_{\text{corr}})_\text{I}$ is plotted over the change $\Delta\sigma$ of the surface tension of mercury taken at $\varepsilon^* = \varepsilon^*_{\text{corr}}$, referred to the potential of zero charge of mercury and of iron, respectively. (From Antropov)

adsorption (see Fig. 7.10). Instead, there probably is covalent bonding of iodide to the surface by electrons shared between substrate and adsorbate.

Adsorption on liquid mercury is easily recognized by measuring surface tension. The question therefore is whether adsorption data readily obtained for mercury can be used to predict adsorption, and thus inhibition, for solid metals. Considering the role of the charge in the metal surface in processes of adsorption from solutions, one suggestion is to change the scale of potentials from values, $\varepsilon$, referred to SHE, to values, $\varepsilon^* = \varepsilon - \varepsilon_{\text{pzc}}$, referred to the potential of zero charge in the respective electrolytic solution. There then should be hope to find a correlation between surface tension data for mercury and corrosion inhibition data for, e.g., iron in acids, taken at equal potentials $\varepsilon^*$. As shown in Fig. 7.14, the factor $(i_{\text{corr}})_\text{I}/(i_{\text{corr}})_\circ$ for iron in sulfuric acid does, in fact, linearly correlate with the decrement $\Delta\sigma$ of the surface tension of mercury, when the latter was determined in the same solutions at potentials $\varepsilon^*$ equal to the respective corrosion potential, $\varepsilon^*_{\text{corr}}$ of iron [22].

As already noted in Sect. 5.3, in-situ scanning tunnel microscopy can be used to obtain images of copper surfaces undergoing anodic dissolution. In this context the influence of adsorbed organic material such as benzotriazole on dissolution morphology in acid sulfate solutions has been studied. The substance is a well known inhibitor of copper corrosion, and it was found to block Cu dissolution. The interesting observation was that the aromatic $C_6$-ring is flat on the surface below −0.15 V, but packed upright above this potential. In the latter state the adsorbed layer contains $Cu^+$ ions providing bonding between adsorbed molecules, and this then is the efficiently inhibiting layer [23].

## References

1. a) Symp. Europ. Inhibiteurs de Corrosion, Ferrara 1960: C.r. Univ. Ferrara, Sez. V, Suppl. n.3, 1961; Proc. Europ. Symp. on Corrosion Inhibitors, Ferrara: ibid.; b) 1966; c) 1971; d) 1976; e) 1980; f) 1985; g) 1990
2. Rosenfeld, I.L.: Corrosion Inhibitors. McGraw-Hill, New York 1981
3. Shastri, V.S.: Corrosion Inhibitors. Principles and Applications. John Wiley & Sons, Chichester etc., 1998
4. Fischer, H.: loc. cit. [1c], p. 3; Fischer, H., Seiler, W.: loc. cit. [1b], p. 19
5. a) Loc. cit. Chap. 5 [1–3]; b) Parsons, R.: in: loc. cit. [1c], p. 3; Damaskin, B., Petrii, O., Batrakov, V.: Adsorption of Organic Compounds at Electrodes. New York, Plenum Press, 1971
6. Akstinat, M.H.: Werkstoffe u. Korr.*21* (1970) 273
7. Hoar, T.P. and Hollyday, R.D.: J. Appl. Chem. *3* (1953) 502
8. a) Kaesche, H.and Hackerman, N.: J. Electrochem. Soc. *105* (1958) 191; b) Kaesche, H.; loc. cit. [1a], p. 137; c) id.; loc. cit. Chap. 6 [4,6]; d) id., in: Dechema Monographien Bd. 39. Verlag Chemie, Weinheim 1961; f) Heidemeyer, J. and Kaesche, H.: Corr. Sci. *8* (1961) 55
9. Heusler, K.E. and Cartledge, G.H.: J. Electrochem. Soc. *108* (1961) 732
10. a) Müller, W., and Lorenz, W.: Z. phys. Chemie N.F. *27* (1961) 23; b) Lorenz, W.: Z. Elektrochemie Ber. Bunsen-Ges. phys. Chemie *62* (1958) 193
11. Szklarska-Smialowska, Z. and Wieczorek, G.: loc. cit.[1c]
12. Hackerman, N., Justice, D., and McCafferty, E.; Corr. NACE *31* (1975) 240
13. Lorenz, W., Yamaoka, H., and Fischer, H.: Z. Elektrochemie Ber. Bunsenges. phys. Chemie *67* (1963) 932
14. Makrides, A.C. and Hackerman, N.: Ind. Engng. Chem. *47* (1953) 1773
15. Cavallaro, L., Felloni, L., and Trabanelli, G.: loc. cit. [1b]
16. Frumkin, A.: in: Advances in Electrochemistry and Electrochemical Engineering, Vol. 1, p. 65, Vol. 2, p. 287, Interscience Publishers, New York, London (1961)
17. Gouy, M.: Ann. chim. phys. 1903, 1906, see the treatises on electrochemistry.
18. Kolotyrkin, Ya.M.: Trans. Farad. Soc. *55* (1959) 455
19. Nikolajewa, N.W., Frumkin, A.N., and Jofa, S.A.; Zhur. Fiz. Khim. SSSR *26* (1952) 1326
20. Jofa, S. and Roshdestwenskaja, G.: ibid. *91* (1953) 1159; see also Frumkin, A.N.: Z. Elektrochemie *59* (1955) 807
21. Ajasjan, E.: Dokl. Akad. Nauk. SSSR 100 (1955) 473
22. Antropov, L.J.; Proc. 1st Int. Conf. Metallic Corrosion. Butterworth, London 1962, p. 147
23. Magnussen, O.M. et al.: loc.cit. Chap. 5 [27]

# 8 Corrosion of Homogeneous Alloys

## 8.1 Introduction

Concerning ambient-temperature corrosion of solid alloys, a preliminary note in Chap. 3 on thermodynamics was that bulk/surface equilibria of alloy components cannot normally be expected, because solid-state diffusion usually is too slow to counteract selective dissolution rates of components. The situation is different for liquid amalgams, as they can be mechanically stirred. Also, liquid amalgam surfaces are uniform down to atomic distances. For solid alloys, slow bulk-to-surface diffusion usually will instead leave the surface depleted with respect to 'fast' components, and enriched with respect to 'slow' components, and the result is an overall decrease in corrosion rate. The crystallografic structure of solid surfaces suggests that the main effect is blocking of kink sites active in metal dissolution, and this supposition will further below be seen to account for what may be described as fine-tuning of concepts, which we do not, however, immediately need to approach.

In practice, the effect of corrosion protection by surface depletion of fast components is, of course, useful especially when the resulting overall corrosion rate of the alloy is negligible, and further means of protection can be dispensed with. The obvious example are dental alloys, if they belong to the noble-metal alloy type, as for instance the gold alloy type. The typical question then is what the mole fraction of gold must be in an alloy with otherwise non-noble components to have it behave like pure gold, which in saliva is thermodynamically stable. Further below, we shall see that this type of alloys, with components having widely different standard electrode potentials, is indeed one of the central objects of investigations. Sometimes, detrimental corrosion of this type of alloys may still occur, if for some reason the surface protection by slow components fails and fast dissolution of non-noble components continues, leaving spongy material without mechanical strength. Well known incidents of this type are *dezincification of brass* and *dealuminization of aluminum bronzes*, both prominent examples of what is termed severe *selective dissolution* and *dealloying*. A somewhat different case is *spongiosis* of cast iron, which is preferential dissolution of ferritic iron leaving a sponge of undissolved carbide or graphite particles.

In the present chapter, we do not yet discuss the influence of alloy composition on passivity by ultra-thin oxide films, as described in Chap. 10. Instead, the alloy surface is assumed to be bare, free of deposits, including passivating films, and thus assumed to undergo 'active' corrosion. Still it will be seen presently that alloys may show quasi-passive behaviour, when the surface is sufficiently enriched with respect to 'slow' components.

Commercial alloys usually are heterogeneous either because they contain inclusions such as graphite and carbide in cast iron, or because they contain different crystallografic phases such as ferrite and austenite in duplex stainless steels, etc.. However, to set out the essentials of alloy corrosion we attend, in the first place, to single-phase alloys, and especially to homogeneous solid solutions.

As for pure metals, corrosion of alloys is governed by the kinetics of both anodic metal dissolution and cathodic reduction of some oxidizing species, determining the corrosion potential. Further below, it will be seen that the position of the corrosion potential relative to a critical potential for dealloying may be decisive for success, or else failure of corrosion protection. Therefore, cathodic reaction kinetics are as important for overall corrosion behaviour as anodic kinetics. In everyday service of metals, corrosion usually is oxygen corrosion. Owing to the high overpotential of oxygen reduction, corrosion potentials do not normally exceed ca. 0.2 V, and this, then, may ease dangers of dealloying often restricted to much more positive potentials.

Regardless of this importance of cathodic reaction kinetics, we shall concentrate in the following on anodic metal dissolution kinetics, as revealed by galvanostatic or potentiostatic measurements as described before. Also, where not noted otherwise, symbols have the same meaning as in previous chapters. Also where not noted otherwise, numerical values of alloy composition are given in atom%.

Electrochemical measurements are in-situ investigations. Parallel to these, the enormous possibilities of surface inspection by transmission electron microscopy (TEM), transmission electron diffraction (TED), scanning electron microscopy (SEM), Auger electron spectroscopy (AES), X-ray surface spectroscopy (XPS), low energy electron diffraction spectroscopy (LEED), surface X-ray diffraction (SXRD), as well as low-angle incidence X-ray diffraction (XRD) increasingly contribute to alloy corrosion science. Mostly, these are ex-situ methods and some, requiring transfer of specimen into vacuum chambers, or surface sputtering, or both, are not always beyond doubt as to producing spurious artefacts. At the same time, although usually allowing only lower magnification, much progress has been gained using in-situ scanning tunneling microscopy (STM) and small angle neutron scattering (SANS). The possibilities offered by new experimental techniques have caused a considerable increase in papers contributing to the subject, of which only few can be quoted in the following.

## 8.2 Uniform Dissolution

At some variance with the use of the term as defined in Chap. 6, anodic alloy dissolution is described as *uniform*, if the dissolution rates of the components are all directly proportional to there bulk concentration. In this case, anodic dissolution leaves the alloy surface plane, and the sub-surface non-porous. As a source of detrimental corrosion, non-uniform metal dissolution from alloys usually is more important. Nevertheless, the principles of uniform alloy dissolution should be understood in the first place, and insofar we follow a path set out especially in [5].

Let the alloy be a binary solid solution of components A and B with bulk mole fractions $\gamma_A$ and $\gamma_B = 1 - \gamma_A$. The interesting electrode reactions are $A \rightarrow A^{x+} + xe^-$, and $B \rightarrow B^{y-} + ye^-$. To simplify matters, assume negligible rates of both reverse reactions of metal re-deposition.

Let $v_A$ and $v_B$ (mol m$^{-2}$ s$^{-2}$) be the rates of dissolution of A and B. Anodic alloy dissolution then is uniform, if $v_A/v_B = \gamma_A/\gamma_B$. Regardless of absolute values, the ratio of rates is, thus, imposed by the bulk composition of the alloy. At the same time these rates obviously are governed by the composition of the surface, i.e. by surface mole fractions $\gamma_A^s$ and $\gamma_B^s = 1 - \gamma_A^s$. We do not yet, at this point, consider details such as the role of kink, ledge or terrace sites in the surface. Instead, we summarily expect that for any given set of relevant parameters, such as temperature, potential and electrolyte solution composition, the rates obey relations of the form

$$a) \quad v_A = k_A \gamma_A^s \qquad b) \quad v_B = k_B \gamma_B^s \, . \tag{8.1}$$

Let the specific kinetics of the partial reactions be independent of surface composition, so that values of the rate constants can be taken from measurements for pure A and pure B. It then follows that [4,5]:

$$a) \quad \gamma_A^s = \frac{\gamma_A}{k_A} K \, , \qquad b) \quad \gamma_B^s = \frac{\gamma_B}{k_B} K \, , \qquad c) \quad K = \frac{k_A k_B}{\gamma_A k_B + \gamma_B k_A} \, , \tag{8.2}$$

and hence

$$a) \quad v_A - \gamma_A K, \qquad b) \quad v_B = \gamma_B K \, . \tag{8.3}$$

The current densities off A and B dissolution are $i_A = x_A F v_A$ and $i_B = y_B F v_B$, respectively; the total current density is $i = i_A + i_B$. From Eqs. (8.2) and (8.3) it further follows that K and both $v_A$ and $v_B$ tend to zero, whenever one of the constants, $k_A$ or $k_B$ tends to zero, and this holds for any mole fraction $\gamma_A$, however little it may be. For less extreme conditions, provided that in the presence of a 'slow' component A, the ratio $k_A/k_B \ll 1$, the rate of dissolution of a 'fast' component B is still decreased far below what might be estimated intuitively, but erroneously, judging from the magnitude of $\gamma_B$.

Concerning the influence of the electrode potential we expect, summarily assuming charge-transfer rate control for dissolution of both A and B, that

$$a) \quad k_A = k'_A \exp(\varepsilon / b'_A) \qquad b) \quad k_B = k'_B \exp(\varepsilon / b'_B) \, . \tag{8.4}$$

We also expect that constants $k'$ are related to the concentrations reactants, and, especially, related to the surface coverage by metal atoms in kink sites, $\theta_{kink}$, through functions of the general form $f\{\theta_{kink}\}$. Therefore, we substitute constants $k'$ as follows:

$$a) \quad k_A = k''_A f_A\{\theta_{kink}\} \exp(\varepsilon / b'_A), \qquad b) \quad k_B = k''_B f_B\{\theta_{kink}\} \exp(\varepsilon / b'_B) \, . \tag{8.5}$$

Using the preceding set of equations, the total anodic current density, which is

$$i = i_A + i_B = FK\left(x\gamma_A + y(1-\gamma_A)\right) , \qquad (8.6)$$

can be expressed as function of the potential. For convenience, we directly consider the special case that a) $f_X\{\theta_{kink}\} = \theta_{kink}$, b) $x = y = z$, and c) $b'_A = b'_B = b'_{AB}$. This leads to

$$i = zF\,\theta_{kink}\,\frac{k''_A k''_B}{k''_A + k''_B}\,\exp(\varepsilon/b'_{AB}) . \qquad (8.7)$$

We assume constant coverage, $\theta_{kink}$. In this case, a semi-logarithmic ln $i$-$\varepsilon$ plot of this relation is a set of parallel Tafel lines. Changing to common logarithms, the Tafel-line slope is $\partial\varepsilon/\partial\log i = 2.303\,b'_{AB} = b_{AB}$. When the composition changes from $\gamma_A = 1$ to $\gamma_B = 1$, these lines shift from

$$\log i = \log\left(k''_A\;z\,F\,\theta_{kink}\right) + \varepsilon/b_{AB} \qquad (8.8)$$

to

$$\log i = \log\left(k''_B\;z\,F\,\theta_{kink}\right) + \varepsilon/b_{AB} . \qquad (8.9)$$

The range of potentials covered by the set of curves broadens in proportion to the difference $|k_A - k_B|$. At the same time, the shift of the lines accompanying changes in $\gamma$ is highly asymmetric. As an example, take A as the slow component, and assume $k_A/k_B = 0.01$. For the pair of compositions, $\gamma_A = 0.5$ and 0.0, the ratio $K_{0.5}/K_{0.0} \simeq 0.01$: At equal concentrations of A and B in the alloy, the overall dissolution rate at constant potential is only 1% of the dissolution rate of pure B.

The separate-determination of parallel anodic metal dissolution reactions is tedious, and has only rarely been attempted. Full experimental evidence is available from radiotracer measurements of uniform dissolution of homogeneous iron/chromium alloys in acid sulfuric solutions [6]. As shown in Fig. 8.1, the partial current-potential curves of dissolution of Fe to $Fe^{2+}$ and of Cr to $Cr^{2+}$ are Tafel lines, indicating charge-transfer reaction rate control, and the description given above, suitably adapted, should apply. In this case, the slow component is iron, and the concentrations 0.85 and 13.2% are comparatively low. Several observations are interesting:

a) As explained in a re-inspection of this work in [5], data for -0.3 V confirm that the ratio $i_{Fe}/i_{Cr} = i_{Fe}/i_{Cr}$ is equal to $\gamma_{Fe}/\gamma_{Cr}$ for both alloys, as required for uniform alloy dissolution. The very low rates of Cr dissolution are the consequence of the $k''_A/k''_B$-ratio, and not, as apparently previously believed, irregular.

b) For Fe-0.85% Cr, the Tafel slope $b_{Cr}$ is low, as for iron in the same alloy (ca. 40 mV). At the same, and similar to iron, chromium dissolution is accelerated by increasing pH. For Fe-13.2% Cr, $b_{Cr}$ his high, as for pure chromium (ca. 100 mV), and as also for iron in the same alloy. At the same time, iron dissolution and chromium dissolution are now independent of pH, as is chromium dissolution from pure chromium, and this was confirmed for Fe-27.9%Cr, while for Fe-6.95%Cr the dependence on pH already was weak but still distinct. On the whole, these observations show that the mechanism of iron

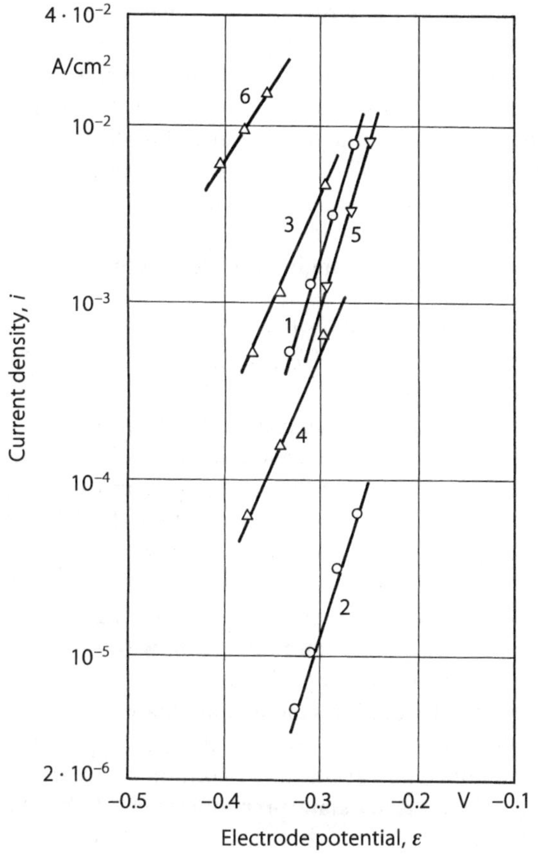


**Fig. 8.1.** Anodic partial current-potential curves for the dissolution of iron and chromium from the pure components, and from two homogeneous iron/chromium alloys in acid sulfate solutions. Partial current densities determined by radiotracer measurements. Concentrations in mass%, approximately equal to atom%. 1) Fe-0.85 Cr: $i_{Fe}$; 2) Fe-0.85 Cr: $i_{Cr}$; 3) Fe-13.2 Cr: $i_{Fe}$; 4) Fe-13.2 Cr: $i_{Cr}$; 5) pure Fe; 6) pure Cr. (From (Kolotyrkin)

dissolution from the alloy can become similar to that of chromium dissolution, and vice versa, depending on alloy composition [5].

At this point we recall that a change in Tafel slopes alone does not necessarily indicate a change in reaction kinetics. If conditions are simple, the Tafel slope is

$$b_{Me} = \partial\varepsilon/\partial \log i_{Me} = RT/(\alpha_{Me}\, n_{Me}\, F)\,, \tag{8.10}$$

where $\alpha_{Me}$ and $n_{Me}$ are transfer coefficient and valence of the rate-determining step of the electrode reaction, respectively. We also recall from Chap. 5 that $n_{Me}$ is not necessarily equal to the electrode reaction valence $z$, that it usually is taken as 1, but sometimes assumed to be 2. A change in the Tafel slope may well be caused by a change in the transfer coefficient only, without a change in reaction kinetics, due to a change in the structure of the electrical double layer at the metal/solution interface depending on alloy composition. Since alloy component dissolution occurs across the same electrical double layer, it follows that the Tafel slope should be the same for dissolution of both A and B, if both are charge-transfer controlled [5]. The situation is more complicated if, as for acid dissolution of iron, the charge-transfer reaction involves intermediate hydroxyl compounds in concentrations obeying a Nernst-type dependence on potential, changing the expression for $b_{\mathrm{Me}}$, as described previously. We pass more details, especially because the issue of changing transfer coefficients does not seem to be very clear.

c) Following the principles of uniform alloy dissolution, the rate of iron dissolution at constant potential is expected to decrease from an initial value, $v_{\circ}$, with increasing chromium content. Instead, it increases to $2.1 \times v_{\circ}$ and $4.7 \times v_{\circ}$ for 0.85 and 13.2% Cr, respectively. This has been explained by reconsidering the functions $f\{\theta_{\mathrm{kink}}\}$ introduced above, assuming the generation of additional kink sites in straight steps and perhaps in terraces of the crystalline surface caused by removal of atoms of the fast components at high overpotentials. The rate of kink generation would be comparatively small and without much influence on the rate of the dissolution of the fast component, while the influence on the rate of dissolution of the slow component is expected to be relatively large, thus explaining the observed increase of the dissolution rate of iron with increasing Cr alloy content [4]. The point is probably open to discussion.

In all of the forgoing, steady-state dissolution kinetics and, therefore, steady-state values of the surface mole fractions, $\gamma^{s}_{\mathrm{Me}}$, were tacitly presupposed. In practice, the change of surface composition from initial conditions created by some surface preparation to the final steady-state preponderance of the slow component may be sluggish. The total current, governed by dissolution of the fast component as long as it is present at high surface concentration, then will decay, at constant potential, only slowly. When the potential is scanned by 'potentiodynamic' recording from more negative to more positive values at a rate of, typically, some mV $s^{-1}$, the resulting curve is not a steady-state current-potential curve as calculated by. e.g., Eq. (8.7) Instead, it is a section through a set of parallel Tafel lines as described for increasing values of $\gamma_{\mathrm{A}}$. Only when, at high overpotential the exponential increase in current densities sufficiently accelerates the electrode reactions, the steady state is reached. Alloy dissolution then is uniform, while before there was transient preferential dissolution of the fast component B. Also before, the recorded curve depends on the scan rate. Fig. 8.2 shows an example of this type of experimental evidence [7a].

Since in steady state the slow component is enriched and the fast component depleted in the surface, there necessarily is a gradient in concentrations across an *interdiffusion zone* below the surface. Again, this is a topic more frequently considered for non-uniform alloy corrosion, and the discussion will be postponed. Suffice it to note that calculating the decay of non-uniform dissolution currents with time, as observed above, is similar to calculating the decay with time of the current of selective dissolution of a fast component only. The case

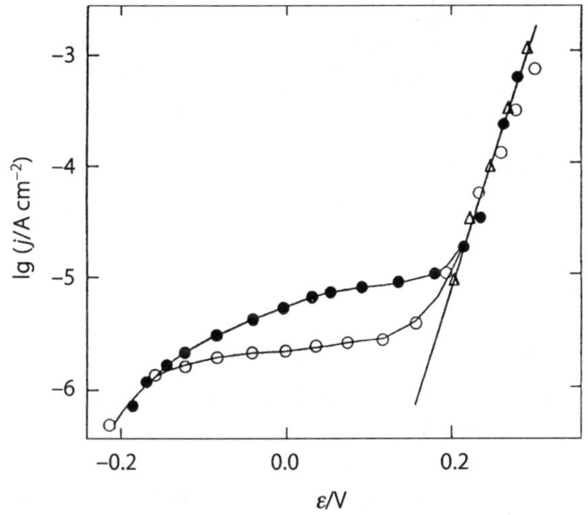


**Fig. 8.2.** Apparent current-potential curves for anodic dissolution of Cu-30Ni in 0.01 M $NaHSO_4$ solution. *Solid circles:* Scan rate 5 mV $s^{-1}$. *Open circles:* Scan rate 10 mV $s^{-1}$. *Triangles:* Steady-state current-potential curve of uniform anodic dissolution. (From Zolotarev, Pchelnikov, Skutarik, Khoklov, and Losev)

has been closely analysed for non-steady state and steady state co-dissolution of Cu and Zn from Cu-30Zn in acid chloride solution, modeling the mechanism by the volume diffusion concept [18].

Finally we note that some ternary glassy alloys containing iron, chromium, and boron are systems which, in terms sense of the rules stated above, show close-to-ideal behaviour, with no coupling of A dissolution and B dissolution kinetics [5]. Glassy alloys will, however, be considered only further below in Chap. 10. For more on coupling of partial reactions during alloy dissolution, as evidenced by an increase of component dissolution rates in alloys as compared to dissolution of pure components, see [7b].

## 8.3 Selective Dissolution and Dealloying

We continue to consider homogeneous binary AB-alloys, but now for the case that one component, B, is thermodynamically stable with respect to corrosion. Systems particularly suitable for investigations are alloys with a noble component such as gold, or palladium, and others in the high end of the standard potential series of the elements, accompanied in solid solution by a less noble component, A, such as copper. Experiments usually are carried out at potentials more positive than the equilibrium potential, $E_{A/A^{x+}}$, of the non-noble component. This condition requires keeping the concentration, $c_{A^{x+}}$ (or, more precisely, the activity $a_{A^{x+}}$), low or even close to zero. At the same time, the potential should be far below the standard equilibrium potential $E_{B/B^{y+}}$ of the noble component. This may seem difficult, since the solution must not contain dissolved $B^{y+}$ ions, which would plate out onto the alloy surface. However, it usually suffices to keep the working potential well below the standard equilibrium potential, $E^{\circ}_{B/B^{y+}}$, because then the equilibrium concentration of $B^{y+}$ is in any case virtually zero.

We recall that, disregarding differences between concentrations and activities, the equilibrium potential of a component Me in an alloy at mole fraction $\gamma_M$ obeys the Nernst equation

$$E_{Me/Me^{z+}} = E^{\circ}_{Me/Me^{z+}} + (RT/z_{Me}F)\ \ln(c_{Me^{z+}}/\gamma_{Me})\ . \tag{8.11}$$

To this we add that, if not otherwise stated, the equilibrium potential is calculated from thermodynamic data for bulk phases with, in principle, infinite extension and plane surfaces. For surfaces curved with a characteristic radius $\rho$ (or 'curvature' $1/\rho$), the equilibrium potential changes by excess energy due to surface tension, or *surface Gibbs energy*, $\sigma$, as described below. Also below, this will cause us to distinguish between equilibrium potentials $E_\infty$ and $E_\rho$ of otherwise identical electrode reactions.

Given a sufficient content of B, anodic A dissolution from the alloy should decrease because the surface becomes covered with undissolved B atoms. Also, it is intuitive to expect that continued dissolution of A requires continuous paths through the alloy occupied by A atoms only. Then, a threshold concentration of B is expected, beyond which such continuous paths along A-sites through the bulk of the alloy do no longer exist, and this then would be the *resistance limit* or *parting limit* of the alloy, which, accordingly, also would be expected to be independent of the composition of the aggressive solution. Tammann's early attempts to explain observations on the basis of this hypothesis have failed owing to oversimplified assumptions [8]. We shall see further below that using modern *percolation theory* quantitative predictions along similar lines have been revived. Before, more detailed experimental facts will be presented in the first place.

As an early and path-breaking example, Fig. 8.3 shows anodic current-potential curves observed by steady-state galvanostatic polarization of CuAu solid-solution homogeneous alloys in an acid copper sulfate solution [9]. For alloys with between 0 and 30% Au, there is Cu-dissolution only all over the range of potentials, with a steep surge in current density beyond a critical potential threshold, $\varepsilon_{crit}$, shifting to more positive values as the gold content of the alloys increases. This threshold potential is a point of major interest in the following. As a general warning against attempting to understand dealloying kinetics from metallurgical principles only we note that $\varepsilon_{crit}$ depends on solution composition: For Cu-30Au in acid sulfate solution it is ca. 1.1 V, for the same alloy in 12 M LiCl solution it is lowered to ca. 0.7 V, which is indeed a very substantial change [10]. Apparently, this is the effect of chloride adsorption on the surface promoting metal dissolution, possibly by decreasing surface tension, or else by metal/chloride complex chemical bonding.

At high anodic current densities, the curves turn into an irregularly wavy anodic limiting current density. These branches of the curves are probably caused

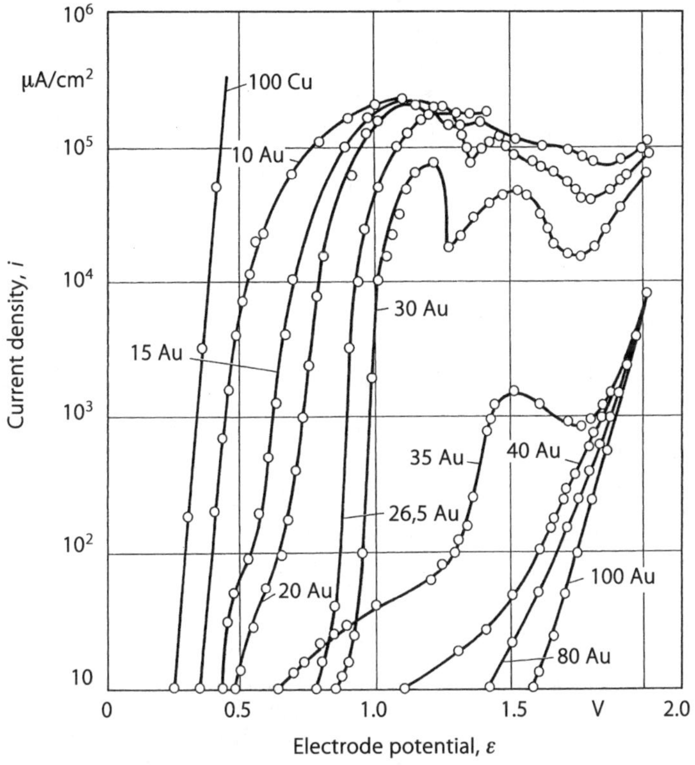


**Fig. 8.3.** Stationary anodic current-potential curves of CuAu solid solution alloys in 1 M $Na_2SO_4$ solution with added 0.005 mol $L^{-1}$ $H_2SO_4$ and 0.1 mol $L^{-1}$ $CuSO_4$. (From Bergmann and Gerischer)

by porous solid copper sulfate deposits on the surface appearing whenever the solubility product of the salt is exceeded in the hydrodynamic boundary layer adjacent to the metal surface. We shall not further discuss this probably trivial effect and discuss, instead, currents only up to $10^4$ μA $cm^{-2}$.

When the Au content of the alloys exceeds 40 at.%, the current-potential curves more and more resemble that of pure gold. These curves do not, however, reflect anodic gold dissolution, but anodic oxygen evolution at oxide-covered gold, which, in fact, is gold passivated by an oxide film. We pass further consideration of this of phenomenon, and also pass considering the intermediate curve type observed for 35 at.% Au as particularly complicated.

The really interesting observation is the threshold potential, $\varepsilon_{crit}$, which, though not very sharp, is a distinctly narrow critical range of potentials. The effect is by no means unique and has been observed for AgAu, CuPd, AgPd, and NiPd alloys as well [9–14]. In this later work, the subcritical range of potentials, disregarded in [9], was a point of particular interest. Figure 8.4 gives quasi-stationary current-potential curves for CuAu alloys in acid sulfate solutions,

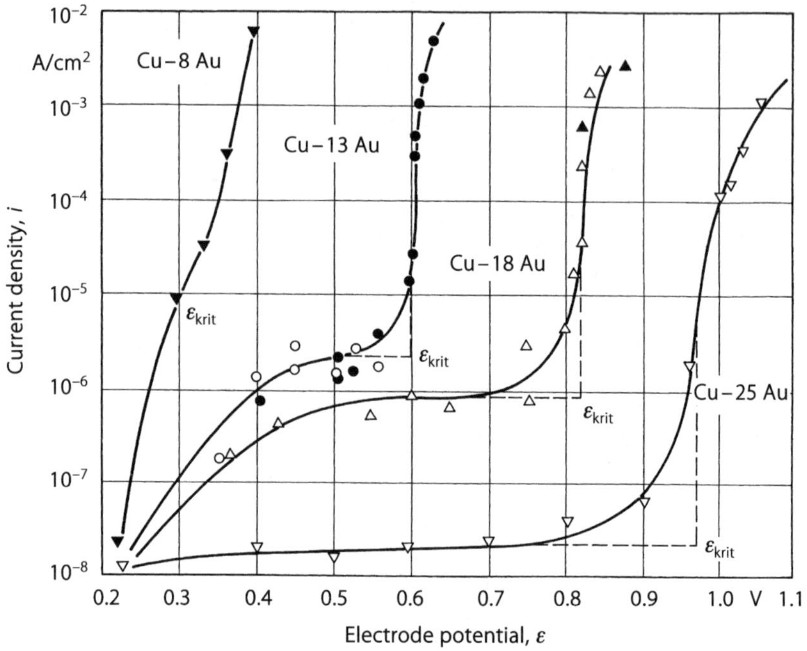


**Fig. 8.4.** Quasi-stationary anodic current-potential cur-ves for CuAu solid solution alloys in 1 M $Na_2SO_4$/0.005 M $H_2SO_4$. *Open circles* for Cu-13Au show anodic current densities determined by chemical analysis of dissolved copper; other symbols are galvanometer readings. (Data for Cu-13Au and Cu-18Au from Pickering and Byrne, for Cu-8Au and Cu-25Au from Popp and Kaiser)

with all current densities equal to the partial current densities of Cu dissolution [11b,14b]. A second example is presented in Fig. 8.5 for CuPd alloys in acid sulfate solutions [12a,b].

From the shape of these curves it appears that below the breakthrough threshold a residual current actually does exist. This, however, is a point of controversy because, typically, the subcritical current often slowly decays with time apparently without limit, as exemplified in Fig. 8.6 for solid-solution Cu-10Pd in acid sulfate solution [12a,b]. As seen in Fig. 8.5, the critical potential was +0.56 V, and only beyond this value the current apparently levels to an approximately stable and relatively high value. Below, the logarithm of the current appears to decay indefinitely. The slope of the curves is ca. −0.7. Literature values of such slopes usually are between ca. −0.6 and −1.2.

Whenever current decays are slow compared to the rate of potentiodynamic scanning, then, similar to what has been explained above for simultaneous dissolution of alloy components, recorded curves may be a section through steady-state curves, each valid for a constant concentration $\gamma_A^s$, in the direction of decreasing surface mole fraction of the non-noble component. Whether the slow decay of dissolution currents always tends to zero, or else whether for

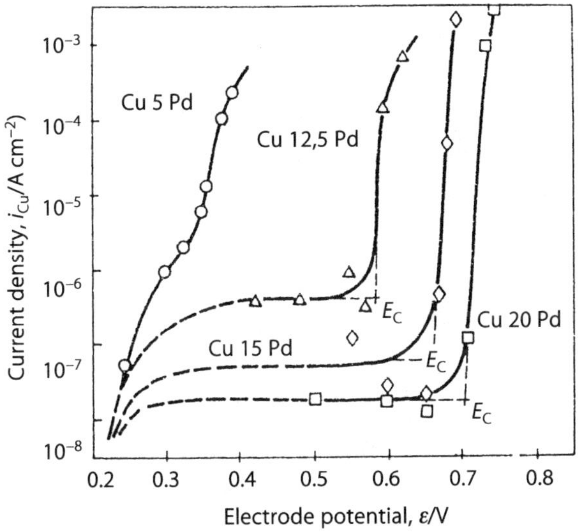


**Fig. 8.5.** Quasi-stationary current-potential curves for homogeneous CuPd alloys in 0.05 M $H_2SO_4$ solutions, pH 2. (From Kabius)

some alloy systems, or perhaps always, a stationary residual current eventually persists, remains a point open to discussion.

Originally, the opinion was that the currents below the critical threshold range in potential are currents of non-noble component dissolution from kink sites in ledges on the crystalline metal surface [9]. While on pure metal surfaces, kink sites are not exhausted during anodic dissolution, dissolution of dissolution leaves kink sites occupied by B atoms, and A atoms on kink sites are soon only rarely available. For anodic dissolution to proceed, the overpotential must be increased to more positive values to eventually supply sufficient energy to remove A atoms out of terrace sites between ledges. Greatly simplifying the real course of events, the situation is schematically shown in Fig. 8.7. Later, residual currents have been considered to indicate persistent below-threshold residual currents, viz., anodic limiting currents [15]. It is then not difficult to formulate equations modeling the observed shape of curves observed in the subcritical potential range. Whenever, instead, residual steady-state currents do not really exist, the following derivation is, of course, invalid. For this derivation, Fig. 8.8 schematically explains the situation, and the argument proceeds as follows:

Assuming a very simple charge-transfer controlled mechanism of anodic dissolution of A, the sum of currents from kink sites and from terrace sites, $i_{\text{kink}} + i_{\text{terr}}$, with coverage $\theta_{\text{kink}}$ and $\theta_{\text{kink}}$, is

$$i = i_{\text{kink}} + i_{\text{terr}} = (k_1\,\theta_{\text{kink}} + k_2\,\theta_{\text{terr}}) \exp(\alpha\, n_e\, F\,\varepsilon/RT) \ , \tag{8.12}$$

also assuming that the exponential term is the same for both reaction paths. Considering the higher binding force of atoms in terrace sites as compared to

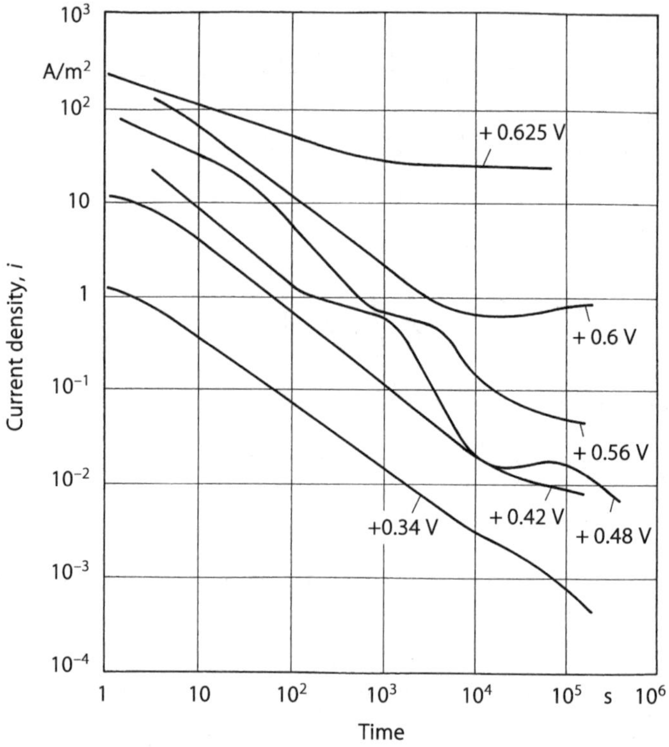


**Fig. 8.6.** Potentiostatic current transients for copper dissolution from an homogeneous Cu-10Pd alloy in acid sulfate solution at pH 2, obtained at different values of electrode polarization beyond the open circuit corrosion potential. The critical potential +0.56 V. (From Kabius, Kaiser, and Kaesche)

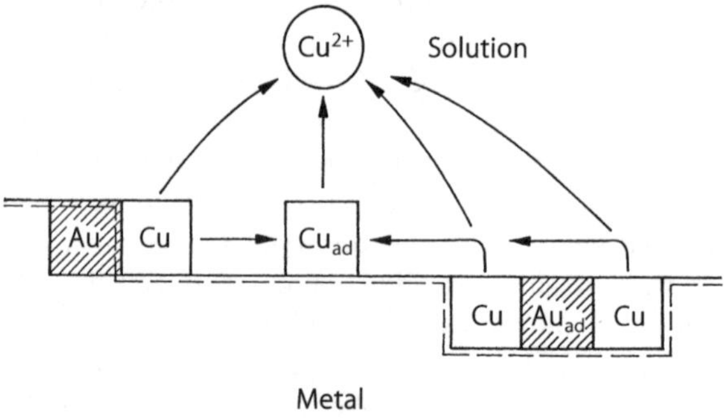


**Fig. 8.7.** Simplified model of the dissolution of copper from homogeneous CuAu alloys via adsorbed atoms or directly via either kink sites or terrace sites. The dashed line schematically represents the solid metal surface, indicating that adsorbed atoms diffuse freely on the surface while kink site atoms have only poor thermal mobility

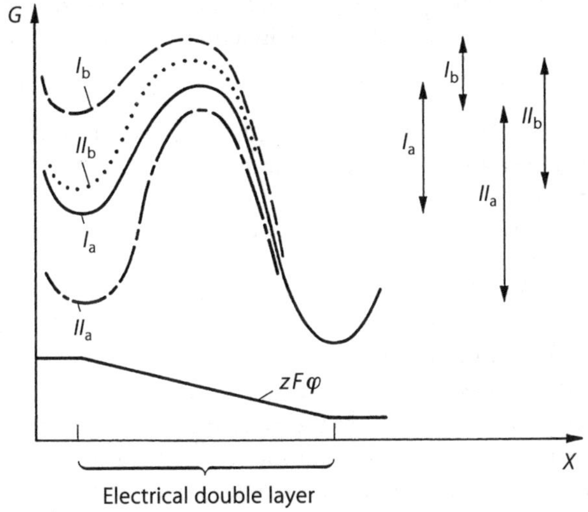


**Fig. 8.8.** Energy barrier in the metal/solution interface across the electrical double layer. *Curves* $I_a$, *and* $I_b$: Energy change for the transfer of metal ions from kink sites with and without super-imposed electrostatic energy. *Curves* $II_a$ *and* $II_b$: Corresponding curves for transfer from terrace sites recognizing the higher binding force caused by higher coordination of atoms in the surface lattice as compared to kink site atoms. (From Gerischer)

atoms in kink sites in ledges, we expect $k_1 \ll k_2$. The first term in the bracket can be adapted to experimental evidence by introducing two more assumptions:

a) While in ledges kink sites are predominantly occupied by noble B atoms, some thermal mobility remains, liberating kinks from time to time from occupying B atoms, with ensuing A dissolution. Essentially, this is an effect of *surface diffusion*.

b) The rate of dissolution is co-determined by crystallisation overpotential. This is due to exhaustion of liberated kink sites by a current-dependent decrease in coverage, from an initial value, $\theta^{\circ}_{\text{kink}}$, to smaller values, $\theta_{\text{kink}}$, as function of $i_l$, eventually to $\theta_{\text{kink}} \to 0$. Formally, this anodic effect is similar to the cathodic diffusion overpotential described in Chap. 5. By this analogy we write, considering, e.g., Eq. (5.45), without further detailed derivation the implicit relation

$$i_{\text{kink}} = k_1\,\theta^{\circ}_{\text{kink}} \left(1 - \frac{i_{\text{kink}}}{i_{\text{limit}}}\right) \exp\left(\alpha\, n_e\, F\, \varepsilon / RT\right) , \tag{8.13}$$

which, solved for $i_{\text{kink}}$, does model the observed current-potential curves in the subcritical range. The relation offers three adjustable parameters to fit experimental observations. It should be noted that relations between surface and bulk quantities previously discussed for uniform alloy dissolution are not useful presently, because for selective dissolution they only predict $\gamma_A \simeq 0$. Therefore, we do not know how the quantities $\theta^{\circ}_{\text{kink}}$ and $i_{\text{limit}}$ depend on the mole fraction $\gamma$. For want of better evidence, empirical and essentially non-physical relations,

$$a)\ \theta^{\circ}_{\text{kink}} = k_3\,\gamma^m, \quad b)\ i_{\text{limit}} = k_4\,\gamma^n\ , \tag{8.14}$$

have previously been offered [15], but better founded relations introducing fewer arbitrary fit parameters certainly would be desirable.

The $i_{\text{terr}}$-component of the total current produced by dissolution of A from terrace sites also needs closer inspection. As more fully discussed further below, dissolution beyond the critical potential is not as simple as so far assumed in Eq. (8.12). Since, at least probably, the critical potential really is a steep increase rather than a sharp step, events typical for critical dissolution already start below the critical potential. In fact, it is not probable that terrace-site dissolution leaves only isolated vacancies in the surface, although vacancy injection and ensuing vacancy bulk diffusion does play a major role in the mechanistic interpretation of events. It also stands to reason that surface vacancies will tend to cluster, forming what are *dissolution nuclei* [1,16] in the sense of negative crystallisation nuclei [12c,17].

Introducing some aspects of nucleation theory, the present treatment follows [20b] by considering three-dimensional ('3D') nuclei. For direct application to the case of selective alloy dissolution producing surface roughness, the nuclei are assumed, also as in [20b], to be either hemispherical cavities in the surface, called 'pits', or hemispherical elevations, called 'bumps'. 'Pits' as defined and described in this chapter must not be confused with the much larger pits extensively discussed in Chap. 12. The assumed shapes for 'pits' and 'bumps' greatly simplify real crystallografic pits and bumps, and numerical values of energy contributions involved in the growth of pits are thus estimated with a wide margin of error. Experimental data on metal/solution interfacial tension, important for the present case, are also usually not readily available.

Consider, then, in the first place, an elementary pure metal Me and its anodic dissolution to $Me^{z+}$ producing pits, or its cathodic deposition producing bumps. Both bumps and pits are thermodynamically unstable with respect to flat surfaces: For uniform dissolution or deposition of Me on a flat surface, the equilibrium potential is $E_\infty$, as given by Eq. (8.11). For hemispheres concave to the metal on the flat surface, with radius $\rho$, the equilibrium potential, $E_\rho$, is, however, more negative, because the thermodynamic tendency is to flatten the hemispheres in favor of the smooth surface by what essentially is *Ostwald ripening*. Whether this occurs by surface diffusion, or else by a dissolution/redeposition process will depend on circumstances. By analogy, the equilibrium potential in cavities is more positive than $E_\infty$. For bumps of Me sitting on the flat substrate also of Me, with a total volume equal to the molar volume of Me, $V_{\text{Me}}$, it follows from the *Thomson equation* describing the vapor pressure of small droplets with radius $\rho$ and surface tension $\sigma$, that the additional Gibbs surface energy is $-\sigma V_{\text{Me}}/\rho$. For bumps the '$-$' sign changes to the '$+$'sign. Therefore, the equilibrium potential of metal dissolution and redeposition for bumps and pits is

$$E_\rho = E_\infty \mp \frac{\sigma V_{Me}}{z F \rho} . \tag{8.15}$$

In other words: At some value $\varepsilon > E_\infty$, cavities can continue to grow only if their radius exceeds a critical value

$$\rho_c = \frac{\sigma V_{Me}}{z F |\eta^*|} ; \quad \eta^* = \varepsilon - E_\infty . \tag{8.16}$$

Smaller cavities disappear by spontaneous deposition of Me from the solution, or alternatively by surface diffusion. It therefore is necessary to assume, regardless of electrode kinetics, that as a result of statistical fluctuations and owing to vacancy agglomeration a spectrum of pits of variable size always exists. By analogy, the same is expected for bumps. By anodic metal dissolution, which is our case presently, only pits with $\rho \geq \rho_c$, as determined by $\eta^*$, have a survival and growth probability $\geq 50\%$ .

Surface tensions or surface energies have values of the order of magnitude 1 J m$^{-2}$. $V_{Me}$ is, for instance, $10^{-5}$ m$^3$ mol$^{-1}$. For z=2 and $\eta^* = 0.25$ V, we find from Eq. (8.16) that $\rho_c \simeq 2 \times 10^{-10}$ m. As radius of a well defined hemisphere consisting of atoms this is rather too small, but it does show that we are considering plausible dimensions. For bumps instead of pits the result is the same.

Aiming at calculating the current-potential curve for $i_{terr}$ as function of the electrode potential, the next step is to estimate the rate, $N$, of nucleation of pits with a radius of at least $\rho_c$. Using Boltzmann statistics we find, analogous to the Arrhenius equation :

$$N = N_\circ \exp(-A_c/kT) ; \quad A_c = 8\pi\sigma^3 V_{Me}/3z^2F^2\eta^2 . \tag{8.17}$$

$A_c$ is the reversible work required to form a pit; it thus is not a molar quantity and the denominator is $kT$, with the Boltzmann constant $k$, and not $RT$. For the pre-exponential frequency factor, $N_\circ$, we expect, by analogy to two-dimensinal nucleation, a relation, $N_\circ = \text{const} \times \exp(zF\beta\eta/RT)$.

To finally arrive at the desired expression for $i_{terr}$, we need to calculate the momentary current contributions from all growing pits. In the first place, we recognize that inside crystallografic pits kinked ledges again will appear, and that metal dissolution via these sites will proceed as before from ledges on the flat surface. In terms of electrode kinetics, pit formation thus essentially is a process creating more kinked ledges. Kinetics governing $i_{terr}$, therefore, are essentially the same as for $i_{kink}$. When a pit has grown to a size large compared with the thickness of the electrical double layer, regular electrode kinetics therefore should once more apply. The contribution $i'$ of any one growing cavity to the current density, $i_{terr}$, then should be proportional to $2\pi\rho^2 \exp(\alpha n_e F \eta/RT)$. We do not attempt to explicitly evaluate the sum of contributions from pits or bumps growing during the total time, $t$, for varying periods $(t-\tau)$. Instead, we approximate the sum by an adjustable constant multiplied with the exponential factor typical for charge-transfer reaction control. Doing so, we also disregard that for small values of $\rho$ the true overpotential $\eta = \varepsilon - E_\rho$ differs from $\eta =$

$\varepsilon - E_\infty$. Contracting constants we arrive, also using Eq. (8.17) at the approximate expression:

$$i_t = \pm K \exp\left(\pm K'\eta/RT\right) \exp\left(-K''/|\eta|^2\right) . \tag{8.18}$$

The '+' sign holds for anodic pit formation, the '–' sign for cathodic bump formation, in both cases neglecting the respective reverse electrode reaction.

A third type of events may substantially contribute to the dissolution current. This is production of vacancies on terrace sites, agglomerating to two-dimensional ('2D') flat nuclei rather than '3D' pits, bordered by ledges of one-atomic depth. The geometrical model would be a round flat disk. This additional path of A dissolution would have to pass an energy barrier intermediate between dissolution from pre-existing kinks and '3D'-nucleation, and it would add a term proportional to exp(const $/\eta$) to the total current density [17b].

As pointed out in [12c], the structure of equations describing current-potential relations for pitting of passive metals is similar to what has been derived above. Whether this similarity is incidental, or else points to similar mechanisms, is disputable. In this context, the question is not, however, whether morphological instability of flat surfaces always starts with the nucleation of pits, as this is demonstrably obvious. Instead, it is whether nucleation can be described by a Volmer-type theory, as attempted above.

When concerned with selective dissolution of a component A from a homogeneous AB alloy, we have to account for the remaining B. After prolonged anodic selective dissolution, a porous layer of B will appear, as described below, but this does not answer the question for the onset of nucleation in an initially flat surface. At this point, percolation theory, as applied to the problem, assumes that in a high-melting disordered alloy there is, at ambient temperatures, considerable clustering of A and B atoms quenched into the material during cooling after solidifaction from the melt [20]. An interesting conjecture is that the statistically predominant diameter of A clusters also is the characteristic diameter of surface nuclei and approximately equal to $\rho$. The latter is estimated as $\xi_A \simeq (1+\gamma_A)a/(1-\gamma_A)$, where $a$ is a characteristic lattice parameter. Again, we are dealing with at most sub-nanomicroscopic dimensions. We note that in this concept the dissolution of A clusters leaves cavities, isolating B clusters, which then tend to coarsen to bumps. We also note that statistical clustering of A and B does not occur in ordered alloys, such as ordered $Cu_3Au$.

From rapid selective dissolution accompanied by surface roughening, we return to below-critical potentials, where selective dissolution is slow and decaying either to zero following a power law, $i \sim t^{-m}$, or to a small residual current density. Owing to the decrease in surface concentration of A and the increase in surface concentration of B, there then exists a gradient of concentrations below the surface extending into the bulk alloy, i.e., the subsurface interdiffusion zone introduced above. In the long run, a layer of pure or almost pure B will cover the surface as a *reacted zone* of thickness $\delta_r$, on a surface presumably at constant composition, $\gamma_B^s \simeq 1$ except for pores, or, as percolation

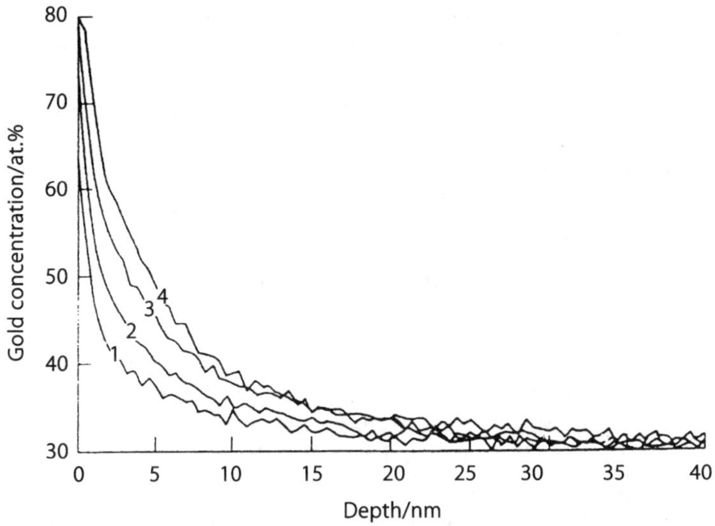


**Fig. 8.9.** AES concentration-depth profiles measured after potentiostatic polarization of Cu30Au in 12 M LiCl solution at 350 mV. Polarization time 1: 1 s, 2: 10 s, 3: $10^3$ s, 4: $10^4$ s. (From Laurent and Landolt)

theory formulates, a 'two-phase interpenetrating composite' with B and pores as components, and the interdiffusion zone only underneath. The dealloying reaction then occurs at the reaction front, which is the interface between the alloy and the reacted zone. As a precursor to this state, the agglomeration of isolated B to islands on the alloy surface is expected and observed.

Initially, and especially at low potentials, the reacted zone is negligible, and anodic dissolution occurs at the uppermost atom layer of the alloy. The surface remains plane, although the volume of the alloy shrinks owing to the loss of Cu dissolution. For this situation, two models have been proposed based on two different sets of suppositions. The first uses evidence from potentiostatic current transients observed for for CuAu alloys in 12 M LiCl solution showing that the $i$-$t^{-m}$ relation holds unequivocally between $10^{-2}$ and $10^4$ s at 0.35 V, with $0.8 \leq m \leq 1.0$, and also uses concentration-depth profiles for gold obtained by AES [10b]. Figure 8.9 presents data showing an interdiffusion zone growing with time up to a depth of ca. 20 nm, also showing a slow decrease of the Au surface concentration $\gamma^s_{\mathrm{Au}}$ from 0.65 after 1 s to 0.95 after $10^4$ s of polarization. In this time interval, the current density of Cu dissolution dropped from $10^{-2}$ to below $10^{-7}$ A cm$^{-2}$, with no tendency to level off. In the following, calculations are directly concerning $\gamma_{\mathrm{Cu}}$, which we briefly designate, temporarily, as $\gamma$. We note that volume concentrations, $c_i$(mol m$^{-3}$) and mole fractions are related by $c_i = \varrho\gamma_i$, where $\varrho$ (mol m$^{-3}$) is the molar density of the alloy.

The data presented in Fig. 8.9 certainly indicate that volume diffusion is involved in the dealloying process. Therefore, Fick's first and second differential equations describing fluxes, $J_i$ (mol m$^{-2}$ s$^{-1}$), of diffusing species, labeled $i$, have to be solved. The problem is one-dimensional, and we need to introduce only an $x$-axis perpendicular to the surface. A complication is that by the loss in alloy volume the surface recedes, so that a moving-boundary diffusion problem arises.

The surface recession rate, $v$, is evaluated from the flux, $J^s$, of copper arriving at and passing the alloy/solution interface: $v = c_{Cu} J^s / \varrho$ (m s$^{-1}$). Volume diffusion towards the surface then obeys the equations

$$a) \quad J_{Cu} = -\varrho D \frac{\partial \gamma}{\partial x} + \varrho \gamma v \,, \quad b) \quad \frac{\partial \gamma}{\partial t} = \frac{\partial}{\partial x}\left( D \frac{\partial^2 \gamma}{\partial x^2} - \gamma v \right) . \qquad (8.19)$$

$D$ is the interdiffusion coefficient. Diffusion is expected to proceed via vacancies in the alloy lattice, and $D$ is related to the vacancy mobility and to the vacancy mole fraction by $D = D_v \gamma_v$. Since we are dealing with Au and Cu interdiffusion, we also have $D_v = \gamma D_{Au,v} + (1 - \gamma) D_{Cu,v}$. The molar density, $\varrho$, is assumed to be constant, although the atomic volumes of Cu and Au are different. While probably unimportant for the numerical calculation, the point merits interest in view of the major role assigned to vacancies: Owing to changes in $\varrho$, stresses may develop in the dealloyed surface, influencing vacancy production or vacancy annihilation [5].

Two major further assumptions are used in [10]: The first is to treat the surface vacancy mol fraction at $x = 0$ as an equilibrium property determined by a formation reaction involving a Gibbs reaction energy, $\Delta_r G = \Delta_r H - T \Delta_r S$. The overpotential, $\eta$, of anodic Cu dissolution is assumed to enter the reaction enthalpy: $\Delta_r H = \Delta_r H^\circ - k_\eta F \eta$. This leads to $D = D_\circ \exp(k_\eta F \eta)$.

The assumption of an equilibrium concentration of vacancies is arbitrary. Also, an influence of electric surface charge rather than of overpotential may merit alternative consideration, since charge reduces surface tension. The point is, however, not crucial, since eventually the overpotential, $\eta$, is replaced by the polarization, $\pi = \varepsilon - \varepsilon_{corr}$, apparently without much influence on the result of the calculations. It is more important that vacancy penetration into the bulk by far exceeds Cu and Au diffusion, so that over the range of the Cu/Au-interdiffusion zone $\gamma_v$ will be nearly constant.

The second assumption is that Cu atoms arriving by volume diffusion in the alloy surface cannot directly dissolve. Instead, they must diffuse on the surface to what is termed 'dissolution sites' existing on the surface at an average distance $\Delta$. The surface diffusion coefficient is assumed to be proportional to the volume diffusion coefficient: $D_s = \text{const} \times D$. For a given surface mole fraction, $\gamma^s$, the diffusion surface flux of Cu atoms is estimated to be $J_{Cu} = -(\text{const}/\Delta) D \varrho \gamma^s$.

The last boundary condition is that, far away from the surface, $\gamma$ has the initial bulk value. This completes the system of equations, which, however, cannot be solved analytically; instead, a numerical finite-difference calculation is carried out, fitting four adjustable parameters for optimal modeling. Since the solution is non-analytical, it is difficult to judge the influence of the surface recession term in the original system of equations. Experimental in-depth concentration profiles are well modeled, including the surface composition at $x = 0$; and so are current decay curves.

A different approach is offered in [11d]. Again, dissolution of Cu from CuAu alloys below the critical potential is considered, also again assuming a subsurface interdiffusion zone in the alloy. As explained by the authors, analogous calculations would also apply when a distinct dealloyed layer already exists, with the interdiffusion zone within this layer rather than in the bulk alloy.

In this derivation, the moving-boundary contribution to volume diffusion is ignored, and the interdiffusion coefficient, $D$, is considered to be constant. In terms of volume concentration, $c$, of copper in the alloy, Fick's second law then reads $\partial c/\partial t = D(\partial^2 c/\partial x^2)$. The current density is equal to the electrical equivalent of the flux through the surface according to Fick's first law. The current density of forward Cu dissolution is assumed to obey a Tafel relation. Thus, two additional equations are valid:

$$a) \quad i = i^\circ (c/c^\circ) \exp(\alpha n_e F\eta)\,, \quad b) \quad i = -zFD(\partial c/\partial x)_{x=0}\,. \tag{8.20}$$

Here, $c^\circ$ is the bulk Cu concentration. For Cu $\rightarrow$ $Cu^{2+} + 2e^-$, z = 2, and it is assumed that $n_e = z$. The set of equations has an analytical solution characteristic for the superposition of charge-transfer and transient concentration overpotential [17b]. It is

$$\frac{c(x,t)}{c^\circ} = erf\left(\frac{x}{2\sqrt{Dt}}\right) + \exp\left(\frac{kx}{D} + \frac{k^2 t}{D}\right) \mathrm{erfc}\left(\frac{x}{2\sqrt{Dt}} + k\sqrt{\frac{t}{D}}\right),$$

$$k = (i^\circ/zFc^\circ) \exp(\alpha n_e \eta/b)\,. \tag{8.21}$$

In this equation, $\mathrm{erf}(w) = \int_0^w \exp(-u^2)du$ is the *error function* with tabular values increasing from erf(0) = 0 to erf(2) = 0.96 and erf($\geq$ 3) $\simeq$ 1. The error function complement is $\mathrm{erfc}(w) = 1 - \mathrm{erf}(w)$.

The inspection of Eq. (8.21) reveals that the exponent in the $t^{-m}$-relation describing potentiostatic current decay is expected to be 1/2 for large values of $k\sqrt{t/D}$, in fact at variance with many observations, but confirmed by evidence quoted by the authors. For small values of $k\sqrt{t/D}$, the current decays only very slowly. As calculated from Eq. (8.21), the surface concentration, $c^s$, decreases only slowly with time, confirming evidence as, e.g., shown in Fig. 8.9. The authors claim that the calculation does not require fitting parameters, apparently because, in principle, both $D$ and $k$ can be determined from independent measurements.

So far, the focus has been on high-melting alloys, where volume diffusion is slow at ambient temperature. Volume diffusion is much easier in low-melting alloys such as SnIn alloys. These have been the object of earlier work described in [18], which has prompted more recent investigations [19,3b]. The result, obtained for an alloy with $\beta$-Sn structure is presented in Fig. 8.10, clearly showing an interdiffusion zone as indicated by the decrease of the In concentration form the bulk value 5 *at.*% to zero surface concentration. The solid curve was calculated using the solution, $c = c^\circ\, \mathrm{erf}(x/2\sqrt{Dt})$, of Fick's second law valid

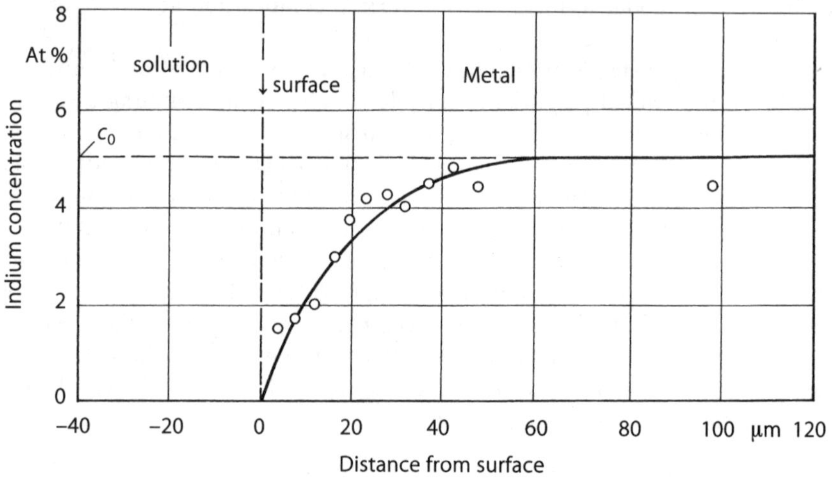


**Fig. 8.10.** In-depth concentration profile of In in Sn-5 at%In after polarization for 105 h at −0.47 V in 3 M NaCl/0.01 M HCl solution, obtained by electron microprobe measurements at metallographic sections. The solid curve was calculated with a diffusion coefficient $D = 5 \times 10^{-12}$ cm$^2$ s$^{-1}$. (From Kaiser)

for diffusion from a semi-infinite medium to the surface where the concentration is kept at zero for all values of the time, $t$. Again the moving-boundary contribution to diffusion was ignored. As seen from the figure, the calculated concentration gradient models experimental evidence very well. The value of $D$ used to fit the observed measurements compares very well with a value 4.3 $10^{-12}$ cm$^2$ s$^{-1}$ reported elsewhere [18c]. It may be noted that InSn alloys are particularly well-behaved as concerns molar density, since it is unaffected by composition. Concerning phase changes, we shall return to the InSn system further below.

As the potential comes closer to the critical potential threshold there is more and more evidence of non-uniform dissolution events changing surface morphology. Earlier evidence from investigations into selective Cu dissolution in CuPd alloys by TEM and TED, trimmed to particularly high resolution [12], has later been confirmed for CuAu alloys by TEM [11d], and for AgAu alloys by in-situ STM [20c]. In this context, recent results of investigations into the dealloying of ordered $Cu_3Au$ and of disordered $Au_3Cu$ in 0.1 M $H_2SO_4$ solutions are interesting. In this work, the lateral resolution of STM was below 1 nm [21]. Monocrystalline specimens were prepared with either (001), or (101), or (111) surfaces. By a combination of high-vacuum STM, LEED, and SXRD, the occupation of the three uppermost atomic layers was established. As an example we quote results obtained for $Cu_3Au(001)$ surfaces: Here, the uppermost atomic layer has 50 at.% Cu, the second is pure copper, the third again 50% Cu. To summarily characterize electrode kinetics, current-potential curves were rapidly scanned in 0.05 M $H_2SO_4$ solution at a rate of 20 mV s$^{-1}$.

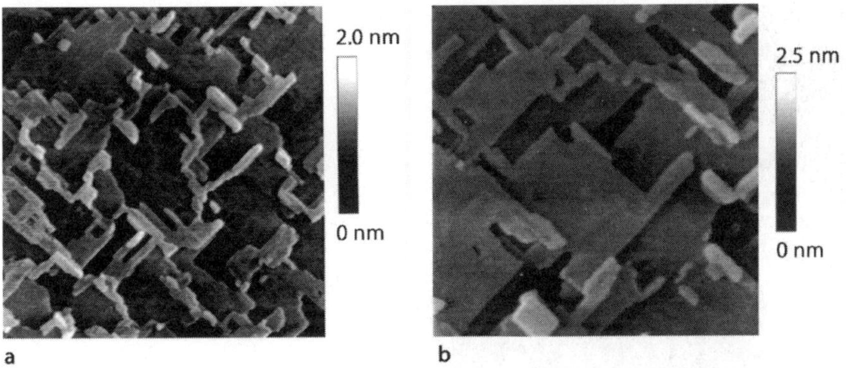


**Fig. 8.11a,b.** Typical in-situ STM images of (001) faces of monocrystalline $Cu_3Au$ during stepwise potentiostatic polarization. The section area is 350 × 350 nm in **a** and 300 × 300 nm in **b**. On the depth scales, darker shades indicate deeper cavities. **a** Potential $\varepsilon = 0.49$ V reached in the 240th minute of stepping the potential, **b** 0.61 V in the 295th minute. (From Eckstein)

This is far from recording quasi-stationary curves, and the critical potential, which showed up at ca. 1.1 V, probably is more positive than the stationary value. Still, the curves qualitatively had the shape shown in Figs. 8.4 and 8.5. For in-situ STM the polarization program changed to stepping the potential in 20 mV-steps with time intervals of up to 30 min. This program still overruns processes of slow current decay, and it leaves the question of residual currents unanswered. The more interesting point is that already at 0.10 V, which is nearly 1 V below $\varepsilon_{crit}$, terrace-site metal dissolution is observed producing clustered arrays of vacancies, accompanied by formation of islands of Au (or Au-rich alloy), unfortunately to small to be shown in reduced-scale print. Also at this pre-breakthrough stage, surface realloying is also detected.

At 0.49 V and at 0.61 V, which still is definitely below $\varepsilon_{crit}$, terrace surface attack becomes severe, as seen in Fig. 8.11. The morphology of surface attack shown in these figures is very similar to what is observed, though on a larger scale, during pitting in passive aluminum (see Chap. 12). The pictures also show the degree of simplification ventured further above by describing such structures as hemispherical pits and bumps.

Closer to $\varepsilon_{crit}$, nucleation events adopt a curious path, with cavities turning into tunnels [12a]. The effect is shown in Fig. 8.12. Again, the similarity with larger-scale events typical for pitting of passive aluminum is striking. Tunneling during dealloying may be typical for low-diffusivity alloys, where volume diffusion to the surface is rate-determining, or at least co-determining [19]. Tunnels, in fact, may be regarded as inverse dendrites, typical for volume-diffusion controlled solidification from molten alloys. Severe roughening during dealloying beyond $\varepsilon_{crit}$ has been explained using similar arguments, albeit in contrast to percolation concepts discussed further below. On the alloy surface, the onset of coarsening by the formation of a porous layer of the noble metal, or of a

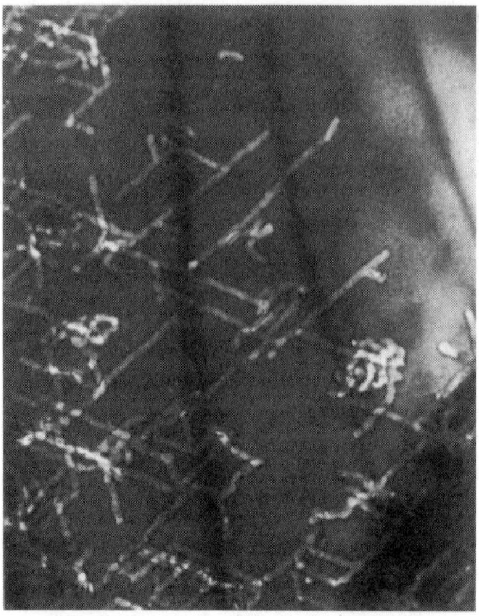

**Fig. 8.12.** Crystallografic tunneling in thin Cu-10Pd foils during potentiostatic dealloying in sulfuric acid solution, only little below the critical potential (see Fig. 8.5). TEM magnification is 60 000. (From Kabius, Kaiser, and Kaesche)

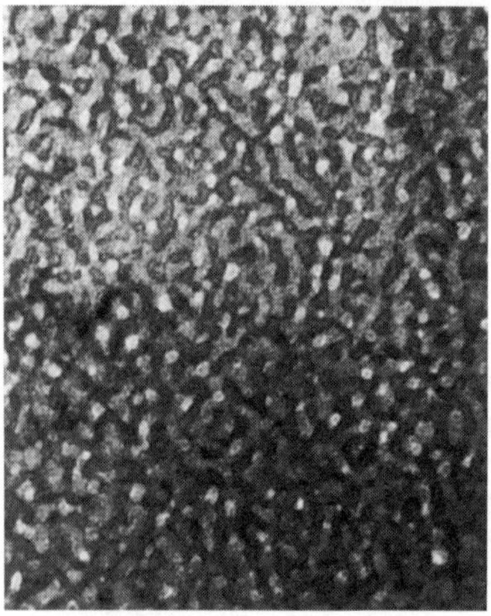

**Fig. 8.13.** Cu-15Pd thin foil after polarization in sulfuric acid, inspected by ex-situ TEM, magnification 300 000, showing the the onset of formation of a porous Pd layer as a residual from dealloying close to but still below the critical potential. (From Kabius, Kaiser, and Kaesche)

layer rich in the noble metal, is shown by Fig. 8.13. This type of morphology has been observed elsewhere as well [11c,22].

Figure 8.14 schematically explains the surface island formation as the result of recrystallazitaion of adsorbed Au atoms: The thermodynamic activity of adatoms left by dealloying is probably excessive, promoting spontaneous '2D' or '3D' nucleation of solid gold particles. The reaction path is either direct surface diffusion, or else a dissolution/redeposition sequence. For gold alloys, the latter mechanism does not seem likely, but the situation will further below be seen to be different for dealloying of brasses.

Below the breakthrough potential, $\varepsilon_{crit}$, an AB alloy with a noble component B is practically stable with respect to continued detrimental corrosion, regardless

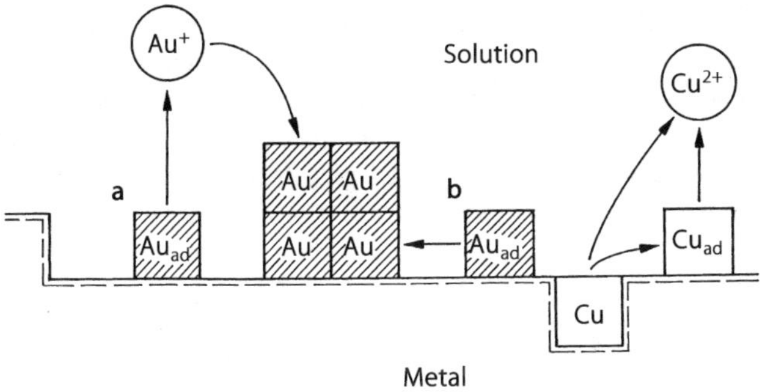


**Fig. 8.14.** Two possible paths of recrystallisation of gold islands from adsorbed gold atoms present at high thermodynamic activity on the surface, isolated by selective copper dissolution: Surface diffusion and the underpotential dissolution-redeposition sequence

of composition. The alloy, therefore, is practically safely protected, as long as the corrosion potential is more negative than the breakthrough potential.

A different type of criticality is the *parting limit* of the alloy. This is the minimal concentration of a thermodynamically stable, i.e. noble component B, beyond which the alloy assumes almost true noble-metal behaviour, regardless of the composition of the aqueous solution, because A-atoms completely cease to come in contact with the solution. These composition limits are the main topic of percolation theory as applied to alloy dissolution [20,23]. As noted above, the first supposition is that an influence of volume diffusion of A on dissolution kinetics is ruled out because, instead, A dissolution proceeds only along preexisting continuous A paths into the alloy. When, as in high-diffusivity alloys, volume diffusion is, in fact, obvious, the following considerations are not applicable.

Short-range order clustering of both A and B atoms, with cluster diameters scattered around a most probable value, $\xi$, was noted previously. Typically, the size of clusters is that of several atoms, and possible dissolution channels following such clusters are much too narrow to allow the entrance of electrolytic solution. As an answer to this problem, surface diffusion in the channel system is assumed to coarsen the interior structure of dealloyed material by Ostwald ripening. This coarsening has, in fact, been demonstrated using in-situ SANS [20d], revealing a two-phase alloy/pores composite with suitable pore diameters. The method has been applied to $Ag_{0.7}Au_{0.3}$, prepolarized for 18.7 h in 1 M $HClO_4$/1 M $AgClO_4$ solution to a beyond-critical potential producing severe attack, and switching the potential back to a below-critical value, where practically no further attack occurred [20d]. During 20.1 h, the characteristic length parameter considerably increased, indicating the growth of a porous channeled structure, and this slow process may control the ingress of electrolytic solution. As the porous channeled structure grows into the alloy, conditions once more resemble those encountered in pitting of passive metals insofar as a local enrichment of dissolved $AgClO_4$ is expected, and also an ohmic voltage drop along deep pores. These effects might well cause a decay in current density with time, which in fact is observed in [20a], but apparently disclaimed in [20c], and not incorporated into the model. Modeling is by computer simulation involving a number of adjustable parameters. The main principle implemented in the algorithm used to produce either '2D' or '3D' pictures of sections through superficially dealloyed material is that, apart from the comparatively simple geometry of site-percolation conditions, surface diffusion is the decisive factor in deciding between resistance to, or else failure by dealloying. In this model, dealloyed structures have a fractal appearance, the roughness being determined by the rules adopted in the simulation. An initial decay of dissolution currents with time is allowed for by the nature of the initial structure of clusters emerging in the metal/solution interface, turning to steady-state conditions inside the material, if the residual current differs from zero. The images produced by the computer illustrate what further above has been described as the reacted zone left by the dealloying process on top the bulk alloy.

In the early version, percolation theory, as applied to dealloying, was at a loss to explain critical potentials which, as shown above, are rather well understood, although by far not yet accessible to quantitative prediction, from nucleation theory. Later, using arguments close to those of the nucleation concept, an explanation was offered, disregarding the complete current-potential curve $i_A(\eta)$, but regarding the particular value critical for the onset of severe surface roughening, $i_A(\eta_{crit})$[20b]. Criticality is seen as the result of '3D'-roughening overtaking surface diffusion tending to smooth the surface. Charge-transfer control is assumed for A dissolution and the Butler-Volmer equation (see Chap. 5) is used in the simplified, symmetrical form $i = i^\circ\{\exp(a\eta) + \exp(-a\eta)\} = 2i^\circ \sinh(a\eta)$. As for nucleation, the formation of pits with average radius $\rho = \xi_A$, in this case equal to the most probable A-cluster diameter, is considered decisive. Pits are thought to be directly formed by anodic dissolution with an overpotential $\eta = \varepsilon - E_\varrho$, rather than as the result of statistical vacancy fluctuation. Therefore,

$$i = 2i^\circ \sinh\{\varepsilon - \left(E_\infty + (\sigma V_B/zF\xi)\right)\} . \tag{8.22}$$

The critical current density is estimated by considering a surface only slightly perturbed into a shallow sinusoidal profile with average amplitude $\xi$, inspecting the competition between profile flattening by surface diffusion and profile-deepening by anodic metal dissolution. Surface diffusion is curvature-driven, as suggested by the influence of curvature on the chemical potentials of surface A and B atoms.

Let $\mu(\xi)$ be the average chemical potential as function of $\xi$ and assume $\mu(\xi) = \mu_\infty + \sigma V/\xi_A$, where $V$ is the molar alloy volume. The gradient of $\mu(\xi)$ along a surface element, $ds$, with curvature $1/\varrho$, is $\partial\mu/\partial s = \sigma V(\partial(1/\varrho)/\partial s$. The rate of surface diffusion then is $v_s = -(D_s\sigma V/RT)\partial\mu/\partial s$ [24]. This relation is used to estimate the rate of profile-leveling as compared to the rate of surface roughening. Using additional simplifications, the estimate eventually is that roughening overtakes leveling, whenever the current density of metal dissolution exceeds the critical value

$$i_c = 4\pi^2 N_s D_s / \xi_A^2 \,. \tag{8.23}$$

$N_s$ is the number of atoms per unit area. Substituting this value for $i$ in Eq. (8.22) and solving for $\varepsilon$, the result should be the critical potential, $\varepsilon_{crit}$. Considering selective dissolution of Ag from AgAu alloys, the values $i_\circ \simeq 1\,\mathrm{Acm^{-2}}$, $\sigma \simeq 0.5\,\mathrm{Jm^{-2}}$, and $D_s \simeq 10^{-10}$ cm$^2$ s$^{-1}$ were used, as suggested from in-situ STM measurements of smoothening of Au(111) surfaces following dealloying [20e]. The result compares well with previous investigations into dealloying of AgAu [9b].

From subcritical and critical dealloying we now turn to beyond-critical rapid selective dissolution of homogeneous CuAu alloys in acid sulfate solutions. For this case, the original opinion was that volume diffusion to the alloy surface is rate-determining [11a]. The mobility of single vacancies was considered insufficient for rapid diffusional transport of Cu and Au to and from the surface. Instead, data indicating a high mobility of divacancies in copper were used [25].

Considering, in particular, dealloying of Cu-3Au, the mobility of divacancies was assumed to be $D_{vv} = 1.3 \times 10^{-12}$ cm$^2$ s$^{-1}$. In the interdiffusion zone beneath the alloy surface, both Cu and Au move via divacancies. Again, the penetration of divacancies into the bulk alloy by far exceeds the depth of the interdiffusion zone. Therefore, the divacancy mole fraction, $\gamma_{vv} \simeq 0.01$ is practically constant within this zone. The diffusion equation was solved assuming zero surface concentration of Cu, aiming at an approximation only, but taking the moving-boundary contribution to the problem into account. The result was that the depth of the interdiffusion zone increases as $\delta \simeq \{2(1-\gamma_{Cu}^{\circ})Dt\}^{1/2}$. The interdiffusion coefficient is $D = \gamma_{vv} D_{vv} \simeq 10^{-14}$ cm$^2$ s$^{-1}$, and the current density of Cu dissolution was calculated to possibly reach $i \simeq 2 \times 10^{-4}$ Acm$^{-2}$. Current decay was disregarded as being of minor interest.

For a a typical polarization time of $10^3$ s, the estimated depth of the interdiffusion zone, $\delta$, is of the order of order of 0.1 μm. At the time of this work, this was a difficulty, because the aim was to use X-ray diffraction to study the interdiffusion zone, but the X-ray penetration depth by far exceeds $\delta$. The solution of the problem was to use the effect of folding of the interdiffusion zone during continued dealloying, caused by the instability of a plane surface with respect to roughening when volume diffusion is rate-determining [25]. To see this, consider a plane surface on an interdiffusion zone of thickness $\delta$, and an indentation in the surface of depth $\Delta\delta$. Below the bottom of this cavity the depth of the interdiffusion zone is smaller than $\delta$ by $\Delta\delta$, but the difference of concentrations across the zone is the same as elsewhere. Therefore, volume diffusion to the bottom of the cavity is faster than to the plane surface, and if volume diffusion determines the rate of Cu-dissolution, the tendency is to deepen the cavity. The volume diffusion mechanism thus tends to increase the amplitude of surface perturbations, which is equivalent to roughening. Considering, as depicted in Fig. 8.15, that by the same principle the curved flanks of the surface again will unstable with respect to smaller-scale instability, a more and more corrugated structure will develop, similar to what is described to occur during dealloying by the percolation mechanism. The thickness, $\delta_r$, of the

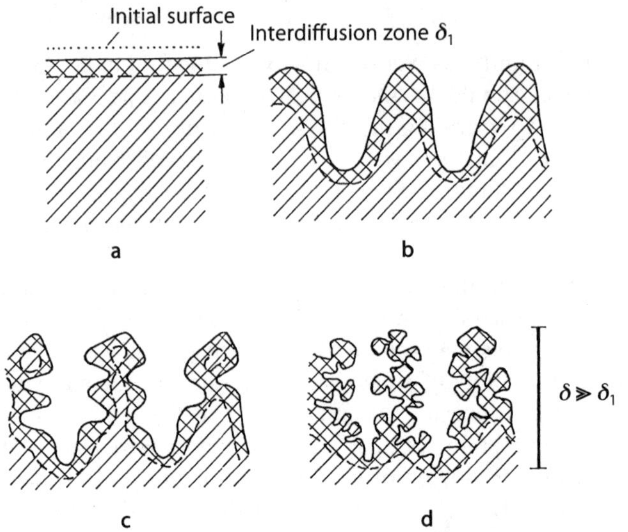


**Fig. 8.15a–d.** Surface roughening creating a reacted porous layer consisting of Au, or of an alloy rich in Au, on a CuAu alloy by selective dissolution of Cu with rate-determining volume diffusion

corrugated reacted zone now is orders of magnitude higher than the thickness of the interdiffusion zone, and accessible to X-ray diffraction analysis.

As shown in Fig. 8.16, the intensity of the Cu–3Au(111) reflex decreases with dealloying time, and a broadening range in reflected intensity shifting to higher values of the refracting angle, 2Θ, is observed. This was taken as indicating the presence of a continuous range of increasing Au concentrations in the reacted zone, tending to but not reaching pure Au. The broad range in reflected intensity can, however, also be explained by reflection from homogeneous but very small crystallites. This was found to apply to dealloying of NiPd alloys as shown in Fig, 8.17, where, as expected for the formation of a reacted zone consisting essentially of pure, but microcrystalline Pd, the Pd(200) reflex is clearly visible but considerably broadened by the size effect [26a].

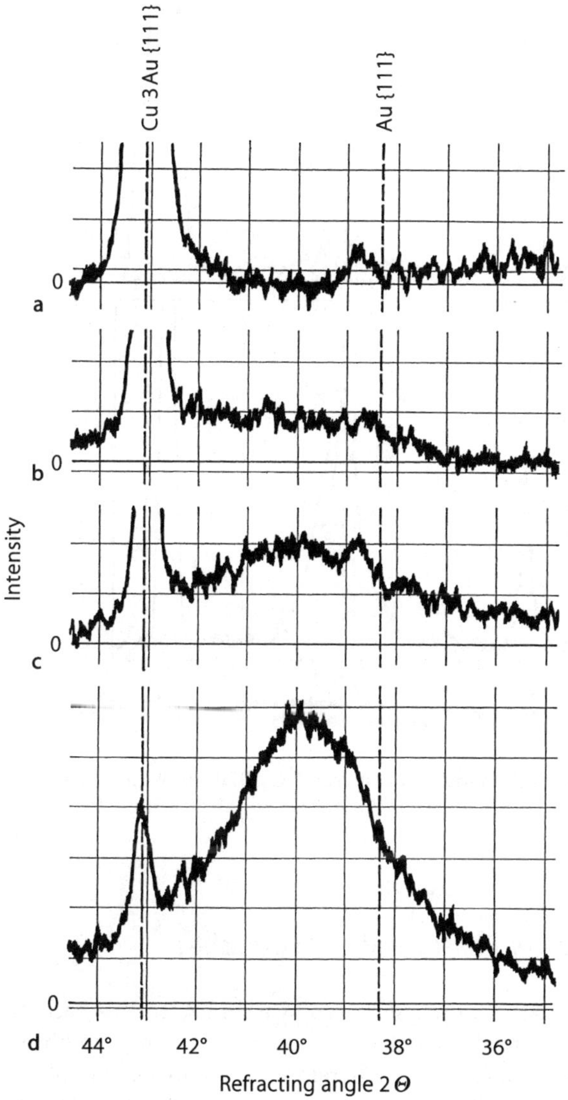


**Fig. 8.16a–d.** Surface X-ray diffraction diagrams taken from Cu-3Au specimens after galvanostatic polarization in 0.05 *M* $H_2SO_4$ solution. The current density was 5 mA $cm^{-2}$. **a** is the initial state, **b**, **c**, and **d** were obtained after passage of 3.0, 6.0, and 24.0 C $cm^{-2}$. 2Θ is the refracting angle. (From Pickering)

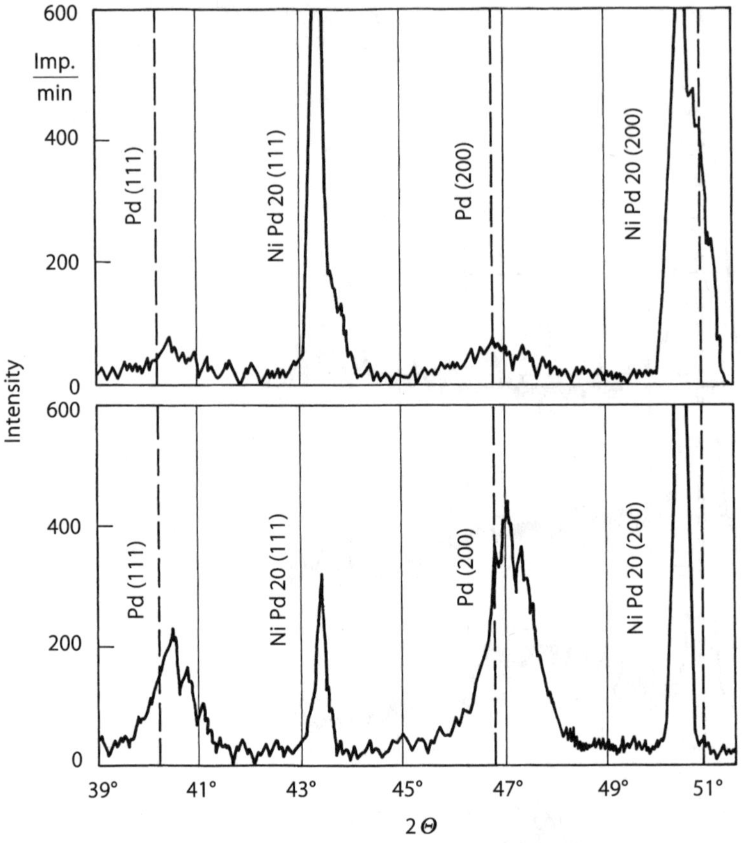


**Fig. 8.17.** Surface X-ray diffraction diagrams taken from Ni-20Pd alloy specimens after anodic polarization at 0.6 V in 0.05 N $H_2SO_4$ solution causing dealloying, after **a** 10, **b** 20 h. (From Kaiser and Kaesche)

## 8.4 Dealloying of Hume-Rothery Phases and of Martensites

A particularly interesting binary alloy system is ZnCu, brass. In this case, the corrosion potential can easily reach values more positive than the equilibrium potential of the more noble component Cu. Therefore, simultaneous dissolution of atoms of widely different nobility now is an interesting possibility. Also, we now encounter a series of Hume-Rothery phases instead of a series of solid solutions, and this raises the question of phase changes induced by dealloying.

In the first place, we have to consider that even below the $Cu/Cu^{2+}$-equilibrium Cu may recrystallize on the dealloyed surface by *underpotential dissolution* caused by a high surface activity of adsorbed Au atoms left by Zn dissolution, followed by redeposition. The tool to distinguish surface diffusion

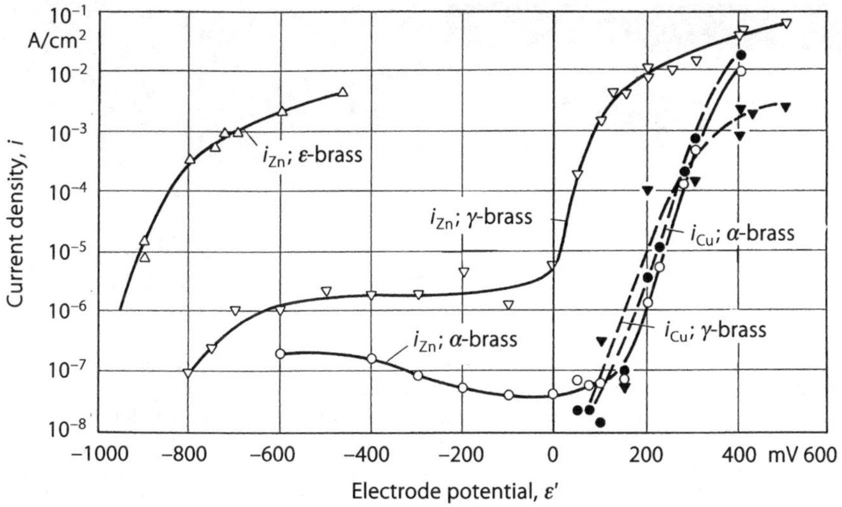


**Fig. 8.18.** Partial current-potential curves for zinc dissolution, $i_{Zn}$, and for copper dissolution, $i_{Cu}$, from $\alpha$-, $\gamma$-, and $\epsilon$- brass, i.e. Cu-30Zn, Cu-65Zn, and Cu-68Zn, in buffered $Na_2SO_4$ solution, pH 5. (From Pickering and Byrne)

and the dissolution/redeposition is a *rotating ring-disk electrode*, which is a rotating disc equipped with an outer ring, and controlled by two potentiostats. The center disc is made from the alloy, the surrounding inert platinum or gold ring is concentric and coplanar with the disk. The idea is that by the solution flux created by the rotation of the disk, dissolved $Au^{3+}$, if there is any, is transported past the electrode surface without complete redeposition, more redeposition being enforced at the ring by suitably adjusting the ring potential. For Cu-13Au no such ring current was observed [11a]. For brass electrodes the ring current was, however, well defined, indicating operation or cooperation of the dissolution/redeposition mechanism [11a,27].

Figure 8.18 shows anodic current potential curves for different Hume-Rothery phases of brass observed in slightly acid sulfate solutions, studied separating the contributions from zinc and copper dissolution, [11b]. As previously, the residual current below the critical potential is very low and perhaps an artefact. For the $\varepsilon$-phase, the curve suggests that $\varepsilon_{crit}$ is below the range of potentials covered by the measurements; at the same time these negative values of $\varepsilon$ rule out anodic Cu dissolution, including underpotential dissolution followed by redeposition. For the $\alpha$-phase, the case apparently is one as described in Sect. 8.1, indicating a steep surge of uniform alloy dissolution, caused by the exponential dependence of current densities on potential, rather than a critical potential for dealloying by Zn dissolution. The behaviour of the $\gamma$-phase is particularly interesting: While beyond $\varepsilon_{crit}$ both alloy components dissolve, Zn dissolution is by far the more rapid process, indicating non-uniform dissolution and, thus, dealloying of a special type.

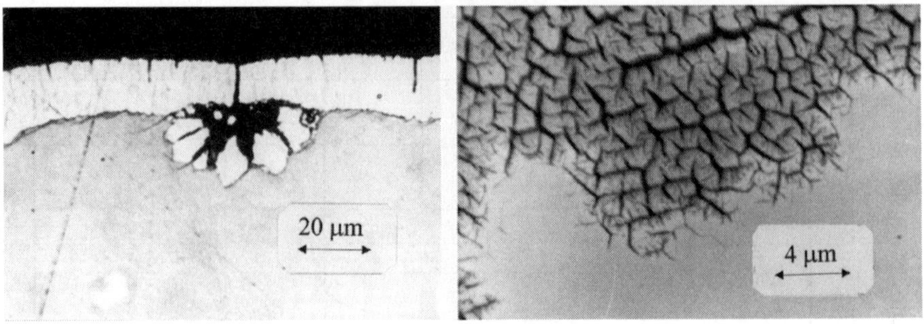


a b

**Fig. 8.19.** **a** Phase transformation ε-CuZn → γ- CuZn caused by 42 h anodic polarization of ε-brass at −075 V in 1 N $Na_2SO_4$ solution, pH 5. Optical micrograph of cross section etched in thiosulfate/bisulphate solution. Note formation of a virtually non-porous reacted zone consisting of columnar γ-crystals. **b** Reacted zone consisting of fissured copper on ϵ-brass after 10 min anodic polarization at −0.25 V in the same solution. SEM micrograph; atomic number contrast. Note propagation of reacted zone by oriented tunnels or slots. (From Gladen, Kaiser, and Kaesche)

The inspection of the relatively thick reacted zone developed during continued dealloying by X-ray diffraction was described above. This method has been applied to dealloying of both ϵ- and γ-brass [11b]. The shift of broadening reflected intensity in the direction of increased solid-solution Zn concentrations, indicating volume diffusion, was again observed. At the same time, reflexes indicating the presence of α-brass appeared in the dealloyed zone on γ-brass, and reflexes from γ-brass in the zone on ϵ-brass gave strong support for the volume diffusion mechanism. This mechanism has been confirmed by further investigations into the dealloying of brass, also clearly showing phase changes owing to dealloying [3b,26d]. Figure 8.19a shows a section through ϵ-brass dealloyed at low potential and current density. A non-porous layer of γ-brass has developed owing to transformation caused by dezincifation. An important point is that the thickness of this reacted layer increases proportional to $t^{-0.5}$, which is strong evidence for rate-determining zinc volume diffusion across the γ-layer to the layer/solution interface. A complication is the preferential advance of the reaction along and at the tip of fissures. At high potential and current density, the situation changes, as seen in Fig. 8.19b: Dealloying now produces a fissured sponge of copper, and the reaction proceeds by tunneling metal dissolution at the bulk/sponge interface. Apparently, ϵ-brass is an alloy of intermediate diffusivity, and the course of dealloying by either the volume-diffusion typical for high-diffusivity alloys, or tunneling corrosion typical for low-diffusivity alloys depends on the ratio of the possible diffusion flux and the required rate of metal dissolution.

The phase diagram of the alloy system FeZn again shows a sequence of Hume-Rothery phases. Expecting, for conditions where oxide and hydroxide films are not formed, a behaviour similar to that of the brasses, dealloying of the δ-phase in acid sulfate solutions has been studied [26e]. Unusual observations

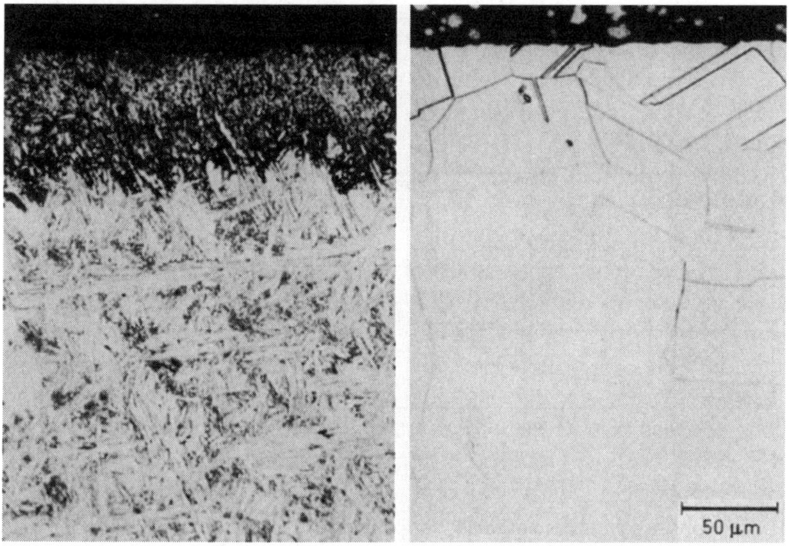


**Fig. 8.20.** Metallographic section through the surface of Cu-10wt.% Al after 48 h potentiostatic anodic polarization at 0.19+V in 0.5 M $H_2SO_4$ solution. *Left:* Rapid dealloying of $\beta'$-martensite; Heat tempered 2 h at 1000°C followed and quenched in water. *Right:* Resistant $\alpha$-solid solution. Tempered 72 h 600°C and quenched in water. (From Langer, Kaiser, and Kaesche)

were made, apparently mainly due to surface cracks developing during the formation of the dealloyed reacted zone, resulting from volume changes in the alloy as a consequence of changes in the lattice spacing of the dealloyed material, as revealed by broadening of X-ray reflexes.

When, as in the case of brasses, composition and structure are varied simultaneously, it is difficult to separate the influences of the two parameters. Therefore, alloys undergoing constant-composition martensitic transformations are interesting. Also, martensites, while of particular practical importance, generally tend to rapid selective corrosion. As shown in Fig. 8.20, martensitic Cu-10wt.%Al rapidly dissolves in sulfuric acid, while at the same composition the solid-solution homogeneous structure is virtually resistant to corrosion [14c]. Here again, the thickness of the reacted zone in the dealloyed martensitic surface is well defined. In this case the rate law for the growth of the reacted zone is linear, as shown in Fig. 8.21, apparently because the dealloying reaction proceeds at the interface between reacted zone and bulk alloy, where reaction conditions remain independent of time [26c]. This requires a continuous ingress of electrolytic solution through the porous reacted zone. As also shown in Fig. 8.21, dealloying of NiPd at beyond-critical potentials also proceeds, although more slowly, at constant rate, probably also controlled by reaction kinetics at the interior surface of the reacted layer [26b].

Another typical example is rapid corrosion of martensitic plates in otherwise austenitic stainless steel [28], or in NiCoPd alloys [26b]. Frequently, rapid

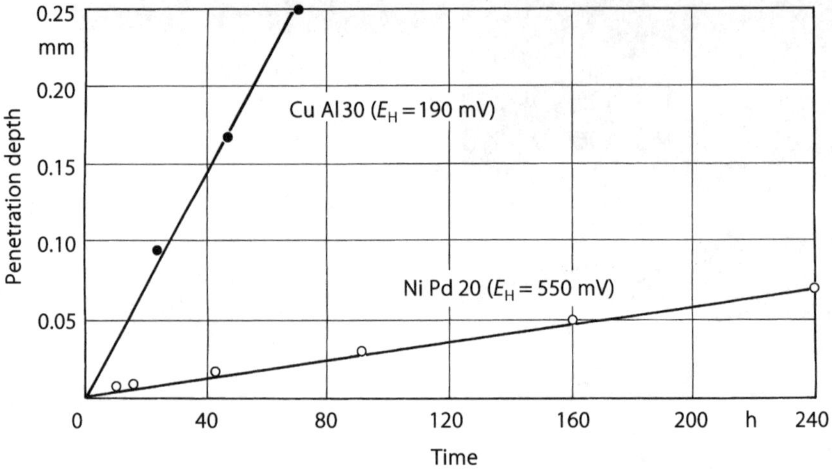


**Fig. 8.21.** Linear advance with time of dealloying during potentiostatic polarization in acid solutions. *Solid circles:* Martensitic Cu-30Al, 0.5 M $H_2SO_4$, 0.19 V (from Langer, Kaiser, and Kaesche). *Open circles:* Ni-20Pd, 0.05 M $H_2SO_4$, 0.56 V (From Kaiser)

corrosion of martensites and *intercrystalline corrosion* (see Chap. 13) occur simultaneously, and probably for the same reasons, which is high density of dislocations, and especially of dissociated dislocations, i.e., stacking faults.

In this chapter, the metal surface in contact with an aggressive electrolytic solution was understood to be free of deposited corrosion product. In actual service, corrosion protection of structural materials such as brasses, or bronzes usually owes more to thin protective layers of corrosion products than to inherent resistance due to composition. The reason is that usually the aggressive solution is near-neutral, and that poorly soluble hydroxides and oxides are soon deposited. We recall from Chap. 3 that corrosion protection by corrosion products is the main topic when dealing with potential-pH diagrams. At the same time, corrosion protection by relatively thick corrosion product layers also depends on other circumstances, especially on the presence of minor alloy components influencing the kinetics of corrosion processes including events such as dezincification and dealuminization. These details, actually of outstanding practical importance, have been and will further be left disregarded, with reference to the literature quoted in Chap. 1.

## References

1. Gerischer, H.: in: Korrosion 14. Ber. Diskussionstaging Korrosionsschutz durch Legieren. Frankfurt 1961. Verl. Chemie, Weinheim, 1962
2. Pickering, H.W.: Corr. Sci. *23* (1983) 1107
3. a) Kaiser, H.: Alloy Dissolution. In: Corrosion Mechanisms (Mansfeld, F., ed.). Marcel Decker Inc., New York and Basel, 1986, p. 85; b) Kaiser, H.: Corr.

Sci. 34 (1993) 683; Kaiser, H. and Eckstein, G.A.: Dealloying. In: Bard, J. and Stratmann, M., eds.: Encyclopedia of Electrochemistry (in preparation)

4. Seo, M. and Sato, N.: Langmuir 1987 *3* (1987) 917
5. Heusler, K.E.: Corr. Sci. *39* (1997) 1177
6. Kolotyrkin, Y.M.; Electrochim. acta *25* (1980) 89
7. a) Zolotarev, E.I., Pchelnikov, A.P., Skuratnik, Ya.B., Khoklov, N.I., and Losev, V.V.: Elektrokhimia *25* (1989) 208, quoted from [5]; b) Zartsyn, I. D., Vvedenskii, A. V., and Marshakov, I. K.: Russian J. Electrochem. (translated from Elektrokhimiya) 30 (1994) 492
8. Tammann, G.: Z. Anorg. Chemie 112 (1920) 233, 114 (1920) 281, 118 (1921) 48; for comments see Dehlinger, U. and Glocker, R.: Ann. Physik 16 (1933) 156
9. a) Gerischer, H.: loc. cit. [1]; b) Tischer, R.P. and Gerischer, H.: Z. Elektrochemie *62* (1958) 50; c) Gerischer, H. and Rickert, H.: Z. Metallkunde *46* (1955) 681; d) Bergmann, M., quoted in loc. cit. [1], see also [9b]
10. a) Rambert, S. and Landolt, D.: Electrochim. acta 31 (1986) 1421, 1433; b) Laurent, J. and Landolt, D.: ibid. 36 (1991) 49
11. a) Pickering, H.W. and Wagner, C.: J. Electrochem. Soc. *114* (1967) 698; b) Pickering, H.W. and Byrne, P. J.: ibid. 118 (1971) 209; c) Pickering, H.W.: Corr. Sci. *23* (1983) 1107; d) Ateya, B.G. and Pickering, H.W.; ibid. *38* (1996) 1245; e) Ateya, B.G., Fritz, J.D., and Pickering, H.W.; J. Electrochem. Soc. *144* (1997) 2606
12. a) Kabius, B., Kaiser, H., and Kaesche, H.: in: Proc. Int. Symp. Honoring N. Hackerman, San Diego 1986. (McCafferty, E. and Brodd, R. J., eds.). Electrochem. Soc. Vol. 86-7, Pennington, 1986; b) Kabius, B.: Doctoral thesis, Erlangen 1987; c) Kaesche, H.: in: Proc. Int. Conf. Localized Corrosion, Orlando 1987. NACE, Houston, 1990
13. Gnieweck, J., Pezy, J., Baker, B.G., and Bockris, J. O'm.: J. Electrochem. Soc. *125* (1978) 17
14. a) Popp, W., Kaiser, H., Kaesche, H., Brämer, W., and Sperner, F.: in Proc. 8th Int. Conf. Metallic Corrosion, Mainz 1981, Vol. 1. Dechema, Frankfurt 1981, p. 76; b) Popp, W. and Kaiser, H.; Unpublished work, 1982; c) Langer, R., Kaiser, H., and Kaesche, H.: Werkstoffe u. Korr. *29* (1978) 409
15. a) Kaesche, H.: in previous editions of this book; b) loc. cit. [3a]
16. Kaiser, H.: private communication 1982
17. a) Volmer, M.: Kinetik der Phasenbildung. Th. Steinkopf, Leipzig and Dresden, 1939; b) Vetter, K.J.: loc. cit. Chap. 5 [1]; c) Budevsky, W.B.; Electrocrystallisation. In loc. cit. [3.7]; d) Kaiser, H. and Kaesche, H.: Jahrestagung der DGM, Erlangen 1983
18. a) Marshakov, I.K., Pchelnikokov, A.P., Losev, V.V., and Kolotyrkin, Ya.M.: Elektrokhimia *17* (1981) 725; b) Marshakov, I., Vvednenskii, K., Kondrashin, W.Ya., and Vonov, G.A.: Anodnoe Rastvorenie i Celektiivnaja Korrosia Cplabob. Voronesh, Isd-wo WGU 1988; c) Verezhnaya, A.G. and Ekilik, V.V.: Elektrokhimia 29 (1993) 910; d) Pchelnikov, A.P., Sitnikov, A.D., Marshakov; I.K., and Losev, V.V.: Electrochim. acta *26* (1981) 591; e) Pchelnikov, A.P.,

Skuratnik, Y.B., Sitnikov, A.D.: in Proc. 3rd Japan-USSR Seminar on Electrochemistry, Kyoto 1978. Electrochem. Soc. Japan, Academy of Science USSR
19. a) Kaiser, H.: Werkstoffe u. Korrosion *40* (1989) 1; b) Kaiser, H. and Kaesche, H.: Ext. Abstracts. Vol. 88-2 Electrochem. Soc. Fall Meeting, Chicago 1988, p. 138
20. a) Sieradsdzki, K., Corderman, R.R., Shukla, K., and Newman, R.C.: Philos. Mag. *A 59* (1989) 713; b) Sieradszki, K.: J. Electrochem. Soc. *140* (1993) 2868; c) Wagner, K., Brankovic, S.R., Dimitrov, N., and Sieradszki, K.: ibid. *144* (1997) 3545; d) Newmann, R.C., Corcoran, S.G., Erlebacher, J., Aziz, M.J., and Sieradski, K.: MRS Bulletin July 1999, p. 24; e) Oppenheim, I.C., Trevor, D.J., Chidsey, C.E.D., Trevor, P.L., and Sieradzki, K.: *254* (1991) 687
21. a) Eckstein, G.A.: Doctoral thesis, Erlangen 2000; b) Eckstein, G.A., Dakkouri, A.S., and Kaiser, H.: 2000 MRS Spring Meeting, Symp. Corr. Metals and Alloys, San Francisco 2000
22. a) Forty, A.J. and Durkin, P.: Phil. Mag. 42 (1980) 295; b) Forty, A.J. and Rowlands, G.: ibid. 43 (1981) 171
23. Stauffer, D.: Introduction to Percolation Theory. Taylor and Francis, London 1985
24. Mullins, W.W.: in Metal Surfaces: Structure, Energetics and Kinetics. Am. Soc. Metals, Metals Park, 1963, p. 17
25. Wagner, C.: J. Electrochem. Soc. *103* (1956) 571
26. a) Kaiser, H. and Kaesche, H.: Werkstoffe u. Korr. 32 (1980) 347; b) Kaiser, H.: Doctoral thesis, Erlangen 1976; c) Kaiser, H., Lenz, E., and Kaesche, H.: in Int. Conf. Mechanism of Environment Sensitive Cracking of Materials. Guildford 1977; d) Gladen, H., Kaiser, H., and Kaesche, H.: Corr. Sci. *30* (1990) 737; e) Kaiser, H.: Materials and Corr. *47* (1996) 34
27. Feller, G.H.: Corr. Sci. *8* (1968) 698
28. Honkasolo, A.: Scand. J. Metallurgy *2* (1973) 156, Corr. NACE *29* (1973) 13

# 9 Rusting of Iron and Steel

Iron and unalloyed or low-alloy steels are the predominant metallic structural materials governing technology and by far out-weighing all other structural metals. Loss of iron and steel by corrosion, and costs of protecting iron and steel against corrosion by coatings, lacquers, and varnishes therefore have outstanding economic importance. Since corrosion of iron and steel usually is rusting, a closer look at the mechanism of iron reacting with oxygen dissolved in approximately neutral solutions appears appropriate.

The sequence of reaction steps producing, $Fe^{2+}$, $Fe^{3+}$, $Fe(OH)_3$ and, finally, iron rust has already been discussed in general, and with particular respect to the influence of solution pH, in Chap. 3. In Chap. 5, the electrode kinetics of oxygen corrosion have been treated, disregarding the influence of solid rust, except for the case of very thin layers, for which the influence was incorporated into a corrected term $D/\delta_{\mathrm{Nernst}}$ in the equations of diffusion-limited cathodic kinetics. Rust was considered to be an inert porous layer with electrode reactions occurring at the metal/solution interface only. Many complicating details of the process of rusting were, thus, so far disregarded [1].

Generally, the rust barrier will, after some time, slow further corrosion. Whether rust is considered protective depends, however, on circumstances. In the case of 'weathering' low alloy steels, to be described further below, rusting during a number of years is considered acceptable, if the rust layer eventually becomes dense, so that additional coatings can be dispensed with. In domestic installations, rust is, on the other hand, considered objectionable whenever it colours water carried in steel pipes. In prestressed reinforced concrete structures, even slightly prerusted high-strength tendons are considered unfit for use owing to the danger of hydrogen embrittlement. Also, the requirements concerning useful life of iron and steel parts vary very much, from 'practically indefinitely' for bridges, to 'many decades' for domestic appliances, to 'several years' for coatings, and to 'months' for phosphated parts.

An example of massive use of steel parts requiring long-time safe corrosion protection is the case of pipes transporting tap water from the water works to the users. Essentially, tap water is a dilute electrolytic solution containing some millimoles per liter of sulfates, chlorides, carbonates and bicarbonates. It is approximately neutral, and it contains dissolved oxygen, which makes oxygen corrosion the expected event. For protection, the interior of large-diameter pipes often is lined with thick layer of cement, apparently aiming at passivation of the steel surface (see Chap. 10), while pipes for domestic installations usually are hot-dip galvanized, especially for the distribution of warm water. Very many water-carrying pipes are, however, left to rust without further protection, expecting, as it is, that the rust layer eventually will become protective. An elaborated and widespread, though not convincingly founded theory had it

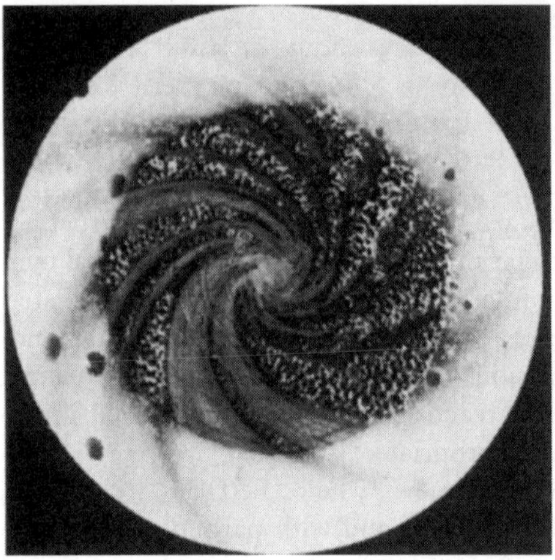

**Fig. 9.1.** Rust deposit on disk of unalloyed steel rotated for three days with a rotation rate of 1 $s^{-1}$ in a solution modeling tap water: $3 \times 10^{-4}$ M $Na_2CO_3$ solution, saturated with respect to both solid $CaCO_3$, and air containing 1% $CO_2$. (From Bohnenkamp)

that a crucial point is the carbonate/bicarbonate/calcium thermodynamic equilibrium in tap water [2]. The opinion was that the pipe interior is protected by a lime/rust mixture slowly developing only if the water is in equilibrium with respect to lime deposition. Closer inspection of old pipes has, however, shown that the protective layer consists mainly of iron oxide, and that, apart from carbon dioxide, other dissolved substances are important: Chlorides appear to be detrimental, while phosphates, silicates, and aluminum hydroxide are useful. An interesting point in a matter otherwise hardly very well accessible to theory is that, since the protective deposit is mainly the corrosion product rust, deaeration of the water is not desirable, in contrast to what on first sight would seem obvious for oxygen corrosion. For the same reason, water flow should be reasonably rapid, or reasonably frequent [3a]. In this context, it has been supposed that in stagnant water large-scale Evans differential aeration cells – for more on these, see Chap. 11 – tend to develop [4c]. More precisely, galvanic cells of this type may be expected to have dimensions not exceeding the thickness of the hydrodynamic boundary layer, and this thickness should grow to the pipe diameter when the solution stagnates. Other effects, such as initial conditions for the growth of deposits on the steel surface, may depend on flow rate as well [3b]. As an example of the influence of slow solution flow on the non-uniform deposition of rust, Fig. 9.1 shows rust deposited on a slowly rotating steel disk in what essentially is harmless model tap water. The figure also clearly shows the nature of solution flow induced by the rotation of the disc.

For domestic installations, hot-dip galvanized pipes are the preferred choice [5]. The protective coating is formed by stacked, interlocked layers of FeZn alloys. The layer sequence is, from the outside to the steel surface, $\alpha$-, $\Gamma$-, $\delta_1$-, $\varsigma$-, and $\eta$-FeZn, with 100-90, 28-21, 12-7, 6, and ca. 0 wt.% Zn, respectively.

The coating thickness does not exceed ca. 0.1 mm. This very thin coating is not likely to survive many decades, though this certainly is the lifetime of the piping expected by users. FeZn alloys have been briefly noted in Chap. 8 in the context of dealloying corrosion, but his is not the point interesting presently. In tap water, the corrosion rate of the zinc layer appears to be controlled by an outer layer of $Zn(OH)_2$ which, in turn, dissolves with rate determining diffusion into the tap water [6]. Under these circumstances, the useful effect of hot-dip galvanizing is not directly obvious, especially because the interior protective layer formed within decades in water apparently does not contain much zinc [7]. Possibly, the main influence of galvanizing is to initially slow down iron rusting by exposing iron only slowly, but in increasing percentage, as corrosion proceeds through the layered FeZn alloys [3].

A second topic of paramount economic importance is atmospheric corrosion of iron and steel, which in Chap. 1 already has been described as oxygen corrosion under thin films of condensed moisture. This latter conjecture is easily confirmed by experimenting with air at controlled relative moisture content, i.e. controlled *relative humidity*, defined as

$$H_{\mathrm{rel}} = \left(p_{\mathrm{H_2O}}/p_{\mathrm{H_2O}}^{(s)}\right) \times 100\% \,, \tag{9.1}$$

where $p_{\mathrm{H_2O}}$ and $p_{\mathrm{H_2O}}^{(s)}$ are water vapour pressure in the atmosphere, and saturation water pressure at the given temperature, respectively. For $H_{\mathrm{rel}} < 100\%$, liquid pure water is unstable with respect to evaporation. Over solutions, including electrolytic solutions, the saturation pressure is reduced: $(p_{\mathrm{H_2O}}^{(s)})_{\mathrm{sln}} < (p_{\mathrm{H_2O}}^{(s)})_{\mathrm{H_2O}}$, and the effect is greatest for a solution saturated with respect to the dissolved substance. This leads to the definition of the *critical relative humidity* which is

$$H_{\mathrm{i,crit}} = \left(p_{\mathrm{i,crit}}/p_{\mathrm{H_2O}}^{(s)}\right) \times 100\% \,, \tag{9.2}$$

where $p_{\mathrm{i,crit}}$ is the equilibrium water pressure over a solution saturated, at a given temperature, with respect to a substance labeled $i$. Values of saturation pressures are tabulated, and values of $H_{\mathrm{i,crit}}$ are thus available, provided that the nature of $i$ is known. Consider, then, a metal surface exposed to air at ambient temperature, and let the relative humidity be $H_{\mathrm{rel}} < 100\%$. Assume contamination of the metal surface by a well defined salt species, $i$, for which $H_{\mathrm{i,crit}}$ is known. When $H_{\mathrm{rel}} < H_{\mathrm{i,crit}}$, the contamination will remain dry, no condensed salt solution will form on the metal surface and no corrosion will occur. When, instead, $H_{\mathrm{rel}} \geq H_{\mathrm{i,crit}}$, the salt is hygroscopic and will attract water, forming an aqueous salt solution, normally causing corrosion. When the metal is unalloyed steel, rusting will be seen to readily set in at the moistened surface.

Table 9.1 shows a series of experiments with various salt contaminations on steel exposed to constant-temperature air at different controlled values of the relative humidity [9]. As expected, rusting starts only if the contamination forms a condensed solution. $NaNO_2$ is a notable exception: In this case, salt solution formation is readily observed, but the substance is a passivating inhibitor (see Chap. 10).

**Table 9.1.** The influence of hygroscopic contaminants on the onset of atmospheric corrosion of mild steel in air at 20°C. (From Buckowiecki)

| Salt | $H_{i,crit}$% | $H_{rel}$% 100 | 90 | 80 | 70 | 60 | 50 |
|---|---|---|---|---|---|---|---|
| $Na_2SO_4 \cdot 10H_2O$ | 93 | ● | ○ | ○ | ○ | ○ | ○ |
| KCl | 86 | ● | ● | × | ○ | ○ | ○ |
| NaCl | 78 | ● | ● | ● | × | ○ | ○ |
| $NaNO_3$ | 77 | ● | ● | ● | ○ | ○ | ○ |
| $NaNO_2$ | 66 | + | + | + | + | ○ | ○ |
| $NaBr \cdot 2H_2O$ | 59 | ● | ● | ● | ● | ● | ○ |
| $NaJ \cdot 2H_2O$ | 77 | ● | ● | ● | ● | ● | ● |
| $LiCl\dot{H}_2O$ | 77 | ● | ● | ● | ● | ● | ● |

● : salt wet, steel rusting underneath; ×: salt discoloured, slight rusting; + : salt wet, no rust; ○ : salt dry, no rust

When the presence of a potentially hygroscopic contamination cannot be avoided, moist air can be dried by air-conditioning until salt solution condensation can no longer occur. Instead of drying, air can be heated at constant water vapor pressure, as this reduces the relative humidity. By the same principle, acid moisture condensation from the $CO_2/H_2O$ combustion gases of power plants contaminated by sulfur dioxide, $SO_2$, is prevented by not allowing the gas temperature to drop below a critical value, which is the 'dew point' of the gas.

Safe exposure of iron and steel to the open atmosphere requires the application of protective coatings. These, however are permeated by both water vapor and oxygen, and it therefore is of outstanding importance to apply coatings to precleaned surfaces only. Otherwise, under-rusting of coatings is difficult to prevent. We only briefly note the countermeasures offered by anti-corrosion pigments such as red lead, which, again, probably is a passivating inhibitor.

An often quoted practical rule is that atmospheric corrosion requires relative humidity to exceed ca. 50-60%. Form the data presented in Table 9.1 it is, however, obvious that, depending on the nature of contaminants, this rule occasionally may fail. A contamination easily encountered in sea-side atmospheres is, e.g., $MgCl_2$ with an ambient-temperature critical humidity of only 35%.

So far, the description of the influence of air humidity was concerned only with the presence of condensed moisture films. Obviously, the duration of wetting then is an important factor for the extent of corrosion. Often, as in the case of rain, but not at all always, the consideration is trivial. The point is important for rules of *anti-corrosion constructing*, which is planning metallic structures such that any condensed moisture can easily evaporate. The wetting time does not, however, directly bear on the rate of corrosion during the duration of wet parts. In this context it might appear, on first sight, that a decrease in relative humidity should promote corrosion, as long as the contamination is hygroscopic, because the salt concentration in condensed moisture is increased. Instead, as shown in Fig. 9.2, the rate of corrosion of unalloyed, unprotected

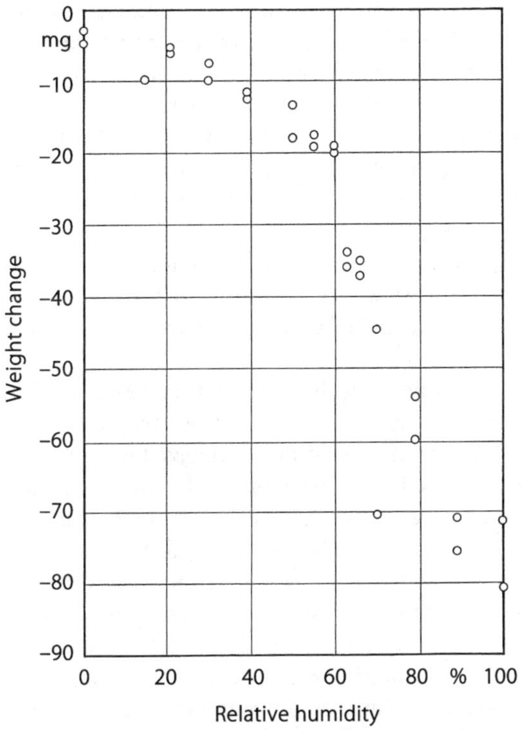


**Fig. 9.2.** Weight loss of 10 $cm^2$-sheet specimens of unalloyed steel contaminated wit $MgCl_2$ during 8 days of exposure to air with controlled relative humidity at 25°C. Data obtained after removal of rust. Note increase of the corrosion rate with increasing relative humidity. (From Evans and Taylor)

steel increases with increasing relative humidity. This has several reasons: a) Corrosion rates are not directly related to salt concentration. b) The concentration of dissolved oxygen increases as the salt concentration decreases. c) When the water vapor pressure in air is $p_{H_2O}$, exceeding the pressure $p_i^{(s)}$ of the saturated solution of $i$, the hygroscopic contamination will produce an aqueous solution of a well defined concentration, $c_i$, with a vapor pressure $p_i = p_{H_2O}$. As $p_{H_2O}$ approaches $p_{H_2O}^{(s)}$, $c_i$ tends to zero, and the volume of condensed moisture tends to infinity. Thus, the amount of condensed moisture increases with increasing relative humidity, and this may well enter the rate of corrosion by increasing the thoroughly wetted part of the steel surface. We note that the observed behaviour requires that even in contact with practically pure water the air-formed oxide film on the specimens was destroyed.

A striking example of the influence of relative humidity is *filiform corrosion*. This is the under-coating spreading of threads of corrosion products, such as threads of rust on unalloyed steel. Experiments with particles of hygroscopic salts under a thin film of lacquer have shown that at the head of rust threads a liquid drop of salt solution of constant volume is moving on the metal surface locally causing corrosion [4b]. The volume is fixed by the nature and the amount of dissolved salt together with the relative humidity of the surrounding atmosphere. Water and oxygen permeate the lacquer film. The mechanism is interesting as well: At the bottom of the drop, iron dissolves to $Fe^{2+}$; it then is oxidized to $Fe^{3+}$ at the ferric oxide membrane separating the drop and the rust thread, which grows by further ferric oxide deposition and pushes the drop ahead. Oxygen reduction occurs at both the the bottom of the drop and at the membrane. Thread growth at constant rate, without loss of volume-determining salt in the drop, has been observed during weeks. On the whole, the course of events again is a variant of the mechanism of an Evans differential aeration cells.

Atmospheric corrosion is greatly influenced by air pollution, especially by $SO_2$ [1,8,11]. Apparently, the substance is not, as previously presumed [12], directly acting as an oxidant. Instead, oxidation to $SO_3$, catalyzed by iron, and condensation to aqueous sulfuric acid solution probably is decisive [13]. As a result, $FeSO_4$ is present in the rust layer. This, however, is very detrimental because the salt is hygroscopic and, by further reaction of ferrous to ferric cations in the moisture thus produced hydrolizes, producing local acidity. Once formed, the salt therefore catalysis further corrosive attack, and this explains the observation that per $SO_2$ molecule between 15 and 150 atoms of Fe are turned into rust [14]. Acid moisture constantly produced in the rust layer will tend to keep the rust layer porous and, thus, to prevent the formation of a dense, protective coating. In this context, Fig. 9.3 shows the results of corrosion testing unalloyed and unprotected sheet steel in routine corrosion tests modeling attack in clean, and also in $SO_2$-bearing wet air [4b]: In clean air corrosion decays owing to a thickening dense layer of rust, while in contaminated air corrosion proceeds at constant rate apparently due to the formation of porous rust. It should be noted that routine corrosion testing in wet air with substantial additions of $SO_2$ may severely exaggerate conditions. In practice, there is no overall wetting of metal surfaces in industrial atmospheres by dilute sulfuric acid, instead, sulfates appear to be nested locally [15].

Important details concerning the role of rust in the mechanism of atmospheric corrosion have been revealed by admirably simple experiments [1a]: It is easily shown that freshly prepared ferric oxide, metallic iron and a slightly acidic sulfate solution form a galvanic cell with iron as the anode and the oxide as the cathode. The short-circuited cell produces a current which anodically dissolves iron and cathodically reduces ferric oxide. The process can be cycled by reexposing the reduced oxide to air causing reoxidation, and so forth. From this observation, a mechanism involving catalysis of rusting by rust is suggested as described in Fig. 9.4: In contact with metallic iron, the innermost oxide should be magnetite, $Fe_3O_4$, as predicted by thermodynamics. The layer may be very thin, and it is assumed porous, with the pores filling with moisture acidified by hydrolysis of $Fe^{2+}$ and/or $Fe^{3+}$, whenever atmospheric humidity is sufficiently high. Iron dissolves as $Fe^{2+}$ into the pores, and further reacts to more $Fe_3O_4$. A

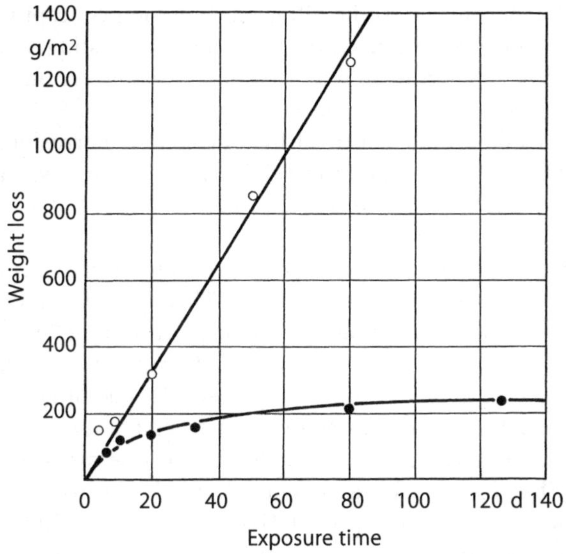


**Fig. 9.3.** Specific weight loss of deep-drawn sheet steel in an industrial-air test chamber. Testing was by water-saturated air alternating between 20 and 40°C. Solid circles: Clean air, essentially modeling tropical conditions. Open circles: Air contaminated with $SO_2$, modeling industrial atmospheres. Data were obtained after removal of rust. (From Kaesche)

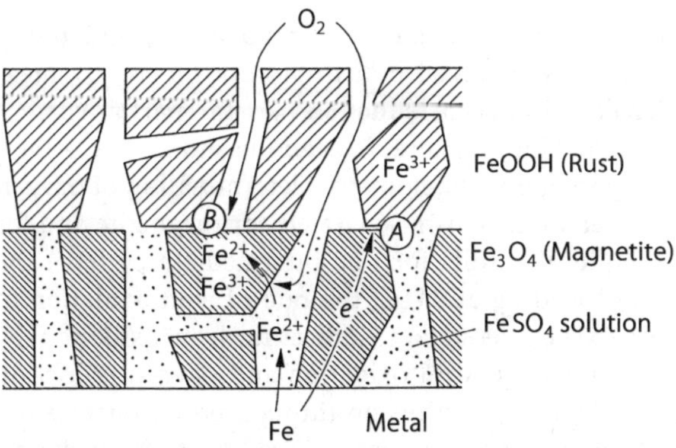


**Fig. 9.4.** Evan's model of the mechanism of atmospheric corrosion of iron in the presence of $SO_2$

porous layer of FeOOH covers the magnetite layer. Cathodic oxygen reduction proceeds by transfer of electrons from the metallic phase through the semiconducting magnetite to the $Fe_3O_4$/FeOOH-interface marked 'A', reducing FeOOH to $Fe_3O_4$. Elsewhere, e.g. at 'B', oxygen oxydizes $Fe_3O_4$ to FeOOH. According to this mechanism, diffusion of dissolved oxygen to the iron surface through the liquid filling pores in the oxide layer is not necessary and oxygen corrosion is not diffusion-limited. Since, however, the diffusion-limitation concept is well confirmed by experimental evidence concerning early stages of rusting, the mechanism described presently apparently applies for thicker rust layers only. Also, long aging may result in loss of reactivity of rust layers.

Reduction of $\gamma$-FeOOH (lepidokrokite), formed by atmospheric corrosion, to magnetite has been observed by other authors as well [16,17]. However, when produced by anodic electrolysis of iron, the same substance is further reduced to dissolved $Fe^{2+}$ ions [16-18].

In more recent work, rust layers were produced by exposing an iron film deposited on a gold substrate to $SO_2$-containing air [19]. This layer was then reduced and reoxidized by external polarization, continuously recording ferromagnetism as an indication of the presence of magnetite. The result was that, initially, the rust layer is $\gamma$-FeOOH (lepidokrokite) and $\alpha$-FeOOH (goethite). Of these modifications, only $\gamma$-FeOOH can be reduced to an intermediate substance FeOHOH, which in turn is easily reoxidized to lepidokrokite, or else dissolves, in acid solutions, to $Fe^{2+}$. Electron transfer, as important for the Evans cycle described above, is attributed to the reaction: $\gamma\text{-FeOOH} + H^+ + e^- \rightarrow \text{FeOHOH}$, which is thought to be reversible. If the potential is made more negative, the intermediate substance reacts, in pores of the rust layer where pH increases, to magnetite: $\text{FeOHOH} + 2\gamma\text{-FeOOH} \rightarrow Fe_3O_4 + 2H_2O$. Reoxidation of magnetite in alkaline solution again leads to $\gamma$-$Fe_2O_3$, also contributing to electron transfer across the rust layer. The rather ill defined intermediate FeOHOH, incidentally, may well really be 'green rust' which is a an $Fe_{II}Fe_{III}$ oxide containing chloride and/or sulfate [20].

Still more refined information was obtained, for cyclic moistening and drying of rust films on an iron substrate in artificial industrial atmosphere, by measuring the partial current density of oxygen reduction, $i_{O_2}$, the partial current density of iron dissolution into the film moisture, $i_{Fe}$, and the sum $i = i_{Fe} - |i_{O_2}|$. This was achieved by combining weighing, manometric determination of oxygen consumption, and recording of ferromagnetism separately for magnetite and for iron. Figure 9.5 gives an example of results for one intermediate cycle of the procedure: During the period of wetting at 24° C, the rate of iron oxidation at first greatly surpasses that of oxygen reduction, indicating iron oxidation by ferric oxide reduction. In a second stage, iron oxidation and oxygen reduction are equivalent, so that, apparently, oxygen diffusion through pore moisture determines the reaction rate. If, now, wetting is discontinued by heating the air at constant water vapor pressure to 40° C, the rates of both iron oxidation and oxygen reduction surge, probably owing to the increase of the diffusion rate of oxygen caused by increasing evaporation of moisture from the pores in the rust layer. Eventually, the cycle ends with a negligible corrosion rate when the rust layer is dry. Metal loss during corrosion apparently is greatest during the initial period of drying rather than during wetting.

*Weathering steels* have been briefly noted in Chap. 5 as essentially very low-alloy steels containing small amounts of components such as copper or chromium [21]. They are known to develop, though usually only during years

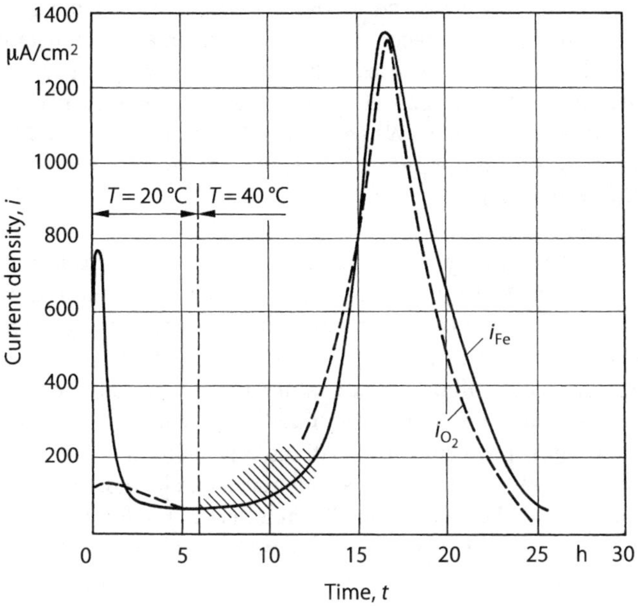


**Fig. 9.5.** Corrosion of iron, precorroded in humid air with additions of $SO_2$, subsequently cycled in pure humid air without sulfur dioxide. Cycling was moistening during 6 h followed by drying during 24 h. The figure shows, for the 4th such cycle, the typical change with time of the partial current densities of anodic iron dissolution into pore moisture and of cathodic oxygen reduction, $i_{Fe}$, and $i_{O_2}$, respectively. The hatched region indicates scattered values of $i_{O_2}$. (From Stratmann, Bohnenkamp, and Ramchandran)

of alternating wetting and drying in atmospheres of only moderate agressivity, dense and protective rust layers. In an attempt to find the origin of this useful behaviour, experiments as described above were continued, replacing pure iron by Fe-0.5wt%Cu. The test material behaved similar to pure iron during the periods of wetting in water-saturated air. In contrast to pure iron the surge in corrosion rate caused by the onset of drying did not appear. Whether this does explain the merits of weathering steels remains to be seen.

Any closer study of electrode kinetics requires current-potential measurements, but for atmospheric corrosion under very thin moisture films it is difficult to set up the three-electrode experiments. A possible alternative is to turn from measuring electrode potentials, $\varepsilon$, to measuring Volta potentials, as introduced in Sect. 3.1. As a useful example, consider iron in metallic contact with gold. The Volta potential, $\Delta\psi = \psi_x - \psi_y$, is the difference of Galvani potentials in air between points directly adjacent to the iron surface and to the gold surface, respectively. It is equal to the sum, $\phi_{air,Au}$, $\phi_{Au,Fe}$, and $\phi_{Fe,air}$, of three Galvani potential differences at the interfaces between air and gold, gold and iron, and iron and air, respectively. Also, it is a measurable quantity and can be determined, among other possible methods, by using *Kelvin probes* [22]. For this purpose, a capacitor with an iron plate and a gold plate is used, and the gold plate is vibrated, thus changing the capacitance of the arrangement at constant Volta potential between the plates. This causes a charging/discharging capacitive current flow. The current can be compensated by an external voltage source, and the Volta potential thus determined. When the iron electrode is covered by a conducting moisture film, the component $\phi_{Fe,air}$ splits into to $\phi_{Fe,l} + \phi_{l,air}$. In this sum, the second term is probably of the order of ca. 0.1$V$, and it should be constant for a given set of conditions. Then, if the iron electrode is polarized by currents, $\phi_{Fe,l}$ and also $\Delta\psi$ change, and the changes in the two properties should be equal. Therefore, measuring functions $\Delta\psi(i)$ essentially amounts to measuring current-potential curves $\varepsilon(i)$, albeit with a different reference voltage.

The method has been applied to measuring passivation current densities of iron under very thin moisture films [19c,d]. More recently, it has become a refined tool to obtain information on the delamination of coatings and thus on the durability of corrosion protection by coatings especially on mild steel. Delamination is serious if, starting from an accidental scratch, moisture can creep between the coating and the coated metal. The effect is readily recorded by scanning the surrounding of such scratches with a miniaturized version of the Kelvin probe, because $\Delta\psi$ changes owing to the appearance of a component $\phi_{Fe,l}$ which is absent as long as the coating tightly adheres to the metal surface. By now, this line of investigations is continued especially concerning the strength of coating adhesion, as depending on the molecular structure of the metal/coating interface. The subject is beyond the scope of this book, but well referenced elsewhere [19e].

## References


1. a) Evans, U. R.: Inst. Metal Finish *37* (1960) 1; b) id.; Corr. Sci. *9* (1969) 813; c) Evans, U.R. and Taylor, C.A.: ibid. *12* (1973) 227; d) Bohnenkamp, K.: Archive Eisenhüttenwesen *47* (1976) 751; e) Forker, W., Worch, H., Rahner, D., and Schatt, W.; Sitzungsber. Akad. Wiss. DDR 24 N, 1982; Worch, H., Forker, W., and Rahner, D.: Werkstoffe u. Korr. *34* (1983) 402
2. See for instance Klas, H., and Steinrath, H.: Die Korrosion des Eisens und ihre Verhütung. 2nd ed. Verlag Stahleisen, Düsseldorf 1974
3. a) Schwenk, W.: Werkstoffe u. Korr. *30* (1979) 34, *31* (1980) 611; b) id.; private communication, 1997
4. a) Kaesche, H.: Werkstoffe u. Korr. *10* (1959) 668b); b) *15* (1964) 379; c) *26* (1975) 175
5. See for instance: Proc. Conf. Korrosion in Kalt- und Warmwassersystemen der Hausinstallationen, Bad Nauheim 1974. Deutsche Ges. Metallkunde, Oberursel, 1975
6. Werner, G., Wurster, E., and Sontheimer, H.: gwf, Wasser, Abwasser *114* (1973) 105; Kruse, O.L.: Werkstoffe u. Korr. *26* (1975) 454; Schwenk, W.: ibid. *27* (1976) 157
7. Friehe, W., and Schwenk, W.: Heizung, Lueftung, Haustechnik *24* (1973) 13

8. See for instance Barton, K.: Schutz gegen atmophärische Korrosion. Weinheim, Verlag Chemie 1973
9. Buckowiecki, A.: Schweizer Archiv Angew. Wiss. u. Technik *23* (1957) 97
10. Evans, U.R. and Taylor, C.A.: British Corr. J. *1974*, 26
11. Bohnenkamp, K.: Tribune du CEBEDEAU Nr.324 (1970) 1
12. Rosenfeld, I.L.: in 1st Int. Conf. Metallic Corrosion. Butterworths, London 1962, p. 243
13. Vannerberg, N.-G. and Syderberger, T.: Corr. sci. *10* (1970) 43
14. Schikorr, G.: Werkstoffe u. Korr. *14* (1963) 69
15. Schwarz, H.: Werkstoffe u, Korr. *23* (1972) 648, *16* (1965) 93, 210
16. Kunze, E.: Neue Hütte *19* (1974) 295
17. Suzuki, I., Masuko, N., and Hisamatsu, Y.: Corr. sci. *19* (1979) 521
18. Cohen, M. and Hashimoto, K.: J. Electrochem. Soc. *121* (1974) 42 648
19. a) Stratmann, M., Bohnenkamp, K., and Engell, H.-J.: Corr. sci. *23* (1983) 969; Werkstoffe u. Korr. *34* (1983) 604; b) Stratmann, M., Bohnenkamp, K., and Ramchandran, T.: Corr. sci. *27* (1987) 905; c) Stratmann, M.: Corr. sci. *27* (1987) 869; d) Stratmann, M., and Streckel, H.: Ber. Bunsenges. Phys. Chemie *92* (1988) 1244; e) Stratmann, M., Fürbeth, W., Grundmeier, G., Lösch, R., and Reinartz, C.R.: in loc. cit. [5.4], p. 373
20. Ernst, K.: Doctoral thesis, Erlangen 1985
21. a) Misawa, T., Kyuno, T., Suetaka, W., and Shimodaira, S.: Corr. sci.*11* (1971) 35; b) Misawa, T., Asami, K., Hashimoto, K., and Shimodaira, S.: ibid. *14* (1974) 279; c) Schwitter, H., and Böhni, H.: Werkstoffe u. Korr. *1131* (1980) 703; d) Scharnweber, D., Forker, W., Rahner, D., and Worch, H.: Corr. sci. *24* (1984) 67
22. a) Kohlrausch, F.: Praktische Physik, Vol. 2, p. 230, Stuttgart 1968; b) Rath, D.L., and Kolb, D.M.: Surface sci. *109* (1979) 493

# 10 Passivity

## 10.1 Introduction

When iron is immersed in concentrated nitric acid, it is seen to immediately react with the solution, as evident by a burst of evolution of gaseous nitrous oxides, but the reaction stops almost immediately, and henceforth the metal remains bright and uncorroded indefinitely. It is 'passivated' with respect to the thermodynamically strongly favored corrosion reaction, thus showing 'similar-to-noble metal behaviour'. The explanation – already proposed by Faraday – is that the oxidizing power of $HNO_3$ is sufficient to produce an thin, nonporous oxide film which, while easily soluble, happens to dissolve into the acid at only a negligible rate. The film thickness is below the wavelength of visible light and the film, therefore, transparent.

Passivity of iron is also observed in neutral and, especially, in weakly alkaline solutions, again as an effect providing near-perfect corrosion protection. In both neutral and alkaline solutions passivating oxide films usually are thermodynamically stable owing to the low solubility of the oxides at sufficiently high pH, and conditions are, thus, on first sight rather less complicated than for passivity in acid solutions where the film is unstable. In non-acid solutions, the mainly interesting kinetics of passivity are, however, often obscured by side effects of secondary oxide deposition on the passivating film.

Local activation of an iron wire passivated in concentrated nitric acid is followed by an activation wave traveling along the wire, leaving the wire, in the wake of the wave, repassivated and temporarily unfit for renewed activation. This behaviour makes the system a surprisingly perfect inorganic model for excitations waves traveling along nerve fibers in organic, living tissues. It was the interest in inorganic modeling of some effects of biophysical chemistry which which was at the root of work on passivity of metals and, by consequence, on general electrode kinetics, by K.-F Bonhoeffer and coworkers in Göttingen, in contrast to the earlier work of U. R. Evans and the Cambridge school mainly concerned with the practical and, in fact, more mundane aspects of the matter.

We shall, in the first place, introduce basic principles of passivity considering iron in acid solutions as the predominant example. It will later be seen that passivity of stainless chromium-nickel steels as well as of nickel and nickel-base alloys are well understood on the grounds of what is known about passivity from experience obtained from iron. The reason is that in every case the passivating oxide films are nonporous semiconductors. These are what might be termed classical examples of passivity, since at the oxide/solution interface electrode reactions essentially follow the same basic laws than at metal/solution interfaces. The situation is different for metals passivated by insulating oxides, as especially aluminum and its alloys, or titanium and its alloys, passivated by valve oxides with asymmetric conductivity. We shall not, however, hesitate to call these metal/solution systems passive, as, again from a practical point of view, they do show the similar-to-noble behaviour which was the typical observation made for iron in nitric acid. If the passivating oxide is an insulator, a very important additional advantage is often available, and typical again especially for aluminum: Under the action of an external high voltage the insulating oxide can be made to grow to considerable thickness. This, then, is 'anodized' aluminum, a structural material of outstanding importance. Anodizing will be considered in the following only as a method of corrosion protection. We note, however, that, since insulating oxides usually are hard minerals, continued anodizing producing, by 'hard anodizing', oxide films of up to several 10 microns of film thickness results in wear-resistant coatings.

We do not give a precise definition of passivity, which indeed is not available. In the present context we use the term to rather loosely describe the similar-to-noble behaviour of metals under conditions where from thermodynamics rapid corrosion would seem the expected event, when the cause is an invisible, or practically invisible oxide film. The film thickness sometimes is very low, indeed, of the order of only few or even only one molecular oxide layers, and the onset of passivity may, in fact, be the result of monomolecular hydroxide adsorption inducing oxide growth instead of metal dissolution.

The importance of passivity of metals for corrosion protection offered without additional means such as coatings is very considerable. A warning may therefore be directly appropriate: It is that most of the really dangerous incidences of corrosion failure typically occur with passive metals. The reason is that owing to the oxide film the corrosion potential tends to comparatively high values. Therefore, local breakdown of the passivating film brings the base metal in contact with the solution at a correspondingly high value of anodic overpotential for metal dissolution. Frequently, the result then is localized rapid corrosion, such as pitting, or intercrystalline corrosion, or stress corrosion cracking, proceeding at high rates, often excessive in comparison to what is observed for non-passive metals under the same conditions. This will become the topic of several chapters following further below.

## 10.2 Iron in Acid Solutions

For iron in concentrated nitric acid, the cathodic partial reaction of the reduction of oxidizing species is particularly complicated, because the undissociated $HNO_3$ molecule is taking part in the reaction, passing through a number of successive reduction states. It is much easier to study passivity of iron in 'non-oxidizing' acids where, if the solutions are deaerated, the cathodic partial reaction of open-circuit corrosion is hydrogen deposition. In 1N $H_2SO_4$ solution, passivity of iron is not spontaneous but can be achieved by external anodic polarization. The experimental advantage is that passivity sets on only

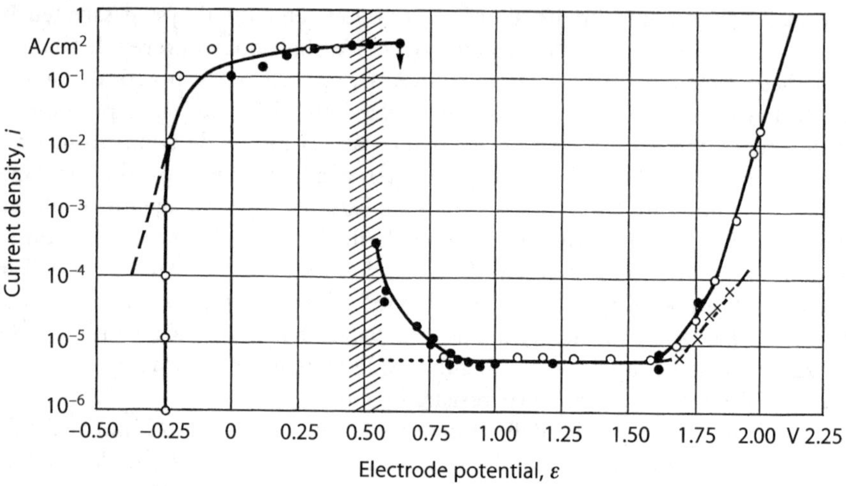


**Fig. 10.1.** Stationary anodic current-potential curve for iron in deaerated 1 N $H_2SO_4$ solution at 25°C. Data from Franck (○), Weil and Franck (×), and Herbsleb and Engell (●)

far beyond the hydrogen equilibrium potential in this solution, and that a wide range of potentials is available for electrochemical studies of passivity kinetics undisturbed by electrode reactions other than anodic iron dissolution.

Figure 10.1 shows the typical stationary current-potential curve [1–3]: Starting from the open-circuit corrosion potential, $\varepsilon_{corr}$, of acid corrosion, the current density of metal dissolution increases exponentially with the potential. The broken line indicates the partial current-potential curve of anodic iron dissolution, which we note as $i_{Fe}(\varepsilon)$, dropping the '$\rightharpoonup$' superscript. Beyond −0.2 V, the measured total current, $i$, is equal to $i_{Fe}$. The current density is limited by a plateau at ca. 200 to 300 mA cm$^{-2}$. This limiting current increases with increasing rate of solution stirring. Beyond a threshold value at ca. 0.5 V, the current density of iron dissolution then suddenly drops by orders of magnitude, and this marks the onset of passivity. Originally, the dotted curve was assumed to represent the curve. Also originally, the drop in current density of iron dissolution was often accompanied by oscillations in the shaded current-potential range. This effect was, however, a property of the electrode/potentiostat system as a whole rather than of the electrode alone, caused by the current-potential curve – essentially corresponding to the characteristic of a tunnel diode – in the active/passive transition range, where the differential polarization resistance (see Chap. 4) is negative. An experimental improvement was the observation of a steady decrease in current density as the potential changes from 0.5 to 0.8 V. Improved measuring equipment revealed that, as indicated in Fig. 10.2, the current density of metal dissolution continuously decreases between a *passivation potential,* $\varepsilon_{p1}$, and a *activation potential,* $\varepsilon_{p2}$, [1c,1e,3-6]. We note that for iron in 1 N $H_2SO_4$, but much less distinct in less acid solutions, the drop of the current density

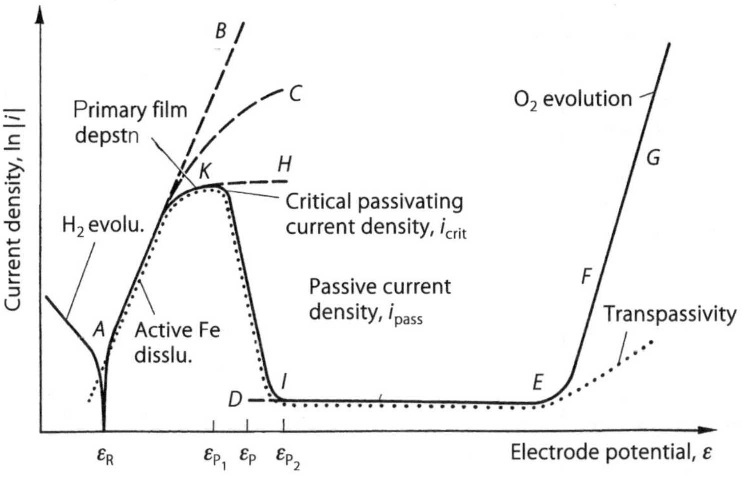


**Fig. 10.2.** Schematic current-potential curve for an electrode undergoing open-circuit acid corrosion, passivated by an semiconducting oxide film formed at the passivation potential $\varepsilon_p$, also showing anodic oxygen evolution beyond the equilibrium potential $E_{O_2}$ accompanied by transpassive increase in metal dissolution rate. Solid curve: Total current-potential curve. Dotted: Partial current-potential curve for metal dissolution. For curves AB, AC, and AH see the text

of metal dissolution at $\varepsilon_{p1}$ is, in fact almost abrupt. Imprecisely, we shall use the term 'passivation potential' sometimes for an average intermediate value $\varepsilon_p$ between $\varepsilon_{p1}$ and $\varepsilon_{p2}$. We also note, that $\varepsilon_{p2}$ is the *Flade potential*, observed as a temporary arrest of the potential, when an electrode passivated previously by external polarization is left to activate spontaneously on opening the external polarizing circuit.

The generally characteristic current-potential curve schematically shown in Fig. 10.2 is explained as follows: From the corrosion potential the current density of iron dissolution increases with increasing potential. When charge-transfer kinetics are rate-controlling, the increase is expected to be proportional to $\exp(\alpha\, n_e F\varepsilon/RT)$, until $\varepsilon = \varepsilon_p$, producing a prepassivation current-potential curve as AB in Fig. 10.2. In the case of low-pH acid corrosion of iron, the current density of metal dissolution close to $\varepsilon_p$ thus would exceed ca. 50 A cm$^{-2}$, and as a result the observed curve would be distorted into AC by an ohmic component of the measured potential. In reality, the current density of metal dissolution is limited to a much lower value owing to the precipitation of a layer of ferrous sulfate on the iron surface, changing the current-potential curve into AH. The limiting current density brings the electrode to the passivation potential and therefore is the *critical passivating current density*, $i_{crit}$. Correspondingly, the current density of metal dissolution in the passive state or, more briefly, the *passive current density* is denoted as $i_{pass}$.

As a typical example of what is termed 'galvanostatic charging curve', and well suited to show essential kinetic events, Fig. 10.3 shows the transient change

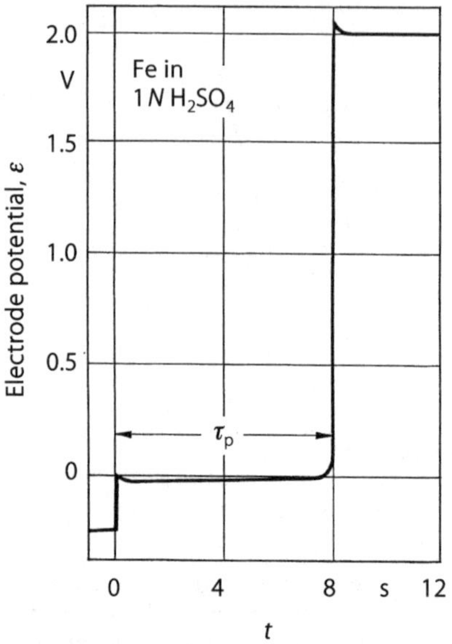


**Fig. 10.3.** Galvanostatic anodic charging curve for iron in 1 N $H_2SO_4$ solution at 25°C showing the potential transient, observed at 1.75 A $cm^{-2}$, from the open-circuit corrosion potential, $\varepsilon_{\text{corr}} \simeq -0.2\ V$ to the potential $\varepsilon_{p1} \simeq 0.0\ V$ of sulfate precipitation, and, after the passivation time, $\tau_p = 8$s, to the potential of rapid oxygen evolution at 2.0 V. (Data from Franck)

in potential for iron in sulfuric acid observed after switching on a polarizing current $i$ distinctly exceeding the critical passivating current density [1d]. Eventually, the electrode is forced into the region of rapid anodic oxygen evolution, but only after a well defined 'passivation time', $\tau_p$. During this period, the sulfate film is precipitated onto the iron surface at a potential which remains constant until all pores in the layer are blocked by salt deposition, and then forcing the potential to jump to oxygen evolution at the passivated surface from which the salt layer has disappeared by dissolution into the acid after the onset of passivation. Assuming a salt dissolution rate corresponding to the current density $i_{\text{crit}}$, the amount of salt deposited after a time $t$ is proportional to $(i - i_{\text{crit}})t$. Since the passivating amount of salt precipitated on the metal surface should not depend on $i$, a relation $(i - i_{\text{crit}})\tau_p = \text{const}$ should hold and is, in fact, confirmed by experimental evidence [1d].

A complication arises, when by vigorous stirring the passivating current density, $i_{\text{crit}}$, is further increased, causing an increase in the ohmic distortion of the current-potential curve [8–10]. In this case, the apparent electrode potential measured at the active/passive transition is more positive than the true electrode potential by the ohmic voltage between metal surface and the tip of the Haber-Luggin capillary. Since this voltage disappears when the current drops to the passive current density, the electrode potential returns through a Z-shaped curve back to a more negative value. For instrumentation suitable to control conditions of this type, see [8].

If passivity is caused by a protective, though invisible, oxide film of composition $Fe_nO_m$, the intuitive suggestion would be that the passivation potential, $\varepsilon_p$, should be equal to the equilibrium potential of oxide formation. If so, a relation

$$\varepsilon_p = E^\circ_{Fe/Fe_nO_m} + (RT/F) \ln a_{H^+} \tag{10.1}$$

should hold (see Chap. 3). But experimental evidence is that, instead [1],

$$\varepsilon_p = 0.58 - 0.059 \text{ pH}, \quad \textit{for} \quad 0 \le \text{pH} \le 6, \tag{10.2}$$

and comparison with data listed in Chap. 3 shows that this is ca. 0.6 V more positive than any equilibrium potential $E_{Fe/Fe_nO_m}$. To answer this problem, the alternative supposition is that initially magnetite, $Fe_3O_4$, is formed in pores of the primary salt film, as it should, according to thermodynamics, at $E_{Fe/Fe_3O_4}$, but that magnetite is non-passivating as it is known to rapidly dissolve in acids. It is also known that an unstable modification of ferric oxide, cubic maghemite, $\gamma$-$Fe_2O_3$, while also easily soluble, dissolves in acids only very slowly, except by galvanic reductive dissolution [11–13a,14]. Therefore, we tentatively assume that for acid solutions the passivation potential marks the onset of oxidation of $Fe_3O_4$ to $Fe_2O_3$. If the ferric oxide were thermodynamically stable, rhomboedric hematite, $\alpha$-$Fe_2O_3$, this would make $\varepsilon_p = 0.20-0.059$ pH, which is closer but not very close to the experimental value for $\varepsilon_p$. Also, hematite again dissolves rapidly in acids. Cubic maghemite, as the preferred choice owing to its slow dissolution in acids, is a highly disordered magnetite, from whose crystal lattice the $Fe^{2+}$ cations have been removed, so the transition from magnetite to maghemite is easily possible, and passivation by the magnetite/maghemite transition is an attractive supposition, indeed. Unfortunately, the Gibbs standard energy of formation, $\Delta G^\circ_{\gamma\text{-}Fe_3O_4}$, is not known. In order to satisfy Eq. (10.2), it should be $-718.2$ kJ mol$^{-1}$, which is quite plausible since (see Table 3.1) for the stable hematite the value is $-742$ kJ mol$^{-1}$ [14].

According to this model, the passive oxide film on iron in acid solutions is a 'sandwich' with $\gamma$-$Fe_2O_3$ on $Fe_3O_4$. Also, this is not mere speculation, since the two oxides have been identified during anodic passivating polarization using the method of intermittent galvanostatic charging. Starting from the open-circuit corrosion potential of acid corrsion and, thus, starting with a bare iron surface, an anodic current of a density exceeding $i_{crit}$ is applied to the electrode and interrupted for short intervals. During intervals, the current and, therefore, the overpotential of the reaction occurring at the respective time is zero, so that the electrode potential drops to the equilibrium potential of the respective electrode reaction. This is shown in Fig. 10.4 for iron in weakly acid solution, and the equilibrium potentials for the $Fe/Fe_3O_4$ and the $Fe_3O_4/\gamma$-$Fe_2O_3$ reactions are seen as successive shoulders in the charging curve [14c]. As the change from the magnetite to maghemite lattice is continuous, the two species form homogeneous solutions with a continuously changing ratio of activities, $a_{\gamma\text{-}Fe_2O_3}/a_{Fe_3O_4}$.

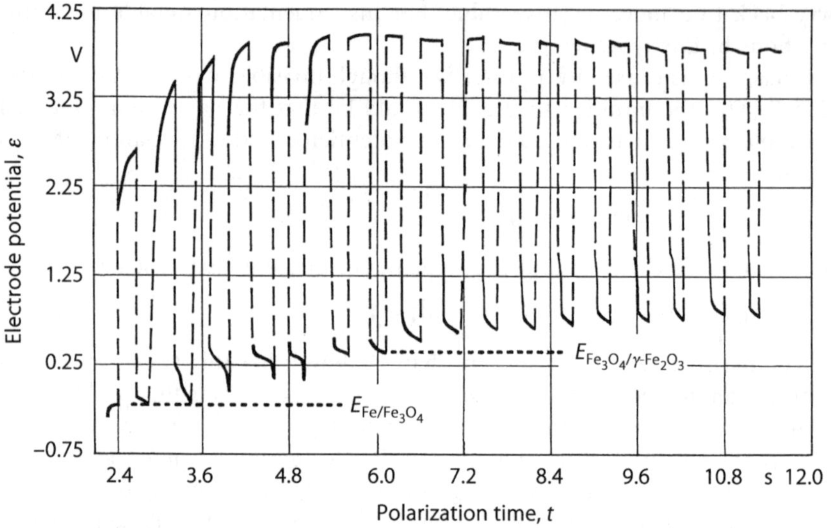


**Fig. 10.4.** Intermittent anodic charging curve for iron in 0.07 M $KNO_3$/0.001 N $H_2SO_4$ solution, pH 2.3, 20°C, obtained with i = 300 mA cm$^{-2}$. (Data from Lange and Weidinger)

The equation of the equilibrium potential for the oxide oxydation, therefore, has to account for this continuous change in degree of oxidation. The resulting relation, which should represent the passivation potential, then reads

$$\begin{aligned}\varepsilon_p &= E_{Fe_3O_4/\gamma\text{-}Fe_2O_3} \\ &= E^{\circ}_{Fe_3O_4/\gamma\text{-}Fe_2O_3} + \frac{RT}{2F} \ln \frac{a^3_{\gamma\text{-}Fe_2O_3}}{a^2_{Fe_3O_4}} + \frac{RT}{F} \ln a_{H^+} \,. \end{aligned} \tag{10.3}$$

This also rationalizes the transition from $\varepsilon_{p1}$ to $\varepsilon_{p2}$ already assumed in Fig. 10.2. We shall see further below that more fine-tuning of the model is required to account for evidence as the thickness and optical properties of the film. The main point in explaining passivity of iron is, however, contained in Eq. (10.3).

For iron in acid solutions, $i_p$ remains constant and small up to the equilibrium potential of anodic oxygen evolution. Beyond $E_{H_2O/O_2,H^+}$ or, more briefly, '$E_{O_2}$', oxygen evolution sets in, with anodic current densities, $i_{O_2}$, increasing exponentially with the electrode potential. This proves electron transport across the film and regular electrode kinetics at the oxide film/solution interface. Therefore, the film certainly is a semiconductor. Oxygen evolution kinetics now involve, instead of the total Galvani potential difference $\phi_{m,l}$ between metal and solution, the Galvani potential difference $\phi_{ox,l}$ only, where $\phi_{ox}$ is the potential in the oxide facing the solution. Concerning kinetics at semiconducting oxide/solution interfaces, more will follow further below. Here, suffice it to note that, as indicated in Figs. 10.1 and 10.2, the rate of anodic iron dissolution also increases, parallel to the onset of oxygen evolution. This effect is designated as *transpassity*.

Up to the passivation potential, iron dissolves to $Fe^{2+}$. Beyond, metal migration through the passivating oxide film and dissolution into the solution is to both $Fe^{2+}$ and $Fe^{3+}$. Corresponding to the increase in degree of oxidation, the fraction of migrating $Fe^{3+}$ increases with increasing potential. Beyond the Flade potential the dissolution into the solution is to $Fe^{3+}$only. At least beyond the Flade potential, the semiconducting oxide film must be nonporous, since an active pore bottom in contact with the surrounding oxide would be a short-circuited galvanic cell, and the cell current would either repassivate the pore bottom, or else, if the bottom were large, cause complete activation of the system [15a,16a].

As a nonporous film, the oxide film is a *barrier layer* with properties distinctly different from those of the precursing salt layer, and also distinctly different from those of the nonprotective rust layers described in Chap. 2. The passive film is not formed by metal dissolution and following oxide redeposition, since this sequence of events, typical for the formation of nonprotective rust layers, is impossible as long as the acid is not saturated with respect to iron oxides or hydroxides. Instead, the mechanism of barrier film formation must be similar to the mechanism of direct growth of high-temperature oxide films, and this makes passivation an event of *low-temperature scaling.*

The thickness, $d$, of passive films on iron in acid solutions has early been estimated from polarization data to be ca. 5 nm [15,16]. It can be determined using *ellipsometry*, i.e. by measuring the change in polarization of monochromatic light reflected in-situ from the passive surface [17,18]. The evaluation is somewhat unsafe, because optical properties of thin films probably differ from those of bulk oxides, but the error can be reasonably well corrected by comparison with polarization data. Values thus obtained in 1 N $H_2SO_4$ solution increase from 3.0 nm at 1.08 V to 4.0 nm at 1.45 V and 3.49 nm at 1.53 V. In 0.01 N $H_2SO_4$/0.5 M $Na_2SO_4$ solution, $d$ was $3.0 \pm 0.15$ nm at 0.90 V and $4.25 \pm 0.25$ nm at 1.40 V [18c]. As very clearly shown in Fig. 10.5, the film thickness increases linearly with time on galvanostatic polarization with currents $i > i_{pass}$ exceeding the stationary passive current density. In steady state, $i_{pass}$ is found to be 7 $\mu$A cm$^{-2}$ regardless of the potential, and $d$ is constant at any given potential. Apparently, the oxide film becomes very thin as the potential approaches $\varepsilon_{p2}$. If the film is crystalline, we also expect that on approaching $\varepsilon_{p2}$ the concept of a film of uniform thickness may become incorrect owing to important differences in the size of the last layer of crystallites. The effect may contribute to the lack in sharpness of the active/passsive transition [16c].

Even considering a continuous change from magnetite to maghemite, it is difficult to reconcile the notion of a distinctly sandwiched oxide film with values of $d$ as low as 3 nm or less. Since the unit cell constant of magnetite is only 0.8 nm, it appears more reasonable to describe the film as being disordered magnetite, $Fe_{(3-\Delta)}O_4$, with an iron deficit, $\Delta$, varying from $\Delta = 0$ at the iron/oxide interface, where the oxide is magnetite, to $\Delta = 0.33$ at the oxide/solution interface, where it is maghemite [19]. We then arrive at a description of the electrode kinetics of iron, passivated by potentiostatic polarization in acid solutions, as

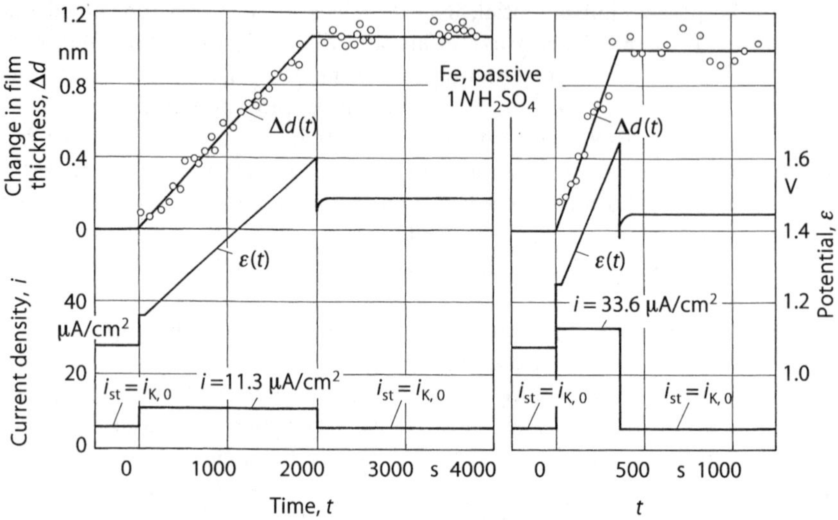


**Fig. 10.5.** Experimenting with passive iron in 1 N $H_2SO_4$ solution at 25°C: Response of electrode potential, $\varepsilon$, and of film thickness (open circles), $d$, with time to galvanostatic anodic polarization in the range of passivity beyond the Flade potential, $\varepsilon_{p2}$. Current density stepped at $t = 0$ from stationary passive current density, $i_{\text{pass}} = 7 \mu A\ cm^{-2}$, to 11.3 $\mu A\ cm^{-2}$ (*left*), resp. 33.6 $\mu A\ cm^{-2}$ (*right*), and stepped back to $i_{\text{pass}}$ after 2000, or 400 s, respectively. Values of $d$ obtained by ellipsometry. (From Vetter and Gorn)

schematically shown in Fig. 10.6. In what follows, Vetters arguments, published in [20a-c], are simplified; for more comprehensive reviews see [20d,6b]. To simplify notation, we write $FeO_n$ for any iron oxide. Thus, we replace $Fe_3O_4$ by $FeO_{4/3}$ and $Fe_2O_3$ by $FeO_{3/2}$, also allowing for any intermediate composition, $FeO_n$, with $4/3 \leq n \leq 2/3$. Further subscripts are '0' and '$d$', indicating positions in the oxide at the metal/oxide interface and the oxide/solution interface, respectively. In the figure, equilibrium of the passage of $Fe^{2+}$,$Fe^{3+}$, and $e^-$ across the metal/oxide interface is assumed. The assumption is not trivial; it has come to be preferred to an alternative assumption of rate-determining transfer reactions across this interface, recently again suggested in [20h], only on account of the success of further reasoning concerning passivity of iron in acid solutions, as described presently. By this equilibrium, the electrochemical potentials of metallic iron, iron ions with average valence $z = 2n_0$, and electrons, respectively, obey the relations

$$a)\ (\mu_{\text{Fe}})_{\text{Fe}} = (\mu_{\text{Fe}})_0$$

$$b)\ (\tilde{\mu}_{\text{Fe}^{z+}})_0 = (\mu_{\text{Fe}^{z+}})_0 + zF\phi_0 \quad c)\ (\tilde{\mu}_{\text{e}^-})_0 = (\mu_{\text{e}^-})_0 - F\phi_0\,. \tag{10.4}$$

Cations $Fe^{2+}$ and $Fe^{3+}$ migrate through the oxide with a rate equivalent, on the whole, to $i_{\text{pass}}$. The $Fe^{2+}/Fe^{3+}$ activity ratio decreases from left to right owing to the change in composition of the oxide as described above. When the electrode

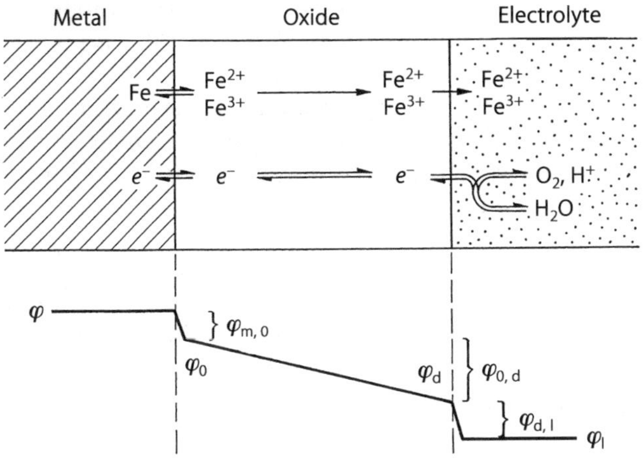


**Fig. 10.6.** Schematic phase system for a semiconducting passivating oxide film on iron, in contact with acid electrolytic solutions for the case that the oxide is soluble, but oxide solution is slow, with a rate equivalent to $i_{\text{pass}}$. Double arrows indicate reactions in equilibrium, single arrows reactions out of equilibrium. For indices see the text

potential is more positive than the Flade potential, $\varepsilon_{p2}$, the concentration and activity of $Fe^{2+}$ in the oxide adjacent to the solution is virtually zero.

As to migration, we shall see presently that in the oxide film the driving force is the electric field, $d\phi/dx$, is rather than the concentration gradient, $dc_{\text{Fe}^{z+}}/dx$. In any case, there is a chemical potential gradient, $d\tilde{\mu}_{\text{Fe}^{z+}}/dx$, so there is, across the film, non-equilibrium with respect to iron ions.

Everywhere in the semiconducting oxide, the reaction $Fe^{2+} = Fe^{3+} + e^-$ is assumed to be in local equilibrium, determining the local activity, and thus the chemical potential, $(\mu_e^-)_{ox}$, of electrons. At the same time, the mobility of electrons (or holes) in the semiconducting oxide certainly is much higher than that of ions. Therefore, it makes sense to assume electronic equilibrium throughout the oxide film regardless of the simultaneous irreversible migration of iron ions. By this assumption, the electrochemical potential, $\tilde{\mu_{\text{e}^-}} = \mu_{\text{e}^-} - F\phi$ of electrons is the same everywhere in the oxide film, and equal to that in the metal. Concerning the film we see that, owing to the local equilibrium between electrons and iron ions, the change in Galvani potential through the oxide film is interrelated with the change in oxide composition.

When the reaction is stationary, the thickness, $d$, of the oxide film is constant. Formally, the overall reaction then is oxide dissolution accompanied by re-formation of oxide. The net rate of oxygen dissolution from the oxide into the solution and of redeposition of oxygen into the oxide is zero, and the net sequence of events more simply is the dissolution of metallic iron to dissolved cations. Also, this is more than an abstract reasoning: The dissolution of ionic compounds in general and of oxides such as iron oxides in particular is, in fact, to be thought of as proceeding, rather than by dissolution of oxide molecules, by the superposition of statistically independent partial currents of cation and anion dissolution [21,22]. We therefore have a passage of iron ions across the oxide film into the solution as the net electrode reaction.

The next step is to consider that, although there is no net oxygen anion transfer through the oxide/solution interface, we have no reason to assume that the oxide/solution interface is blocked with respect to this ionic transfer. Instead, it is legitimate to suppose that the electrode reaction

$$(O^{2-})_d + 2(H^+)_l = (H_2O)_l \tag{10.5}$$

between oxygen anions in the oxide at the oxide/solution interface, hydrogen ions in the solution and water molecules is in equilibrium. From this, an expression for the Galvani potential difference $\phi_{ox,l} = \phi_d - \phi_l$ across the oxide solution interface is available. In the first place, we have

$$(\tilde{\mu}_{O^{2-}})_d + 2(\tilde{\mu}_{H^+})_l = (\mu_{H_2O})_l \, . \tag{10.6}$$

Splitting electrochemical potentials, $\tilde{\mu}_i = \mu_i + zF\phi$, and rearranging terms we arrive at

$$\phi_{ox,l} = (1/2F)\left\{(\mu_{O^{2-}})_d - (\mu_{H_2O})_l + 2(\mu_{H^+})_l\right\} \, . \tag{10.7}$$

At this point we add that, owing to the equilibrium described in Eq. (10.4c), we also know that

$$\phi_{m,0} = (1/F)\left\{(\mu_{e^-})_m - (\mu_{e^-})_0\right\} \, . \tag{10.8}$$

We consider a cell

$$\text{Pt}|\text{Fe}|\text{acid sln. I}(a_{H^+})||\text{acid sln. II } (a_{H^+}{=}1 \text{ mol L}^{-1}),H_2(p_{H_2}{=}1 \text{ bar})|\text{Pt},$$

and we assume it equipped with a potentiostat controlling the cell potential, which is

$$\varepsilon = \phi_{m,0} + \phi_{0,d} + \phi_{d,l} + \phi_{l,\text{Pt}} + \phi_{\text{Pt},m} \, . \tag{10.9}$$

Substituting Eqs. (10.7) and (10.8), Eq. (10.9) changes into

$$\varepsilon = \frac{1}{F}\left\{\frac{1}{2}(\mu_{O^{2-}})_d - (\mu_{e^-})_d\right\}$$
$$+\frac{1}{F}(\mu_{H^+})_l - \frac{1}{2F}(\mu_{H_2O})_l + \phi_{0,d} + \frac{1}{F}\left\{(\mu_{H_2})_g - (\mu_{H^+})_l\right\} \, . \tag{10.10}$$

The sum in the first bracket is related to the chemical potential of molecular oxygen dissolved in the oxide at $d$:

$$\frac{1}{2}(\mu_{O^{2-}})_d - (\mu_{e^-})_d = \frac{1}{4}(\mu_{O_2})_d = \frac{1}{4}\left\{\mu^\circ_{O_2} + RT\ \ln(p_{O_2})_d\right\} \, . \tag{10.11}$$

We remember from Chap. 3 that sums $\Sigma\nu_i\mu_i$ can be replaced by sums of Gibbs energies of formation, $\Sigma\nu_i\Delta G^\circ_i$, and that $\Delta G^\circ_i = 0$ for elements and for $H^+$. Combining Eqs. (10.10) and (10.11) we thus find that

$$\varepsilon = -\frac{\Delta G^{\circ}_{H_2O}}{2F} + \frac{RT}{4F}\ln(p_{O_2})_d + \frac{RT}{F} ln a_{H^+} + \phi_{o,d} \,. \tag{10.12}$$

Except for $\phi_{0,d}$, this is the equation for the equilibrium potential of an oxygen electrode, as determined by the oxygen pressure $(p_{O_2})_d$. An interesting point is that this equation holds regardless of the nature of the metal, which does not need to be iron as long as electronic equilibrium between oxide and metal is established.

The concentration of oxygen dissolved in the oxide, and also $(p_{O_2})_d$, will be very small, indeed, but this does not impede thermodynamic reasoning. For $n_d = n_\circ = 4/3$ the latter pressure simply would be the dissociation pressure of $O_2$ in equilibrium with both iron and magnetite, but this cannot apply to iron in acid solutions. For the non-equilibrium composition $FeO_{n_d}$, the corresponding oxygen pressure is obtained by considering the Gibbs reaction energy for the equilibrium of the reaction

$$\nu\, FeO_n + O_2 = \nu\, FeO_{n+2/\nu} \,. \tag{10.13}$$

The Gibbs reaction enthalpy is calculated either as the infinitesimal change in free enthalpy for a system of finite size, or as the finite change in free energy for a system of infinite size, always referred to the formula reaction. Therefore, we can calculate $\Delta_r G$ by increasing $\nu \to \infty$ and decreasing $1/\nu = dn \to 0$:

$$\Delta_r G = 0 = \lim_{\nu\to\infty, dn\to 0}\left[\nu(\Delta G^{\circ}_{FeO_{n+2/\nu}} - \Delta G^{\circ}_{FeO_n})\right] - RT\ln(p_{O_2})_d \,,$$

$$RT\,\ln(p_{O_2})_d = \frac{d\Delta G^{\circ}_{FeO_n}}{dn} \,. \tag{10.14}$$

In the lower part of Fig. 10.6 a linear decrease of the Galvani potential $\phi(x)$ along $x$ between $x = 0$ and $x = d$ is assumed. This certainly is a simplification; however, when – beyond the Flade potential, and except close to the metal/oxide interface – the oxide mostly is $\gamma$-$Fe_2O_3$, the indicated constant slope $d\phi/dx$ and thus a constant strength of the electric field is probably approximately correct.

Ions migrate through the film by successive jumps across the periodic array of energy barriers imposed by the crystalline structure of the oxide. At ambient temperature, migration by diffusion caused by gradients in concentrations will be negligible compared to migration caused by the electric field. Since the oxide film is very thin, and since the voltage, $\phi_{0,d}$, in the film may easily be of the order of 0.1 V or more, very high field strengths, of the order $10^6$ V cm$^{-1}$ or more, are expected. If so, Ohm's law does no longer hold, and a *high-field mechanism* of ionic migration must be assumed [20]. We simplify the situation by considering migration of $Fe^{3+}$ ions (or the equivalent reverse migration of vacancies) only. According to the high-field mechanism, the current density of ion migration is an exponential function of the field strength and is expressed as

$$i_{Fe^{3+}} = K\exp\left(\frac{3\beta F}{RT} a \frac{\phi_{o,d}}{d}\right) , \tag{10.15}$$

where $0 \leq \beta \leq 1$ is a transfer coefficient, and $a$ is the jump distance between lattice defects used by the migrating species. When, instead, the current density, or else the electrode potential is stepped to more positive values, the oxide grows. At constant values of $(i - i_{\text{pass}})$, which was the case in the experiments producing the data shown in Fig. 10.5, the rate of oxide growth is constant. At constant electrode potential, and thus also constant $\phi_{\circ,d}$, the current density decays corresponding to the decay in $\phi_{0,d}/d$. The integration of the film growth law is difficult, but an approximate solution is the *inverse logarithmic law*: $(i - i_{\text{pass}}) = A - B \ln t$.

When experimenting with the oxide film thickness by pulsing the polarization current, or else the potential, both $\phi_{m,ox}$ and $\phi_{ox,l}$ may temporarily deviate from their equilibrium values by some overpotential, $\Delta\eta$. In this context it is interesting to remember that, originally, the high-field mechanism was deduced from the decay of currents during barrier-layer anodizing of such metals as aluminum and tantalum [20e,f]. Experimenting with voltages as high as 120 V, phase boundary potential differences, $\phi_{m,ox}$ and $\phi_{ox,l}$ and contributions from a reference electrode were probably negligible, so that $\phi_{0,d} \simeq \varepsilon$. When, instead, low-voltage polarization is studied, the potential differences at the metal/oxide and oxide/solution interfaces may considerably contribute to the electrode potential. For intermediate voltage steps of, e.g., several volts, the non-equilibrium overpotentials at the phase boundaries may temporarily co-determine the rate of oxide growth.

One variant of explaining passive oxide film growth has it that the field strength in the film actually is, if at all, only a weak function of the film thickness [20h–j]. In this so-called 'point defect model' the growth of the oxide film is controlled by the overpotential of oxygen anion vacancy injection into the oxide at the metal/oxide interface. The decay of current density observed during potentiostatic polarization experiments is explained by a corresponding decay of this overpotential with increasing film thickness. Of late, the model has been applied, rather than to acid passivity of iron, predominantly though not exclusively, to such metals as zirconium, tungsten and tantalum. Excellent experimental evidence is claimed to support the concept. Concerning the details of this approach, the reader is referred to the literature quoted above.

For the transfer of $Fe^{3+}$ across the oxide/solution interface we assume usual charge-transfer control kinetics and a negligible rate of the reverse reaction. Once more limiting consideration to the stationary state, we therefore have

$$i_{\text{pass}} = K_2\, \theta \exp\left(\frac{\alpha\, n_e\, F}{RT} \phi_{\text{ox,l}}\right) . \tag{10.16}$$

In this equation, $\theta$ is the concentration of $Fe^{3+}$ in the oxide surface, determined by the degree of oxidation of the film material. $n_e$ is the valence of the charge transfer reaction which may be the passage of trivalent ferric ions, which would make $n_e = 3$. As $i_{\text{pass}}$ remains constant regardless of an increase in potential, the conclusion is that $\phi_{ox,l}$ also remains constant, and that the change in $\varepsilon$ is equal to the change in $\phi_{\circ,d}$ necessary to maintain the flux of iron ions through the film of increasing thickness.

Using the interrelation of degree of oxidation, electronic equilibrium, oxygen transfer equilibrium, and $Fe^{3+}$ transfer kinetics, an expression for the potential drop at the oxide/solution phase boundary, and for the dependence of the passive current density on pH is obtained [19]. Passing the derivation, we note only the result:

$$a)\quad \phi_{ox,l} = \text{const} + (RT/F)\ln a_{H^+}\,,\qquad b)\quad i_{pass} = K'\theta\,(a_{H^+})^{\alpha n_e}\,. \qquad (10.17)$$

An experimental value reported in the literature is $\partial(\log i_{pass})/\partial\,\mathrm{pH} = n_e\alpha = 0.84$.

Beyond the oxygen equilibrium potential, oxygen evolution begins to increase with increasing potential. This proves an increase of the Galvani potential difference, $\phi_{ox,l}$, governing charge-transfer across the interface. Consequently, the current density $i_{Fe^{3+}}$ and, thus, the current density, $i_{Fe}$, of iron dissolution also rises, since this electrode reaction occurs through the same gradient in Galvani potential. The details of this transpassive iron dissolution accompanying oxygen evolution, which we also pass, are described in [5,15c]; for a review see [5b]. For details of the mechanism of oxygen evolution at semiconductor surfaces see further below.

## 10.3
## Iron in Weakly Acid, Neutral, and Alkaline Solutions

The current density of passive iron dissolution, $i_{pass}$, decreases with increasing pH. The same is true for the critical passivating current density, $i_{crit}$: In acid phtalate buffer solutions at pH 3.4, and 6, $i_{crit}$ was found to be 0.05, 0.007, and 0.002 A cm$^{-2}$, respectively [16a]. In these solutions, the critical current density for passivation apparently was no longer governed by the precipitation of a salt layer, but rather by adsorption of $OH^-$ ions [23]. The point is of major interest because, as described in Chap. 5, adsorption of $OH^-$ also is an essential step in the catalysis of anodic iron dissolution. At intermediate values of pH, a branching point apparently exists, where on further increase of the electrode potential iron dissolution kinetics involving $OH^-$ can change from catalysis to prepassivation probably by layers of $Fe(OH)_2$ [24,25a]. This change from hydroxide precursors to passive oxide films is at the root of most of what is to be reported presently on passivity of iron in weakly acid, neutral, and alkaline solutions. It is important to remember that the alternative always is rusting, producing, instead of protective barrier oxide films, nonprotective porous layers of porous rust. Which way the iron/solution system will take at any pH very much depends on solution composition, especially on the nature of the anions present. Figure 10.7 shows stationary current-potential curves observed in phosphate/phosphoric acid and in borate/boric acid solutions [26a]. These are solutions free of aggressive anions, such as chloride. At the same time, phosphate is not inert with respect to iron corrosion; on the contrary, it favors inhibition owing to the low solubility of iron phosphate. The figure once more shows the decrease of $i_{pass}$, for values of pH $\geq 7$ it also once more shows the

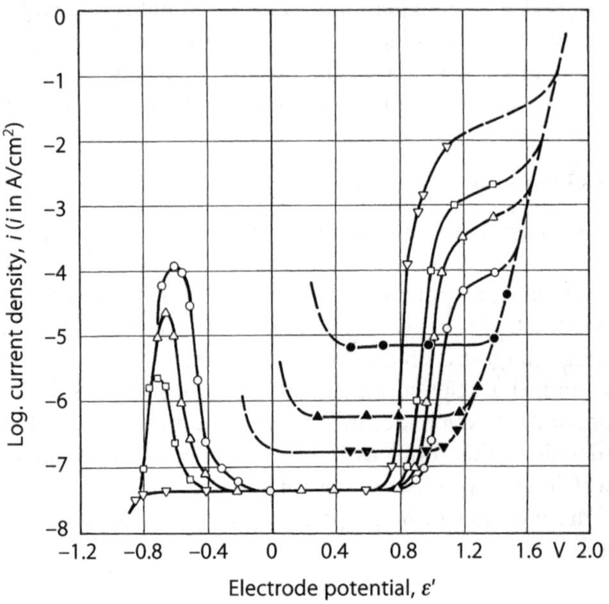


**Fig. 10.7.** Stationary anodic current-potential curves for iron in solutions buffered by non-aggressive anion/acid couples. Sodium phosphate/phophoric acid solutions: pH values: ● 1.85, ▲ 3.02; ▼ 3.90. Sodium hydroxide/ sodium borate/ boric acid solutions: pH values: ○ 7.4; △ 8.42, □ 9.37, ▽ 11.50. Solutions deaerated, 20 °C. $\varepsilon'$ referred to sat. cal. electrode. (From Sato, Noda, and Kudo)

decrease of $i_{crit}$. We leave the broken curves, indicating some kind of *secondary passivation,* without discussion.

Concerning the current density of passive iron dissolution in neutral and alkaline solutions, caution is recommended: When the solution is saturated with respect to the passivating oxide, the passive current density nominally should be zero, but the decay of polarizing currents caused by slow growth of the oxide layer becomes very slow, and steady states are difficult to obtain [16a]. This situation also tends to make comparisons of data from different sources difficult.

An important practical aspect of the matter is that in a solution of pH 11.5 passivation is spontaneous; and this applies to solutions of higher pH as well. One such solution is saturated $Ca(OH)_2$ solution, which is the liquid filling pores in moist concrete, because concrete stores free CaO, buffering any moisture to pH 12.6, which is well within the range where passivation is spontaneous. As a result, low-alloy steel reinforcements in concrete are safe with respect to corrosion owing to passivity, as long as chloride is absent and as long as there is calcium oxide available in the concrete bed [27]. Details of this subject, though of outstanding importance, are beyond the scope of this book. We only note that reinforced concrete structures are not indefinitely safe, because carbon dioxide in the atmosphere slowly reacts with calcium oxide to calcium carbonate: $CaO + CO_2 \rightarrow CaCO_3$. Once this reaction reaches the steel surface, local pH decreases to ca. 7, because calcium carbonate is scarcely soluble, and rusting will set on. For reviews of recent work see on steel in concrete see [27c,d]; for pitting, stress corrosion cracking and hydrogen embrittlement of steel in concrete, see following chapters.

For weakly acid to neutral solutions, Fig. 10.7 does not show the active/passive transition. In this context, we recall Fig. 5.23, presenting current-potential curves of iron dissolution in weakly acid solutions apparently changing from a Tafel line with a 35 mV-slope to a Tafel line with a higher slope and a lesser dependence on pH, passing through an intermediate minimum of the current density of iron dissolution. The tentative explanation was a change from rate-determining catalysis of dissolution by adsorbed FeOH to rate-determining dissolution via $Fe(OH)_2$. Obviously, the first maximum can be seen as a primary quasi-passivation caused by a layer of $Fe(OH)_2$ which fails as a protective film by a re-increase of the dissolution current density. It needs a second true passivation process to arrive at 'safe' passivity. Apparently, this stepwise arrival at passivity is obscured in high-acidity solutions. It is, however, readily observed for instance in weakly acid phosphate buffer solutions [28]. As a secondary effect, an influence of both phosphate concentration and solution stirring was obvious, indicating the participation of phosphate anions in the passivation processes. As a second example, Fig. 10.8 presents a series of near-stationary current-potential curves recorded in solutions similar to Theorell buffer solutions but with chloride replaced by sulfate [29]. Sulfate, though inert in acid solutions, turned out to be antagonistic to passivation: At pH 7.1 an increase in sulfate concentration from 0.1 to 1.5 M, caused the current maximum of the second peak to increase from 0.5 to 10 mA $cm^{-2}$. Again, stirring of the solution also was important, indicating that sulfate transport to the metal surface enters the rate-determining step of the reaction.

Spontaneous passivation requires that the metal/solution system spontaneously passes over the prepassivation and passivation peaks in the respective current-potential curve, and the peak heights determine the critical passivating current density. When $i_{crit}$ is high by the influence of such nominally harmless ions as sulfate, spontaneous passivation fails. This then explains why in approximately neutral solutions, i.e. for ordinary everyday corrosion, rusting of iron and low-alloy steel is normally to be expected rather than passivation [30a].

Double peaks in the quasi-stationary current-potential curve have been observed in more systems, and apparently they represent the normal course of events between active iron dissolution and passivity [31,32]. To explain this behaviour, the complicated interplay of reactions between the many possible species: metallic iron, iron hydroxides, iron oxides, and oxyhydroxides, has frequently been modeled, also assuming appropriate adsorption isotherms, and usually introducing a considerable number of non-experimental factors in the role of adjustable parameters. Detailed information has also been obtained by

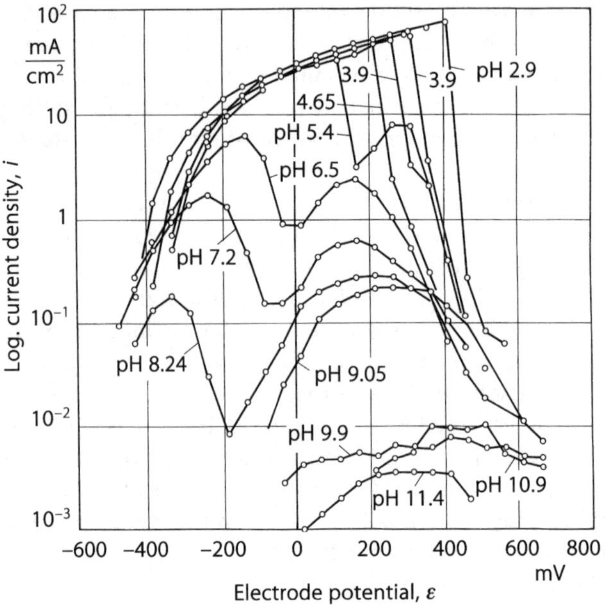


**Fig. 10.8.** Quasi-stationary current-potential curves for iron disc electrodes rotated at 400 rvls. $min^{-1}$ in deaerated buffer solutions prepared from citric, phosphoric and boric acid with additions of sulfuric acid and 0.5 M $Na_SO_4$. Potentiostatic scan rate 20 mV $min^{-1}$. (From Moeller)

impedance measurements (see Appendix), and by surface spectroscopy, as described below.

Values of the oxide film thickness as a function of both electrode potential and solution pH in probably inert buffer solutions, as determined by ellipsometry, have been reported in [26a]. For these measurements it was important to avoid prepassive dissolution of $Fe^{2+}$, and this was achieved by prepassivating the specimens in an alkaline solution where immediate passivation is spontaneous, and only then to change to the desired lower pH. As seen in Fig. 10.9, films thus obtained have have low values of $d$ similar to those found in 1 N $H_2SO_4$ solution, increasing linearly with increasing potential, and also increasing with increasing pH.

If during prepassivation $Fe^{2+}$ dissolution is allowed to occur, passivation kinetics tend to complicate, because the passive film grows by an additional oxidative reaction of initially dissolved iron to iron oxide [13b, 26b,c]. By this reaction water enters the oxide film according to a variant of reaction Eq. (3.99):

$$2Fe^{2+} + xH_2O \rightarrow Fe_2O_3 \cdot (x - 3)H_2O + 6H^+ + 2e^- \ . \tag{10.18}$$

The water content of the film has been confirmed by comparing ellipsometric values for $d$ with values obtained by analysing galvanostatic potential-time curves for oxide film reduction, or by recording the tritium content of oxide

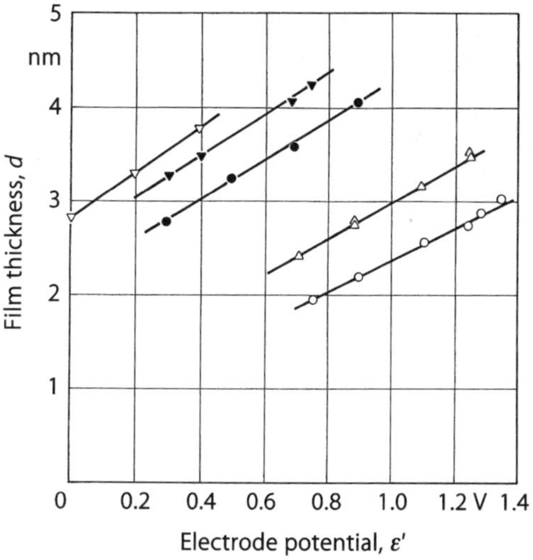


**Fig. 10.9.** Thickness of passive oxide films on iron, determined by ellipsometry, in sodium phosphate/phosphoric acid solutions, as function of the electrode potential with solution pH as parameter.
pH 1.85 (○); pH 3.02 (△); pH 6.15 (●); pH 9.12 (▼); 11.50 (△). Polarization time at the indicated potential was 1 h; all electrodes prepassivated at −0.2 V in boric acid/sodium hydroxide solutions, pH 11.50. Potential values are referred to sat. cal. electrode. (From Sato, Noda, and Kudo)

films formed in tritiated water [33a], by Mössbauer spectrometry [25b], and by Auger electron spectrometry [25c,34]. Auger spectrometry is, however, an ex-situ method requiring transfer of the specimens form the solution into the high vacuum chamber of the spectrometer. As discussed further below, this and other ex-situ procedures raise questions as to possible changes in composition and/or structure of passive oxide films during transfer from the solution to the spectrometer. In solutions containing $Fe^{2+}$, the passive oxide film may, in principle, grow to any thickness. As shown in Fig. 10.10 the increase in film thickness also masks the influence of pH reported previously as observed when prepassivation anodic iron dissolution was prevented [26]. By reaction with dissolved iron-II-salt the passive oxide film then may grow indefinitly.

For iron spontaneously passivated in alkaline solutions, the open-circuit corrosion potential is governed by the reaction Eq. (10.18) [13a,30]. Assuming, as usual, that the outer oxide layer is $\gamma$-$Fe_2O_3$, the equation relating potential, pH and activity of dissolved iron is similar to Eq. (3.100):

$$\varepsilon_{corr} = E_{Fe^{2+}/\gamma\text{-}Fe_2O_3} + \frac{RT}{F} \ln \frac{a^3_{H^+}}{a_{Fe^{2+}}}, \qquad (10.19)$$

and the dependence on pH and on $a_{Fe^{2+}}$ is confirmed by experimental evidence. The activation potential is more negative: A potential scan, shown in Fig. 10.11, starting from an electrode potential more positive than $\varepsilon_{corr}$ in the first place

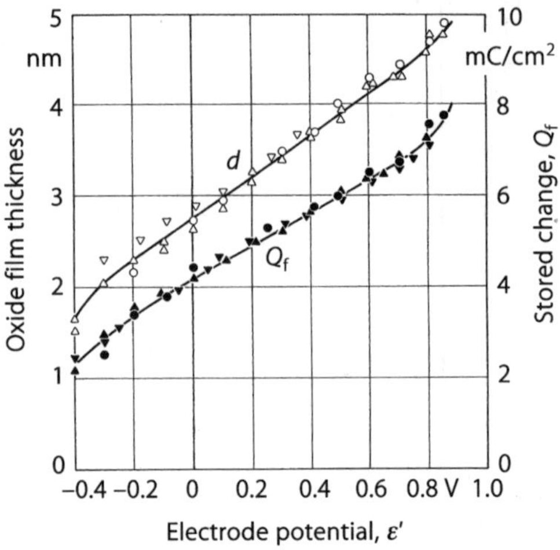


**Fig. 10.10.** Open symbols: Thickness of passivating oxide films, as obtained from ellipsometry, formed (without preceding rapid prepassivation) in sodium hydroxide/sodium borate/ boric acid buffer solutions, 25°C, by anodic polarization during 1 h at the indicated potential. Solid symbols: Amount of Coulomb charge stored in the oxide layer. pH 7.45 (circles); pH+8.42 (upright triangles); pH 9.65 (squares); pH 11.50 (inverted triangles). Potential referred to sat. cal. electrode. (From Sato, Noda, and Kudo)

shows anodic currents of oxidative $Fe^{2+}$ precipitation. The current then passes through zero when the equilibrium potential of this reaction is reached; it turns to cathodic values of reductive dissolution of the oxide below the equilibrium potential, always remaining in the range of passivity of the electrode. At still more negative potentials, the current then again is reversed to positive values, when the activation potential, $\varepsilon_{p2}$ is passed by the potential scan. The much higher passivation potential, $\varepsilon_{p1}$, is not shown. When the scan is reversed to the direction of increasing potentials, passivation and activation potentials are the same as before.

Concerning the passivation potential, we once more note that in neutral and alkaline solutions conditions are more simple than in low-pH solutions: While in the latter iron passivates in spite of the thermodynamic instability of oxides by the incidental kinetic effect of slow dissolution of $\gamma$-$Fe_2O_3$, iron oxides are scarcely soluble in neutral and alkaline solutions, and any oxide may suffice to passivate the metal if only the mechanism of formation is such that the film is non-porous as the result of growth by a low-temperature scaling mechanism instead of a dissolution/redeposition mechanism producing a porous deposit.

In this context, we may inspect Fig. 10.11, which for 50 °C clearly shows the active/passive transition between −0.5 and −0.3 V. Neglecting, for a change in temperature of only 25 K, changes in standard potentials (see Chap. 3), the supposed equilibrium potential for the magnetite/maghemite transition would be $E_{Fe_3O_4/\gamma\text{-}Fe_2O_3} \simeq 0.58 - 0.064$ pH $\simeq 0.0$ V, and this is, in fact, definitely

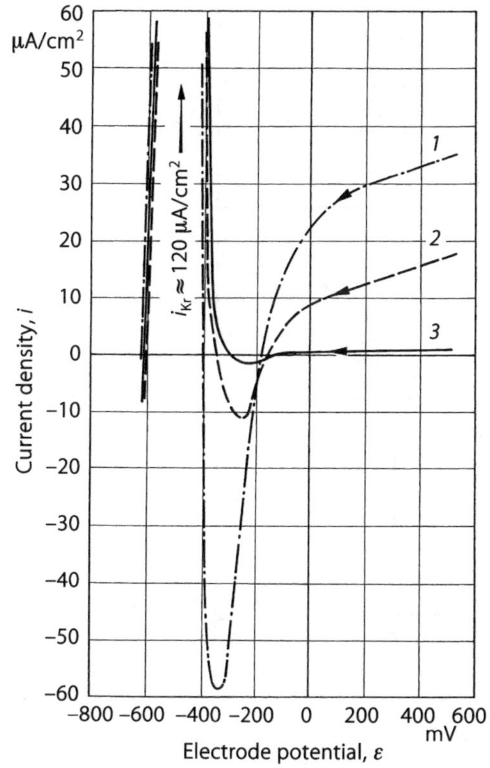


**Fig. 10.11.** Potentiodynamic current-potential curves for iron in sodium hydroxide/glycocoll buffer solutions at pH 9.0, 50°C, recorded in the direction of decreasing potentials, showing the transient currents during activation of prepassivated electrodes. Scan rate was 6 mV $s^{-1}$. The concentration of dissolved $Fe^{2+}$ was $1.45 \times 10^{-3}$ for curve (1), $0.85 \times 10^{-3}$ for curve (2), and approximately 0 mol $L^{-1}$ for curve (3). (From Oelsner)

more positive than the observed passivation potential. We note that, nevertheless, the magnetite/maghemite transition is readily observed, whenever the transition potential is passed [16d]. At the same time, the iron/magnetite transition potential, $E_{Fe/Fe_3O_4}$, or else every other iron/oxide equilibrium potential, is more positive than the passivation potential. One might, therefore, speculate that in this solution the magnetite/hematite equilibrium is decisive, with $E_{Fe_3O_4/\alpha\text{-}Fe_2O_3} \simeq 0.20 - 0.064\ \text{pH} \simeq -0.38$ V. In any case, passivation kinetics in weakly alkaline solutions appear to be more complicated than expected on first sight, and much dependent on circumstances such as prepassivation history, foreign anion content of solutions, polarization program, etc. The simple mechanism of direct passivation at some $Fe/Fe_nO_m$ equilibrium apparently is effective, if at all, for still more alkaline solutions only.

For a wide range of pH, Fig. 10.12 summarizes experiences concerning the nature of passivating films on iron obtained in solutions of always similar com-

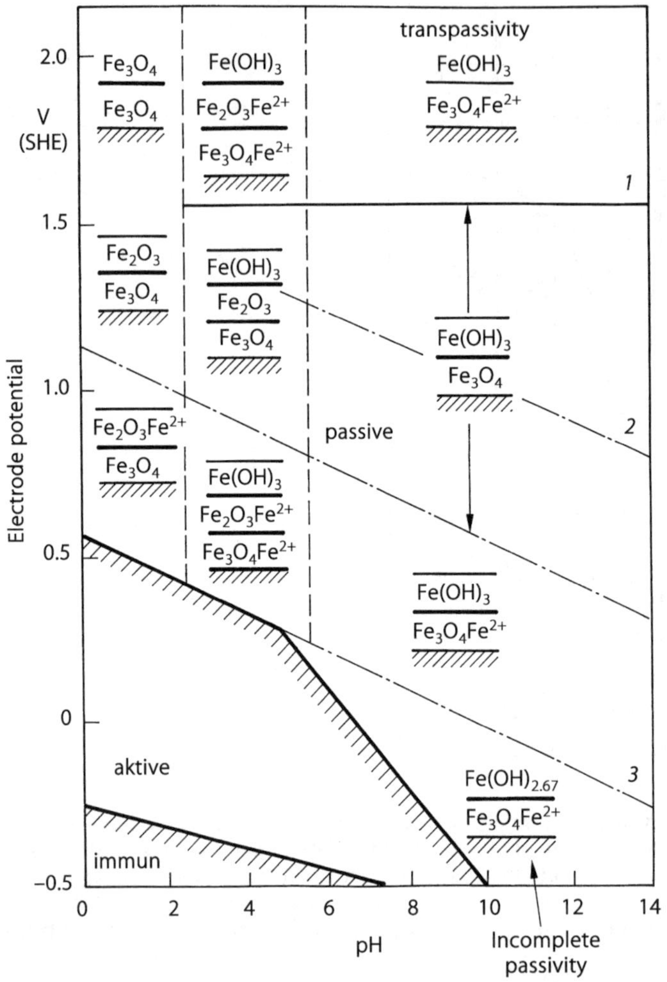


**Fig. 10.12.** Experimental potential-pH diagram for iron in phosphate solutions, showing domains of existence of layered oxide phases forming passivating barrier films and non-passivating hydroxide deposit films. (1) Potential threshold for onset of transpassivity; (2) oxygen equilibrium potential; (3) activation ('Flade') potential.(From Nishimura and Sato)

position [26d]. These and more data were obtained by a combination of ellipsometry, cathodic galvanostatic discharge curves, and chemical analysis. Briefly, circumstances are as follows: In acid solutions, the film is duplex, $Fe_3O_4/Fe_2O_3$, and purely barrier-type. In weakly acid solutions the film still is duplex but thickened by an outer porous hydroxide layer. In neutral and alkaline solutions a non-duplex barrier film with an additional outer hydroxide layer is assumed. At steady-state the thickness of the barrier film increases linearly with the electrode potential regardless of pH. Compared at equal values of the overpotential of oxide formation, the the thickness of barrier films is independent of pH. The deposit layer on top of the barrier layer does not much depend on the potential and increases with pH up to ca. pH 8. In the deposit layer, the valence of iron is three, in the barrier layer, it varies between 2.33 and 3.00, with an average value changing in dependence on pH, solution composition, potential and passivation procedure. Apparently, the film also is ion-selective.

On the whole, this compilation of data appears quite consistent with the vast body of experience concerning the passivity of iron. There rather is an abundance of relevant investigations, because oxide films on iron in neutral and alkaline solutions have been and still are a favorite for spectroscopic surface studies by various methods. As an example, we quote ex-situ XPS (X-ray Excited Photon Emission Spectroscopy) investigations into the behaviour of iron in NaOH solutions carried out with particular care to avoid any phase change in the film except, perhaps, evaporation of water [35a–c]. Figure 10.13 presents the type of information thus obtained: In the first place, the insert shows a rapid potential scan summarily characterizing the electrode kinetics of the system, which is passive throughout the range of potentials studied except at the most negative potentials. A small first peak appearing at the beginning of the scan is due to the oxydation of traces of hydrogen deposited at very low potential, and of no particular concern. The solution was checked for dissolved traces of iron, and the valence of the species thus analyzed is also indicated. The main assumption is that a twovalent oxide or hydroxide forms at low potential and is then oxidized to a threevalent oxide. The lower part of the figure gives film thickness values showing that, on the whole, the overall film thickness, *d*, is an approximately linear function of the electrode potential. From angular XPS data not shown it is concluded that the film is duplex rather than continuous, with FeO under $Fe_2O_3$, but the conclusion is perhaps not safe because the thickening outer layer obscures the signal from the bottom layer. Also, a continuous change in the two/three valence relation through the film appears likely in spite of the supposed duplex structure.

We once more note that secondary deposits of ferric hydroxide tend to complicate the situation without, however, much changing the essentials of passive electrode behaviour. Still it is possible that secondary films offer additional protection with respect to passivity breakdown by pitting.

For more results of spectroscopic investigations, including SIMS (Secondary Ionic Mass Spectrometry), Mössbauer spectrometry, surface enhanced Raman spectroscopy, and other methods, into the nature of passive films on iron especially in neutral and alkaline solutions, see, e.g., the review in [33b]. On the the whole, the picture of the structure, composition and electrochemical behaviour of these films obtained by the impressive plethora of experimental methods is quite similar to what has been described further above for the films present in acid solutions. We pass over most of this work quoting only some more results of in-situ investigations such as Mössbauer spectrometry revealing that passivating oxide films may not be well crystallized in the spinel structure of

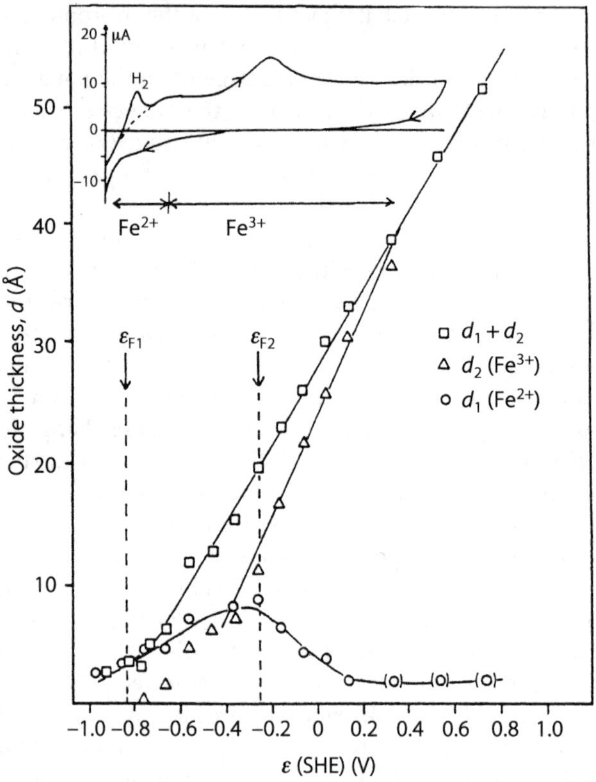


**Fig. 10.13.** Investigating the passivity of iron in 1N NaOH solution with XPS. Insert: Rapidly scanned potentiodynamic triangular wave current-potential curve. According to the model, the passivating film is a sandwich, with FeO or $Fe(OH)_2$ as inner layer of thickness $d_1$ and $Fe_2O_3$ as outer layer of thickness $d_2$. The plot shows $d_1$ and $d_2$ observed after polarization during 300 s at each value of potential.(From Haupt, Calinski, Collisi, Hoppe, Speckmann, and Strehblow)

magnetite and maghemite but tend to be at least in part amorphous, and that they contain water [36]. Much interest has been and is being paid to in-situ transmission XAS (X-ray Absorption Spectroscopy) applied to thin oxide films on equally thin metallic iron films. Absorption of electromagnetic radiation in the range of X-rays, available from synchroton white radiation by adjustable monochromators, and investigated especially in the vicinity of the energy edge of adsorption, is a useful tool to obtain information about the valence and crystal coordination of iron atoms in the metal and ions in the oxide film. The method was introduced as EXAFS (Extended X-ray Absorption Fine Structure) [33b–d]. Comparison of in-situ and ex-situ EXAFS inspection again shows that the spinel structure observed ex situ apparently is a secondary product produced by drying an initially more glassy or amorphous structure. Apparently, amorphous passivating films are particularly protective. A disadvantage of XAS

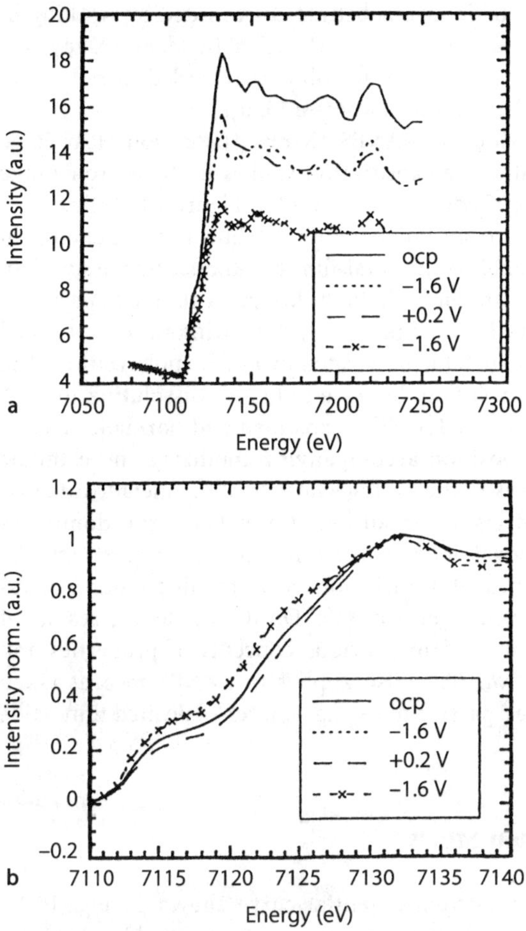


**Fig. 10.14a,b.** In-situ transmission XAS near the Fe K edge of iron in borate buffer solutions at pH 8.4. The iron electrode was a metallic film of 10 nm thickness; the electrode potential was stepped by a potentiostat; spectra were acquired after 5 min at each potential. **a** 'Raw' data showing that by reduction at −1.6 V the film formed at open circuit potential (OCP) is dissolved by reductive dissolution, as seen from the shift in edge height, while it is formed without change in thickness of the specimen, i.e. without loss owing to anodic dissolution, on reanodizing at +0.2 V. **b** Expanded near-edge spectra normalized to unity edge height after background subtraction, indicating by a negative shift a decrease in average valence caused by cathodic removal of the film, and and by a positive shift an increase in valence by anodic polarization. Note that this shift is positive with respect to the open-circuit spectrum. (From Virtanen, Schmuki, Büchler, and Isaacs)

is the strong signal from the metallic iron substrate co-recorded by working in transmission. As an alternative we note GIXAS (Grazing Incidence X-ray Absorption) where X-ray absorption from surface films is recorded in reflection, undisturbed by absorption in the metallic substrate [35d].

A special type of EXAFS, which is XANES (X-ray Absorption Near Edge Spectroscopy) emphasizes in-situ X-ray absorption studies in the narrow range of X-ray energies close to the Fe K edge at 7112 eV [37]. Figure 10.14 illustrates the type of information available from this method: As shown by the change in edge height, which is proportional to the total film thickness, the passive film formed in borate dissolves by cathodic reduction, leaving a thinned specimen. It is formed without change in film thickness and, thus, without loss of metal by dissolution to $Fe^{2+}$, when the metal is passivated by anodic polarization. The situation changes in 1 N NaOH solution, because reductive dissolution to $Fe^{2+}$ is replaced by reduction to solid $Fe(OH)_2$. The expanded and normalized spectra show, by the shifts in edge position accompanying anodizing and cathodic reduction, corresponding changes in the average valence of the metal/oxide system. Although the observed effects are small because metallic iron dominates the spectrum, the complete removal of oxide or hydroxide species after cathodic polarization appears clearly indicated. Conclusions as to the distribution of valence of iron through the film would seem unsafe. On the whole, the results of this type of investigations again confirm previous concepts of properties and behaviour of passive films on iron. This also applies to XANES records taken during the reduction of 'artificial' passive films, e.g. sputter-deposited thin oxide films [37b].

## 10.4 Chromium and Iron-Chromium Steels

As obvious from the stationary current-potential curve shown in Fig. 10.15, passive chromium is similar to passive iron in many respects [38a]: As for iron, the active/passive transition between potentials $\varepsilon_{p1}$ and $\varepsilon_{p2}$ is distinctly marked, followed by a potential-independent range of constant passive current density, and eventually followed by a steep transpassive increase of the current density of metal dissolution. In view of practical applications, the chromium/acid system has, however, important advantages as compared to iron: In 1 N $H_2SO_4$ solutions, the critical passivating current density is much lower than for iron, and passivation is complete at ca. 0.0 V, which is more negative than the hydrogen equilibrium potential in the same solution. Therefore, chromium can passivate spontaneously in deaerated solutions, in contrast to iron. The passive current density also is lower than that for iron. Transpassivity sets in at potentials more negative than in the case of iron, but this disadvantage is not usually very serious because open-circuit potentials are not normally close to values as positive as 1 V (see Chap. 11).

Commercial pure chromium is brittle and unimportant as a structural material. Also, the previous fashion of galvanizing car bumpers and the like with

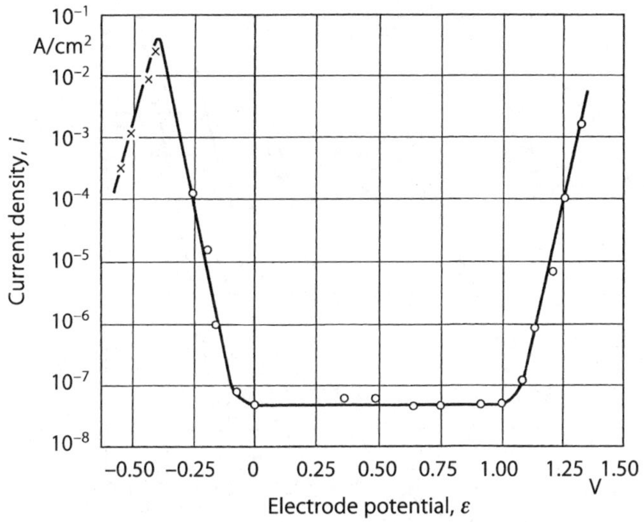


**Fig. 10.15.** Stationary anodic partial current-potential curve for metal dissolution from pure chromium in 1 N $H_2SO_4$ solution. Note well defined critical passivating current density, $i_{crit} = 40$ mA cm$^{-2}$, passive current density $i_{pass} < 140$ $\mu$A cm$^{-2}$, passivation potential $\varepsilon_{p1} = -0.4$ V, activation potential, $\varepsilon_{p2} = -0.1$ V, and onset of transpassivity at $+1.1$ V. (From Kolotyrkin)

nickel, topped with some chromium, has become obsolete. The main use of the pure metal nowadays is for hard (i.e. thick) chromizing of hydraulic pistons and similar applications. The much more important aspect is the use of chromium in chromium-iron and chromium-iron-nickel alloys which, if sufficiently rich in chromium, are spontaneously passive in most aerated solutions.

Concerning the properties of passive chromium already noted we add that, again similar to iron, the active/passive transition shifts to more negative potentials with increasing pH. In earlier work the slope was reported to be ca. −012 V/pH [39], but the more likely value −0.06 V/pH appears equally well confirmed [6]. As usual this indicates passivation by a reaction involving oxide formation or oxide/oxide transformation. Since the passive current density, $i_{pass}$, although very small, is well defined, the passive oxide film again is not in equilibrium with the acid solution. At pH 1.7 and pH 2.2 the passivating current density, $i_{crit}$, drops to 1 and 0.4 mA cm$^{-2}$, respectively [40]. From ellipsometry and from impedance measurements, the oxide film is estimated to be very thin with, in acid solutions, 1 nm at 0.2 V, and 2 nm at 1.2 V, respectively [41,5c]. Before the onset of transpassivity, chromium dissolves to $Cr^{3+}$ and this probably reflects the oxide composition throughout the film. Therefore, the oxide is supposed to be $Cr_2O_3$ or perhaps a hydrated hydroxide $Cr(OH)_3 \cdot 0.3H_2O$ [42]. Similar to iron, oxide dissolution is thought to proceed by transfer of $Cr^{3+}$ alone, rather than by transfer of oxide molecules and subsequent molecule dissociation. Still similar to iron, passivation is complete at potentials considerably

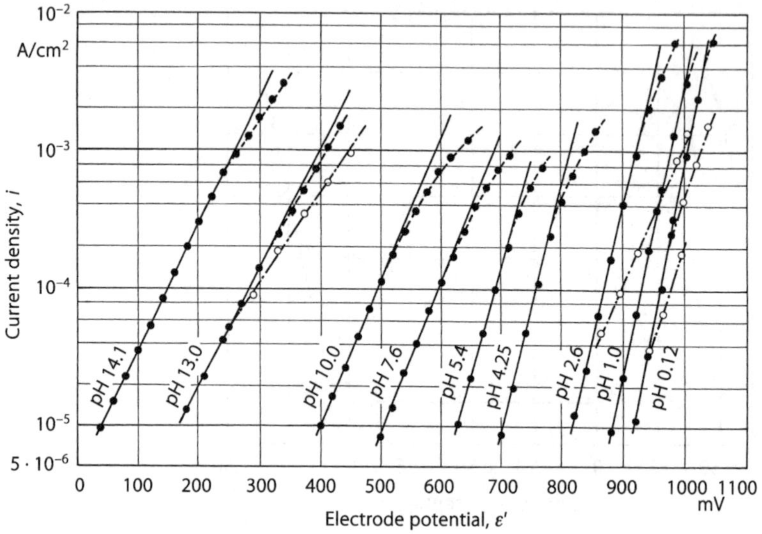


**Fig. 10.16.** Stationary (*solid circles*) and non-stationary (*open circles*) current-potential curves for transpassive dissolution of chromium to chromate ions as a function of pH in NaCl/HCl/NaOH/citrate/borate buffer solutions at 23°C. Electrode potential referred to sat. cal. electrode. (From Knoedler and Heusler)

more positive than the $Cr/Cr_2O_3$ equilibrium, so that, apparently, passivation is achieved by the oxidation of an oxidic precursor of lower valence, similar to the behaviour of iron in acid solutions, where the precursor is magnetite [43].

On pure chromium, anodic oxygen evolution cannot be observed, because of a very early onset of transpassive rapid dissolution of the oxide to hexavalent chromate ions: $2Cr + 7H_2O \rightarrow Cr_2O_7^{2-} + 14H^+ + 12e^-$. Probably this is a two-step process via $Cr^{3+}$, because chromate is an unlikely component of the oxide film [44]. The apparent onset of transpassivity probably is caused by an exponential current-potential law, governing chromate dissolution, thus leaving the corresponding current density negligible at low potentials, but overtaking $Cr^{3+}$ dissolution currents at sufficiently high overpotentials. At any rate the steep increase in current density observed at and beyond the onset of transpassivity must not be confused with threshold potentials marking the onset of some morphological instability such as, especially, pitting. The same holds for transpassive FeCr alloys as well as for transpassive FeCrNi alloys described further below.

The equilibrium potential of a $Cr/Cr_2O_7^{2-}$ electrode is $E = 0.28 - 0.069$ pH (V). Considering, instead, a $Cr/CrO_4^{2-}$ electrode, it is $E = 0.37 - 0.079$ pH (V). Rapif transpassive dissolution is observed only at considerably more positive potentials, viz. at pH 0 only beyond ca. 1 V (see Fig. 10.15). Still the expected strong dependence of the current increase due to chromate dissolution solution on pH is well confirmed by experimental evidence, as seen from Fig. 10.16 [5c].

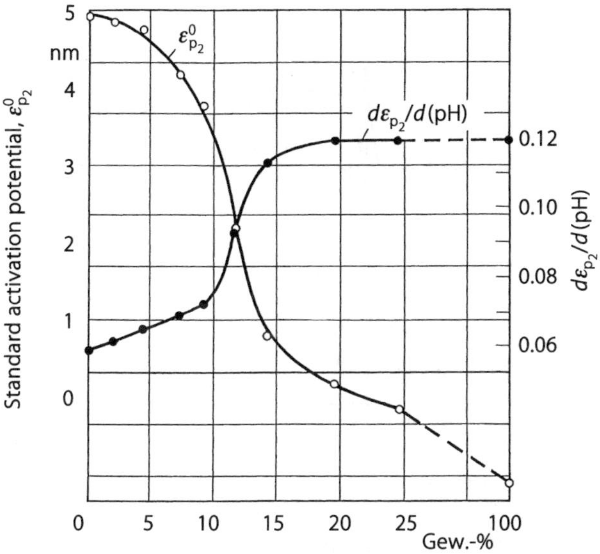


**Fig. 10.17.** The change of the 'standard' activation potential, i.e. $\varepsilon_{p2}$ in $H_2SO_4$ solution at pH 0 with the chromium content of FeCr alloys. Also showing the slope $d\varepsilon_{p2}/d(\mathrm{pH})$. (From Rocha and Lenartz)

Turning attention to FeCr alloys, we first show in Fig. 10.17 early data on the change of the activation potential with composition observed in acid solutions [39]. Data of this type have long been taken as signifying an important change of structural properties of the alloys at ca. 10 wt.% Cr. For practical purposes, this is, in fact, approximately the minimal Cr content necessary for safe passivity of, e.g. the drums of washing machines. Discussions as to conjectured important physical changes in the structure of the alloys at this apparently critical composition are, however, misleading, since there is no important structural change at this composition.

For a better understanding of corrosion kinetics, consider the current-potential curves shown in Fig. 10.18, as recorded in deaerated 1 N $H_2SO_4$ solution at ambient temperature for the mainly interesting range of compositions of binary FeCr alloys and for the pure components [43a]. This work confirms previous measurements in 1 N $H_2SO_4$ [45]. As seen from the figure, there is a gradual shift of the passivation potential to more negative values with increasing chromium content, as well as a gradual shift of the passive current density to smaller values. For the alloys, chromium-type transpassivity is observed at potentials again shifting to more negative values, coinciding with pure-chromium transpassivity behaviour for chromium contents exceeding 6 wt.%. Different from what has been described for pure chromium, the chromium-type transpassivity changes to iron-type transpassivity by passing through a range of more elevated current densities which depend only little on

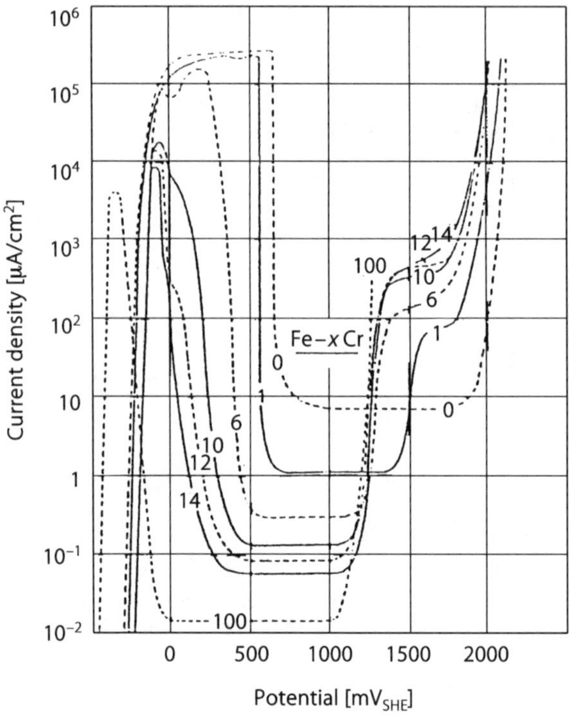


**Fig. 10.18.** Stationary current-potential curves for iron/chromium alloys and for pure iron and pure chromium in 1 N $H_2SO_4$ solutions at 24°C. Chromium contents are in wt.%. (From Kirchheim, Heine, Fischmeister, Hofmann, Knote, and Stolz)

potential. Disregarding, for the moment, the previous argument that in a practical sense 'passivity' really means negligible anodic metal dissolution rate, this may be seen as some kind of *secondary passivation* which we do not attempt to discuss.

The decrease of the passive current density is much more than proportional to the chromium content of the alloys. This effect will be readily explained, as previously for alloys in Chap. 8, as the consequence of steady-state requirements for uniform simultaneous dissolution of ionic species from oxides with widely different rate constants for 'slow' and 'fast' components [46a]. Before, we quote some observations concerning the structure of the passivating film and its chromium content, as obtained by various methods of surface spectroscopy.

The thickness of the passivating film, as revealed by in-situ ellipsometry in acid solutions, is between several nm and 1 nm [41,47]. As for pure chromium, Cr always dissolves to $Cr^{3+}$ in the range of primary passivity, and to hexavalent chromate in the range of secondary passivity. At the same time, as revealed by XPS, iron dissolves to $Fe^{2+}$ below the Flade potential calculated for pure iron, and to $Fe^{3+}$ beyond this threshold potential [35e]. From angular-resolved XPS investigations a duplex structure with and Fe-rich outer and an Cr-rich inner part is concluded [35b].

An interesting point is the crystallinity of the passive oxide as revealed by electron diffraction of films grown in acid solutions [48]. Up to 5 wt.% Cr, the film is a well defined spinel; at 12% Cr the spinel lattice is less ordered; at 19% Cr the film is predominantly, and at 24% Cr it is completely amorphous or glassy. The decrease of the passive current density is attributed to the increase in glassiness, on the supposition that ionic migration in non-crystalline films, as in glassy enamel, is slower than in crystalline material. We shall meet this opinion again when discussing the properties of glassy alloys further below.

Most of the investigations quoted above describe the chromium content of the interior of the oxide film, including depth profiles. This method has been expanded to more precisely measuring, again with XPS, the solution-side surface composition, and by determining, with the aid of high-resolution atomic absorption spectroscopy of the solution composition, the partial currents of iron and chromium dissolution to $Fe^{3+}$ and $Cr^{3+}$ [43a]. In this work, an important point was to always wait for stationary states to be attained, if necessary during several days.

In steady state at constant film thickness, and thus at zero net transfer of oxygen across the oxide/solution interface, the dissolution rates of iron and of chromium from the oxide must obey the continuity laws already introduced for the uniform steady-state corrosion of homogeneous alloys in Chap. 8. Thus, the oxide must behave like a homogeneous alloy, with the oxygen lattice of the crystalline film in the role of a solvent only. We describe the consequences using derivations described in Chap. 8, appropriately adapted to the present purpose. Writing symbols we follow the authors of [43a], and denote mole fractions a) in the bulk of the alloy as $x_{Cr,a}, x_{Fe,a}$, b) in the oxide as $x_{Cr,f}, x_{Fe,f}$, and c) in the solution-side surface of the oxide film as $x_{Cr,e}, x_{Fe,e}$, respectively. When the valence of iron and chromium ions passing the interfaces and the film is the same everywhere, we can compare migration and transfer fluxes directly as current densities, $i_{Cr,f}, i_{Fe,f}, i_{Cr,e}$, and $i_{Fe,e}$. We shall follow this suggestion, although the flux of iron ions more precisely is the sum of fluxes of $Fe^{2+}$ and $Fe^{3+}$, as governed by the change in oxygen deficit, $\Delta$, across the film as described previously. By this simplification we replace the more exact sum by an equivalent $Fe^{3+}$ flux without important detriment to the arguments following presently. Continuity then requires that

$$i_{Cr}/i_{Fe} = x_{Cr,a}/x_{Fe,a} = i_{Cr,f}/i_{Fe,f} = i_{Cr,e}/i_{Fe,e} \; . \qquad (10.20)$$

We expect that the dissolution rates across the metal/solution interface are proportional to the component mole fractions in the surface. Since we also expect charge-transfer control at the oxide/solution interface for both iron and chromium dissolution, we directly write, assuming with the authors $n_e = 3$,

$$a) \quad i_{Fe,e} = k_{Fe}\, x_{Fe,e} \exp(3\,\alpha_{Fe}\, F\, \phi_{ox,l}/RT))$$

$$b) \quad i_{Cr,e} = k_{Cr}\, x_{Cr,e} \exp(3\,\alpha_{Cr}\, F\, \phi_{ox,l}/RT) \; . \qquad (10.21)$$

Considering that everywhere $x_{Fe} = (1 - x_{Cr})$, we can calculate the mole fraction of chromium in the solution-side oxide surface:

$$\frac{x_{Cr,e}}{1 - x_{Cr,e}} = \frac{k_{Fe}}{k_{Cr}} \exp\left(\frac{3F}{RT}(\alpha_{Fe} - \alpha_{Cr})\right) \frac{x_{Cr,a}}{1 - x_{Cr,a}} . \tag{10.22}$$

Distinguishing between transfer coefficients $\alpha_{Fe}$ and $\alpha_{Cr}$ is perhaps overly correct in view of other simplifications, and we shall drop the exponential term in the following. The relation between the chromium mole fractions in the oxide surface and in the alloy then is:

$$x_{Cr,e} = \frac{(k_{Fe}/k_{Cr})\, x_{Cr,a}}{1 - x_{Cr,a} + (k_{Fe}/k_{Cr})\, x_{Cr,a}} , \tag{10.23}$$

Ideally, rate constants are independent of concentrations. If this holds for both chromium and iron dissolution from the oxide, the ratio of rate constants can be calculated from the passive current densities for pure chromium and pure iron (0.014 and 7 $\mu$A cm$^{-2}$, respectively):

$$k_{Fe}/k_{Cr} = (i_{pass})_{Fe}/(i_{pass})_{Cr} = 500 . \tag{10.24}$$

From this it follows that, ideally, the chromium enrichment in the oxide surface is very pronounced already for small bulk chromium contents; it is, in fact, theoretically 98% when $x_{Cr,a} = 1 at.\%$. For comparison, Fig. 10.19 shows a depth profile of the ratio, in %, Cr/(Fe+Cr) obtained for a Fe-10wt.% Cr specimen passivated in the range of primary passivation. The abscissa is film thickness in nm as defined by the usual comparison with a $Ta_2O_5$ film of known thickness. The curve is interpreted as indicating a very high chromium concentration (of unknown absolute value) in the solution-side oxide surface, immediately dropping to a considerably lower value within the oxide, and more or less constant throughout the film of ca. 2 nm thickness, eventually dropping to the bulk alloy composition.

The lower chromium enrichment observed within the film is a further point of considerable interest: The continuity condition requires that the ratio of fluxes or current densities, $i_{Cr,f}/i_{Fe,f}$ through any geometrical surface in the film at a given distance from the alloy surface again must be related to the bulk alloy composition by Eq. (10.20). Inside the film, the mechanism is not interfacial charge transfer, it is, instead, migration in the electrical field. Introducing, in analogy to the migration of ions in solution described in Chap. 3, the *ionic mobilities*, $u_{Fe^{3+}}$ and $u_{Cr^{3+}}$, we now expect that

$$\frac{i_{Cr,f}}{i_{Fe,f}} = \frac{x_{Cr,f}\, u_{Cr^{3+}}}{x_{Fe,f}\, u_{Fe^{3+}}} \tag{10.25}$$

and

$$x_{Cr,f} = \frac{(u_{Fe^{3+}}/u_{Cr^{3+}})\, x_{Cr,a}}{1 - x_{Cr,a} - (u_{Fe^{3+}}/u_{Cr^{3+}})\, x_{Cr,a}} . \tag{10.26}$$

The observed enrichment of Cr within the film indicates that in the electric field across the film $Cr^{3+}$ moves slower than $Fe^{3+}$. The result is compatible with the

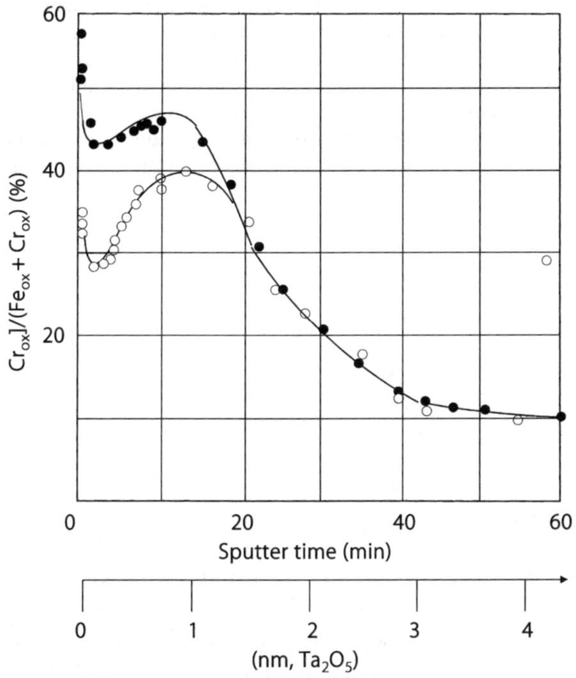


**Fig. 10.19.** Concentration profiles of the relative chromium content of the stationary passive oxide film on an Fe-10wt.% Cr alloy passivated in 1 N $H_2SO_4$. ● measured at ca. 0.7 V (range of primary passivity), ○ measured at 1.5 V (secondary passivity). (Figure redrawn using data from Kirchheim et al.)

high field mechanism postulated for ionic migration in passive oxide films; it would be equally well compatible with migration of ions obeying Ohm's law instead. In any case, the field has to be calculated from the potential drop, $\phi_{0,d}$, across the film, and not from the electrode potential $\varepsilon$. During non-stationary states of oxide film growth, the passive current density, $i_{\text{pass}}$, is found to exceed the stationary value, indicating a transient increase in the interfacial potential $\phi_{ox,l}$.

The calculation of ionic mobilities in the film would have required investigations into the chemistry, structure, disorder, and distribution of ionic concentrations beyond the scope of the work presently described.

Table 10.1 lists data for the amounts of dissolved iron and chromium as related to the composition of the bulk alloy. From these data, the fractions of chromium and iron contents, Cr/(Cr + Fe), in the solution are calculated. These fractions are equal to the fraction of the current densities, $i_{\text{Cr}}/(i_{\text{Cr}} + i_{\text{Fe}})$ and these should be equal to the corresponding component fractions in the alloys. The condition is well satisfied, and the underlying theory of uniform dissolution of chromium and iron from a chromium/iron oxide thus appears unusually well founded. From continued calculations, which we pass, the relation

**Table 10.1.** Checking the continuity condition for potentiostatic stationary uniform dissolution of iron and chromium from passive FeCr alloys in slowly flowing 1 N $H_2SO_4$ at 25°C: Observation time was 96 h in every case. Cr(el.), Fe(el.): Dissolved chromium and iron, determined by AAS. $X$(al),X(el): Cr/(Fe+Cr) ratio in the bulk alloy and in the solution. $i_{pass}$: Stationary passive current density, i.e. ($i_{Cr}+i_{Fe}$). For each Cr concentration in the alloy, the potential was chosen in the middle between the activation potential and the potential of onset of transpassivity. (Data selected from tables reported by Kirchheim et al.)

| Cr(el.) $\mu$g mL$^{-1}$ | Fe(el.) mg L$^{-1}$ | $X$(al) % | $X$(el) % | $i_{pass}$ $\mu$A cm$^{-2}$ |
|---|---|---|---|---|
| $4.6 \times 10^{-3}$ | 0.41 | 1.01 | 1.12 | 1.1 |
| $6.8 \times 10^{-3}$ | 0.11 | 5.89 | 5.83 | 0.30 |
| $5.0 \times 10^{-3}$ | $4.35 \times 10^{-2}$ | 10.4 | 10.4 | 0.13 |
| $3.8 \times 10^{-3}$ | $2.64 \times 10^{-2}$ | 13.0 | 12.6 | 0.081 |
| $3.4 \times 10^{-3}$ | $1.88 \times 10^{-2}$ | 15.2 | 15.5 | 0.060 |

$$i_{pass} = i_{Cr} + i_{Fe} = k_{Cr}\, x_{Cr,e} + k_{Fe}\, x_{Fe,e} \qquad (10.27)$$

is also found to be very well satisfied, proving that the simultaneous chromium and iron dissolution through the iron/chromium oxide shows ‘ideal' behaviour.

A further test of the theory of metal dissolution from uniform binary oxides is offered by the phenomenon of secondary passivity. In this range of potentials, the passive current density is increased by orders of magnitude, as shown in Fig. 10.18. The cause is an increase in the current density of chromium dissolution, reversing the ratio $(i_{pass})_{Fe}/(i_{pass})_{Cr}$. As also shown in Fig. 10.19 for Fe-10Cr, the surface concentration, $x_{Cr,e}$, is indeed found to be distinctly lowered as compared to the case of primary passivity, very well confirming the proposition. As secondary passivity cannot be studied with pure chromium, it is unknown whether the behaviour is ideal in the sense defined by Eq. (10.27). It merits particular attention that, at the same time, the oxide composition, $x_{Cr,f}$, close to the metal surface apparently remains largely unchanged, disproving, incidentally, previous conjectures in [50,51]. Apparently, metal dissolution from the oxide still is simultaneous chromate and iron dissolution, and not exclusively chromate dissolution.

As a last item in this section we note that so far nothing was said concerning the alloy composition at the metal/oxide interface. Complications may arise when by active dissolution before the onset of passivation the surface metal composition is changed by selective corrosion. Hence, a further source of discrepancy of results from different contributors. With respect to reports from other sources concerning the composition of passivating films on FeCr alloys, the situation is generally quite similar to what has been described above for iron: The flow of data supplied by the various methods of advanced surface spectroscopy is considerable, but a coherent picture is difficult to obtain and will not attempted in this text. For more information, reference is suggested to, e.g., [35,37,49–52].

## 10.5 Nickel, Molybdenum, and Stainless Steels

With respect to both passivation kinetics and passive properties, *nickel* is again similar to both iron and chromium. The typical N-shaped current-potential curve, recorded in 1 N $H_2SO_4$, and presented in Fig. 10.20 shows a critical passivating current density, $i_{crit}$, lower, and a passive current density, $i_{pass}$, higher than that of iron [44b,46b]. The passivation potential, $\varepsilon_p$, is close to that of iron. MA is the anodic current-potential curve for active nickel dissolution, followed through BC by passivation apparently involving a precursor, assumed to be NiO or some other low-valence oxide/hydroxide, which transforms into the passive oxide $Ni_3O_4$ or some other higher-valence oxide/hydroxide, at the respective equilibrium potential. Also, a continuous change in valence, similar to what has been described above for iron, has also been reported for passive nickel on the grounds of modulated reflection spectroscopy [5d].

With increasing solution pH, the active/passive transition potential shifts to more positive potentials by 0.06 V/pH in acid solutions, and by more than this in alkaline solutions. At constant sulfate content of the solution, 0.5 M, and at pH 0.3, 3.1 and 11.7, $i_{crit}$ is $7 \times 10^{-2}$, ca. $10^{-5}$, and ca. $10^{-8}$ A cm$^{-2}$, respectively. Between D and E $i_{pass}$ drops from $10^{-6}$ A cm$^{-2}$ at pH 0.3 to ca. $10^{-8}$ A cm$^{-2}$ at pH 5, remaining approximately constant up to pH 10, increasing again in more alkaline solutions.

Various explanations have been offered for the current-potential curves in the range EFG, where the electrode shows an increase in metal dissolution

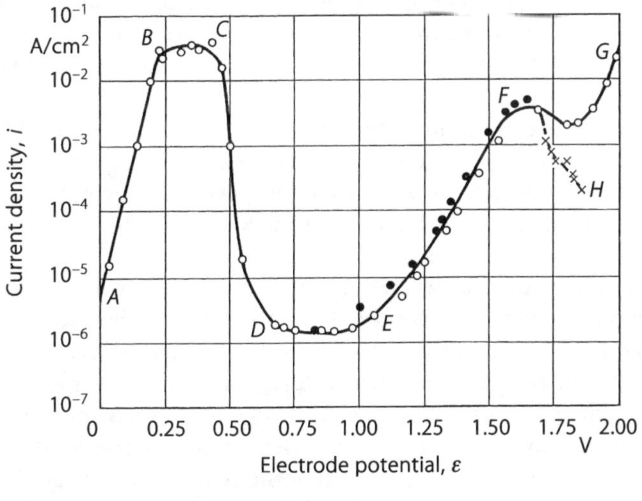


**Fig. 10.20.** The stationary current-potential curve for pure nickel in 1 N $H_2SO_4$ at 25°C. **ABCDEFH:** Partial current density of nickel dissolution; **FH:** partial current density of oxygen evolution. (*Open circles* and *crosses* from Sato and Okamoto, *solid circles* from Vetter and Arnold)

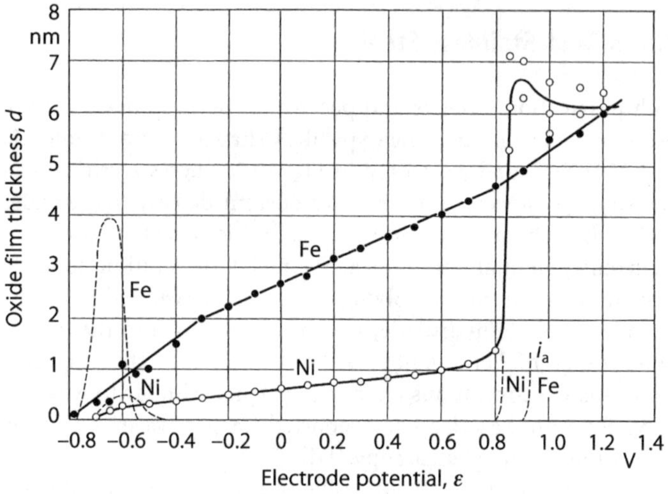


**Fig. 10.21.** The ellipsometric thickness of passive films on nickel (open circles) and on iron (solid circles) in boric acid/ borate solutions at pH 8.42, 25° C as function of the electrode potential. $\varepsilon'$ referred to sat. cal. electrode. Each value of $d$ was obtained after 1 h potentiostatic polarization. Broken curves schematically show the current-potential curves for iron and nickel with current densities in arbitrary units. (From Sato and Kudo)

current density similar to transpassivity. An early suggestion was that beyond ca. 1 V the oxide film recrystallizes and that it then may become porous and allow active nickel dissolution through the pores [46c]. A perhaps more likely supposition would be that an increase in passive current density is, in principle, easily explained by a corresponding increase in the interfacial Galvani potential difference $\phi_{ox,l}$ .

The thickness of the passivating films has been studied in buffered weakly alkaline solutions. [46f]. Figure 10.21 shows that nickel behaves distinctly different from iron, where the film thickness increases in proportion to the potential: On nickel, the film is very thin, and the change in thickness with the potential is small up to the onset of an apparent transpassivity.

*Nickel-base alloys* will be described later as important engineering materials especially for service in high-temperature, high-pressure waters. *Iron-base iron-chromium-nickel alloys* are more directly interesting presently: When the chromium and nickel contents are sufficiently high, these are the *stainless steels* famous for excellent corrosion resistance owing to spontaneous passivity in most aerated solutions even at low pH. Depending on composition, these steels are face-centered austenitic, or body-centered ferritic, or duplex austenitic/ferritic materials. In the following, we concentrate on the prototype containing 18wt.% Cr and 8% Ni, with or without several % Mo. Also, we concentrate in this chapter on uniform corrosion, postponing the discussion of pitting, intercrystalline corrosion, and stress corrosion cracking to later chapters, thus also postponing the discussion of the influences of carbon and titanium as well as the influence of heat treatment.

Concerning pure *molybdenum*, some brief remarks may suffice, as the pure metal is not important in our context. As structural materials, pure molybdenum or Mo-based alloys still are somewhat exotic, although, apparently owing to good high-temperature properties, an increased use for missile and aerospace applications is notable. Concerning the behaviour in aqueous solutions, investigations into the electrode kinetics of molybdenum corrosion are comparatively rare [53,54]. The metal is readily passivated in a range of potentials more negative than that for iron and nickel. The passive current density of metal dissolution is small up to the onset of transpassivity which, however, occurs far below the onset of transpasivity of either iron, or chromium, or nickel. In 1 N $H_2SO_4$ at ambient temperature, transpassivity is observed already at 0.3 V.

In FeCrMo *alloys* exposed to sulfuric acid, molybdenum increases the overpotential of active anodic metal dissolution but decreases the critical passivating current density by orders of magnitude [49,53,55]. In concentrated hydrochloric acid, 11 at.% of Mo added to Fe-24 at.% Cr lower the passive current density by between two and three orders of magnitude. Early transpassivity, which might be expected due to the molybdenum content is not observed; instead, the alloy shows chromium-type transpassivity apparently caused by a predominant chromium content in either the metal surface and/or the oxide film [49,54,56].

As evident from Fig. 10.22 the main effect of constituents on the passivity of austenitic stainless steels certainly is that of chromium [57b]: With increasing chromium content, the passivation potential shifts to considerably more negative values, as in FeCr alloys. At the same time, the passive current density decreases to very low values in the range of potentials below the onset of transpassivity, which also is the range of practical use of the passive alloys. Beyond the onset of transpassivity, secondary passivity is observed, increasingly similar to pure chromium and eventually shifting to anodic oxygen evolution beyond the oxygen equilibrium potential. From this it appears that the role of nickel is more important with respect to mechanical properties such as ductility and toughness than with respect to corrosion. In comparison to FeCr alloys, nickel does, however, have some beneficial effects on passivity. The critical passivating current density and the passive current density are much smaller in the presence as in the absence of 8% Ni in the alloys. The effect is not due to the face centered austenitic lattice of the alloys: FeCrMnNi alloys with only 2% Ni but 18% Mn are austenitic just as well, and they are also readily passivatable, but manganese leaves both $i_{crit}$ and $i_{pass}$ similar to that found for body centered ferritic FeCr alloys of the same Cr content [57a].

The composition of passive oxide films on stainless steel is a matter of much speculation, but also of much experimental endeavor. The difficulties of determining, by preference including in-depth profiling, simultaneously the iron, chromium, nickel and molybdenum content of barrier layers only some few nm thick are obvious, and only some token results will be described presently. For more comprehensive reviews reference is made to, e.g., [58,59] As for FeCr

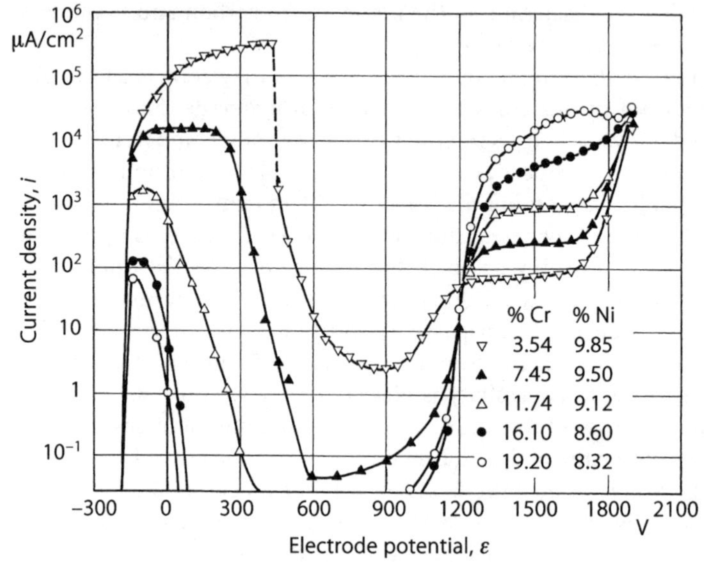


**Fig. 10.22.** Stationary current-potential curves for FeCrNi alloys with practically constant nick-el content as a function of the chromium content in 1 M $H_2SO_4$ at 25°C. (From Osozawa and Engell)

alloys, the oxide film generally is thought to be duplex, with a barrier layer grown by low-temperature scaling and particularly rich in chromium, topped by an secondary more hydroxidic and more porous, precipitated film. An interesting conjecture is that the film is a bipolar ion exchanging membrane, and that the barrier film is anion selective due to a positive charge, while the outer film is cation selective due to a negative charge, with a bipolar rectifying boundary in between [46d,g]. The point will reappear in the next section. Of various other contributions to the field in question, Fig. 10.23 shows the presumable composition of a passive film on Fe-19Cr-11Ni-2Mo passivated in 0.1 M HCl/ 0.4 M NaCl solution at 0.45 V.

Further above, the successful application of the rules of alloy dissolution to the case of steady-state metal dissolution from binary oxides at constant film thickness has been described. An extension of this work to more than binary oxides would have been possible, but was not pursued except for some remarks, as follows [43a]: Considering the barrier layer on FeCrNi alloys the first clue is that the mobility of nickel ions in an electric field is probably similar to that of iron. This then readily explains why in the passivating oxide nickel is depleted, as is iron, owing to the higher mobility of chromium ions. The oxide surface again would be expected to be predominantly chromium oxide, perhaps hydroxylated, in the range of primary passivity, while in the range of secondary passivity, where the chromium dissolution rate constant is high, the surface chromium concentration should be low. The observation (see Fig. 10.23, and

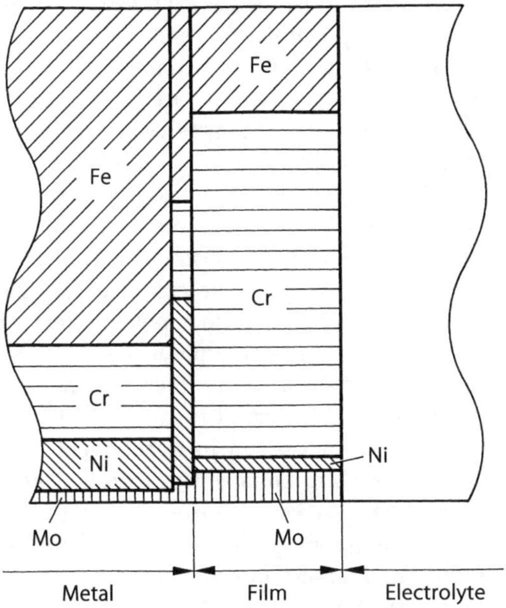


**Fig. 10.23.** Composition of the pasive oxide film on the austenitic alloy Fe-19Cr-11Ni-2Mo passivated in (0.1 M HCl + 0.4 M NaCl) solution at 0.45 V. Also shown: Surface composition of the alloy as caused by selective corrosion.(From Olefjord and Elfström)

[50]) that nickel is enriched in the surface of the alloy can be understood as resulting from a low transfer rate of nickel from the metal into the alloy, which may be caused by the lower thermodynamic stability of NiO as compared to the oxides of molybdenum, chromium, and iron, although this conclusion from thermodynamics to kinetics is not cogent. Concerning molybdenum, it has been remarked above that the pure metal is transpassive in the potential range of passivity of FeCr alloys. Therefore, passivation of FeCrMo alloys, and probably FeCrNiMo alloys as well, would be expected to begin with by growth of an oxide practically free of Mo. In the thickening film, Mo should later be incorporated. As shown in Fig. 10.23, a slight enrichment in Mo is observed, which points to a comparatively low mobility of Mo in the electric field across the film.

## 10.6 Amorphous and Nanocrystalline Alloys

Amorphous metal ribbons can be produced e.g. by melt-spinning of one or more of the transition metals, Fe, Ni, Cr, and Mo, alloyed with one or more of the metalloids, B, C, Si, P, and Ge. They are interesting due to sometimes exceptional mechanical, magnetic, thermal, and electrical properties. Some also have outstanding corrosion resistance owing to unusually stable passivity, and this has caused much interest and many investigations, as reviewed in [60–63]. Some interesting systems are metallurgically simple, as Fe-20 at.% B, others are distinctly complex, such as Fe-36Ni-12P-6B. For multi-component systems, difficulties in elucidating corrosion mechanisms obviously are to be expected. Also, it is not always quite clear whether investigators have been dealing with truly amorphous, or else with nanocrystalline material.

As note previously, increased stability of passivity, as expressed by low values of passive current densities has been tentatively attributed to amorphous oxide films. Expecting that oxide films grown on amorphous substrates may also tend to be amorphous, improved corrosion resistance may be expected for amorphous metals, provided that the metalloid component does not interfere with passivity. At the same time, active dissolution from amorphous metals should, if anything, be favored rather than inhibited, because the thermodynamic activity of atoms in the surface certainly will be increased in comparison to the crystalline material. This has been verified for Co-25B in weakly acid sulfate solution, comparing amorphous and crystalline homogeneous $Co_3B$, whereas crystalline two-phase mixtures of $Co_2B$ and Co were inferior, probably due to local cell action between different crystallografic phases [64]. As reported in [65,66], decreased rates of acid corrosion of amorphous metals, as compared to the crystalline pure metals can be observed, if the cathodic partial reaction of reduction of the oxidizing agent present in the electrolytic solution is inhibited.

As to passivity, alloys containing Fe, or Co, or Ni and one or two metalloids apparently are inferior to the respective pure metals [61]. An example is Fe-13P-7C, which cannot be passivated in 1 N $H_2SO_4$. Some multi-component systems containing Cr or Mo as well as iron are, however, unusually well behaved, when amorphous: The passive current density of metal dissolution in sulfuric acid for Fe-5Cr-13P-7C is lower than that for stainless steel [67].

A striking example of the merits of amorphous alloys is presented in Fig. 10.24 for FeCrMoPC alloys [67c]. Anticipating what will be more fully described in Chap. 12, we note that concentrated hydrochloric acid is a medium normally very difficult to handle even for molybdenum-bearing steels, because of pitting corrosion. Apparently, these amorphous alloys are immune against this danger, and this certainly is a conspicuous advantage. We note that most of the passive current densities, recorded in Fig. 10.24 below the onset of transpassivity, are much too high for a useful engineering material. This, however, probably is only the effect of rapid potential scanning; still it remains to ask what would have been the values observed in steady state.

Resistance to concentrated acid chloride solutions does not necessarily require amorphous substrates. An interesting example are NiCrTa alloys. In the crystalline state, passivity of 60 Ni-10Cr-30Ta apparently is safe even in boiling 6 N HCl, and amorphous 58Ni-10Cr-30Ta-2P is not better behaved. In the same solution, however, 60Ni-20Cr-20Ta fails by rapid dissolution, whereas 58Ni-20Cr-30Ta-2P is passive [67d]. These and other observations raise the question of the composition of passive films on amorphous alloys. Apparently, there is considerable evidence showing that the metalloids phosphorous and boron are

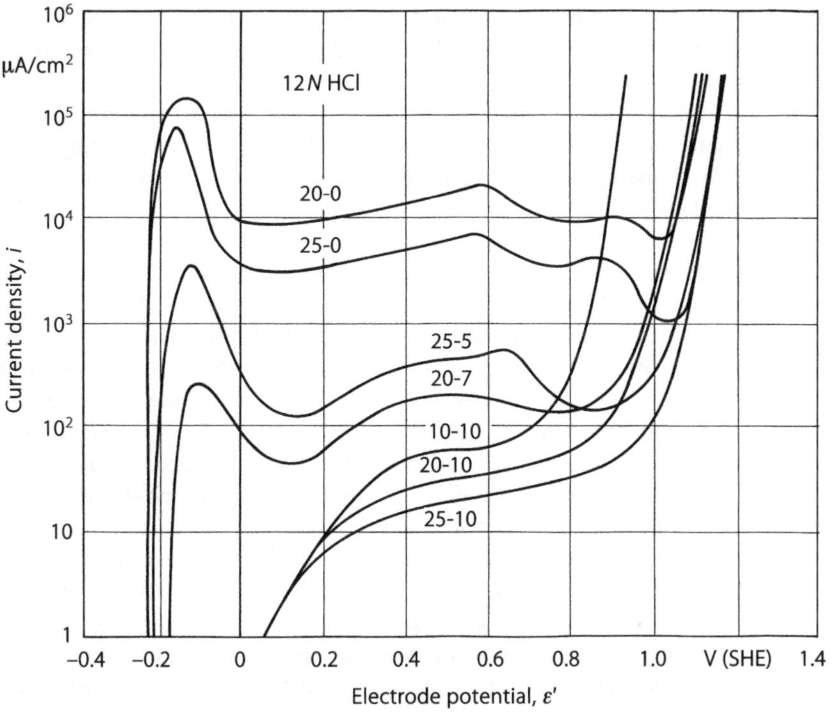


**Fig. 10.24.** Non-stationary current-potential curves for glassy/amorphous Fe-xCr-yMo-13P-7C alloys in 12 M HCl solution at ambient temperature. Cr content $x$ and Mo content $y$ in at.%. Note that for 10 Mo alloys are self-passivating. Potential referred to sat. cal. electrode. (From Kobayashi, Asami, and Hashimoto)

incorporated in the films. For a review of the literature, see [63]. As an example, we quote results from current-potential and impedance measurements in $H_2SO_4$,HCl, and NaCl solutions, comparing crystalline Fe-17Cr with amorphous Fe-10Cr alloys containing B, P, and C in varying amounts between 7 and 13 at.%. In this case, the precursor of passivation apparently is not an adsorbed hydroxide, but a porous film of iron phosphate. After exposure to the solution, this precursor film is rapidly deposited, reducing the critical passivating current density, and it also is an active cathode for hydrogen deposition or oxygen reduction, thus shifting the corrosion potential into the range of spontaneous passivation, as explained in the following section. In the passive state, the main effect apparently is the incorporation of phosphate into the passive oxide film, counteracting any incorporation of chloride from the solution and, thus, counteracting pitting (see Chap. 12). On the whole, this and other evidence supports the view that it is the peculiar composition of passive films on amorphous alloys rather than the amorphous or glassy structure which improves corrosion resistance. This may also explain why amorphous alloys sometimes fail contrary to expectations based on the composition of the material. Failures of this type have been reported for 40Fe-40Ni-20B, 78Ni-8Si-14B, and 58Co-10Ni-5Fe-11Si-16B exposed to sulfuric acid and to weakly alkaline boric acid solutions [69].

For practical applications it would be attractive to precipitate amorphous protective metal films onto engineering metals and alloys with the aim of arriving at exceptionally corrosion resistant passive oxide films. Technically, it will be difficult to precipitate metal/metalloid layers, and the aim rather is the deposition of amorphous alloys without metalloids. On first sight, this does not seem very promising, since it is the metalloids which stabilize the amorphous state. The deposition of thin amorphous metallic alloy layers has however been achieved by direct-current magnetron sputtering of the desired alloy components from small disks onto glass substrates. This work has been reported in a series of papers since 1993, reviewed in [67e]. Binary surface alloys, either amorphous or nanocrystalline, of Cr or Mo or W, with Ti, Zr, Nb, and Ta were produced and found to be passive and resistant to pitting in concentrated hydrochloric acid. The passive oxide films are described as composed of double oxyhydroxides of the alloy components, and they are notable because frequently the passive current density of metal dissolution is lower than that of either component. A distinct advantage of the method is that by sputter-depositing practically any binary alloy can be obtained regardless of its thermodynamic stability. Sputtering onto glass substrates is, of course, not what is sought for practical applications. However, the interior of austenitic stainless steel tubes has also been used as a target for sputtering from a central rod of tantalum, producing amorphous FeNiCrTa layers, showing good corrosion resistance in boiling concentrated nitric acid. Apparently, resistance to pitting was not yet the aim of this application.

The procedure of surface alloying of stainless steel with tantalum may also be seen as an extension of previous and more conventional demonstrations that mild steel surfaces can be alloyed with chromium by rapid laser melting of electrodeposited or sputter deposited chromium into the the surface [70]. Still in the same context it should, then, also be noted that already rapid laser-remelting of mild steel surfaces sometimes definitely improves passivatability of the material [71].

## 10.7 Semiconducting Oxide Films, Spontaneous Passivation, and Passivating Inhibitors

In this section, the interesting point is spontaneous passivation of iron and iron-base alloys by semiconducting passive oxide films, and the action of passivating inhibitors. The latter are reducible substances shifting, by the superposition of the cathodic reduction current, the free corrosion potential of a metal into the range of passivity. Typical oxidizing redox couples interesting in this context are $Fe^{3+}/Fe^{2+}$, or $Ce^{4+}/Ce^{3+}$, and many others, as described further below. Figure 10.25 explains the action for the case of acid corrosion below the hydrogen

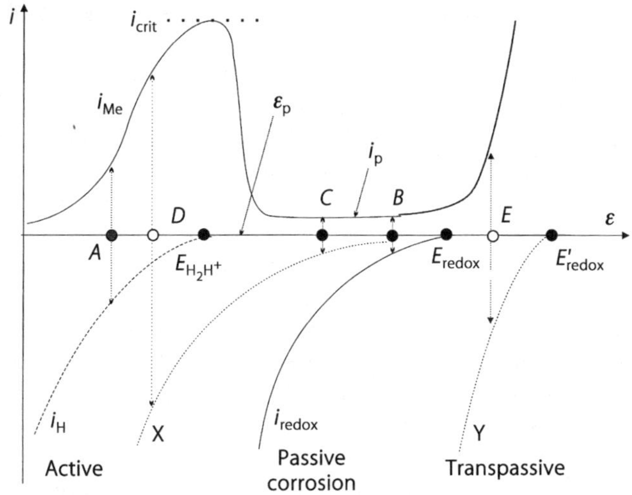


**Fig. 10.25.** Current-potential diagram explaining passivating inhibition for metals with active/passive and passive/transpassive transition, and semiconducting passivating films. $i_{\text{crit}}$, $i_p$ : Critical passivating and passive current density. $\varepsilon_p$ : Average active/passive transition potential. Equilibrium potentials (●): $E_{H_2H^+}$ for hydrogen deposition, $E_{\text{redox}}$ for regular, and $E'_{\text{redox}}$ for excessive inhibition. For corrosion potentials (○) A...E, and for current-potential curves $i_H$,$i_{\text{redox}}$, X and Y, see the text. Vertical arrows indicate active and transpassive corrosion rates

equilibrium potential. Without inhibition, the open-circuit corrosion potential is A, established by the superposition of the current-potential curve $i_H$ for hydrogen deposition. Superposing, instead, the current-potential curve for inhibitor reduction, $i_{\text{redox}}$, the corrosion potential shifts to position B, and the metal is passive. To achieve this, two conditions must be met: In the first place, $E_{\text{redox}}$ must be well beyond the passivation potential, $\varepsilon_p$; at the same time, the system must overcome the current peak before the active/passive transition, and this requires, that $|i_{\text{redox}}(\varepsilon_p)| > i_{\text{crit}}$.

The diagram also shows a possible, unsafe situation where by superposition of the inhibitor-reduction curve in a position, such as X, the steady state corrosion potential can be established either at C in the range of stable passivity, or at D in the range of active, and even accelerated corrosion. A previously passivated metal, activated by some accident, will then not return to passivity; and inhibition is both unstable and dangerous.

The critical behaviour just described must not be confused with the danger of action of oxidizing inhibitors in cases where pitting or stress corrosion cracking is imminent beyond some critical potential threshold beyond $\varepsilon_p$. For details, we refer to later chapters; here we only note that it is this for this reason that passivating inhibitors sometimes are generally classified as 'dangerous' in the sense that there application then does require expert knowledge of circumstances.

A further undesirable case, also shown in the figure, is excessive inhibition, when the inhibitor-reduction equilibrium is in position $E'_{\text{redox}}$ rather than $E_{\text{redox}}$, and when the inhibitor-reduction overpotential is low, drawing the steady-state corrosion potential into the range of rapid transpassive dissolution.

Supplementing what has been described in Chap. 5, we need to consider the transfer of both metal cations and electrons across the semiconductor/solution interface. For metal dissolution from the oxide surface into the electrolytic solution we expect, in principle as before, an exponential function relating current density and interfacial Galvani potential difference, $\phi_{ox,l}$. For electron transfer across the interface, the situation is, however, different, though this was not yet noted further above, when anodic oxygen evolution at passive electrodes was briefly described. The important point, so far disregarded, is the concentration of electrons in the oxide at the film/solution interface. Concerning this topic, the following is a brief outline only; for more information, competent reviewing literature is available [72a,73]. For special details, see also [72b,c,74e,20i].

A well investigated semiconducting film is the oxide on iron passivated by external potentiostatic polarization to an electrode potential in the range of the potential-independent and constant passive current density, $i_{\text{pass}}$, of metal dissolution. We discuss this case to illustrate the concepts and notions involved in semiconductor theory applied to passivity, rather than deriving theoretical principles in general. For experiments as well as theory, we directly refer to investigations into the behaviour of passive iron in deaerated nitrate/borate solutions at pH 8.4, where $i_{\text{pass}}$ is practically negligible [74a,b]. Therefore, we may disregard the electric field driving metal cations through the film. We remember that electronic equilibrium prevails throughout the metal and the oxide film, and that the electrochemical potential of the electrons, $\tilde{\mu}_{e^-}$, is the same throughout the metal and the film. We also recall from Chap. 3 that the electrochemical potential is the differential reversible work, referred to 1 mol, to take electrons from the uncharged infinity into the metal, and thus is equal to the Fermi energy, $E_F$.

Figure 10.26 schematically shows the situation: The film is near-stoichiometric $Fe_3O_4$ close to the metal and near-stoichiometric $\gamma$-$Fe_2O_3$ close to the solution. Following the authors of [74a,b], we simplify matters by disregarding the probably continuous change in cation valence across the film. Instead, the film is thought to be sandwiched, with a thin layer of $Fe_3O_4$ on the metal under a layer of $\gamma$-$Fe_2O_3$. At the same time, the film is barrier-type throughout, and any additional deposit on top of the film would complicate the picture. Finally, the oxide is assumed to be crystalline, so that conventional semiconductor theory is applicable.

Concerning $Fe_3O_4$ the situation greatly simplifies because this oxide has excellent, almost metallic conductivity, and we do not need to explicitly consider semiconductor properties. Therefore, the thickness of the $Fe_3O_4$ layer does not matter. This is due to the massive doping of the substance with $Fe^{2+}$ in the role

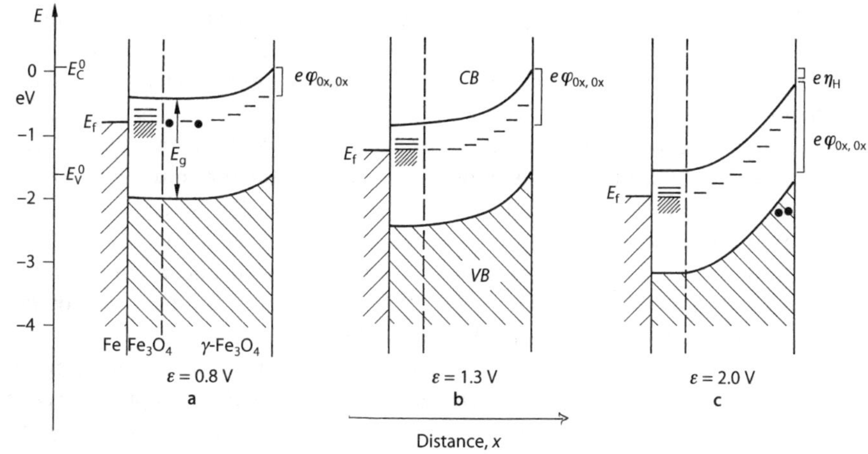


**Fig. 10.26a–c.** Semiconductor model of the barrier $Fe_3O_4/\gamma$-$Fe_2O_3$-film on passive iron. Semiconduction is n-type due to doping with $Fe^{2+}$. Doping of $Fe_3O_4$ is abundant, and conductivity in this oxide is near-metallic. Doping of $\gamma$-$Fe_2O_3$ is by residual $Fe^{2+}$ in near-stoichiometric oxide. Abscissa: Distance in the oxide film from the metal surface. Ordinate: Gibbs energies, of electrons in the metal and in the two oxides layers, noted as '$E$'. $E_F$: Fermi energy of electrons, assumed constant throughout the metal and both oxide layers. In the oxides, $E_F$ is within the gap between the nearly completely occupied valence band and the nearly empty conduction band, but close to the conduction band owing to the presence of the donor states. Gibbs energies are arbitrarily referred to the Gibbs energy of electrons in the solution-side oxide surface, $E_c^\circ = 0$. **a,b,c**: Occupied (-•-) and empty ( – ) donor states, and band bending at three characteristic values of the electrode potential, $\varepsilon$ (SHE). (From Stimming and Schultze)

of donors of excess electrons. In $\gamma$-$Fe_2O_3$, however, the concentration of these donors is small, where 'small' means ca. $10^{20}$ $cm^{-3}$ donor states. Therefore, the composition is near-stoichiometric. For the layer of this oxide of constant thickness, $d$, the figure shows the almost completely filled valence band, the bandgap of 1.6 eV, the almost empty valence band, and the Fermi level. Conduction is either caused by delocalized mobile excess electrons in the conduction band or delocalized and also mobile defect electrons (holes) in the valence band. In our case, the substance is an n-type semiconductor due to the existence of localized donor states only some 0.1 eV below the lower edge of the conduction band. These are the immobile, or more precisely nearly immobile residual $Fe^{2+}$ ions in the role of occupied donor states and $Fe^{3+}$ ions on the same lattice places as empty donor states. We note that an empty donor state also is an immobile, trapped defect electron. We also note that owing to the equilibrium between donors and excess electrons the overall electrochemical potential of electrons remains unchanged. The probabilities, $W_c$ resp. $W_v$, for electrons to enter the conduction band, resp. for holes to enter the valence band, is proportional to the differences in energy between the lower edge of the conduction band, $E_c$, resp. the upper edge of the valence band, $E_v$, and the Fermi energy,

$E_F$ : $W_c \sim \exp(-(E_c - E_F))$, rsp. $W_v \sim \exp(-(E_F - E_v))$ . From this it follows that in our case, except for special effects of excessive band bending described below, semiconduction in $\gamma$-$Fe_2O_3$, is mainly due to excess electrons in the conduction band.

So far, the influence of the adjacent electrolytic solution has not entered the picture. The conditions prevailing at the semiconductor/solution interface are, however, decisive for the kinetics of electron transfer reactions, and the important point is that, owing to the low conductivity in the oxide, the situation is very different from that for a contact between a metal and a concentrated elctrolytic solution. Now, the rigid Helmholtz part in the solution side of the double layer normally is but a small part of the total potential difference, and the greater part is in a space charge layer in the semiconductor adjacent to the solution. We denote this space charge part of the total potential difference as $\phi_{ox,d} = \phi_{ox} - \phi_d$, and the Helmholtz part in the solution as $\phi_{d,l} = \phi_d - \phi_l$. Considering this we see that what enters the Butler-Volmer equations describing the kinetics of ionic charge transfer reactions at the semiconductor/solution interface is the Helmholtz potential difference, $\phi_{d,l}$, and this quantity replaces what imprecisely has been noted as $\phi_{ox,l}$ and $\phi_{d,l}$ further above. Fortunately, the point is not crucial for what has been derived previously. In the present context it is more important to note that in the range of electrode potentials where, at constant pH, $i_{\text{pass}}$ and, therefore, $\phi_{d,l}$ are constant, any change $\Delta\varepsilon$ of the electrode potential is caused by a change in the potential drop inside the oxide, $\phi_{0,d}$.

The space charge locally changes the Galvani potential in the oxide, and thus also changes $-e_\circ\phi$ or, referred to 1 mol, $-F\phi$, which enters the expression for the local Gibbs electronic energy. This, then, is *band bending*, schematically shown in Fig. 10.26 for three characteristic values of the electrode potential.

Bending is referred to the *flatband* lower band edge, and the Galvani potential in the oxide beyond the space charge layer is the *flatband potential*, which we refer to SHE and denote as $\varepsilon_{fb}$. At the flatband potential, the surface charge in the oxide is zero, so is the surface charge in the electrolytic solution, and $E_{fb}$ is equal to the potential of zero charge, already introduced in Chap. 5. For the film on passive iron, $E_{fb}$ is $-0.1$ V.

The band structure schematically shown in the figure has been deduced from experimental values of the capacity $C$(μF cm$^{-2}$), of the oxide film [74a]. For this purpose, $C$ was not determined from alternating-current impedance data (see Appendix), but from current transients following potential pulses of 2 ms duration. The method is equivalent to measuring the a.c. capacity at frequencies exceeding 10 kHz. In several series of experiments, the potential was slowly stepped, between fast pulses, from 0 to 1.2 V. By this method, values of the capacitance, $C$, were available as function of the potential. In Fig. 10.27, $1/C^2$ is shown as a function of the electrode potential in a *Mott-Schottky* plot. This is convenient, because for low values of $(\varepsilon - \varepsilon_{fb})$ the capacity of the oxide film is dominated by the capacity of the space charge layer, and described by the *Mott-Schottky equation*

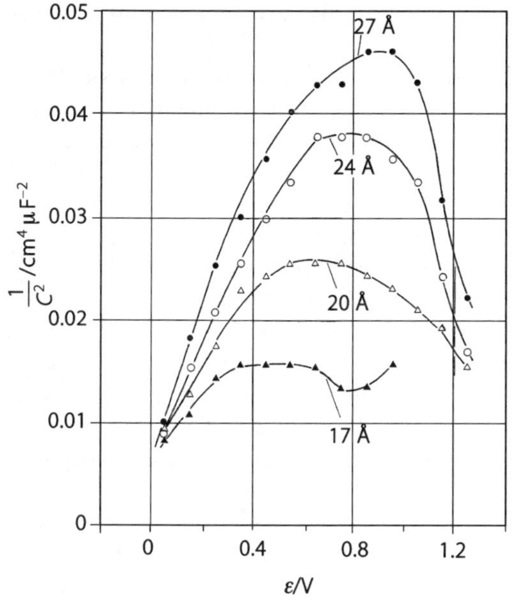


**Fig. 10.27.** Mott-Schottky plot of the capacitance of passive iron in 1 M nitrate solutions buffered to pH 8.4 by borate. Parameter of the group of curves is the thickness of the $\gamma$-$Fe_2O_3$ film, practically equal to the total thickness of the passive film. (From Stimming and Schultze)

$$\left(\frac{1}{C}\right)^2 = \frac{2}{\epsilon\,\epsilon_\circ e_\circ N}\left(\varepsilon \quad c_{fb} - \frac{kT}{e_\circ}\right) . \tag{10.28}$$

Here, $\epsilon_\circ$ is the vacuum permittivity, $\epsilon \simeq 12$ is the dielectric constant (or relative permittivity) of the oxide; the other quantities have been defined before. The relation predicts a linear increase of $(1/C)^2$ with increasing potential, intersecting the abscissa at the flatband potential. As seen in Fig. 10.27, a relation of this type holds as expected for low values of $\varepsilon$ up to 0.4 V. We disregard the changes of the initial slope of the curves with the oxide film thickness, $d$. Instead, we note that in the vicinity of 0.8 V the capacity, $C$, tends to level at a value which decreases with increasing $d$. As indicated in Fig. 10.26b, this is explained by the exhaustion of donor states. In this range of potentials, the film essentially behaves as an insulator, with a capacity inversely proportional to film thickness: $C = \epsilon\epsilon^\circ/d$. Beyond this range of approximate constancy, $C$ increases with increasing potential: As indicated in Fig. 10.26c, band bending in this range is such that in the surface the upper edge of the valence exceeds $E_F$. This indicates that the semiconducting oxide is now *degenerate*; and also indicates that the valence band becomes charged by mobile defect electrons turning semiconduction to p-type. The following final increase of $C$ at potentials exceeding 1.3 V is accompanied by the onset of anodic oxygen evolution, filling holes in the valence band with electrons from oxygen ions, also accompanied by a change in the Helmholtz potential difference, $\phi_{d,l}$.

Concerning the kinetics of cathodic reduction currents at the oxide/solution interface, we once more restrict the consideration to potentials below the onset of oxygen evolution and transpassivity, i.e. to potentials where changes in $\varepsilon$ are equal to changes in $\phi_{ox,d}$. We now note that, owing to the negative charge, the space charge layer in the n-type semiconducting oxide is depleted with respect to mobile electrons moving into the interior flatband region where the local electronic energy is lower. By this effect, the concentration of electrons in the conduction band decreases with increasing electrode potential. It is this change in surface availability of electrons which now enters the rate of cathodic reduction of, for instance, a passivating inhibitor at the oxide surface. Electron transfer proceeds by 'adiabatic' tunneling through the Helmholtz layer to energy states of the redox couple at equal height. Without proof we note that the contribution, $i_c$, of the conduction band to the (cathodic) current density is

$$i_c = -i_c^\circ \left(1 - \exp\left(-F\,\eta/RT\right)\right) , \tag{10.29}$$

whereas the contribution from the valence band, negligible in the present case of passive iron, is

$$i_c = -i_c^\circ \left(\exp\left(F\,\eta/RT\right) - 1\right) . \tag{10.30}$$

The overpotential of the redox electrode reaction is $\eta = \varepsilon - E_{\text{redox}}$. From these relations it follows that the semiconductor surface is blocked for anodic currents, while at sufficiently negative values of the overpotential the current-potential relation is a Tafel line. We can, therefore, continue describing cathodic current-potential curves in the Butler-Volmer terminology: $i_{\text{red}} = i^\circ_{\text{redox}} \exp(\alpha\, F\, \eta / RT)$, although close to the equilibrium potential, $E_{\text{redox}}$, the curves are highly asymmetrical.

Two points merit at least cursory comments. The first is that, when the thickness, $d$, of the passivating film does not exceed two or three nm, electron transfer from the metal to the redox couple in the solution by direct tunneling may considerably contribute to the total cathodic current. The second is the possible role of amorphous or glassy passive oxide films. Crystallinity is no prerequisite of semiconduction in oxides. When, however, semiconducting material is partially or completely amorphous, the sharp band edges turn into tails of decreasing density of electronic states, and localized states increasingly populate the bandgap. One of the consequences is that resonance tunneling of electrons via donor states in the bandgap further contributes to the total current. In this context, photoelectrochemistry should be quoted as a useful tool for in-situ investigations into the energies of electron distribution in bands and bandgaps. For reviews and general comments see [74c,75a]. The essential effect is that by illuminating the oxide film with light of frequency $\nu$, and with energy $h\nu$ exceeding the bandgap energy, excess electrons are injected into the conduction bands from the valence band, leaving defect electrons. In the case of an n-type semiconductor with a negative space charge, the defect electrons then concentrate at the oxide surface, but the excess electrons move away as before, and the cathodic reductive current remains unaffected, although a photocurrent might still be observed as different from the background reductive current, owing to other faradaic processes such as valence changes in the oxide. In the absence of redox couples in the solution, the latter current would be directly observable. This type of measuring photocurrents, always with the aim of analysing the band structure of semiconductors including amorphous materials, has been used frequently. As an example we quote investigations where the photoresponse, together with the capacity and with light absorption, of iron in borate solutions without redox couple content was studied [37i]. The results were compared to that observed for 500-nm layers of $Fe_3O_4$, doped by partial oxidation, on gold substrates. It was found that these certainly crystalline artificial layers were quite useful as artificial models of the passive films, showing that whatever the contribution to the photoresponse from intra-bandgap electronic states might have been does not need not be due to a glassy structure. From this it appears that assuming crystalline true passive films is not as doubtful as on first sight it might seem considering the properties of ultra-thin films.

As described above, the passivating oxide film on iron can turn to p-type semiconduction by inversion, and anodic currents, as in anodic oxygen evolution, can be observed. This similarly holds for passive nickel as well. Not the passive oxide films, but other oxides including NiO, are p-type semiconductors allowing only positive currents to flow. In other cases the bandgap is too large to allow for noticeable conductivity. Oxides of this type, as especially $Al_2O_3$, technically are *insulators*, with important consequences for the corrosion behaviour, as described further below.

Turning from metals and oxides with near-metallic conductivity to semiconductors and, finally, to insulators as passivating films, we naturally expect a decrease in the value of exchange current densities for redox reactions. This is illustrated in Fig. 10.28 by a compilation of data from different sources for reductive and/or oxidative $Fe^{3+}/Fe^{2+}$ currents at various electrodes [74d]. For better comparison, all current densities have been divided by the respective concentration of the redox couple in the electrolytic solutions. The data for non-passive gold and platinum exemplify the normal high values of exchange current densities of the redox system. So-called passive platinum is covered with an adsorbed layer of oxygen, only little impeding electrode reactions. As an oxide with almost metallic conductivity, $RuO_2$ behaves similarly to bare metals. NiO, as a p-type semiconductor, is seen to allow only much lower rates of the electrode reaction. Also, while at passive nickel anodic and cathodic currents are possible, cathodic electron transfer is blocked at massive NiO. Otherwise, passive nickel is comparable in behaviour to NiO, with exchange current densities lower by several orders of magnitude as compared to bare metals. At passive iron, the reaction rate is still more depressed, although electron transfer is again possible in both directions. Massive $SnO_2$ is an n-semiconductor allowing cathodic currents only. At the same time, currents are much lower than at NiO owing to an increase in the bandgap from 2.3 to 3.6 eV. Titanium oxides, which will become a point of special interest in the next section, as well as tantalum

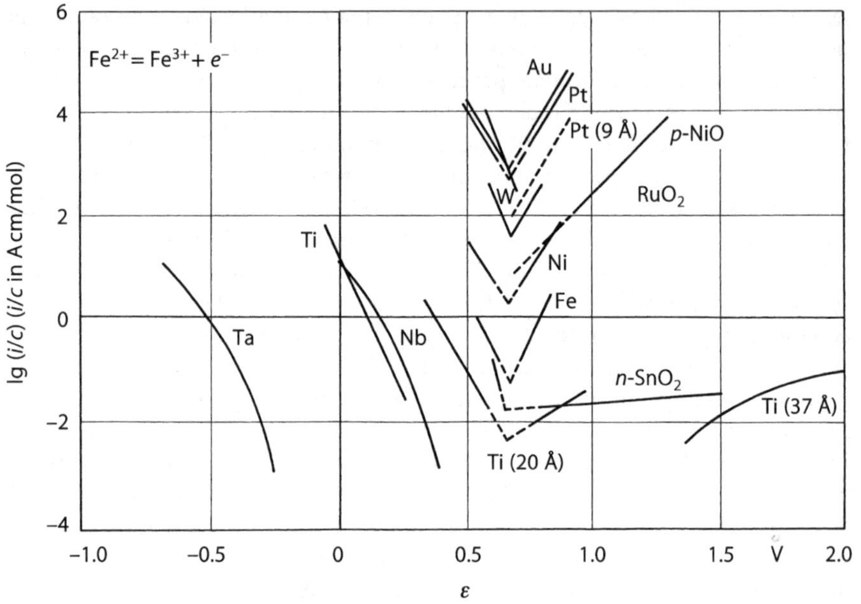


**Fig. 10.28.** Logarithmic current-potential curves for the redox reaction $Fe^{3+} \leftrightarrows Fe^{2+}$ at bare platinum and gold, at massive ruthenium dioxide, nickel oxide and tin dioxide, and at passive platinum, tungsten, nickel, iron, titanium, niobium and tantalum. The current densities are normalized by division by the concentrantion of the dissolved redox couple. (Data from various sources compiled by Schultze and Mohr)

and niobium oxides are, in principle, n-type semiconductors, and technically insulators.

As noted above, passivating inhibitors shift the free corrosion potential from the range of active corrosion into that of passivity, and redox couples are particularly useful substances of this type owing to there normally high exchange current densities. In this context, then, the obvious practical question concerns the role of the omnipresent dissolved oxygen as a typical oxidant. By the very positive value of the equilibrium potential, briefly noted as $E_{O_2} = 1.21 - 0.059\,\text{pH}$ (V), the first condition for spontaneous passivation, $E_{O_2} > \varepsilon_p$, will almost always be satisfied. At the same time, as described in Chap. 5, exchange current densities for oxygen reduction usually are very small due to the inherent stability of the oxygen molecule, as compared with that of, e.g., the hydronium ion. According to the rate equation: $\frac{1}{2}(O_2)_l + (H_2O)_l + 2(e^-)_m \rightarrow 2(OH^-)_l$, the reaction involves the passage of electrons from the metal into the solution. In the case of passivation by semiconducting films, semiconductor electrode reaction kinetics will apply to the reaction step of electron transfer, and overpotentials required for oxygen reduction will tend to very high values, indeed. The second condition for spontaneous passivation, $|i_{\text{cath}}(\varepsilon_p)| > i_{\text{crit}}$, will then often be the obstacle to passivation. In this context, the very low values of both $i_{\text{crit}}$ and $i_{\text{pass}}$ typical

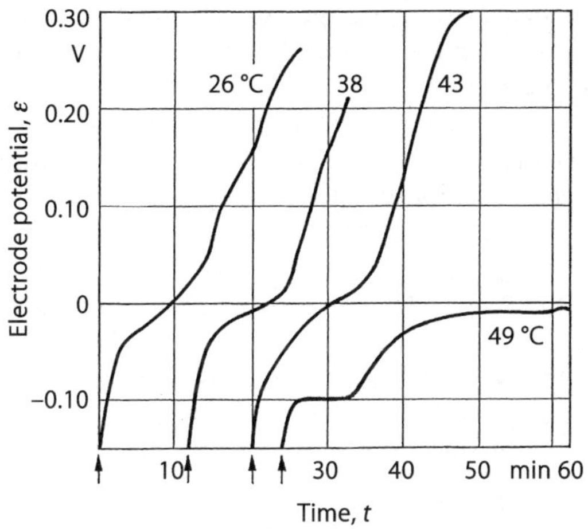


**Fig. 10.29.** Spontaneous passivation of an 18 wt.%Cr-9wt.%Ni-1.5wt%Mo-steel in 20% $H_2SO_4$ after activation by cathodic prepolarization, recorded for increasing temperatures. Note non-passivation at 49°C. (From Rocha and Lennartz)

for stainless steels are interesting, as they suggest spontaneous passivation of these alloys even in acid solutions, if aerated. Early experiments confirming this conjecture are shown in Fig. 10.29 [39]. We do not comment on the obviously complicated kinetics of spontaneous passivation of alloys previously activated by cathodic polarization involving oxide film reduction, and only note that, apparently, spontaneous passivation fails in hot acid solutions. In this work, the stationary value of $\varepsilon_{corr}$ is not recorded, but from what has been said for the cathodic oxygen overpotential, it probably is clear that, eventually, $\varepsilon_{corr}$ will be observed to remain far below $E_{O_2}$. In fact, typical corrosion potential values for passive metals such as stainless steels in electrolytic solutions rarely exceed ca. 0.1 to 0.2 V. We shall return to this point in Chap. 11.

Concerning practical aspects of corrosion protection we note that passivation can, of course, also be achieved by external polarization, and this method does have important applications, as described in the Appendix. We also note that, since passivity finds its limit of useful application at the threshold potential of transpassivity, careful control of the potential is important. This is illustrated by case Y in Fig. 10.25. A special case is concentrated nitric acid, where the very effective, strongly oxidizing species is the undissociated $HNO_3$. Exposed to hot concentrated nitric acid, stainless steel has a corrosion potential close to the transpassivity threshold, and this would have been a point of some concern for the abandoned techniques of chemical reprocessing of radioactive waste. We do not comment on the complicated kinetics of $HNO_3$ reduction reactions; we also do not comment on such processes of practical interest as prepassivation of stainless steels again by nitric acid. Instead, we consider the more straightforward model cathodic kinetics of such systems as the $Ce^{4+}/Ce^{3+}$ or

$Fe^{3+}/Fe^{2+}$ couples. These also are strong oxidizing agents, with standard equilibrium potentials 1.44 and 0.77 V, respectively. Figure 10.30 presents the results of a well made investigation into the passivation of chromium steels, showing the partial current-potential curves for metal dissolution and passivation in acid solution, also for passivator reduction, and their superposition to the total current-potential curve for the metal in inhibited solution [76]. The shift of the open-circuit corrosion potential from ca. −0.44 V indicating active acid corrosion, to ca. −0.25 V indicating stable passivity is clearly visible. As seen from Fig. 10.30b, the cathodic current for $Fe^{3+}$ reduction ends in a plateau region, and this also is the current effective at the peak passivation potential of the steel, $\varepsilon_{p1}$. Apparently, this is the diffusion-limited current density for inhibitor reduction, $i_{Fe^{3+},D}$, which from Chap. 5 we know to be $-3\,F\,D_{Fe^{3+}}\,c_{Fe^{3+}}/\delta_{Nernst}$. In the work presently reported, the Nernst diffusion layer thickness, $\delta_{Nernst}$, was well defined by using cylinders axially rotating at controlled speeds, $v$. Passivation could easily be observed by recording the change of the corrosion potential into the range of passivity. As shown in Fig. 10.31a, this change occurs at decreasing rotational speeds, when, at constant acid concentration, the concentration of the passivator $Fe^{3+}$ is increased, confirming expectations concerning the influence of the ratio $c_{Fe^{3+}}/\delta_{Nernst}$. The situation is explained by the schematic current-potential diagram in Fig. 10.32. It should be noted that in this diagram a possible increase of the critical passivating current density, $i_{crit}$, owing to an acceleration of the dissolution of a salt film preceding passivation in acid solutions has been neglected.

Figure 10.31b shows a second series of measurements with constant speed of specimen rotation and increasing acid concentration, shifting the initial potential for active corrosion to increasingly more positive values owing to a decrease in hydrogen deposition overpotential, and an increase in $i_{crit}$. To match, at constant $\delta_{Nernst}$, this obstacle against passivation, $i_{Fe^{3+},D}$ had to be increased by increasing $c_{Fe^{3+}}$.

When, as assumed in the case just described, the presence of the dissolved passivator leaves the current-potential curve for anodic metal dissolution unchanged, the passivator acts as a kind of chemical potentiostat. This is illustrated by Fig. 10.33 for the case of nickel in acid solutions, showing the result of a series of measurements a) of the current-potential curve in pure acid, and b) of open-circuit potentials and open-circuit corrosion rates observed in inhibited acid solutions [38b], As seen from the figure, experimental pairs of values of the corrosion potential and the corresponding current density, determined by chemical analysis of the solution, perfectly match current-potential curve through the complete range of active corrosion, passivation, passivity and transpassivity.

As already described, chromium in FeCr alloys lowers both the passivation potential and the critical passivating current density. Figure 10.34 illustrates that, for a given and constant cathodic partial current-potential curve, an important shift of $\varepsilon_{corr}$ in the direction to more noble values is expected at a critical threshold chromium composition. The schematic figure neglects a possible, and in fact probable, increase of the overpotential for oxygen evolution on passive

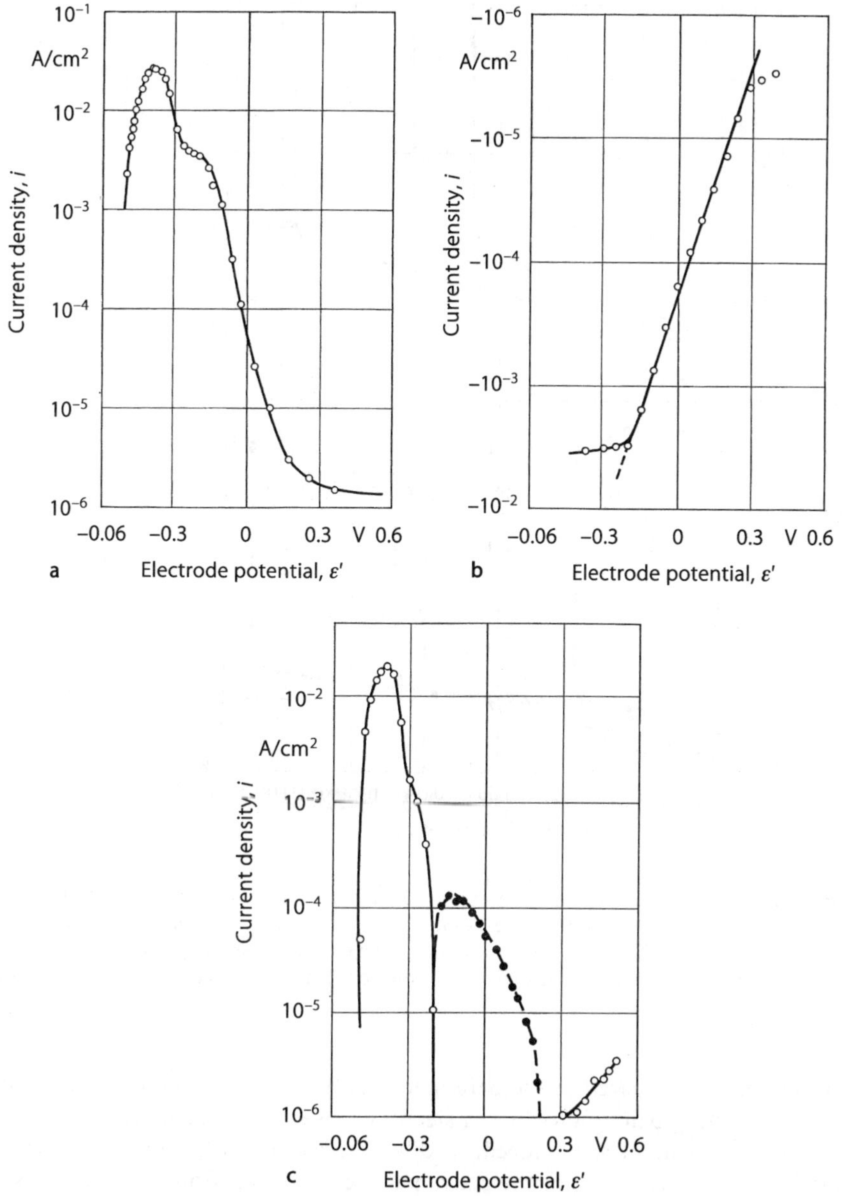


**Fig. 10.30a–c.** Passivation of Fe-12wt.%Cr steel (AISI 410) in deaerated 0.52 N $H_2SO_4$/0.08 N $Na_SO_4$ solution at 30° C. **a** Anodic partial current-potential curve for metal dissolution, $i_{Me}(\varepsilon)$; **b** cathodic partial current-potential curve for reduction of $Fe^{3+}$ in the same solution with added 0.014 mol kg$^{-1}$ $Fe(SO_4)_2$. **c** Total current-potential curve, $i_{Me}(\varepsilon) + i_{cath}(\varepsilon)$. In **c**, open circles are anodic, solid circles, cathodic currents. Electrode potentials are referred to sat.cal.electrode. (From Makrides and Stern)

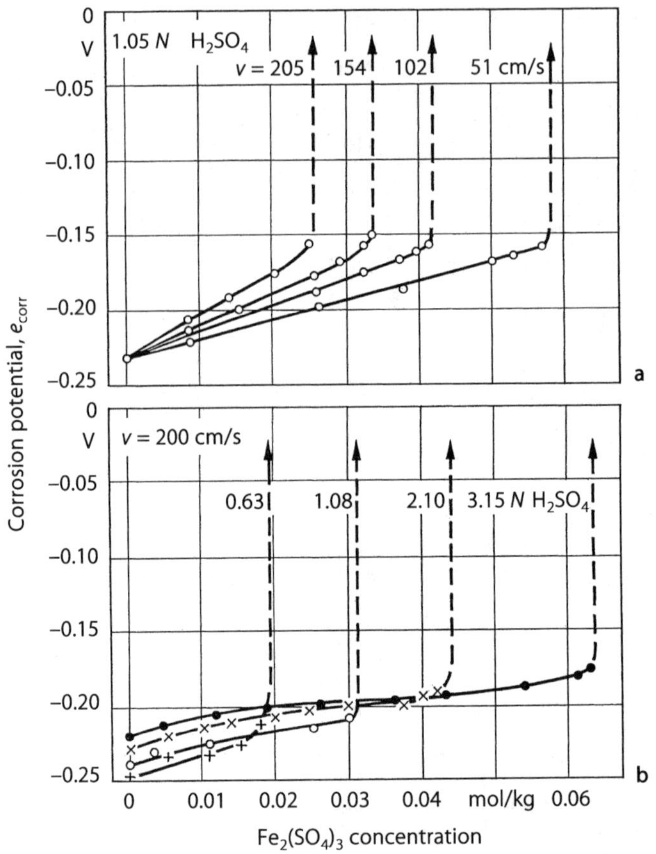


**Fig. 10.31a,b.** Spontaneous passivation of Fe-12 wt.%Cr steel in deaerated $H_2SO_4/Fe(SO_4)_2$ solution at 30° C as function of the $Fe(SO_4)_2$ concentration. **a** Constant concentrations 0.52 N $H_2SO_4$/0.08 N $Fe(SO_4)_2$ and variation of the circumferential speed $\upsilon$ cm s$^{-1}$; **b** constant circumferential speed and variation of the concentration of $H_2SO_4$. (From Makrides and Stern)

chromium as compared to the bare metal surface owing to the presence of the semiconducting oxide film. Such a steep increase of the corrosion potential at a certain Cr content was, indeed, readily observed in early work, as shown in Fig. 10.35 [77]. The shift of the active/passive transition from 13 to 10 wt.%Cr should be noted as particularly instructive, although not surprising after what has been explained further above: Adding hydrogen peroxide to the solution has the same effect as greatly increasing the oxygen content, and thus lowers the chromium content required for spontaneous passivation.

The critical passivating current density, $i_{crit}$, decreases with increasing solution pH (see Figs. 10.7 and 10.8). This, then, readily explains the change to spontaneous passivity with increasing pH shown in Fig. 6.12 in Chap. 6.

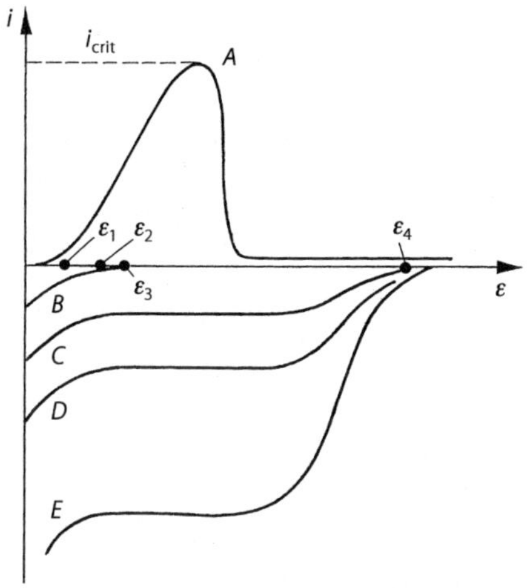


**Fig. 10.32.** Current-potential diagram schematically explaining the active/passive transition as achieved by decreasing $\delta_{\mathrm{Nernst}}$. **A** Anodic currents of metal dissolution; **B** cathodic currents of hydrogen evolution; $\varepsilon_1$: $\varepsilon_{\mathrm{corr}}$ for acid corrosion in deaerated solution. **C,D,E** Cathodic currents of oxygen reduction with increasing values of the diffusion-limited current density owing to decreasing $\delta_{\mathrm{Nernst}}$ $v$. $\varepsilon_2,\varepsilon_3,\varepsilon_4$: Accompanying shift of $\varepsilon_{\mathrm{corr}}$

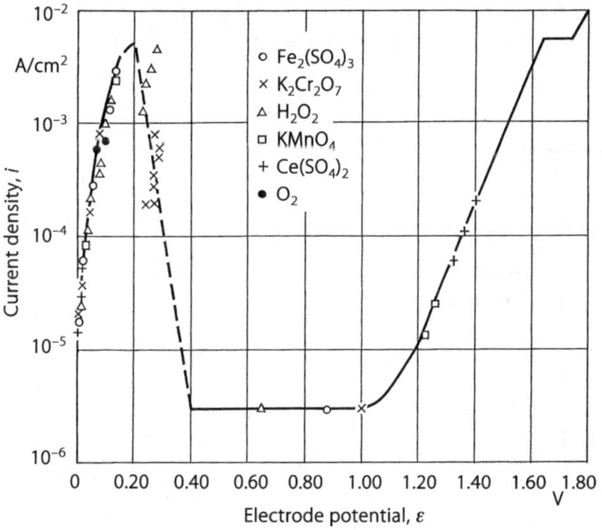


**Fig. 10.33.** The anodic current-potential curve for nickel in $1\mathrm{N}\,H_2SO_4$ in the absence of passivators ( — - -), recorded by external polarization, and experimental points obtained for the corrosion potential and the anodic current density by adding various oxidizing species to the solution, and by chemical analysis of dissolved nickel. (From Kolotyrkin and Bune)

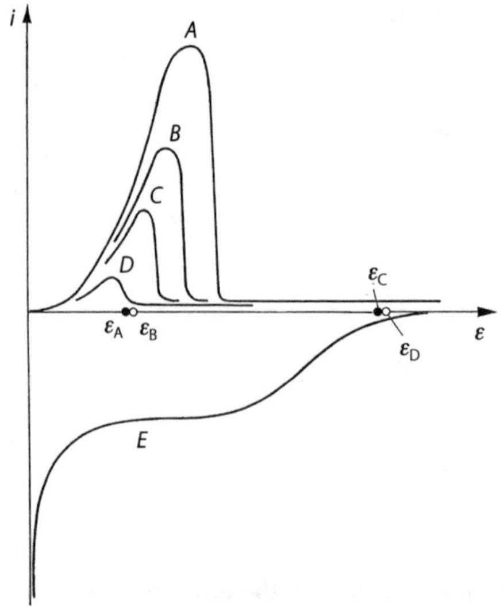


**Fig. 10.34.** Schematic current-potential diagram showing conditions for the spontaneous passivation of FeCr alloys in aerated acid solutions. **A,B,C,D:** Partial current-potential curves for anodic metal dissolution as determined by the chromium content. **E:** Cathodic partial current-potential curve for oxygen reduction from aerated solutions, and superimposed cathodic hydrogen deposition. $\varepsilon_A$,$\varepsilon_B$,$\varepsilon_C$,$\varepsilon_D$: Corrosion potential as determined by the Cr content. Note change to spontaneous passivation, when $|i_{O_2,d}| > i_{crit}$

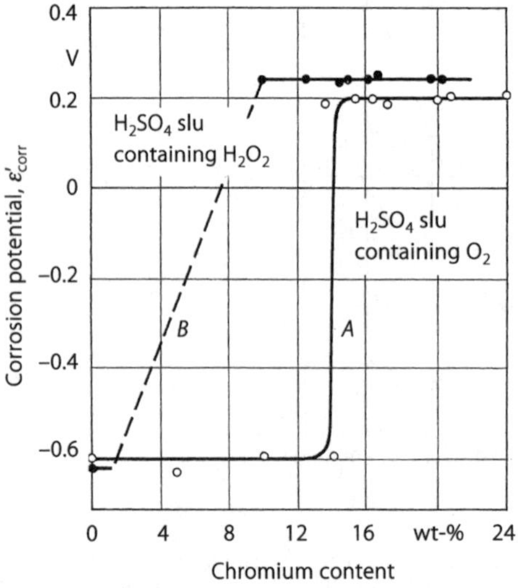


**Fig. 10.35.** Stationary corrosion potentials of FeCr alloys in dilute sulfuric acid solution, referred to N/10 KCl cal.electrode. Open circles, A: aerated solution. Solid circles, B: $H_2O_2$ added to the solution. (From Strauss)

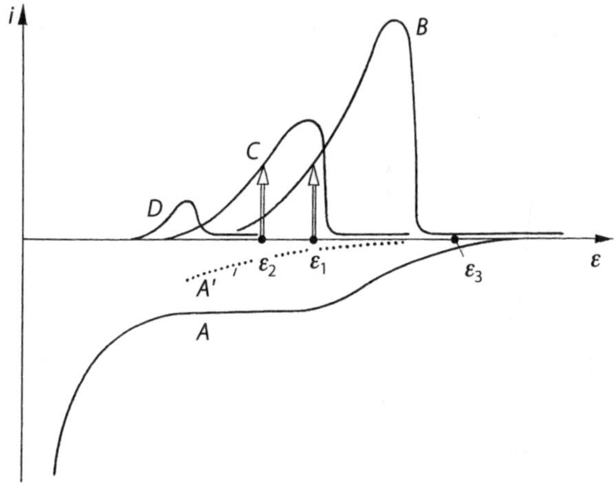


**Fig. 10.36.** Current-potential diagram for corrosion of iron with rate-determining cathodic oxygen reduction in the range of neutral to weakly alkaline solutions. **A:** Partial cathodic current-potential curve for oxygen dissolution with superimposed hydrogen deposition. **B,C.D:** Partial anodic curves for metal dissolution, assuming acceleration of active dissolution and decrease of the critical passivating current density with increasing pH. $\varepsilon_1$,$\varepsilon_2$: Corrosion potential for active metal dissolution. $\varepsilon_3$: Corrosion potential for stable passivity in alkaline solutions. Note that, when the situation is as depicted in the figure, $\varepsilon_3$ also is a possible corrosion potential for 'unsafe' passivity in neutral solutions. A′: Cathodic partial curve indicating an important decline in current densities of oxygen reduction due to the presence of a relatively thick semiconducting passive oxide film

As shown in Fig. 6.13, the spontaneous passivation of plain steels in sufficiently alkaline solutions occurs without an distinct shift of the corrosion potential. In this case, other than for high-alloy chromium steels, the schematic current-potential diagram apparently must be corrected, in order to account for an important increase in cathodic activation overpotential for oxygen reduction caused by the presence of the oxide film. In Fig. 10.36, this is indicated by curve A′.

An inexpensive agent to increase the pH of aqueous solutions with the aim of passivating plain steels is sodium carbonate, soda, which hydrolyses acording to: $CO_3^{2-} + H_2O \rightarrow HCO_3^- + OH^-$. Soda can be replaced by organic bases. The oxidizing species now is dissolved oxygen, and this also holds for other non-oxidizing substances commercially used as passivating inhibitors, such as phosphates or silicates, which also decrease the critical passivating current density, although not through an increase in pH, but through a protective layer of scarcely soluble iron compounds such as phosphates and the like. As an instructive example, Fig. 10.37 in the first place shows that passivation of plain steel can already be achieved by sufficiently increasing the oxygen content of the water [78a]. The critical oxygen content is greatly reduced by the addition of polyphosphates to the water, and this clearly indicates a decrease in the critical passivating current, probably caused by inhibitor adsorption. Once pas-

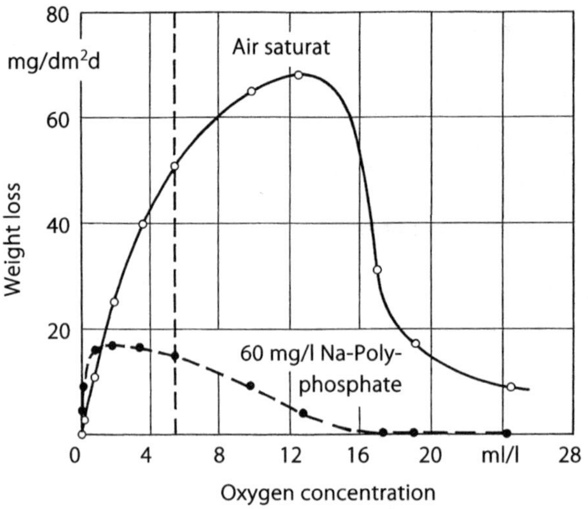


**Fig. 10.37.** The average corrosion rate of mild steel in oxygenated water as a function of the oxygen content. Solid curve: pure water. Broken curve: Water with added sodium polyphosphate as an inhibitor. (From Uhlig, Triadis, and Stern)

sivation is achieved, the oxide film is expected to have the usual composition. For more on passivating inhibition by phosphates, see [79a,b]. It should once more be noted that safe passivation requires that the danger of pitting corrosion is negligible.

Other well known passivating inhibitors, such as $NaNO_2$, $Na_2CrO_4$, $Na_2MoO_4$, and $Na_2WO_4$ are themselves oxidizing. Passivation of iron by chromate has been closely investigated, with the main result that the substance clearly is incorporated into the passivating oxide film [78b, 79b,c,d]. For more detailed systematic investigations into the action of chromate, permanganate, tungstate, pertechnate, and the surprisingly ineffective perrhenate see [79e].

So far, spontaneous passivation has been treated as a desirable effect requiring the passage through the activ/passive transition. In practical operation, the situation usually is different: The important point then rather is how to assure passivity starting from a surface already carrying an air-formed oxide film, i.e. how, if necessary, to change this into a truly passivating film. One well known method suitable for passivation by semiconducting oxide films is prepassivation in oxidizing acids such as nitric acid. In the case of insulating oxide films to be described further below, the useful answer often is high-voltage anodizing.

In aerated, but highly concentrated sulfuric acid, ordinary stainless steels fail, as documented by weight loss data presented in Fig. 10.38 [80], but small additions of only between ca. 0.1 and 1 wt.% of of percent Cu, or Pt, or Pd drastically improve the behaviour. A similar effect, again on first sight surprising, is observed for passive titanium, and the explanation offered in more detail in the next section probably is, mutatis mutandis, the same in both cases: Apparently, Pt or Pd, and in the case of stainless steels also Cu, plate out on the metal sur-

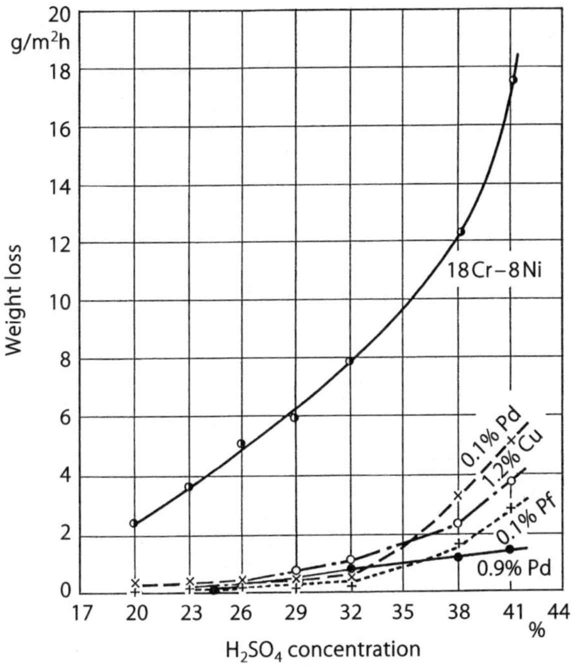


**Fig. 10.38.** The influence of small additions of Cu, or Pd, or Pt to 18Cr-8Ni steels on the rate of weight loss in concentrated sulfuric acid solutions of increasing acid concentration, at 20°C. (From Tomashov)

face by selective alloy dissolution, decreasing the overpotential of the respective cathodic partial reaction, which for stainless steels is oxygen reduction, and thus shifting the corrosion potential beyond the active/pasive transition threshold.

## 10.8 Titanium, Aluminum

The so-called *valve metals*, including titanium, aluminum, tantalum, niobium, and zirconium, are examples of materials with often outstanding corrosion resistance, in every case due to passivity. In this group, passivating oxide films typically have asymmetric conductivity, blocking anodic reactions except, at high applied voltages, oxide growth by anodization. Of this group, we discuss titanium and aluminum only.

Concerning *titanium*, the standard potential for the $Ti/Ti^{2+}$ couple is listed in Table 3.1 as $= -1.63$ V. For the $Ti^{2+}/Ti^{3+}$ redox couple, the tabulated value usually is $-0.37$ V, which gives a value $-1.21$ for the standard potential of the $Ti/Ti^{3+}$ electrode. These numbers are, however, highly controversial, since the $E^{\circ}_{Ti^{2+}/Ti^{3+}}$-value is almost certainly erroneous. It might be as low as $-2.3$ V [81], making the $Ti^{2+}$ an extremely unstable ion, so that active titanium dissolution

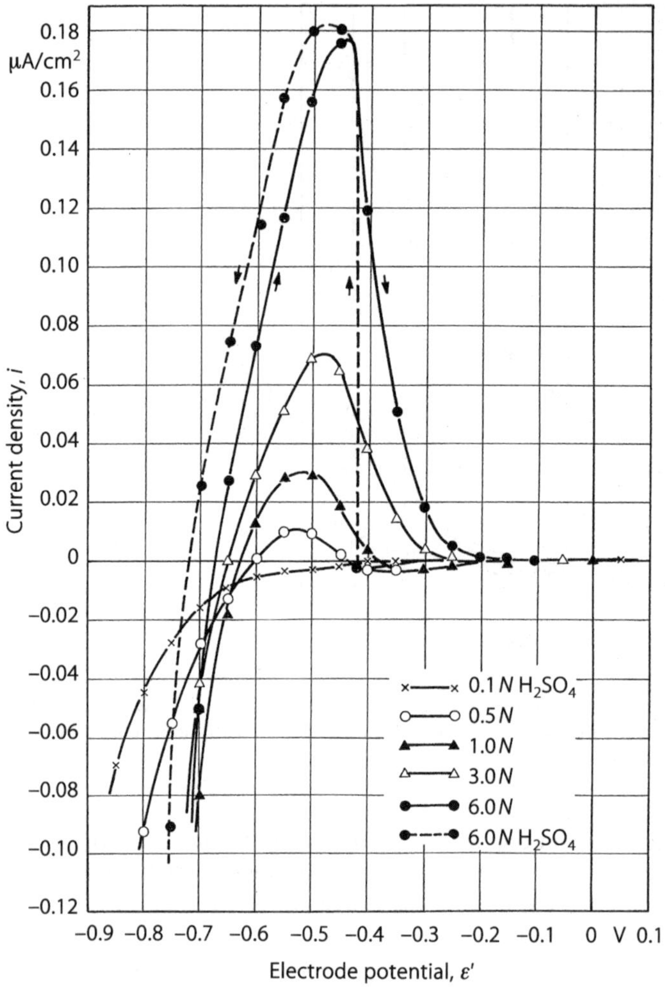


**Fig. 10.39.** Stationary current-potential curves for titanium of commercial purity at 25°C in sulfuric acid solutions.Potentials referred to sat. cal. electrode. (From Brauer and Nann)

probably proceeds directly to dissolved $Ti^{3+}$. In any case, active dissolution, accompanied by cathodic hydrogen evolution, is thermodynamically favored under most conditions, so that passivation is the prerequisite for any practical use of the metal. Considering the role of titanium alloys, especially Ti-6Al-4V (wt.%), as bone implant material, the importance of stable passivity is then quite obvious. With respect to the kinetics of passivation and passivity we shall not, however, comment on alloys, as detailed investigations are rare. As a warning we note, however, that for bone implants mechanical wear of the passive film in moving joints is a problem [83]. Ion implantation into the passive film, as described for pure titanium further below, may improve wear resistance of implant materials [83b,c], but this again is beyond the scope of this text. Other titanium alloys, especially Ti-8Al-1Mo-1V for supersonic aircraft, will be considered in the context of stress corrosion cracking, but again detailed knowledge as to the kinetics of electrode processes involved will be seen to be scarce. Even pure titanium is capricious, for instance concerning the role of the hydride, $TiH_2$, and the relevant discussions are controversial. In the following, we concentrate on essentials, referring the reader for more details to [82].

As an engineering material, commercial titanium is difficult to handle owing to impurity. This should not greatly affect essential electrode kinetics, but we note that the electrodes used in investigations reported below usually are of commercial purity only. Typical current-potential curves are shown in Fig. 10.39 [84], and the typical sequence of events in passing from the active to the passive state are recognized: Active acid corrosion followed by an increase of the current of metal dissolution to a critical passivating density, decreasing with decreasing acid concentration, followed by the active/passive transition between potentials $\varepsilon_{p1}$ and $\varepsilon_{p2}$, followed by a range of low passive current densities. In 20%$H_2SO_4$, the critical current density $i_{crit}$ is several $\mu A\,cm^{-2}$ [76c]; in 1N$H_2SO_4$, it has dropped to $0.1 \mu A\,cm^{-2}$ [30b].

Figure 10.40 gives more precise values of the very low passive current densities, which may, however, still be higher than true steady-state current densities [85]. The weak increase of anodic currents, observed at high values of the electrode potential, should be due to some slow and perhaps only transient oxygen evolution in spite of the valve metal properties of the metal.

A useful aspect is the very negative active/passive transition potential, indicating spontaneous passivation already by the superposition of cathodic hydrogen evolution to the anodic metal dissolution. The overpotential of cathodic hydrogen deposition is, nevertheless, high. In deaerated 0.1 N $H_2SO_4$ the metal is spontaneously passive, but at higher acid concentrations of sulfuric as well as in hydrochloric and hydrofluoric acids titanium remains active [86]. As expected, the metal is passiv in oxidizing acids and in most neutral solutions. Unless polarized to high anodic potentials, titanium is immune to pitting corrosion in chloride solutions and thus resistant to sea-water, but crevice corrosion is dangerous. Fluorides in acid solutions accelerate acid corrosion until a visible dark salt layer eventually becomes protective. The corrosion potential then is unusually negative at ca. −0.95 V [86].

The passivating oxide is anatase or rutile, $TiO_2$. The mechanism of passivation is difficult to investigate owing to the high reactivity of the bare metal. One suitable method is recording galvanostatic intermittent charging curves obtained from electrodes activated by mechanical scraping [14e]. Figure 10.41 illustrates the three states of pseudo-activity, prepassivation and passivation of the metal, as conjectured on the basis of such experiments. The similarity of the supposed sequence of events with that previously described for iron is obvious. As for iron, assuming a less distinctly sandwiched passivating oxide layer, with $TiO_2$ as the main constituent, is probably more realistic. Also similar to iron, there is

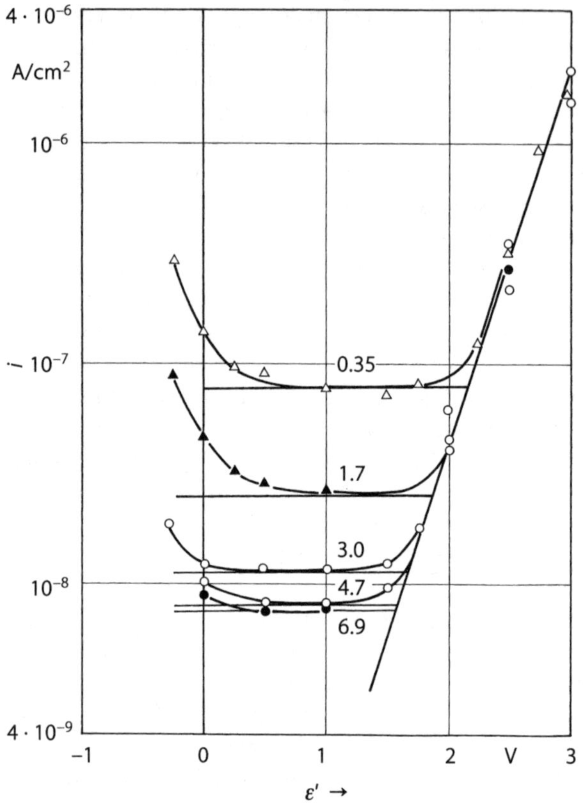


**Fig. 10.40.** Quasi-stationary current-potential curves for titanium (99.6 wt.%) at 25°C in deaerated solutions containing $H_2SO_4$, $NaHSO_4$, $CH_3COONa$, and $NH_4COONa$ at different values of pH. Potentials referred to sat. cal. electrode. (From Hurlen and Wilhelmsen)

a stage in the kinetics where after preliminary hydroxylation of the surface by intermediate products, according to: $Ti \rightarrow Ti(OH)_{ad} \rightarrow Ti(OH)^{+}_{ad} \rightarrow Ti(OH)^{2+}_{ad}$, the reaction can proceed, depending on circumstances, either to dissolved $Ti^{3+}$, indicating active dissolution, or else to intermediate oxides and, finally, to $TiO_2$, causing passivation [82]. A quite different model of the electrode kinetics of titanium has been proposed in [87], based on the assumption that allegedly 'bare' titanium always is hydrided. The presence of some hydride, surviving anodic passivation, has indeed been established by reflection electron diffraction recorded from a titanium surface passivated in acid solution by anodic polarization [84]. Table 10.1 lists the reflexes identified, by comparison with ASTM tables, as indicating the presence of $TiH_2$. Besides the hydride, nonstoichiometric TiO compounds were observed at this potential. When, however, initially passivated titanium was polarized to potentials only slightly below the active/passive transition potential, the metal activated without re-formation of hydride. Therefore, the hydride probably was a remnant from a cathodic pre-

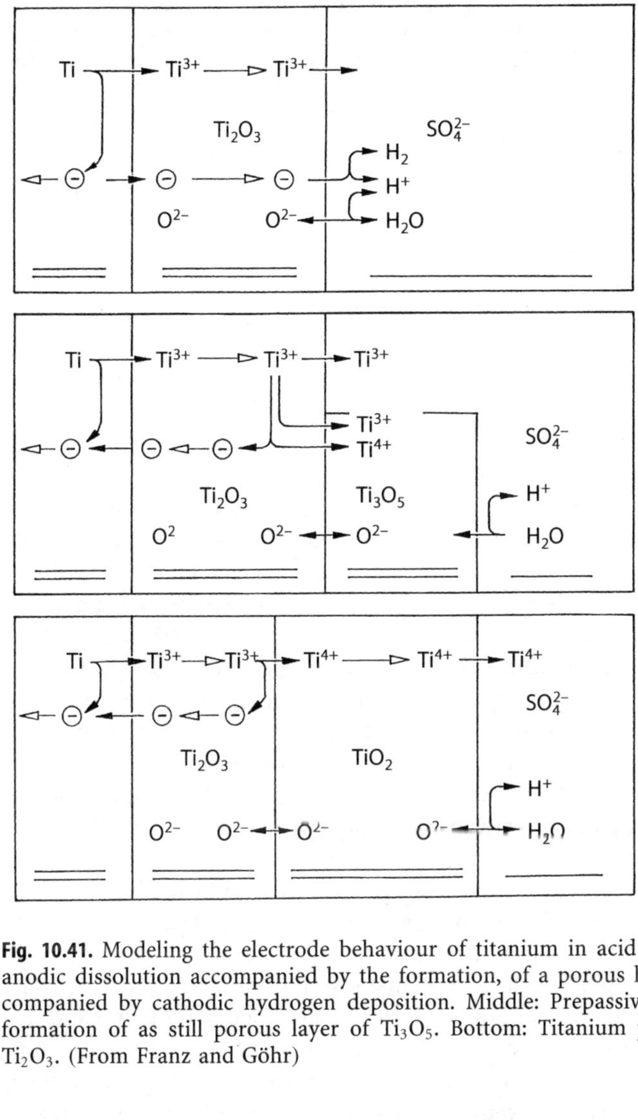


**Fig. 10.41.** Modeling the electrode behaviour of titanium in acid solutions. Top: Active anodic dissolution accompanied by the formation, of a porous layer of $Ti_2O_3$, also accompanied by cathodic hydrogen deposition. Middle: Prepassivation by the transient formation of as still porous layer of $Ti_3O_5$. Bottom: Titanium passivated by $TiO_2$ on $Ti_2O_3$. (From Franz and Göhr)

treatment of the metal, and in fact it reappeared below −0.72 V. Beyond the active/passive transition, i.e. beyond $\varepsilon_{p_2}$, the passivating oxide was $TiO_2$. Elsewhere, it has been claimed that the film rather is $TiO_2(H_2O)_{1.4}$ or $TiO_{0.6}(OH_{2.8})$ [46], and some hydroxylation and hydratation of passive films does in fact generally seem probable at least close to the active/passive transition. At potentials distinctly more positive than this threshold, the presence of $TiO_2$ probably cannot well be contested.

Since $TiO_2$ is an n-type semiconductor with oxygen vacancies in the role of donors, anodic polarization of the metal causes the film to grow, and films of up to ca. 100 nm can thus be produced, giving rise to interference colours. Oxide

**Table 10.2.** Proof of the presence of titanium hydride on titanium polarized to −0.45 V (sat.cal.el.) during 15 hours in 6 N $H_2SO_4$: Comparison of lattice spacings calculated from electron diffraction peaks, and peak intensities, with tabulated ASTM values for $TiH_2$. (Data form Brauer and Nann)

| Experimental values | | ASTM $TiH_2$ | |
|---|---|---|---|
| d(nm) | Intensity | d(nm) | Intensity |
| 0.256 | strong | 0.256 | 100 |
| 0.221 | weak | 0.221 | 40 |
| 0.156 | middle | 0.156 | 60 |
| 0.133 | strong | 0.133 | 60 |
| 0.128 | very weak | 0.127 | 20 |
| 0.109 | very weak | 0.110 | 5 |
| 0.102 | middle | 0.102 | 30 |
| 0.099 | middle | 0.099 | 30 |

growth is, however, accompanied by both transient anodic oxygen evolution and anodic $Ti^{4+}$ dissolution [64d,e,]. From ellipsometry it is known that the film thickens linearly with applied potential by between 2 and 3 nm/V [46h]. Recently, the differences in behaviour of oxide crystallites grown on different substrate grain surfaces has been investigated using a combination of micromethods for electrochemical, ellipsometric, and photochemical techniques, and atomic force imaging suitable to elucidate the semiconductor properties as a function of substrate grain orientation [74f,g]. The most conspicuous result was that oxide grains grown on close-packed (0001) faces of titanium allowed considerable rates of anodic oxygen evolution, while oxide grains grown on lower-density crystal faces proved to be purely n-type, allowing oxide growth but not electron transfer across the oxide/solution interface. Oxide growth on lower-density substrate faces was comparatively rapid, producing homogeneous films apparently with a low density of oxygen vacancies, while growth on close-packed faces was slower resulting in highly doped material.

Titanium is an important structural material for the chemical industry. It also was a candidate for the containment and as electrode material for the electrochemical Purex process planned to recover uranium and plutonium form radioactive waste, and it still is a candidate for the final disposal of high-radioactivity waste. Therefore, radiation damage, especially by $\gamma$-radiation, i.e. hard X-ray radiation, is a much investigated topic, and one aspect is the photoeffect exerted by radiation on the semiconducting oxide film on titanium. In this respect, long-term low-energy $\gamma$-irradiation can be simulated by short-term, high-energy (ultraviolet)-laser irradiation [74h].

To understand irradiation effects, we assume that the charge in the oxide surface is negative, and that electron-state band bending is such that the space charge region is enriched with respect to holes and depleted with respect to electrons. Three effects are, then, possible: The first is that electron holes, noted as $h^+$, may react with water molecules to oxygen: $4h^+ + 2H_2O \rightarrow O_2 + 4H^+$. The second is that holes may also react with the oxide to dissolved titanium ions and, again, oxygen: $4h^+ + TiO_2 \rightarrow Ti^{4+} + O_2$. Finally, the rate of overall oxide formation according to: $Ti + 2H_2O \rightarrow TiO_2 + 4H^+ + 4e^-$, may either increase or decrease, depending on other circumstances.

Using similar micromethods as described above, and focusing the laser beam to a diameter of ca. 10 μm, far below that of the oxide crystallites, oxide films grown in sulfuric acid on titanium to a thickness of up to 50 nm, at potentials up to 15 V, were irradiated with Ar laser light of 257 nm wavelength and a power density in the kW $cm^{-2}$-range. The main effect was an acceleration of oxide growth, producing oxide highly disordered by large amounts of defects in the bandgap. Growth acceleration was higher for smaller crystallites, probably because photo-effects on the high-field growth mechanism are more intensive where the oxide film is thinner. These observations imply that focused laser illumination did not cause local oxide film thinning, which would amount to photo-induced pitting corrosion. Apparently, some anodic oxide evolution was also observed, which is typical for transient oxide growth processes on passive titanium.

In similar context, implantation of palladium, iron and xenon ions into the passive oxide film on titanium (and on hafnium) had been investigated earlier, combining records of implantation by Rutherford backscattering with oxide film analysis by XPS, photochemical, and electrochemical measurements [74i]. Passivation was carried out in sulfuric acid at potentials up to 20 V, producing $TiO_2$ films of up to 50 nm thickness. The effect of ion implantation into the oxide was an increase in the dielectric constant and in the donor state density, as well as an increase in electron states within the bandgap, indicating an amorphization of the oxide film.

The passivation potential of titanium is more negative than the hydrogen equilibrium potential, but in concentrated non-oxidizing solutions the metal still is not self-passivating, apparently owing to the high overpotential of hydrogen deposition. Looking for possibilities to lower the hydrogen overpotential to achieve spontaneous passivation, the deposition of a porous layer of such metals as palladium or platinum on the titanium surface appears suggested, because these are metals useful as hydrogen equilibrium electrodes owing to high values of exchange current densities, $i_H^o$, and, thus, generally low hydrogen overpotential. As it turns out, a practicable method is to alloy titanium with several tenths of percent of either Pd or Pt [76c, 88–90], because the desired porous layer formation then is achieved by initial selective titanium dissolution from the alloy. The schematic diagram Fig. 10.42 and its legend explains the effect, and Fig. 10.43 presents experimental evidence for the spontaneous passivation small of titanium 0.1 wt% platinum in hot sulfuric acid, which otherwise is difficult to handle [89a]. The same effect might be achieved by adding a platinum or palladium salt to the solution, as the noble metals would plate out onto the titanium substrate, but this technique would hardly be practical. Short-circuiting of pure titanium as anode with a noble metal electrode as cathode would serve the same purpose, and this, then, would be an example of *anodic protection* (see Appendix). The mechanism assumed above to be effective for

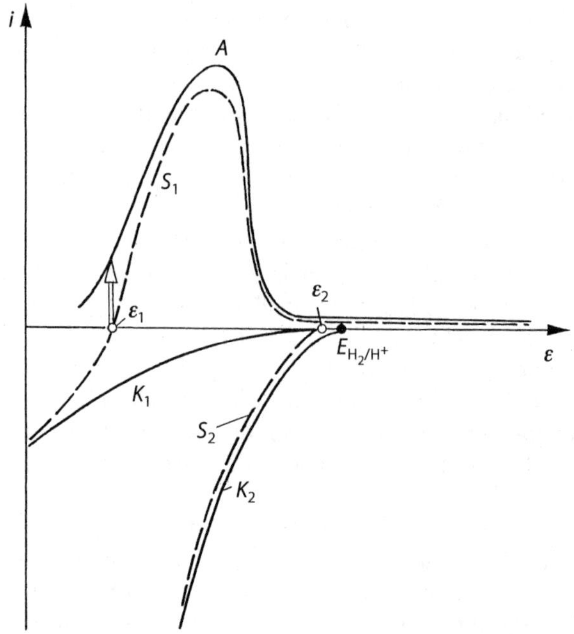


**Fig. 10.42.** Spontaneous passivation of titanium alloyed with small amounts of palladium or platinum. $E_{H_2/H_+}$ : Hydrogen equilibrium potential, more positive than the passivation potential. **A:** Partial Current-potential curve for Ti dissolution. $K_1$,$K_2$ : Partial current-potential curves for hydrogen deposition on pure Ti, and on platinum or palladium, respectively. $S_1$, $\varepsilon_1$ and *open arrow:* Total current-potential curve, corrosion potential, and corrosion rate of active titanium. $S_2$ and $\varepsilon_2$ : Total current-potential curve, and corrosion potential for TiPd or TiPt alloys after formation of porous noble-metal surface film

self-passivation of TiPd alloys in fact is essentially the same on a microscopic scale: As sketched in Fig. 10.44, noble-metal crystallites are supposed to play the role of local cathodes, where hydrogen is deposited at low overpotential, short-circuited to regions of passivated titanium as local anodes. In this model, complications possibly arising from the presence of intermetallic phases such as $Ti_2Pd$ are disregarded [90]. The ohmic voltage between local electrodes is negligible in acid solutions. According to this model, local cathodes out of metallic contact with the alloy surface would be ineffective. Therefore, it is interesting to note that, when the $TiO_2$ film is sufficiently thin, gold plating onto the oxide film also accelerates electron transfer trough the oxide, although there is no metallic contact between the gold deposit and the titanium substrate. Apparently, this effect is due to changes in electron states in the bands of the semiconducting film induced by gold deposition onto the film [74j]. Ion implantation effective in the same direction was already mentioned further above. At this point, the use of platinized titanium electrodes for installations of cathodic protection may be noted as an item of considerable practical interest (see Appendix).

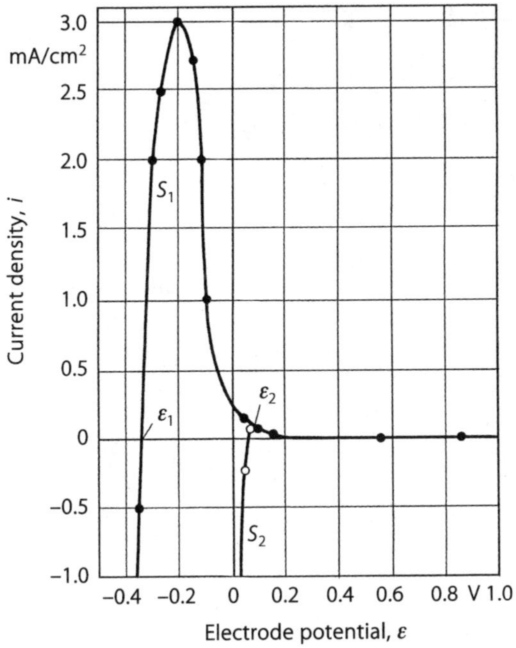


**Fig. 10.43.** Total current-potential curves of commercially pure titanium (•) and of titanium alloyed with 1 wt.% Pt(○) in 40% $H_2SO_4$ at 50°C demonstrating self-passivation of alloyed material. (From Tomashov, Tschernova, and Altovski)

*Aluminum* is the second prominent valve metal owing corrosion resistance to passivity only. For increased strength, the metal is alloyed with several wt.% of Mg, Zn, Cu, and, more recently, Li, producing a large number of commercial alloys, all usually inferior in corrosion resistance as compared to the pure metal. Therefore, plating of Al alloys with pure Al is standard procedure at least in the aerospace industry. Obvious difficulties from the holes in the plated material, necessary for placing rivets, are counteracted by rivets 'swimming' in inhibitive greases; or else the problem is evaded by preferring glued structures. Many of the adverse consequences of alloying will be more fully discussed further below in chapters dealing with intracrystalline and intercrystalline corrosion and with stress corrosion cracking. For pure aluminum, the typical danger is pitting, as extensively described in Chap. 12, including counteraction by inhibitors especially of the 'conversion' type, i.e. by inhibitors such as chromate reacting with aluminum and/or its oxide.

Passivation is always spontaneous, since in this case the passivating oxide film is identical with the air-formed oxide film. At the same time, aluminum is *amphoteric*, i.e. the oxide, or else the hydroxide formed from the oxide during dissolution, is easily soluble in alkaline as well as in acid solutions by the reactions

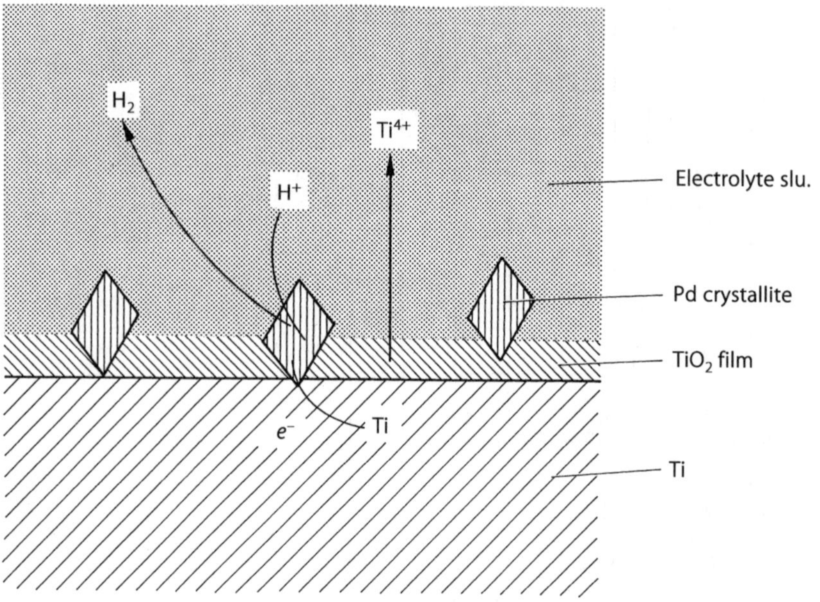


**Fig. 10.44.** Modeling spontaneous passivation of a Ti-1 wt% Pt alloy in deaerated acid solution by selective titanium dissolution leaving Pd crystallites in metallic contact with the alloy surface: Short-circuited galvanic cells formed by passivated titanium local anodes and Pd crystallites as local cathodes

$$\text{a)}\quad Al_2O_3 + 3H_2O \rightarrow 2Al(OH)_3 \tag{10.31}$$

$$\text{b)}\quad Al(OH)_3 + OH^- \rightarrow Al(OH)_4^-$$

$$\text{c)}\quad Al(OH)_3 + 3H^+ \rightarrow Al^{3+} + 3H_2O\,.$$

This makes the metal extremely useful only for everyday corrosion conditions, where aggressive solutions normally are more or less neutral. Outdoor aluminum facades are typical examples of suitable applications, although once fashionable aluminum panels rolled to mirror gloss have not always met expectations. Facades are typical in other respect as well, since as a rule they are *anodized* and, in addition, *vapor sealed* for increased protection.

Anodizing, by preference but not necessarily of pure aluminum, with the aim of thickening the oxide film without concurrent anodic metal dissolution and oxygen evolution requires that the oxide has no electronic conductivity, and $Al_2O_3$, is, indeed, a typical insulator. Oxide thickening is achieved by applying a high voltage across the film in suitable electrolytic solutions, making aluminum the anode. Actually, as described further above, the high-field mechanism of ionic migration across oxide films was deduced from anodizing experiments with direct current voltages typically in the 100 V-range. During oxidation, aluminum ions migrate towards the oxide/solution interface, and oxygen ions migrate to the oxide/metal interface, so that oxide growth occurs at both phase boundaries. Concerning the techniques, we have to distinguish between anodizing in nearly neutral solutions such as boric acid, and anodizing in sulfuric acid. In the first case very thin barrier-type oxide films are obtained, less interesting for corrosion protection than as dielectric material for electrolytic capacitors. The thin air-formed oxide film on aluminum existing without additional anodizing also is of the barrier type. Anodizing in acids, which on first sight may appear strange, since the oxide tends to dissolve, has in fact outstanding merits: By this process, porous oxide grows on top of a bottom barrier layer to films of – in the case of high quality treatment – up to 20 μm thickness. The pores can be sealed by treating the anodized sheets in hot water or in water vapor, as noted above, plugging the pores with hydroxide. Outdoor panels forming facades of buildings thus treated are very durable in clean atmospheres; they may still last decades without repair when air pollution is not more than average. The porous oxide film can be impregnated with colours, which is, however, not our point of interest. For the technical details of anodizing processes, see [21].

Porous oxide films with a thickness of several tens of micrometers, as produced by acid anodizing of aluminum, and especially if sealed by additional treatment, probably should be listed under the heading of corrosion protection by inert coatings rather than by passivation. The same may be claimed for non-porous films grown by anodizing to only tenths of a micrometer. At the same time, the properties of these thickened non-porous oxide films should be very similar to that of the non-porous air-formed film, or the non-porous film on non-anodized aluminum in aqueous solutions. These natural films probably are boehmite, but perhaps partly amorphous as well, with a thickness of some nanometers, and they also are of the barrier-type passivating oxides.

For barrier films grown by anodic oxidation, simultaneous cation and anion migration has been confirmed by more recent investigation [91a,92a,b]. They do not only grow on pure aluminum, but on alloys as well, and this, then, makes for interesting comparisons with investigations into the composition of passive semiconducting films on alloys such as stainless steels. As also described further above, analysing the composition of the ultra-thin semiconducting films on stainless steel is no easy task, and the much thicker insulating oxide films on aluminum alloys are much easier accessible for investigation. In this direction, we quote work on anodic oxidation of AlCu alloys [35f,g,91b,c]. As might be expected, copper is enriched in the alloy surface, entering the oxide film only after considerable accumulation. During anodization of AlCr alloys, Cr is enriched in the surface, until simultaneous migration of both $Al^{3+}$ and $Cr^{3+}$ sets in [91d]. Apparently, chromium oxidation is sometimes accompanied by the production of gaseous oxygen leading to film rupture. A special case was anodic oxidation of an amorphous AlMo film sputter-deposited on anodized aluminum [91g]. The procedure resulted in the development of a film depleted in molybdenum in the outer regions and enriched in molybdenum in the inner regions.

The observation of, on the whole, preferential aluminum oxidation in all of these cases summarily corresponds to expectations considering the very negative

Gibbs energy of formation of $Al_2O_3$. An earlier observation concerning the commercial alloy Al-2.5 wt.%Mg-5.5 wt.% Zn fits into this picture: The oxide film formed in dry air was found to be MgO, because this oxide is even more stable, with $Al_2O_3$ appearing only after access of moisture. ZnO, in fact much less stable, was not found [93].

Corrosion of passive aluminum has been reviewed in more detail in [94a]. The main point is that whenever the passivating oxide film is stable in approximately neutral solutions and not subject to localized attack, the rate of corrosion is close to negligible. For $Na_2SO_4$ solutions, a value of ca. 0.1 $\mu A\ cm^{-2}$ has been reported and may be considered typical [94b,c]. Almost completely blocked for electrode reactions of any kind at the film/solution interface, the interface essentially is an electrolytic capacitor, adjusting to any accidental changes in the voltage between metal and solution, $\phi_{m,l}$, and thus to any change in electrode potential, $\varepsilon$, by a slight change in film thickness. In the range of passivity, the current-potential curve practically falls on the potential axis, with almost zero current density of Al dissolution. Slight residual currents of electrode reactions mostly flow through defects in the oxide film. These are not atomic defects in the sense of crystal disorder theory, but flaws more similar to pores. Further above it was noted that pores in conducting and semiconducting films on passive metals in solution cannot be stable, owing to galvanic current flow between pore bottom and surrounding oxide film. In insulating films, pores are, instead, stable indefinitely. An early estimate was that about between 0.1 and 0.001% of the surface of passive Al are pores, ore flaws equivalent to pores as far as electronic conductivity is concerned [95]. A very instructive early experiment was to simply try to deposit copper onto aluminum in a solution of a copper salt: Cathodic copper deposition produced isolated copper spots only, obviously decorating pores or similar flaws [96]. From recent investigations it appears that these residual flaws in the oxide film are due to impurities present even in 99.99% Al, enriched for instance in grain boundaries or cellular boundaries intersecting the metal surface [91f,g]. For the probable role of flaws during pitting corrosion see Chap. 12.

Considering flaws in the oxide with some enhanced electronic conductivity, the capacitor model introduced further above needs a correction by adding parallel resistance representing a 'leak' to the equivalent model circuit. The leak becomes more important when we turn from pure aluminum to alloys. For comparison, Fig. 10.45 shows cathodic currents for oxygen reduction on pure aluminum, on an Al-4 wt.% Cu alloy and on the intermetallic phase $Al_2Cu$ [97]. Whether the effect of accelerated oxygen reduction at a passive AlCu alloy is a semiconductor effect due to doping of the oxide film remains open to discussion. Experimenting with the commercial alloy Al-1.5 wt.% Mg-4.5 wt.% Cu, it was shown that rapid oxygen reduction does not really occur at the passive oxide but at intermetallic phase particles penetrating, as local cathodes, the oxide film [94e].

As already noted, pure aluminum together with the insulating oxide film and the adjacent neutral solution layer is a capacitor with almost no leaking

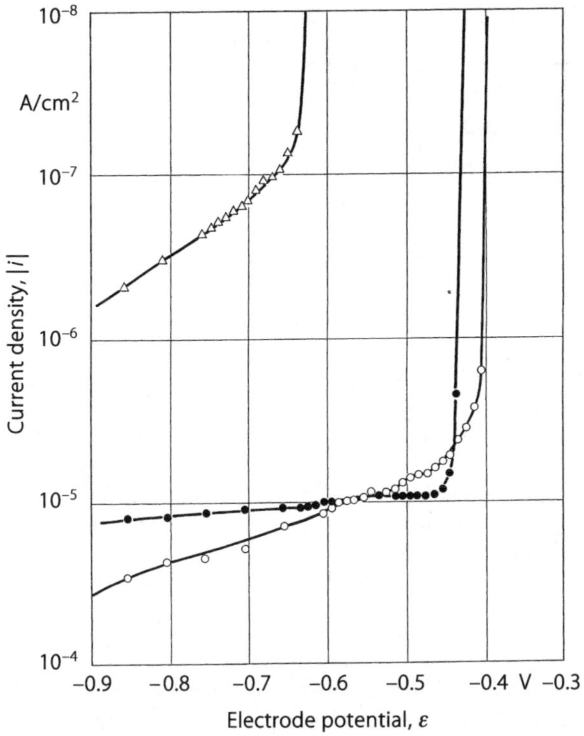


**Fig. 10.45.** Cathodic branches of the total current-potential curves, essentially equal to the partial current potential curves for oxygen reduction from an aerated NaCl solution on pure aluminum (△), on homogeneous Al-4 wt.%Cu (□) and on the intermetallic phase $Al_2Cu$. (From Galvele and Demicheli)

current. This is in contrast to systems such as iron in acids, or iron passivated by a semiconducting oxide. The consequence is that the picture of superposition of anodic and cathodic reactions controlling the free corrosion potential, while in principle certainly still applicable, is no longer very helpful. Instead, the picture of a capacitor with a potential more or less accidentally floating under the influence of electric charges on the plates is rather more appropriate. We shall come back to this point when discussing contact corrosion in Chap. 11.

There is, of course, no obstacle against controlling the potential of pure aluminum electrodes by a potentiostat, and measuring the current-potential curve. Anodic polarization in neutral or in acid solutions other than halide solutions always results in a transient thickening of the oxide film. In alkaline solutions, the situation is different: While the dissolution mechanism of the oxide in neutral and in acid solutions apparently is not well understood, there is little doubt that in alkaline solutions the electrode reaction is: $Al(OH)_3 + OH^- \rightarrow Al(OH)_4^-$, and that it is transport-controlled by the diffusion of either $OH^-$ or $(Al(OH)_4^-$ [94a-c]. As long as cathodic hydrogen evolution does not interfere, the

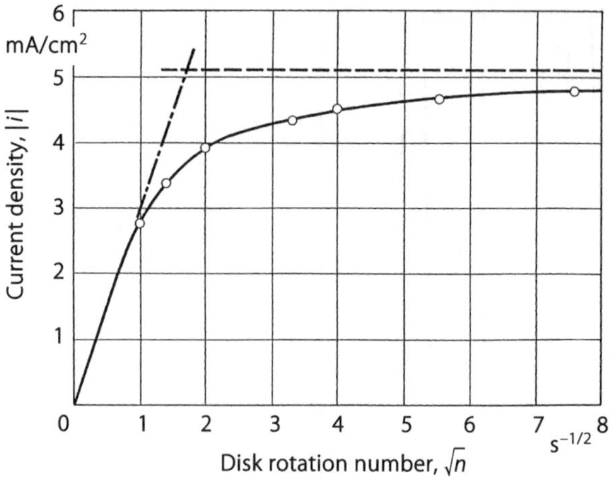


**Fig. 10.46.** Investigating anodic dissolution of Al from a disk rotating with rate $n$ $s^{-1}$. Solid curve: experimental; dash-dotted line: Anodic diffusion-limited current density, inversely proportional to $\sqrt{n}$, indicating rate-determining diffusion; dashed line: reaction-limited current density, indicating charge-transfer control overtaking diffusion rate control. (From Heusler and Allgeier)

current-potential curve is an anodic diffusion-controlled potential-independent plateau, proportional to solution pH: $i_{Al}$ = const ×pH. This has been more accurately investigated using aluminum disks rotating in alkaline solutions, and measuring the rate of anodic metal dissolution by the recording the change of electric eddy currents, which very sensitively respond to changes in the thickness of the disk [64f]. The rate of transport through the diffusion layer should be inversely proportional to $\delta_{Nernst}$ and thus proportional to the square root, $\sqrt{n}$, of the rotation number (see Chap. 5 and Appendix). As shown in Fig. 10.46, this is confirmed for comparatively slow disk rotation. Increasing the rotation number, charge-transfer control takes over, and the current density depends on the potential rather than on the speed of rotation.

Since passivation is due to an oxide film, the pasive/active transition by reduction of the film at a passivation potential, $\varepsilon_p$, should be observable. Possibly, $\varepsilon_p$, is the equilibrium potential of the $Al/Al_2O_3$ or of the $Al/Al(OH)_3$ electrode, which is $E_{Al/Al(OH)_3} = -1.49-0.059(V)$. This potential is, however, very far below the hydrogen equilibrium potential in the same solution, and the consequence is that at this potential considerable overpotential energy for hydrogen evolution is made available by cathodic polarization. Therefore, whenever activation of aluminum by cathodic polarization is achieved, vigorous cathodic hydrogen evolution is expected to set in. By the reaction $2H_2O + 2e^- \rightarrow H_2 + 2OH^-$ hydroxyl ions are produced, resulting in what in terms of the techniques of cathodic protection (see Appendix) is 'wall alkalisation'. In the present case the effect is intensely detrimental, since local enrichment in $OH^-$ catalyses the dissolution of the oxide to $Al(OH)_4^-$. The effect is documented by the data

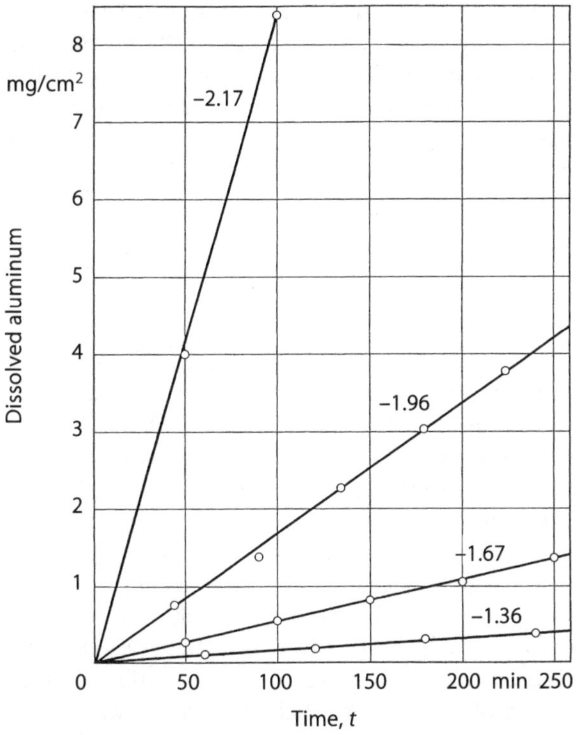


**Fig. 10.47.** Illustrating the cathodic corrosion of aluminum: Weight loss of pure aluminum from 1 M $Na_2SO_4$/NaOH solution at pH+11, 25°C at various values of the potentiostatic electrode potential, $\varepsilon$. Note that the accompanying cathodic hydrogen evolution is not shown. (From Kaesche)

presented in Fig. 10.47 showing that at any very negative potential the rate of this *cathodic corrosion* is excessively high, increasing apparently indefinitely with decreasing potential [94a], Since these are data for alkaline solutions it should be noted that a similar effect has been observed in neutral solutions as well [94a,c]. In view of this data there is no doubt that by some mechanism, however complicated it may be, cathodic breakdown of the passivating oxide is possible. Conducting flaws in the film may serve as starting points for hydrogen evolution and, by local increase in pH, of oxide dissolution, but detailed knowledge of such processes apparently is not available. Still it is interesting to note that in this case local alkalinity appears important, while during pitting a decisive factor is, on the contrary, local acidification, as described in Chap. 12.

The surface state of aluminum during rapid cathodic corrosion is not very clear. Considering these circumstances, there is interest to create, with the aim of studying passivation rather than activation, bare aluminum surfaces by other means. As compiled in [98], widely different methods such as cracking, scratching, or guillotining aluminum samples have been used. In [98], breaking a thin metallic film, deposited on an isolating substrate and coated with $SiO_2$, under potentiostatic control in an electrolytic solution is advocated. By this method, only the cross section of the sample is exposed to the solution, and the coating prevents capacitive contributions to the transient current. A direct logarithmic time law is claimed to fit current density transients, recorded during passivation, better than the inverse logarithmic law, contradicting the usually supposed high-field mechanism of film-growth.

## 10.9 Zinc, Magnesium

Similar to aluminum, *zinc* is amphoteric owing to the solubility of the hydroxide ZnO or the oxide $Zn(OH)_2$ in both alkaline and acid solutions. Also similar to Al, the dissolved species is a complex zincate ion, probably $Zn(OH)_4^{2-}$, at high, and the ion $Zn^{2+}$ at low pH, respectively. In the range of intermediate values of pH, the oxide or hydroxide is scarcely soluble. As described in Chap. 2, zinc very well resists corrosion in contact with solutions of intermediate pH, owing to the formation of thin layers of oxides, or oxycarbonates, qualitatively corresponding to what is expected from the solubility of these compounds.

Zinc in contact with alkaline solutions is an important and much investigated electrode material for batteries such as the zinc/silver oxide battery. While it is not passive in the practical sense of negligible dissolution rate, it still does show the typical active/passive N-type of current-potential curve, albeit with a high passive current density. As an early example of measurements, Fig. 10.48 shows a quasi-stationary curve observed in 4 M NaOH at 25°C [99]. For information about many more earlier contributions to the topic, see [100a–d].

As shown below, 'passivation' of zinc in alkaline solutions at concentrations of up to several Mol $L^{-1}$ goes through a complicated sequence of steps. In less alkaline solutions, more relevant to considerations concerning corrosion rather than batteries, the mechanism is less complicated, as obvious form investigations into the passivity in 1 M $Na_2CO_3$, pH 12, 25°C, described in [94d]. In

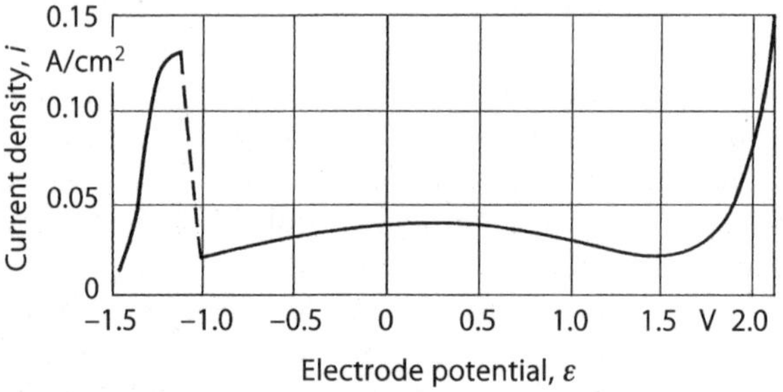


**Fig. 10.48.** Quasi-stationary anodic current-potential curve for zinc in 4 M NaOH solution. The current increase beyond 1.6 V is probably due to transient oxygen evolution. (From Franck and Lüdering)

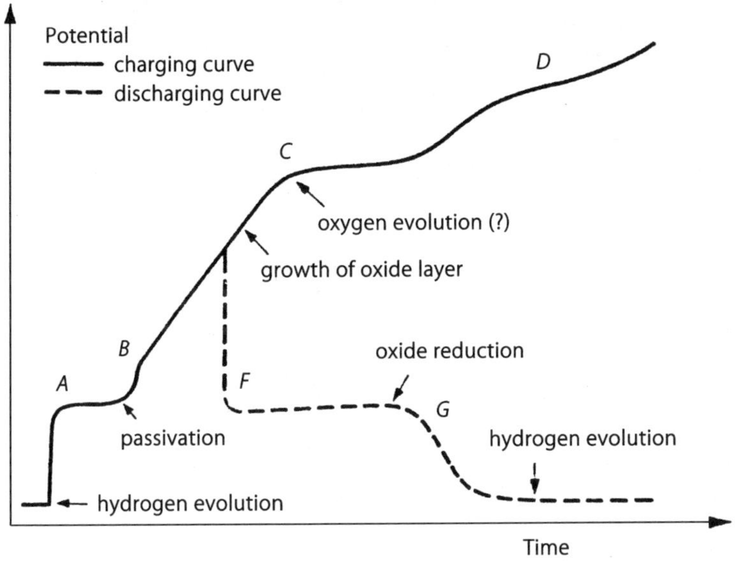


**Fig. 10.49.** Potential-time curves for zinc in 1 M $Na_2CO_3$ solution, 60°C, during galvanostatic polarization. Electrodes initially kept at -10 mA $cm^{-2}$, then switched to +10 mA $cm^{-2}$, causing ZnO formation **(A)** and growth **(BC)** with superimposed oxygen evolution **(CD),** and finally re-switched to -10 mA $cm^{-2}$ causing reduction of ZnO (EF) and, eventually, return to hydrogen evolution (FG). (From Kaesche)

this work, galvanostatic polarization was used to study the behaviour of the electrodes, and Fig. 10.49 gives an example of the observations: On changing the current from initial cathodic protection to anodic polarization, the potential at first goes to a value slightly beyond the Zn/ZnO equilibrium potential, which at 60°C and pH 12 is −1.2 V. AB is the passivation time, $\tau_p$. When the first layer of oxide is complete, oxide thickening sets in until at high values of the potential oxide growth is slowed by the onset of oxygen evolution. The amount of oxide produced is easily controlled by switching the current to cathodic oxide reduction followed, finally, by hydrogen evolution indicating re-activation of the metal.

The much more complicated electrode behaviour observed in more alkaline solutions is obvious from potentiodynamic measurements, accompanied by recording, as in Fig. 10.50, the appearance of the electrodes while the potential was scanned from the corrosion potential to increasingly more noble values [100e]. Here, the current density at first rises to very high values, owing to rapid metal dissolution to zincate ions: $Zn + 4OH^- \rightarrow Zn(OH)_4^{2-} + 2e^-$. When the diffusion layer adhering to the metal surface becomes saturated with respect to ZnO or $Zn(OH)_2$, by depletion with respect to $OH^-$, oxide or hydroxide precipitates in the boundary layern: $Zn(OH)_4^{2-} \rightarrow Zn(OH)_2 + 2OH^-$, but not directly on the metal surface and thus not passivating the metal. As

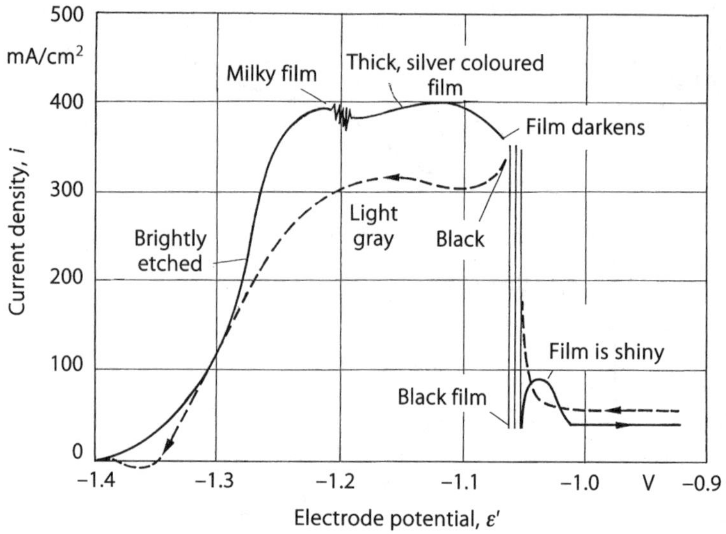


**Fig. 10.50.** Potentiokinetic current-potential curve for a zinc wire in quiescent 5 M KOH solution at ambient temperature. The scan rate was 1.1 mV $s^{-1}$; the potential is referred to an Hg/HgO electrode in the same solution (From Hull, Ellison, and Itoni)

the potential is shifted to more positive values, oxide starts to grow directly on the metal surface, and passivation sets in, with oxide growth probably continuing by the reaction $Zn + 2OH^- \rightarrow ZnO + H_2O + 2e^-$. This sequence of events has been observed by other authors as well [94e,101]. As expected, current densities increase with increasing rate of stirring of the solution. The more intriguing point is the reproducibility of the active/passive transition potential, which clearly is not the Zn/ZnO equilibrium potential calculated for the bulk pH. In some analogy to the case of iron, an oxidation of ZnO to the peroxide $ZnO_2$ has been proposed, but apparently on weak grounds [101d]. So far, the phenomenon is not well understood. One more likely supposition is that a critical lowering of the local pH at the metal surface is necessary for the growth of oxide to start [101c]. It should be noted that the apparent reversibility of the active/passive transition suggested by the observations reported in Fig. 10.50 is somewhat misleading, because the transition shifts to more negative values of the potential when the passivating oxide is allowed to grow to considerable thickness. As revealed by electron microscopy, the re-dissolution of the oxide film begins with a pitting-type of local attack, and this suggests that reactivation sets in when the oxide film thickness drops below a critical value [94e]. In some respect, the reactivation phenomenon is reminiscent of the behaviour of lead in chromate solutions which will be described in Chap. 12 as a model for the mechanism of pitting corrosion.

Bulk ZnO is an n-type semiconductor with a bandgap energy of 3.2 eV and excess-electron conductivity owing to the dissociation of interstitial excess Zn atoms as donor states [75b]. Upon photoanodic dissolution by the reaction of oxide and holes in the valence band according to $ZnO + 2h^+ \rightarrow Zn^{2+} + (1/2)O_2$ is observed [101e]. Since sharp semiconductor band-edge characteristics are lacking, the passivating film probably is amorphous. More information as to semiconductor properties has been obtained from Mott-Schottky investigations revealing that the donor concentration decreases with increasing formation voltage. At the same time, the space charge region appears to always occupy the entire thickness of the barrier layer [101f]. These findings are claimed by the authors to further confirm the point defect model for passive oxide films as opposed to the more classical high field model, as described further above.

As noted previously, zinc hot-dip galvanizing is very effective for long-term protection of unalloyed steel especially against atmospheric corrosion. The effect was described as being due to dense and thin corrosion product layers, which are inert and non-conducting, but not really in the role of passivating or at least quasi-passivating films. An interesting application is hot-dip galvanizing of rebar in moist concrete, i.e. at pH 12.3 or above, when in marine atmospheres safe passivity of iron is threatened by pitting corrosion (see Chap. 12). Zinc is, in fact, much better resistant to pitting corrosion, regardless of the comparatively high pH of the pore moisture in concrete. Furthermore, a defect in the zinc layer exposing bare iron is normally safe due to local cathodic protection (see Appendix). A warning is, however, appropriate concerning hydrogen embrittlement of galvanized prestressed high-strength tendons in concrete (see Chap. 14).

A third field of wide use of hot-galvanized steel is mild steel pipes for the distribution of warm water in domestic installations. Since this is fresh water containing dissolved oxygen, bare mild steel pipes tend to rust, while galvanized steel tends to develop dense layers of iron and zinc corrosion products. Again, when initially these are not yet effective as inert barriers separating water and metal, the bottom of accidental defects in the zinc layer is protected by the short-circuit with surrounding metallic zinc (see Chap. 11). A curious effect then is sometimes observed in water at temperatures above ca. 60°C, containing bicarbonate (and/or nitrate, which, however is highly undesirable for other reasons) and only little amounts of sulfate and chloride [102]: Zinc sometimes spontaneously really passivates under conditions where iron does not. By this effect, the corrosion potential of zinc becomes more positive than that of iron; hence the description as *potential reversal of zinc*. Using dilute $NaHCO_3$ solution as a model of domestic water, the case has been studied more closely [94f], and the result was that the effect of cathodic protection of bare iron at the bottom of defects in the zinc layer fails. However, the opposite and detrimental reversal in galvanic action, which would accelerate iron corrosion at defect bottoms, also does not occur. The reason is that the passivating, practically insulating ZnO layer does not allow cathodic oxygen reduction. When, however, traces of copper where dissolved in the model solution, short-circuited iron/zinc couples did produce currents of considerable strength by anodic iron dissolution accompanied by rapid oxygen reduction at the passivated zinc surface. Apparently,

ZnO now was doped with copper, increasing the conductivity of the film. The origin of copper traces in domestic installations, incidentally, is copper tubing used to heat the water to the desired temperature. The effect often causes massive pitting ending, often within the first months of life of new installations, in perforating thick pipe walls.

To put zinc into a coherent picture, we remember from Chap. 3 that zinc in zinc sulfate solution is the anode in the Daniell cell because the open circuit potential is equal to the $Zn/Zn^{2+}$ equilibrium as a consequence of a high value of the exchange current density of zinc dissolution and redeposition, $i^{\circ}_{Zn}$. No passivation and in fact only little change in potential at all occurs on anodic polarization, indicating a very low polarization resistance. Similar conditions occur in chloride solutions including sea-water. When, as in sea-water, the activity of dissolved zinc is virtually zero (or, by the Pourbaix convention, ca. $10^{-6}\,\mathrm{mol\,L^{-1}}$), the free corrosion potential is very negative ($\lesssim -0.8$ V). Electrode polarizability still is low, and this explains the use of zinc as material suitable for 'sacrificial' anodes for cathodic protection of the hulls of sea-going ships, as described in the Appendix. The useful effect now is that general hull corrosion is replaced by the consumption of the easily replaceable anodes.

*Magnesium* is a further example of a structural material usually protected to some degree against long-term corrosion by oxide or hydroxide layers. The resistance against corrosion is, however, not very satisfactory, and corresponds even less to what is understood as regular passivity or at least quasi-passivity, as that of zinc. In the following, we only briefly comment on this behaviour, referring for details to the review in [103a].

The metal has the most negative standard potential, $E^{\circ}_{Me/Me^{z+}}$, of the engineering metallic materials (see Table 3.2), and readily reacts with either dry or wet oxygen to $Mg(OH)_2$ and/or MgO, but the protective quality of this air-formed oxide film is much poorer than that of films on aluminum. Magnesium alloys, mostly with aluminum and with zinc, such as Mg-9wt.%Al-1wt.%Zn, are equally unsatisfactory, but they are increasingly used especially in the aircraft and transport industries because of their light weight and their high weight-to-strength ratio. Parts made from magnesium alloys usually are castings with high wall thickness, so that superficial corrosion can be tolerated, as for years it does not seriously affect service conditions.

The origin of the difference between well-behaved oxide films on aluminum and comparatively ill-behaved such films on magnesium probably is porosity. As an explanation, recurs can tentatively be taken to the old *Pilling and Bedworth rule* of 1923, stating that an oxide film on a metallic substrate is dense and protective only if the oxide/metal volume ratio exceeds the value 1, so that the film is squeezed onto the metal surface: For aluminum, the ratio is 2.0, while for magnesium it is only 0.79. Of course, when monomolecular adsorption of hydroxide layers or epitaxial growth of similarly thin oxide films is important, the rule is obsolete. When, however, it comes to relatively thick films, as on aluminum and magnesium, it may offer a useful possibility to explain simple observations. Film porosity may also explain an effect which previously has experienced attention under the name of *negative difference effect* [102b]. This is the observation, paradoxical on first sight, that anodic polarization of magnesium in chloride solutions does not only cause anodic metal dissolution, but also increased cathodic hydrogen evolution. A variant of the effect has already been described above for the case of cathodic corrosion of aluminum, where cathodic polarization was seen to accelerate not only cathodic hydrogen evolution but anodic metal dissolution as well. The observation was explained as being due to rapid alkaline corrosion setting in whenever aluminum, bared by oxide reduction, comes in direct contact with the electrolytic solution. The case of anodic polarization of magnesium is analogous to that of anodic polarization of aluminum in chloride solutions causing pitting, which again is accompanied by local hydrogen evolution, as described in Chap. 12. For magnesium, the difference probably only is that in this case, owing to the preexisting highly defective oxide film, the pitting-type of well defined breakdown of passivity of an otherwise perfectly passivating oxide is not a necessary prerequisite of the phenomenon.

Details of magnesium electrode reaction kinetics are not well known. A point of more practical interest is the detrimental influence of impurities, and the effects of iron, nickel, copper, and others are well documented. As an example we refer to iron, which in magnesium is without effect up to a content of 0.018 wt.% but greatly increases the rate of corrosion in salt water beyond this 'tolerance limit' [102c]. The mechanism at work apparently is acceleration of corrosion due to heterogeneous inclusions of metallic iron particles acting as local cathodes for hydrogen deposition. It was by the suppression of beyond-tolerance amounts of alloy components that commercial magnesium alloys have much gained in corrosion resistance.

In contrast to aluminum and zinc, magnesium is not amphoteric. Therefore, it is quite stable in alkaline solutions owing to the insolubility of the oxide layer, but there is not much technical interest in this merit, or else, considering possible alkaline batteries, disadvantage. The metal is widely used as sacrificial anode material for cathodic protection of water boilers (see Appendix) for two reasons: In the first place, the free corrosion potential usually is very negative ($\lesssim -1.4\ V$), though considerably more positive than the standard $Mg/Mg^{2+}$ equilibrium potential. At this negative potential, cathodic hydrogen evolution certainly should be much more important than oxygen reduction from aerated solutions, and magnesium in hot water is, in fact, seen to slowly develop gaseous hydrogen accompanying metal dissolution. In the second place, the change in potential caused by anodic polarization is relatively low, indicating a low metal dissolution overpotential and thus a high value of the exchange current density of magnesium dissolution and redeposition, $i^{\circ}_{Mg}$. Both circumstances again point to a state of electrode kinetics governed by an inert but highly defective oxide layer, leaving parts of the metal surface uncovered and available for active metal dissolution and cathodic hydrogen deposition.

## 10.10 Oxide Films on Steels in High-Temperature Aqueous Solutions

Turning to high-temperature, mostly high-pressure aqueous corrosion of steels, we are once more concerned not with passivity in the ideal sense of the concept, but still with efficient corrosion protection by oxide films formed on the metal surfaces by reaction with the surrounding solution [104]. Again these films are no longer, or else not necessarily, ultra-thin and transparent but, instead, visible, with a thickness in the micrometer range. The prominent case is black magnetite, $Fe_3O_4$, for the safe corrosion protection of immense lengths of mild steel or low-alloy steel tubes installed in fossil-fuel power plant loops carrying water from ambient temperature to the pressurized elevated boiling point. The rate of uniform corrosion, which is the passive, resp. quasi-passive, current density is low, with films growing in neutral solutions to several micrometer thickness only within thousands of service hours. To avoid salt deposition, the circulating water is kept almost pure by deionisation; to minimize corrosion it also is deaerated, leaving a situation which, in the limiting case, is expected to be one of slow non-electrolytic reaction of iron with water to a scarcely soluble oxide film by the overall reaction $3Fe + 4H_2O \rightarrow Fe_3O_4 + 4H_2$. Concerning essential reaction kinetics this process probably is more similar to high-temperature dry scaling than to electrolytic corrosion with superposition of distinct anodic ad cathodic partial reactions. Since the reaction product films are relatively thick, growth is slow and the growth law, parabolic or other, is one of the main points of interest. For low-temperature passivity by nanometer oxide films, film growth kinetics, mostly disregarded in the preceding sections, kinetics certainly can be studied just as well, particularly with the aim of identifying the character of ionic transport across the film, but the point then is academic rather than of practical interest.

Within the narrow limits imposed by plant operation conditions, some slight chemical water treatment is always applied, varying according to the type of plant, and varying according to tradition. Since water and steam also pass through the turbines and the condensors and auxiliary equipment, demands concerning water treatment are complicated. Typical low-concentration additions are ammonia, $NH_3$, and hydracene, $N_2H_4$. Ammonia increases pH by dissociation: $NH_3 + H_2O \rightarrow NH_4^+ + OH^-$, while hydracene removes oxygen by the reaction: $N_2H_4 + O_2 \rightarrow N_2 + 2H_2O$. Increasing pH lowers the solubility of iron oxides, so an 'all volatile' treatment with both ammonia and hydracene makes obvious sense, if low thermodynamic oxide solubility is decisive. By long experience, a 'neutral' water treatment also is established practice, with pH left unchanged, and with oxygen purposely added to a concentration of, typically, 200 ppb or $6 \times 10^{-6}$mol $L^{-1}$. This appears to point in two directions: The first is that minimizing oxide solubility is not all that crucial. Still we note in this connection that the calculation of high-temperature equilibrium solubility data, by preference plotted as Pourbaix diagrams, is frequently considered important. The second is that oxide-film passivation is favored by slightly increasing

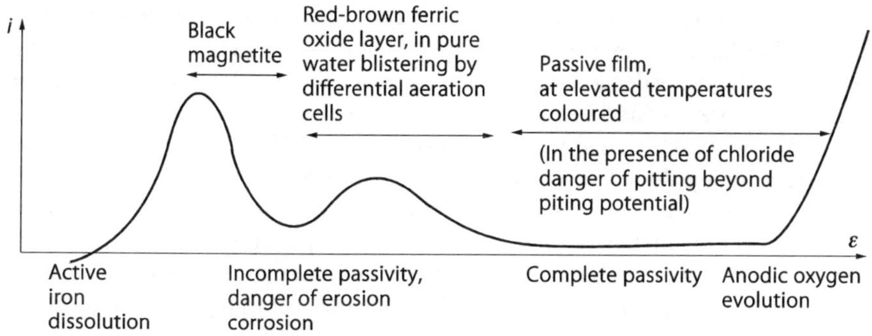


**Fig. 10.51.** Schematic anodic current-potential curve for iron dissolution and iron oxidation from a low-alloy preheater steel exposed to continuously flowing nearly high-purity water at 180°C, indicating typical regions of precursors to passivity and regular passivity. Also shown: Possible range in potentials where pitting is expected in chloride-contaminated water. (From Rohlfs and Kaesche)

the oxygen content of the water, i.e.presumably by shifting the electrode potential to more positive values. If so, the advent of a third, 'combined' water treatment, combining a slight increasing both pH and oxygen content, is not surprising [105a].

If the electrode potential is a point of concern, investigations into potential-dependent metal/oxide/solution transfer reactions are advised and, thus, current-potential measurements. In very dilute solutions the simple capacitor model of the electrical double layer, as described in Chap. 5, will certainly need correction for an extended diffuse part of the potential drop, but this is of minor importance as long as the critical issue only is the formation of thick protective oxide films. The more difficult point is the low conductivity of the solutions. The order of magnitude of feed-water conductivity is 1 $\mu S\ cm^{-1}$ or $(\mu\Omega\ cm)^{-1}$. Therefore, the order of the ohmic voltage included in the electrode potential measured at a current density of e.g. 10 $\mu A\ cm^{-2}$ with a reference electrode capillary tip at a distance of, e.g., 0.05 cm from the metal surface is 0.5 V. By proper choice of electrode geometry, and using ohmic voltage feedback to the potentiostat, the situation is alleviated, but the number of current-potential curves reported in the literature is small. For current-potential measurements, solution conductivity usually is increased by changing to a borate buffer solution at equal pH, thus leaving overall thermodynamics unchanged and ohmic voltages minimized, but not necessarily without interfering with reaction kinetics. We ignore details; instead, Fig. 10.51 qualitatively summarizes the experiences concerning the anodic partial current-potential curve for quasi-stationary anodic iron dissolution from a typical 0.1% C-steel preheater tube containing Si, Mn, Cr, Mo, Ni, and Cu, in nearly high-purity water at 180°C [105e].

The figure compares well with expectations extended from ambient temperature behaviour of low-alloy steel, with electrode states passing through what further above was described as a precursor state of passivity, to 'true' passivity.

In contrast to ambient-temperature behaviour, where precursor films and stable passivating films are invisible, there now is always magnetite distinctly visible, passing to higher-valence oxides again growing to considerable thickness and, finally, to 'true' passive hematite films with colors similar to colored annealing films. In practical service, the latter stage would be expected to develop in waters highly doped with dissolved oxygen. Concerning heat transfer, and also concerning hydrodynamics of water flow, it might on first sight appear desirable. Unfortunately, this also is the range of potentials where in chloride-contaminated water pitting is imminent. We note that, in the intermediate state, localized attack by formation of small differential aeration cells is typical and also undesirable.

Service conditions usually are such that in the water circulation system, low-alloy steels are in the range of protective magnetite layer formation, and 'passive' enough in a practical sense. It is this same range, where investigations into film structure, chemistry and growth have frequently been carried out . Magnetite usually is found to be deposited as a two-layered scale with a dense inner and a less dense, often coarse, outer part. Therefore, scale growth presumably is determined by the kinetics of mass transfer through the inner layer. Over the years, a number of different mechanisms have been suggested, as reviewed in [106]. The underlying main concept is that inner-layer growth, occurring at the metal/oxide interface, requires migration of water to this interface. Water transport by solid state diffusion being much too slow, water vapor was thought to diffuse through pores in the inner layer. Outer-layer growth originally was thought to proceed by transport of $Fe(OH)_2$ to the outer layer through solution-filled pores [107a]. However, experimental pore diameters were found too small to allow rapid outward iron migration [107b]. Therefore, a combination of outward solid state cation diffusion and inward migration of water, again through pores, appeares better suggested, or even a mechanism involving solid state diffusion in both directions, governed by a change from n- to p-type semiconduction across the oxide film. [107b,c]. More recently, concepts have been updated by postulating inward transport of water through nanometer-pores and outward short-circuit diffusion of cations along grain boundaries [107d].

In nuclear power stations, where again nearly pure water is the coolant, water and steam are usually transported through austenitic stainless steel pipes, and passivity, or else quasi-passivity, once more is responsible for safe corrosion protection. In Germany, a steel with 32/35 Ni, 20/23 Cr (wt.%), stabilized against intercrystalline corrosion (see Chap. 13) by adding titanium, with $\%Ti \geq 8 \times (\%N + \%C)$, is the preferred choice.

In pressurized water reactors (PWR), the primary loop coolant is a dilute solution containing LiOH and boric acid, also containing dissolved hydrogen. In the secondary loop, conditions are essentially the same as in the water/vapor circle of conventional power plants, and similar methods of water treatment are applied. In the single coolant loop of boiling water reactors (BWR), more highly specialized conditioning is required, which we pass.

On stainless steels, passive oxide layers are thinner than under similar conditions on low-alloy steel by about one order of magnitude, probably indicating substantially lower steady-state corrosion rates. The situation thus is quite similar to ambient-temperature passivity of stainless steels in relation to low-alloy steels. Also, the analogy extends to the nature of the protective oxide films as well: Similar to bi-layers on low-alloy steels, films on austenitic steels are duplex with an inner and an outer oxide layer of different morphology but similar crystal structure [108]. Formed in deaerated water, both layers are inverted spinels as represented by magnetite but, in this case, with an inner composition close to $(Fe,Cr)_3O_4$ and an outer composition close to $Fe_3O_4$. Depending on formation conditions, the spinel structure is able to accommodate some nickel cations replacing either iron or chromium ions. When the water contains some dissolved oxygen, $Fe_2O_3$ and/or FeOOH can be detected in the oxide film's outer surface [110]. Oxygen or hydrogen peroxide are possible products of water radiolysis, but these contaminants are suppressed by adding hydrogen in the primary loop coolant of pressurized water reactors.

Concerning growth kinetics, some authors claim that 'regular' passivity, as typical for ambient temperature, is destroyed within hours by some change in film properties, if passive specimens are taken to temperatures beyond 185/195°C [109b]. The change is indicated by an accompanying increase in pitting susceptibility wich is not our concern presently. The effect is described as confirmed in [111], where it is explained on the assumption that initially amorphous films, as they have been described further above for chromium steels, recrystallize at elevated temperatures, and that the resulting microcrystalline films are less protective than vitreous films.

Regardless of possible intermediate changes in the quality of oxide films, which in the long run certainly are crystalline, suggested mechanisms of high-temperature film growth on austenites are scaled-down versions of what has been described for low-alloy steel [107c]. As before, the inner magnetite layer is thought to grow by the access of water through narrow pores, while the outer layer grows by short-circuit diffusion of iron cations along grain boundaries.

## References


1. 1a) Franck, U.F.: Habilitation thesis, Göttingen 1954; b) Weil, K.G. and Franck, U.F.: Z. Elektrochemie 56 (1952) 814; c) Franck, U.F and Fitzhugh, R.: Z. phys. Chemie N.F. *3* (1955) 183; d) Franck, U.F.: Z. Naturforschung *4a* (1951) 378; e) id.: Werkstoffe u. Korr. *11* (1960) 401
2. Herbsleb, G. and Engell, H.J.: ibid. 65 (1961) 881
3. Osterwald, J.: Z. Elektrochemie *66* (1962) 401
4. Ord, J.L. and Bartlett, J.H.: J. Electrochem. Soc. *112* (1965) 1260
5. a) Heusler, K.E.: Ber. Bunsenges.Phys. Chemie *72* (1968) 1197; b) loc. cit. Chap. 5 [16b]; c) Knoedler, R. and Heusler, K.E.: Electrochim. acta *17* (1972) 197; d) Heusler, K.E. and Schoner, K.: Ber. Bunsen Ges. phys. Chemie *20* (1964) 35

6. a) Vetter, K.J.: loc. cit. Chap. 5 [1]; b) Heusler, K.E.: loc. cit. Chap. 5 [16b]
7. a) Proc. 4th Int. Symp. Passivity, Airlie, VA, 1977 (Frankenthal, R.P. and Kruger, J., eds.) The Electrochem. Soc., Princeton, 1978;b) Proc. 9th Int. Congr. Metallic Corrosion, Toronto 1984. National Research Council Canada; c) Proc. 5th Int. Symp. Passivity of Metals and Semiconductors, Bombanne 1983. (Froment, M., ed.). Elsevier, Amsterdam etc. 1983; d) Proc. Int. Symp. Surfaces, Inhibition, and Passivation (Honor. N. Hackerman), San Diego 1986 (McCafferty, E. and Brodd, R.J., eds.). The Electrochemical Society, Pennington, 1986; e) Proc. Symp. Passivity and its Breakdown, Paris 1997 (Natishan, P.M., Isaacs, H.S., Janik-Czachor, M., Macagno, V.A., Marcus, P., and Seo, M., eds.), The Electrochem. Soc., Pennington, 1995; f) Corrosion and Environmental Degradation, Vol. I, II (Schütze, M., ed.), in: Materials Science and Technology (Cahn, R.W., Haasen, P., and Kramer, E.J., eds.). Wiley VCH, Weinheim etc., 2000; g) Proc. Eur. Symp. Modifications of Passive Films, Paris 1993 (Marcus, P., Baroux, B., and Keddam, M., eds.). The Institute of Materials, 1994; h) Proc. 4th Int. Conf. Metallic Corrosion, Amsterdam 1969. NACE, Houston, 1972,
8. a) Epelboin, I., Gabrielli, C., Keddam, M., and Takenouti, H.: The Study of the Passivation Process by Electrode Impedance Analysis. In: loc. cit. Chap. 3 [6], Vol. 4; b) Epelboin, I. and Keddam, M., in: loc. cit. [7a], p. 184
9. Russel, P.P. and Newman, J.J.: J. Electrochem. Soc. *130* (1983) 547
10. Lorbeer, P., Jüttener, K., and Lorenz, W.J.: Werkstoffe u. Korr. *34* (1983) 290
11. Pryor, M.J.and Evans, U.R.: J. Chem. Soc. *1949*, 3330; *1950*, 1259, 1267, 1274
12. a) Mayne, J.E.O., Menter, J.W., and Pryor, M.: J. Chem. Soc. *1950*, 3229; *1954*, 103; b) Mayne, J E.O. and Pryor, M.: J. Electrochem. Soc. *98* (1961) 263; *100* (1953) 203
13. a) Nagayama, M. and Cohen, M.: ibid. *109* (1962) 781; *110* (1963) 670; b) ibid. *110* (1963) 670
14. a) Göhr, H. and Lange, E.: Naturwiss. *43* (1956) 12; b) ibid.: Z. Elektrochemie *61* (1957) 1291; c) Lange, E. and Weidinger, H.: Naturwiss. *43* (1956) 12; d) Lange, E. and Göhr, H., in: loc. cit. Chap. 3 [5]; e) Franz, D. and Göhr, H.: Ber. Bunsen Ges. phys. Chemie *67* (1963) 680
15. a) Vetter, K.J.: Z. phys. Chemie *202* (1953) 1; b) Z. Elektrochemie *58* (1954) 230; c) Vetter, K.J. and Gorn, F.: Electrochim. acta *18* (1973) 321
16. a) Weil, K.G. and Bonhoeffer, K.-F.: Z. phys. Chemie N. F. *4* (1955) 175; b) Weil, K.G.: Z. Elektrochemie 59 (1955) 711; c) id.: ibid. 110 (1963) 640; d) Heusler, K.J., Weil, K.G., and Bonhoeffer, K.-F.: Z. phys. Chemie NF *15* (1958) 149
17. Proc. Symp. Ellipsometry in the Measurement of Surfaces and Thin Films. ( Passaglia, E., Stromberg, R.R., and Kruger, J., eds.) NBS Misc. Publ. 256, US Government Printing Office, Washinton, 1964; Proc. Symp. Recent De-

velopments in Ellipsometry. (Bashara, N.M., Buchman, A.B., and Hall, A.C., eds.). Surface Sci. *16* , (1969)

18. a) Gorn, F. and Vetter, K.J.: Z.phys. Chemie N.F. *77* (1972) 317; b) Gorn, F.: Optik *35* (1972) 536; c) Vetter, K.J. and Gorn, F.: Z.phys. Chemie N.F. *86* (1973) 113
19. Wagner, C.: Ber. Bunsenges. phys. Chemie *77* (1973) 1090
20. Vetter, K.J.: a) Z. Elektrochemie Ber. Bunsen Ges. Phys. Chemie *66* (1962) 577; b) ibid. *69* (1965) 589; c) J. Electrochem. Soc. *110* (1963) 5597; d) Electrochim. acta *16* (1971) 1923; e) Cabrera, N. and Mott, N.F.: Rep. Prog. Phys. *12* (163) 1948; f) Günterschultze, A. and Betz, H.: Z. Physik *12* (1934) 367; g) Vermilyea, D.A.: Acta met. *1* (1953) 282; h) Chao, C.Y., Lin, L.F., and Macdonald, D.D.: J. Electrochem. Soc. *128* (1981) 1194; i) Macdonald, D.D., Urquidi-Macdonald, M., and Lenhardt, S.J., in: loc. cit. [7d], p. 402; j) Macdonald, D.D., Zhang, L., Sikora, E., and Sikora, J., in: loc. cit. [7e]
21. a) Young, L.: Anodic Oxide Films. Academic Press, London, New York, 1961; b) Oxides and Oxide Films (Diggle, J.W., ed.). Marcel Dekker, New York, 1973; c) Reschke, L., in: Korrosion u. Korrosionsschutz (Tödt, F., ed.). DeGruyter, Berlin, 1961
22. a) Jaenicke, W.: Z. Elektrochemie *56* (1952) 473; b) Jaenicke, W. and Haase, M.: ibid. *63* (1959) 521; c) Engell, H.J.: Z. phys. Chemie N. F.: *7* (1956) 158; d) Valverde, N. and Wagner, C.: Ber. Bunsenges. phys. Chemie *80* (1976) 330 473; e) loc. cit. [21]
23. a) Hoar, T.P.: J. Electrochem. Soc. *117* (1970) 17C; b) Evans, U.R.: Electrochim. acta *16* (1971) 1825
24. Worch, H., Forker, W., and Rahner, D.: Werkstoffe u. Korr. 34 (1983) 402
25. a) Wroblowa, H., Brusic, V., and Bockris, J.O'M: J. Phys. Chem. 75 (1971) 2823; b) Grady, O. and Bockris, J.O'M.: Chem. Physics Letters *5* (1970) 116; Surface sci. *38* (1973) 249; c) Revie, R.W., Baker, B.G., and Bockris, J.O'M.: J. Electrochem. Soc. *122* (1975) 1460
26. a) Sato, N., Noda, T., and Kudo, K.: Electrochim. acta *19* (1974) 471; b) Sato, N. and Kudo, K.: ibid. *16* (1971) 447; c) Sato, N., Kudo, K., and Noda, T.: ibid.*16* (1971) 1909; Corr. sci.*10* (1970) 785; d) Nishimura, R. and Sato, N., in: loc. cit. [7b] Vol.1, p. 96
27. a) Bäumel, A. and Engell, H.J.: Zement, Kalk, Gips *13* (1960) 2370; b) Kaesche, H.: ibid. *13* (1960) 240; c) Böhni, H. et al.: Dokumentation D 020, 021 Schweizerischer Ingenieur- u. Architekten-Verein, Zürich 1988; d) Elsener, B., in: loc. cit. [7f], Vol. II, p. 389
28. a) Florianovitch, G.M., Kolotyrkin, Ya.M., and Kononova, M.D.: loc. cit. [7h] p. 694
29. Moeller, F.: Diploma thesis, Erlangen 1974
30. a) Oelsner, G.: Doctoral thesis, Berlin 1974; b) Diploma thesis, Berlin 1961
31. a) Geana, D., El Miligy, A.A., and Lorenz, W.J.: Corr. sci. *13* (1973) 505; b) Bessone, J., Karayaka, L., Lorbeer, P., and Lorenz, W.J.: Electrochimica acta *22* (1977) 1147; c) Lorbeer, P., and Lorenz, W.J., in: loc. cit. [7], p. 607; d) iid.: Electrochim. acta *25* (1980) 375; see also loc. cit. Chap. 5 [16c]

32. a) Bech-Nielsen G.: Electrochim. acta *18* (1973) 671; *19* (1974) 821; *20* (1975) 619; *21* (1976) 627; loc. cit [7], p. 614
33. a) Yolken, H.T., Kruger, J., and Calvert, J.P.: Corr. sci. *8* (1968) 103; b) Kruger, J., in: loc. cit. [7b], Vol. 1, p. 96; c) Kruger, J., Long, G.G., Kuriyama, M., and Goldmann, A.: ibid. p. 163; d) Kruger, J., Long, G.G., in: loc. cit. [7d], p. 210
34. a) Lumsden, J.B.and Staehle, R.W.: Scripta met. *6* (1972) 1205; b) Seo, M., Lumsden, J.B., and Staehle, R.W.: Surface sci. *42* (1974) 337; *50* (1975) 541
35. a) Haupt, S., Calinski, C., Collisi, U., Hoppe, H.W., Speckmann, H.-D., and Strehblow, H.-H.: Surface and Interface Analysis *9* (1986) 357; b) Haupt, S. and Strehblow, H.-H.: Langmuir *3* (1987) 873; c) Strehblow, H.-H.: ibid. *12* (1988) 363; d) Strehblow, H.-H., Borthen, P., and Lützenkirchen-Hecht, D.: Synchroton Radiation News *11* (1998) 12; e) Strehblow, H.-H., Calinski, C., Simson, S., Druska, P., Hoppe, and Rossi, A.: loc. cit. [7g], quoting previous publications; f) Strehblow, H.-H., Melliar-Smith, C.M., and Augustyniak, W.M.: J. Electrochem. Soc. *125* (1978) 915; g) Strehblow, H.-H. and Malm, D.L.: Corr. sci. *19* (1979) 469
36. O'Grady, W.E.: J. Electrochem. Soc. *127* (1980) 555
37. a) Davenport, A.J. and Samsone, M.: ibid. *142* (1995) 725; b) Schmuki, P., Virtanen, S., Davenport, A.J., and Vitus, C.M.: ibid. *143* (1996) 574; see also Schmuki, P., Virtanen, S., Isaacs, H.S., Ryan M.P., Oblonsky, L.J., and Böhni , H.: loc. cit. [7e], p. 183; c) Büchler, M., Schmuki, P., and Böhni, H.: ibid. *144* (1997) 2307; d) Oblonsky,, Davenport, A. J., Ryan, M. P., Isaacs, H.S., and Newman, R.C.: ibid. *144* (1997) 2398; e) Büchler, M., Schmuki, P., and Böhni, H.: ibid. *145* (1998) 609; see also Büchler, M., Schmuki, P., Virtanen, S, Böhni, H., Stenberg, T., Vippola, M., and Mäntylä, T.: loc. cit. [7e], p. 195; f) Virtanen, S., Schmuki, P., Büchler, and Böhni, H.: ibid. *146* (1999) 4089; g) Schmuki, P., Virtanen, S., Davenport, A.J., and Vitus, C.M.: ibid. *143* (1996) 3997; h) Virtanen, S., Schmuki, P., Isaacs, H.S., Ryan, M.P., Oblonsky, L.J., Vippola, M., and Böhni, H., in: loc. cit. [7e]; i) Büchler,M., Schmuki, Böhni, H., Stenberg, T., and Mäntylä, T.: J. Electrochem. Soc. *145* (1998) 378
38. a) Kolotyrkin, Ya.M.: Z. Elektrochemie *62* (1958) 700; b) Kolotyrkin, Ya.M. and Bune, N.Ya.: Z.phs. Chemie *214* (1960)274
39. Rocha, H.-J. and Lennartz, G.: Archiv Eisenhüttenwesen 26 (1955) 117
40. Heumann, T.W. and Rösener, W.: Z. Elektrochemie 59 (1955) 722
41. Andreeva, V.V.: Corr. NACE 20 (1964) 35
42. Seo, M., Furuichi, R., Kamoto, G.O., and Sato, N.: Trans. Jap. Inst. Met. 16 (1975) 519
43. a) Kirchheim, R., Heine, B., Fischmeister, H., Hofmann, S., Knote, H., and Stolz, U.: Corr. Sci. *29* (1989) 899; b) Kirchheim, R.: Electrochim. acta *32* (1987) 1619
44. a) Plieth, W.J. and Vetter, K.J.: Ber. Bunsenges. Phys. Chemie 73 (1969) 1077; b) Vetter, K.J. and Arnold, K.: Z. Elektrochemie *64* (1962) 240,407
45. Olivier, R.: Thesis, Leyden, 1955. See Vetter, K.J.: loc. cit. Chap. 5 [1]

46. a) Seo, M. and Sato, N.: loc. cit. Chap. 8 [4]; b) Sato, N. and Okamoto, G.: J. Electrochem Soc. *110* (1963) 703; c) Okamoto, G., Kobayashi,H., Nagayama, M., and Sato, N.: Z. Elektrochemie *62* (1958) 775; d) Sato, N.: J. Electrochem. Soc. *129* (1982) 255; e) id.: Corr. Sci. *31* (1990) 1; f) Sato, N. and Kudo, K.: Electrochim. acta 19 (1974) 461; g) Sato, N.: Corr. NACE *45* (1989) 354; h) Ohtsuka, T., Masuda, M., and Sato, N.: J. Electrochem. Soc. *132* (1985) 787
47. a) Tronstadt, L.: Trans. Farad. Soc. *29* (1933) 502; b) Okamoto, G. and Shibata, T.: Corr. sci.*10* (1970) 371:; c) iid., in: loc. cit. [7a], p.646; d) Goswami, K. and Staehle, R.W.: Electrochim. acta *16* (1971) 1895; e) Bulman, G.M. and Tseung, A.C.C.: Corr. sci. *13* (1973) 531
48. a) Kruger, J., in: loc. cit. [7b], Vol. 1, p. 79; b) Revesz, A.G. and Kruger, J., in: loc. cit. [7a], p. 137; c) McBee, C.L. and Kruger, J.: Electrochim. acta. *17* (1972) 1337
49. Kolotyrkin, Ya. M.: Electrochim. acta *25* (1980) 89
50. Fischmeister, H. and Roll, U.: Z. Anal. Chemie *319* (1984) 639
51. Brüesch, P., Müller,K., Zeller, H.R., and Bennemann, K.H.: Surf. sci. *169* (1986) L327
52. Asami, K., Hashimoto, K., and Shimoidara, S.: Corr. sci. *18* (1978) 151
53. Goetz, R. and Landolt, D.: ibid. *27* (1982) 1061
54. a) Olefjord, L. and Elfströem, B.-O.: Corr. NACE *38* (1982) 46; b) Clayton, C. and Olefjord, I., in: loc. cit. Chap. 5 [4], p. 176
55. Bockel, M.B.: Corr. NACE *29* (1973) 393
56. a) Pozdejeva, A.A., Antonowskaja, E.I., Suchotin, A.M.: Zash. Metal. *1* (1965) 20; b) Neidmann, N.S., Kolotyrkin, Ya.M., Knyasheva, V.M., Plaskejew, A.W., Dembrowski, M.A.: Dokl. an SSSR *216* (1974) 1331; Leygraf, C., Hultquist, G., Olefjord, I., Elfström, B.-O., Knyasheva, V.M., Plaskeyev, A.V., Kolotyrkin, Ya.M.: Corr. sci. *19* (1979) 343
57. a) Engell, H.-J. and Ramchandran, T.: Z. phys. Chemie *215* (1960) 176; b) Osozawa, K. and Engell, H.-J.: Corr. sci. *6* (1966) 389
58. a) Atrens, A., Baroux, B., and Mantel, M.: J. Electrochem. Soc. *144* (1997) 3697
59. Lu, Y.C., Clayton,, and Brooks, A.R.: Corr. sci. *29* (1989) 863
60. a) Hashimoto, K.: Chemical Properties. In: Amorphous Metallic Alloys (Luborsky, F.E., ed.). Butterworths, London, etc., 1983, p. 472; b) Proc. 4th Int. Conf. Rapidly Quenched Metals. (Masumoto, T. and Suzuki, K., eds.) Jap. Inst. of Metals, Sendai, 1982
61. Archer, M.D., Corke, C.C., and Harij, B.H.: Electrochim. acta *32* (1987) 13
62. Proc. Summer School on Amorphous Metals, Wilga 1985. (Matya, H. and Zielinski, P., eds.). World Scientifique, Singapore, 1985
63. Janik-Czachor, M.: Corr. sci. *31* (1990) 325
64. a) Kapusta, S.and Heusler, K.E.: Z. Metallkunde *72* (1981) 785; b) Heusler, K.E. and Huerta, D.J.: Non-cryst. Solids *56* (1983) 261; c) iid.: loc. cit. [62], p.302; d) Allard, K.D. and Heusler, K.E.: J. Electroanal. Chem. *77* (1977)

35; e) Allard, K.D., Ahrens, M., and Heusler K.E.: Werkstoffe u. Korr. *26* (1975) 694; f) Heusler, K.E. and Allgeier, W.: ibid. *22* (1971) 297
65. Kovacs, P., Farkas, J., Tacs, L., Awad, M.Z., Vertes, A., Kiss, L., and Lovas, A.: J. Electrochem. Soc. *129* (1982) 695
66. Linker, U. and Plieth, W.: Werkstoffe u. Korr. *34* (1983) 391
67. a) Naka, M., Hashimoto, K., and Masumoto, T.: Corr. NACE *32* (1976) 146; b) iid.: J. Non-cryst. Solids *31* (1979) 355; c) Kobayashi, K., Asami, K., and Hashimoto, K.: loc. cit. [60b]; d) Shimamura, K., Kawashima, A., Asami, K.,and Hashimoto, K.: Sci. Rep. Tohoku Univ., Ser. A. *33* (1986) 196; e) Hashimoto, K., Bhattarai, J., Li, X.-Y., Park, J.-H., Kim, J.-H., Akyama, E., Habazaki, H., Kawashima, A., Asami, K., Shimamura, K.: loc. cit. [7e], p.369
68. Virtanen, S. and Böhni, H.: Corr. sci. *31* (1990) 333
69. Linker, U. and Plieth, W.: Werkstoffe u. Korr. *34* (1983) 391
70. McCafferty, E and Moore, P.G.: J. Electrochem. Soc. *133* (1986)1090
71. Peter, R., Löschau, W., Pompe, W., and Forker, W.: Werkstoffe u. Korr. *37* (1986) 621
72. a) Schmickler, W.: loc. cit. Chap. 5 [3]; b) id.: loc. cit.[7d], p. 102; c) id.: loc. cit. [7c], p. 23
73. a) Gerischer, H., in: Advances in Electrochemistry and Electrochemical Engineering. (Delahay, P. and Tobias, C.W., eds.) Interscience Publ., New York 1961; b) Many, A., Goldstein, Y., and Grover, N.B.: Semiconductor Surfaces. North Holland Publ., Amsterdam, 1965; c) Myalin, V.A. and Pleskov, Yu.V.: Electrochemistry of Semiconductors. Plenum Press, New York, 1967
74. a) Stimming, U. and Schultze, J.W.: Ber. Bunsen Ges. phys. Chemie *80* (1976) 1299; b) ibid.: Electrochim. acta *24* (1979) 859; c) Stimming, U.: ibid. *31* (1986) 415; d) Schultze, J.W. and Mohr, S.: Dechema Monographien *90* (1981) 231; e) Searson, P.C., Stimming, U., and Latanision, R.M.: loc. cit. [7d], p.175; f) Schultze, J.W. and Kudelka, S.: The Electrochemical Society *Interface*, Summer 1997, p. 28; g) Schweinsberg, M., Kudelka, S., Michaelis, A., and Schultze, J.W.: Materials and Corr. *49* (1998) 161; ; h) Michaelis, A. and Schultze, J.W., in: loc. cit. [7d], p. 44; i) Schultze, J.W., Danzfuss, B., Meyer,O., and Stimming, U.: Materials Sci. and Engng. *69* (1985) 273; j) Bartels, C., Danzfuss, B., and Schultze, J.W., in: loc. cit. [7c], p. 35
75. a) Gerischer, H.: Corr. sci. *31* (1990) 81; b) id.: J. electrochem. Soc. *113* (1966) 1174
76. a) Makrides, A.C. and Stern, M.: J. Electrochem. Soc. *107* (1960) 877; b) Makrides, M.C.: ibid. *108* (1961) 412; *111* (1964) 392, 401; *113* (1966) 1158; c) Stern, M. and Wissemberg, W.J.: J. Electrochem. Soc. *106* (1959) 751, 756
77. Strauss, B.: Z. Elektrochemie *33* (1928) 717
78. a) Uhlig, H.H., Triadis, D.N., and Stern, M.: J. Electrochem. Soc. *102* (1955) 59; b) Uhlig, H.H. and King, P.: ibid. 106 (1959) 1

79. a) Pryor, M.J. and Cohen,J.: J. Electrochem Soc.*98* (1961) 263 and earlier work quoted therein; b) Szklarska-Smialowska, S. and Staehle, R.W.: ibid. *121* (1974)1146, 1393; c) Brasher, D.M., Kingsbury, A.H.: Trans. Farad. Soc. 54 (1958) 1214; d) Cohen, M. and Beck, F.: Z. Elektrochemie 62 (1958) 696; e) Cartledge, G.H.: J. Electrochem. Soc. 114 (1967) 39 and earlier work quoted therein
80. Tomashov, N.D.: Z. Elektrochemie *62* (1958) 717; Corr.sci. *4* (1964) 315; Werkstoffe u. Korr. *27* (1976) 636
81. Data from Chap. 3 [9], see also [82]
82. Kelly, E.J.: Electrochemical Behaviour of Titanium. In: Modern Aspects of Electrochemistry, No.14. ( Bockris, J.O'M., Conway, B. E., and White, R.E., eds.). Plenum Press, New York and London, 1982
83. a) Thull, R. and Schaldach, M.: Biomed. Technik *20* (1975) 51,111; b) Schmidt, H., Stechemesser, G., Witte, J., and Soltani-Farsh, M.: Corr. Sci. *40* (1998) 1545; Schmidt, H., Exner, H.E., Rüeck, D.M., Angert, N., and Fink, U., in: Advances in Material Science and Implant Orthopedic Surgery, Dordrecht, 1995
84. Brauer E. and Nann, E.: Werkstoffe u. Korr. *20* (1969) 676; ibid. *25* (1974) 309, 481
85. Hurlen, T. and Wilhelmsen, W.: Electrocheim. acta *31* (1986) 1139 676
86. a) Straumanis, E.M., Shim, S. T., and Schlechten, A.W.: J. Phys. Chem. *59* (1955) 317; b) Straumanis, E.M. and Gill, C.B.: J. Electrochem. Soc. *101* (1954) 10
87. Sukhotin, A.M. and Tungusova, L.I.: Prot. Met. (English translation) *4* (1968) 5, quoted from [81]
88. a) Rüdiger, O., Fischer, W.R., and Knorr. W.: Z.Metallkunde *47* (1956) 599; b) Fischer, W.R., Ilschner-Gench, C., Knorr, W.: Werkstoffe u. Korr. *12* (1961) 597; c) Rüdiger, O.: ibid. *16* (1965) 109
89. a) Tomashov, N.D., Chernova, G.P., and Altovsky, R.M.: Z. phys. Chem. *214* (1960) 321; Zhur. Fiz. Khim.*35* (1961) 1068; b) Tomashov, N.D., Chernova, G.P., Modestova, V.N., Chukalovskaya, T.V., Volkov, L.N., and Vasiyeva, R.P.: Proc. 4th Int. Conf. Metallic Corrosion, Amsterdam 1969. NACE, Houston, 1972, p. 694
90. Cotton, J.B.: Platinum Met. Rev. *11* (1967) 50
91. a) Sheldon, P., Shimigu, K.,Thomson, G.E., and Wood, G.C.: Phil. Trans. R. Soc. London A *348* (1994) 295; b) Habazaki, H., Paez, M. A., Shimizu, P., Sheldon, P., Thomson, G. E., Wood, G. C., and Zhou, X.: Surf. Interface Anal. *23* (1996) 892; c) ibid.: Corr. sci. *38* (1996) 1033; d) Habazaki, H., Sheldon, P., Thompson, G.E., Wan, J., Wood, G.C., Zhou, X., and DeLaet, J.: J. Electrochem. Soc. *144* (1997) 4217; e) Habazaki, H., Sheldon, P., Shimizu, K. Thompson, G.E., Wood, G.C.: Corr. sci. *37* (1995) 1497; f) Richardson, J. A. and Wood, G.C.: ibid. *10* (1970) 313; g) Brown, G. M., Shimizu, K., Kobayashi, K., Thompson, G.E., and Wood, G.C.: ibid. *34* (1993) 1045, 2099
92. a) Brown, F. and Mackintosh, W.D.: *120* (1973) 1096; b) Kalil, N. and Leach, J.S.: Electrochim. acta *31* (1986) 1279

93. Viswanadham, R.K., Sun, T.S., and Green, J.A.S.: Corr. NACE *36* (1980) 275
94. a) Kaesche, H, in: loc. cit. [7a], p. 935; b) id.: Z. phys. Chem. N. F. *34* (1962) 87; c) id.: Werkstoffe u. Korr. *14* (1963) 557; e) Nikol, T. and Kaesche, H.: Preprints Eurocorr '87, Karlsruhe. Dechema, Frankfurt, 1987, p. 505; d) Kaesche, H.: Electrochim. acta *9* (1964) 383; e) Grauer, W. and Kaesche, H.: Corr. sci. *12* (1972) 617; f) Kaesche, H.: Heizung, Lüftung, Haustechnik *13* (1962) 332
95. Ergang, R. and Masing, G.: Z. Metallkunde *41* (1950) 272
96. Pryor, M.J. and Keir, D.S.: J. Electrochem. Soc. *102* (1955) 605
97. Galvele, J.R. and DeMicheli, S.M., in: loc. cit. [7h], p. 439
98. Frankel, G.S., Johnes, C.V., Brusic, V., and Davenport, A.J.: J. Electrochem. Soc. *142* (1995) 2290
99. Franck, U.F. and Lüdering, W., quoted from Franck, U.F.: Werkstoffe u. Korr. *9* (1958) 504
100. a) Huber, K.: Z. Elektrochemie *62* (1958) 675; b) Schwabe, K.: Z. phys. Chemie *205* (1956) 304; c) Bartelt, H. and Landsberg, R.: ibid. *222* (1963) 217; d) Kabanov, B.N.: Electrochim. acta *6* (1962) 253; e) Hull, M.N., Ellison, B.T., and Itoni, J.E.: J. Electrochem. Soc. *117* (1970) 192
101. a) Powers, R.W. and Breiter, M.W.: J.Electrochem. Soc. *116* (1969) 719:; b) Powers, R.W.: ibid. *116* (1969) 719; *116* (1969) 1652; c) McKubre, M.C.H. and Macdonald, D.D.: ibid. *128* (1981) 524; d) Prentice, G.A., Chang, Y., and Shan, X.: ibid. *138* (1991) 890; e) Scholl, P., Shan, X., Bonham, D., and Prentice, G.A.: ibid. p. 895; f) Macdonald, D.D., Ismail, K.M., and Sikora, E.: ibid. *145* (1998) 3141
102. a) Kroenig, W.O. and Pawlow, S.F.: Metallschutz *9* (1933) 268; b) Schikorr, G.: Gas u. Wasserfach *82* (1939) 834; c) Roters, H. and Eisenstecken, F.: Archiv Eisenhüttenwesen *15* (1941) 59; d) Gilbert, P.T., in: Pittsburgh Int. Conf. Surface Reactions. Corr. Publ. Comp. 1948, p. 127; e) Hoxeng, R.B.and Prutton, C.F.: Corr. NACE *14* (1958) 585; f) Shuldiner, H. and Lehrmann, L.: ibid. *14* (1962) 332
103. a) Song, G.L. and Atrens, A.: Advanced Engng. Materials *1* (1999) 1; b) See for instance Polmaer, I.J., in: Physical Metallurgy of Magnesium Alloys. DGM Informationsgesellschaft, Oberursel, 1992, p. 201; c) Nisancioglu, K., Lunder, O., and Aune, T.K.: Proc. 47th Congr. World Magnesium Assocoation, 1990, p. 43
104. a) Freier, R.K.: Chemie des Wassers in Thermischen Kraftanlagen. W.de Gruyter, Berlin,New York, 1984.; b) Proc. Int. Conf. High-Temperature High-Pressure Electrochemistry in Aqueous Solutions, Surrey 1973. (Jones, D.,deG. and Staehle, R.W., eds.). NACE Houston 1974 with Water and Steam, Heidelberg 1992. (Dooley, B. and Bursik, A., eds.). EPRI, Pleasant Hill, 1993
105. a) Effertz, P.-H., Hickling, J., Heinz, A., and Mohr, G.: Allianz Berichte Nr.23 (1985); b) Fuji, C.T. and Meussner, R.A.: J. Electrochem. Soc. *111*(1964)1215; c) Schoch, W., Richter, R. and Effertz, P.-H.: Der Maschi-

nenschaden *43* (1970) 65; d) Effertz, P.-H., in: Proc. 5th Int. Conf. Metallic Corrosion, Tokyo 1972. NACE, Houston 1974, p. 920; e) Rohlfs, U. and Kaesche, H.: Der Maschinenschaden *57* (1984) 11
106. a) Mann, G.M.W., in: loc. cit. 104b, p. 34; b) Wright, I.G. and Agraval, A.K., in: loc. cit. [104c].
107. a) Castle, J.E. and Mann, G.M.W.: Corr. sci. *6* (1966) 253; b) Tomlinson, L.: Corr. NACE *37* (1981) 591; c) Robertson, J.: Corr. sci. *29* (1989) 275; d) id.: ibid. *32* (1991) 443
108. Lister, D.H., Davidson, R.D., and McAlpine, E.: Corr. sci. *27* (1987) 113
109. a) Winkler, R. and Lehmann, H.: VGB Kraftwerkstechnik *65* (1985) 421; b) Winkler, R., Hüttner, F., and Michel, F.: ibid. *69* (1989) 527
110. Pick, M.E. and Segal, M.G.: Nucl. Engng. 22 (1983) 433
111. a) Stellwag, B.: Corr. NACE *53* (1997) 120; b) id.: Corr. sci. *40* (1998) 337

# 11 Galvanic Corrosion Cells

## 11.1 Dissimilar Metal Contact Corrosion

### 11.1.1 Introduction

Two dissimilar metals, A and C, subject to corrosion in the same electrolyte, normally are at different corrosion potentials, $(\varepsilon_a)_{\text{corr}}$ and $(\varepsilon_c)_{\text{corr}}$. The electrode system metal C/solution/metal A then is a galvanic cell with the open-circuit voltage

$$\Delta\varepsilon = (\varepsilon_c)_{\text{corr}} - (\varepsilon_a)_{\text{corr}} \tag{11.1}$$

We choose indices such that $(\varepsilon_c)_{\text{corr}}$ is more positive than $(\varepsilon_a)_{\text{corr}}$, so that A is the anode and C the cathode in the cell. When short circuited by an external metallic connection between A and C, a cell current, $j$, will flow, moving positive charge carriers in the solution, and electrons in the metalllic connection, in the direction from the anode to the cathode. The current passing the anode/solution interface has the positive strength $+|j|$; a current of the same strength but opposite sign, $-|j|$, passes the solution/cathode interface. At both electrodes the total current, $j$, is, in principle, the sum of anodic and cathodic partial currents. Often, circumstances are simpler, when at the anode the cathodic partial current of the reduction of an oxidizing species, and/or when at the cathode the partial current of anodic metal dissolution are non-existent or negligible, but this is not the general case. The early assumption that, instead, the action of galvanic cells is the general principle of corrosion in fact has been a source of much confusion.

In the special case of the titanium anode/platinum cathode cell described in Chap. 10, the anode is passivated by the contact with the cathode, but this and similar cases again are exceptions. The rule is that anodic polarization of the anode increases the rate of anode metal dissolution, thus accelerating corrosion; and this effect defines *contact corrosion.*

Some aspects of the matter have already been described in Chap. 4, more will follow presently, but some complications will be further disregarded, including the possibility that the system under consideration is a galvanic chain of the type metal A /solution I/metal C/solution II/metal A, with an electrolytic instead of a metallic short circuit. Normally, A directly contacts C, as for steel tubing welded to copper tubing, and in many other cases. This then creates conditions different from what so far has been assumed in this text, which was uniform distribution of all partial current densities, including the current density of metal dissolution, everywhere on the surface of the respective electrode. When, instead, direct contact between dissimilar metals in one and the same solution causes short-circuit galvanic current flow, the electrode potentials and, by consequence, the densities of partial currents on both metals, normally will be functions of the distance from the three-phase common boundary. The main problem then is to evaluate the distribution of current densities on the two metal surfaces, and especially the distribution of the current densities of metal dissolution on the anode.

The mathematical formulation of the problem is straightforward, set out by the following simultaneous 6 conditions :

1) Let $j_r$ be the current strength of the $r$th of $m$ partial electrode reaction. Electroneutrality of the system then requires that

$$\sum_1^m j_r = 0 \,. \tag{11.2}$$

2) Owing to the high conductivity of metals, ohmic potential differences in metals are negligible and the Galvani potentials, $\phi_a$ and $\phi_c$, therefore constant in the interior of A and C, although $\phi_a \neq \phi_c$.

$$\phi_a = \text{const} \,, \quad \phi_c = \text{const}' \,, \quad \text{const} \neq \text{const}' \,; \quad \text{in metals.} \tag{11.3}$$

In very large metallic structures, ohmic voltages along the metal may become considerable, but this is a complication arising only when considering cathodic protection of, e.g., extended pipeline systems and the like.

3) In the electrolytic solution, ohmic voltages often will be considerable owing to the much lower conductivity even of concentrated electrolytes. The net electrical charge of the bulk of the solution is, however, zero, and *Laplace*'s differential equation must be satisfied, prescribing that the divergence of the gradient of the Galvani potential is zero:

$$\operatorname{div}\operatorname{grad}\phi = \frac{\partial^2\phi}{\partial x^2} + \frac{\partial^2\phi}{\partial y^2} + \frac{\partial^2\phi}{\partial z^2} = 0 \,; \quad \text{in electrolytic solutions.} \tag{11.4}$$

More briefly, this is expressed as $\Delta\phi = 0$. It is useful to remember at this point that in any homogeneous conducting phase, including electrolyric solutions, measurable differences of Galvani potentials are ohmic voltages.

4) At insulating surfaces, such as solution container walls, the component of the potential gradient in the direction of the normal vector, $\vec{n}$, disappears, since no current enters the insulator:

$$\frac{\partial\phi}{\partial\vec{n}} = 0 \,; \quad \text{at insulator surfaces.} \tag{11.5}$$

5) Any total current, resulting from the superposition of partial currents, passing a metal/solution interface governed by electrode kinetics, also must satisfy

Ohm's law: The total current density, $i_s$, is the product of solution conductivity, $\sigma$, and the local gradient of the potential normal to the surface. Conductivity often is expressed as reciprocal resistance, $\sigma/(\Omega\,\text{cm})^{-1}$ in 'Siemens' units.

$$i_s = \sigma \frac{\partial \phi}{\partial \vec{n}}\,; \quad \text{at electrode surfaces.} \tag{11.6}$$

6) At the same time, total current densities, $i_s$, are determined by the kinetics of the reactions at the respective electrode as expressed by current-potential functions, $i(\varepsilon)$, or in the inverse form, $\varepsilon(i)$, introduced by Eq. (4.11). For the present purpose it is convenient to summarily write, for anodes and cathodes, respectively,

$$a) \quad \varepsilon_a = (\varepsilon_a)_{\text{corr}} + |\pi_a|\,, \qquad b) \quad \varepsilon_c = (\varepsilon_c)_{\text{corr}} - |\pi_c\,.| \tag{11.7}$$

In principle, this set of equations always completely describes the problem to be solved. Unfortunately, the current-potential relations, which enter the system of equations as boundary conditions, usually are non-linear, as for instance exponential functions, and this then normally prevents analytical solutions of the problem under consideration. To illustrate the basic circumstances, we therefore discuss simple, or else simplified cases only.

A typical attempt is to overcome complications by linearizing current-potential relations. An example is described in [1a]. The problem was to estimate the conditions for safe cathodic protection (see Appendix) of steel in a neutral aerated salt solution as achieved by contact with a magnesium sacrificial anode. The anode was supposed to be practically unpolarizable, since, as described in Chap. 10, its polarizability is at least low. Therefore, the function $\varepsilon_a(i)$ is replaced by the simple relation: $\varepsilon_a \simeq$ const. Corrosion of mild steel in neutral aerated solutions has been described in Chap. 6 as being controlled by cathodic oxygen reduction, more or less in the range of potential-independent cathodic limiting current densities. Inspecting Fig. 6.6 (case $A_3$), it is then seen that corrosion is negligible when the total polarizing current is equal to the current density of oxygen reduction. In this protective range of potentials, the current-potential curve for iron reduces to the expression $i_c = i_{O_2,D}$. Thus, both current-potential relations are linearized. The problem is further simplified when the anode is placed far away from the cathode, since there then is no three-phase common boundary. We do not continue the calculation, as it applies to cathodic protection rather than contact corrosion. The set-up of boundary conditions is, however, remarkable not only for simplicity but for its origin in electrode kinetics as well. Usually, linearization is achieved rather more arbitrarily, as seen further below.

In [1b], oxygen corrosion of heterogeneous metal surfaces, representing coplanar arrays of microscopic anodes and cathodes, i.e. systems of *local cells,* has been analyzed, again assuming unpolarizable anodes, and diffusion-controlled oxygen reduction at cathodes,providing rules of thumb suitable to estimate the degree of danger of contact corrosion.

When in a galvanic cell the anode and the cathode are in direct contact, and a three-phase boundary does exist, the problem requires the application of Fourier series or Fourier integrals even in the relatively simple case of linear current-potential curves [2]. Computer-supported iterative finite element solutions have been obtained even for exponential current-potential functions as boundary conditions [3]. These calculations will be ignored in the following in favor of the consideration of very simple cases showing some essentials of contact corrosion events.

### 11.1.2
### Cells with Homogeneous Distribution of Current and Potential

To illustrate the influence of solution conductivity and electrode polarization on the behaviour of contact corrosion galvanic cells, we once more return to an electrode arrangement without a three-phase common boundary. Instead, a flat anode and a flat cathode of equal, for instance rectangular, contour form the parallel endplates of an electrolytic cell of length $l$. The area of both electrodes is $A_{met}$. As indicated in Fig. 11.1, the short circuit by an external metallic wire can be interrupted by a switch. Two reference electrodes equipped with Haber-Luggin capillaries are used to measure electrode potentials. For this geometry, the solution of Laplace‘s equation is trivial: Equipotential surfaces (or more briefly: potential surfaces) are equidistant planes parallel to the electrodes, including the electrode surfaces. Current lines are equidistant straight lines orthogonal to the set of potential planes. Anode current density, $(i_s)_{an}$, and cathode current density, $(i_s)_{cath}$, are constant on the anode and the cathode, respectively, and they also are equal in value for anode and cathode:

$$(i_s)_a = +|i_s|\,, \quad (i_s)_c = -|i_s|\,. \tag{11.8}$$

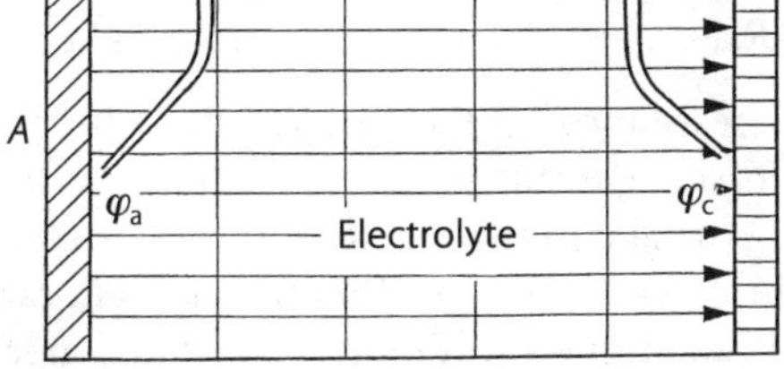


**Fig. 11.1.** Schematic section through a galvanic cell with plane-parallel electrodes as endplates of the solution container, equipped with reference electrodes and an external switch

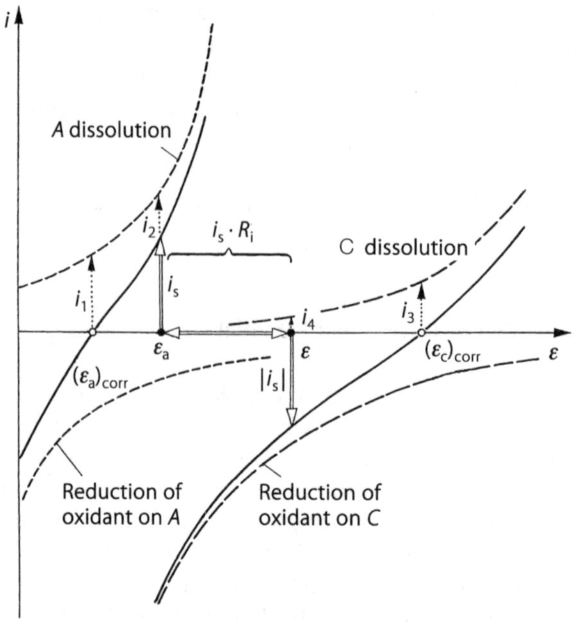


**Fig. 11.2.** Schematic Current density-potential diagram for a galvanic cell with equal-area anode and cathode as in Fig. 11.1, showing anodic and cathodic partial current density -potential curves (- - - -)and total current-potential curves (—-) for both electrodes. Reverse partial reactions are assumed to be negligible. $(\varepsilon_a)_{corr}$,$(\varepsilon_c)_{corr}$,$i_1$,$i_3$ : Open-circuit anode and cathode potentials and metal dissolution current densities. $\varepsilon_a$,$\varepsilon_c$,$i_2$,$i_4$ : Short-circuit potentials and current densities. $i_s$: Cell current density

Concerning potentials we have:

$$\varepsilon_a \leq \varepsilon_c\,, \quad \varepsilon_a \geq (\varepsilon_a)_{\text{corr}}\,, \quad \varepsilon_c \leq (\varepsilon_a)_{\text{corr}} \tag{11.9}$$

The flow of ions in the solution causes an ohmic Galvani potential difference, $\phi_a - \phi_c = \phi_{a,c}$, which is positive, since cations migrate from the anode to the cathode. This potential difference is measured with the two reference electrodes as the electrode potential difference: $\phi_{a,c} = \varepsilon_c - \varepsilon_a$. Therefore, current flow in the solution obeys Ohm's law in the form

$$\phi_{a,c} = \varepsilon_c - \varepsilon_a = (+|i_s|\,l)/\sigma = +|j_s|\,R_i\,, \tag{11.10}$$

where $R_i = (\sigma l/A_{\text{met}})$ is the internal resistance of the electrolytic solution. Figure 11.2 schematically shows the complete current-potential diagram for the short-circuited cell, assuming negligible partial current densities of reverse electrode reactions at both the anode and the cathode. We note that the diagram can be adapted to the more realistic case of anodes and cathodes of unequal area, $A_a \neq A_c$, by replacing current densities, $i$, by current strengths, $j$. In any case it is seen that the intensity of galvanic cell action, which is contact corrosion of the anode accompanied by cathodic protection of the cathode, is determined by the current-potential curves and by the internal resistance of the solution. Using Eq. (11.7), we can relate the short-circuit current to the open-circuit potential difference of the cell, $\Delta\varepsilon_{oc} = (\varepsilon_c)_{corr} - (\varepsilon_a)_{corr}$, and to the anode and cathode polarizations, $\pi_c$,$\pi_a$:

$$i_s = \{\Delta\varepsilon_{oc} - |\pi_c(i_s)| + |\pi_a(i_s)|\}\,(\sigma/l)\ . \tag{11.11}$$

This may not seem to make particular sense because, for any electrode, $\varepsilon_{corr}$ and $\pi$ are interrelated by the superposition of the same partial current-potential relations, and these themselves depend on solution composition and, therefore, conductivity. Still, we may further proceed by introducing integral polarization resistances as defined by Eq. (4.14). Dropping 'integral' subscripts and counting cell currents as positiv, we introduce

$$\text{a) } (R_\pi)_c = |\pi_c|/i_s\,; \quad \text{b) } (R_\pi)_a = |\pi_a|/i_s\ . \tag{11.12}$$

Substitution into Eq. (11.11) results in the expression

$$i_s = \frac{\Delta\varepsilon_{oc}}{(l/\sigma) + (R_\pi)_a + (R_\pi)_c} \lesssim \frac{\Delta\varepsilon_{oc}}{(R_\pi)_a + (R_\pi)_c}\ . \tag{11.13}$$

This is analogous to Ohm's law, though integral resistances are not ohmic, except in the rare case of linear current-potential relations. Equation (11.13) is useful, as it does make it very clear that for an estimate of the danger of dissimilar metals contacts in any geometric configuration at least an approximate knowledge of the current-potential curves is necessary, considering at least the open-circuit corrosion potentials of anodes and cathodes, and approximate magnitude of polarizations, $\pi$, or integral resistances, $R_\pi$ [4a].

Integral polarization resistance values or, more briefly, polarizability data, are not, however, usually available even in cases where open-circuit corrosion potential data are recorded in what is termed 'practical series' of electrode potentials for metals and alloys in actual service. As an exception to this rule, we quote corrosion potential values as well as polarizability data listed for a number of commercial metals and alloys immersed in two types of model electrolytes in [5]. To simulate almost pure moisture, as encountered for instance in tropical climates, a phtalate/hydroxide buffer solution was used, expecting that the phtalate ion does not much interfere with corrosion kinetics. To model aggressive marine surroundings, artificial seawater was prepared following a German standard. In this latter test solution, chloride is the most aggressive component, frequently causing pitting corrosion. Table 11.1 lists a selection from the much more comprehensive data, all applying to commercial metals and alloys as fully specified in the original publication. The accompanying compendium of polarization data probably is not easily available [5b]. Commenting on the contents of the table, the first note is that, in the long run, contact corrosion conditions may change owing to the growth of layers of corrosion products. Also, for every metal widely different polarizability data may apply, whenever the solution composition changes importantly. An example is tin which, as a consequence of complex ion formation in fruit juice tends tends to become anodic to carbon steel, which is interesting for the huge amounts of tinned steel fruit cans [6a].

**Table 11.1.** Values of quasi-stationary corrosion potentials of some commercial metals and alloys in two solutions at 25°C, modeling typical corrosion conditions: Aerated phtatalate/hydroxide solution, pH 6.0, modeling moisture without aggressive electrolytes ('water'), and aerated artificial seawater according to DIN 50907, adjusted to pH 7.5, modeling a solution particularly aggressive due to chloride content ('seawater'). Values selected, and rounded to tenth of volts, from more comprehensive data reported by Elze and Oelsner. Data in brackets indicate potentials drifting with time

| metal | 'water' | 'seawater' | metal | 'water' | 'seawater' |
|---|---|---|---|---|---|
| gold | (+0.31) | (+0.24) | aluminum | −0.17 | −0.67 |
| silver | +0.19 | +0.15 | tin | −0.28 | −0.81 |
| titanium | +0.18 | (−0.11) | lead | (−0.28) | −0.26 |
| Monel[1]) | (+0.13) | +0.01 | carbon steel | −0.35 | −0.34 |
| brass[2]) | +0.15 | +0.13 | cadmium | −0.57 | −0.52 |
| copper | +0.14 | +0.01 | zinc | −0.79 | −0.79 |
| stainl.steel[3]) | +0.08 | −0.13 | magnesium | −1.15 | −1.14 |

[1]) Ni-38.8Cu, [2]) Cu-33Zn, [3]) Fe-18Cr-8Ni (Wt.%)

Another example, already described in Chap. 10, is the potential reversal of zinc, referred to iron, in warm water. In the same chapter it also was explained that, on the other hand, any contact of aluminum in non-pitting, such as sulfate solutions, with any cathode is tolerable, because the metal is passivated by an insulating oxide film. The polarizability then tends to infinity: $(R_\pi)_{an} \rightarrow \infty$. Contact between two dissimilar passive metals, as between stainless steel and titanium, is without danger due to negligible metal dissolution rates, always provided that pitting does not occur.

Owing to the origin of the corrosion potential by the superposition of partial current-potential curves, a discussion of the practical range of $\varepsilon_{corr}$ values is difficult. Trivial exceptions are cases where, as for zinc in zinc sulfate solutions, or copper in copper sulfate solutions, $\varepsilon_{corr}$ equals $E_{Me/Me^{z+}}$. A large group of other, irreversible, potentials is, however, also rather easily explained: Inspecting Table 11.1, it is conspicuous that passive metal corrosion potentials in neutral aerated solutions usually fall into a comparatively narrow range, $\varepsilon_{corr} = 0.1 \pm 0.2V$, and that they tend to drift with time. The reason is that these metals have corrosion potentials more positive than the hydrogen equilibrium potential in the same solutions, and that owing to high anodic overpotential of metal dissolution they should tend to settle close to the oxygen equilibrium potential, if this were not prevented by the even higher overpotential for oxygen reduction. In a way, the role of most passive metals for contact corrosion is that of a poor oxygen electrode. It is typical that even noble metals, such as gold and platinum, spontaneously adopt an irreversible potential considerably more negative than the $E_{O_2}$. Whether this is due to a well-defined superposition of partial current-potential curves, or else due to spurious other electrode effects, may as well remain open to discussion as the role of oxygen adsorption similar to incipient passivity on noble metals.

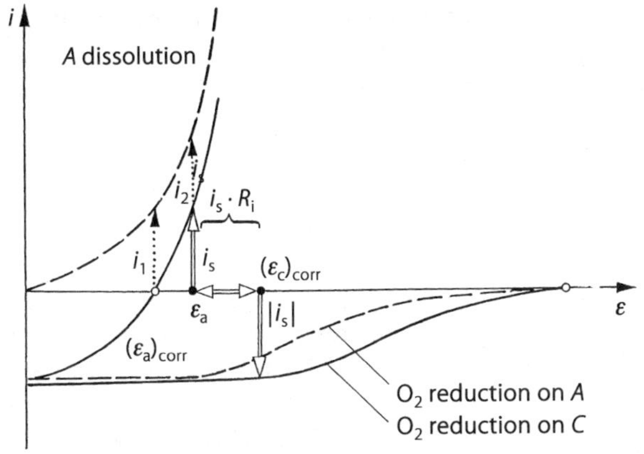


**Fig. 11.3.** Schematic Current density-potential diagram for an iron electrode in contact, without common boundary, with a platinum cathode of equal surface area in the same aerated, stirred, neutral solution. Hydrodynamic conditions are supposed to be equal at the electrode surfaces, and the short-circuit potentials of both anode and cathode to be in the range of the cathodic diffusion-limited current density of oxygen reduction. Indexing of potentials and currents is as in Fig. 11.2

For non-passive electrodes, especially for plain steel in neutral salt solutions, the distance to the more positive oxygen equilibrium potential in the same solutions, i.e. from $\varepsilon_{corr}$ to $E_{O_2}$, easily is of the order of 1 V. This, then, often brings the corrosion potential into the range of the cathodic diffusion-limited current density of oxygen reduction, $i_{O_2,D}$. An interesting case then is contact corrosion of an iron electrode caused by a noble-metal cathode such as platinum, where cathodic oxygen reduction is the only electrode reaction occurring. For the simple case of equal-area, plane-parallel anode and cathode, Fig. 11.3 illustrates the situation, assuming a cathode potential in the range of $i_{O_2,D}$. It is seen that under the conditions assumed the rate of anodic iron dissolution, $i_{Fe} = i_2$, which also is the rate of iron contact corrosion, is twice the open-circuit iron dissolution rate: $i_2 = 2i_1$, and $i_1 = |i_{O_2,D}|$.

For a system with different anode and cathode areas, $A_{an}$ and $A_{cath}$, but with both short-circuit electrode potentials still in the range of $i_{O_2,D}$, we conclude from Eq. (11.2) that

$$i_{Fe}A_{Fe} = -i_{O_2,D}(A_{Fe} + A_{Pt}) . \qquad (11.14)$$

Here, the interesting case is that of a small anode, as for instance a pore in a protective coating, in contact with a large cathode, i.e. $A_{Fe} \ll A_{Pt}$. Then

$$i_{Me} \simeq -i_{O_2,D}(A_{Pt}/A_{Fe}) , \qquad (11.15)$$

which for a small anode of constant area essentially reduces to $i_{Me} \sim A_{Pt}$. This expresses the *oxygen catching area principle*, stating that the rate of corrosion is proportional to the cathodic area 'catching' diffusing oxygen [6b]. For a small anode within a large cathode, the density of anodic metal dissolution will increase with decreasing distance to the anode/cathode boundary line. The overall effect of contact corrosion may then still be formulated for the average metal dissolution current density, defined as $\bar{i}_{Fe} = \int i_{Fe} dA_{Fe}$. With increasing cathode area, the principle eventually fails, when the ohmic voltage between the anode and distant points on the cathode is so large that the cathode potential shifts in the positive direction into the range beyond $i_{O_2,D}$, where oxygen reduction is charge-transfer controlled.

A notorious case of an unfavourable anode/cathode area ration was pores in the copper/nickel/chromium coating of car bumpers, exposing anodic iron to cathodic coating metals. The useful counter-example still is zinc-coated steel, because, except for the rare event of 'potential reversal' (see Chap. 10), a pore in the coating always is cathodically protected by zinc rather than anodically attacked by contact corrosion.

### 11.1.3
### Contact Corrosion Cells with Coplanar Electrodes

In most practical cases, a contact corrosion cells has a common three-phase anode/cathode/solution boundary, and we have to inspect the consequences at least semi-quantitatively. For simplicity, we assume, as in Fig. 11.4, flat electrodes of infinite extension under an equally infinitely large volume of electrolytic solution, with no insulator surfaces anywhere. For this geometry, Laplace's differential equation reduces to: $\partial^2\phi/\partial x^2 + \partial^2\phi/\partial y^2 = 0$.

When anode and cathode polarizabilities are small, it is useful to inspect, in a first step, the so-called 'primary distribution of potentials and currents', neglecting terms $\pi_a$ and $\pi_c$, i.e. setting

$$\varepsilon_a = (\varepsilon_a)_{corr}\,, \quad \varepsilon_c = (\varepsilon_c)_{corr}\,; \quad \text{for vanishing polarization,.} \qquad (11.16)$$

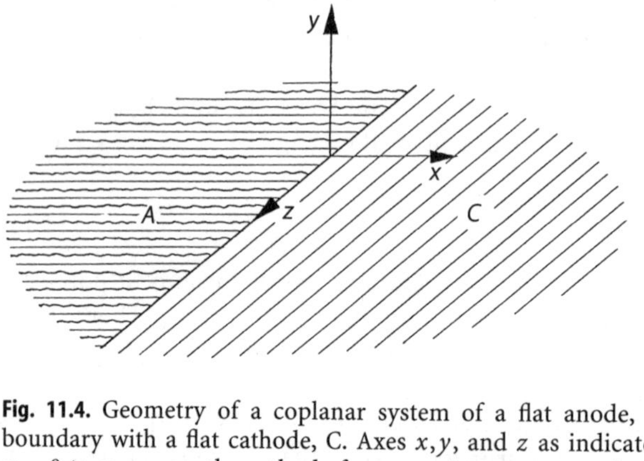


**Fig. 11.4.** Geometry of a coplanar system of a flat anode, A, in contact by a straight boundary with a flat cathode, C. Axes $x$,$y$, and $z$ as indicated. The anode extends from $x = 0$ to $x = -\infty$, the cathode from $x = 0$ to $x = +\infty$, the electrolytic solution volume from $z, x = \pm\infty$ to $0 \leq y \leq \infty$

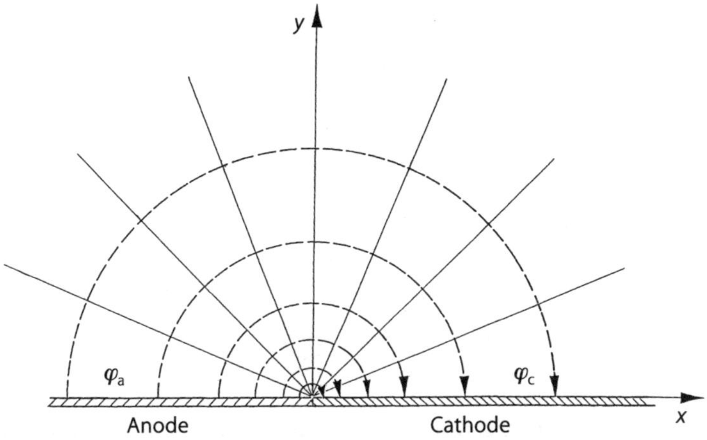


**Fig. 11.5.** Section at $z$ = const through the current lines and the potential field over a flat anode in contact with a flat cathode under an electrolytic solution in the case of negligible polarizibities.polarizability of both anode and cathode. *Solid lines:* sections through equipotential surfaces; *broken lines:* current lines

As a consequence of this boundary condition, both electrode surfaces are equipotential surfaces at $\phi_a$ and $\phi_c$, respectively. As shown in Fig. 11.5, the sections through the equipotential surfaces at any constant value of $z$ form a set of concentric straight lines with the boundary line, $x,y = 0$, as origin. The field of current density lines is orthogonal to the field of potential surfaces, and each line is a semi-circle originating at some point, $-x_n$, on the anode and ending at the corresponding point, $+x_n$, on the cathode at constant $z$. Each pair of current lines maps the section through a channel through the solution volume, and in each such channel the same current flows from anode to cathode. The reciprocal channel width is a measure of the local current density including, for primary conditions, the current density $i_s$ at the respective electrode surface. Since the length of each current line is $\pi x$, we have

$$i_s\,(\mp x) = \pm\,\sigma\phi_{a,c}/\pi x = \pm\,\sigma\,\Delta\varepsilon_{oc}/\pi x\;. \tag{11.17}$$

According to this equation, the cell current density tends to infinity when the distance from the anode/cathode boundary tends to zero. This, of course, is impossible owing to the finite polarizability of both anode and cathode in every realistic case, and Fig. 11.6 qualitatively explains the influence of small and approximately equal constant values of $(R_\pi)_a$ and $(R_\pi)_c$, especially showing changes in anode and cathode electrode potentials increasing with decreasing distance from the anode/cathode boundary. In this figure, we approach what is termed the 'secondary distribution of currents and potentials'. What has been explained concerning current channels still applies, but for the evaluation of current densities at the electrode surfaces the $\vec{n}$-component of the current has to be considered only. A special case, which can still be handled relatively easily, is $(R_\pi)_a = (R_\pi)_c = R = \text{const}$, assuming linear and symmetrical current-

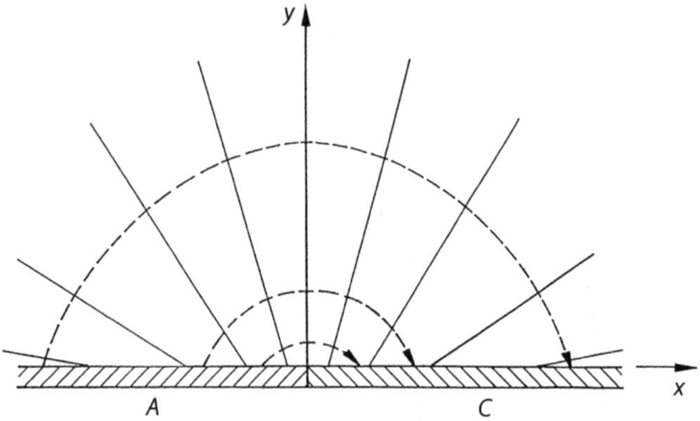


**Fig. 11.6.** Section through the current-potential field of a short-circuited contact corrosion cell as in Fig. 11.4, but allowing for finite, but small and approximately symmetrical polarizability of anode and cathode

potential curves for anode and cathode. If, furthermore, $R$ is small, the current lines still are approximately complete semicircles, and the instructive relation arrived at is [7a]:

$$i_s(\mp x) = \pm\Delta\varepsilon_{oc}/(\pi x/\sigma + 2R)\ . \tag{11.18}$$

The influence of changes of R and $\sigma$ on the local cell current density at the anode calculated with this relation is shown in Fig. 11.7. On the cathode, the current distribution is exactly anti-symmetric to this field. Equation (11.18) reflects the increase of the anode current density of metal dissolution, i.e. the anode corrosion rate, with decreasing distance from the anode/cathode boundary. We also note that, as judged from the same equation, the intensity of contact corrosion of the anode directly adjacent to the cathode is a function of the polarizability only, and the only effect of solution conductivity is to spread contact corrosion to a distance from the boundary increasing with increasing conductivity. We remember, of course, that in any real case it may be difficult to change the conductivity of the solution without changing electrode kinetics, thus changing polarizability as well.

As seen in Fig. 11.7, the assumption of infinitely large electrodes does not seriously affect the calculated results, since the effect of contact corrosion vanishes with increasing distance from the anode/cathode boundary. The assumption of small, equal and constant integral polarization resistance values replacing actual current-potential curves certainly is much more serious. We do not, however, enter more realistic but not very transparent numerical calculations. Instead, the description of experimental evidence concerning a model contact corrosion cell is preferred to illustrate the functioning of a real system.

As described in [8], a model cell was constructed from a large rectangular nickel-plated steel sheet, 30.5 × 81.3 cm, from whose length the nickel coating was removed on one end, leaving a length of 20.3 cm of bare mild steel; and

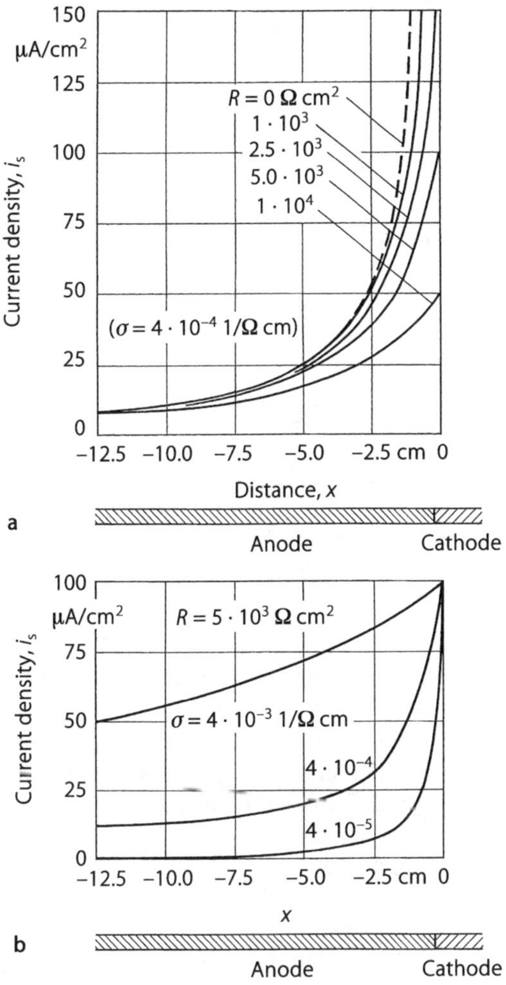


**Fig. 11.7a,b.** The dependence of the local density of the short-circuit cell current in a plane anode/plane cathode contact corrosion cell on the anode, calculated from Eq. (11.18). **a** Influence of changing values of the polarization resistance $R$ in solutions of constant conductivity. **b** Influence of changing values of the solution conductivity,$\sigma$, for constant polarization resistance. (From Ilschner-Gensch)

this sheet was then immersed in tap water ($\sigma = 7 \times 10^{-5}\,\Omega^{-1}\ \text{cm}^{-1}$) open to air. The case is similar to corrosion of iron in contact with a noble-metal oxygen electrode, because nickel is passivated in tap water by a sufficiently semiconducting oxide film, and is acting as catching area for oxygen reduction, in addition to oxygen reduction accompanying active metal dissolution at the iron surface. After 12 weeks exposure, the electrostatic Galvani potential field

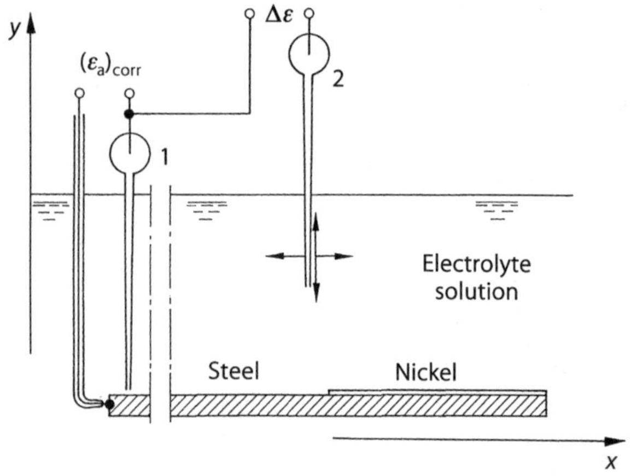


**Fig. 11.8.** Setup for the determination of the ohmic potential field in tap water over a flat iron-nickel contact corrosion cell. Fixed and moving reference electrodes, '1' and '2', resp., equipped with capillary probes. Electrode '1' is placed far away from the iron/nickel boundary, with the probe tip close to the metal surface, and also used to measure the iron potential, $(\varepsilon_a)_{\text{corr}}$, outside the range of short-circuit action

in the test vessel was scanned, measuring the potential difference, $\Delta\varepsilon$, between a moving and a fixed reference electrode, as explained in Fig. 11.8.

A typical result is shown in Fig. 11.9 for the interesting vicinity of the contact area only. From this field, the orthogonal field of current lines can be constructed by simple conformal mapping procedures, and from this the density, $i_s$, of the cell current at the metal surfaces is also available. The local current density of iron dissolution was then calculated using a consideration which we cut short by noting that in this case $i_{\text{Fe}} = (i_{\text{corr}})_\infty + i_s$, where the first term is the corrosion rate of iron at infinite distance from the anode/cathode boundary, measurable from depth of corrosion attack far away from the boundary, using Faraday's law. The simple relation holds according to Eq. (6.11), applicable to the present case. The depth of attack was independently determined, after one year exposure, on the iron surface, and the result is shown in Fig. 11.10 together with the values obtained from applying Eq. (6.11). The correspondence of the two curves is reasonable, but most of it is due to the use of the same experimental $(i_{\text{corr}})_\infty$ value for both sets of data. We note that, at the time of this work, an alternative method to arrive at the same purpose apparently was not recognized: It would have been to measure a) the partial current-potential curve for iron dissolution, $i_{\text{Fe}}(\varepsilon)$, by suitable independent polarization experiments, b) the electrode potential as a function of the distance from the anode/cathode boundary, $\varepsilon(x)$, scanning the potential at $y = 0$, and c) combining the results to obtain, as desired and probably with similar accuracy, $i_{\text{Fe}}(x)$.

With a view on contact corrosion by local microcell action, an interesting intermediate case is that of a flat strip of cathodic metal imbedded, as shown

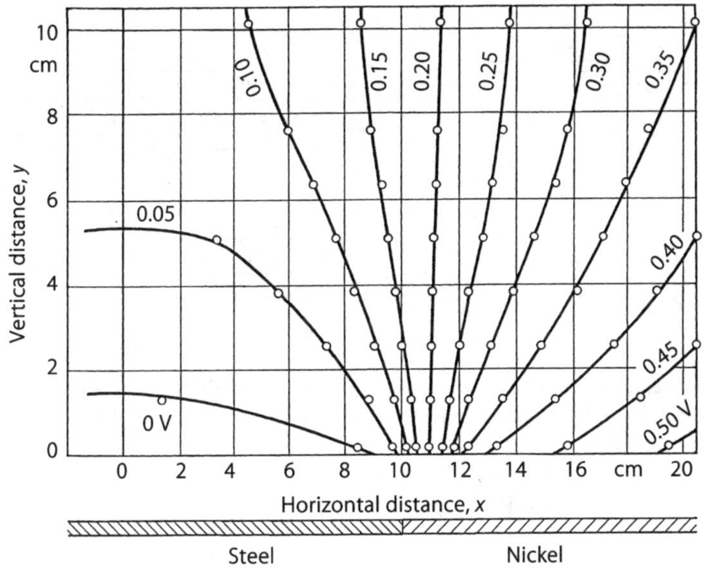


**Fig. 11.9.** Section through the field of equipotential surfaces in the electrolytic solution covering an large iron/nickel contact corrosion cell, measure after 12 weeks test duration. Potential values referred to the free corrosion potential of the iron anode unaffected by contact corrosion. (From Copson)

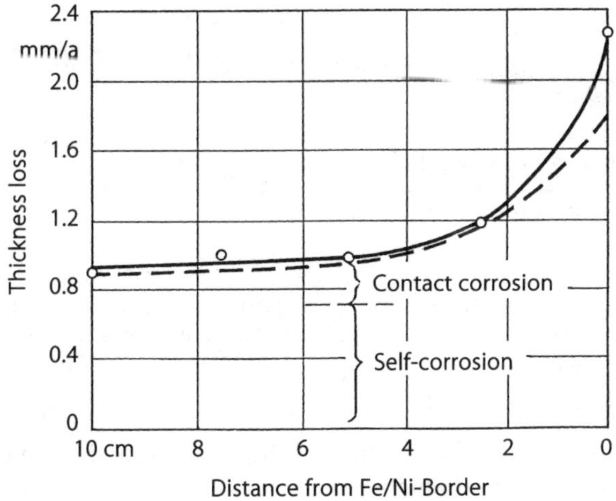


**Fig. 11.10.** Depth of metal loss after 1 year exposure observed for the iron anode in the iron/nickel contact corrosion cell. *Open circles:* Direct observation of iron surface recession. *Broken curve:* Recession calculated from the current density of anodic iron dissolution, as evaluated from the electrostatic potential field shown in Fig. 11.9. (From Copson)

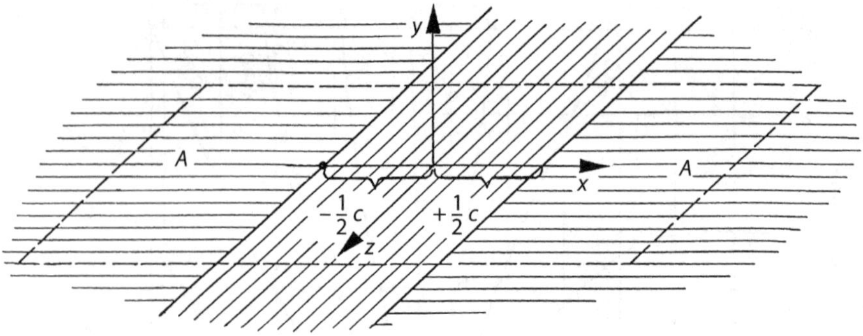


**Fig. 11.11.** Coplanar arrangement of a strip of cathodic metal of width *c* embedded in an infinite plane of anodic metal under an electrolytic solution filling the volume over the cell surface

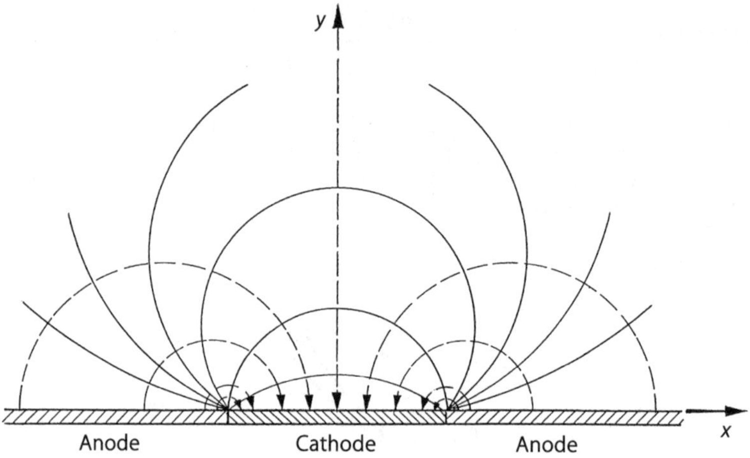


**Fig. 11.12.** Section through the current-potential field of a short-circuited contact corrosion cell as in Fig. 11.11, calculated assuming vanishing polarizability of anode and cathode metal in the given solution

in Fig. 11.11, in a large anode. Again, the solution of the problem is the same for every value of *z*, and Fig. 11.12 schematically shows the section through the field of equipotential surfaces at for any constant *z*-value, together with the orthogonal field of current lines. Close to both anode/cathode boundaries, the situation is the same as in Fig. 11.5, and in the vicinity of the boundaries the effect of finite anode and cathode polarizability again is to spread equipotential surface foot points as in Fig. 11.6. A similar system is the pencil-shaped cell shown in Fig. 11.13, which has been used to investigate acid corrosion of zinc in contact with platinum [9a]. The electrode was a zinc rod, platinized in the

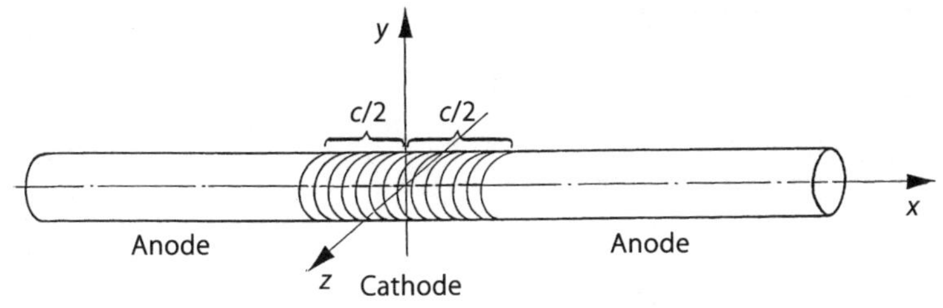


**Fig. 11.13.** Pencil shaped zinc-platinum contact corrosion cell, indicating cartesian and cylinder axes, $x,y,z$ and $x,r,\alpha$

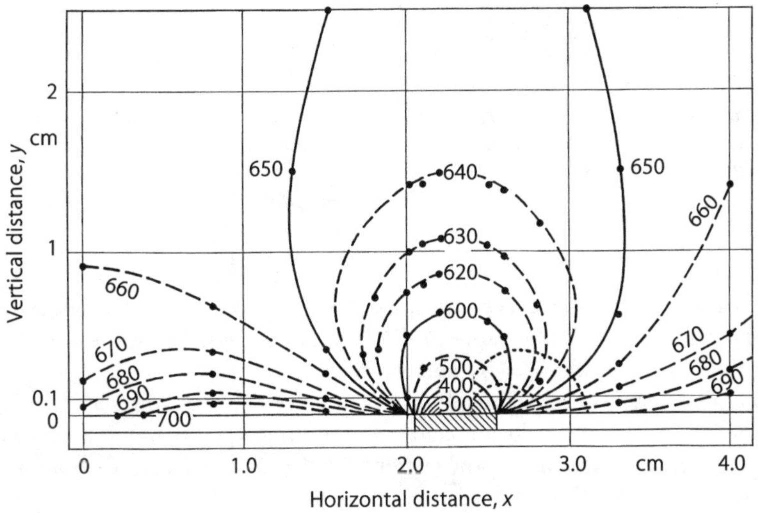


**Fig. 11.14.** One-side section through the potential field of the zinc/platinum pencil electrode in 0.05 N HCl solution. The platinized part of the rod extends from 2.1 to 2.6 cm. Values of the Galvani potential are referred to SHE. Of he set of orthogonal currrent lines only two are shown dotted. (From Jaenicke and Bonhoeffer)

centre on a length $d$. When fully immersed in an electrolytic solution, suitable coordinates are $x$ in the axis of the pencil, radius $r$ and angle $\alpha$. Owing to the cylindric symmetry of the system the solution of the Laplace differential equation is the same for every plane, $\alpha = \text{const}$, and it is symmetric in $\pm r$. For any plane $\alpha = \text{const}, r \geq 0$, the primary distribution of potentials and currents is as in Fig. 11.12. A real distribution, observed in 0.05 M HCl using an experimental set-up much like that described in Fig. 11.8, is shown in Fig. 11.14, presenting experimental data rather close to what is expected from Fig. 11.12. As explained further above, the directions and density of lines of current flowing in the solution, and from these the density of electrolytic currents, $i_s$, can be evaluated from the potential field. It should be remembered from Chap. 5 that, owing to the high hydrogen overpotential typical for this metal, the anode

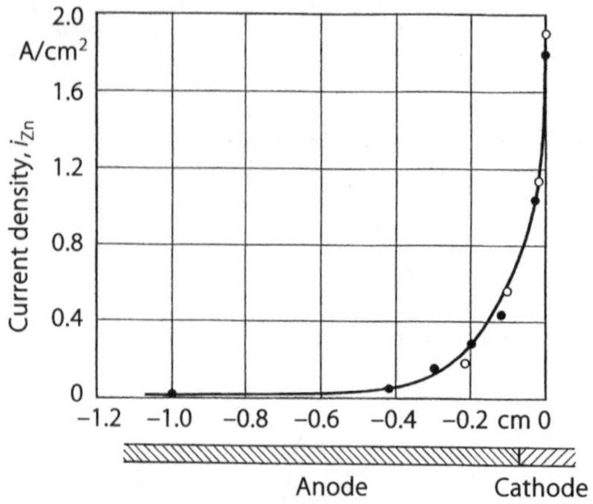


**Fig. 11.15.** The distribution of the current density of anodic zinc dissolution, practically equal to the total short-circuit current density, $i_s$, as function of the distance from the zinc/platinum boundary, evaluated from experimental data on surface recession in 1 N HCl solution. Open and solid points are from two different experimental runs. (From Jaenicke and Bonhoeffer)

corrosion rate is negligible except in contact with a low-overpotential metal such as platinum. Since, at the same time, platinum is immune in the solution it follows that at the anode $i_s = i_{Zn}$ and at the cathode $i_s = i_H$.

In a parallel series of experiments, the zinc surface recession was directly measured, albeit in 1 N HCl solution, and not assuming equality but at least similarity of results. Values for the anodic current density, $i_{Zn}$, thus directly obtained are shown in Fig. 11.15. Introducing a common reduction factor accounting for the different acid concentrations, reasonable congruence between theoretical and experimental current density data was found. Closer inspection revealed higher-order differences: Close to zinc/platinum boundaries, hydrogen deposition apparently depleted the solution with respect to hydrogen ions, causing an increase in pH difficult to account for in the reduction factor. In the limiting case of very small cathodes for hydrogen evolution, diffusion control is certainly probable to predominate anyway [10].

Further approaching the case of micro-sized local cells, the cathode length of the pencil model cell was varied from 3 cm down to 0.1 cm. The ensuing change in the distribution of electrode potentials measured at the distance $r = 0$, very large for broad cathodes, was found, as seen in Fig. 11.16, to reduce for very short cathodes to very small anode/cathode potential differences.

On the whole, the action of contact corrosion cells is dominated by three parameters: Solution conductivity, $\sigma$, a characteristic length, such as in the present case $d$, and a some polarization parameter. Following procedures typical for similarity theory, a dimensionless number $d/\sigma R'_\pi$, with a suitably defined average polarization resistance, $R'_\pi$, has been proposed to characterize the danger

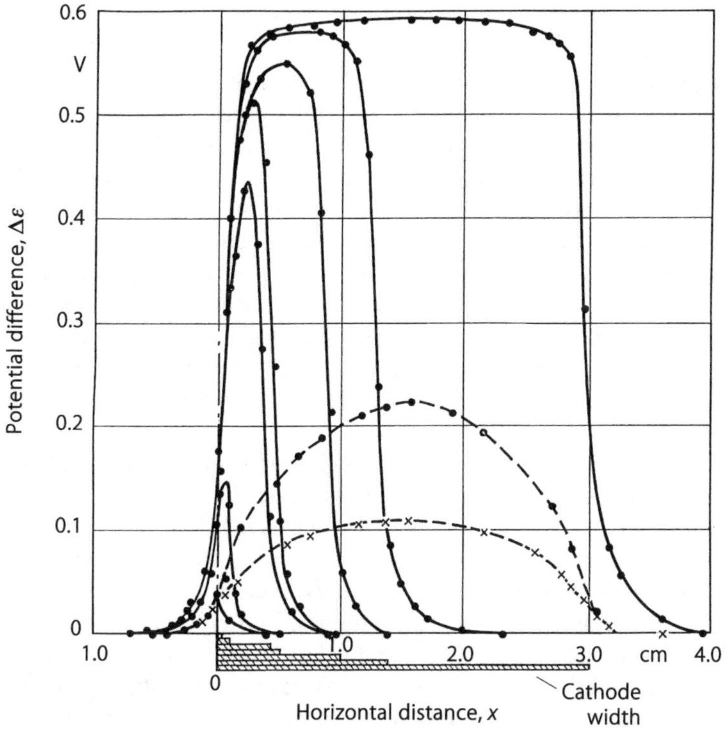


**Fig. 11.16.** Electrode potential profiles for a zinc/platinum/zinc contact corrosion cell. Solid curves: Profiles observed in 0.05 N HCl for varying widths of the platinum cathode. Broken curves: Two profiles observed in 0.05 N HCl/1 M KCl solution for the cathode length 3 cm. (From Jaenicke and Bonhoeffer)

of contact corrosion effects, but this concept apparently has not met much interest [1c,d].

By definition, local cells are microcopic anode/cathode systems on metal surfaces. From measurements as described above it follows that for local electrodes with, e.g., a diameter not exceeding 0.1 mm, differences between anode and cathode potentials drop into the range of millivolts, or, in well conducting solutions, even less. For routine potential measurements with an accuracy of some mV only, this is negligible. Therefore a metal surface, where local cells are actually present, nevertheless has a practically uniform potential, and the microscopic non-uniformity does not show up.

From this it also follows that during current-voltage curve measurements with the usual limited accuracy, local cells remain undetected, and that their influence on electrode kinetics is tacitly incorporated into the values of rate constants evaluated on the assumption of uniform electrode reactions. Local anodes and cathodes certainly will differ in corrosion rate, and this will often cause slow changes in surface composition, together with effects of selective corrosion as described in Chap. 8.

When the purpose of experiments is to detect local cell action on corroded surfaces, a very fine tip of the Haber-Luggin capillary probe of the reference electrode is necessary. Also, the use of low-conductivity solutions is recommended, since it tends to cause an increase in measurable ohmic voltages. As an example we quote investigations into the behaviour of cadmium/bismuth alloy surfaces in 0.05 N HCl where capillary tips with diameter between 5 and 9 μm [9b] have been used. The difference of the free corrosion potentials of Cd and Bi in this solution was 635 mV. Heterogeneous model alloys were used, which contained Cd and CdBi eutectic, or else Bi and the same eutectic. In every case, the detectable difference of potentials along the surface was less than 1 mV. On impure zinc in the same solution differences were even smaller, of the order of 0.1 mV.

We shall describe later the role of heterogeneities in stainless steel welds for non-uniform corrosion. In this context, the possibilities of supporting metallographic evidence by electrochemical potential scans have been investigated [7]. To increase ohmic voltages between microscopic neighboring zones of different alloy composition, the conductivity of the solution can be much lowered by turning to alcoholic instead of aqueous acids. Solutions of this type have always been typical tools of metallographic surface etching preceding microscopic inspection. Optical metallography is, in fact, largely based on differentiating surface heterogeneities by their differences in corrosion rates, producing differences in colour and reflectance of crystallites.

Probing ohmic voltages in solutions over heterogeneous electrodes as described above can be greatly improved by using a reference electrode which vibrates in the electrostatic field in the solution. By the vibration of the probe tip in the field, a small a.c. voltage enters the recording instrument instead of a static potential. The sensitivity of this method is quite considerable, and the principle has been successfully applied to probing corroded surfaces. As described in it is possible to supplement the method by scanning the surface with the vibrating electrode, and the purpose of mapping the complete state of a surface, including its slow change with time, is well documented [12].

Even if local cell action on heterogeneous metal surfaces can be directly detected measuring potentials with the aid of fine capillary tips, the the probe tip must be very close to the surface. Removing the tip to a distance large compared with the dimensions of local cells, the recorded potential again will be practically uniform. Whether this situation then creates difficulties for current-potential curve measurements depends on circumstances as well as purposes of the experiment [13a].

So far, the solution volume over the anode/cathode arrangement was assumed large. A different situation of outstanding practical importance is contact corrosion under thin moisture films, as they are typical especially for atmospheric corrosion. Obviously, the range of cell currents between anodes and cathodes decreases with decreasing thickness of the solution layer covering the electrodes. Experimentally, the evaluation of the ohmic potential field in thin films is difficult owing to the difficulty of placing suitable reference electrodes. At this point, the technique of recording Volta potentials instead of electrode potentials, using a scanning Kelvin probe as described in Chap. 9, is an intelligent solution to the problem. It has been shown that metal surfaces under moisture films as thin as 2 μm can be studied, and that even Volta potential-current curves are experimentally accessible, with the Volta potential proportional to the electrode potential [14]. The method has been used to study the corrosion of zinc anodes in contact with iron cathodes under thin films of salt solutions, modeling cathodic protection of bare defects in an otherwise intact zinc layer of galvanized mild steel [15]. In this work, the Volta potential field was probed under pure water and under NaCl solution, using moisture films with a thickness of 400 and of 100 μm. In contact with iron, the zinc potential remained almost constant, while the iron potential markedly decreased close to the zinc boundary. This corresponds to expectations: Iron now is an oxygen-catching cathode probably in the range of the cathodic limiting reduction current density, with $R_\pi$ tending to large values, while zinc is an active anode dissolving without much overpotential and, thus, with low $R_\pi$. On the whole, this behaviour is similar to what has been described above for a large iron-nickel contact corrosion cell, with nickel replaced by iron and iron, by zinc. The lateral range of cell currents, which in this case is the range of useful cathodic protection of iron, of course was much lower: Under films of 400 and 100 μm thickness, the current range was ca. 8 and ca. 3 mm in water, and >15 mm and ca.10 mm in 0.5% NaCl solution, respectively [15]. To judge the effectiveness of a cell with a view on cathodic protection of the cathode, the authors propose the use of a characteristic length, $\xi =$ , defined by $\xi^2 = R'_\pi / \delta\, \sigma$. In this case, an average polarization resistance, $R'_\pi$($\Omega$ cm is used, but the assessment of this property needs some insight into the character of the current-potential curves involved, which we pass.

## 11.2 Differential Aeration Cells

In Chap. 4, the basic model of a differential aeration cell was described, with two iron electrodes separated by a diaphragm between solutions of different oxygen concentration but otherwise constant composition. A particularly interesting variant of this effect is the spontaneous establishment of differences in oxygen concentration in stagnant aerated solutions due to differing lengths of the diffusion path from the solution surface to different parts of the metal electrode. The phenomenon is at the root of *crevice corrosion*, and also important for many more aspects especially of the corrosion behaviour of iron, carbon steels and low-alloy steels.

We continue considering oxygen corrosion in the range of cathodic diffusion control, i.e. in the range of electrode potentials where every oxygen molecule arriving at the metal surface by diffusion is directly reduced. We cannot, however,

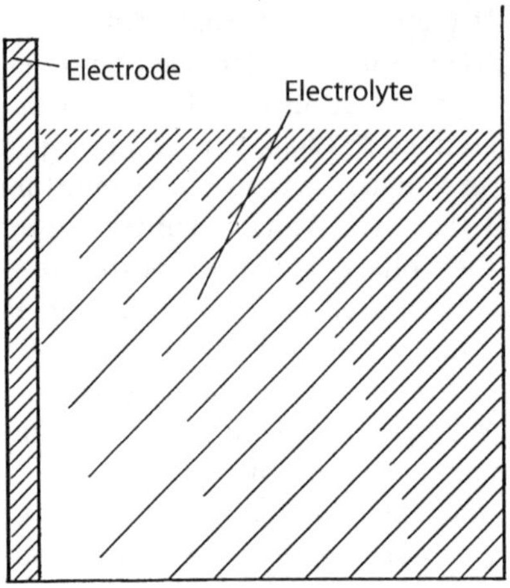


**Fig. 11.17.** A flat iron electrode as one wall of a rectangular vessel filled with well conducting salt solution in contact with the atmosphere. The density of shading schematically shows the distribution of the oxygen concentrations, $c_{O_2}$, in the solution for the case that at the boundary with the atmosphere the solution always is saturated with respect to dissolved oxygen, while everywhere at the metal/solution interface the concentration is zero

continue calculating the cathodic limiting current density of oxygen reduction using the simple Nernst model suitable for a well stirred solution. Instead, the situation is as described in Fig. 11.17, schematically illustrating that two oxygen equiconcentration surfaces exist: the interface between solution and atmosphere, and the interface between solution and iron surface. Between the two, the distribution of oxygen concentration is as schematically indicated by the shaded areas. A more quantitative description will follow below, but it is obvious that the length of diffusion paths through the solution to the iron surface is short for points on the metal close to the solution/atmosphere interface, and long for points on the metal close to the bottom of the container. Thus, the gradient of oxygen concentration normal to the metal surface decreases from from top to bottom of the iron electrode. As a consequence, the well aerated top part is cathodic with respect to the deaerated bottom part, and the system has spontaneously dehomogenized, in fact to a short-circuited galvanic cell.

The main properties of the cell behaviour can be discussed using the two-compartment model system described in Fig. 4.3, where deaeration of the anode and aeration of the cathode compartment are enforced by flushing with nitrogen and air, respectively. The difference is that in this model system anode and cathode have equal area, $A_{\text{met}}$, and they each are at constant electrode potential, $\varepsilon_a$ and $\varepsilon_c$. Also, current densities of electrode reactions are constant on the anode and on the cathode; and the internal solution resistance $R_i$, is essentially equal to the resistance of the diaphragm separating anode and cathode compartments. Thus, a schematic current-potential diagram can be unambiguously constructed,

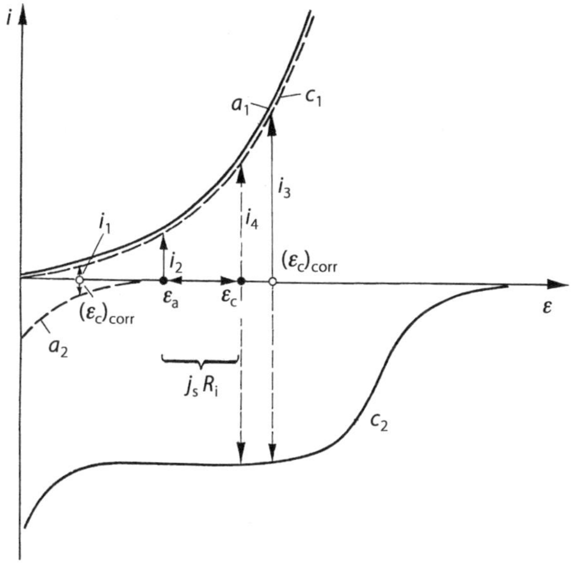


**Fig. 11.18.** Current-potential diagram for two iron electrodes of equal area in a model differential aeration cell as in Fig. 4.3. $a_1$,$c_1$. Identical current-potential curves $i_{\mathrm{Fe}}(\varepsilon)$ at both electrodes. $a_2$,$c_2$ : current-potential curve $i_{\mathrm{H}(\varepsilon)}$ at the anode, and for $i_{\mathrm{H+i_{O_2}}}$t at the cathode, resp. $(\varepsilon_a)_{\mathrm{corr}}$ ,$(\varepsilon_c)_{\mathrm{corr}}$ ,$\varepsilon_a$,$\varepsilon_c$ : Open-circuit and short-circuit anode and cathode potentials. $i_1, i_3$ $i_2$ $i_4$: Partial current densities of anode and cathode iron dissolution on open-circuit and on short-circuit, resp. $j_s$ cell current, $R_i$ solution resistance

as shown in Fig. 11.18. By its legend, the diagram should be easily understood. For simplicity, the total anode and cathode current-potential curves have been omitted. We therefore note that the short-circuit cell current is given by $j_s = A_{\mathrm{met}}(i_c - i_a)$. We also note that from experience it is known that the corrosion rate, $(i_a)_{\mathrm{corr}}$, of iron in neutral completely deaerated solutions is negligible. We further remember that, since oxygen reduction at the cathode is diffusion-controlled, we also have $(i_c)_{\mathrm{corr}} = |i_{\mathrm{O_2,D}}|$. Inspection of the diagram then reveals that the short-circuit current densities for anode and cathode iron dissolution are related as follows [4]:

$$a)\quad i_a \le i_c\,,\quad b)\quad 0 \le i_a \le \frac{1}{2}\,(i_c)_{\mathrm{corr}}\,,\quad c)\quad \frac{1}{2}\,(i_c)_{\mathrm{corr}} \le i_c \le (i_c)_{\mathrm{corr}}\,. \tag{11.19}$$

In Eq. (11.19a) the ‘equals’ sign applies for negligible values of the ohmic potential drop, $j_s R_i$. In this limiting case, anode and cathode contact corrosion rates are equal, otherwise $i_a < i_c$.

An earlier, erroneous assumption was that, generally, the cathode of a differential aeration cell is catching oxygen only, and the iron dissolution current of the cathode was expected to be negligible. This supposition was, in fact, sometimes confirmed by experimental evidence, because sometimes the cathode is passive, owing to an increase in pH of the catholyte as already described in Chap. 4, by consequence of oxygen reduction producing excess $OH^-$ ions. In

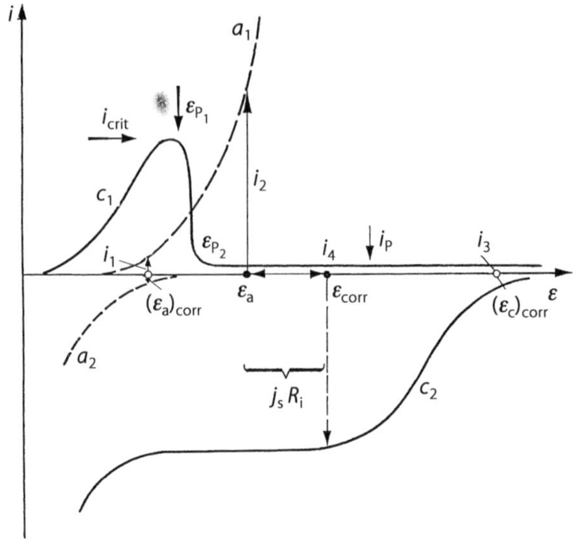


**Fig. 11.19.** Current-potential diagram for two iron electrodes of equal area in a model differential aeration cell as in Fig. 4.3, assuming passivation of the iron cathode in a solution at increased pH owing to oxygen reduction producing excess $OH^-$ ions. Notation of symbols as in Fig. 11.18

this case, the role of the passive iron cathode essentially is similar to that of the platinum cathode in the cell in Fig. 4.3. Figure 11.19 schematically shows the corresponding current-potential diagram, again for the case that anode and cathode have the same area. The anode contact corrosion rate is now greatly increased, since we have the relation

$$\text{a)}\quad i_c = 0\,, \qquad \text{b)}\quad i_a = -(i_{O_2})_{\text{cathode}} = -i_{O_2,D}\,, \tag{11.20}$$

while the open-circuit corrosion rate now is negligible for both electrodes. From this we can easily return to cathodes of varying area acting as oxygen catching surfaces already considered further above.

It is interesting to inspect the behaviour of a iron or zinc sheet vertically immersed, as in Fig. 11.17, in a stagnant neutral salt solution open to air, considering a quasi-passivating air-formed oxide film initially covering and protecting all of the electrode surface. When the solution is pure water there is much hope that nothing will happen whatsoever. When it is a salt solution, local breakdown of passivity must be expected, obeying mechanisms described in Chap. 12, and in the course of time the system dehomogenizes into a macroscopic differential aeration cell with an active anode and a non-activated quasi-passive or passive cathode. An important difference to the behaviour of anodes and cathodes kept in separate solution compartments then is that the path of metal cation migration from the anode to the cathode now is short, and that, therefore, in an interdiffusion zone between anolyte and catholyte metal hydroxides are precipitated. As an example of investigations into the spontaneous formation of large

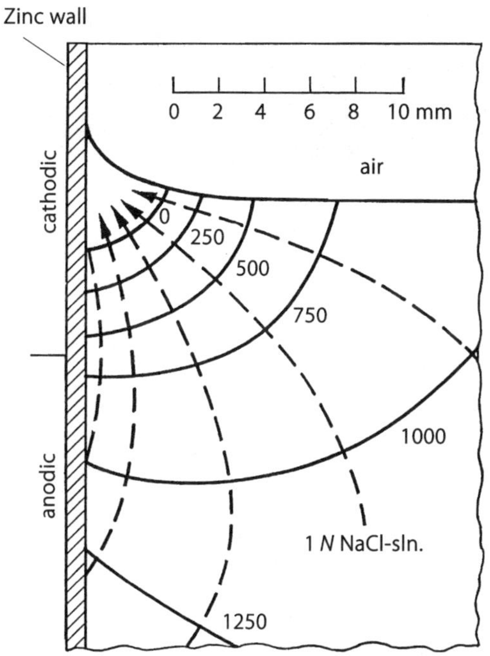


**Fig. 11.20.** Experimental equipotential surfaces and current lines in stagnant 1 N NaCl solution open to air bevor a vertically immersed zinc sheet. Potentials are in mV on an arbitrary scale, essentially referred to the solution meniscus. (From Agar and Evans)

scale differential aeration cells we present in Fig. 11.20 the electrostatic potential field established after some time in a salt solution in contact with a vertical flat zinc electrode, as it was observed by scanning the solution with a reference probe, together with the orthogonal field of electrolytic migration currents in the solution from the anodic areas deep in the solution volume to the cathodic areas close to the solution surface where oxygen supply is abundant [6c].

The well known 'Evans drop' experiment is an instructive demonstration of the role of macroscopic, rate-determining diffusion in stagnant solutions, overtaking initial microscopic metallurgical influences in establishing the stable pattern of macroscopic anodic and cathodic areas in a differential aeration cell. The cell develops on a horizontal sheet of iron on which a drop of NaCl solution has been placed [6d]. The solution contains traces of colour indicators responding to dissolved $Fe^{2+}$ and to pH, respectively. The sequence of events is described in Fig. 11.21: After placing the drop on the bare steel surface, which has been ground previously, corrosion starts by events of local pitting probably at defective places in the air-formed oxide film, as indicated by blue patches from the reaction of ferrous ions with ferrocyanide. These events are accompanied by oxygen reduction at neighboring surface sites, increasing local pH and causing the pH indicator to turn to red. Since the solution is stagnant,

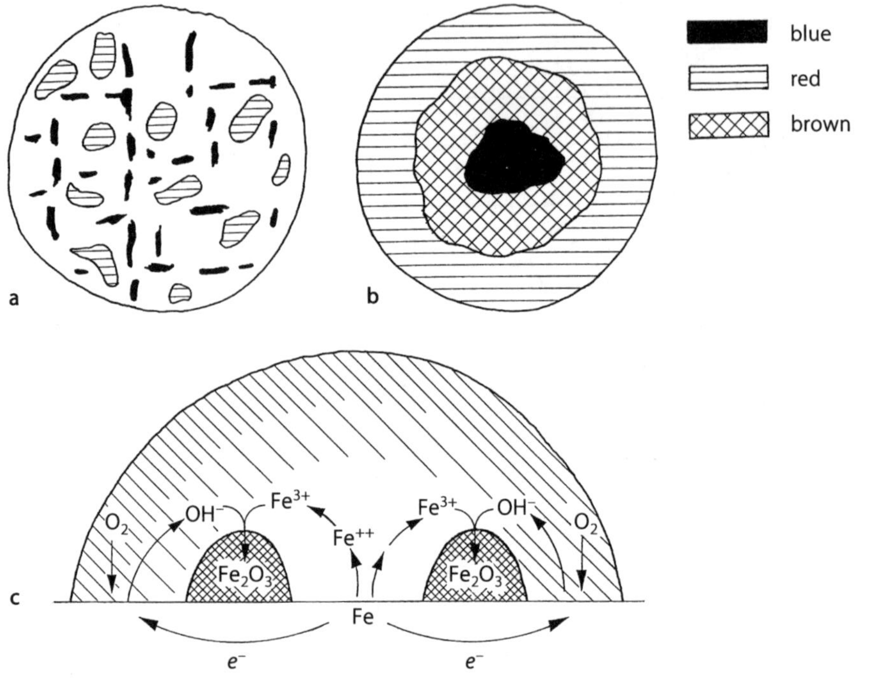


**Fig. 11.21a–c.** Schematic illustration of the Evans experiment with a drop of NaCl solution on a mild steel surface previously ground. The solution contains phenolphtaleine as an indicator of changes in pH to alkaline values by changing in colour to red, and potassium ferrocyanide indicating the presence of dissolved ferrous ions by the formation of prussian blue. **a** The initial distribution of anodic and cathodic areas after local breakdown of quasi-passivity. **b** The quasi stationary state established arrived at within minutes. **c** Schematic description of electrode reactions in quasi-steady state

it will become depleted in dissolved oxygen locally, according to the length of diffusion paths from the air/solution interface to the steel surface; and oxygen reduction will thus practically cease in the center of the drop. Therefore, oxygen reduction will concentrate in a ring-shaped area close to the air/steel/solution three-phase boundary, causing an increase in pH over the same surface ring range and, by consequence, causing repassivation of the outer ring area. The pH indicator turns the colour of this part of the solution red. At the same time, active iron dissolution recedes to the center of the drop, increasing in current density by the action of the differential aeration cell now established, and the center part of the solution turns to blue due to the colour change of the iron indicator. Diffusion and electrolytic migration of $Fe^{2+}$ from the center outwards considerably contributes to this dehomogenization of the system: As on their way to the outer cathode these ions encounter dissolved oxygen, they react to $Fe^{3+}$ ions which in turn react with $OH^-$ ions in the anode/cathode interdiffusion zone of the solution to insoluble ferric hydroxide and oxide, i.e.

rust, forming a brown deposit as an intermediate between the red catholyte and the blue anolyte, further stabilizing the separation of the system into different electrodes.

In an active/passive iron differential aeration cell, cathodic partial currents at the anode and anodic partial currents at the cathode have practically disappeared. The corrosion rate of the anode must, therefore, be equivalent to the short circuit cell current. This effect has early been used to settle the dispute concerning the electrolytic mechanism of corrosion, by measuring currents flowing between the anode and the cathode of differential aeration cells developing on vertical steel sheets immersed in stagnant chloride solutions and proving that the loss in anode metal weight was quantitatively the same as that calculated from the amount of electricity passed between the electrodes through an external connection [6e]. The proof succeeded because it was possible to cut the sheet in two pieces along the anode/cathode boundary which happened to be reproducible, although this boundary would have been difficult to predict a priori. For a recent review of this work, see [4c]. Unfortunately, this success has given support to the misleading opinion that the fundamental law of electrolytic corrosion is the action galvanic cell, either macroscopic or microscopic. For the counterproof, see Chap. 4.

## 11.3 Modeling Concentration Fields in Galvanic Cells

Concentration differences caused by galvanic cell action normally should reach a steady state. If so, changes in concentration of dissolved species, labeled $i$, with time disappear: $dc_i/dt = 0$, and Fick's first law applies, or else the second law describing non-stationary diffusion processes simplifies to $\Delta c_i = div\, grad\, c_i = 0$. At the same time, boundary conditions for any such problem are that at walls, where no electrode reaction involves the species $i$, the gradient in concentration normal to the surface disappears: $dc_i/d\vec{n} = 0$. For the case of a differential aeration cell developing on an iron sheet in a stagnant salt solution, this applies to non-metallic walls of the trough containing the solution. Our interest now is in the distribution of values of the oxygen concentration, assuming oxygen reduction at cathodic areas, in the range of the cathodic diffusion-limited current density. For a system as described in Fig. 11.17 we further have the condition, that the solution/air and the iron/solution interface both are equiconcentration surfaces: At the solution/air interface the concentration of dissolved oxygen is equal to he saturation concentration, $c_{O_2} = (c_{O_2})_s$, whereas at the iron surface $c_{O_2} = 0$. From this, a resulting field of equiconcentration surfaces, and a corresponding orthogonal field of diffusion paths, as schematically shown in Fig. 11.22a, is intuitive.

In this simple case, the quantitative solution of the problem of calculating equiconcentration surfaces and diffusion lines should not be difficult. Many more complex problems of similar kind are treated in the literature especially

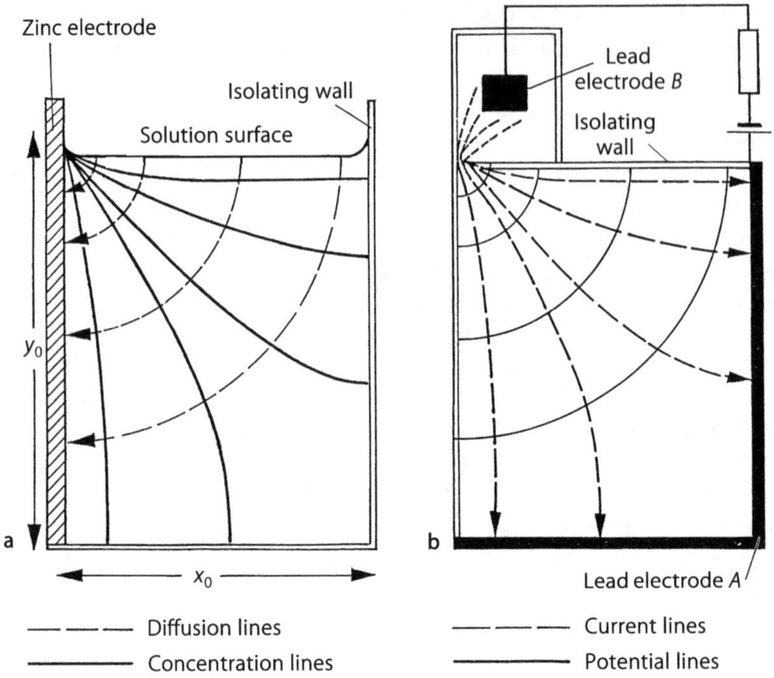


**Fig. 11.22. a** The field of oxygen equiconcentration surfaces (—-), and of oxygen diffusion lines (- - -) through the solution, for a an electrolytic trough with an iron wall as in Fig. 11.17. **b** A conjugate electrical system obtained by replacing the iron sheet and the solution surface by inert walls and replacing vessel walls by lead walls. Equipotential surfaces (—-) equivalent to diffusion current lines; - - - electrolytic current lines equivalent to equiconcentration surfaces. (From Bianchi)

for the conditions in narrow cracks, and this will become a point of interest in Chap. 15. For a review, the reader may refer to [17].

Here, we consider the experimental rather than the theoretical proof of the supposed diffusion field [13b]. The interesting problem is the distribution of dissolved oxygen down narrow slots open to the atmosphere. In principle, this requires an experimental determination of the distribution of oxygen concentrations throughout the electrolytic trough, which is difficult. At this point, the helpful suggestion is that, formally, Fick's second law, as applied to cases of steady state diffusion in a space with no sources or sinks for oxygen except at surfaces, is analogous to Laplace's equation for the potential field in a conducting space with no electrical charges, also except at surfaces. By suitable choice of boundary conditions, a field of equipotential surfaces and current lines can be created which is equivalent to the diffusion system under consideration, and can be probed relatively easily with the aid of scanning reference electrodes. In the present case this would require to replace the air/solution interface as well as the iron wall by two flat nonpolarizable but otherwise arbitrary equi-

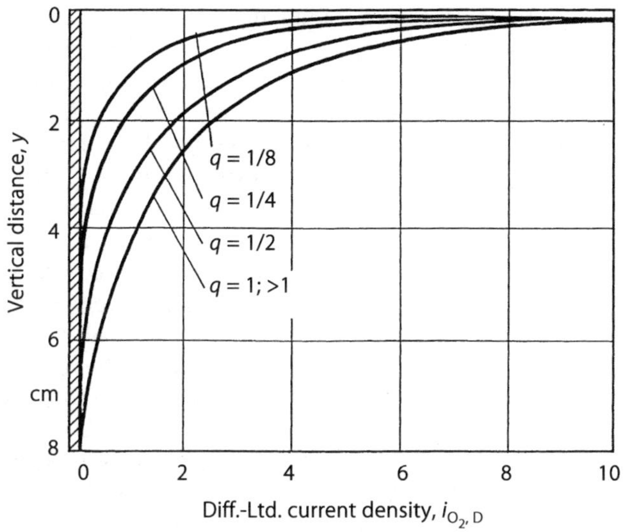


**Fig. 11.23.** The result of monitoring the conjugate-analogue Galvani potential field in an electrolytic trough: The distribution of densities of oxygen reduction in the range of limiting diffusion control as a function of the depth $y$ on a vertical electrode in contact with a stagnant solution. The parameter of the set of curves is the ratio of width, $x_\circ$, of the trough, and length, $y_\circ$, of the electrode: $q = x_\circ / y_\circ$. The abscissa is in arbitrary units of current density. (From Bianchi)

librium electrodes at different electrode potential. The two electrodes can be identical, since the anode/cathode difference of potentials can be simulated by an external voltage source introduced into the short circuit. The aerated solution then is replaced by an electrolyte containing the ions of the reversible reaction occurring at the metal electrodes. In the present case, lead as electrode material and lead sulphamate as electrolyte were found useful. The difficulty still was that at the anode/cathode boundary local polarization of the lead/lead sulphamate electrodes could not be avoided. The final solution was to change to the mathematically equivalent conjugate system as shown in Fig. 11.22b: The metal electrodes were replaced by inert glass walls, the previous glass walls were replaced by lead/lead sulphamate electrodes, and an auxiliary electrode of the same kind was placed beyond a slot between the glass walls indicated in the figure. The figure schematically shows the field of equipotential surfaces and current lines for this system, as obtained from actual probing the solution volume. The final aim is to arrive at the distribution of current densities of oxygen reduction at the original iron surface as a function of the depth in the original aerated salt solution. This, then, can be evaluated from the density of equipotential surfaces as they arrive at right angles at the glass wall, which has replaced the iron electrode, by conformal mapping procedures. As seen in Fig. 11.23, the current density of oxygen reduction, limited by diffusion, drops to low values within a few mm below the solution surface, and it does so increasingly rapidly as the width of the solution vessel decreases. Recalling that for a differential aeration cell this essentially applies to the depth of the cathodic zone on a vertical electrode, this result qualitatively compares well with the experimental Galvani potential field reported in Fig. 11.20.

We note that, even for the smallest width/length ration, the slot is far from being as narrow as a crack, so that down a typical crack, acting as a differential aeration cell, oxygen access is expected to be confined to a narrow region at the crack mouth. Illustrating complications to be expected in many such cases, it may be noted that the assumption of overall diffusion control of oxygen reduction at zinc electrodes did not always apply in this work [13b]. Instead, it took pre-etched electrodes to find the supposition verified. Otherwise, charge-transfer reaction control for oxygen reduction on zinc apparently was predominant, similar to the very high hydrogen deposition overpotential observed for zinc in acids. As a consequence, oxygen reduction then essentially is confined to heterogeneous inclusions, for instance of FeZn compounds, intersecting the surface. In this case, the width of the slot under consideration becomes unimportant, owing to bottle-neck crowding of diffusion lines in the vicinity of inclusions. Acid pre-etching probably has the effect of enriching inclusions until their diffusion zones in the solution overlap, leaving an essentially uniform surface as an effective oxygen cathode.

## References

1. a) Wagner, C.: J. Electrochem. Soc. *98* (1951) 116; b) id., in: Handbuch der Metallphysik, Vol. 1.2. Akad. Verlasge. 1940, p. 165; c) J. Electrochem. Soc. *99* (1952) 1; *107* (1960) 447; d) Werkstoffe u.Korr. *11* (1960) 673
2. a) Waber, J.T.: J. Electrochem. Soc. *101* (1954) 27, *102* (1955) 420, *103* (1956) 567; b) Waber, J.T. and Rosenbluth, M.: ibid. *102* (1955) 354; c) Waber, J.T. and Fagan, B.: ibid. *103* (1956) 64; d) Waber, J.T., Morrissey, J., and Ruth, J.: ibid. *103* (1956) 13; e) Kennard, E. and Waber, J.T.; ibid. *117* (1970) 880; f) Melville, P.H.: ibid. *126* (1979) 208; g) Parrish, W.R. and Newman, J.: ibid.*117* (1970) 43; h) Newmann, J. and Harrar, J.E.: ibid. *1120* (1973) 104
3. a) Klingert, J.A., Lynn, S., and Tobias, C.W.: Electrochim. acta *9* (1964) 297; b) Helle, H.P.E., Beel, G.H.M., and Ligtelijn, J.T.: Corr. NACE *37* (1981) 522; c) Smyrl, W. and Newman, J.T.: J. Electrochem. Soc. *123* (1976) 1423; d) Doig, P. and Flewitt, P.E.J.: Phil. Mag. B *38* (19789) 27; e) ibid.: British Corr. J. *13* (1987) 118; f) ibid.: J. Electrochem. Soc. *126* (1979) 2057; g) McCafferty, E.: Corr. sci. *16* (1976) 183; h) id.: J. Electrochem. Soc. *123* (1976) 1869
4. a) Kaesche, H.: Werkstoffe u. Korr. *14* (1963) 557; b) ibid. *15* (1964) 379; c) in: Symp. Honor. R. Frankenthal. (Frankel, G.S. and Isaacs, H.S., eds.). Spring Meeting Electrochem. Soc., Philadelphia 2002. The Electrochem. Soc., Pennington (in the press)
5. a) Elze, H. and Oelsner, G.: Metalloberfläche *12* (1958) 129; b) Strom-Spannungs-Diagramme Wichtiger Gebrauchsmetalle. Bundesanstalt f. Materialprüfung, Berlin, 1958

6. a) Evans, U.R.; Corrosion and Oxidation of Metals. Arnolds, London, 1960, p. 648; b) id.: J. Chem. Ind. *47* (1928) 731; c) Agar, J. N. and Evans, U. R., quoted from Evans, U.R.: J. Iron and Steel Inst. *141* (1940) 219; d) Evans, U.R.: Metal Ind. *29* (1926) 481; e) Evans, U.R. and Hoar., T.P.: Proc. Roy. Soc. A *137* (1932) 343
7. Ilschner-Gensch, C.: a) Z. Elektrochemie *64* (1960) 275; b) Techn. Mitt. Krupp *18* (1960) 29
8. Copson, H.R.: Trans. Electrochem. Soc. *84* (1943) 71
9. a) Jaenicke, W. and Bonhoeffer, K.F.: Z. phys. Chemie A *193* (1944) 301; b) Jaenicke, W.: ibid. *191* (1942) 350
10. Levich, B. and Frumkin, A.: Acta physicochim. SSSR *18* (1943) 325
11. Proc. Symp. Electrochemical Techniques. Corrosion NACE, Boston, 1985. NACE Houston, 1986
12. Isaacs, H.S. and Ishikawa, Y.: loc. cit. [11], p. 17
13. a) Bianchi, G. and Caprioglio, G.: Werkstoffe u. Korr. *9* (1958) 440; b) Bianchi, G.: Metallurgia ital. *45* (1953) 123, 323
14. a) Stratmann, M. and Streckel, H.: Corr. sci. *30* (1990) 681, 697, 715
15. Tahara, A. and Kodama, T.: ibid. 42 (2000) 655
16. Proc. NATO Advanced Research Workshop on Modelling Aqueous Corrosion, Plymouth 1993 (Threthewey, K.R. and Roberge, P.R., eds.). NATO Advanced Science Institute Series E: Applied Science, Vol. 266 Kluwer Academic Publ., Dordrecht etc., 1994
17. Turnbull, A., in: loc. cit. [16], p. 29

# 12 Pitting Corrosion

## 12.1 General Aspects

In this chapter, the subject is *pitting corrosion* as a dangerous process causing deep localized attack in metal surfaces, including perforations of thin-walled structures. In this context, we do not consider trivial cases of localized corrosion caused by accidental defects in coatings. Instead, the matter of serious practical concern is unexpected pitting of metals passivated by oxide films. Attending to this type of pitting only, two other types of attack also resulting in pitting are disregarded: One is pitting of iron after prolonged exposure to acid solutions containing an adsorption inhibitor [2a]. The other is pitting of GaAs [3a], which is interesting, as apparently a space charge layer in the semiconductor surface is now in the role of the oxide film on passive metals. The reason to concentrate on pitting of passive metals is its outstanding practical importance for safety considerations. In contrast to uniform corrosion, whose rate will normally be predictable from vast experience, pitting corrosion tends to proceed at an unexpected and high rate of localize metal dissolution. This major aspect of the matter is, at least in principle, easily understood considering that passive metals usually are at a corrosion potential, $\varepsilon_{corr}$, which by the presence of the oxide film is shifted far beyond the equilibrium potential of most of the metals alloys components, $E_{Me/Me^{z+}}$. Whenever in pore of the film bare metal is exposed to the electrolytic solution, the overpotential, $\eta_{Me} = \varepsilon_{corr} - E_{Me/Me^{z+}}$, available for the acceleration of anodic metal dissolution thus tends to be very high.

When the oxide is semiconducting, as for iron-group passive metals, the oxide film area surrounding a pore will become a large cathode in contact with the small anodic pore, and short-circuit galvanic cell action will be the result, with an unfavorably high cathode/anode area ratio sustaining rapid pit growth. When the oxide is an insulator, as in the case of aluminum, the danger of pitting certainly is much less acute, although aluminum otherwise is a favorite for investigations into the mechanism of pitting. For iron group metals, we may consider the action of differential aeration cells, as described previously, as a simple model for the mechanism of coarse pitting. In the Evans drop experiment, the center anode of such a cell can be seen as a very large pit, with true pitting arrived at when the center anode shrinks to microscopic dimensions. Describing the behaviour of the Evans drop we have, so far, failed to fully recognize one essential property of the mechanism, and this is the specific action of certain aggressive anions causing local breakdown of passivity. In practice, the dangerous species usually are chloride ions present in the electrolytic solutions, but we shall see that other anions are effective as well.

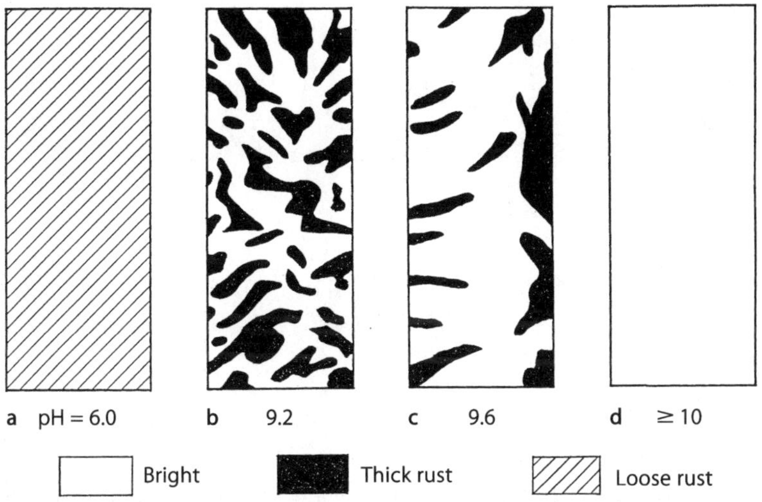


**Fig. 12.1a–d.** The influence of solution pH on the corrosion of plain steel in aerated, stirred $10^{-3}$ M NaCl solutions during 24h exposure at ambient temperature. Note the change from overall rusting to overall passivity via intermediate stages of localized corrosion with decreasing anodic area. (Schematic, from photographs by Resch and Odenthal)

The typical danger of localized corrosion by the action of differential aeration cells, in comparison to uniform corrosion, is illustrated by the influence of a gradual shift of pH of a NaCl solution to higher values [4]. Figure 12.1 schematically shows the typical distribution of rust on mild steel specimens exposed to dilute chloride solutions during 24 hours: At pH 6, which is near-neutral, ordinary rusting is observed, causing near-uniform surface attack of the steel at a rate probably equal to the diffusion-limited cathodic current density of oxygen reduction through the porous rust film. Beyond pH 10, the specimen remains passive in spite of the presence of dissolved chloride. We shall see further below that this is an important observation, as it proves inhibition of chloride attack by adsorption of hydroxyl anions on the passive film. Between pH 9 and 10, the steel surface separates into bright and unattacked passive regions, and regions which are covered by thick rust layers which afterwards are found to cover areas of corrosive attack. Obviously, these are regions of anodic iron dissolution followed by rust deposition. The rust layers impede oxygen access to these anodic areas, leaving the bright passive parts of the surface as cathodic oxygen catching areas. When the solution is made slightly more alkaline, the anodic regions shrink, eventually producing a state of the surface where the passive cathodic area is close to the total area, while the active anodic area is very small. Since the electrolytic solution is well conducting, we expect that oxygen reduction continues to be diffusion-controlled, and that the true local rate of anodic metal dissolution now by far exceeds the overall rate of general rusting in neutral solutions. We particularly note that this effect also illustrates 'dangerous' inhibition, in this case by hydroxyl ions.

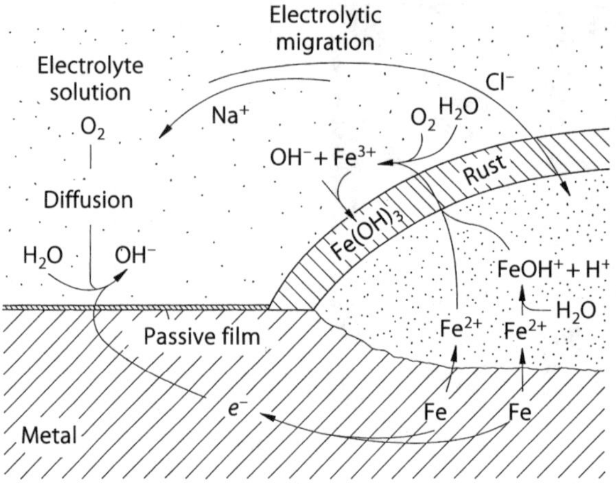


**Fig. 12.2.** Mechanism of a differential aeration cell spontaneously developed on iron in aerated, weakly alkaline solution containing dissolved chloride

Figure 12.2 schematically illustrates the functioning of differential aeration cells of this type, showing (similar to Fig. 11.21) anodic dissolution of iron to $Fe^{2+}$, accompanied by cathodic oxygen reduction at the surrounding passive iron surface. As described previously, the predominance of oxygen reduction at the passive oxide film will tend to increase local pH, stabilizing passivity. At the same time, hydrolysis of ferrous ions will tend to slightly lower pH over anodic areas, stabilizing activity of the iron surface. This effect, of minor importance for the present case, will be met in a major role especially for aluminum further below. The separation of the surface into passive cathodic and active anodic areas is further stabilized by the formation of a rust membrane as the result of the ferrous-to-ferric ion reaction already described for the Evans drop experiment. As also already noted, the local activation of the iron surface starts at defects such as mechanical scratches apparently covered by highly defective, or else particularly thin oxide.

The passivity of unalloyed steel reinforcements, or rebar, in moist concrete has been described as an effect of outstanding importance in Chap. 10. It is equally important that rebar is subject to pitting corrosion, when the surrounding moist concrete contains chloride. Previously, before regulations made it illegal, calcium chloride was frequently used as an inexpensive anti-freeze additive to concrete, and pitting of steel reinforcements was the frequent consequence, remaining a cause of trouble even nowadays for old concrete structures [5a,2b]. In the present context, the mainly important point is the existence of a well defined threshold potential for the onset of pitting corrosion [2b]. This is clearly seen in Fig. 12.3, where stationary current-potential curves observed for mild steel in saturated $Ca(OH)_2$ solution are shown for different concentrations of added sodium chloride. The solution models concrete pore moisture. Within

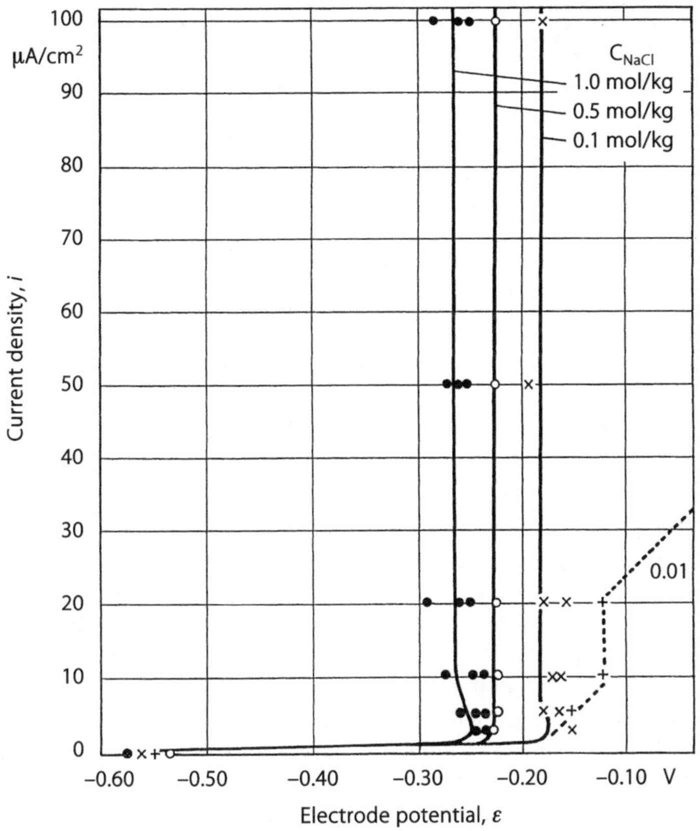


**Fig. 12.3.** Stationary anodic current-potential curves for heat-treated mild steel in deaerated sat. $Ca(OH)_2$ solution, pH 12.6, 25°C, with varying additions of NaCl. Note that experimental points are galvanostatic steady state values observed after decay of initial overshoot. Note also persistent overshoot observed at low chloride concentration. (From Kaesche)

plus/minus several 10 mV, the threshold is well defined, and it is seen to shift to more noble values as the concentration in dissolved chloride is decreased.

Here and everywhere in this text, the threshold potential for the onset of pitting corrosion is designated as the *pitting potential*, $\varepsilon_{pit}$. Unfortunately, there is much controversy concerning the definition of this potential, mainly as a consequence of the convenience of rapid recording non-stationary current-potential curves as cyclic voltammograms, i.e. by recording potentiodynamic triangular potential sweep curves. Therefore, it should be noted that the curves presented in Fig. 12.3 are true stationary curves, obtained by stepwise galvanostatic polarization, and that the meaning of the threshold is as follows: Up to potentials, $\varepsilon$, more negative than $\varepsilon_{pit}$ by some small difference, $\Delta\varepsilon$, the anodic current density of iron dissolution is the negligible passive current density, $i_{pass}$. For the magnitude of $\Delta\varepsilon$, see further below. At potentials more positive than $\varepsilon_{\text{pit}}$, the current density of metal dissolution then increases with time almost indefinitely by quasi-continuous pit nucleation and pit growth. At $\varepsilon_{\text{pit}}$, the current-potential curves become practically vertical, making the electrode essentially non-polarizable. As explained further below, this observation, on first sight surprising, is readily understood: The total area of active corrosion, $A_{\text{pit}}$, is proportional to the apparent current density of metal dissolution, $i_{\text{Fe}} = j_{\text{Fe}}/A_{\text{geom}}$, where $A_{\text{geom}}$ is the total geometric surface of the electrode, i.e. the sum of areas $F_{\text{pit}}$ undergoing corrosion, and areas $F_{\text{pass}}$ still passive. Thus, the true current density of metal dissolution referred to the activated surface area, $j_{\text{Fe}}/A_{\text{pit}}$, is constant. These sharp bends of current-potential curves are typical for pitting corrosion in general, and have been observed much earlier for aluminum in chloride solutions [6a,b].

Concerning techniques of current-potential measurements, we add here that the potentiostat certainly is the ideal instrument to record relatively 'flat' current-potential curves with high values of the polarization resistance $R_\pi = \partial\varepsilon/\partial i$. However, as $R_\pi$, decreases the potentiostat becomes less suitable, and for our case, where $R_\pi \to 0$, the galvanostat certainly is the better choice. For this reason, curves as presented in the figure, and others to be presented further below, have been obtained by stepwise galvanostatic polarization. At each step of the polarizing current density, i.e. $j/A_{geom}$, the potential normally will tend to temporarily overshoot the stationary value, $\varepsilon_{\text{pit}}$, reflecting 'slow' kinetics of slow increase in activated surface. Also usually, these kinetics will become 'slower' when the concentration of activating ions, such as chloride, in the electrolyte decreases, until, as seen in the figure for 0.01 $\text{mol}\,\text{L}^{-1}\,\text{Cl}^-$, the expected stationary pitting potential is no longer observed.

Since the existence of a stationary, or steady-state pitting potential is important for many serious safety considerations, we note two higher-order corrections: The first is that, whenever, among other contributions, an ohmic component in deep recesses is co-determining the electrode potential, $\varepsilon_{\text{pit}}$ may slowly drift to more negative values, if increasing pit depth causes this component to increase. This should not, however, impede first-order considerations aiming at a quantitative description of the pitting potential as a function of important parameters. One parameter of generally outstanding practical importance is bulk solution pH, but in the case of iron and mild steel, and other cases as well, $\varepsilon_{\text{pit}}$ is independent of pH, in contrast to the kinetics of pit nucleation. The second correction is that, as also described below, isolated, rare events of pitting will are observed definitely below, and even far below, the steady-state pitting potential.

Again concerning experimental techniques, we further note that the transient overshoot in potential observed during stepped galvanostatic polarization can be analyzed with respect to the kinetics of pit initiation, including the kinetics of local breakdown of passivity. The same purpose can be served by recording the transient current increase observed during stepped potentiostatic polarization beyond the pitting potential. In principle, an analysis of the kinetics of the electrode events under consideration is possible from potentiodynamic current-potential curves as well, and this even is the usual approach to the matters in question. When the potential is scanned at a sufficiently slow rate, the stationary pitting potential may, in fact, be observed with almost the same accuracy as by galvanostatic polarization. More rapid scanning may also yield useful, though usually only summary information. For comparatively rapid scanning it then makes sense to distinguish a lower range in potential, where transient current spikes indicate the transient life of small, unstable pits, from a higher range of the potential where a surge of the current indicates the onset of massive pit initiation. If, by appropriate choice of the potential scan rate, repassivatable and non-repassivatable pitting can be safely distinguished, it then makes sense to define a threshold potential, $\varepsilon_{\mathrm{pit1}}$, up to which pits are microscopically small and only shortlived, and a higher threshold potential, $\varepsilon_{\mathrm{pit2}}$, indicating the surge to more or less unlimited pitting. Unfortunately, these scanning experiments rather often are not completed by scan reversal, although without recording the hysteresis of forward and backward scanning the information obtained tends to remain obscure. During reverse scans, the metal dissolution current often remains high below both $\varepsilon_{p2}$ and $\varepsilon_{\mathrm{pit1}}$, dropping to the original passive current density only when the potential passes a still lower threshold, $\varepsilon_{\mathrm{pit3}}$, designated as *repassivation potential.* Since at $\varepsilon_{\mathrm{pit3}}$ preexisting pits cease to grow, this repassivation potential should coincide with the pitting potential, $\varepsilon_{\mathrm{pit}}$, defined above. On closer inspection, higher-order corrections again may apply owing to reverse overshoot during rapid scanning, but the first-order alleged equality is acceptable. The threshold $\varepsilon_{\mathrm{pit2}}$ is widely registered under the name of 'pitting potential', which is unfortunate since, depending on conditions, such as scan rate, solution pH, and others, which leave $\varepsilon_{\mathrm{pit}}$ unchanged, make for scatter of experimental values for $\varepsilon_{\mathrm{pit2}}$ and, for that matter, $\varepsilon_{\mathrm{pit1}}$ as well, sometimes up to hundreds of mV for one and the same system [7a].

Pitting potentials, as defined presently, can be observed for passive metals only at potentials more positive than the respective passivation potential. But pitting potentials, as defined above, are independent of pH, again except probably for second order effects, while passivation potentials invariably increase with decreasing pH. In the case of iron and mild steel, this has the consequence that in neutral and acid solutions the pitting potential cannot be observed, and pitting occurs over the whole range of potential of passivity.

Figure 12.4 shows the pitting attack observed after continuous long-time galvanostatic polarization of steel in a moist bed of mortar soaked with calcium hydroxide and chloride [2b]. This also is to show what is meant by dangerous pitting, in this case of concrete rebar. Impressive damage cases are also reported for copper tubes [8a], or galvanized steel tubes [8b,c], all perforated by gross pitting corrosion in domestic water supply systems.

Without external polarization, i.e. in actual service, the occurrence, or non-occurrence of pitting will depend on the position of the free corrosion potential, $\varepsilon_{\mathrm{corr}}$, relative to the pitting potential. $\varepsilon_{\mathrm{corr}}$, in turn, will be established by the superposition of the anodic partial voltage curves for metal dissolution and for,

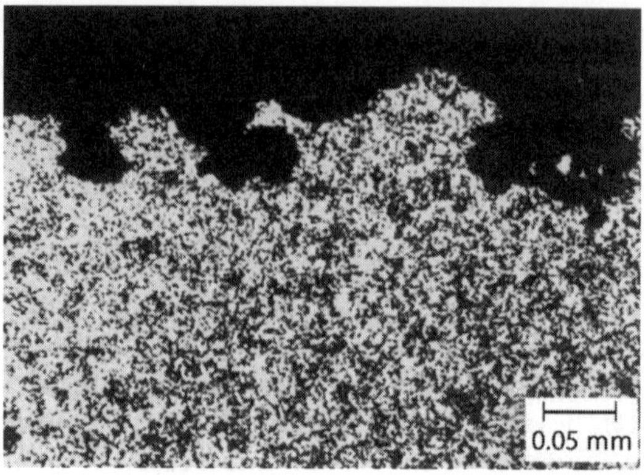


**Fig. 12.4.** Etched cross section of the surface of mild steel after prolonged galvanostatic anodic polarization in mortar soaked with $Ca(OH)_2/CaCl_2$ solution. The rust layer covering the surface has been removed. (From Kaesche)

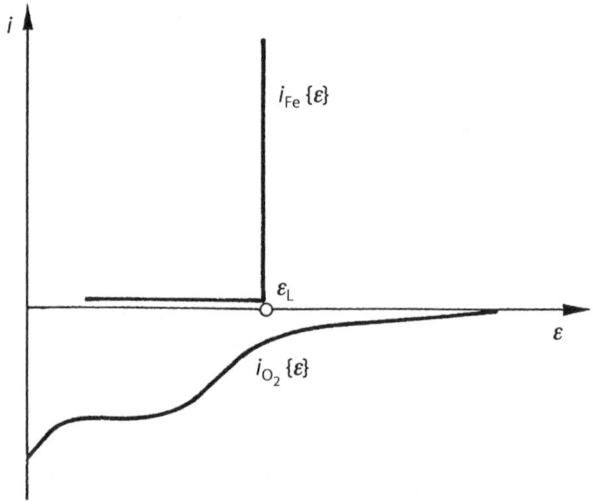


**Fig. 12.5.** Schematic current-potential diagram for iron in aerated weakly alkaline chloride solution. Cathodic oxygen reduction is assumed to reach the diffusion limited rate only at potentials more negative than the pitting potential

e.g., oxygen reduction. In Fig. 12.5 schematically shows the situation encountered for iron and mild steel in aerated weakly alkaline chloride-containing solutions. The passive oxide film being semiconducting, oxygen reduction is possible everywhere, although, as indicated, probably requiring high values of cathodic overpotential. From this figure it should be obvious that the corrosion potential is expected to be equal to the pitting potential, $\varepsilon_{corr} = \varepsilon_{pit}$, and the metal surface subject to spontaneous pitting corrosion.

Owing to a rust membrane forming above the anodic area, the behaviour of typical differential aeration cells producing pitting is rather complicated. In

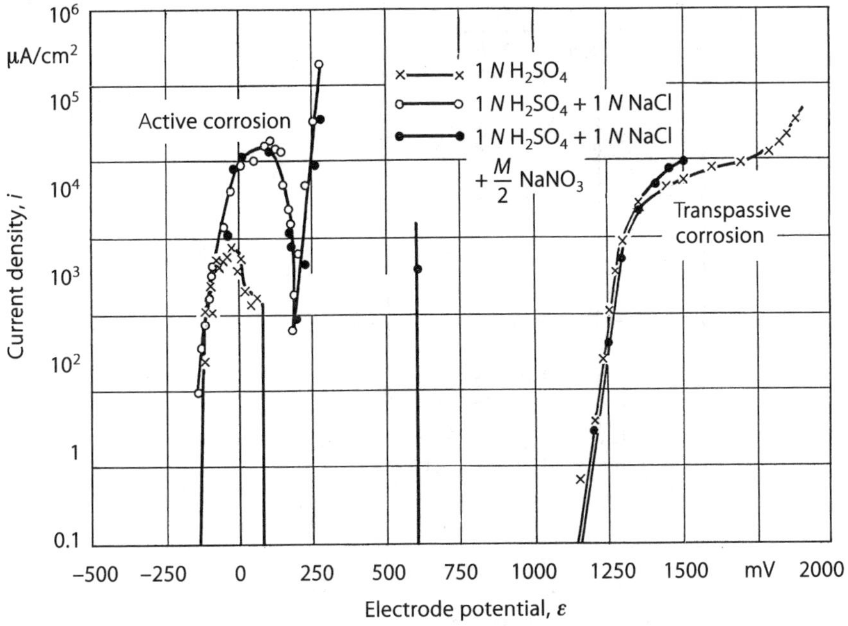


**Fig. 12.6.** Stationary current-potential curves for 18 Cr 10 Ni stainless steel in deaerated 1 N $H_2SO_4$ solution with and without additions of NaCl and $NaNO_3$. (From Brauns and Schwenk)

acid solutions, where rust deposits cannot develop, pitting of passive iron still is possible, and has been the subject of much research, as shown further below. For the moment we turn to stainless steel, and Fig. 12.6 in the first place shows the typical current-potential curve for the material in sulfuric acid, including the onset of transpassivity, which must not be confused with pitting [9a]. It also shows the twofold effect of high additions of NaCl to the solution: In the presence of 1 mol $L^{-1}$ of chloride, the critical passivating current density is greatly increased for reasons which we leave unexplained. More important is the onset of pitting corrosion at a well defined pitting potential, $\varepsilon_{pit}$, which in this case happens to be close to the passivation potential. We note, that $\varepsilon_{pit}$ moves to more positive values when the chloride content of the solution is decreased. In the presence of high concentrations of chloride, transpassivity and anodic oxygen evolution are no longer observable. This situation changes, however, by the addition of a nitrate salt, $NaNO_3$, to the solution: In the presence of this substance, there appears an upper critical potential, limiting pitting, in this case to $-0.2 \lesssim \varepsilon \lesssim 0.6$ V. The effect, curious and singular, apparently never has experienced further and closer inspection.

As seen in Fig. 12.6, the stationary current-potential curve indicating the onset pitting of stainless is very steep, again pointing to an apparent very low polarizability. In fact, this behaviour is quite typical for pitting corrosion in.

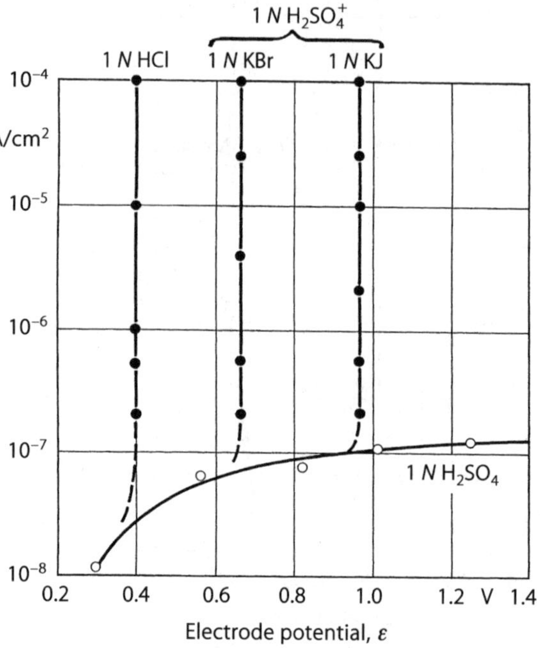


**Fig. 12.7.** Stationary current-potential curves for zirconium in 1 N $H_2SO_4$ solutions with one molar additions of chloride, bromide and iodide. (From Kolotyrkin)

general. For comparison, Fig. 12.7 shows stationary current-potential curves for pitting of passive zirconium in sulfuric acid solutions containing chloride, or bromide, or iodide [10a]. In every case the stationary curves indicating pitting are practically vertical. Aluminum and its alloys behave very similarly, including the change of the pitting potential with a change of the halogen from chloride to bromide and iodide. Figure 12.8 presents the stationary current-potential curve for pure aluminum in a weakly alkaline solution containing chloride, and Fig. 12.9 shows a cross section of a typical pit grown in this solution [2d]. These pits are open to the alkaline solution, because aluminum dissolves to complex aluminate cations, as described below. The growth of open pits can easily be observed under a low-magnification microspcope, but the observation is blurred by gas bubbles, in fact hydrogen bubbles, escaping from the pit interior, although in alkaline solutions the pits are growing at a pitting potential which is anodic to the hydrogen equilibrium potential. This apparent paradox will be explained further below. Furthermore, it is important to note that well defined growth of roughly hemispherical pits shape is characteristic for pitting of aluminum in alkaline solutions only, whereas pitting in lower-pH solutions, and especially pitting in hot hydrochloric acid, produces a different morphology. In any case, aluminum is very susceptible to pitting corrosion, and pitting is easily studied using external polarization. The situation is different for open-circuit free corrosion, and thus for the danger of severe pitting in practical service:

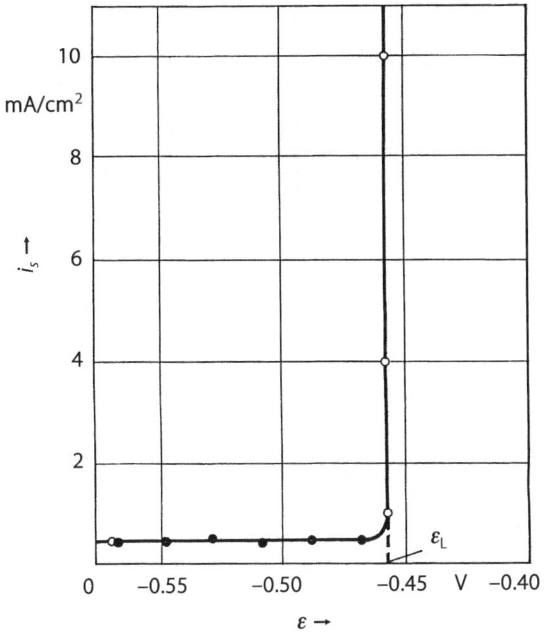


**Fig. 12.8.** Stationary current-potential curve for pure aluminum in deaerated NaOH/0.5 M NaCl solution, pH 11, 25°C. Solid circles potentiostatic, open circles galvanostatic. (From Kaesche)

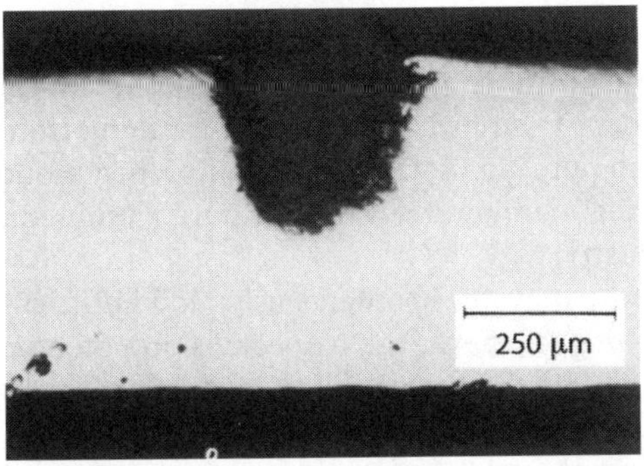


**Fig. 12.9.** Cross-section of a typical pit grown during galvanostatic anodic polarization in pure aluminum in weakly alkaline chloride solution. The metal was 99.99 wt.% Al and cold rolled. Note the near-hemispherical shape of the pit, characteristic for alkaline solutions, and the roughness of the inner pit surfaces. (From Kaesche)

Since the passivating oxide on aluminum is an insulator, galvanic cell action between anodic pits and cathodic passive surroundings is virtually impossible, owing to the internal resistance of the cell tending to infinity. At the same time, cathodic reduction of oxidizing species inside the pit, requiring transport of, e.g., dissolved oxygen into the pit, is much to slow to sustain pit growth. This

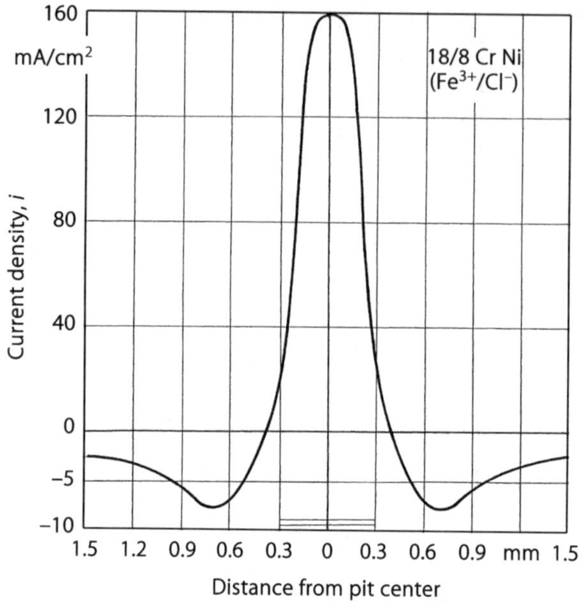


**Fig. 12.10.** The distribution of current densities flowing in $FeCl_3$ solution from a pit actively growing, without external polarization, in a 18 Cr 10 Ni stainless steel surface. Values of current densities have been obtained from the field of ohmic voltages recorded in the solution. Note different scales for anodic and cathodic currents. (From Rozenfeld and Danilov)

holds regardless of some cathodic deposition of molecular hydrogen always occurring inside growing pits, as evident from hydrogen bubbles escaping from pits. Therefore, the danger of pitting in pure aluminum is, in fact, low; hence one of the reasons for the merits of plating aluminum alloys with pure aluminum, as used especially in the aircraft industry.

In neutral solutions stainless steel passivates spontaneously, and sufficient chloride additions will then cause pitting whenever the free corrosion potential shifts beyond $\varepsilon_{pit}$. An agent suitable to test the susceptibility of stainless steels, as well as other passive iron base alloys, therefore is dissolved ferric chloride, $FeCl_3$, since it is at the same time strongly oxidizing by the action of $Fe^{3+}$ and pit-inducing by the action of $Cl^-$. In these, and other acid solutions, pits on stainless steel grow to considerable depth without being clogged in the opening by oxide redeposition. Therefore, it is possible to investigate the short-circuit galvanic cell action between the anodic pit opening and the cathodic passive surrounding scanning the ohmic potential field in the solution above a pit, using experimental methods as described in Chap. 11. Fig. 12.10 shows the result for a pit, which at the time of recording the potential field had an opening diameter of 0.6 mm. Clearly, the data confirm the supposed pit propagation by short-circuit galvanic cell action involving cathodic $Fe^{3+}$ reduction [10d]. From these

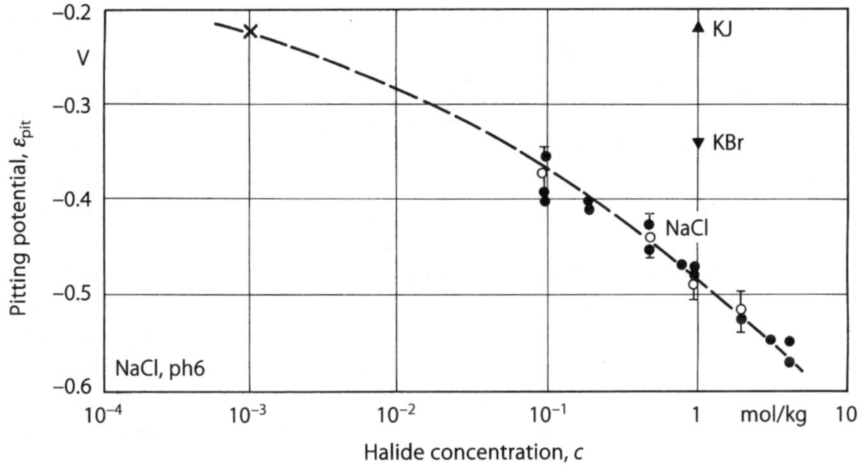


**Fig. 12.11.** Steady-state pitting potentials, $\varepsilon_{pit}$, for pure aluminum in solutions of chloride, bromide, and iodide, respectively, as determined by galvanostatic anodic polarization. Solid symbols: pH 11, ×: pH 6 (From Kaesche). Open circles: Open-circuit corrosion potentials, $\varepsilon_{corr}$, for pure aluminum in neutral chloride solutions containing 5 ppm dissolved $Cu^{2+}$. (From Böhni and Uhlig)

measurements, the current density of metal dissolution, referred to the surface of the pit opening, was found to exceed 0.16 A $cm^{-2}$.

As galvanic coupling between cathodic passive oxide and anodic active pit is inefficient in the case of aluminum, the free corrosion potential is not the pitting potential; instead, $\varepsilon_{corr}$ floats below $\varepsilon_{pit}$. If, however, a copper salt is dissolved in an aerated chloride solution contacting aluminum, metallic copper plates out onto local defects in the oxide film, as described in Chap. 10, creating local cathodes. At these, oxygen is easily reduced, $\varepsilon_{corr}$ shifts to the pitting potential, and pitting sets in.

In this context, data presented in Fig. 12.11 are interesting. Here, solid symbols show pitting potentials of pure aluminum, as determined by galvanostatic polarization under conditions where the corrosion potential is more negative in the range of uniform dissolution of passive aluminum. We recall that uniform dissolution is negligible at pH 6 or less is negligible, but rapid at pH 11 [2d,e]. The change in $\varepsilon_{pit}$, observed for bromide and iodide solutions is similar to that observed for zirconium cited above, and will become a point of interest when discussing the nature of the pitting potential. The figure also presents values for $\varepsilon_{corr}$ observed for pure aluminum in neutral chloride solutions containing traces of dissolved cuprous ions, and the coincidence of these values with $\varepsilon_{pit}$ is obvious [11a]. Incidentally, this also explains the occurrence of pitting of aluminum cooking pots in copper-containing supply water, as observed earlier [8b,c]. The effect of deposition of metallic copper onto defects of the oxide film can be replaced by short-circuiting pure aluminum electrodes with large solid cathodes such as copper, or nickel, or platinum. Obviously, the ratio of the corrosion rate of passive aluminum dissolution, negligible in neutral solutions, and the considerable rate of pitting in contact with oxygen catching cathodic areas makes for a particularly important effect of contact corrosion [2f,12b].

As noted above, as implicit in Fig. 12.11, and as well confirmed by other more direct measurements [2f], the pitting potential of aluminum does not change with solution pH. In acid chloride solutions, for pH $\lesssim 4$, the corrosion potential of pure aluminum is no longer more negative than, but equal to $\varepsilon_{pit}$ and well defined [2f,g], due to onset of pitting corrosion without the aid of external polarization or contact corrosion, probably owing to the thinning of the oxide film. Under these conditions, the type of attack changes from the growth of well defined approximately hemispherical pits to an irregular roughening of the surface accompanied by redeposition of non-protective aluminum oxide/hydroxide layers. In more acid solutions, where the oxides/hydroxides again are easily soluble, the result of pitting is a very fine roughening used for the production of electrolytic capacitors, as described below.

To complete this preliminary introduction we add that aluminum in fluoride solutions does not fit onto the picture: Here, the air-formed oxide film on aluminum changes to an oxyfluoride film and, while the rate of uniform aluminum dissolution apparently increases, pitting is not observed. For a review of previous and recent work on the subject, see [13].

## 12.2 Breakdown of Passivity and Pit Nucleation

Spontaneous pitting of passive metals usually starts at sharp edges, burrs, and the like, where the passivating oxide film is thin or otherwise defective. Also, the edges of varnishes used to coat electrodes are often preferred starting points of localized attack, but this is crevice corrosion (see below), which is an effect closely related to, but different from pitting. In every case, pitting of aluminum probably starts at preexisting pore-like flaws in the isolating oxide film, as they have been described in Chap. 10, and pores with thin oxide bottoms will reappear as the model for pit nucleation in aluminum further below. Much more extensive research work has been devoted to pit nucleation in iron group metals, and especially to stainless iron-chromium, and iron-chromium-nickel steels. Very often, and not surprisingly, manganese sulfide inclusions in CrNi steels or chromium oxide inclusions in FeCr steels are the sites of pit nucleation [14]. In the case of MnS inclusions, pit nucleation appears to be a microscopic variant of crevice corrosion, with sulfide dissolution as the first step, and a critical concentration of manganese chloride in the resulting cavity required for the onset of pitting of the steel matrix [14f]. It is, in fact, difficult to study passivity breakdown and pit nucleation in steel surfaces undisturbed by these and other irregularities in the oxide film. Nevertheless, studies of pitting in pure surfaces have produced and are producing a wealth of data sometimes difficult to review. The reason apparently is that, at least when the frequency of pit initiation events is high and the distribution of such events over the surface

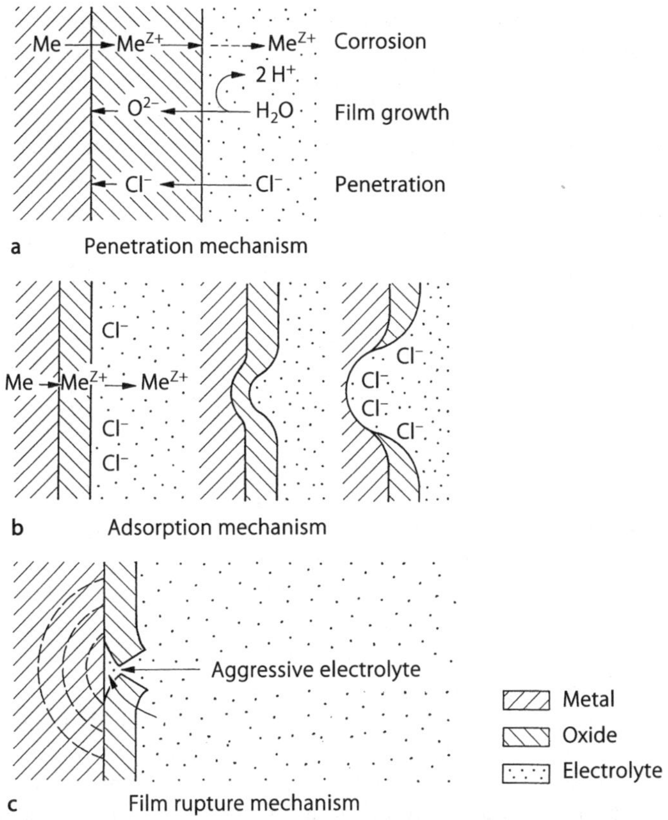


**Fig. 12.12a–c.** Three types of pit initiation processes, excluding the case of preexisting flaws as in aluminum oxide films: **a** Penetration of chloride ions through the film, **b** island adsorption of chloride ions on the film, and **c** mechanical failure due to internal stresses in the film. (From Strehblow)

is random, the subject lends itself to all methods of advanced surface analysis and impedance spectroscopy. This also includes photoeffects: As described in [3b,15a] for both passive iron and passive stainless steel, pit initiation is impeded by illuminating the electrodes with red light. Apparently, the effect is due to some change in the defect structure of the semiconducting oxide films caused by suitable irradiation.

In principle, three different types of initiation mechanisms appear possible, a shown in Fig. 12.12 [16a] These are the *penetration mechanism* involving, as the main step, the incorporation of chloride (or other pit promoting ions) into the film [17a], the *adsorption mechanism* assuming island adsorption at the oxide/solution interface only [10b,17b], and the *film breaking mechanism* involving stresses in the oxide film on the substrate causing local film rupture [16a,b,18a].

The film breaking mechanism hypothesis considers stresses between oxide film and metal substrate as unavoidable in view of various influences such as the difference in specific molar volume of metal and oxide, changing with hydration, or else dehydration, of the oxide; or interfacial tension between oxide and substrate; or even electrostriction caused by the Galvani potential field. It has been argued that the probability of local rupture processes, as predicted from the influence of electrostriction, corresponds to observations concerning the probability of pit nucleation [18a]. More data confirming the film rupture mechanism have been gained for pitting of both iron and nickel [16b]: For iron in acid solutions we recall from Chap. 10 that active metal dissolves to $Fe^{2+}$, while the passive oxide slowly dissolves to $Fe^{3+}$. Dissolution to $Fe^{2+}$ can be identified using a rotating ring-disk arrangement as described in Chap. 8, setting the ring potential to a value where ferrous ions are reduced to iron. If now the potential of the central iron disk is suddenly stepped from steady state passive $Fe^{3+}$ dissolution at +1.3 V down to +0.7 V, which is still in the range of passivity, there is a momentary onset of co-dissolution to $Fe^{2+}$. Apparently, this is due to local ruptures in the oxide film caused by changing the field strength across the film. In the absence of dissolved chloride, immediate repassivation blocks further dissolution of $Fe^{2+}$, while in the presence of chloride pitting starts. This clearly shows that chloride action was not required for film rupture, but only for continued pit growth. Still, ion adsorption may also have an influence on film breaking processes by changing surface stresses, as claimed in [19a].

An earlier concept, described in [17], is that pit initiation is caused by adsorption of chloride, or other pit promoting ions, on the oxide surface. The further supposition is that clusters of adsorbed ions locally lower the strength of chemical bonds between iron and oxygen atoms in the film, causing local film thinning of the film and a local increase of the current density of passage of iron ions from the metal through the film into the solution. As a consequence, pit-like dimples are expected to appear and to grow into the iron surface, eventually turning into the bare bottom of an active pit. Island adsorption of halide, viz. bromide ions onto passive iron, and of dissolved chloride ions onto passive aluminum has in fact been observed using radioactive ion isotopes [20]. Whether this proves island adsorption, or else clustering of ions migrating into the oxide films, is probably open to discussion.

For iron in neutral solutions, a considerably refined model of pit initiation by the adsorption mechanism is available [21]. Figure 12.13 schematically shows the supposed sequence of events: Again, island adsorption of chloride ions is the first step, leading to a local increase in the rate of dissolution of the oxide by the simultaneous dissolution of $Fe^{3+}$ and $O^{2-}$, owing to a weakening of the chemical bonds in the oxide surface. This process is non-faradaic and leaves the passive current density, $i_{pass}$, of transport of $Fe^{3+}$ from the metal into the solution unchanged, although it causes local thinning of the oxide film. The overall increase of the $Fe^{3+}$ dissolution rate is detected with the aid of a rotating ring-disk electrode, with the passive iron disk at the desired potential in the range of pitting, and the ring serving as cathode for the reduction of ferric ions.

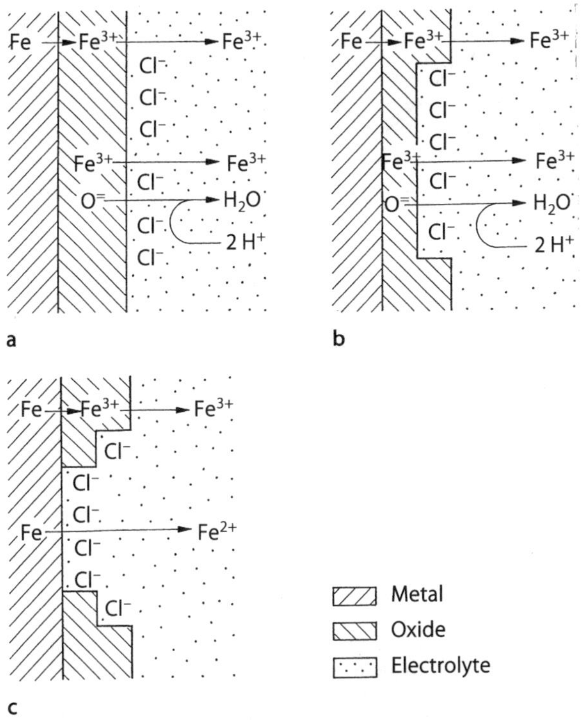


**Fig. 12.13a–c.** Pit initiation by adsorption-enhanced oxide dissolution, according to Heusler

Figure 12.14 shows the observed effects caused by the addition of chloride to a neutral borate buffer solution. A long induction period of accelerated chemical oxide dissolution owing to adsorbed chloride is clearly indicated, eventually followed by a rapid increase of the current of anodic iron dissolution once the oxide film has been thinned locally down to the iron surface.

In this picture, entry of chloride into the oxide film is excluded, and it is useful to note that this supposition is corroborated by XPS spectroscopy for pitting of iron as well as of nickel [16q]. The model has been further elaborated by checking the statistics of initiation times, aiming at determining the average value as well as the statistical variance as a function of the electrode potential. It was found that the important deterministic property governing the stochastic distribution of initiation times is a difference $\Delta\varepsilon = \varepsilon - \varepsilon^{\star}_{\text{pit}}$. The asterisk is added, since $\varepsilon^{\star}_{\text{pit}}$ is not the pitting potential as defined above, but a 'pit initiation potential', described as being reproducible to within few mV. It should be related to the potentials $\varepsilon_{\text{pit1}}$ and $\varepsilon_{\text{pit2}}$ described further above, but the point remains to be investigated [21]. Apparently, $\varepsilon^{\star}_{\text{pit}}$ depends on the defect structure of the iron substrate, such as screw dislocations emerging in the surface: Monocrystalline iron whiskers, grown without dislocations, are immune to pitting up to at least

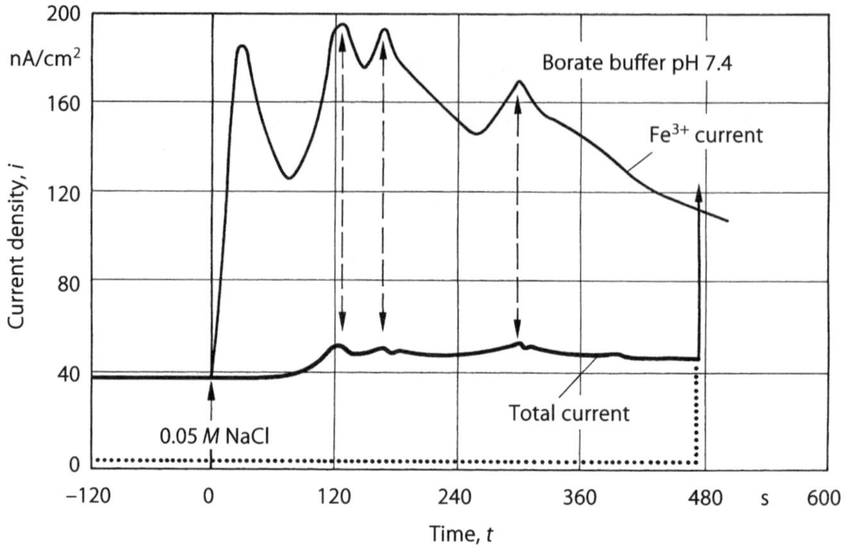


**Fig. 12.14.** Testing the adsorption theory of pit nucleation in passive iron surfaces in borate buffer solution at pH 7.4: At time zero, 0.05 mol $L^{-1}$ NaCl are added to the solution. While the passive current density, $i_{pass}$, at the iron center disk remains constant, the rate of $Fe^{3+}$ dissolution immediately rises, as detected by the reductive current at the surrounding platinum ring. After an induction time of, in this case, ca. 470 s the disk current rises owing to the onset of $Fe^{2+}$ dissolution from the bottom of the first pit nucleus reaching the iron surface. The ring-disk electrode was rotated at 2000 $\text{min}^{-1}$, and the potentiostatic disc potential set to 0.245 V. (From Heusler and Nachstedt)

+1 V. For less exotic material, an average initiation time, $\bar{\tau}$, was shown in [21b] to depend on the chloride concentration, $c_{Cl^-}$, on pH, and on $\Delta\varepsilon$ as

$$\ln \frac{\bar{\tau}}{\bar{\tau}^\star} = \left\{ k\,ln \left( \frac{c_{Cl^-}}{c^\star_{Cl^-}} \frac{\text{pH}}{\text{pH}^\star} \Delta\varepsilon \right) \right\}^{-1} . \tag{12.1}$$

In a further series of experiments, the occurrence or non-occurrence of pit initiation within a given time, $t$, from the moment of addition of chloride to the solution (or from the moment of switching the potential to a value $\varepsilon > \varepsilon^\star_{\text{pit}}$ in chloride containing solutions), was counted for a large number of specimens. Counting the relative number, $P(t)$, of pitted electrodes, the probability of non-pitted survival, $1 - P(t)$, was found to obey the relation:

$$1 - P(t) = \exp(-At) ; \tag{12.2}$$

$$A = \{1 - \exp(-t/b_1)\} \{b_2 + b_3 \exp(-t/b_4)\} .$$

In this equation, all coefficients, $b_i$, are functions of $\Delta\varepsilon$ in the form $\ln(b_i/b_i^\star) \sim 1/\Delta\varepsilon$. From a computerized evaluation of these data a model of pit initiation was quantified, describing the statistics of processes of local oxide film thinning under adsorbed fluctuating patches of chloride alternating with film thickening.

The important detail is that re-adsorption of chloride patches can occur, where after preceding thinning film repair is still incomplete, thus deepening what already is a recess, and the statistics of these overlapping processes allows for the occasional thinning down to the metal surface, where true pitting then starts. It should be recognized that the overall degree of thinning of the passive film accompanying these processes is only slight, in fact less than 1%.

During potentiostatic experiments, the fluctuating processes of oxide thinning and oxide repair should cause corresponding, though very small, fluctuations of the polarizing current. This, then, is *electric noise* (see Appendix), and the frequency distribution has been studied with respect to pit-precursor processes below $\varepsilon^{\star}$. For reviews of this work, see [21d].

On the whole, this mechanism of passivity breakdown of iron in chloride solutions appears well established. Judging from the wealth of literature accumulating constantly [1], a coherent picture of pit initiation in general is not, however, easily available. Mechanisms may, in fact, change from case to case if the same end of breakdown of passivity is accessible on different parallel ways [16c]. This also applies to possible processes of pit initiation by ions penetrating through the oxide film. In the first place, this mechanism may be ruled out, whenever pit initiation is due, or else is considered to be due to adsorption of large ions, as for pitting of iron in sulfate or perchlorate solutions [16d,10c,22], and aluminum in nitrate solutions [22]. Concerning perchlorate, it is now believed that pit initiation is preceded by decomposition of the ion producing chloride ions as the true promoters of pitting [16h].

It is suggestive to use in-depth profiling of the chloride concentration, using argon sputtering during Auger surface spectroscopy, or other spectroscopic methods, to show the absence, or else presence of chloride in the oxide. Reports on the results of such methods are, however, controversial. As reported for iron in neutral solution in [21], initiation times for the onset of pitting are much to short to allow for chloride diffusion through the passive film. This has been confirmed in [24], contested in [25], and reconfirmed in [26]. In all of these papers, the discussion is based on the classical concepts as described in Chap. 10, assuming equilibrium of metal ion transfer across the metal/oxide interface, and high-field ion migration across the film, with complications arising from influences such as film glassiness. As also noted previously, there is, at variance with this approach, the suggestion that the rate-determining step is the transfer of metal ions at the metal solution interface. This 'point defect model' has been extended to also include a concept of pit nucleation by cation vacancy agglomeration, apparently following a Kirkendall mechanism, at the same interface [15b,c]. Chloride is supposed to occupy oxygen vacancies, thereby accelerating outward diffusion of cations. Considering, instead, the passivating oxide as predominantly amorphous, with the water content preventing crystallization, a different model results [26]: Now, incorporated chloride is thought to replace water molecules, increasing the overall rate of oxide dissolution.

Pitting of passive nickel in acid solutions has been reinvestigated recently [28]. The method was novel, using in-situ backside-recording of the intensity of

$\beta^-$ radiation from radioactive $^{36}Cl^-$, adsorbed on the front face of thin nickel electrodes in dilute sulfuric acid containing dissolved chloride during both potentiodynamic and potentiostatic polarization. Measurements where complemented by ex-situ recording of in-depth XPS profiles of the composition of the oxide film in the absence and in the presence of chloride. Results where explained on the basis of a two-layer model of the passive film, which had a total thickness of ca. 1 nm, assuming the presence of an outer nickel hydroxide film and an inner nickel oxide film. For electrodes prepassivated in chloride-free solutions it was observed that below what presumably is the pit nucleation potential, chloride did enter the outer film but did not migrate into the inner film. Supposedly, it is only beyond the pit nucleation potential, that chloride also enters the inner film. This picture, while certainly interesting, does not in itself explain the onset of pitting. Instead, the inner film apparently does endure uniform saturation with respect to dissolved chloride, and pit nucleation requires some local effect of chloride over-saturation which, however, remains unknown. Important overall film thinning preceding pitting again is excluded.

For aluminum it has been claimed that chloride ions migrate into the oxide film in spite of the very low crystallografic disorder of the film, and that the penetration mechanism again is operative [29]. The view is maintained in [30], where localized dissolution of aluminum at the metal/oxide interface is postulated, followed by blistering due to local hydrogen evolution, and eventual film rupture opening the pit. As already noted, the more likely supposition for pit nucleation in insulating oxide films on aluminum is that the film has pores, and that nucleation occurs by breakdown of the still oxydic pore bottom after chloride adsorption [31]. This will be discussed in more detail further below.

## 12.3 Growth of Pit Nuclei

Pit initiation by local breakthrough of the passive oxide film is followed by growth of pit nuclei. Predominantly, these nuclei are crystallografic, bounded by low-index crystal faces. Figure 12.15 shows an early stage of growth [19b]. Apparently, attack is mostly crystallografic, but some non-crystallografic rounded faces are visible as well. A somewhat later stage, with all growing pit nuclei now showing regular crystallografic faces, is seen in Fig. 12.16 [16a]. Many more cases of crystallografic pit growth are described in the literature [1]. At the same time it does appear, although the conclusion is perhaps somewhat unsafe, that on increasing the electrode potential far beyond the pitting potential, pit nuclei tend to adopt structureless hemispherical shapes with bright, electropolished interior. This was observed at least for aluminum [31d,2h], for nickel [32], and iron [33].

Severe corrosive attack is the later-stage result of overlapping of growing pit nuclei. For iron, Fig. 12.17 illustrates such presumably later stages of pit growth caused by coalescence of growing crystallografic nuclei turning into an electropolished hemispherical pit [19b]. This electropolishing is interpreted as

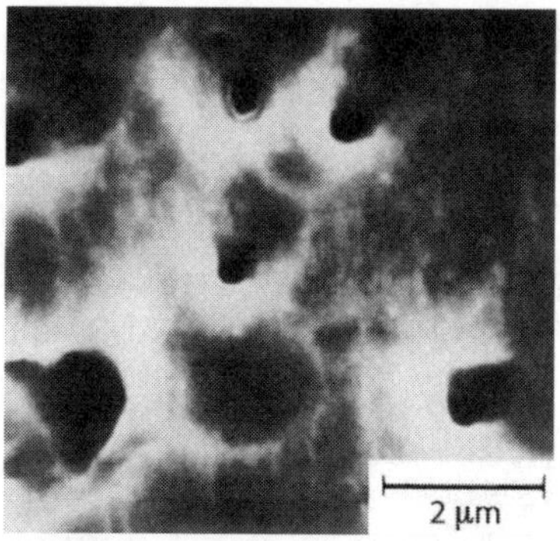


**Fig. 12.15.** Polygonal etch pits in the surface of passive iron showing the onset of pitting in chloride, sulfate, and perchlorate solutions, observed by scanning electron microscopy. (From Pickering and Frankenthal)

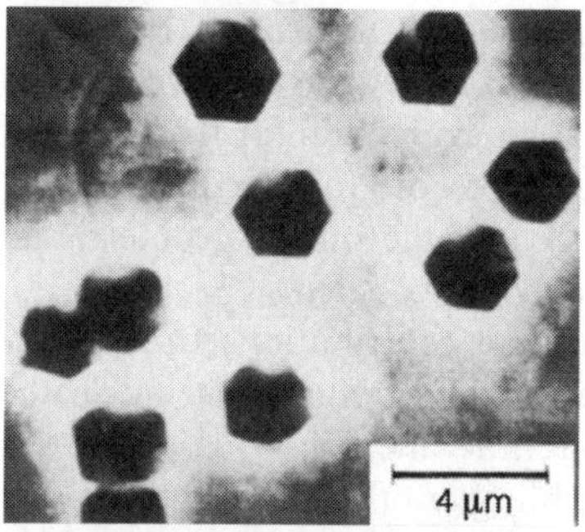


**Fig. 12.16.** Polygonal etch pits in the surface of passive iron observed after 3 s at 1.18 V in phtalate buffer solution, pH 5.0, and 0.01 mol $L^{-1}$ $Cl^-$ and 0.5 mol $L^{-1}$ $SO_4^{2-}$. (From Vetter and Strehblow)

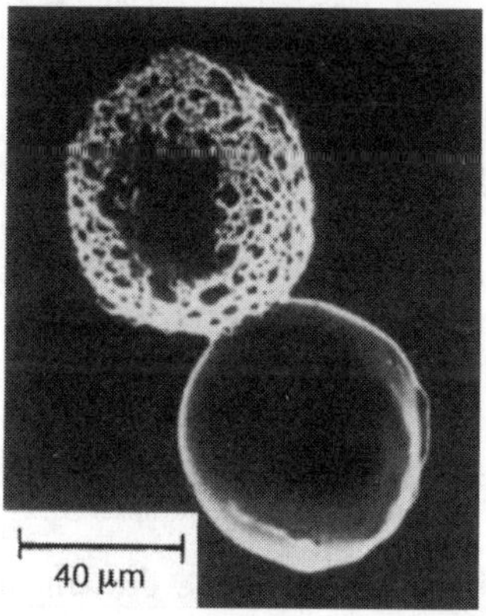


**Fig. 12.17.** Two stages of pit growth in the surface of passive iron in 0.5 M $H_2SO_4$/3 M NaCl solution at +.85 V. *Above left:* Already hemispherical pit still showing crystallografic facets presumably from overlap of many initial nuclei. *Bottom right:* Completely polished pit interior. (From Pickering and Frankenthal)

occurring under a gelatinous salt film, in this case supposed to be supersaturated iron sulfate devoid of hydrating water. The size of these pits is that of visible pinpricks, which can be inspected with respect to the local electrode potential introducing sufficiently fine Haber-Luggin capillaries. In one such investigation, the electrode potential at the pit bottom was found to be ca. 1 V more negative than the potentiostatic potential impressed on the electrode, revealing a very considerable voltage drop across the electropolishing film [5e]. Similar observations have been reported by several authors [10e,19d,34].

When pits grow as hemispheres, the anodic current density of metal dissolution is the same everywhere inside the pit. Let $j'$ be the contribution to the anodic current from a single pit of radius $r$, so that the density of the current passing the pit opening is $I' = j'/\pi r^2$. This is the current density referred to the pit opening surface, while for the physically more interesting current density of metal dissolution referred to the interior pit surface we have $I = I'/2$, since this surface is $2\pi r^2$. The rate of loss in volume of the metal is $2\pi r^2\,(dr/dt)$. From Faraday's law we also have

$$j' = \pi r^2 I' = 2\pi r^2 \frac{zF}{V_{\mathrm{Me}}} \frac{\partial r}{\partial t} \,, \tag{12.3}$$

where $V_{\mathrm{Me}}$ is the atomic volume of the metal, and z is the valence of the dissolving cation. When, at any constant potential, the pit nucleation rate, $v$, is constant, and when in every pit, regardless of its size, the true pit current density, $I$, and, thus also $I'$, is the same, we arrive at an expression for the apparent current density of metal dissolution, $i_{\mathrm{Me}}$, referred to the total geometric surface, $A_{\mathrm{Me}}$, of the electrode, as follows [5b]: Let $A_{\mathrm{pit}}$ be the sum of geometric pit opening surfaces, and $A_{\mathrm{pass}}$ the remaining passive area, let $a_{\mathrm{pit}} = A_{\mathrm{pit}}/A_{\mathrm{geom}}$ and $a_{\mathrm{pass}} = A_{\mathrm{pass}}/A_{\mathrm{geom}}$, respectively. Summing all contributions from single pits we have $\Sigma j' = a_{\mathrm{pit}} I'$. At any time, $t$, a particular pit which had nucleated at some previous time $t'$, has grown during the time interval $t - t' = \tau$ to the radius $r = (V_{\mathrm{Me}}/zF)(I'/2)\tau$. In the time $dt$, the number of pits nucleated is $v dt$ or $v d\tau$, and $a_{\mathrm{pit}}$ as a function of $t$ is calculated by integrating as

$$a_p = v \int_o^t 2\pi \left( \frac{V_{\mathrm{Me}}}{zF} \frac{I'}{2} \tau \right)^2 d\tau \,. \tag{12.4}$$

For the apparent current density of metal dissolution, referred to the total geometric surface of the electrode, we have $i_{\mathrm{Me}} = (1 - a_{\mathrm{pit}}) i_{\mathrm{pass}} + a_{\mathrm{pit}} I'$. Considering that for short times of observation $a_{\mathrm{pit}} \ll a_{\mathrm{pass}}$, we can approximate:

$$i_{\mathrm{Me}} = i_{\mathrm{pass}} + \frac{\pi}{12} \left( \frac{V_{\mathrm{Me}}}{zF} \right)^2 v (I')^3 t^3 \,. \tag{12.5}$$

This $t^3$-law was very well confirmed by experimental evidence, thus confirming the presuppositions as to the constancy of both $v$ and $I'$ [5].

$I'$ was determined separately by microscopic observation of the growth of single pits. It was found to increase with the chloride concentration of the sulfuric acid solution from ca. 0.1 A cm$^{-2}$ in $10^{-3}$ M $Cl^-$ to ca. 5 A cm$^{-2}$ in 0.1 M $Cl^-$. At the same time, $I'$ was the same for potentials between 0.85 and 1.65 V, which points to a constant electrode potential underneath the electropolishing salt film adjusting its voltage to changes in external potential.

Intuitively, one might expect that the nucleation rate, $v$, increases with increasing potential, and this is what is observed, e.g., for aluminum, as described below. For iron in acid chloride solution, the experimental evidence, shown in Fig. 12.18 is, however, different [5d]: In this case, the pit nucleation rate, $v$, greatly decreased when the potential was made more positive. We remember

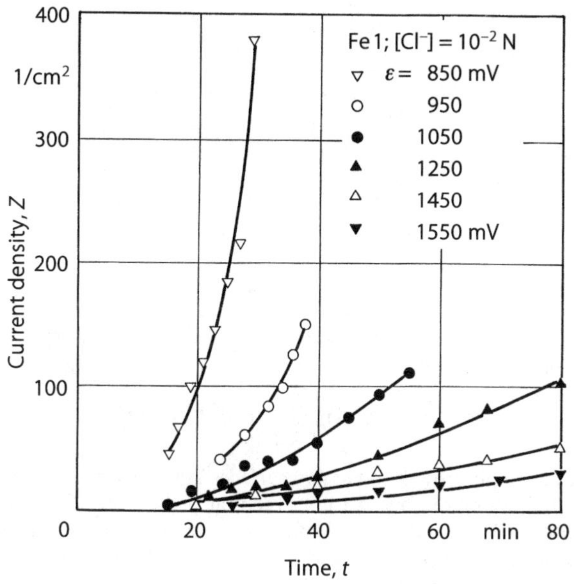


**Fig. 12.18.** The number of pits with a radius exceeding 0.02 mm, observed in the surface of passive pure iron, in 1 *N* $H_2SO_4$/0.01 *N* HCl solution as a function of time during potentiostatic polarization, as function of the potential. (From Herbsleb and Engell)

that an increase in potential is accompanied by an increase in passive oxide film thickness, and this apparently is effective in increasing the resistance to pitting. The figure also shows that the assumption of a potential-independent pit nucleation rate is a simplification and that $v$ tends to increase with time.

If we assume the inner pit surface to be covered by a low-conductivity, gelatinous or otherwise largely dehydrated salt layer, we may further assume that, eventually, as the pit grows, the layer recrystallizes to a hydrated crystalline porous film. When the solution is sulfuric acid, and the metal, iron, this state would correspond to that at the passivating potential $\varepsilon_{p1}$, and the critical passivating current density, $i_{crit}$, should be close to the pit current density (see Chap. 10). In this model, schematically shown as Case a in Fig. 12.19, a considerable ohmic voltage drop exists in the pit across the surface layer. The model fails to explain observed pit current densities sometimes exceeding $i_{crit}$ by an order of magnitude, and conditions under polishing films may indeed differ from those effective at the passivation potential. As a variant of the model we note the proposition that the quantity mainly requiring consideration is a large ohmic voltage drop not across a film at the pit bottom, but in the pit electrolyte [19]. One possibility among others is clogging of narrow pits by hydrogen bubbles [19b]. According to this concept, stable pit growth only requires that the ohmic potential drop always keeps the pit bottom below the passivating potential in the pit electrolyte. This supposition evades the initial origin of this potential drop and, thus, also evades the mechanism of pit initiation.

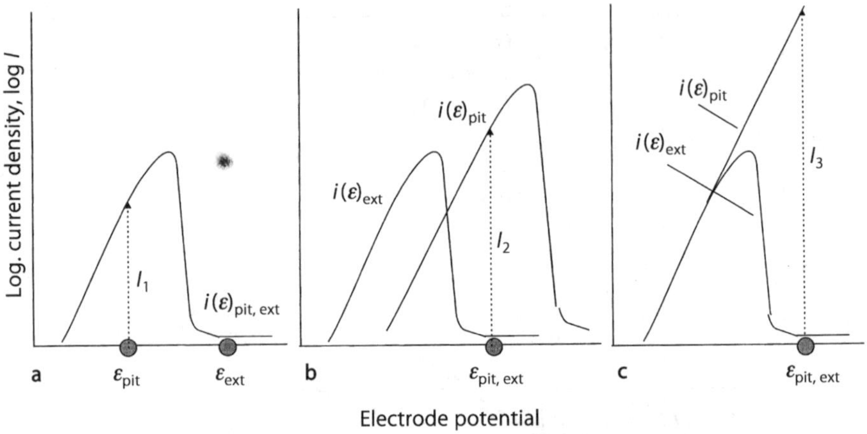


**Fig. 12.19a–c.** Three models for active pit growth in passive surfaces at the external potential $\varepsilon_{ext}$, and an internal potential below the passivation potential. $I_1$,$I_2$, and $I_3$ indicate true pit growth current densities. **a** Bulk and pit electrolyte composition are the same, and the $i(\varepsilon)$ curve is the same for both pit exterior and pit interior, but the pit-bottom potential, $\varepsilon_{pit}$, is more negative by a large ohmic drop, $\Delta\varepsilon_{Ohm}$. **b** Neglect $\Delta\varepsilon_{Ohm}$, but consider a large decrease, ΔpH, of the pit electrolyte due to acidifying, shifting the pit passivation potential to a more positive, the critical pit passivating current density to a higher value. **c** applies for both negligible ΔpH and $\Delta\varepsilon_{Ohm}$, which is possible for a brief time interval after pit nucleation, when a pinprick pit is still small and shallow

The second model ascribes the main effect in pitting to an increase in acidity due to the hydrolysis of dissolved cations of the pitted metal, Me, by the reaction $Me^{z+}+H_2O \rightarrow Me(OH)^{z-1}+H^+$. The importance of hydrolysis in pit electrolytes was pointed out in [17] for the obvious reason that sufficient acidity of the pit electrolyte readily explains that repassivation does not occur. In Fig. 12.19, this is Case b, and we note that again the pit current density cannot exceed the critical passivating current density, though it may be higher in the acidified pit than in the bulk solution. Composition and, thus, pH of the pit electrolyte are governed by diffusion and migration of dissolved metal salt out of the pit, as will be more closely discussed below. In a simple model based on this concept, transport processes are treated as one-dimensional, considering the pit as a passive slot or hole with an active bottom, and it is argued that the product of pit depth, $x$, and current density, $I$, must exceed some critical value, $(xI)^*$, to allow stable pit growth, as otherwise repassivation is expected [23b].

Case c in Fig. 12.19 considers the very onset of pitting, where pinprick pits still are shallow, and both $\Delta\varepsilon_{Ohm}$ and ΔpH still are negligible. The pit bottom potential is on the active branch of $i_{Me}(\varepsilon)$, extrapolated to beyond the passivation potential. Thus, very high initial pit current densities are predicted, as they are indeed observed, as noted below.

Regardless of a specific model, it is obvious that pit electrolytes should be enriched with respect to cations of the dissolved metal and, by compensation, in anions migrating into the pit. Quantifying pit growth kinetics therefore requires the calculation of changes in composition in the pit electrolyte as compared to the bulk electrolyte. In principle, mathematics are straightforward: Four contributions to mass transport processes are to be considered: a) Diffusion governed by Fick's first law; b) electrolytic migration governed by Ohm's law; c) influx of bulk electrolyte into the growing pit, and d) convection caused by gas bubbles evolving in the pit and escaping through the pit mouth, if hydrogen evolution is occurring. The slow movement of ions carried by the flux of bulk solution into the growing pit normally is negligible in the sense that it does not much affect the gradients in concentrations and in potential, and term (c) will therefore be neglected in the following. This has a major consequence: If the pit volume is taken as constant, and concentrations in the pit are in steady state, then the fluxes of ions not taking part in electrode reactions at the pit bottom formally vanish. In pitting, this applies to all ions except the cations, $Me^{z+}$, dissolving by anodic dissolution from the active pit surface.

Let $X_m^{z_m}$ be any one such ionic species, and $c_m$ its concentration. The flux, $J_m (\mathrm{mol\,m^{-2}\,s^{-1}})$, is given by the *Nernst-Planck* equation

$$J_m = -D_m \nabla c_m - \frac{z_m}{|z_m|} c_m u_m \nabla \phi \text{ , for dissolving metal cations ;} \tag{12.6}$$

$$J_m = 0 \text{ , for other ions in the pit electrolyte .}$$

The Nabla operator is $\nabla = \partial/\partial x + \partial/\partial y + \partial/\partial z\ (\mathrm{m^{-1}})$, the diffusion coefficient, $D_m$ ($\mathrm{m^2\,s^{-1}}$) and the mobility, $u_m$ ($\mathrm{m^2\,s^{-1}\,V^{-1}}$). The intricacy is that the separation of diffusion and migration terms is strictly correct only in dilute solutions, whereas pit electrolytes typically are concentrated. For high concentrations, the hydration shells of ions overlap, and the effect is mutual drag of cations and anions. This, then, makes the transport equations, as applied to pitting problems, approximate in any case. Quite the same applies to the additional *Einstein* relation $D_m/RT = u_m/|z_m|F$, which brings Eq. (12.6) into the often used form:

$$J_m = -D_m \left( \nabla c_m + c_m \frac{z_m F}{RT} \nabla \phi \right) . \tag{12.7}$$

In concentrated solutions, ionic activities, $a_m = c_m f_m$, may differ very much from concentrations. According to thermodynamics, diffusion tends to equalize electrochemical potentials, i.e. $\tilde{\mu}_m = \mu_m^\circ + RT \ln c_m f_m, + zF\phi$, rather than concentrations. Therefore, Fick's first law can be reformulated into the non-empirical, rather axiomatic form $J_m = -(RTD_m/c)\nabla \tilde{\mu}_m$. Since, however, diffusion kinetics continue to be governed by collisions and, thus, numbers of particles rather than activities, the improvement of numerical results obtained for highly concentrated solutions is questionable, and even more so, as the factors, $f_m$, are single-ion activity coefficients, inaccessible to precise experimental determination. If adopted nevertheless, the procedure adds a term $c_m \nabla \ln f_m$ to the sum in the brackets in Eq. (12.7). The point is interesting in the context of attempts to estimate single-ion activities for ions such as aluminum ions in highly concentrated solutions [36]. A predicted sharp increase in the activity coefficient of the aluminum ion in highly concentrated solutions, where hydration shells of

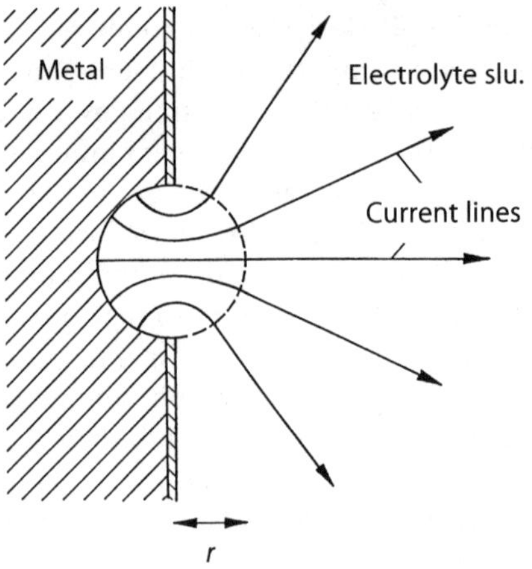


**Fig. 12.20.** Anodic current lines of metal dissolution out of a hemispherical pit of radius $r$ in an otherwise passive metal surface. (From Vetter and Strehblow)

the ions bind most of the available water molecules, plays a role in a particular model of pitting of aluminum in [37a].

The change in composition of pit electrolytes, referred to the surrounding bulk electrolyte, has been considered for the initial stage of still very small pits, with a characteristic size of no more than ca. 1 μm, in [16d]. For simplicity, the bulk electrolyte is a dilute solution of the cations, $Me^{x+}$, of the pitted metal, and chloride anions, $Cl^{y-}$, where $y = 1$. The geometry is as shown in Fig. 12.20. Outside the dotted contour, the radial-symmetric distribution of anodic current lines leaving the pit interior and entering the bulk electrolyte is obvious. From this current distribution, the increase, $\Delta c$, in concentration of $MeCl_x$ at the area bordered by the dotted contour, referred to the distant interior of the bulk solution at $r = \infty$, is calculated as

$$\Delta c_r = \frac{rI}{xFD_{Me^{x+}}(1 + x/y)} . \tag{12.8}$$

To calculate from this the more interesting change of $\Delta c$ in the pit interior, at $(-r)$ is, in principle, a matter of conformal mapping [16i]. Instead, an intuitive estimate $\Delta c_{(-r)} \simeq 3\Delta c_r$ was used originally. For the particular case of pitting of iron (see below), the experimental value for the pit current density was $I = 0.9\ \text{Acm}^{-2}$. Since the diffusion coefficient is $D_{Fe^{2+}} \simeq 5 \times 10^{-6}\ \text{cm}^2\text{s}^{-1}$, it followed that $\Delta c_{(-r)} \simeq 0.9\ \text{mol L}^{-1}$. This was considered to be far below the saturation concentration, which is 4.4 mol $\text{L}^{-1}$, and the conclusion was that, while the interior pit surface certainly must be chlorided directly from the start of pit life, this surface iron chloride, or possibly a chloride adsorption layer, was not in equilibrium with a solution saturated with respect to the metal chloride. In this context, it is useful to recall that, as shown by the passivity of iron in acids, films covering electrode may very well be thermodynamically unstable.

In view of the approximations entering the reasoning, the estimate may not seem quite safe as proving, even for very small pits, the undersaturation of the pit electrolyte with respect to dissolved metal chloride. Also, when the bulk electrolyte is rich in, e.g., dissolved NaCl, the precipitation of $MeCl_x$ is expected even earlier. Two further conclusions, gained by reasoning along similar lines, probably are safe: The first is that an ohmic potential drop, $\Delta\varepsilon_{\mathrm{Ohm}}$, in very small pits does not exceed several 10 mV, and the second is that the diffusion potential between pit bottom and bulk solution is of the order of $RT/F$, i.e., also of the order of several 10 mV. All this will have to be reconsidered for cases such as nickel, where the initial pit current density is very high, viz. ca. 100 $\mathrm{Acm}^{-2}$.

## 12.4 More on Aluminum and Aluminum Alloys

### 12.4.1 Weakly Alkaline Solutions

In the work described in [2c,d,s], which was prompted by a case of severe pitting of an aluminum can, the solution was weakly alkaline at pH 11, and hydrogen bubbles were escaping from the pits at a pitting potential distinctly more positiv than the hydrogen equilibrium potential, $E_{\mathrm{H_2/H^+}}$. From this it either followed that the pit bottom was at a correspondingly more negative potential, which for wide open pits was not, however, considered likely, or else that the pit electrolyte was sufficiently acid owing to the hydrolysis of $Al^{3+}$. We recall that the growth of gross hemispherical, open pits is characteristic for weakly alkaline solutions owing to inhibition of pit nucleation by hydroxyl adsorption on the passivating oxide film. The effect should not be confused with cathodic corrosion of aluminum in more alkaline solutions, caused by aluminate dissolution of the oxide film, as described in Chap. 10.

Adopting the supposition that, essentially, pit growth in aluminum is due to hydrolytic acidity of the pit electrolyte, the situation is as described and explained in Fig. 12.21. The mechanism, explained in the figure caption, includes cathodic hydrogen evolution in the pit, and by this effect the net current density of Al dissolution exceeds the faradaic equivalent of the current current flowing out of the pit by ca. 10%. This effect is neglected further below in the calculation of transport processes. Also, we dismiss the more accurate description of the superposition of anodic aluminum dissolution and cathodic hydrogen deposition. In this context, an interesting conjecture, alternative to the concept of independent superposition of partial reactions of aluminum dissolution and hydrogen evolution, is that part of the anodic aluminum dissolution proceeds to monovalent $Al^{+}$ or divalent $Al^{2+}$ only, followed by a chemical oxidation to $Al^{3+}$ by reduction of water producing hydrogen [38].

Hydrogen bubbles, escaping from the acid pit interior, passing the outer oxide film in an alkaline solution at a potential anodic to the equilibrium po-

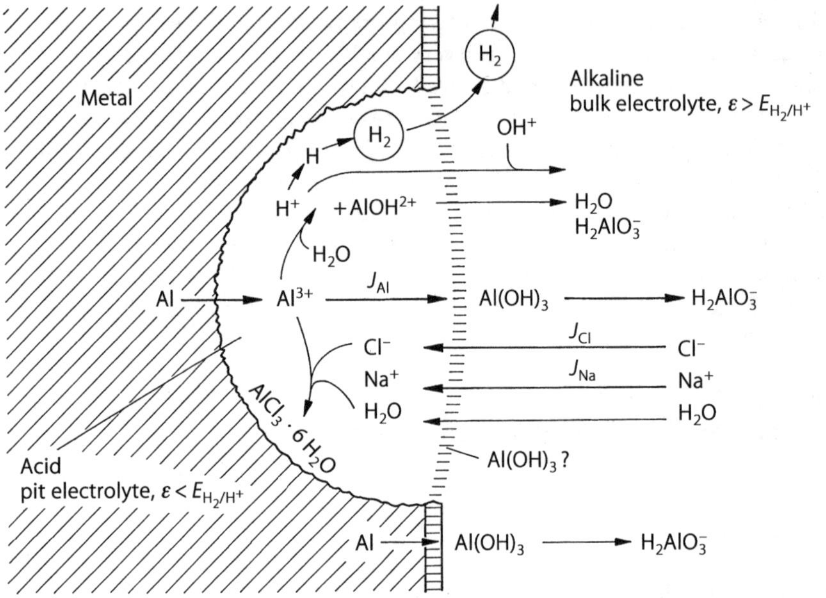


**Fig. 12.21.** The growth mechanism of hemispherical pits in aluminum, observed in NaCl solutions at pH 11 during galvanostatic anodic polarization at the pitting potential. Outside the pit, the metal is passive, but owing to high pH the rate of uniform dissolution from Al via $Al_2O_3$ and $AlOH_3$ to $AlOH_4^-$ or $H_2AlO_3^-$ is well defined. In the pit, aluminum dissolves anodically into the pit electrolyte to hydrated $Al^{3+}$. Hydrolysis of a fraction of these cations to $AlOH^{2+}$ acidifys the pit electrolyte, causing cathodic hydrogen evolution in the pit, producing hydrogen bubbles which stir the pit electrolyte before they escape from the pit interior. During transport of $Al^{3+}$, through the pit mouth into the bulk electrolyte, an intermediate deposition of a veil of solid hydroxide may occur, followed by aluminate redissolution. The acidity of the pit electrolyte is such that repassivation of the pit interior is excluded at least when the pit electrolyte is saturated with respect to hydrated aluminum chloride. (From Kaesche)

tential, should be reoxidized, but this reaction is blocked owing to the lack of electronic conductivity in the passivating oxide on aluminum.

Concerning pitting kinetics, more information was obtained by potentiostatic polarization experiments at potentials positive to $\varepsilon_{pit}$, where the current continuously increases with time, owing to continuous nucleation and growth of pinprick open pits. The current increase obeys the $t^3$-law, as derived above indicating, at any rate approximately, constant values of the pit nucleation rate, $v$, and of the pit current density, $I'$ and $I = I'/2$. Values of $v$ where obtained by ex-situ counting of pinprick pits visible at binocular low magnification, because hydrogen bubbling out of every pit prevented the in-situ observation of pit growth. Using these values, $I'$ was evaluated from Eq. (12.5). Figure 12.22 shows the dependence of both $v$ and $I'$ on the deviation $\Delta\varepsilon = \varepsilon - \varepsilon_{pit}$ of the potential from the stationary pitting potential and on the chloride concentration (see Fig. 12.10).

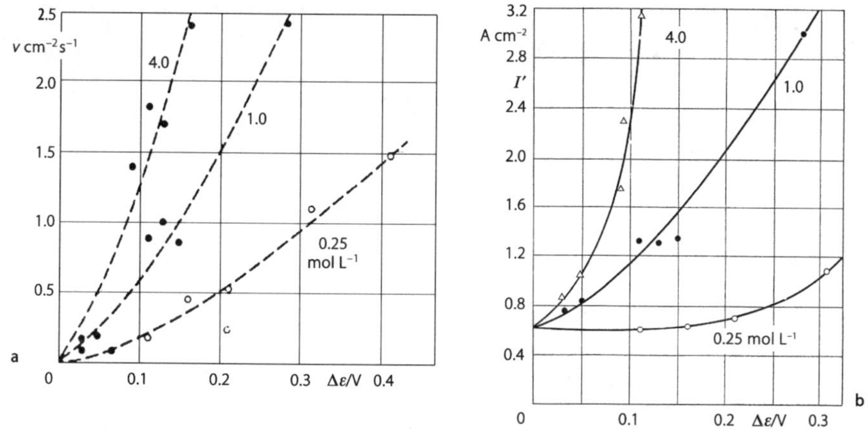


**Fig. 12.22a,b.** Pitting of pure aluminum in NaCl solutions, pH 11, during potentiostatic polarization, as a function of the deviation $\Delta\varepsilon$ of the electrode potential from the pitting potential, $\varepsilon_{\text{pit}}$ in solutions of varying chloride concentration. **a** Pit nucleation rate, $v$; **b** pit dissolution current density, $I'$, referred to the pit mouth. (From Kaesche)

The pit nucleation rate increases not only with increasing chloride concentration of the solution, but also with increasing potential difference, referred to the pitting potential. Also, the reason for choosing the variable $\Delta\varepsilon$ is obvious from this plot: In every case, $v = 0$, when $\Delta\varepsilon = 0$. Hence, the pitting potential, though changing with the chloride concentration, always is the potential of vanishing pit nucleation rate, and we remember that it also is the potential below which preexisting pits repassivate. At the same time, $\Delta\varepsilon = 0$ marks the limiting condition of stable pit growth with $(I)_{\varepsilon_{\text{pit}}} = 0.3\,\text{Acm}^{-2}$, regardless of the bulk solution chloride concentration. Clearly, this indicates a steady state with respect to pH and/or chloride concentration in the pit electrolyte. From this the most obvious supposition was to assume that at $\varepsilon_{\text{pit}}$ the pit electrolyte is saturated with respect to aluminum chloride.

Similar records of pit nucleation rates and continuous current increase at potential beyond the the pitting potential, performed in 1 M KBr solution, established a value $(I)_{\varepsilon_{\text{pit}}} = 0.2$ A $\text{cm}^{-2}$. While this is not exactly the same as before, it does prove that the increase in $\varepsilon_{\text{pit}}$ observed in bromide as compared to chloride solution was not due to an increase of current densities with potential.

As pits grow to macroscopic depth, the current density, $I$, apparently decreases and pit growth slows. This was observed by recording the time of perforation of aluminum foils during potentiostatic polarization in alkaline chloride solution [11]. Figure 12.23 shows that the decay of $I$ with increasing time, $t$, is approximately parabolic, and this was taken as pointing to a rate-determining potential gradient, $\Delta\varepsilon_{\text{Ohm}}$, in the pit electrolyte, which flattens with increasing pit depth. A parabolic change of $I$ with time would, however, also be expected

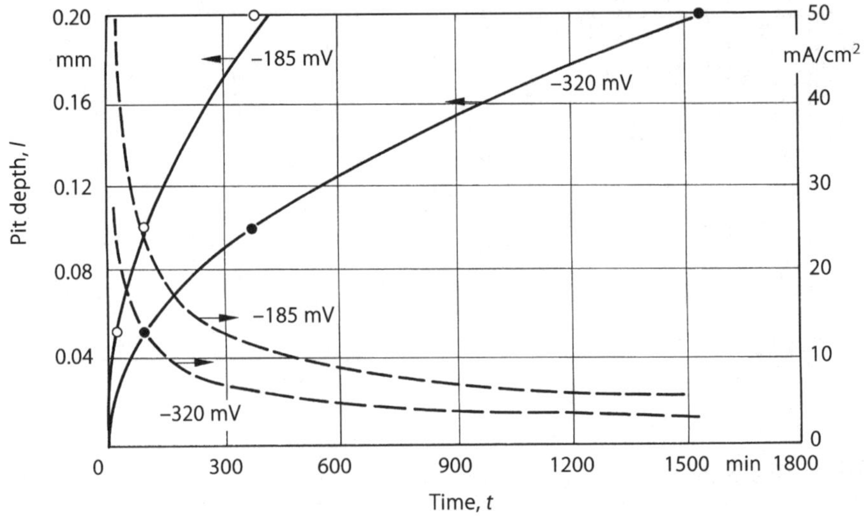


**Fig. 12.23.** Depth (—-) and local pit current density (- - -), referred to the interior pit surface, observed for pitting in pure aluminum foils during prolonged potentiostatic polarization in 0.01 M NaCl solution at pH+11. (From Hunkeler and Böhni)

in the case of rate-determining diffusion driven by concentration gradients. These points of long-time pit growths will not further be considered in the following.

Concerning the salt deposit at the inner pit surface, the original simple assumption was to consider $AlCl_3 \cdot 6H_2O$. The presence of this substance was not, however, substantiated by experimental evidence, and some other critical concentration would suit the model just as well, provided that solution pH is low enough to prevent the deposition of aluminum hydroxide. The composition of the pit electrolyte and of precipitated phases is, indeed, a matter of much speculation. Judging from the literature on aluminum chemistry, the presence of complex aluminum oxychloride cations is described as probable in [35b]. In the same paper, experiments with artificial pit electrolytes are reported, obtained by rapid anodic dissolution of aluminum from 'pencil electrodes', i.e. narrow wires embedded in glass tubes. This and similar devices are frequently used in model pitting experiments, and they probably come rather close to conditions of deep, gross pitting. In the present case, the nuclear magnetic response (NMR) spectrum of model pit electrolytes proved to be similar to that of solutions of $Al(OH)_2Cl$ and $Al(OH)Cl_2$, with mostly monomeric dissolved ions at a concentration close to 3 mol $L^{-1}$, and pH $\simeq$ 3. Elsewhere [39,35c], values pH $< 0$, or pH $\simeq 0$ have been reported for concentrated aluminum chloride solutions, so the assumption of a high solubility of aluminum oxide in the pit electrolyte does seem quite acceptable.

Considering a high hydration number for dissolved aluminum cations, and cation concentrations of up to several mol $L^{-1}$, the availability of water molecules certainly becomes a problem. Even limiting considerations to the case of pits growing at the pitting potential, i.e. with the lowest possible current density, we are, therefore, dealing with highly concentrated solutions, and the application of the Nernst-Planck equations to mass transport through the pit electrolyte certainly is an approximation for reasons explained previously. In view of errors introduced by ignoring these complications, activity coefficients may well remain disregarded within the the scope of what is semi-quantitative reasoning anyway. With these limitations, the deduction procedure was as follows [2e]:

Diffusion and migration are considered for $Al^{3+}$,$Cl^-$, and $Na^+$ ions only, ignoring other ionic aluminum compounds, also ignoring $H^+$. The pit electrolyte is considered well stirred owing to the flushing with hydrogen bubbles and, therefore, homogeneous, as also is the bulk electrolyte. Concentration gradients thus exist in a disk-shaped thin boundary layer in the pit mouth only. Here, the probable presence of a veil of aluminum hydroxide and the disturbance by gas bubbles escaping from the pit are disregarded. We recall that convective flow into the growing pit is considered negligible.

Under theses conditions, diffusion and migration through the boundary layer are one-dimensional, and the Nabla operator in Eqs. (12.6) and (12.7) reduces to $d/dx$. The total faradaic current flowing in the hemispherical pit is $2\pi r^2 I$ or $\pi r^2 I'$. Referred to the area of the pit mouth, the flux densities to be considered are

$$\begin{aligned} &\text{a)}\quad J_{Al^{3+}} = \frac{I'_p}{3F} = -D_{Al^{3+}}\left(\frac{dc_{Al^{3+}}}{dx} + \frac{3F}{RT}c_{Al^{3+}}\frac{d\phi}{dx}\right), \\ &\text{b)}\quad 0 = -D_{Cl^-}\left(\frac{dc_{Cl^-}}{dx} - \frac{F}{RT}c_{Cl^-}\frac{d\phi}{dx}\right), \\ &\text{c)}\quad 0 = -D_{Na^+}\left(\frac{dc_{Na^+}}{dx} + \frac{F}{RT}c_{Na^+}\frac{d\phi}{dx}\right). \end{aligned} \tag{12.9}$$

Together with the electroneutrality condition,

$$3c_{Al^{3+}} + c_{Na^+} = c_{Cl^-}, \tag{12.10}$$

the system can be solved for the concentrations, $c_{Al^{3+}}$,$c_{Cl^-}$, and $c_{Na^+}$, in the pit by integrating from the pit interior, i.e. for $c_{Al^{3+}} = (c_{Al^{3+}})_{pit}$, through the boundary layer of thickness $\delta$, into the bulk solution at $c_{Al^{3+}} = 0$, and by introducing a critical concentration of aluminum chloride. This was chosen to be the saturated solution, with 3.0 mol $L^{-1}$. Inserting reasonable values for $1 \times 10^{-3} \lesssim \delta \lesssim 5 \times 10^{-3}$ cm, and $1 \times 10^{-6} \lesssim D_{Al^{3+}} \lesssim 5 \times 10^{-6}$ $cm^2$ $s^{-1}$, an order of magnitude of several 0.1 A $cm^{-2}$ was estimated for the pit current density, $I$, in accordance with the experimental value observed at the pitting potential. Up to several mol $L^{-1}$ of chloride in the bulk solution, the calculated pit electrolyte composition is almost pure, saturated $AlCl_3 \cdot 6H_2O$ solution. All this is no great surprise, but it does show that the approximate calculations in fact confirm the original suppositions. The observed dependence of the pitting potential of aluminum (and, as described below, of aluminum alloys) on the bulk chloride concentration (or, as also described below, on other halide concentrations) is readily, though only qualitatively explained as the change in the diffusion potential in the pit mouth between pit and bulk electrolyte.

The exact nature of the aluminum salt film supposed to be present on the active pit bottom so far remains a matter of speculation, although the presence of some halided aluminum compound in the respective halide solution is hardly questionable. This is also corroborated by the data in Fig. 12.11 for bromide and iodide solutions: As noted in [2d], the changes of $\varepsilon_{pit}$ from the chloride- to the bromide- and iodide-value surprisingly well parallel the differences in the Gibbs standard energy of formation of $AlCl_3$, $AlBr_3$, and $AlJ_3$. This does not conclusively prove the nature of the film present during active pitting, as other, perhaps more directly important properties of aluminum may be decisive, but these, then, should again correlate with the same Gibbs energies.

The other, equally important point to be made here is that the stationary pitting potential appears to be governed by the pit current density, $I(\varepsilon_{pit})$, as the lowest rate of aluminum dissolution able to support active pit growth [2e]. A change in $\varepsilon_{pit}$ caused by a change in the halide concentration and, thus, by a change in the diffusion potential in the pit mouth, leaves this critical current density unchanged. As to changes in $\varepsilon_{pit}$ caused by changes in the nature of the halide, or in the composition of aluminum alloys, see below. As described above, aluminum electrodes become non-polarizable at the pitting potential, because the total area of active pitting is proportional to the apparent current density, referred to the geometric electrode area.

During galvanostatic pitting at the pitting potential, preexisting pits are growing, and this, then must increase the active pit area, violating the condition of constancy of this same area. Galvanostatically pitted electrodes can, however, easily adapt to this situation: Some pits stop to grow by repassivation, decreasing the active area, and thus increasing the true pit current density. This causes fluctuations of the potential to more positive values, where new pits are nucleated, with following return to $\varepsilon_{pit}$, etc. Such fluctuations in potential have been, in fact, observed already much earlier [6a].

## 12.4.2 Neutral and Acid Solutions

In neutral solutions, the passive current density, $i_{pass}$, of uniform aluminum dissolution is negligible, otherwise the stationary current-potential curve, and especially the stationary pitting potential as defined previously is the same as before.

For completeness, three side-effects, observed in neutral and acid solutions, and not further discussed below, need brief comment:

(a) When, after prolonged periods of galvanostatic anodic polarization causing severe pitting, the potential is decreased to below the pitting potential, $i_{pass}$

is found to be persistently higher than before, although still small, of the order of 1 μA $cm^{-2}$ [2h,i,p]. This is not a pitting current; instead, while preceding pitting was crystallografic, the residual current below the pitting potential gradually rounds crystal facets. Apparently, the residual current is an increased passive current, possibly pointing to a persistent damage to the passive film, possibly as the result of incorporation of chloride into the passive film.

(b) For pitting of very thin foils of aluminum the repassivation potential, $\varepsilon_{pit3}$, as defined previously, depends on foil thickness [7b]. We pass this effect, as it disappears for foils thicker than 10 μm.

(c) A third effect, actually more important with respect to crevice corrosion described below, is the following: In acid solutions, aluminum certainly still is passive owing to an oxide film, although the thermodynamic oxide solubility is high. The passive film is, however, thinned, and film attack is occurring at a noticeable rate definitely below the pitting potential. Depending on the potential, the morphology of this sub-pitting attack changes from irregular etching to the formation of microscopic shallow, and sometimes actually crystallografic dimples, as documented in [2t].

At the time of the work described in [2d,e], instrumentation for a closer inspection of pit walls was not available, but roughness of pit interiors was readily identified, as shown, e.g., in Fig. 12.9. In the meantime, it has become widely recognized that aluminum attack during pitting often is crystallografic, as reported first in [17e], and subsequently documented in many examples in the meeting volumes [1] and, especially, in [2h,t,u].

While all of the following describes work on pitting in either neutral or acid solutions, it is quite probable that a closer inspection of pits grown in weakly alkaline solutions would reveal the same microscopic and submicroscopic morphology, and that insofar the subdivision of the chapter is probably incidental.

As noted before, pit interiors tend to change, at highly anodic potentials, from crystallografic facets to electropolished hemispheres. An incident of this type is shown in Fig. 12.24 for pitting of aluminum polarized to beyond 1.7 V in a nitrate solution [23a]. Since very high anodic polarization potentials are of little concern for corrosion, we do not further consider details of this high-potential pit growth. Further below, it will, however, be seen that crystallografic pitting closer to the pitting potential sometimes also produces vaulted instead of clearly crystallografic faces, but these are either bounding very small and fast-growing nuclei, or else they really are faceted on a lower scale.

Returning to galvanostatic crystallografic pitting directly at the pitting potential, Figs. 12.25 and 12.26 present typical examples of the many different variants of pit morphology observed in neutral salt solutions during galvanostatic polarization of aluminum and homogeneous binary aluminum alloys at the respective pitting potential [23a,2p]. Some are metallographic sections, others SEM micrographs looking into open pits; others show intricate tunneling corrosion following crystallografic orientation into the bulk of electrodes. The latter tunnel structures prepared by anodic oxydation of tunneled specimens followed by dissolution of the metal in methanolic bromine solution leaving oxide replica

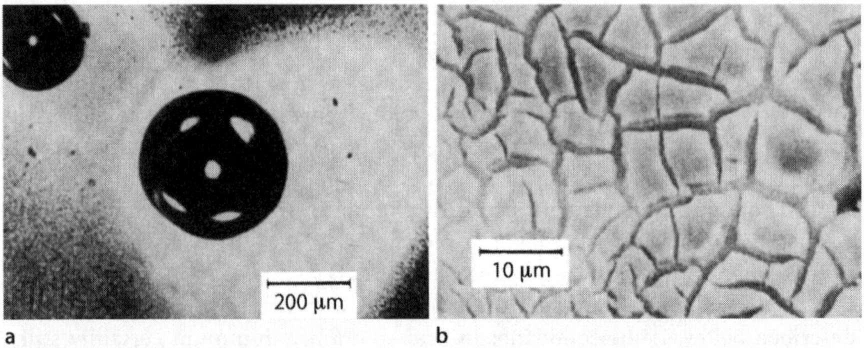


**Fig. 12.24.** **a** Hemispherical electropolished pit grown in 99.99 wt.%Al at >1.7 V in 1 M $NaNO_3$ solution. **b** SEM picture of the pit bottom showing the dried and ruptured electropolishing film. (From Galvele, DeMicheli, Muller and Wexler)

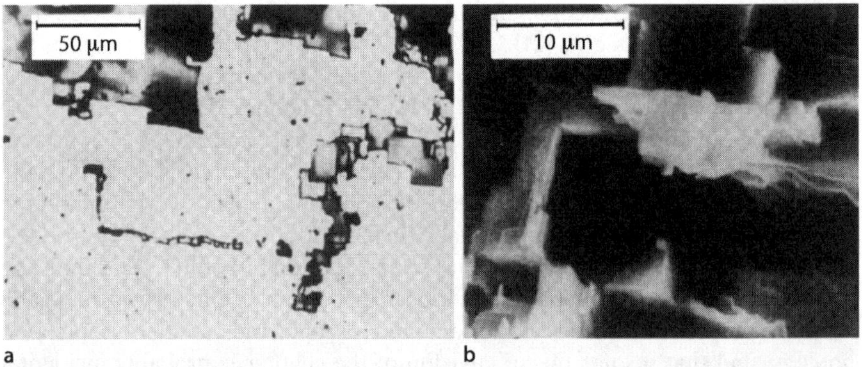


**Fig. 12.25.** **a** Metallographic cross section of a pit bottom in pure aluminum after galvanostatic anodic polarization at the pitting potential in 0.1 M KCl solution. (From Pleva). **b** Scanning electron microscope picture of the bottom of a pit bottom in pure aluminum in 1 M NaCl solution. (From Galvele, DeMicheli, Muller, Wexler, and Alanis)

of the tunnels. The surprising variety in microscopic pit morphology is not as yet well understood, and no coherent explanation can be attempted. We note the appearance of rounded faces and of some (110) faces, while (100) faces certainly are predominant. Pit growth always proceeds in $< 100 >$ directions. In the following, true local current densities, $I$, therefore are rates of propagation of pit fronts in these directions.

For AlCu, AlAg, and AlZn alloys, current-potential curves are not shown. We note, therefore, that they have been recorded in every case, always confirming a perfectly well defined pitting potential marking the potential of onset of non-polarizability of the electrodes [2p]. In any halide solution, the pitting potential increases with decreasing halide concentration, as exemplified by the data in Fig. 12.11. The dependence always approximately obeys a relation $\varepsilon = a - b \ln c$, where the numerical value of $b$ usually is between $RT/F$ and $2\ RT/F$,

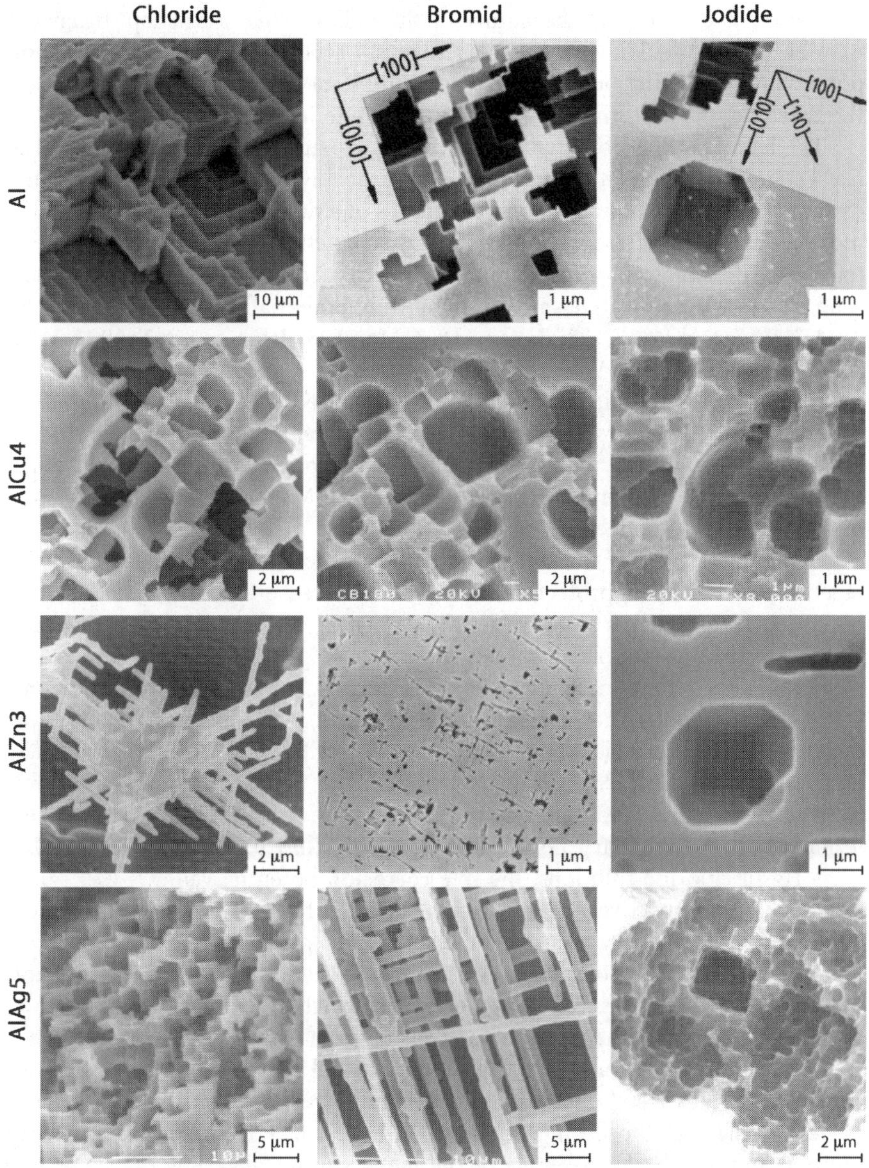


**Fig. 12.26.** Crystallografic pitting in aluminum and homogeneous aluminum alloys observed at the respective pitting potential in halide solutions: **a** Pure aluminum in 0.01 M NaJ; **b** Al-4 wt% Cu in 0.1 M NaBr; **c** Al-3 wt.% Zn in 1 M NaCl (oxide replica); **d** Al-4 wt.% Zn in 1M NaCl (oxide replica); **e** Al-5 wt% Ag in 0.1 M NaBr; **f** Al-3 wt.% Zn in saturated $AlCl_3 \cdot 6H_2O$. (From Knörnschild and Kaesche)

and where $a$ depends on the nature of the halide. This 'Nernstian' behaviour may be purely incidental, but it also suggests a possible dependence of some potential-determining process on some multiple of $(RT/F) \ln a_{X^-}$, where $a_{X^-}$ is the thermodynamic activity, $c_{X^-} f_{X^-}$, of the halide in the solution.

The list of pitting potential values can be expanded to more alloys in 0.1 M NaCl solutions using data reported in [40], probably recorded by slow potentiodynamic scanning of current-potential curves of a number of supersaturated homogeneous alloys. These were prepared by magnetron sputtering of aluminum together with tungsten, molybdenum, tantalum, chromium and zirconium, and they revealed large positive shifts of the pitting potential. We quote −0.43 V for Al, 1.72 V for Al-9%W, 0.35 V for Al-8% Mo, 0.28 V for Al-8% Ta, 0.15 V for Al-4.5% Cr, and −0.23 V for Al-3% Zr (wt.%).

No change of the pitting potential, as compared with pure aluminum, was observed in Al-Li alloys (types 8090 and 2091) regardless of whether the alloy surface was rich or depleted in lithium, although lithium-depleted zones markedly differ in general corrosion kinetics [2j]. Magnesium as an alloy component is equally ineffective.

Generally, a shift of the pitting potential to more positive values is advantageous for corrosion protection, as chances increase that the free corrosion potential, $\varepsilon_{corr}$, remains negative to $\varepsilon_{pit}$. When, however, aluminum materials are to be used as sacrificial anodes for cathodic protection installations (see Appendix), the situation changes. For this purpose, aluminum is effective only if the electrode polarization resistance is low, and if $\varepsilon_{corr}$ is considerably more negative for the sacrificial anode than for the metallic structure to be protected. Both effects are achieved by using technical AlZn alloys containing additional 'activators', such as indium, tin, gallium, and mercury [41a]. Not for these ternary, but for the binary alloys Al-01 wt.% In, Al-0.1/0.2 wt.% Sn, and Al-0.01 wt.% Hg in chloride solutions, the free corrosion potential values −0.9, −0.15, and −1.35 V, respectively, have been reported [41b-e]. These are potentiodynamic, perhaps not very accurate values, but they probably reflect a real trend. At the same time, the polarization resistances indeed are low, and the probable explanation is that these are pitting potentials, which, if we consider these values as real, on the whole then appear to cover a range from −1.35 to +1.72 V.

Concerning the nature of the pitting potential for aluminum, the first assumption, as described above, was to conclude from the constancy of the pit current density at the pitting potential, $I(\varepsilon_{pit}) = 0.3$ A cm$^{-2}$, that at the pit bottom, and at $\varepsilon_{pit}$, the coverage of the metal surface always is the same, and probably a film of aluminum chloride. We recall that it is now understood that deep gross pitting in aluminum observed in weakly alkaline solutions in a way is incidental, owing to the interplay of nucleation inhibition by hydroxyl adsorption and autocatalysis by local acidifying in pits. We repeat this explicitly, as it bears on some conflict of opinion in the literature concerning the decisive role of either low pH or high chloride concentration for stable pit growth. Obviously, both factors have interrelated importance for pitting kinetics, although the pitting potential, $\varepsilon_{pit}$, as defined in this text, is independent of pH.

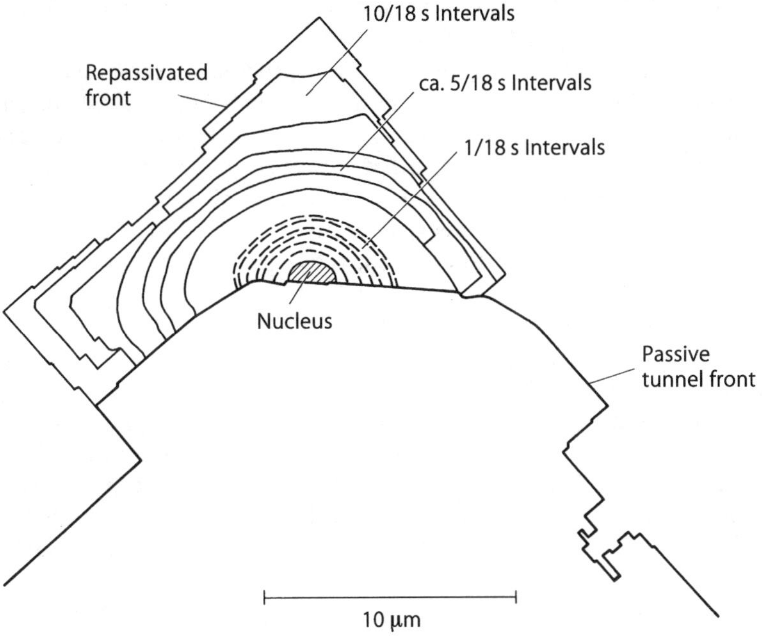


**Fig. 12.27.** The growth of a single crystallografic micropit in a (100) foil of pure aluminum in 1 N NaCl solution at 25°C. The drawing compiles contour pictures taken at the indicated intervals. The original contour is that of a previously grown pit which had been stopped growing by switching the potential to below $\varepsilon_{\text{pit}}$. Note start of growth with a non-crystallografic nucleus, growing at an excessive rate, turning crystallografic and slowing down with increasing time. The 'dead' pit is bounded by (100) faces everywhere. (From Baumgärtner and Kaesche)

A crucial point needing explication is the unusually high rate of local crystallografic attack. For aluminum, this has been closely investigated using thin monocrystalline foils of pure aluminum in neutral chloride solutions, where the progress of pitting from a single hole in the foil could be recorded under a microscope through a transparent coating of the electrode surface [2k–o]. In this case, polarization was potentiostatic at potentials well defined below, identical with and beyond the $\varepsilon_{\text{pit}}$, respectively. Figure 12.27 gives a typical example of successive pit contours, as observed in situ under a microscope during pit growth only slightly beyond $\varepsilon_{\text{pit}}$. The current density of aluminum dissolution calculated from consecutive displacements of the contour varies between several A cm$^{-2}$ and several 10 A cm$^{-2}$. The scatter appears to be due to stepwise growth, indicating that dissolution is pulsating at a frequency much lower than 18 s$^{-1}$. Directly after nucleation, $I$ is much higher, up to ca. 100 A cm$^{-2}$. No solid deposit is visible on the growing pit surfaces. The overall active pit surface narrows as the pit grows deeper, i.e. any such pit tends to taper until its growth comes to a full stop. The initial small nucleus could not be more closely inspected. At a slightly later stage the impression was that apparently vaulted contours really have stepped (100) faces.

Wherever inside a pit active growth is locally proceeding, the rate of propagation always is the same. Hence, this local pit growth rate is constant, until growth comes to an overall full stop. This observation is typical for all cases of pit growth so far closely investigated; and it indicates two important properties of pit growth: a) During growth, the kinetics of anodic dissolution at active pit fronts are the same everywhere, independent of pit depth. b) There is a factor allowing or disallowing, similar to a switch, active pit growth, which becomes critical at a certain pit depth.

Local growth rates are also independent of the bulk chloride concentration, and this certainly is strong argument for constant conditions at advancing surfaces, regardless of bulk solution properties. Obviously, the active pit bottom is in contact with a solution, or else with a film, which always has the same composition. On first sight, this does seem to support the assumption that the pit bottom composition is that of an aluminum chloride solution of critical composition, probably stabilized by an equilibrium with a solid salt deposit. There is, however, an obstacle to this explanation: The next likely supposition would be that the deposit is $AlCl_3 \cdot 6H_2O$, and this suggests to test the effect of replacing the sodium chloride in the bulk solution by aluminum chloride. In fact, as long as the bulk solution is undersaturated with respect to this chloride, nothing new happens as compared with other chlorides at the same concentration. When, however, the bulk solution is saturated with respect to the aluminum chloride hexahydrate, the situation drastically changes: The attack is now similar to what is shown for $AlZn_3$ in Fig. 12.26, and the pit current density, $I$, is only between ca. 0.3 and ca. 3 A $cm^{-2}$, i.e. one order of magnitude lower than in other chloride solutions.

It is now remarkable that these values of $I$, observed for saturated $AlCl_3 \cdot 6H_2O$ solutions by direct microscopic observation of moving fronts of pitting attack, are essentially the same as growth current densities evaluated at the pitting potential in alkaline chloride solutions from the $t^3$-increase of currents with time. Therefore, the previous interpretation of gross pitting, at $\varepsilon_{pit}$, of aluminum in weakly alkaline as due to the presence of an $AlCl_3 \cdot 6H_2O$ deposit, may still be correct. If so, the case is, however, abnormal, with pit growth slowed to comparatively very small values by the presence of this very deposit. The point will once more reappear further below.

Pit growth rates have also been determined for Al-4wt.%Cu and Al-3wt.% Zn [2p–r]. Mostly, experimental techniques were the same as with pure aluminum, and crystallografic recording of pit growth in AlCu4 was quite unambiguous. In contrast to pure aluminum, no indication of a pulsating mechanism was observed. In every case, and regardless of either potential or chloride concentration, the pit current density, $I$, was between 3 and 9 A $cm^{-2}$. Again, the exception is a bulk solution saturated in $AlCl_3 \cdot 6H_2O$, where $I$ once more dropped to 0.3 A $cm^{-2}$. All this is very similar to what was fond for pure aluminum. For pitting in AlZn3, the pit current density at the bottom of the very narrow

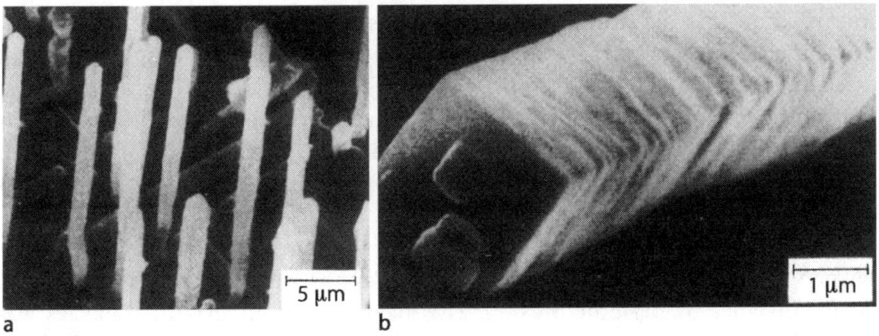


**Fig. 12.28.** **a** Submicroscopic corrosion tunnels grown in <100>-direction into the surface of pure aluminum during galvanostatic anodic polarization in 1 N HCl at 80°C. REM picture of oxide replica obtained by anodic oxidation of the specimen and dissolution of aluminum in methanolic bromine. **b** Further magnification of the end of a tunnel. (From Alwitt, Beck and Hebert)

tunnels shown in Fig. 12.26 was determined by direct microscopic observation, and also by analysing the length of tunnels from oxide replica prepared after the pitting experiment. Again, $I \simeq 5$ A $cm^{-2}$, except for saturated for $AlCl_3 \cdot 6H_2O$ solutions.

We note that the narrow corrosion tunnels observed in AlZn3 do not grow to more than ca. 100 µm length, and this is similar to what is consistently observed for pure aluminum in hot hydrochloric acid. This latter case has been thoroughly investigated, with results reported in a series of publications [35]. The main observation is that under special conditions, i.e. when pretreated in acid solutions (probably with the effect of thinning the oxide film) followed by intensive galvanostatic polarization during, e.g., 2 min, the metal surface is found to have enormously increased in total surface by the growth of ca. $10^5$ $cm^{-2}$ short and narrow tunnels. The well known technical application is the production of electrolytic capacitors, as noted previously. The tunnels can be observed from oxide replica and typical examples are shown in Fig. 12.28. These tunnels originate from microscopic cubic pits in the metal surface, nucleated and growing during a brief potential overshoot following the onset of polarization with an apparent overall current density, $i$, (referred to the geometric metal surface) of ca. 200 mA $cm^{-2}$ [35d,e,f]. The true local overshoot current density, i.e. the initial value of $I$, is up to 100 A $cm^{-2}$. An interesting detail is that during overshoot the dissolution current has a capacitive component owing to the change in charge of the electrical double layer at the passive metal/solution interface.

By the same authors, the change from three-dimensional cubic pitting to one-dimensional tunneling was described in a mathematically model. We ignore this detail and proceed to the further attempts to model the growth of tunnels along <100> directions, with passive walls and active (100) bottoms. One convenient point then is that after the initial overshoot, the potential, $\varepsilon$, as recorded outside the tunnel openings, is constant throughout the anodic polarization experiment. In the terminology used in this text, although not in that used by the authors, this then is the pitting potential, $\varepsilon_{pit}$. The authors also once more clearly state that the sum of actively propagating pit tunnel fronts is constant at constant apparent current density, $i$, and that pit propagation is proceeding at a constant local pit current density, $I$, regardless of the magnitude of $i$. As described further above, this implies that the active front area adapts to the available current density, $i$. In the present work, this was confirmed by observing that tunnels abruptly changed width, when $i$ was changed, as for instance by periodic pulsing. The pit propagation current, $I$, always was up to 20 A $cm^{-2}$, i.e. in the range of current densities typical for crystallografic pitting processes. It was further carefully confirmed that $I$ did not vary with tunnel length.

From experiments carried out between 65 and 95°C, an apparent activation energy of ca. 63 J $mol^{-1}$ was evaluated. While the physical meaning of this property is somewhat difficult to see, it does allow to predict that $I$ drops to ca. 0.2 A $cm^{-2}$ at room temperature, and this is the value of $I$ reported previously for gross pitting in weakly alkaline solutions. Conflicting experience comes, however, from experiments in hot aluminum chloride solutions, where tunneling is observed similar to that in hot hydrochloric acid. Here, the local current density, $I$, remained unaffected by the bulk concentration of aluminum chloride, but the length of tunnels decreased with increasing salt concentration, and was practically zero when the bulk solution was saturated with respect to aluminum chloride. We note that the authors introduce, instead of tunnel length, $l$, a geometry factor convenient to handle the growth of tapered tunnels, which however reduces to $l$ for parallel-wall tunnels.

The authors consider that at the bottom of tunnels passivated walls coexist with active, propagating bottoms, and they conclude that this is possible only at what they describe as a singular 'repassivation potential'. To avoid confusion, we note this as $\varepsilon^*_{repass}$, in contrast to the pitting potential, $\varepsilon_{pit}$, and a differently defined repassivation potential, previously introduced as $\varepsilon_{pit3}$. We recall that both $\varepsilon_{pit}$ and $\varepsilon_{pit3}$ are measured at the pit mouth with the aid of a reference electrode, and that a tentative assumption was that these two characteristic potentials essentially are the same. We also recall that, usually, experimental evidence is that $\varepsilon_{pit} = a - b \ln c_{Cl^-}$, and that $b$ usually is between the numerical values of $RT/F$ and $2RT/F$. By the present authors, this external pitting potential, including its approximately logarithmic dependence on bulk chloride concentration, is taken for granted. In principle, this also holds for $\varepsilon^*_{repass}$, but with two additional arguments: Again $\varepsilon{*}_{repass}$ obeys a 'Nernstian' dependence on the $Cl^-$ in the electrolyte at the pit bottom, though with different values for $a$ and $b$. Also, as a tunnel grows in length, this pit bottom chloride concentration increases, shifting $\varepsilon{*}_{repass}$ to increasingly negative values. At the same time, the local $Al^{3+}$ concentration also increases. Eventually, tunnel life comes to its end, when the pit bottom electrolyte becomes saturated with respect to aluminum chloride.

This leaves the nature of active tunnel bottom surfaces open. A tentative suggestion is to assume that, although no solid aluminum chloride deposit should be present, a chlorided surface always exists, because otherwise the required change of $\varepsilon^*_{\text{repass}}$ accompanying chloride enrichment of the tunnel electrolyte would be difficult to explain. Essentially, the question is what $z$ and $x$ are in a deposit, or an adsorbed layer of $AlCl_z \cdot xH_2O$, as already noted above.

For further progress, the mass transport Eqs. (12.6) must be solved for the particular case. This has been achieved in [35g] for both tapering and non-tapering tunnels with passive walls and active bottoms. For parallel-wall tunnels, the geometry is particularly simple, as the anodic current density of aluminum dissolution at the tunnel bottom, $I$, also is the density of the total current passing each section of the tunnel between the bottom and the mouth. To simplify matters, the pit electrolyte was assumed to be pure $AlCl_3$ solution up to the pit mouth, which is correct only when the bulk electrolyte also is an aluminum chloride solution. For bulk concentrated hydrochloric acid the approximation is inaccurate owing to the high concentration in $H^+$. Still, the assumption that $H^+$ is driven out of the pit electrolyte by the anodic current is justified by the argument that, in the presence of concentrated aluminum chloride binding most of the water molecules in hydration shells, $H^+$ is no longer an exceptionally mobile species. When, thus, the pit electrolyte is essentially binary, the system of transport equations reduces accordingly.

The migration terms in Eqs. (12.6) can be replaced by considering that the part of the electric current flowing in the solution carried by $Al^{3+}$ and by $Cl^-$, is $I \times t_{Al^{3+}}$ and $I \times t_{Cl^-}$, respectively, with transference numbers $t_{Cl^-}$ and $t_{Al^{3+}}$. Therefore, the migration terms can be replaced by $(+I\, t_{Al^{3+}}/3F)$ and $(-I\, t_{Cl^-}/F)$. Electroneutrality further requires, that $(dc_{Al^{3+}}/dx) = 3dc_{Cl^-}/dx$. Three more assumptions were important: (a) Convective flow of bulk solution into the deepening tunnel is disregarded. (b) The diffusion coefficients and, therefore, the transference numbers of aluminum and chloride ions are constant along the tunnel length. (c) The tunnel electrolyte is not stirred by hydrogen bubbles, which requires that dissolved hydrogen diffuses in the tunnel electrolyte to the tunnel mouth and desorbs into gas bubbles only outside.

With these simplifications, the change of the concentration of dissolved aluminum chloride from the bulk of the solution, along the depth, $x$, of a tunnel, to its bottom is

$$c_{\text{bottom}} = c_{\text{bulk}} + (I\, t_{Cl^-}/3FD)\, x\,. \tag{12.11}$$

$D$ is the effective coefficient of diffusion for aluminum chloride as calculated from $D_{Al^{3+}}$ and $D_{Cl^-}$. The calculation continues on the argument, that the tunnel-tip potential, as measured referred to a reference electrode outside the tunnel mouth, is the sum off the true tunnel- bottom potential plus an ohmic and also a diffusion potential along the tunnel length:

$$\varepsilon_{\text{pit}} = \varepsilon_{\text{bottom}} + \Delta\varepsilon_{\text{Ohm}} + \Delta\varepsilon_{\text{diff}}\,. \tag{12.12}$$

Considering the geometry of parallel-wall tunnels, the conductivity of the pit electrolyte, $\kappa$, and, for growing tunnels, $\varepsilon_{\text{tip}} = \varepsilon^{\star}_{\text{repass}}$, the potential measured at the tunnel mouth is expressed as

$$\varepsilon_{\text{pit}} = \varepsilon^{\star}_{\text{repass}} + \int_0^l \frac{I}{\kappa} dx + \frac{4RT}{3F} \int_{c_{\text{bulk}}}^{c_{\text{tip}}} t_{\text{Al}^{3+}} d(\ln a_{\text{AlCl}_3}) \,. \quad (12.13)$$

The derivation ends with an estimate of the sum, $\Delta\varepsilon_{\text{Ohm}} + \Delta\varepsilon_{\text{diff}}$, i.e. the sum of the two integrals on the right side of the equation, as a function of tunnel length. The estimated value is 30 mV for a 50 μm tunnel, which is convenient in the context of the next section, as it shows that these contributions to the measured potential are comparatively small.

The simplification to disregard, for the pit or tunnel electrolyte, other species than aluminum ions and chloride would probably also have been good enough in the case of gross hemispherical pits described further above. On the other hand, the limitation of the calculation to a binary salt in the solution can be relaxed in the present case as well [35g], without, however, much change in numerical results.

### 12.4.3
### The Nature of the Pitting Potential

The pitting potential, $\varepsilon_{\text{pit}}$, has been described further above as independent of solution pH. We note that the 1 M NaCl-value, −0.48 V, reported in earlier work [2d,e], was later corrected to −0.51 V [2o]. If the bulk solution is saturated $AlCl_3 \cdot 6H_2O$, $\varepsilon_{\text{pit}} = -0.64$ V. For pH 11, $\varepsilon_{\text{pit}}$ was described as dividing the more negative range of potentials where passivity is safe, from the more positive range of continuous pit nucleation. This was explained as the consequence of the nature of the pitting potential as follows:

For uniform polarization of the entire surface composed of pit openings or pit mouths, and passive surfaces, the potential in any pit mouth, $\varepsilon_{\text{mouth}}$, is the sum of the potential at the interior active pit surface, rsp. the pit bottom, $\varepsilon_{\text{bottom}}$, an ohmic voltage, $\Delta\phi_{\text{Ohm}}$, and a diffusion potential $\Delta\phi_{\text{diff}}$ (or, as previously noted, $\varepsilon_{\text{diff}}$) :

$$\varepsilon = \varepsilon_{\text{bottom}} + \Delta\phi_{\text{Ohm}} + \Delta\phi_{\text{diff}}; . \quad (12.14)$$

Of course, this also applies, as in Eq. (12.12), to the special case that $\varepsilon$ is $\varepsilon_{\text{pit}}$. At the same time, $\varepsilon_{\text{bottom}}$ is the sum of the equilibrium potential, $E_r$, of some potential-determining reaction across the bottom/pit solution interface, and an overpotential, $\eta_r$, of the reaction, which depends on the reaction current density. In our case, the reaction is aluminum dissolution, and the current density is $I$. If now the supposition is that the reaction involves the formation of a solid layer of some aluminum halide, $Al/AlX_z \cdot xH_2O$, the equation describing the potential turns into

$$\varepsilon = E^{\circ}_{\text{Al/AlX}_z \cdot x\text{H}_2\text{O}} - \frac{RT}{F} \ln(a_{\text{X}^-})_{\text{bottom}} + \eta(I) + \Delta\phi_{\text{Ohm}} + \Delta\phi_{\text{diff}} \,. \quad (12.15)$$

Here, $X^-$ is any of the halides except fluoride, and $(a_{X^-})_{\text{bottom}}$ is the activity of dissolved halogen at the active bottom. The unusual case of micropitting caused by fluoride, described in [2k,t], will be passed in the following. For chloride, bromide and iodide, the pitting potential $\varepsilon_{\text{pit}}$ then is the particular position of $\varepsilon$, at and below which the rate of formation of hemispherical pits reduces to zero, and where all active pits grow with the same true current density, $I(\varepsilon_{\text{pit}})$. This particular value of $I$ is explained as the minimal, critical current density of aluminum dissolution, $I(\varepsilon_{\text{pit}}) = I_{\text{crit}}$, required to prevent re-passivation of the growing pit.

The further assumption was that in chloride solutions the hydrated aluminum-halogen compound is $AlCl_3 \cdot 6H_2O$, and $(a_{X^-})_{\text{tip}}$ is the chloride activity in a solution saturated with respect to this compound. This, however, is not a necessary, but rather a conveniently arbitrary choice to express the more basic supposition of a well defined constant critical property of the pit bottom solution, still to be more closely investigated.

An essentially similar proposition is to write

$$\varepsilon_{\text{pit}} = (\varepsilon_{\text{corr}})_{\text{satd.AlCl}_3} + \pi_{\text{Al}}(I)\ , \tag{12.16}$$

where the first term on the right side is the free corrosion potential of active aluminum in saturated aluminum chloride solution, and the second a 'polarization' term in the sense defined in Chap. 5, describing an anodic shift of the potential sufficient to enforce a dissolution current high enough to stabilize the concentrated pit electrolyte [23b]. Ohmic and diffusion contributions are disregarded, making the derivation rather more approximate. Clearly, the sums $E^\circ_{\text{Al/AlX}_z\cdot x\text{H}_2\text{O}} - (RT/F)\ln(a_{X^-}) + \eta_{\text{Al}}(I)$ in Eq. (12.15) and $(\varepsilon_{\text{corr}})_{\text{satAlCl}_3} + \pi_{\text{Al}}(I)$ in Eq. (12.16) are essentially equivalent. In [2l], the further refinement was to interpret $\eta_{\text{Al}}(I)$ as a nucleation overpotential, specifying, more precisely than before, that $\varepsilon_{\text{pit}}$ is the potential, where micropitting, already occurring below $\varepsilon_{\text{pit}}$, surges, so that the current of anodic aluminum dissolution becomes continuous. This requires a sufficient increase in the micropit nucleation rate, $N(\text{s}^{-1})$, and an overlap of lifetimes, $\tau$, of single micropitting events. To substantiate this, Fig. 12.29 shows this surge, occurring at the pitting potential, from isolated microevents of crystallografic pitting to an overall roughening of the aluminum surface by a multitude of such events which overlap with respect to both number and to lifetime. Occasionally, isolated micropits appear far below $\varepsilon_{\text{pit}}$, but we skip this exceptional effect.

The main argument, continued presently, may appear incomplete, as we are now considering fine micropits instead of hemispherical gross pits. Still, the objection can be dropped, although crystallografic pitting was not originally observed inside gross pits. Observations (see Fig. 12.9) only revealed that the pit interior was rough and, thus, far from being electropolished. Gross pits can be produced, however, in neutral as well as in alkaline solutions, by a variant of the inhibition effect already described [2k,t]: It suffices to inhibit pit nucleation by either adding a massive amount of sulfate to the solution, for instance 1M $Na_2SO_4$ to 1 M NaCl, or else to reduce the chloride concentration to, e.g. $10^{-3}$ M

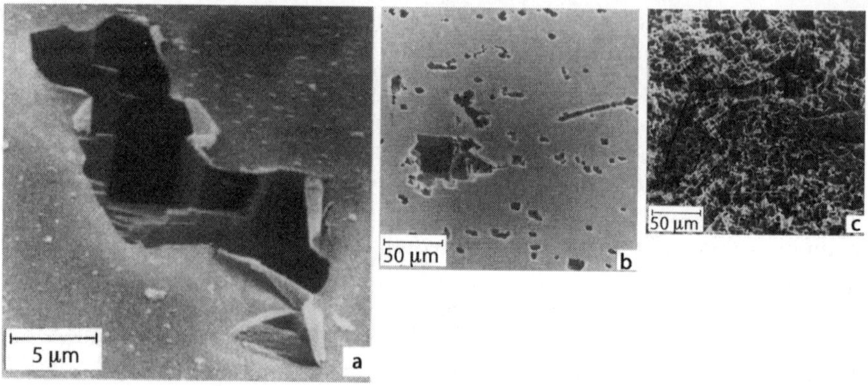


**Fig. 12.29a–c.** Pitted surface of pure aluminum after 24 hours potentiostatic anodic polarization in 1 N NaCl solution at potentials: **a** $\varepsilon_{pit} - 20$ mV; **b** $\varepsilon_{pit}$; **c** $\varepsilon_{pit} + 20$ mV. (From Baumgärtner and Kaesche)

NaCl. Crystallografic micropitting is then easily recognized in the interior of gross pits, and the conclusion that the same holds for alkaline solutions probably is convincing. This, then, would also suggest that gross pitting proceeds by intermittent growth of crystallografic micropits and tunnels, each growing for a limited time at a rate of, e.g., 10 A $cm^{-2}$, causing an overall rate of growth of gross pit at an average much smaller rate of, e.g, 0.3 A $cm^{-2}$. Once more, the situation may be more complicated as apparent on first sight, because we recall that micropitting in saturated $AlCl_3 \cdot 6H_2O$ solutions proceeds at a much lower rate than micropitting in low-concentration chloride solutions,including low-concentration aluminum chloride solutions,) by intermittent growth of both fine and short tunnels at a rate similar to the 0.3 A $cm^{-2}$ recorded for overall growth of gross pits in alkaline solutions. The point remains open to discussion, waiting for further experimental evidence. It will be seen presently that this is not the last of the open points.

We recall that during potentiostatic polarization in alkaline solution beyond $\varepsilon_{pit}$, both $v$ and $I$ were found to increase as approximately exponential functions of the potential. We must, therefore, specify that in principle both the nucleation rate of tunneling micropits and the propagation rate of tunnels are properties essentially different from $v$ and $I$. Nevertheless, we shall continue to label local current densities as '$I$'.

Returning to nucleation phenomena, we repeat that, as for surface roughening and tunneling during selective alloy dissolution, the statement of nucleation of events of massive local dissolution producing morphological instability of surfaces is trivial, and that the question only is how to properly apply nucleation theory. For selective alloy dissolution, this was comparatively easy, since nuclei are small, and the application of the Volmer type of theory therefore was unambiguous. For pit nucleation, the situation is rather more difficult [2i–u], Nuclei of any type of pitting or tunneling corrosion must be small holes through the passive oxide layer reaching into the metallic surface, and able to grow. They must be created by a local accumulation of atomic anodic oxide dissolution and subsequent metal dissolution events. Within the scope of any nucleation theory, details of electrode kinetics of this type of processes do not need consideration. Instead, the important issue is that further growth of a small hole is possible if the accompanying gain in Gibbs volume energy exceeds the increase in surface energy. Every nucleation event is considered accidental and subject, with respect to size and frequency, to Boltzman statistics. To simplify matters we consider, in the following, hemispherical instead of crystallografic nuclei with a radius, $\varrho$, and we refer to nucleation basics as described in Chap. 8. As in Eq. (8.17) The frequency, $N$, of pit nucleation is then proportional to the product of some basic frequency, $N_\circ$, and an an exponential term: $N = N_\circ \exp(-A_{\mathrm{nucl}}/kT)$, where $A_{\mathrm{nucl}}$ is the reversible work to create, at a given overpotential for nucleation, $\eta$, a hemispherical nucleus of critical radius, $\varrho_{\mathrm{crit}}$, still of much less than pinprick size but able to spontaneous further growth. As in Eq. (8.17), $A_{\mathrm{nucl}}$ then should be determined by the molar volume, $V$, of the nucleus, by the specific tension, $\sigma$, of its surface, and by the nucleation overpotential, $\eta$. Whith increasing nucleation overpotential the critical radius of the nucleus decreases, and so does $A_{\mathrm{nucl}} \sim 1/(\eta)^2$. Following similar lines of argument as for the description of the breakthrough potential for selective alloy dissolution, the equation relating the apparent current density $i_{\mathrm{Al}}$, of aluminum dissolution, related to the geometric electrode surface, and the overpotential is analogous to Eq. (8.18). Disregarding the negligible current of passive metal dissolution, it reads $i_{\mathrm{Al}} = const \times \exp(\alpha n_e F\eta/RT)\exp(-K/\eta^2)$. The argument of the second exponential term is $(-A_{\mathrm{nucl}})$. The physical nature of the nucleus is, however, somewhat difficult to see. The view of the authors in [2i,etc.] was that nucleation occurs at the bottom of pore-like flaws in the passive oxide film, as they already have been described as typical for oxide films on aluminum in Sect. 10.8. This would mean that nucleation involves the local breakthrough of the oxidized bottom of a pore, with immediate reaction of the bared aluminum with the halogen causing breakthrough, thus creating a nucleus with a halogenated aluminum bottom and aluminum oxide walls. In this case, the volume to be removed by the nucleation process would be mostly aluminum oxide. Also, the overpotential for nucleation is defined as the difference of the electrode potential and the equilibrium potential of the electrode reaction creating the nucleus: $\eta = \varepsilon - E_{\mathrm{nucl}}$. In [2i,j], the overpotential for micropit nucleation, $\eta_{\mathrm{nucl}}$, was set equal to the overpotential term in Eq. (12.15), since for the growth of single crystallografic micropits all electrode and transport reactions are the same as for multitude of such microevents causing gross pitting. If this is correct procedure, it introduces a change from the notion of $\varepsilon_{\mathrm{pit}}$ as a singular potential value to that of a narrow, but finite potential threshold marking the steep surge of nucleation events. The apparent threshold is then mainly due to the second of the two exponential functions governing $i_{\mathrm{Al}}$.

At this point, pulsed polarization experiments at two potentials, $\varepsilon_{\mathrm{high}} > \varepsilon_{\mathrm{pit}}$ and $\varepsilon_{\mathrm{low}} \leq \varepsilon_{\mathrm{high}}$, are interesting [2k,t], and the experiment documented in

Fig. 12.27 needs reconsideration: In this case, the propagation of the micropit under observation had been completely stopped by stepping the potential down to a value 0.45 V below $\varepsilon_{pit}$. On switching the potential back, the micropit started to grow again by the formation of a nucleus growing at a rate much higher than the rate of subsequent crystallografic steady-state propagation. The observation, thoroughly reproducible, showed that the momentarily 'dead' pit interior was the preferred site for pit re-nucleation. This, then, can have one of two reasons:

Either the entire micropit surface was truly repassivated by an oxide layer, which initially is, however, very thin. At the moment of downstepping the potential, the pit electrolyte, and especially the pit bottom electrolyte, certainly is still rich in dissolved aluminum chloride and, thus, in the pit-inducing chloride ions. Then, on the whole, pit nucleation inside the 'dead' pit is much more probable than outside, provided that the time of interrupted growth was too short for the oxide film to thicken and the pit electrolyte to dilute by diffusion and/or convection.

Or else there exists a state of pseudo-passivity different from regular oxide film passivity. The pseudo-passive surface layer may be an aluminum chloride layer, but other compounds, such as hydroxychlorides, offered by the complicated chemistry of concentrated solutions of aluminum salts, may also well merit attention.

So far, the first variant appears more readily acceptable, but then it will be seen presently that pit bottom conditions really may need considerations unusual within the scope of normal electrode kinetics. Before entering a discussion of this point, we record more observations, made by stepping down the potential to below $\varepsilon_{pit}$, but by only $\leq 0.1$ V. Now, any active micropit continues growth for some time, but the active propagation area shrinks, until after some time pit propagation comes to a complete stop. In the meantime, which is tenths of seconds, any switching back of the potential to values anodic to the pitting potential is accompanied by an increase in the active pit front. This implies that a micropit front previously de-activated can be re-activated without nucleation simply by expanding adjacent areas of active pit propagation. This, then, again points to a state of pseudo-passivation.

The directly observed local rate of crystal face propagation always is the same, irrespective of the potential below, at, or beyond $\varepsilon_{pit}$. We explicitly repeat that in a typical active micropit as shown in Fig. 12.27, where at any moment some crystal faces propagate, while others are inactivated either temporarily or definitely, the propagation rate always is the same everywhere, or else zero, regardless of the electrode potential. At any potential, including potentials positive to $\varepsilon_{pit}$, the lifetime of any such micropit is finite, and during life any micropit tends to taper to smaller overall width. It is, then, this lifetime which decreases, with decreasing potentiostatic potential. Therefore, it is the depth of micropits which limits lifetime, and this certainly points to resistance and diffusion as critical parameters. Since, now, in contrast to the parallel-walled very short microtunnels observed during hot-acid pickling described above, the comparatively large micropits growing in neutral solutions are irregular in shape,

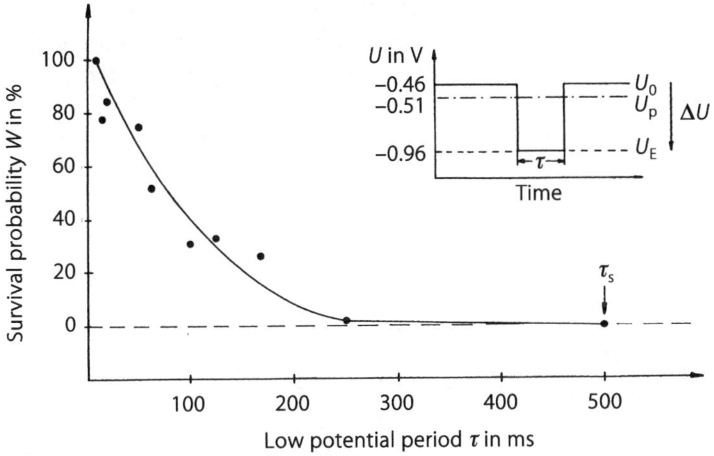


**Fig. 12.30.** The survival probability of microcrystallografic pits in pure aluminum in 1 N NaCl solution during periods of downstepping the potential to 0.45 V below $\varepsilon_{pit}$, as a function of step duration, $\tau$. *Insert:* Potentiostatic polarization program applied to the electrode. (From Baumgärtner and Kaesche)

mass transport is no longer simply one-dimensional, with pit shapes very much changing during pit growth. Therefore, a calculation of mass transport effects was not attempted in this work.

More insight limits of pit life was gained by determining the statistical survival probability as a function of the of the height of potential steps, $\Delta\varepsilon$, to below $\varepsilon_{pit}$ and of the duration, $\tau$, of these steps. This was possible by counting, during numerous single experiments with varying values of $\Delta\varepsilon$ and $\tau$, the number of electrodes with pits surviving the negative potential pulse. As an example, Fig. 12.30 shows the survival probability, $P$, observed after switching the potential to considerably below $\varepsilon_{pit}$, and it is seen that $P$ vanishes only after a limiting step duration value, $\tau_{lim}$, of between 0.3 and 0.4 V. More such experiments revealed that $\tau_{lim}$, while independent of $\varepsilon_{high}$, increases with $\varepsilon_{low}$, reaching ca. 10 s and more as $\varepsilon_{low}$ approaches $\varepsilon_{high}$.

Thus, $\varepsilon_{pit}$ is a threshold not for the surge of the micropit nucleation rate $N$, but for a surge of the survival probability as well. This, then, points to a relation between $N_{nucl}$ and $P$, which remains to be understood. The combination of the two effects causes the steady-state, well defined reproducibility of $\varepsilon_{pit}$.

For further insight, we return to nucleation in chloride solutions. As described previously, an acceptable supposition is that bottoms of rapidly growing shallow nuclei are immediately chlorided, while the pit solution still has the bulk composition, and while the ohmic voltage across in the pit electrolyte is still negligible. For this case the terms $\Delta\varepsilon_{Ohm}$ and $\Delta\varepsilon_{diff}$ in Eq. (12.14) cancel, and the argument $I$ in the overpotential function $\eta(I)$ is very high, as described previously. Also, the pit bottom halide activity now is still the same as the bulk activity. This implies that, while the pit electrolyte is not saturated with respect to the halide deposit on the metal surface, the exchange of the halide ion between surface layer and solution is in equilibrium. Clearly, this is reminiscent of the situation of passive iron in acid solutions, where the oxide is not in equilibrium with the solution,but where equilibrium does exist with respect to transfer of hydroxyl ions.

For completeness we note that the nucleation concept outlined presently does not explain the occasional occurrence of isolated crystallografic pits as far as 1 V below $\varepsilon_{pit}$ [2u]. From what will be described, further below, for pitting of alloys, this rather points to occasional low-potential pitting not of pure aluminum, but of foreign atom clusters in the surface of aluminum.

Concerning absolute values of $\varepsilon_{pit,}$ an attempt to arrive at a plausible value for the threshold potential by estimating the magnitude of $N_{\circ}$, V, and $\sigma$, and by assuming the presence of the regular hexahydrate of aluminum chloride is described in [2i]. This we leave for future re-discussion, turning, instead, to considering the more readily accessible changes in $\varepsilon_{pit}$ accompanying changes from chloride to bromide and iodide solutions, reported further above. In the first place, these are too large to be explained by changes in the surface tension of pit nuclei. Instead, the observations again point, as for pitting in alkaline solutions, to a shift in $E^{\circ}_{AlX_z \cdot xH_2O}$, whatever the values of $z$ and $x$ really are. Also as previously, the shifts in $\varepsilon_{pit}$ with changes in the bulk halide concentration are well explained by the Nernstian term $(RT/F)\ln a_{X^-}$ in the equilibrium potential.

When the very rapid initial growth of nuclei slows to crystallografic propagation rates still of several A $cm^{-2}$, both resistance and diffusion contributions to the electrode potential should increase. For galvanostatic experiments, where the external electrode potential always remains practically constant and equal to $\varepsilon_{pit}$, it follows that the pit bottom potential drifts to values more negative than $\varepsilon_{pit}$ by an increase of the sum $(\Delta\phi_{Ohm} + \Delta\phi_{diff})$. This, then, explains the end of the life of any micropit: Pit growths ceases, whenever the pit bottom potential, $\varepsilon_{bottom}$ drops to below a a critical value.

As described further above, the finite length of tunnels growing during hot-acid etching of aluminum has the same cause: Tunnel life was supposed to end whenever the tunnel bottom (or tunnel tip) potential drops to below a 'repassivation potential, '$\varepsilon^{\star}_{repass}$, again by an increase in the sum of ohmic and diffusion contributions to the external electrode potential, $\varepsilon_{pit}$. In [35g], the concept was that the diffusion boundary layer occupies all of the length of a microtunnel, and that the increase in tunnel length is accompanied by an increase in the concentration of dissolved aluminum chloride at the tunnel botom with respect to dissolved aluminum chloride. By this effect, $\varepsilon^{\star}_{repass}$, drifts to increasingly negative values only until saturation of the tunnel bottom electrolyte is reached. A further increase in $\Delta_{Ohm}$ then causes the tunnel bottom potential to drop below $\varepsilon^{\star}_{repass}$, and tunnel growth comes to its stop.

In [2k] the concept, again applied to chloride solutions, is somewhat different: Again, the bottom chloride concentration, and, thus, $\Delta\phi_{diff}$ as well as well as $\Delta\phi_{Ohm}$ are thought to increase with time, but this increase is irregular owing

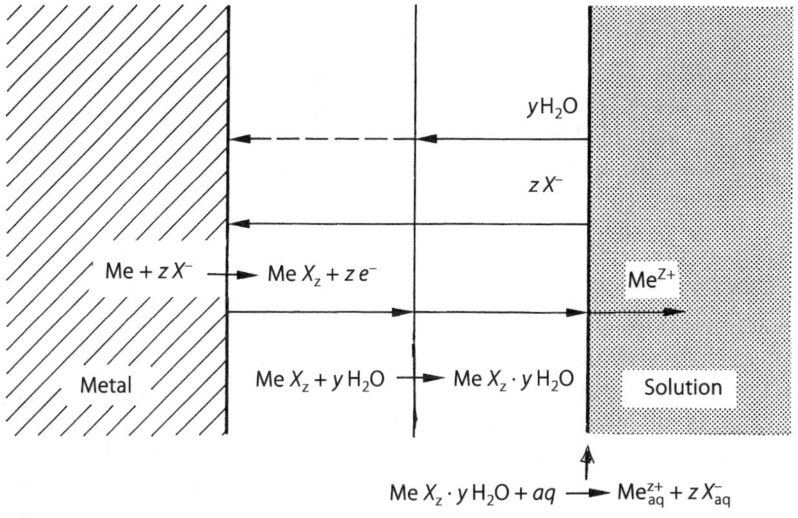


**Fig. 12.31.** Modeling the bottom of an active micropit or microtunnel in a metal such as aluminum subject to pitting in chloride (or other halide) solution, assuming a non-equilibrium metal salt film of varying degree of hydration, also considering cathodic hydrogen evolution at the metal/film interface. (From Beck)

to the irregular micropit shapes. The observation, evident e.g. in Fig. 12.27, of discontinuous processes of crystallografic pit front propagation steps may well be explained by local concentration effects in the micropit electrolyte, but this point is not, at present, well understood. Possibly, occasional events of local crystallization from supersaturated solutions are important. At the same time, the systematic tapering of tunnels, together with observations during potential down-stepping, rather more convincingly point to a critical increase in $\Delta_{\mathrm{Ohm}}$ eventually causing the drop in micropit-tip potential to below what again is the repassivation potential. This, however, now has a definite explanation: It is the potential described in Eq. (12.14), as applied to $\varepsilon_{\mathrm{pit}}$ and $I_{\mathrm{crit}}$.

We already have mentioned 'pencil electrode' experiments as attempts to study macroscopic models of pitting by observing the rapid anodic dissolution of a wire embedded in an inert rod. Experiments of this type have also been reported for high-current anodic aluminum dissolution into chloride solutions in [35h]. From direct observation, and from additional impedance measurements, a picture of the pit bottom behaviour as in Fig. 12.31 was outlined: According to this model, which may well represent the state of an active pit bottom, the active metal surface is covered with a metal salt film assumed to be duplex, dehydrated inside and increasingly hydrated outside. At the metal/film interface, the exchange current density of metal dissolution and metal redeposition into the film is high. Metal ion transfer into the film may even be in equilibrium, similar to the equilibrium at metal/oxide films postulated for passive systems.

If so, the Galvani potential difference, $\phi_{Me/film}$, also has an equilibrium value, and the overpotential so far is zero. Metal ions migrate through the non-porous dehydrated salt layer by a high-field mechanism, as also described for passive metals further above, and through the porous hydrated outer layer by a low-field ohmic mechanism. The overall potential difference across the duplex film is $\phi_{film,fim}$. Finally the metal ions pass the film/solution interface and the Galvani potential difference, $\phi_{film,sln}$. From here, further transport through the pit interior and out of the pit into the bulk electrolyte is by electrolytic migration and diffusion as usual. It remains to integrate, into this picture, the experimental evidence of hydrogen evolution. The proposal is to assume diffusion of hydration water through the film to the surface, followed by cathodic reduction of water at the metal/film interface producing dissolved molecular hydrogen, followed by diffusion of molecular hydrogen back through the film into the pit solution and, eventually desorption into the gas phase. We note that in a pencil electrode experiment evolution of gaseous hydrogen efficiently stirs the pit electrolyte, whereas the evolution of bubbles with a size exceeding that of micropits or tunnels should block pit propagation. Hence the assumption in [35g] that bubble formation occurs only outside the microtunnels growing during hot-acid etching.

If this attractive line of thought is applicable to micropitting of aluminum in chloride solutions, it implies that

$$E_{AlCl_z \cdot xH_2O} - (RT/F)\ln(a_{Cl^-})_{tip} + \eta(I) \tag{12.17}$$

$$= const + \phi_{Me/film} + \phi_{film,film} + \phi_{film,sln} \,,$$

where *const* is the contribution to the electrode potential from the reference electrode. Formally, the overpotential is $\eta = \phi_{film,film} + \phi_{film,sln}$, and it is left waiting for further specification as to origin and magnitude. It is, at any rate, no longer a nucleation overpotential. We therefore have here two distinctly different descriptions of active micropit bottom propagation, and pit nucleation in the passive outer surface, respectively.

Hydrogen evolution accompanying micropitting in neutral solutions has been extensively studied for pure aluminum and for aluminum alloys with copper, zinc, and magnesium [2q]. As already noted further above, the true current density of aluminum dissolution at active fronts observed at the pitting potential, $I(\varepsilon_{pit})$, always has approximately the same value of between several and ca. 10 A cm$^{-2}$, as evaluated from direct observation of moving pit fronts in ca. 1 s intervals. Disregarding possible short pulses of even higher current density in pure aluminum, we therefore have the observation of always essentially the same rate of Al dissolution regardless of alloy composition and chloride concentration of the bulk electrolyte. To this we now add that, as established by manometry of hydrogen evolution during pitting, the ratio $i_H/i_{Al}$ of apparent current densities of hydrogen evolution and aluminum dissolution also always is almost the same, with values varying between ca. 0.1 and ca. 0.2. The conclusion that the same applies to the ratio $I_H/I$ of true local current densities at active pit fronts is probably convincing. It is, then, important to recognize that this implies true local current densities, $I_H$, of up to ca. 1 A cm$^{-2}$ which, for hydrogen evolution, are quite enormous. If this is the rate of an independent cathodic partial reaction superposed to the partial reaction of metal dissolution, a rough estimate of the required cathodic overpotential at the pit bottom, $\eta_H = \varepsilon_{bottom} - E_{H_2/H^+}$, is that, probably, $\varepsilon_{tip} \lesssim -1.7$ V. This, then, also implies a considerable voltage drop across the salt film covering the active pit bottom.

Both $I$ and $I_H$ are approximately constant, or else variations of these current densities are small as compared to what would be normally expected for exponential current-potential functions. Of course, conditions may not be normal in the sense of regular electrode kinetics. Still, it may also be considered that diffusion of some reacting species is rate-determining, because then the influence of potential on kinetics is much less, or in the limiting case even vanishing.

Concerning hydrogen evolution, the supposition then would be that the diffusion of neutral water molecules through the salt film is rate-determining, probably in the limiting case of zero water concentration at the metal/film interface. Minor variations of $I_H$ by a factor of up to two would then be explained by variations in the composition and thickness of the salt film. This also offers an explanation of repassivation: Hydrogen evolution by water decomposition produces hydroxyl ions which, if not rapidly removed by electrolytic migration through the salt film, can react with aluminum to solid passivating hydroxide [35h]. This, by the way, would re-introduce local pH as a decisive parameter in pitting kinetics. As an alternative, we once more point to the possibility that aluminum ions, $Al^{z+}$, enter the film with varying valence, predominantly with z = 3, but also with z = 1 or z = 2, with subsequent chemical reaction with water molecules, and that hydrogen is produced in a subsequent homogeneous reaction of water with low-valency aluminum ions. At present, this aspect remains open to further investigation, and especially it is usual procedure to tentatively continue to describe hydrogen evolution kinetics in terms of potential-dependent transfer reactions. As shown presently, this also does make sense, considering the very negative pitting potentials of some alloys.

Concerning aluminum dissolution, the corresponding supposition is an anodic diffusion-limited current through the salt film. This, however, would require diffusion of an uncharged aluminum species rather than cations, at any rate as long as high-field ion migration through the inner film is operating.

In every case, conditions at the film/solution interface so far are not as well understood as at oxide/solution interfaces. We note, that this is a nearly nonaqueous phase in contact with the aqueous pit electrolyte, and this raises more questions as to electrode kinetics and transport processes. We also note that, similar to electropolishing films, the film can adapt to changing conditions with respect to potential and currents, by changing water content and thickness, all of which again are unknown properties in the present case.

For the few alloys, for which data on pit propagation rates are available, all showing values largely independent of either alloy or bulk electrolyte composition, the supposition of an always similar set of conditions for pit growth certainly makes sense. Apparently, the second component additions, all in the

**Table 12.1.** Galvanostatic stationary pitting potentials, $\varepsilon_{pit}$ (V, SHE), for aluminum and aluminum alloys in 1M halide solutions. (From Knörnschild and Kaesche)

| Metal | Al-3 wt.%Zn | Al-5 wt.%Ag | Al | Al-4wt%C u |
|---|---|---|---|---|
| Solution | | | | |
| NaCl | -075 | -0.59 | -0.51 | -0.36 |
| NaBr | -0.55 | -0.46 | -0.40 | -0.25 |
| NaJ | -0.28 | -0.36 | -0.23 | -0.11 |

range of only 1 or 2 at.%, make no important difference in the later stage of crystallografic micropitting, whereas the influence on the pitting potential, i.e. on the pit nucleation process, is quite pronounced. Concerning later-stage crystallografic micropitting, the tentative assumption then is that in this stage pit propagation characteristics always are essentially the same as for pure aluminum. If this is true, the further assumption is that by the high aluminum dissolution current the relatively very few second-component atoms are dragged along across the metal/film interface by a process more mechanical than electrochemical, irrespective of the nominal Gibbs energies of charge transfer. Thermodynamics then are satisfied by redeposition of these second-component metals in the growing pit, as is easily observed during pitting of AlCu alloys. Assuming usual charge-transfer mechanisms for anodic aluminum dissolution, the assumption also implies that the pit bottom electrode potential always is at essentially the same, and probably very negative level. The value of $\varepsilon_{bottom}$ then should be more negative than the most negative of all pitting potentials observed [2o]. Assuming, instead of rate-determining charge-transfer kinetics, a potential-independent anodic diffusion-limited current density, this constancy of pit-bottom properties would be relaxed; this, however, is highly speculative reasoning.

While for the alloys later-stage crystallografic micropit propagation rates always have similar values, nucleation rates of the same alloys vary very much, as indicated by important shifts of the pitting potential recorded in Table 12.1. We recall that, again in every case, a quasi-Nernstian dependence of $\varepsilon_{pit}$, on the halide concentration is observed, pointing to a nucleation mechanism always involving, as before, metal/halide and/or oxide/halide interaction. To explain the variations of $\varepsilon_{pit}$ with alloy composition we consider thermodynamics: Except for Mg and Li, which both are even less noble than aluminum, and in fact leave $\varepsilon_{pit}$ unchanged, second components of binary aluminum alloys are consistently more noble than the parent metal. During air exposure of the alloys, slow selective oxidation therefore should produce approximately pure aluminum oxide on a metal surface considerably enriched with respect to the more noble second component. This has, in fact, been confirmed by investigations into the electrolytic oxidation of AlCu alloys showing oxidation to pure aluminum oxide with zero copper content [2q,16k,l]. the unexpected complication observed in this work is that, in the alloy, copper atoms agglomerate to clusters, and it then depends on details of oxide growth kinetics whether these are embedded in the oxide, or else only enriched in the alloy surface beneath the oxide. When the oxide film is thinned in acid solutions, such copper clusters can penetrate the electrically isolating oxide, causing local conductor paths sufficient for efficient cathodic oxygen reduction. Apart from this particular effect, noble-metal enrichment in the aluminum alloy surface may well be responsible for the observed changes in $\varepsilon_{pit}$. Afterwards, micropitting or tunneling corrosion rapidly reaches a depth of attack where the alloy has the nominal, low second-component content, insufficient to greatly influence pit nucleation and growth as compared to pure aluminum. Thus an explanation for both the the strong influence of noble alloy components on the onset of pitting and the lack of this influence for later-stage pit propagation is offered.

More information concerning pitting of alloys certainly would be helpful, especially concerning local current densities of initial pit nucleation and of subsequent pit propagation. This applies to both the group of second-component metals such as In, Sn, or Hg, which depress $\varepsilon_{pit}$ to very negative, as well as to second-component metals such as Cr, Mo, W, etc. which shift $\varepsilon_{pit}$ to very positive values. Still one might tentatively again proceed to suppose that, during later-stage crystallografic pit propagation, $\varepsilon_{bottom}$ again always essentially is the same and that, therefore, $\varepsilon_{bottom}$ is more negative than the most negative of the values observed for $\varepsilon_{pit}$ [2q]. Owing to lack of data on details of dissolution mechanisms this is, as yet, inconclusive reasoning.

Concerning initial pit nucleation, the situation may appear less difficult: For aluminum and aluminum alloys, galvanostatic pitting potentials, $\varepsilon_{pit}$, and slowly recorded potentiodynamic pitting potentials, previously indexed as $\varepsilon_{pit2}$, are not much different. Thus, many numerical values for these potentials, for the wide range from AlIn alloys to AlW alloys are available. However, without more information on nucleation rates and propagation current densities, not much safe modeling is as yet possible. In the literature, various suggestions have been offered to explain the observed large shifts of pitting potentials. These include correlations of the pitting potential with the solubility of the second-component oxides [14i], or with the potential of zero point zero charge again of the second-component oxide [30], or with the second-component metallic cohesion [42]. For AlMo, AlCr, and AlTa alloys, a further tentative explanation is incorporation of $MoO_4^{2-}$,CrOOH, and $Ta_2O_5$ in the passive oxide causing reduced suscepitibility with respect to pitting by a variety of alternative reasons such as electrostatic repulsion of $Cl^-$, or oxide structure modifications, or barrier layers of the oxidized solutes at the alloy/film interface. The pronounced effect of tungsten is attributed to an ill defined synergystic interaction with the hydrated aluminum oxide structure [40e]. We leave these correlations without further discussion, referring the reader to the literature quoted above.

For direct practical applications, chromium oxide in aluminum oxide is particularly interesting in connection with the technique of adding sealing to anodic oxidation of aluminum, as briefly described in Chap. 10. By anodizing, the nucleation rate for pitting is depressed, and usual electrochemical experiments fail to lead to pitting. Still, the oxide coating has pore-like flaws, and sealing is

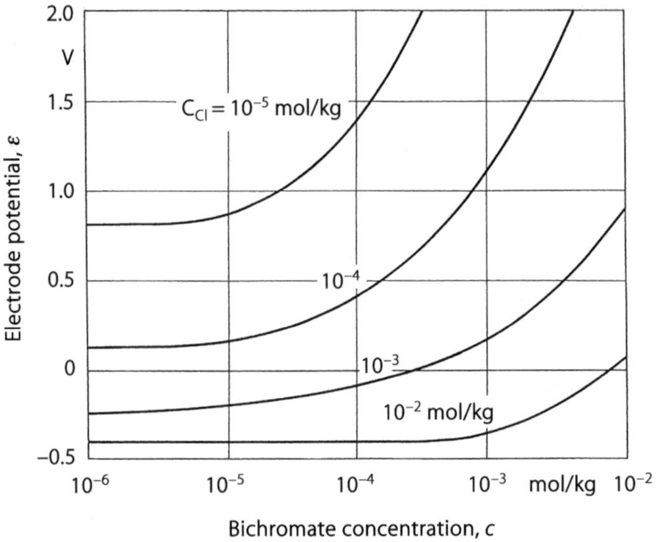


**Fig. 12.32.** The change in the potentiodynamic pitting potential, $\varepsilon_{pit2}$, of aluminum in $NaCl/K_2CrO_4$ solutions as a function of the chloride and the chromate concentration. The potential scan rate was 1 $Vh^{-1}$. (From Anderson and Hocking)

the added procedure to arrive at long-endurance surface protection, as demonstrated by many an aluminum facade of buildings. One particularly efficient method of sealing is 'conversion coating' with chromate solutions. An important application is protection of airplane fuselages. These must be periodically stripped of coatings, and it is this point where chromate has come into disrepute owing to high toxicity. For attempts to replace chromate compounds by, e.g. rare-earth compounds see e.g. [43]. Apparently, chromate is difficult to replace, and this is due to the specific nature of requirements. Inhibition of pitting of aluminum implies blocking the access of chloride through flaws in the oxide film. These also are sites of increased apparent electronic conductivity, and one interesting reaction requiring passage of electrons is reduction of dichromate to solid hydrated chromium oxides. By this effect, pore-like flaws can be decorated and thus made visible. At the same time, flaws are blocked by these processes, and this probably is the main merit of conversion coatings [31f].

We do not further describe details of the wide field of the technical applications of anodizing and sealing for large-scale corrosion protection of aluminum. In the present context, it is more interesting to note that the stationary pitting potential, $\varepsilon_{pit}$, as defined in this chapter, is almost unaffected by chromate additions [2f]. The inhibiting influence on pit nucleation kinetics is, however, readily recognized by rapid potentiodynamic current-potential scanning. An early instructive application of this method to record what is $\varepsilon_{pit2}$ is shown in Fig. 12.32. The scan rate $d\varepsilon/dt = 1\,\mathrm{V\,h^{-1}}$ was high enough to considerably shift the onset of pitting by overshoot of the stationary pitting potential. We note that, in contrast to the minute amounts of inhibitor substances required for inhibition of acid corrosion, the amount of inhibiting chromate necessary for pit nucleation inhibition typically is very large. Apparently, large concentrations of inhibiting chromate are required to replace chloride in an adsorption layer on the passive oxide.

## 12.5 More on Iron, Nickel, and Stainless Steel

For iron in alkaline solutions, the pitting potential was further above shown to be well defined by experimental evidence. Since it is independent of pH, it cannot be observed in acid solutions because it now is more negative than the the Flade potential. This is confirmed in [16o], by data obtained from a combination of galvanostatic and potentiostatic methods, including forward and backward potential scanning. In the same work, a variation of $\varepsilon_{\text{pit}}$ for chloride, bromide, and iodide solutions, quite similar to what already has been reported for aluminum, was observed, once more pointing to an influence of metal/halide bonding on the pitting potential. As for aluminum, fluoride plays a curious role, which we pass. Still with respect to iron, the same authors state that electropolished hemispherical pits are not observed except in the presence of excess sulfate. Otherwise, pits are described as always microscopic and polygonal, i.e. crystallografic. The authors particularly point out that the microscopically sharp border between pits and surroundings, where active and passive areas coexist, excludes any considerable difference of either Galvani potential or solution composition across the border [16d]. This, then, was the reason to exclude, for small and shallow such pits, ohmic voltages as well as concentration gradients, as already discussed previously. An alternative explanation of apparently sharp pit borders is that what really is observed is the edge of an overhanging passive film [37b]. As described below, oxidic covers of small pits are indeed often visible, but not, however, in the present case of pure iron. As also described above, the presence of a metal halide layer in the pit interior was postulated in spite of the non-saturation of the pit electrolyte with respect to the same halide. We now add that this was confirmed by ex-situ microprobing, i.e. X-ray emission analysis. Fig. 12.33 gives an example showing both the electron backscatter picture and the chloride coverage of a polygonal small pit [16m]. The film had a thickness of several nanometers, very similar to the thickness of the passive oxide film. It is from observations of this type that pit grows at this stage can be seen as a special case of 'quasi-passivition' by a non-porous, non-equilibrium salt film. This surface state differs from technically useful passivating oxide films by a 'quasi-passive' dissolution current density exceeding $i_{\text{pass}}$ by a factor $10^7$. Again, migration through the film is thought to occur by a high-field mechanism.

In chloride solutions with an excess sulfate content, the situation is different [16m]. Pits now grow to electropolished hemispheres, and ex-situ microprobing now reveals the presence of a much thicker film of, mostly, iron sulfate,

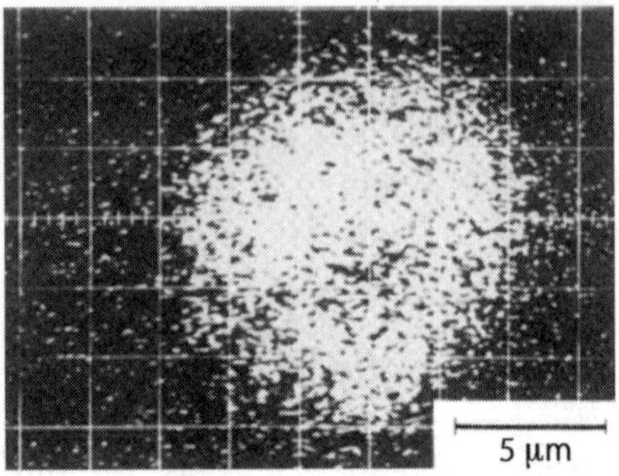


**Fig. 12.33a,b.** Ex-situ investigation of a small polygonal pit grown on pure iron in phtalate buffer/1 M KCl solution, pH 5, at +1.11 V. **a** Direct image obtained with backscatter electrons; **b** detection of chloride coverage by $Cl(K_\alpha)$ X-ray emission. (From Strehblow, Vetter, and Willgallis)

deposited from pit electrolyte saturated with respect to this salt. This, then, probably is the mechanism producing potentiostatic $t^3$-current transients and large voltage drops between pit mouth and metal surface beneath the film, as describe previously.

On both iron and nickel surfaces, the initial growth of small polygonal pits can easily be observed under a microscope [16j,n]. These are the measurements proving extremely high true values of the pit propagation current density, $I$, for the very initial stages of pit growth, referred to further above. The quite conspicuous difference of the values for iron and nickel is not, at present, well understood.

Also for nickel in slightly acid solutions, we note in the first place that the existence of the pitting potential in the sense used in this text has been unambiguously established by galvanostatic polarization experiments [16p]. The apparent non-polarizability of pitted electrodes, caused by a constant value of $I$ and an active pit surface, $A_{pit}$, adapting to the apparent current density, $i$, was confirmed as well.

For a review of many more details of pitting of passive iron and nickel, see [16o]. Here, we only add a general note: Since iron, nickel, and FeCr and FeCrNi alloys as well, all are passivatable in acid solutions, where the passivating oxide films are non-equilibrium species, the stability of active pits cannot be explained by local acidifying causing dissolution of the passivating oxide. Instead, the important parameter apparently is the concentration of the pit electrolyte with respect to metal halide, keeping pit fronts active.

This also applies to stainless steel, whose susceptibility to pitting is one of the deficiencies of an otherwise excellent material. From the many papers describing investigations into this topic, we particularly quote [18c–f]. In this work, emphasis was on pitting of austenitic 18-wt.% 5Cr-8 wt.% Ni steel exposed to chloride solutions. According to the concepts developed in this series of papers, local breakdown of passivity is caused by local oxide film thinning. More

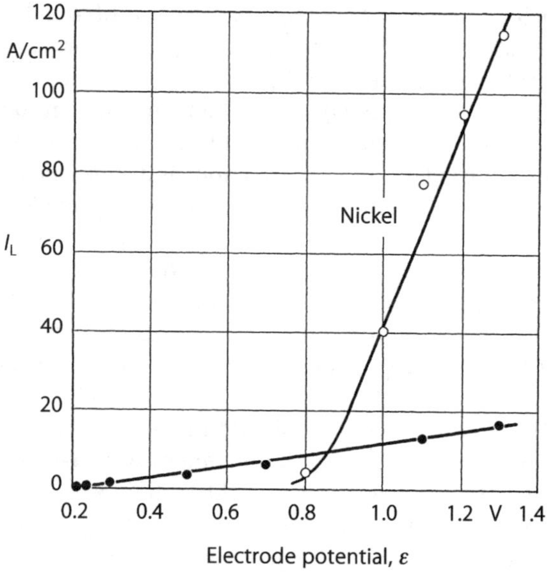


**Fig. 12.34.** Observing the growth of polygonal microscopic small pits on iron and nickel surfaces during the first second after pit initiation in phatalate/KCl solutions at pH 5. The ordinate gives values of the true local current density, $I$, as calculated from the rate of increase in pit diameter. The abscissa is the potentiostatic electrode potential. (From Wenners, Ives, and Vetter)

precisely, thinning proceeds locally by a mechanism similar to that of overall transpassive dissolution at more elevated potentials. The supposition is that the transpassivity threshold potential is locally shifted to more negative values by locally clustered chloride adsorption. The effect is believed to be due to quasi-metallisation of the oxide film surface by injection of electron donor states. Pits grow by the electropolishing mechanism under a solid salt layer deposit, and they are hemispherical. Repassivation occurs, when on lowering the potential the pit dissolution current density decreases, until it is insufficient to maintain the pit electrolyte at its critical composition. This repassivation potential, which in the notation used in this text is $\varepsilon_{pit3}$, decreases with increasing pit depth. At the same time, pit electrolyte pH is the important factor determining the position of the Flade potential of overall active/passive transition. When the Flade potential is more positive than $\varepsilon_{pit3}$, complete repassivation does not occur, but electropolishing pitting changes to what is described as 'active state pitting'. By this type of localized attack deep irregular recesses are growing into the metal, essentially similar to crevice corrosion as described further below. If, instead, $\varepsilon_{pit3}$ is more positive than the Flade potential, true repassivation does occur.

This view on pitting of stainless steel certainly is interesting. Still it may be noted that, normally, it will be 'active-state pitting' rather than elevated-potential electropolishing pitting, which is dangerous to the practical use of stainless materials. As a consequence, lower-potential pitting is the process probably needing most attention in corrosion engineering [9b].

Except for the very initial states of pit growths, it is almost trivial to recognize that the composition of the pit electrolytes is different from that of the bulk solution. Nevertheless, the subject has found much attention, and experimental evidence confirming the supposition is readily available. An early example is described in [9c]. In this work it was shown that pitting of stainless steel in $FeCl_3/NH_4Cl$ solutions can be suppressed by rotating suitably oriented specimens, because the concentrated pit electrolyte has a higher density than the bulk solution, and thus can be centrifuged out of the pit. As expected, growth of pits covered by an oxide lid continued unaffected by the drag on the pit electrolyte.

If pit grow by active-state metal dissolution, it should be useful to directly study electrode kinetics in solutions of the supposed pit composition. Evidence from this type of experiments is that it needs iron chloride solutions of up to 60% or more of saturation to keep the surface of stainless steel electrodes active [14j,45]. While this does confirm the necessity of highly concentrated solutions for maintaining activity, it still leaves open the question whether simple charge transfer kinetics really are rate-determining.

Many have been the attempts to devise macroscopic models of pits and crevices with the aim of insight into the chemistry of pit electrolytes. For a review see [46]. Here, we particularly refer to macroscopic electrode arrangements of the lead-in-pencil type, where the face of a thin wire embedded in an inert rod is completely activated by intense anodic polarization in a chloride bulk solution. After some recess of the face of the wire, the geometry is that of an active flat bottom of a passive tunnel. Experiments of this type have been reported for stainless steel in [37]. We quote interesting investigations into the composition of model pit electrolytes by in-situ energy-dispersive X-ray fluorescence microanalysis, available when a cyclotron white light source is at hand [37d]. The electrode was a thin metal sheet sandwiched between plastic covers. The model pit or, rather, model crevice was 1.6 mm deep, and it was scanned for composition as it passed a 50 μm-beam of high-intensity white X-radiation. A salt deposit was detected on the metal surface and found to be depleted, as compared to the bulk alloy, with respect to both nickel and chromium. While this qualitatively corresponds to expectations, the thickness of the deposit at the bottom of the slot may have been more than in real pits under otherwise similar conditions. Typically, anodic current densities in artificial pits tend to be lower than what is observed in real pits. The same authors also have been able to determine concentration gradients along the slot, again confirming what would be expected for a geometry of this type.

Lead-in-pencil model pits of much smaller dimensions have been studied in [37b]. In this work the aim was to establish, with respect to both electrode potential and pit electrolyte composition, critical conditions for the change from mass-transport controlled pit growth at relatively high, to allegedly charge-transfer controlled pit growth at relatively low potentials. Separating the two ranges in potential, a 'transition potential' threshold, $\varepsilon_{\mathrm{trans}}$, is defined. The desired information was sought by two-step experiments: In the first step, the embedded wire face is anodically polarized to complete activation and into the mass-transport control range, where $\varepsilon > \varepsilon_{\mathrm{trans}}$, e.g. $\varepsilon = 0.5$ V. The diffusion-limited current density, $i_{\mathrm{lim}}$ in the authors' notation, is estimated from Fick's first law to be inversely proportional to the depth, $L$, of the cavity: $i_{\mathrm{lim}} = zFDc_{\mathrm{sat}}/L$, where $D$,$c_{\mathrm{sat}}$, are effective diffusion coefficient and saturation concentration, respectively. In the second step, the potential is decreased at the high rate of 10 mV s$^{-1}$. During this negative potential sweep, the current remains constant and equal to $i_{\mathrm{lim}}$, supposedly as long as $c_{\mathrm{sat}}$ is maintained at the electrode surface. But the available stock of undissolved salt deposit is soon exhausted, because the supply of dissolved metal decays with decaying potential until, at $\varepsilon_{\mathrm{trans}}$, the metal surface is bare. At this moment, the current is still equal to $i_{\mathrm{lim}}$, decaying to lower values directly afterwards. The threshold potential, $\varepsilon_{\mathrm{trans}}$ is explained as the sum of three terms: The first is the free corrosion potential, $\varepsilon_{\mathrm{corr}}$, which would be observed in the saturated metal salt solution; the second is the polarization, $\pi$, i.e. the excess overpotential required to increase the anodic current density from $i_{\mathrm{corr}}$ to $i_{\mathrm{lim}}$, and the third is the ohmic voltage, $\Delta\phi_{\mathrm{Ohm}}$, in the pit electrolyte. Assuming simple charge-transfer control of anodic metal dissolution, the polarization is set at $\pi = const \times \ln(i_{\mathrm{lim}}/i_{\mathrm{corr}})$. The ohmic voltage is taken as approximately constant, independent of pit depth, because the ohmic resistance is the product of the average specific conductivity of the pit electrolyte and pit depth: $R = \varrho L$. Since $\Delta\phi_{\mathrm{Ohm}} = i_{\mathrm{lim}}R$, the factor $L$ cancels. This supposed constancy is confirmed by rapidly re-sweeping the potential in the positive direction, and recording what essentially is the polarization resistance of the active state system. Therefore, we have

$$\varepsilon_{\mathrm{trans}} = \varepsilon_{\mathrm{corr}} + const \times \ln(i_{\mathrm{lim}}/i_{\mathrm{corr}}) + const' , \qquad (12.18)$$

very similar in essence to the contents of Eq. (12.16), adapted to the present case. It follows that $i_{\mathrm{lim}}$ and $\varepsilon_{\mathrm{trans}}$ decrease with increasing depth, $L$, of artificial pits, similar to what has been predicted in [18]. The meaning of this result for real pits growing to irregular shapes in what again is active-state pitting remains to be more closely analysed. Again, below-transition pitting is thought to be inherently metastable, but real pitting may in fact proceed by frequent repetition of this metastable pitting processes, so these should not be underestimated as being of minor importance.

## 12.6 A Note on Crevice Corrosion

Narrow crevices in metallic structures, as well as crevices between metal surfaces and loosely covering deposits, naturally are a disadvantage with respect to corrosion, if only for the simple reason that any trapped moisture will evaporate only slowly. The effect is trivial, although very important as a consideration governing sound constructing techniques.

Considering crevices in fully immersed structures, it is no surprise that the crevice interior, although depleted in, e.g., dissolved oxygen, is not less subject to corrosion than the surrounding outside. Instead, galvanic differential aeration cell action between crevice interior and surroundings, making the crevice interior anodic, is well understood. In fact, *crevice corrosion* often is an example of spontaneous formation of macroscopic active/passive short-circuited galvanic cells. In a loose sense, it resembles pitting, and it is in this context that some comments are appropriate presently. For more general reviews of the subject, see [47].

Here, we mainly comment on differences between crevice and pitting corrosion. The first interesting point is that crevice corrosion is observed at potentials distinctly more negative than the pitting potential valid for the case under consideration, especially more negative than the steady-state pitting potential referred to as $\varepsilon_{pit}$ in this text. For aluminum, it has been shown that what occurs in narrow crevices exposed to neutral chloride bulk solutions is, in fact, not really rapid pitting [21]. Instead, it is a slow and irregular process of enhanced passive aluminum dissolution including the growth of microscopic shallow dimples, briefly described further above, which is dominant for an accelerated but still comparatively slow attack in crevices. These processes are autocatalytic in the sense that they tend, by enhanced aluminum dissolution to cations, which then hydrolyse, to locally acidify the electrolytic solution, so that oxide film thinning spreads inside the crevice. For pure aluminum, this requires external anodic polarization, or else contact corrosion by galvanic cell formation with contacting metals such as copper in an aerated solution. For commercial alloys with electron-conducting precipitates piercing the oxide film, the situation will be different, and spontaneous crevice corrosion is more probably. In any case, the appearance of aluminum surfaces undergoing crevice corrosion is the same as that of flat aluminum anodically polarized to potentials below the pitting potential in $AlCl_3$ solution. For more details, see also [35i,j].

A prominent example of a material susceptible to crevice corrosion is stainless steel. For this alloy, the transient state of alleged active-state pitting has already been described above, including its similarity to crevice corrosion. The subject has been treated in a series of papers emphasizing the properties of active/passive galvanic cells necessarily at work [19c–f]. This cell action causes an ohmic voltage in the crevice electrolyte between crevice bottom and crevice mouth, and the point of interest now is the depth of the crevice, where the potential drop lowers the electrode potential to below the active/passive transition threshold. Below this depth, the crevice is active, above, passive. Local acidification is no major argument in this context, since stainless steel readily passivates in acid solutions. Crevice solution pH nevertheless is important, as it determines the active/passive transition potential. The concept is quantified by appropriate solutions of the set of equations governing diffusion and migration, as presented further above, but we pass these details, which will reappear in the context of stress corrosion cracking further below. We note, however, that in this treatment of the subject the existence of an ohmic voltage is taken for

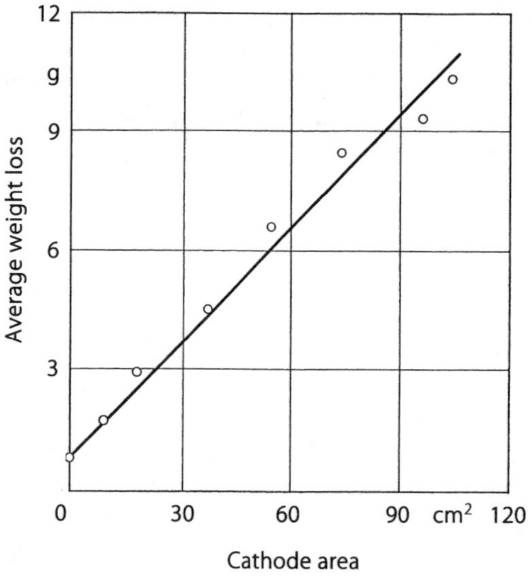


**Fig. 12.35.** Checking the validity of the oxygen catching principle in the case of model crevice corrosion of stainless steel (AISI 430) in seawater: Metal weight loss by corrosion of anodic artificial crevice vs. area of contacting cathodic open surface area. (From Ellis and LaQue)

granted, although initially, bevor local activation in the depth of the crevice has set in, no current flows and no voltage drop exists. Initiation processes, thus, remain outside these considerations.

Because the passivating oxide film on stainless steel is semiconducting, cathodic electrode reactions proceed, if only with increased overpotential, according to charge-transfer kinetics. Stainless passive surfaces outside crevices therefore act as efficient oxygen catching areas in aerated solutions, and crevice corrosion, once started, is sustained by galvanic cell action between anodic crevice interior and passive exterior. As demonstrated by data presented in Fig. 12.35, the weight loss by anodic metal dissolution in an artificial crevice was sometimes found to be directly proportional to the surface area of an adjoining contacting non-creviced portion of the same alloy [48]. There is an interesting statistical aspect of the matter: From many such creviced specimens, crevice corrosion started only on 22% of the total number in the case of steel AISI 316, and 58% in the case of AISI 304. Such differences in behaviour of single specimens are not surprising, in view of the complicated interplay of accidental processes of local passive film thinning in the crevice to occasional build-up of quasi-stationary quasi-active corrosion in the crevices tending to repassivation.

Crevice corrosion detrimental to the chemical process industry is not usually caused by differential aeration cells. A typical different mechanism is corrosion failure of titanium by crevice corrosion, which came as a disappointing surprise in the use of this metal with otherwise excellent pitting resistance. In this case, and especially at elevated temperatures, chlorine containing moisture was the active agent. Chlorine reacts with water to hypochloric acid and chloride, producing an unusually aggressive mixture, which nevertheless does not much attack titanium except in narrow recesses [49].

## References


1. a) Proc. Int. Conf. Localized Corrosion, Williamsburg, 1971 (Brown, B. F., Kruger, J., and Staehle, R. W., eds.). NACE, Houston, 1974; b) Proc. Int. Conf. Advances in Localized Corrosion, Orlando, 1987 (Isaacs, H., Bertocci, U., Kruger, J., and Smialowska, S., eds.). NACE, Houston, 1990; c) Proc. Int. Symp. Pits and Pores: Formation, Properties, and Significance for Advanced Luminescent Materials, Montreal, 1997 (Schmuki, P., Lockwood, D.J., Isaacs, H.S., and Bsiesy, A., eds.). The Electrochemical Society, Penninton, 1997; d) Proc. 4th Int. Symp. Passivity and Breakdown of Passivity, Airlie, 1977 (Frankenthal, R.P. and Kruger, J., eds.) The Electrochem. Soc., Princeton, 1978; e) Corrosion Mechanisms in Theory and Practice (Marcus, P. and Oudar, J., eds.). Marcel Dekker, New York etc., 1998; f) Passivity of Metals and Semiconductors (Froment, M, ed.). Elsevier Science Publ., New York, 1983 *74* (1970) 1024 – g) Proc. US-German Conf. Passivity and Breakdown of Passivity, Nümbrecht 1987. Corr. sci.*29* (1989) Nrs. 2, 3; h) Proc.. 6th Int. Symp. on Passivity, Sapporo, 1989. Corr. sci. *31* (1990); i) Proc. Int. Symp. Corrosion Science and Engineering, Brussels 1989 ( Rapp, R.A., Gokcen, N.A., and Pourbaix, A., eds.). Rapports Cebelcor, vol. 157, RT 297, 1989; j) Proc. Conf. Corrosion Chemistry within Pits, Crevices and Cracks, Teddington 1984. (Turnbull, A., ed..), HMSO,London 1987; k) Corrosion Mechanisms (Mansfeld, F., ed.). Marcel Dekker, New York, Basel, 1987; i) Proc. Symp. Passivity and its Breakdown, Paris 1997 (Natishan, P.M., Isaacs, H.S., Janik-Czachor, M., Macagno, V.A., Marcus, Ph., and Seo, M., eds.). The Electrochem. Society, Pennington, 1998
2. a) Heidemeyer, J. and Kaesche, H.: Corr. sci. *8* (1968) 377; b) Kaesche, H.: loc. cit. Chap. 10 [27b]; c) id.: loc. cit. Chap. 10 [94f]; d) id.: Z. phys. Chemie N.F. *26* (1960) 138 (In (d), exchange the values given for $\varepsilon_{\mathrm{pit}}$ in chloride and in iodide solutions); e) id.: ibid. *34* (1962) 87 (In Eq. (12), replace $z_j$ by $z_j/|z_j|$); f) id.: Werkstoffe u. Korr. *14* (1963) 557; g) id.: loc.cit. [1a], p. 516; h) Pleva, J.: Doctoral thesis, Erlangen, 1976, see also [2g]; i) Kaesche, H., and Baumgärtner, M., in: loc. cit. [1i], p. 439; j) Roth, A. and Kaesche, H., in: Proc. 5th Int. AlLi Conf., Williamsburg 1989. (Sanders, T.H. and Starke, E.A., eds.). Materials and Component Engng., Birmingham, 1989, p. 1207; k) Baumgärtener, M. and Kaesche, H., in: loc. cit. [1j]; l) iid.: Corr. sci. *29* (1989) 363; m) ibid., iid. *31* (1990) 231; n) iid., in: loc. cit. [1b], p. 407; o) iid.: Werkstoffe u. Korr. *42* (1991) 158; p) Knörnschild G. and Kaesche, H.: Extended Abstracts Vol. 91-2, ECS Meeting, Phoenix (1991). The Electrochem. Soc., Pennington; q) iid., in: loc. cit. [1c], p. 163;

r) Knörnschild, G., Heldt, J., Mitterbacher, H.,and Kaesche, H.: Materials and Corr. *46* [1994) 572; s) Kaesche, H.: Habilitation thesis, Berlin, 1962; t) Baumgärtener, M.: Doctoral thesis, Erlangen, 1989; u) Knörnschild, G.: Doctoral thesis, Erlangen 1994

3. a) Schmuki, P., Fraser, J., Vitus, C.M., Graham, M.J., and Isaacs, H.S.: J. Electrochem. Soc. *143* (1996) 3316; b) Schmuki, P. and Böhni, H.: Electochim. acta *40* (1995) 775
4. a) Resch, G and Odenthal, H.: Korrosion 15. Verlag Chemie, Weinheim, 1962, p. 33; b) Hömig, H.E.: Werkstoffe u. Korr. *8* (1957) 321
5. a) Bäumel, A. and Engell, H. J.: loc. cit. Chap. 10 [27a]; b) Engell, H.-J. and Stolica, N.D.: Z. phys. Chemie N.F.: *20* (1959) 113; d) iid.: Archiv Eisenhüttenwesen *30* (1959) 239; d) Herbsleb, G. and Engell, H.-J.: Z. phys. Chemie *215* (1960) 167; e) iid.: Z. Elektrochemie *65* (1961) 881; f) iid.: Werkstoffe u.Korr. *17* (1966) 365
6. a) Ergang, R. and Masing, G.: Nachr. Akad. Wiss. Göttingnen *1* (1946) 62; b) Ergang, R. and Möhling, M.: Z. Elektrochemie em 56 (1952) 9
7. a) Frankel, G.S.: J. Electrochem. Soc. *145* (1998) 2186; b) Frankel, G.S., Scully, J.R., and Jahnes, C.V.: ibid. *143* (1996) 1834
8. a) Campbell, H. S.: loc. cit. [1], p.625; b) Kaesche, H.: Heizung, Lüftung, Haustechnik: *13* (1962) 332; c) Stichel, W., in: Korrosion in Kalt- u. Warmwassersystemen der Hausinstallationen, Bad Nauheim 1974. DDGM, Oberursel, 1975
9. a) Brauns, E. and Schwenk, W.: Archiv Eisenhüttenwesen *32* (1961) 387; b) Schwenk, W.: Corr. NACE *20* (1964) 129t; c) Fischer, W.R.: Techn. Mitteil. Krupp *17* (1959) 137
10. a) Kolotyrkin, Ya.M.: J: Electrochem. Soc. *108* (1961) 209; b) id.: Corr. NACE *19* (1964) 261t; c) Freiman, L.I. and Kolotyrkin, Ya.M.: Dokl. Akad. Nauk SSSR *154* (1963) 886; Corr. sci. *5* (1965) 199; d) Rozenfeld, I.L. and Danilov, I.S.: Corr. sci. *7* (1961) 387
11. a) Böhni, H. and Uhlig, H. H.: J. Electrochem. Soc. *116* (1969) 906; b) Hunkeler, F. and Böhni, H.: a) Werkstoffe u. Korr.*32* (1981) 129; *34* (1983) 593; c) Corr. NACE *37* (1981) 645, *40* (1984) 534; d) Böhni, H.: Localized Corrosion. In loc. cit. [1k]
12. a) Porter, F.C. and Hadden, S.E.: J. appl. Chem. *3* (1953) 385; b) Pryor, M.J. and Keir, D.S.: J. Electrochem. Soc. *102* (1955) 605, *104* (1957) 269
13. Carroll, W.M., Murphy, M., and Breslin, C.B.: Corr. sci. *33* (1992) 1495
14. a) Smialowski, M., Szklarska-Smialowska, Z., Rychik, M., and Szummer, A.: Corr. sci.*9* (1969) 123; b) Szklarska-Smialowska, Z., Forchhammer, P., and Engell, H.J.: Werkstoffe u. Korr. *20* (1969) 1; c) Stoffels, H. and Schwenk, W.: ibid. *12* (1961) 493; d)Herbsleb, G. and Schwenk, W.: ibid. *18* (1967) 685; e) Szummer, A., Szklarska-Smialowska, Z. and Janik-Czachor, M.: Corr. sci.*8* (1968) 827; f) Baker, M.A. and Castle, J.E.: Corr.sci. *98* (1992) 97; g) Ke, R. and Alkire, R.: J. Electrochem. Soc. *142* (1995) 4056; h) Szklarska-Smialowska, Z.: Corr. sci. *33* (1992) 1193; i) Inturi, R.B. and Szklarska-

Smialowska, Z.: Corr. sci. 34 (1993) 705; j) Mankowski, J. and Szklarska-Smialowska, Z.: Corr.sci. *15* (1975)

15. a) Macdonald, D.D., Sikora, E., Balmas, M.W., and Alkire, R.C.: Corr. sci. *38* (1996) 97; b) Chao, C.Y., Lin., L.F., and Macdonald, D.D.: J. Electrochem. Soc. *128* (1981) 1187; c) Macdonald, D.D.: ibid. *128* (1981) 1194
16. a) Strehblow, H.-H.: Werkstoffe u. Korr.*27* (1976) 792; *35* (1984) 437; b) Löchel, B.P. and Strehblow, H.-H.: ibid. *31* (1980) 353; c) Strehblow, H.-H., loc. cit. [1e], p. 201; d) Vetter, K.J. and Strehblow, H.-H.: Ber. Bunsenges. phys. Chemie *74* (1970) 1024; e) Löchel, B.P. and Strehblow, H.-H.: Electrochim. acta *28* (1983) 565; f) Strehblow, H.-H. and Löchel, B.P., in: loc. cit. [1f], p. 379; g) Khalil, W.S., Haupt, S., and Strehblow, H.-H.: Werkstoffe u. Korr.*36* (1985) 16; h) Prinz., H. and Strehblow, H.-H.: Corr. sci. *40* (1998) 1671; i) Newman, J., Hanson, D.N., and Vetter, K.J.: Electrochim. acta *22* (1977) 829; j) Strehblow, H.-H. and Ives, M.B.: Corr. sci. *16* (1976) 317; k) Strehblow, H.-H., Melliar-Smith, C.M., and Augustyniak, W.M.: J. Electrochem. Soc. *125* (1978) 915; l) Strehblow, H.-H. and Malm, D.L.: Corr. sci. *10* (1979) 469; m) Strehblow, H.-H., Vetter, K.J., and Willgallis, A.: Ber. Bunsenges. Phys. Chemie *75* (1971) 822 –n) Wenners, J.: Diploma thesis, FU Berlin 1971, quoted from Vetter, K.J., in: loc. cit. [1a], 240; o) Vetter, K. J. and Strehblow, H.-H.: Ber. Bunsen-Ges. phys. Chemie *74* (1970) 449; p) Strehblow, H.-H. and Wenners, J.: Z.phys. Chemie *98* (1975) 199; q) Strehblow, H.-H., in: loc. cit. [1e], p.
17. a) Hoar, T.P., Mears, D.C., and Rothwell, G.P.: Corr. sci. *5* (1965) 279; b) Hoar, T.P. and Jacob, W.R.: Nature *216* (1967) 1299; c) Hoar, T.P.: Corr. sci. *7* (1967 341); d) id.: Trans. Farad. Soc. *33* (1937) 1152; *45* (1949) 683; Disc. Farad. Soc. *1* (1947) 299; e) Edeleanu, C.: J. Inst. Metals 89 (1960)
18. a) Sato, N.: Electrochim. acta *16* (19717) 83; b) Sato, N., Kudo, K., and Noda, T.: ibid. *16* (1971) 1909; c) Sato, N., in: loc. cit.[1h], p 1; d) id.: J. Electrochem. Soc. *129* (1982) 255, 260; e) Corr. NACE *45* (1989) 354; f) id.: Corr. sci. 31 (1990), 1
19. a) Xu, Y., Wang, M., and Pickering, H.W.: J. Electrochem. Soc. *140* (1993) 3448; b) Pickering, H.W. and Frankenthal, R.P., in: loc. cit. [1a], p. 261; c) iid.: J. Electrochem. Soc. *119* (1972) 1297, 1304; d) Pickering, H.W. and Ateya, B.G.: J. Electrochem. Soc. *122* (1975) 1018; e) Pickering, H.W., in: loc. cit. [1b], p. 77; f) Xu, Y. and Pickering, H.W., in: Critical Factors in Localized Corrosion. (Frankel, G.S. and Newman, R., eds.). The Electrochem. Soc., Pennington, 1992
20. a) Weil. K.G. and Menzel, D.: Z. Elektrochemie *63* (1967)669; b) Verkerk, B.: in: Int. Conf. Radioisotopes in Scient. Research, UNESCO /BS/R1C44, quoted from Pryor, M.J.: loc. cit. [1a], p. 2
21. a) Heusler, K.E. and Fischer, L.: Werkstoffe u. Korr. *27* (1976) 550, 697, 780; b) Dölling, R. and Heusler, K.E.: in Proc. Int. Conf. Metallic Corrosion, Toronto 1984; see also Z. Phys. Chem. N. F.*139* (1984) 39; c) Nachstedt, K. and Heusler, K.E. : Electrochim. acta *33* (1988) 311; d) Heusler, K.E.: in loc. cit [1g], p. 131, [1h], 597

22. Tousek, J.: Corr. sci. *14* (1974) 251
23. a) Galvele, J.R., DeMicheli, S.M., Muller, I.L., DeWexler, S.B., and Alanis, I.L.: in loc. cit. [1a], p. 580; b) Galvele, J.R.: J. Electrochem. Soc. *123* (1976) 464
24. a) Szklarska-Smialowska, Z., Viefhaus, H., and Janik-Czachor, M.: Corr. sci. *16* (1976) 649; b) Janik-Czachor, M., Kaszcyszyn, S.: Werkstoffe u. Korr. *33* (1982) 500
25. a) Painot, J. and Augustinski, J.: Electrochim. acta *20* (1975) 747; b) Augustinski, J. and Painot, J.: J. Electrochem. Soc. *123* (1976) 841; Koudelkova, M. and Augustinski, J.: ibid. *124* (1977) 1165
26. a) Pou, T.E., Murphy, O.J., Young, V., Bockris, J.O'M., and Tongson, L.L.: ibid. *131* (1984) 1244; b) Jovancicevic, V., Bockris, J.O'M., Carbajal, J.L., Zelenay, P., and Mizuno, T.: ibid. *131* (1986) 2219; c) Bockris, J.O'M.: loc. cit. [1g], p. 291
27. a) Okamoto, G, and Shibata,T.: Corr. sci. *10* (1970) 371; b) Saito, H., Shibata, T., and Okamoto, G.: ibid. *19* (1979) 693; c) O'Grady, W.E. and Bockris, J.O'M.: Surf. Sci. *10* (1973) 249; d) Revie, R.W., Baker, B.G., and Bockris, J.O'M.: J. Electrochem. Soc. *122* (1975) 1460
28. a) Herbelin, J.-M., Barbouth, P., and Marcus, P.: J. Electrochem. Soc. *137* (1990) 3410; b) Herbelin, J.-M. and Marcus, P., in: Proc. Conf. The Application of Surface Analysis Methods to Environmental Materials Interactions. (Baer, D.R., Clayton, C.R., and Davis, G.D., eds.). The Electrochemical Society, Pennington, 1991; c) Marcus, P. and Herberlin, J.-M.: Corr. sci. *34* (1993) 1123
29. Pryor, M.J., in. loc. cit. [1a], pp. 2, 540
30. a) Natishan, P.M., McCafferty, E., and Hubler, G.K.: J. Electrochem. Soc. *133* (1986) 1061; b) McCafferty, E.: Corr. sci. *35* (1995) 481
31. a) Wood, G.C., Richardson, J.A., Sutton, W.H., Riley, T.N., and Malherbe, A.G.: in loc. cit. [1a], p. 526; b) Thompson, G.E., Xu, Y., Shimizu, K., Skeldon, P., Skeldon, M., and Wood, G.C.: in loc. cit. [1b], p. 47; c) Wood, G.C., Richardson, J.A., Abd Rabo, M.F., Mapa, L.B., Sutton, W.H.: in loc.cit. [1d], p. 973; d) Richardson, J.A. and Wood, G.C.: Corr. sci. *10* (1970) 313; e) Wood, G.C.: ibid.33 (1992) 1371;f) Brown, G.M., Shimizu, K., Kobayashi, K., Thompson, G.E., and Wood, G.C.: Corr. sci. *34* (1993) 2099
32. Garz, I., Worch, H., and Schatt, W.: ibid. *9* (1969) 71; b) Kesten, M.: ibid. *14* (1974) 665
33. Franck, U.F.: Habilitation thesis, Göttingen 1954; b) id.: Werkstoffe u. Korr. *11* (60) 401; c) Franck, U.F. and Meunier, L.: Comptes rs. *244* (1957) 2610
34. a) Greene, N.D., France, W.D., and Wilde, B.E.: Corr. NACE *21* (1965) 275; b) France, W.D. and Greene, N.D.: ibid. *24* (1968) 247
35. a) Beck, T.R. and Alkire, R.C.: J. Electrochem. Soc. *126* (1979) 1662; b) Wong, K.P. and Alkire, R.C.: J. Electrochem. Soc. *137* (1990) 3010; c) Beck, T.R.: Electrochim. acta *29* (1979) 485; d) Wiersma, B.J. and Hebert, K.R.: J. Electrochem. Soc. *138* (1991) 48; e) Tak, Y. and Hebert, K.R.: J. Electrochem. Soc.*141* (1994) 1455; f) Zhou, Y. and Hebert, K.R.: ibid. *145* (1998) 3100;

g) Hebert, K.R. and Alkire, R.: ibid. *135* (1988) 2146; h) Beck, T.R.: Electrochim. acta; *29* (1984) 485, *30* (1985) 725; i) Hebert, K.R. and Alkire, R.: J. Electrochem. Soc. *130* (1983) 1001, 1007; j) Alkire, R., Thomasson, T., and Hebert, K.R.: ibid. *132* (1985) 1027
36. Edwards, R.A.H.: loc. cit. [1b], p. 381
37. a) Newman, R.C.: Corr. sci. *37* (1995) 527; b) Laycock, N.J. and Newman, R.C.: Corr. sci. *39* (1997) 1771; c) Newman, R.C. and Isaacs, H.S.: J. Electrochem. Soc. *130* (1621) 1983; d) Isaacs, H.S., Cho, J.-H., Rivers, M.L., and Sutton, S.R.: ibid. *142* (1995) 1111
38. Straumanis, M.E.: J. Electrochem.Soc. *105* (1958) 284
39. Basow, V.P., Karapety, Y.A, and Shut'ko, A.P.: Ukrainskii Khimicheskii Zh. *39* (1973) 1096, quoted from [35b]
40. a) Moshier, W.C., Davis, G.D., Ahearn, J.S., and Hough, H.F.: J. Electrochem. Soc. *133* (1986) 1063, *134* (1987) 2677; b) Moshier, W.C., Davis, D.G., and Cote, G.O.: ibid. *136* (1989) 356; c) Moshier, W.C., Davis, G.D., Fritz, T.L., and Cote, G.O.: ibid. *137* (1990) 422; d) Shaw, B,A., Fritz, T.L., Davis, G.D., Moshier, W.C.: ibid. *137* (1990) 1317e); e) Shaw, B.A., Davis, G.D., Fritz, T.L., Rees, B.J., Moshier, W.C., and Moshier, W.C.: ibid. *1138* (1991) 3288
41. a) Keir, D.S., Pryor, M.J., and Sperry, P.R.: J. Electrochem. Soc. *116* (1969) 319; b) DeMicheli, S.M., in: 7th ICMC, Rio de Janeiro, 1978; c) Reboul, M.C., Gimenez, Ph., and Rameau, J.J.: Corr. NACE *40* (1984) 366; d) Tohma, K. and Kudo, H.: Boshoku-Gijutsu *34* (1985) 3; e) Breslin, C.B. and Carroll, W.M.: Corr. sci. *33* (1992) 1161, 1735
42. Vijh, P.M.: Corr. sci. *11* (1971) 161; *36* (1994) 1615
43. a) Mansfeld, F., Chen, C., Breslin, C.B., and Dull, D.: in loc. cit. [1i], p. 602; b) Hughes, A.E., Gorman, J.D., and Paterson, P.J.K.: Corr. sci. *38* (1996) 1957
44. Anderson, P.J. and Hocking, M.E.: J. Appl. Chem. *8* (1964) 352
45. Hakkarainen, T.: in loc.cit. [1j]
46. Fuji, C.T.. in loc. cit. [1a], p. 144
47. a) Rozenfeld, I.L. and Marchakkov, I.K., in: 2nd Int. Conf. Metallic Corrosion, New York 1963; b) Uhlig, H.H.: Mat. Prot. & Performance *12* (1973 42; c) Pourbaix, M., in: loc. cit. [1a], p. 12; d) Wilde, R.E.: in ibid., p. 342; e) Ijsseling, F.P.: British Corr. J. *15* (1980) 51; f) Turnbull, A.: Corr. sci. *23* (1983) 833
48. Ellis, O.B. and LaQue, F.L.: Corr. Nace *7* (1951) 362
49. a) Gleekman, L.W., in: loc. cit. [1a], p. 669; b) Cerquetti, A., Mazza, F., and Vigano, M., in: ibid. p. 661; c) Vincentini, B., Siaglia, B., Taccani, G.: Corr. sci. *15* (1975) 479

# 13 Intercrystalline and Intracrystalline Corrosion

*Intercrystalline corrosion* (IC) is preferential attack along grain boundaries into the interior of polycrystalline metallic materials. *Intracrystalline corrosion* is preferential attack into grains, along subgrain boundaries, or stacking fault planes, or other crystallografic irregularities.

In the context of intercrystalline corrosion, we pass, as not causing serious corrosive damage, theoretically interesting slow changes in crystallografic surface morphology owing to thermodynamic surface tension effects, as formation o grain boundary trenches with, in principle, well defined contact angles between adjacent crystal faces. This subject is reviewed in [1]. We also pass, though in contrast to the case of pitting, the details of nucleation of IC by breakdown of passivity along grain boundaries. Instead, we shall concentrate on corrosion propagation along grain boundaries into the material, owing to chemical and/or structural differences in microscopically thin zones along grain boundaries as compared to bulk grain material. As for pitting, the predominantly interesting cases are caused by active/passive short-circuit cell action, where grain boundary zones are active and grain faces are passive. In serious cases, the effect eventually is that of *grain disintegration* of the metal in question, with important or even total loss in strength caused by, on the whole, very little corrosion.

The term *intracrystalline corrosion* is coined in analogy to that of intercrystalline corrosion to indicate that now crystallografic chemical and/or structural inhomogeneities existing in the interior of grains cause selective attack by corrosion proceeding into grains, typically producing crystallografic attack. The effect must not be confused with selective dealloying as described in Chap. 8. The phenomenon is by far not as dangerous as intercrystalline corrosion, but similar in many respects.

The most notorious example of IC is that observed in stainless steels, i.e. carbon-containing iron base alloys with, typically, 18 wt.%Cr and 8 wt.% Ni. For the main properties of these materials, see Chap. 10. As manufactured, stainless NiCr steels usually have been solution annealed at $> 1000°C$ and cooled to room temperature in order to obtain a homogeneous austenitic, i.e. face centered cubic crystal structure. If afterwards accidentally reheated to, e.g., 600 °C they are found to be 'sensitized' with respect to intercrystalline corrosion. Ferritic stainless chromium steels are susceptible directly after cooling from the preceding annealing treatment. The typical appearance of intercrystalline corrosion is shown in Fig. 13.1 for a molybdenum bearing steel [2].

The influence of composition, heat treatment, and other factors on IC is the subject of a voluminous special literature, referred to in the standard text books quoted in Chap. 1. We also quote some of the original investigations into the subject in [3]. Also, the origin and progress of IC in stainless steels is well understood: Sensitation is the consequence of chromium depletion in

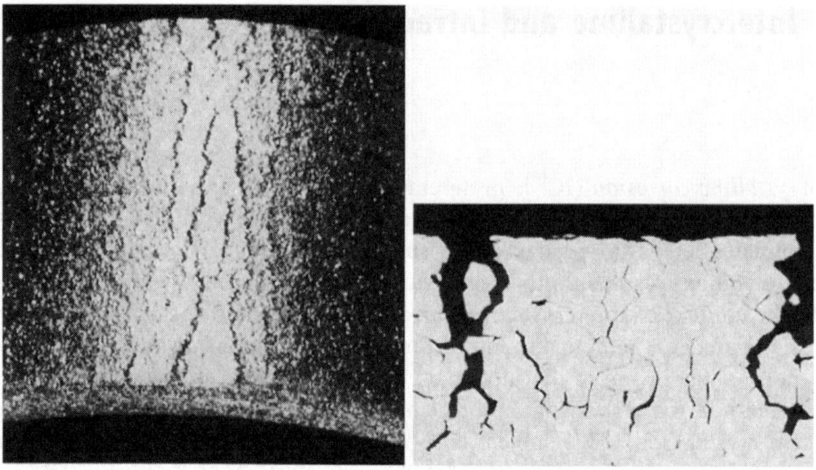

a b

**Fig. 13.1a,b.** Intercrystalline corrosion of sensitized CrNiMo stainless steel exposed to Strauss solution. **a** Macrophoto of a specimen bent after the test to reveal attack. **b** Metallographic section showing intercrystalline trenches. (From Herbsleb)

grain boundary zones caused by local precipitation of the mixed chromium iron carbide $(Cr,Fe)_{23}C_6$. This, in turn, is the consequence of thermodynamic instability of the homogeneous austenitic structure owing to supersaturation at room temperature with respect to this compound. At room temperature, carbide precipitation is negligibly slow, and no sensitizing occurs. When, however, during welding, parallel to weld seams the temperature drifts into the range where carbide solubility is still low, but precipitate nucleation now is rapid, carbide precipitation does set in. The preferred precipitation sites are grain boundaries, because the nucleation energy is lower at crystallografic interfaces than in the bulk of grains. This leaves the grain boundary zone depleted in chromium, which is the alloy component securing passivity.

From this mechanism, the influence of aging time during sensitation at any given temperature is obvious: After short times, chromium depletion remains negligible. After extended times, which is 'overaging', overall equilibrium is reached, with precipitates still present, but the chromium concentration leveled throughout the alloy. Then, depleted zones along grain boundaries no longer exist. Normally, the overall uniform chromium depletion will be low, and the alloy then remains passive as before. Only in an extreme case is the overall content in residual dissolved chromium so low that overall passivity of the alloy is lost. It is only in an intermediate temperature/time domain that IC sensitation is dangerous.

This *chromium depletion theory* explains observations quite sufficiently, and other factors such as stresses around carbide precipitates apparently are not important. The effect is thus the same as that which would be observed if the grain boundary zone depleted in chromium where replaced by a thin sheet of an

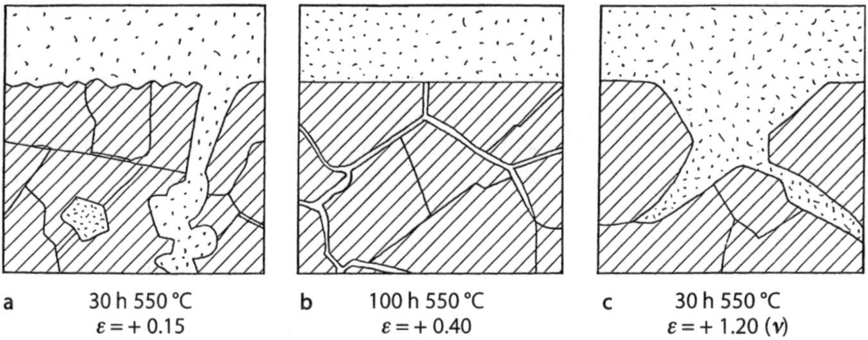


**Fig. 13.2a–c.** Intercrystalline corrosion of an 18 10 CrNi steel sensitized by heating to 550°C. The steel was exposed to boiling 2 N $H_2SO_4$ solution at controlled potentials in the range of active dissolution of both bulk grains and chromium depleted zones, **b** in the range of passive bulk grains and active chromium depleted zones, and **c** in the range of beginning transpassivity of both bulk and grain boundary material. The drawing is schematic, using pictures of metallographic sections obtained by Bäumel, Bühler, Schüller, Schwaab, Schwenk und Zitter

active metal such as zinc. In any case, the result is the action of active/passive short-circuited cells, whose internal ohmic resistance is low owing to the negligible distance between between activated anodic grain boundary zones and passive cathodic grain faces. Under these conditions, the difference in electrode potential between anodic and cathodic zones is negligible. According to this model, the common electrode potential of both grain boundary zones and grain faces should be in the range where chromium depleted steels are active, while grain faces, still having the nominal composition, are passive. In acid solution, such as 2 N $H_2SO_4$, this dangerous range of potentials of sensitized 18/8 CrNi steel is between 0.1 and 0.5 V [31]. This explains, why the common Strauss test is suitabe to detect IC sensitation: For this test, steel specimens suspected to be sensitized are exposed during hours to hot sulfuric acid containing dissolved copper sulfate, and also containing copper turnings. This mixture acts as a chemical potentiostat, fixing the corrosion potential in the critical region, because the specimens practically adopt the $Cu/Cu^{2+}$ equilibrium potential, and this happens to be in the critical region where chromium-depleted grain boundary zones are active and grain faces passive.

On closer inspection, conditions are rather more complicated. Figure 10.18 had already shown that an increase in overall chromium concentration causes a higher rate of anodic metal dissolution at every potential the range of transpassivity, and this includes the range of passivity. Grain-boundary chromium-depleted zones therefore should show some accelerated dissolution in almost any case. This is confirmed by experimental evidence schematically illustrated in Fig. 13.2 [31].

Under these conditions it does not surprise that a variety of tests, other than the Strauss test, are in common use to identify sensitation with respect to IC.

One is the nitric acid/hydrofluoric acid test, where the corrosion potential is ca. 0.1 V, another is the Streicher test with iron-III-sulfate/sulfuric acid at ca. 0.8 V, a third the Huey test in nitric acid between 0.8 and 1.3 V. A curiosity is the oxalic acid test applying anodic polarization up to 2.4 V, which is a range of potentials where any accelerated intercrystalline attack cannot well be due to chromium depletion of grain boundary zones. We do not further comment these testing methods, referring for details to a competent review in [3m].

In the context of pitting corrosion, we had taken occasion to comment on potentiodynamic current-potential scanning as the popular method to conveniently record what are more or less quasi-steady-state current-potential curves. If the aim is winning general characteristics of electrode processes, this type of investigation can be well used to obtain information rapidly, if only without deeper insight into mechanisms, and this includes a standard test suitable to identify IC susceptibility of sensitized stainless steels. The principle is that, when IC is due to chromium depletion in grain boundary zones, such zones still can be passivated by rising the potential beyond their active/passive transition or Flade potential. Sweeping the potential, this active/passive transition range should be detected as a rather high maximum in the quasi-potentiostatic current-potential curve. We recall from Chap. 10 that for non-sensitized stainless steel the passivating current density is almost negligible, so it does not show up much such during potentiodynamic experiments. The method has been used and found successful already in [4a,b]. Over the years, it has been elaborated, with variants such as backwards sweeping the potential from the range of passivity of both chromium-depleted and chromium-rich material, hence the designation as 'electrochemical potentiokinetic reactivation method' (EPR). The test is carried out in hot sulfuric acid, sometimes with additions of potassium rhodanide, either sweeping the potential from more positive to more negative potentials only ('single loop testing'), or else, as would seem recommended in any case, sweeping the potential back from activation of chromium-depleted zones back to their passivation ('double loop testing'). The test has gained widespread recognition since it was used to identify IC susceptibility of heat affected zones of weld seams in stainless steel (AISI 304) used as a structural material is nuclear power plants [4c]. For a review of the history of this test, see [4d].

If chromium depletion in grain boundary zones owing to chromium carbide precipitation is the important issue, then the obvious remedy is avoiding precipitation, and there are a number of possible solutions, all presupposing that an undesirable heat treatment of austenites cannot strictly be avoided. In the first place, sensitized parts, if small, may be re-homogenized by re-heating for a sufficient period to the temperature where the carbide precipitates re-dissolve. The second obvious measure is to change from carbon-containing to carbon-free alloys of otherwise similar composition. This is feasible, since in stainless steel carbon is not required for strength as it is in plain steel. Considering, finally, that it is dissolved carbon only which can react with chromium, the third and most usual method is to add several tenths of percent of titanium, or niobium, or tantalum already to the melt of carbon containing alloy. Because the carbides of these metals are almost insoluble already at the melt temperature, they precipitate in fine dispersion already during solidification, leaving almost no residual dissolved carbon available for detrimental later chromium carbide precipitation. These 'stabilized' austenitic stainless steels, with Ti, Nb, or Ta added in amounts of a small multiple of the nominal carbon content, are widely used whenever IC is imminent, or considered as not altogether improbable.

As for stainless steels, IC is a matter of concern for precipitation hardening aluminum alloys. For these, and different as compared to steels, it is the very hardening effect which is at the root of IC susceptibility. We describe, in the first place, IC in the binary alloy Al-4wt.% Cu, which as a homogeneous alloy is supersaturated with respect to the tetragonale intermetallic phase $Al_2Cu$. The equilibrium precipitation sets in at elevated temperatures, > 300°C, only, and hardening at room temperatures is caused by intermediates of the precipitation process, i.e. Guinier-Preston zones of type GP I, coherent with the host lattice, followed, at temperatures of up to ca. 200°C, by coherent precipitates of the $\Theta''$-phase, also labeled GP II zones, and of the no longer fully coherent $\Theta'$-phase. The details of the hardening mechanism are not our concern, since it is known that IC susceptibility is due to heterogeneous grain boundary precipitates of $Al_2Cu$, or else of a phase of similar composition [5]. Also, it was shown in this work that grain boundary zones are anodic with respect to grain faces. The very simple proof merits attention: The authors applied an isolating coating alternately a) to all grain faces excepting narrow strips along grain boundaries, and b) along such narrow strips only, and they found that, after critical degrees of aging, the corrosion potential in aerated chloride solutions of boundaries anodic was indeed anodic to that of grain faces.

As observed already in [6], $\varepsilon_{corr}$ for homogeneous AlCu alloys in aerated chloride solutions increases to more positive values with increasing copper content, and this probably reflects the increase of the pitting potential described in Chap. 11, although $\varepsilon_{corr}$ may still remain negative to $\varepsilon_{pit}$. For the intermetallic phase $Al_2Cu$, $\varepsilon_{corr}$ in 3% NaCl solution is anodic to the corrosion potential of pure Al by 0.16 V, and also anodic to homogeneous AlCu4 alloys. It follows that the observed polarity of the grainboundary/grain face galvanic cells cannot be due to grain boundary $Al_2Cu$ precipitates. Instead, the conclusion is that these precipitates rich in copper leave copper-depleted grain boundary zones essentially of pure aluminum, and the galvanic cell effect essentially is that of coupling pure Al with homogeneous AlCu4. In fact, this copper depletion of grain boundary zones caused by precipitation of $Al_2Cu$ has been confirmed by TEM combined with measuring electron energy losses [7].

So far we are left with the puzzle to explain why essentially pure aluminum in grain boundary zones should depassivate in contact with either the AlCu4 matrix or $Al_2Cu$ precipitates. This has been explained in [8], where it was shown that the effect is due to pitting of aluminum in grain boundary zones in contact with grain faces rich in copper. By the contact, the grain boundaries are polarized to their pitting potential, which is negative to that of the copper-depleted boundary

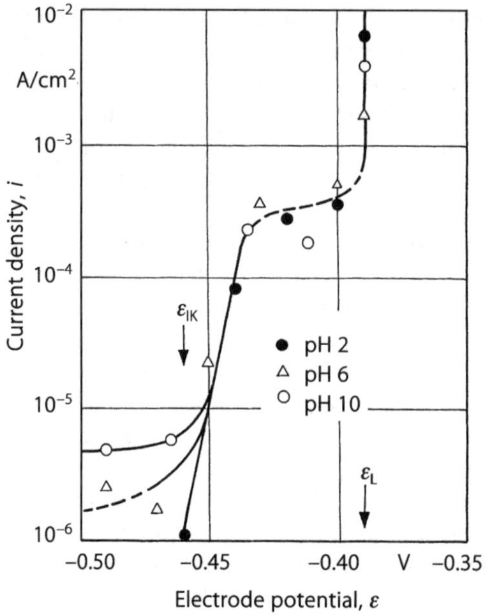


**Fig. 13.3.** Steady-state anodic current-potential curve for AlCu4, aged 3h at 180°C, in 1 M KCl solution. NaOH or HCl was added to adjust pH. Duration of galvanostatic polarization was ca. 20 h at each measuring point. (From Pleva)

zone. At the same time, the copper-rich grain faces are cathodically protected, polarized to a potential negative to their own more positive pitting potential.

Recalling what has been described in Chap. 11 for short-circuited galvanic cells, we need to explain, for IC without external polarization, that the internal resistance of the cells at work is low: In the first place, the ohmic resistance, $R$, between grain boundaries and grain faces is indeed low because the lateral distance between anodes and cathodes is microscopic. In the second place, the polarization resistance, $(R_\pi)_{an}$ of a local electrode dissolving by the pitting mechanism is practically non-existent. This leaves the polarization resistance, $(R_\pi)_{cath}$, of passivated homogeneous AlCu4, but this again is low, as demonstrated in Chap. 10 by the ease of cathodic oxygen reduction at this alloy, in contrast to reduction at pure aluminum.

The mechanism of IC of age hardening AlCu alloys by pitting concentrating in copper-depleted grain boundary zones is, thus, well established. Therefore, we may define a threshold potential for IC of aged AlCu which is practically equal to the pitting potential of pure Al: $(\varepsilon_{IC})_{AlCu} \simeq (\varepsilon_{pit})_{Al}$. If this is true, $(\varepsilon_{IC})_{AlCu}$ should be as independent of pH as $(\varepsilon_{pit})_{Al}$, and this is confirmed by data presented in Fig. 13.3 [9a], showing steady-state current-potential curves for aged AlCu4 in chloride solutions. This figure also shows a shoulder in the curve, similar to an anodic limiting current density, which apparently indicates the spreading of pitting attack over all of the susceptible grain boundaries. On increasing

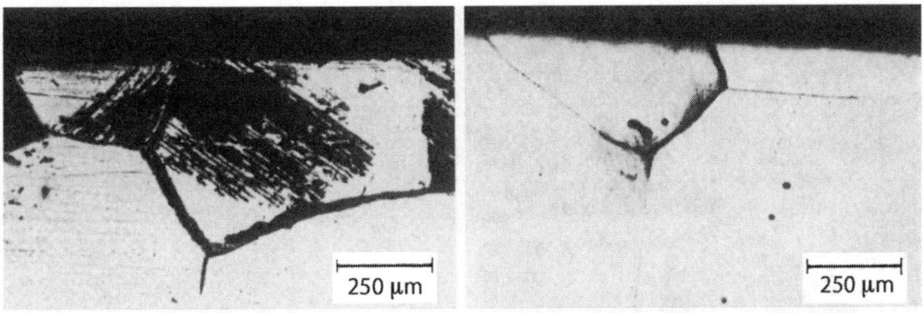


a b

**Fig. 13.4a,b.** Metallographic sections of corrosion attack in AlCu4, aged 3h at 178°C, after 30 min potentiostatic polarization in 1 M NaCl solution. **a** $\varepsilon = -0.35$ V, anodic to the pitting potential of both matrix and grain boundary zones. **b** $\varepsilon = -0.43$ V anodic to pitting potential of grain boundary ones but cathodic to pitting potential of matrix faces. Not shown: Complete absence of corrosive attack at $-0.66$ V, cathodic to the pitting potentials of both matrix and grain boundaries. (From Galvele and DeMicheli)

the galvanostatic current, the potential then passes through a second threshold, which we readily recognize as the more positive pitting potential, $(\varepsilon_{pit})_{AlCu}$, of the copper-rich alloy matrix. The following steep increase in current, i.e. the negligible polarizability, is typical for pitting corrosion as long as the pitting area has not yet spread over the total copper-depleted grain boundary seams.

In Fig. 13.4, the two types of local attack, IC of copper-depleted grain boundary zones and pitting of matrix faces of aged AlCu4, are well documented by metallographic sections of potentiostatically attacked specimens [8a]. Closer inspection of Fig. 13.4a reveals that the matrix attack, superimposed to IC at $\varepsilon_{pit}$, is of the tunneling type described in Chap. 10. At the same time it appears that in an aged AlCu specimen this tunneling attack somewhat differs from what was reported in this chapter for non-aged AlCu alloys. The differences may seem slight, still it is mainly this type of crystallografic attack which is coined 'intracrystalline corrosion'. We also remember structure-dependent selective dealloying described in Chap. 8, and we note that, on the whole, the overlap in mechanisms of pitting, IC, and dealloying, makes for quite a variety of morphological aspects of the matter. To further illustrate this point, Fig. 13.5 shows crystallografic intracrystaline attack of a quenched and air-stored specimen of AlCu4 after 24 h of anodic galvanostatic polarization in 1 M KCl solution [9a]. A further case showing that, essentially, intercrystaline and intracyrstaline corrosion of aged aluminum alloys are both governed by pitting, directed to either matrix faces, or to grain boundary zones, is that of AlZnMg alloys [9b,c]. Aged to equilibrium, the precipitate is the intermetallic $\eta'$-phase $MgZn_2$, and deposition of the precipitate depletes the matrix with respect to both Mg and Zn. As described in Chap. 10, as also reported in [8b], magnesium does not much affect the pitting potential, whereas zinc in solid solution considerably lowers this pitting threshold. Precursors of the equilibrium precipitate, decisive for age hardening, again are, in the first place, coherent Guinier-Preston zones, and in a

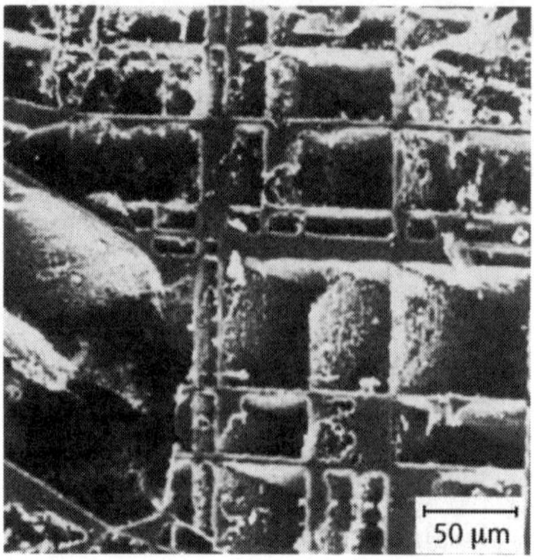


**Fig. 13.5.** Morphology of intracrystaline attack in AlCu4, quenched from solution temperature, aged in air at ambient temperature, after 24 h of anodic galvanostatic polarization in 1 M KCl solution. (From Pleva)

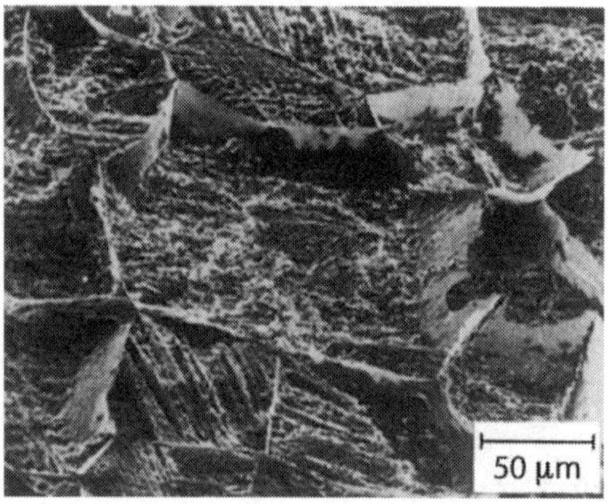


**Fig. 13.6.** Morphology of potentiostatic anodic attack of aged AlZnMg1 in 1 M NaCl solution, 30°C. The potential was −0.67 V. Heat treatment of the specimen: 1 h at 753 K, quenched in water, stored 7 d at room temperature, reheated 12 h to 368 K and 16 h to 18 K. (From Stoll, Hornig, Richter, and Kaesche)

second stage semi-coherent $\eta'$-phase precipitates. By suitable choice of duration of heat treatments stepped to temperatures below and above the Guinier-Preston solvus temperature, the distribution and concentration of the different types of precipitates can be varied with respect to the composition of grains and grain boundary zones, correspondingly varying conditions for intracrystalline and intercrystalline corrosion. The latter effect can be further modified by choosing heat treatment conditions favoring intracrystalline precipitates on certain preferred crystal face orientations. It even is possible to invert conditions such that grain boundary zones are depleted in Zn and Mg, leaving the matrix pitting potential, $\varepsilon_{\text{pit}}$, more negative than $\varepsilon_{\text{IC}}$. Then, as shown in Fig. 13.6 potentiostatic

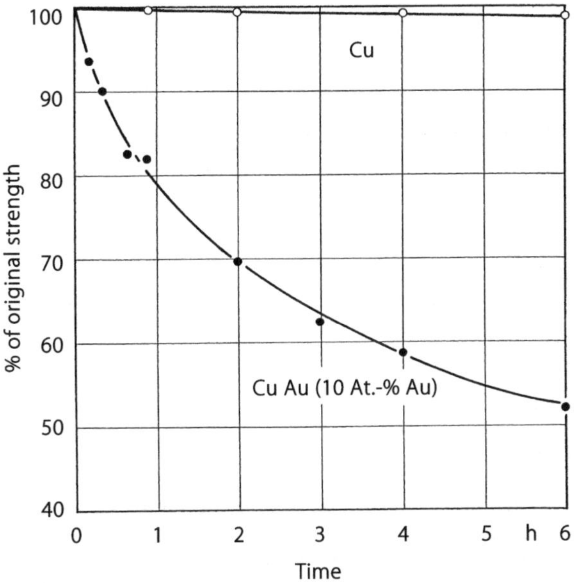


**Fig. 13.7.** Fracture load of tensile specimens after load-free exposure to 2% $FeCl_3$ solution as a function of exposure time. (From Graf)

polarization at a potential $\varepsilon_{IC} \leq \varepsilon < \varepsilon_{pit}$ produces intracrystalline corrosion only, leaving the grain boundary zones unattacked.

For the result of potentiostatic testing, the precipitates themselves are of little interest. This may change, however, for corrosion experiments without external polarization, i.e. for routine corrosion testing. In this case, precipitates may act as electrodes in galvanic microcells. For investigations in this direction, see [10].

IC of homogeneous rather than of precipitation hardening binary alloys has been the object of an extensive series of investigations concerning binary alloys [12] In this work, intercrystalline corrosive attack was characterized by the time to failure by fracture of specimens loaded, after exposure, to 2% $FeCl_3$ solution. The emphasis was on differentiating a 'static solid solution effect', which is IC, and a 'dynamic solid solution effect', which essentially is stress corrosion cracking (see Chap. 15), observed when specimens where exposed to the test solution during loading.

As shown in Fig. 13.7, the 'static' effect is pronounced, indeed, indicating a weakening of grain boundaries due to homogeneous alloying which, however, at the time remained without further explanation. CuAu alloys have also been studied in this context in [13]. A similar weakening of grain boundaries was found in many binary homogeneous alloys with a noble component thermodynamically stable with respect to anodic dissolution [12].

As usual, rather better insight into reaction mechanisms is available from controlled-current or controlled-potential measurements. As an example,

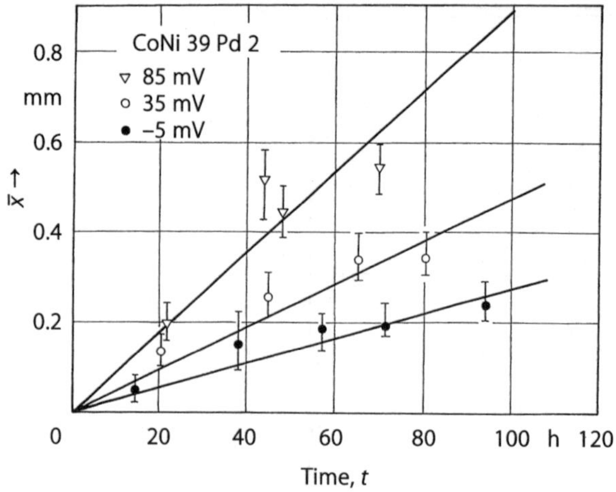


**Fig. 13.8.** Average depth of intercrystalline penetration of the homogeneous alloy Co-39Ni-2Pd (at.%) in 0.1 M NaCl solution as a function of time during anodic potentiostatic polarization to the indicated potentials. (From Kaiser and Kaesche)

Fig. 13.8 shows the increase in average depth of penetration of a homogeneous alloy CoNiPd alloy, where Pd is the thermodynamically immune component, in 0.1 M NaCl solution as a function of time during potentiostatic anodic polarization [9d]. The penetration rate increases with increasing potential, indicating a rate-determining charge-transfer electrode reaction at the tip of deepening intercrystalline trenches. The dependence on time at constant potential is approximately linear, or else at least not obviously parabolic, as would be required for rate-determining transport reactions from the tips of IC trenches to the specimen surface.

To quantify what so far has been loosely termed 'grain boundary weakening' the increase in misfit energy, caused by substitution of, e.g. Cu by Au atoms in solid solutions, was considered to be the important parameter [12c]. Grain boundaries where assumed to be sources of misfit dislocations produced during primary solidification of the alloy, later moving from boundaries into grain interiors, leaving slip steps in grain boundaries, increasing in density with increasing second component content of the alloy. Anodic metal dissolution can be expected to prefer ledges of slip steps, and IC was thus seen as the effect of anodic metal dissolution overpotential lowered in grain boundaries, with a maximum of IC susceptibility predicted at 50% of the noble metal component. For some comments on this concept, see+[14].

The results of investigations into the behaviour of homogeneous NiCoPd alloys rather complicate the picture [9d,e]. The system is useful in the present context, since it is susceptible to IC due to the Pd content, and since it allows to control the stacking fault energy and, hence, the stacking fault density, by changing the Ni/Co ratio: Pure nickel is face centered cubic, but pure cobalt is close packed hexagonal, and on increasing the content in cobalt, stacking

**Fig. 13.9.** TEM image of micropitting attack of stacking faults in homogeneous Ni-69Co-2Pd (at.%) after 39 min of anodic potentiostatic polarization to 0.085 V. (From Kaiser)

faults increase in density, until at ca. 60 at.% Co a plastic deformation suffices to induces a martensitic fcc→cph transition. As already described in Chap. 8, martensitic phases are particularly susceptible to corrosion. In the present case, as illustrated in Fig. 13.9, this is reflected in the preferential micropitting attack of stacking faults, which are precursors of martensite. The reason for the susceptibility of either stacking faults or martensites so far remains to be established, but a suggested mechanism for IC of alloys low in cobalt is preferential attack of stacking faults accumulating in grain boundaries.

## References


1. Gleiter, H.: Mat. sci. and engng. *52* (1982) 131
2. Herbsleb, G.: Schriftenreihe Schweißen und Schneiden *5* (1974) 3
3. a) Strauss, B., and Schottky ,H., and Hinüber, J.: Z. anorg. u. allgem. Chemie *188* (1930) 309; b) Huey, W.R.: Trans. ASST *18* (1930) 1126; c) Schüller, H.J., Schwab, P. and Schwenk, W.: Archiv Eisenhüttenwesen *33* (1962) 853; d) Ternes, H. and Schwenk, H.: ibid. *33* (1965) 99; e) Bäumel, A.: Werkstoffe u. Korr. *18* (1967) 289; f) Herbsleb, G. and Schwenk, W.: ibid. *19* (1968) 103; g) Herbsleb, G.: ibid. *19* (1968) 204, 406; h) Herbsleb, G. and Schwaab, P.: ibid. *19* (1974) 484; i) Herbsleb, G.: Archiv Eisenhüttenwesen *41* (1970) 875; j) Heumann, T. and Rockel, M.: ibid. *42* (1971) 111; k) Herbsleb, G.: Metalloberfläche *24* (1970) 195; l) Bäumel, A., Bühler, H.E., Schüller, H.-J., Schwaab, P., Schwenk, W., and Zitter, H.: Corr. sci. *4* (1964) 89; m) Herbsleb, G. and Schwaab, P.: Werkstoffe u. Korr. *37* (1986) 24
4. a) Čihal, V. and Pražák, M.: Werkstoffe u. Korr. *9* (1958) 517; b) iid.: Corr. NACE *16* (1960) 138; c) Clarke, W.L., Cowan, R.L., and Walker, W.L.: in ASTM STP 656. (Steigerwald, R.F., ed.) ASTM, Philadelphia, 1978, p. 99; d) Čihal, V., Desestret, A., and Wagner, G.H.: Materials and. Corr. *52* (2001) 7

5. a) Mears, R.B., Brown, R.H., and Dix, E.H.: in Symp. Stress Corr. Cracking of Metals. ASTM-AIME, Philadelphia and New York, 1945, p. 323; b) Dix, E.H.: Trans. AIME *137* (1940) 11; c) id.: Trans. ASME *42* (1950) 1057
6. Holub, L.: unpublished work, quoted in: Vosskühler, H.: Werkstoffe u. Korr. *1* (1950) 143, 310; *8* (1957) 463
7. a) Doig, P. and Edington, J.W.: Proc. Roy. Soc. London *A 399* (1974) 37; b) Edington, J.W. and Hibbert, G.J.: J. Microscopy *99* (1973) 125
8. a) Galvele, J.R. and DeMicheli, S.M.: in 4th Int. Conf. Metallic Corrosion. Amsterdam, 1969. NACE, Houston, 1972, p. 439; b) Muller, I.L. and Galvele, J.R.: Corr. sci.*17* (1977) 995
9. a) Pleva, M.: Doctoral thesis, Erlangen, 1976; b) Stoll, F., Hornig, W., Richter, J., and Kaesche, H.: Werkstoffe u. Korr. *29* (1978) 585; c) Richter, J., and Kaesche, H.:ibid. *32* (1981) 174; d) Kaiser, H. and Kaesche, H.: Z. F. Metallkunde *70* (1979) 582; e) Lenz, E., Kaiser, H., and Kaesche, H.: in Int. Conf. Mechanisms of Environment Sensitive Cracking of Materials, Guildford, 1977
10. a) Grauer, R. and Widmer, E.: Werkstoffe u. Korr. *31* (1980) 550; b) Mazurkiewicz, B.: Corr. sci. *32* (1983) 687; c) Mazurkiewicz, B. and Pietrowski, A.: ibid. p. 697
11. a) Proc. NATO Symp. Theory of Stress Corroson Cracking in Alloys (Scully, J.C., ed.). NATO Scientifique Affairs Divison, Brussels, 1971; b) Int. Conf. Surface Effects in Crystal Plasticity (Latanision, R.M., Fourie, J.F., eds.). NATO Advanced Studies Institutes Series E Nr. 17. Nordhoff, Leyden, 1973; c) Stress Corrosion Cracking and Embrittlement (Robertson. W., ed.). New Xork, Wiley; Chapmann & Hall, London, 1965
12. a) Graf, L.: Acta met. *6*(1958) 116; b) Graf, L.: in loc. cit. [11a], p. 399, and previous literature quoted therein; c) Frank, W. and Graf, L.: in loc.cit. [11b], p. 781; d) iid.: Z. Metallkunde *66* (1975) 555; d) Frank,W.: ibid. *71* (1980) 559
13. Robertson, W.D. and Bakish, R.: in loc. cit. [11c], p. 65
14. Engell, H.-J., Staehle, R.W., and Bullock, R.: in loc. cit. [11b], p. 790

# 14 Hydrogen Embrittlement

## 14.1 Introduction

An important danger to structural materials subject to tensile load is unexpected brittle, instead of usually predictable ductile, fracture, and a notorious variety is embrittlement by uptake of atomic hydrogen as a solute in the metal lattice. Since ductility of metals is caused by dislocations moving in stress fields, hydrogen embrittlement usually is an effect of interaction of absorbed hydrogen with dislocations. The subject and its implications are vast, and we shall not go into many details, referring the reader to competent reviews [2]. Hydrogen embrittlement is not, however, always caused by hydrogen/dislocation interaction. Atomic hydrogen, if dissolved to high concentration or high thermodynamic activity, also can recombine at internal surfaces to molecular gaseous hydrogen at high pressure, and the combination of the two effects tends to very much complicate the situation [3]. Therefore, the two limiting cases of pure effects: a) high-pressure internal molecular hydrogen and b) dissolved atomic hydrogen, will be briefly introduced presently, postponing the description of details to later sections. For the limiting cases, the main principles are, in fact, well understood and well documented [3–5].

In the context of electrolytic corrosion, the main issue is the initial uptake of hydrogen in metals, and especially in unalloyed and highly alloyed steels, during wet corrosion processes. An obvious case is acid corrosion of mild steel, proceeding by the superposition of anodic metal dissolution and cathodic hydrogen deposition. Hydrogen uptake then is from the adsorption layer on the metal surface, where adsorbed atomic hydrogen is an intermediate of the reduction of either $H^+$ or $H_2O$. When the acid contains adsorption inhibitors, as described in Chap. 7, the thermodynamic activity of adsorbed hydrogen atoms can reach excessively high values, as compared to metal surfaces in contact with dry hydrogen gas. In mild steel, this leads to the formation of blisters in the interior of the metal, owing to recombination of atomic to molecular hydrogen at a pressure corresponding to this high surface activity. In carbon steels of average strength, blistering tends to change to internal cracking, and this, then, defines the limiting case *hydrogen-induced cracking*, HIC, as a mechanism of cracking not requiring additional mechanical stresses either internal or external. The phenomenon not associated with the pinning of dislocations, or other hydrogen/dislocation interactions, and thus different, but not always recognized as different from other cases of hydrogen damage.

At high strength levels, and especially for high-strength steels, hydrogen embrittlement, now indeed by dislocation/hydrogen interaction, adds to the danger of cracking under the influence of mechanical stresses. The appropriate theoretical treatment then is *fracture mechanics*, as briefly outlined in the Appendix. The point of interest is crack propagation from the tips of stress-raising notches at stress levels below what in the absence of dissolved hydrogen would be the *fracture toughness* of the material. The crucial phenomenon now is the concentration of embrittling dissolved atomic hydrogen, whose chemical potential is decreased in the notch tip vicinity in the strained metal lattice. This thermodynamic effect drives dissolved atomic hydrogen into the very region of peak detrimental action. As very little dissolved hydrogen suffices to embrittle the notch tip region, severe electrolytic corrosion is not required for this damage mechanism, and it is, in fact, observed already in gaseous hydrogen at ambient temperature. A more practical case is that of notched high-strength steels in air, i.e. in the absence of gaseous hydrogen, but undergoing some, if possibly only minute, slight corrosion by reaction with moisture, producing adsorbed atomic hydrogen by a reaction similar to $3H_2O + Fe \rightarrow Fe(OH)_3 + 3H_{ad}$. Within the concepts of fracture mechanics, this is *environment-induced subcritical crack growth* in the special variant of *hydrogen-induced stress corrosion cracking*, HISCC. As other mechanisms of stress corrosion cracking (SCC) may be at work simultaneously, 'hydrogen-assisted' SCC would have been a more appropriate description, but HISCC apparently is the accepted term especially in Germany. At any rate, pure HISCC certainly is the second limiting case of failure of metals influenced by hydrogen in the metal.

Particularly in niobium, tantalum, vanadium, zirconium, hafnium, and titanium, hydrogen uptake can lead to heterogeneous precipitation of hydrides. These are brittle intermetallic phases, and there presence in the host metal is similar in effect to the presence of internal notches. Hence the description as *hydride embrittlement*, whose details we pass with reference to the literature reviewed e.g. in [2f]. The subject has practical interest in connection with the Purex-type reprocessing of nuclear fuels employing electrolysis of uranium and plutonium solutions in contact with Ti and Hf walls [6a]. In this case, hydride embrittlement caused by cathodic hydrogen uptake was found to be restricted to narrow subsurface zones and thus, in fact, of little danger to strength.

## 14.2
## HIC by High-Activity Hydrogen in Low Strength Steels

If there is hydrogen embrittlement, the metal in question obviously was able to dissolve atomic hydrogen, and the hydrogen solubility, i.e. the concentration of absorbed atomic hydrogen, $H_{abs}$, therefore is one of the decisive factors of the matter in question, together with the diffusivity of this dissolved atomic hydrogen. Previous to dissolution, adsorbed hydrogen atoms must be present on the metal surface as reaction intermediates. As described in Chap. 5, cathodic hydrogen deposition from acids proceeds by the Volmer reaction: $H^+ + e^- \rightarrow H_{ad}$, producing adsorbed atomic hydrogen as intermediates, followed by either Tafel recombination: $2H_{ad} \rightarrow H_2$, or Heyrovsky recombination: $H^+ + H_{ad} \rightarrow H_2$, to molecular hydrogen. The surface coverage of adsorbed atomic hydrogen,

$\theta_H$, will depend on pH, current density of hydrogen deposition, temperature, and properties of the metal substrate such as, especially, the Gibbs energy of adsorption, and it is a property difficult to control or to analyse. Still, it obviously is $\theta_H$ which governs the entry of adsorbed hydrogen into the metallic substrate.

The difficulties to precisely describe $\theta_H$ as a function of important parameters will reappear in the description of the uptake reaction: $H_{ad} \rightarrow H_{abs}$. For the moment, we take the presence of dissolved atomic hydrogen for granted, and we consider a metal with internal flaws such as microscopic voids or microcracks. Part of the hydrogen diffusing from the surface into the metal, encountering internal surfaces of these flaws, will react to atomic hydrogen adsorbed at such surfaces, and eventually recombine to molecular gaseous hydrogen entering the void. The interesting point then is the pressure, $p^{(i)}$, of gaseous hydrogen in voids. When hydrogen entry into the metal is from a gaseous surrounding phase instead of by cathodic deposition, the limiting case, which is equilibrium between molecular hydrogen in voids and in the surrounding gas phase, can easily be stated: Consider a static metal/hydrogen system, where the metal is not being deformed. If the surrounding gas phase is the origin of hydrogen entering the metal, then the pressure in voids, $p^{(i)}$, cannot exceed the pressure in the surrounding gas phase, $p^{(a)}$, and we have $p^{(i)} \leq p^{(a)}$. More precisely, the limiting condition is equality of the thermodynamic activities or fugacities, rather than pressures, of hydrogen in the two gas phases, but then hydrogen gas behaves practically as an ideal gas up to very high pressures. Concerning concentrations, $c^{(i)}$, of dissolved atomic hydrogen in the metal, it follows that in the limiting case this is the saturation concentration, which normally obeys Sievert's law: $c^{(i)}_{sat} = \text{const} \times \sqrt{p^{(a)}}$.

When the the metal is undergoing plastic deformation, conditions change. Now, the interesting case is deformation-assisted hydrogen uptake, for instance during tensile testing in hydrogen atmosphere. The action of the straining machine supplies excess energy to the system, and beyond-equilibrium states can thus be reached. One possibility then is dislocation transport of excess atomic hydrogen into the metal increasing the concentration of dissolved hydrogen to values beyond saturation, also causing $p^{(i)}$ to exceed $p^{(a)}$.

A second possibility to arrive at excessive hydrogen gas pressures in internal voids is cathodic hydrogen deposition at electrode potentials below the equilibrium potential, $E_{H_2/H^+}$, at a cathodic overpotential, $\eta_H$, supplying extra Gibbs energy, $F\eta_H$, to the system. The prominent case is formation of subsurface blisters in mild steel during acid sulfide corrosion. To explain this, Fig. 14.1 shows typical cathodic current-potential curves for iron in dilute sulfuric acid, and in saturated hydrogen sulfide solutions of the same pH $\simeq$ 4 [7]. Similar results have been reported in [6c].

The figure shows, for a sulfuric acid solution, the plateau of the cathodic diffusion-limited current, $i_{H,D}$ for hydrogen deposition by $H^+$ reduction, as predicted by Eq. (5.9). From what has been described in Chap. 6, it follows that at the corrosion potential, $\varepsilon_{corr} \simeq -045$ V, the corrosion current density, $i_{corr}$, is close to, or less than this limiting current density, although on further lowering

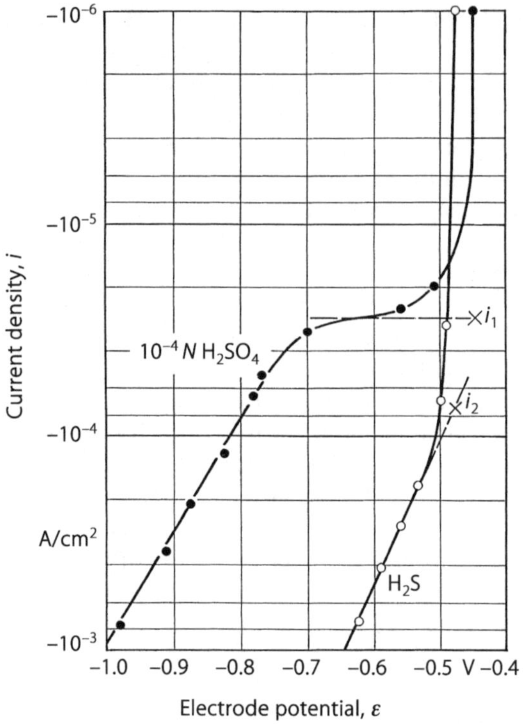


**Fig. 14.1.** Steady-state total cathodic total current-potential curves for Armco iron in deaerated, stagnant $10^{-4}$ N $H_2SO_4/Na_2SO_4$ and saturated $H_2S/Na_2SO_4$ solutions. (From Naumann and Carius)

the potential the cathodic current once more increases by superposition of $H_2O$ reduction. It is more interesting that in the acid sulfide solution $i_{H,D}$ has disappeared, and that, therefore, $i_{corr}$ gas greatly increased. The reason is that the sulfide solution is buffered with respect to pH by the excess content in dissolved undissociated $H_2S$ at a concentration $c^{\circ}_{H_2S} \simeq 0.1M$. It is this concentration rather than $c_{H^+} \simeq 10^{-4}$which now enters Eq. (5.9). A are more realistic formulation of the Volmer reaction step now is: $H_2S + e^- \rightarrow H_{ad} + HS^-$.

As reported in [8] anodic iron dissolution is catalyzed by $HS^-$ ions acting similarly as $OH^-$. This, however, is of minor importance in the present context. The important point, illustrated in Fig. 14.2, is that adsorbed $(HS^-)_{ad}$ acts on the mechanism of hydrogen deposition by increasing the coverage of the metal surface with adsorbed atomic hydrogen, $H_{ad}$. By this effect, the rate of uptake of absorbed hydrogen in the metal is increased, subsurface hydrogen blisters are formed inside the metal by recombination at internal surfaces, proving that the pressure in internal voids, $p^{(i)}$, certainly by far exceeded 1 bar, also proving non-equilibrium between molecular hydrogen in voids, and in bubbles simultaneously escaping from the acid solution. This, then, further proves that

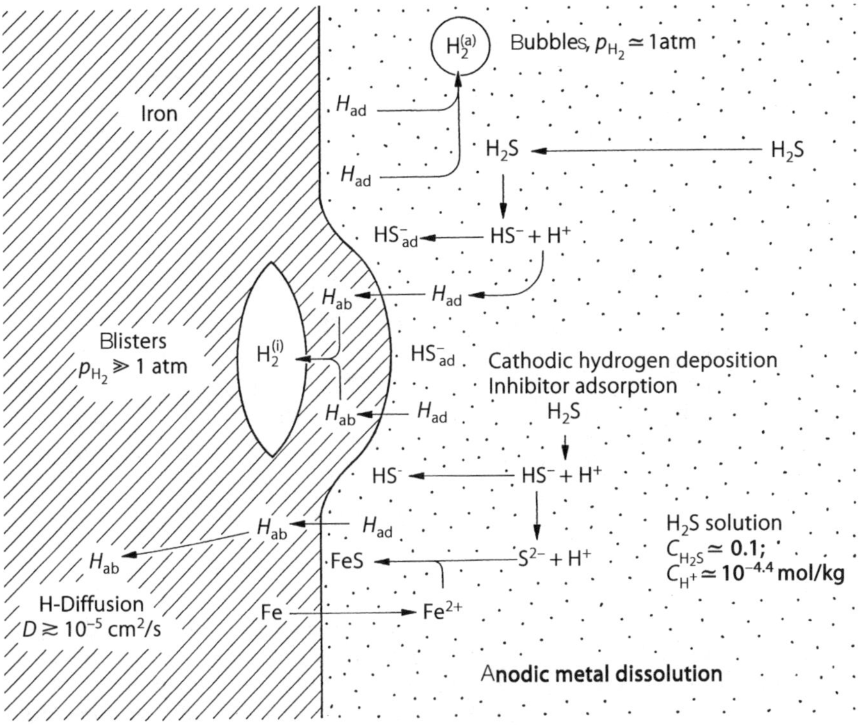


**Fig. 14.2.** The mechanism of corrosion of iron in $H_2S$ solutions, explaining the formation of subsurface blisters filled with high-pressure hydrogen, assuming the Tafel-Volmer sequence of reaction steps. Not that a change to the Tafel-Heyrovsky mechanism, probable for high values of surface coverage by atomic hydrogen, leads to the same blistering

the coverage, $\theta_H$ of the metal surface was much higher than in equilibrium with gaseous hydrogen in the bubbles with 1 atm hydrogen pressure.

For this, there are two possible causes: Adsorbed $(HS^-)_{ad}$ can catalyse the Volmer step of cathodic hydrogen deposition, and this may well cause an increase in $\theta_H$. Or else, $(HS^-)_{ad}$ inhibits the recombination of adsorbed atomic hydrogen to desorbing molecular hydrogen through either the Tafel or the Heyrovsky mechanism, piling up adsorbed hydrogen atoms and, thus, again increasing $\theta_H$. The same holds for other so called *promotors of hydrogen embrittlement,* such as traces of arsenic acid, or cyanide compounds. Together with $HS^-$, these species also are designated as 'poisons of electrode reactions', which is synonymous with undesirable inhibition.

The second possibility, poisoning of hydrogen recombination reactions, allows an estimate of the pressure build-up in voids in an extreme, limiting case: Assume equilibrium between adsorbed atomic hydrogen, $(H)_{ad}$, gaseous hydrogen in internal voids, $(H_2^{(i)}$, and hydrogen ions, $H^+$, in the sulfide solution, but non-equilibrium between adsorbed atomic hydrogen and hydrogen escap-

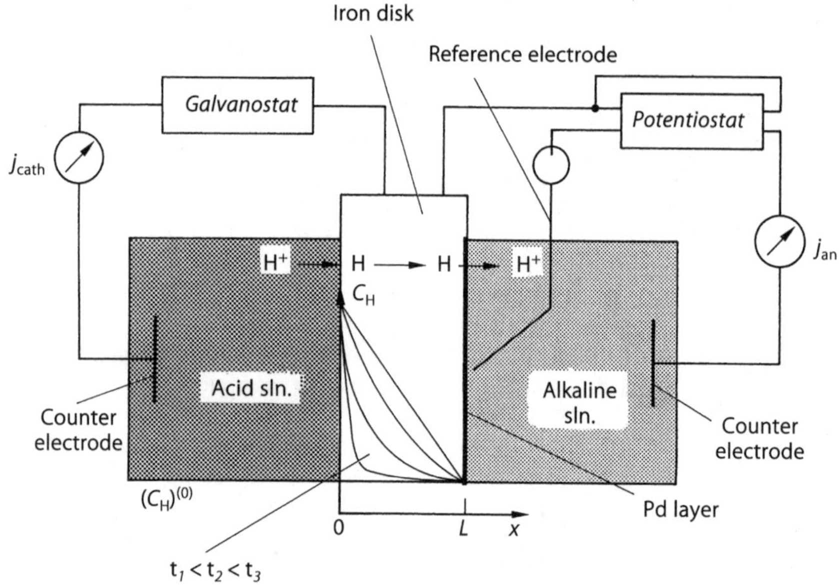


**Fig. 14.3.** Schematic experimental setup for measuring hydrogen permeation through a thin iron foil, observed by galvanostatic cathodic deposition on the front, and potentiostatic anodic hydrogen ionisation at the reverse side of the membrane. Also shown: The change in the concentration profile of diffusing atomic hydrogen, provided that at the front side the coverage in adsorbed hydrogen is kept constant, while at the reverse side it is kept at zero throughout the experiment

ing in bubbles from the solution, $H_2^{(a)}$. This means assuming that the hydrogen overpotential, $\eta_H$, is Tafel recombination overpotential only, in conflict with experimental evidence as described in Chap. 5. On this assumption, the overpotential simply would be the difference in hydrogen equilibrium potentials, $E_{H_2/H^+}$, calculated for the two pressures $p^{(a)}$ and $p^{(it)}$ :

$$\eta_H = E(p^{(i)}) - E(p^{(a)}) = -(RT/2F)\, ln\,(p^{(i)}/p^{(a)})\,, \tag{14.1}$$

where $p^{(a)} = 1$ bar. From this, the estimate for the void pressure $p^{(i)}$, for the easily accessible overpotential $\eta_H = 290$ mV would be $10^{10}$ bar, which, of course, is irrealistic, though instructive.

For more realistic calculations, the coverage, $\theta_H$, essentially is an important parameter, accessible, however, only with difficulties [1a]. In this situation, it is very useful that, more directly, the concentration of atomic hydrogen dissolved in the metal can well be determined, and even without recourse to such tedious methods as hot hydrogen extraction. The solution is recording the diffusion of dissolved hydrogen permeating a thin iron disk. The experiment is explained in Fig. 14.3 [1a,9].

In this setup, hydrogen is deposited from an acid solution onto the front of the disk, causing atomic hydrogen to enter the metal phase and to diffuse to the reverse side, where it leaves the metal phase. To ease hydrogen egress, this side

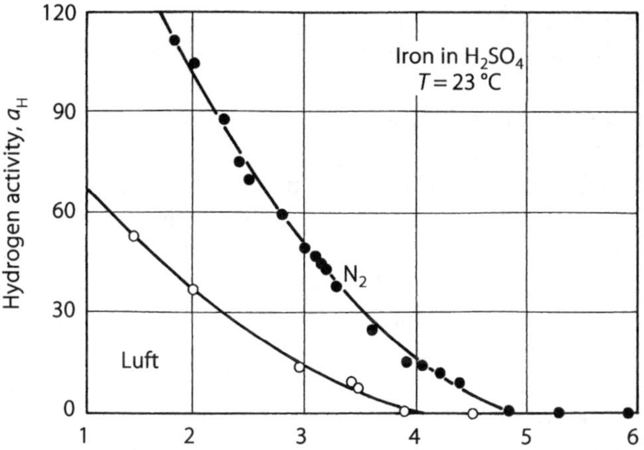


**Fig. 14.4.** The thermodynamic activity of atomic hydrogen dissolved in iron at 23° $C$ in dilute $H_2SO_4$ solutions as a function of pH, in aerated and in deaerated solutions. (From Riecke and Johnen)

has a thin galvanic palladium layer, and it is kept at a potential where hydrogen emerging from the disk is continually reoxidized to $H^+$, which dissolves into an alkaline solution. The current density of this anodic hydrogen oxidation is a measure of the rate of arrival of diffusing hydrogen at the reverse disc side. The method is surprisingly sensible mainly because the diffusion coefficient of hydrogen in iron is comparatively very high: $D_H \simeq 10^{-5}$ cm$^2$ s$^{-1}$.

In practice, the transient increase, $i(t)$, of the anodic hydrogen ionisation current density is recorded together with the stationary current observed when, at time $t \to \infty$, a constant gradient in hydrogen concentration is established across the disk of thickness L. Setting x = 0 at the disk front, and x = L at the reverse side, we have, in this case, $dc_H/dx = (c\,(0) - c\,(L))/L$, with $c\,(L) = 0$, owing to the continuous hydrogen oxidation. c(0) is the concentration at the interior front surface in the disk, and assumed to be equal to the saturation concentration, $(c_H)_{sat}$ of hydrogen dissolved in the metal at the given current density of cathodic hydrogen deposition, i.e. at some fixed , though unknown, value of $\theta_H$., Without proof we note that for the transient increase of the oxidation current density the expression:

$$i(t)/i(\infty) = \left(2L/\sqrt{\pi D_H t}\right) \exp(-L^2/4D_H t)\,, \tag{14.2}$$

is a good approximation (see Appendix). Thus, one and the same record of the transient increase of $i(t)$ up to the steady-state plateau allows the simultaneous determination of the saturation concentration of hydrogen dissolved at a given cathodic current density, and of the hydrogen diffusion coefficient in the metallic phase. As an example, Fig. 14.4 presents data on the saturation activity of dissolved atomic hydrogen in iron at 23°C in dilute sulfuric acid as a function of pH. Here, the 'activity', $a_H$ is somewhat unusually defined as the ratio $a^2_{H^2} = p^{(i)}/(1bar)$. The observed increase of $a_H$ with increasing solution acidity is intuitive. The decrease in aerated solutions, compared to nitrogen-flushed solutions apparently is due to a decrease in $\theta_H$ caused by oxidation of some of the $H_{ad}$ atoms to water. The interesting observation is that hydrogen uptake by wet corrosion, sufficient to cause delayed cracking by HISCC (see further below) is noticeable already at pH 5.

We have referred to blistering of mild steel and the possible change to HIC in acid sulfide solutions as one of two limiting cases. The details of hydrogen-induced cracking under the influence of internal pressure build-up, starting at stress-raising oxide inclusions, or at hard segregation zones, has been the object of very many extensive investigations into the details, which are beyond the scope of this text [1c,10]. We only describe the possibility to identify, by repeated permeation experiments, traps in crystal lattices [11]. The method does not allow to discern the nature of different types of traps, such as inclusions, dislocations, grain boundaries, pores, twins, and the like. Instead, only two groups of traps are distinguished according to their hydrogen-bonding energies either below 0.8 eV (77 kJ $mol^{-1}$), or distinctly above this value. In the first case, traps are designated as 'shallow', and hydrogen is assumed to be trapped reversibly, in variable equilibrium with diffusing atomic hydrogen. In the second case, traps are 'deep', and bonding is irreversible. Extracting hydrogen from deep traps requires heating to ca. 600°C.

When hydrogen is trapped in the metal membranes used for permeation experiments, diffusion is retarded, and this distorts the transients of the ionisation current, $i(t)$. When the permeation cycle is repeated, deep traps remain occupied, and are no longer noticed as delaying hydrogen diffusion. Provided that the activity of dissolved atomic hydrogen is kept low enough to prevent recombination to molecular hydrogen and, thus, blistering and cracking, repeated cycling thus is useful to study trapping. In Fig. 14.5 an experiment of this type is schematically described, using experimental curves observed with membranes of a steel containing 0.2 C, 1.4 Mn, 0–3.5 Ni and some Cr, Mo, and Cu [11]. Curve A is observed during the first permeation cycle starting with hydrogen-free steel. On switching off the cathodic hydrogen-loading current, the anodic ionisation current decays as curve B. From the initial tangent of curve B, $(di/dt)_o$, the diffusion coefficient, $D_H$, is easily calculated, and from this value, together with the plateau current of steady- state diffusion through the membrane, the saturation concentration, $(c_H)_{sat}$, is also available. Using these data, curve C is calculated as the charging curve expected for the case of the same steel in the absence of any traps. In a second charging cycle, a curve similar to D is observed instead. The areas between curves D and A, and between curves C and D, allows an estimate of hydrogen in 'deep' and in 'shallow' traps. In a further run the cathodic charging current is increased until the corresponding increase in internal hydrogen activity induces subsurface cracking of the membrane at sulfide inclusions. This then is revealed by curve E, indicating an excessive uptake of hydrogen. From E, a new value of the product $D_H \times (c_H)_{sat}$ is calculated. Since cracking should not affect the trap population, the hydrogen concentration distribution can be re-evaluated. Finally, metallographic sections of the membrane are prepared,

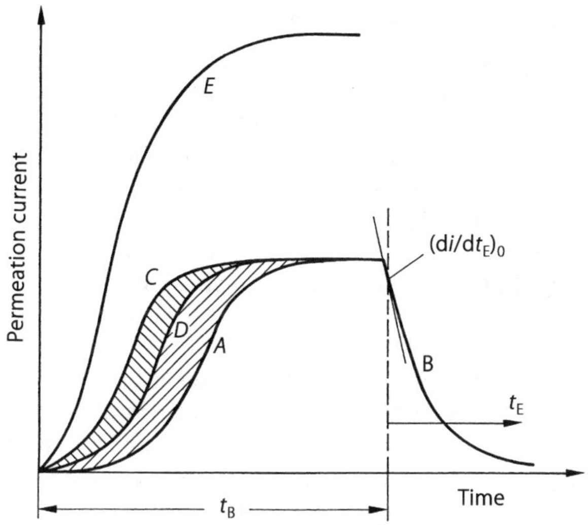


**Fig. 14.5.** Demonstrating the procedure used to determine, by repeated permeation, the hydrogen content in mild steel in 'shallow' low-energy, and in 'deep' high-energy traps, respectively. The membrane was steel with 0.2% C, 1.4% Mn, some Cr, Mo, and Cu, and Ni varied between 0 and 3.5 wt.%. For the details, see the text. (From Coudreuse and Charles)

and the distance of subsurface cracks from the surface is determined, and from this, the minimum hydrogen concentration necessary to trigger HIC is obtained. It turns out to vary, depending on steel composition, between 0.2 and 0.7 ppm, which compares well with values obtained from hot hydrogen extraction.

## 14.3 HISCC by Low-Activity Hydrogen in High-Strength Steels

For high-strength, thick-wall structural components, fracture toughness, $K_C$, as defined in fracture mechanics (see Appendix), is more important than ultimate tensile strength, $R_m$. $K_C$ is the critical value of the *stress intensity factor*, $K\,(\mathrm{N\,m^{-3/2}})$, beyond which a preexisting notch in the surface is unstable with respect to propagation as a crack. $K$ itself is the characteristic factor in the function describing the increase in the crack-opening component of the stress tensor, $\vec{\sigma}$, close to the crack tip. Essentially, this is the result of a three-axial state of stress in the vicinity of notches in high-strength components, which prevents plastic deformation and thus causes inherent brittleness. The net effect is an increase of the local stress to up to tree times the proportionality limit, $R_p$, as defined by the conventional tensile test. As a result, very little uptake of dissolved hydrogen can suffice to further increase brittleness until crack extension is possible below $K_C$, and this is HISCC as one variant of subcritical crack growth.

Under these circumstances, an equally very little reaction of high strength steels with moisture sometimes suffices to produce critical degrees of hydrogen uptake, far below saturation or even supersaturation as in acid sulfide solutions. For reproducible testing, careful control of stress distribution in notched tensile specimens is required. For convenience, test specimen dimensions should allow the application of *linear fracture mechanics*. The test specimens are *double cantilever beam* (DCB) specimens in general, as described in the Appendix and there shown in Fig. 17.2, and in particular *compact tension* (CT) specimens with standardized dimensions. In the depth of these, the stress state is three-dimensional (and the strain state, therefore, 'plane'), except for a narrow zone of plastic deformation ahead of the crack tip. The radius of the plastic zone moving with the growing crack narrows with increasing inherent brittleness of the tested material; it disappears in ideally brittle materials. Since in the surface of the specimens the perpendicular stress component disappears, we have, here, a state of two-dimensional stress (or a state of 'plane stress'). Furthermore, when the total tensile stress is acting perpendicularly on the crack faces, we are dealing with 'Mode I' loading, in contrast to shear or torsion loading of the fracture faces ('Mode II', or 'Mode III', respectively). For Mode I loading, the stress intensity factor is designated as $K_{\mathrm{I}}$, and the fracture toughness as $K_{\mathrm{IC}}$.

In the case of ideal brittleness, the crack-opening stress component, $\sigma_{x,y}$, at a distance $x$ from the crack tip in the $xz$-plane is $\sigma_{yy} = K_{\mathrm{I}}/\sqrt{2\pi x}$. In real materials, the singularity in $x = 0$ disappears. In the Appendix, Fig. 17.2, shows $\sigma_{yy}$, as a function of $x$ in the plastic zone, as derived from approximate calculations in [12]. For a total tensile stress $\sigma$ acting on a crack of depth $a$, the stress intensity factor is $K_{\mathrm{I}} = \sigma\sqrt{\pi a}\beta$, where $\beta$ is a correcting factor calculated from the geometry of the specimen. In the depth of the specimen, the radius of the plastic zone ahead of the crack is $r_{\mathrm{plast}} \simeq (1/6\pi)(K_{\mathrm{I}}/R_p)^2$, and the plastic crack tip opening is $\delta_{\mathrm{tip}} \simeq 0.5K_I^2/(R_pE)$, where $E$ is Young's modulus of elasticity. The distance of the maximum in $\sigma_{yy}$ from the crack tip is of the order of $2\delta_{\mathrm{tip}}$, which is in the µm range. An important point is that an increase in $K_{\mathrm{I}}$ causes the diameter of the plastic zone to increase, leaving the maximum in local stress in this zone unchanged. We note that for a running crack, i.e. for non-stationary states, the calculations need corrections which we pass, referring the reader to [12].

In vacuum, a preexisting crack is, in principle, able to propagate when the stress intensity factor exceeds the fracture toughness, $K_{\mathrm{IC}}$. This is explained by a relation between $K_{\mathrm{I}}$ and the energy release, $G_{\mathrm{I}}$, accompanying crack extension, which is the energy gain by relaxation of the elastic strain in the specimen. The estimated value is $G_{\mathrm{I}} = (1 - \nu)K_{\mathrm{I}}^2/E$, where $\nu$ is Poisson's number of contraction. Beyond a critical value, either $G_{\mathrm{IC}}$ or $K_{\mathrm{IC}}$, the energy gain exceeds the energy loss resulting from moving the plastic zone, and from creating additional fracture surface. Clearly, this is thermodynamic reasoning, and without non-thermodynamic additional assumptions not directly bearing on kinetics. Nevertheless, it is standard procedure in fracture mechanics to plot crack ex-

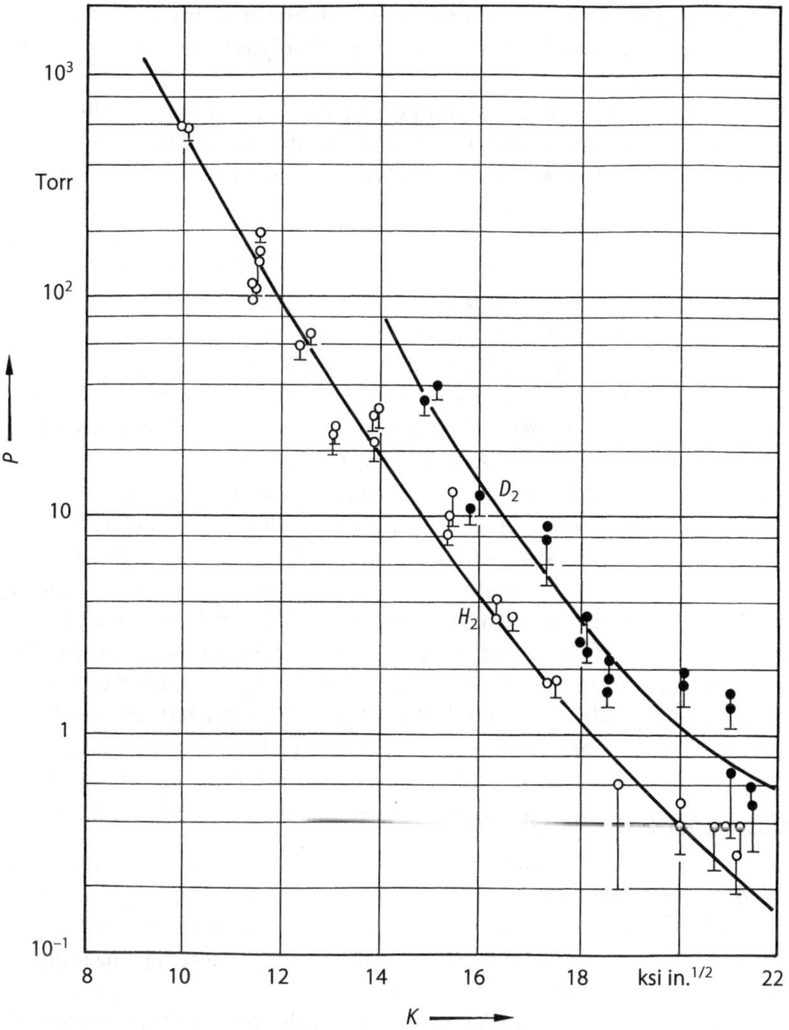


**Fig. 14.6.** Demonstrating the lowering of the critical value of the stress intensity factor, $K_{\mathrm{I}}$ for the onset of the crack extension in a notched steel specimen loaded in hydrogen and in deuterium atmosphere in tension. The high-strength steel was maraging 18Ni AISI 4340, with a yield strength of 1700 MPa. Hydrogen pressure in mm Hg. 1 torr = 133 Pa; 1 ksi $(\text{inch})^{-1/2}$ = 1.099 MPa $m^{1/2}$. (From Oriani and Josephic)

tension rates, $v$, as a function of $K_{\mathrm{I}}$, and results of such investigations will be reported later in this section and in the following chapter.

A direct method to check the influence of hydrogen uptake is to determine critical pairs of values of the hydrogen pressure, $p^{(a)}$, in pure gaseous hydrogen and corresponding values of the stress intensity factor, $K_{\mathrm{I}}$, required for the onset of crack extension in DCB specimens. Figure 14.6 shows data obtained for a high-strength steel strained in hydrogen and in deuterium, and the lowering in critical stress intensity values by what are remarkably low values of hydrogen activity is striking [13,2b].

Since the crack opening stress component has a maximum in the plastic zone at some small distance ahead of the crack tip, one obvious supposition is that elementary crack-opening events also occur at this distance, creating a crack nucleus ahead of the main crack tip, which then runs back into the main crack. This causes a discontinuous growth of cracks. Crack nucleation in the plastic zone should be favored at inhomogeneities such as matrix/precipitate interfaces. For high-activity hydrogen, crack nucleation ahead of the crack is easy to understand. In the case of low-activity hydrogen the mechanism is less obvious, but then we may consider that, during plastic deformation ahead of the running crack, dislocation transport of hydrogen is able to increase the activity of dissolved hydrogen to beyond-equilibrium values. More important effects are directly caused by the dislocation/hydrogen interaction itself. This interaction may favor planar slip and, thus, also favor dislocation pile-up, increasing local stresses, also increasing apparent brittleness. Pinning of dislocations by dissolved hydrogen should also contribute in this direction. The opposite possibility is local softening in the plastic zone owing to *hydrogen-enhanced localized plasticity* (HELP). According to this concept, so-called brittle fracture proceeds by microscopic steps of shear fracture processes, and brittle crack propagating is continuous. We do not go into these details, referring the reader to [14].

The increase of hydrogen solubility in strained crystal lattices was briefly noted further above. More precisely, this is the effect of the hydrostatic component, $\sigma_h$, of the stress tensor, and the increase o solubility is approximately:

$$(c_\mathrm{H})^*_\mathrm{sat} = (c_\mathrm{H})_\mathrm{sat} \exp\left((\sigma_h V_\mathrm{H} - \mathrm{const}\ \sigma_h^2)/RT\right) . \tag{14.3}$$

Here, $V_\mathrm{H} = \partial V / \partial n_\mathrm{H}$ is the partial molar volume of dissolved hydrogen, and const is adjustable. The derivation uses the argument that the sum $\mu_\mathrm{H} = \mu_\mathrm{H}^\circ + RT \ln c_\mathrm{H} - \sigma_h V_\mathrm{H}$ is the chemical potential of hydrogen in a body under Hookean linear-elastic strain. The additional term, const $\sigma_h^2$, approximately accounts for non-Hookean behaviour in the plastic zone

At this point, the *decohesion hypothesis* offers the concept that in steels the chemical bond between metal ions on lattice sites is weakened by interstitial hydrogen atoms [13,16]. More precisely, the suposed effect is due to interstitial protons, as the electrons are likely to enter the electron gas of the metal. The force of bonding is a function of the distance between lattice metal atoms, and in a strained lattice this distance is increased as compared to unstrained material. Consider, now, two neighboring lattice planes emerging in the crack tip. The distance between the two planes is strained, and the bond strength, $F_\mathrm{chem}$, weakened already in the absence of interstitial protons. To separate the faces, and thus to propagate the crack, the stress opening component, $\sigma_{yy}$, must exceed $NF_\mathrm{chem}$, where N is the number of metal ions per unit surface. The supposed influence of dissolved hydrogen on the bond strength is described by the arbitrary relation $F^* = F_\mathrm{chem} - const \times c_\mathrm{H}$. Eventually, the authors arrive at calculating a critical stress intensity factor, $K_\mathrm{I}^* < K_\mathrm{IC}$ :

$$K_I^* = const' \times \sqrt{\varrho}\, N(F_{chem} - const \times c_H) \,, \qquad (14.4)$$

where $\varrho$ is the crack tip radius. In this relation, $c_H$ is substituted by $(c_H)^*_{sat}$ from Eq. (14.3), and the hydrodynamic stress component, $\sigma_h$, is expressed as a function of $K_I/\varrho$. We omit this final calculation, and only note that, with constants appropriately fitted, the solid curves in Fig. 14.6 have been calculated, well representing experimental evidence. $K_I^*$ marks the onset of subcritical crack growth and essentially is the same property as $K_{ISCC}$ introduced in the Appendix and in the following chapter on stress corrosion cracking.

The semi-quantitative success of this concept leaves the details of decohesion processes within the plastic zone ahead of the crack tip open to further investigation. A distinct influence of heterogeneities such as precipitate surfaces an grain boundaries certainly merits attention. Also, macroscopic loading in Mode I may change to microspcopic loading in Mode II in the plastic zone. One obviously important item is the hydrogen concentration at the crack tip. For this, a very large value: $(c_H)^*_{sat}/(c_H)_{sat} \simeq 10^4$, is estimated in [13b]; other estimates vary between 0.2 [15a] and 70–700 mol $m^{-3}$ [15b]. This particular point has been cleared for the steel 90 MnV8, tempered to values of $R_p$ varying between 1600 and 2000 MPa [6e,f]. The experimental procedure is described in Fig. 14.7: A CT specimen is stressed in air to an intensity factor below $K_{IC}$, i.e. able to sustain the stress indefinitely, and the constant length of the precrack is verified constantly. To do this, a 'potential probe' is used, continually recording the constant-current ohmic voltage across the geometric plane in the specimen containing the crack. The specimen then is loaded with dissolved atomic hydrogen by cathodic deposition onto the flat backside of the specimen. The diffusion coefficient, $D_H$, and the saturation concentration, $(c_H)_{sat}$ at the backside are known from previous permeation experiments. Hydrogen now diffuses through the specimen from the back towards the front side, eventually arriving at the crack tip. As soon as at the crack tip $c_H$ is sufficiently high, subcritical crack propagation sets in, as directly monitored with the potential probe. The specimen fails shortly afterwards by fracture. Fractografy then reveals that crack faces always are previous austenite grain boundaries, probably because the crack follows impurities segregated to previous grain boundaries during austenite tempering.

The calculation of the increase in concentration of dissolved hydrogen at the crack tip with time requires of solving the differential equation of non-stationary diffusion. As explained in Chap. 12, the useful approach is to consider the gradient of the chemical potential, $grad\, \mu_H$, rather than the concentration gradient, as the thermodynamic force driving the diffusion flux of hydrogen atoms, $J_H$. The chemical potential of hydrogen in unstressed material, $\mu_{H^+\circ} = RT \ln c_H$. As previously, but neglecting higher-order terms, lattice strain is accounted for by an additional term $-\sigma_H V_H$. Since diffusion is non-stationary, the flux has sinks everywhere due to slow local build-up of the hydrogen concentration, and the divergence of the flux, $div J_H$, is not zero. Adding, finally, the calculation of the hydrodynamic stress component as appropriate to the problem, we arrive at a system of equations as follows:

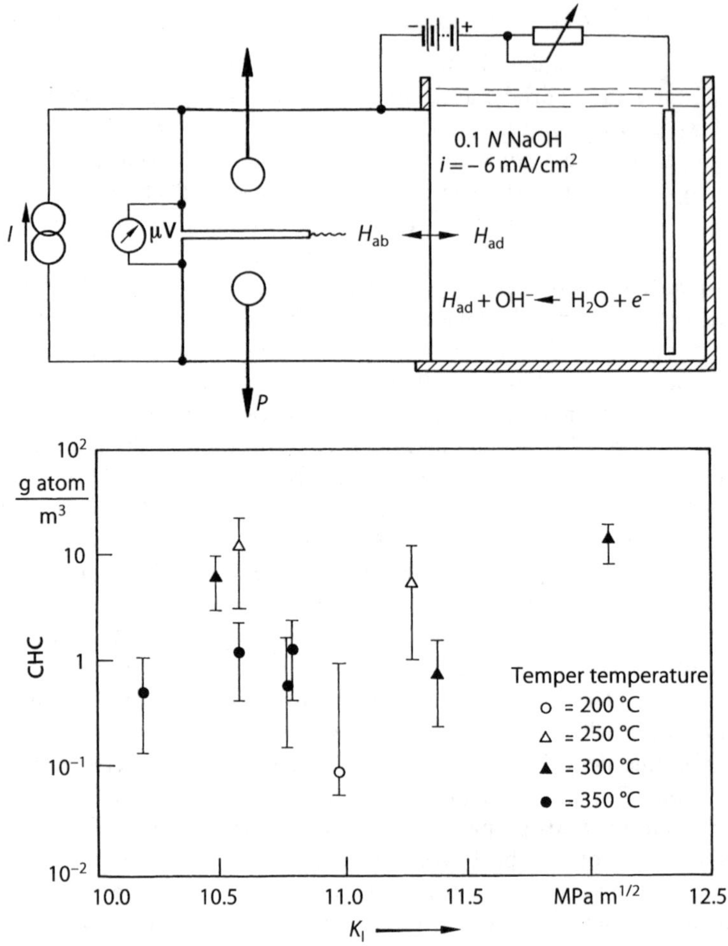


**Fig. 14.7.** The measurement of the critical concentration, $c_{\mathrm{H,crit}}$, of atomic hydrogen dissolved directly ahead of the notch tip in a CT specimen of hardened steel 90 MnV8. **a** Experimental set-up for cathodic charging the specimen with hydrogen from the backside, and for recording crack propagation by the potential probe method. **b** $c_{\mathrm{H,crit}}$, as determined at different values of the stress intensity factor, $K_{\mathrm{I}}$. (From Maier, Popp, and Kaesche)

$$\text{a)}\quad J_{\mathrm{H}} = -\frac{D_{\mathrm{H}}}{RT}\, c_{\mathrm{H}}\, grad\, \mu_{\mathrm{H}}; \qquad \text{b)}\quad \mu_{\mathrm{H}} = \mu_{\mathrm{H}}^{\circ} + RT \ln c_{\mathrm{H}} - \sigma_h V_{\mathrm{H}} \qquad (14.5)$$

$$\text{c)}\quad \frac{\partial c_{\mathrm{H}}}{\partial t} = -\, div J_{\mathrm{H}}; \qquad \text{d)}\quad \sigma_h = \frac{2}{3}(1+\nu)\frac{K_{\mathrm{I}}}{\sqrt{2\pi r}} \cos\frac{\theta}{2}\,.$$

Here, the abscissa, $x$, is pointing into the specimen from the backside; $r$ and $\theta$ are cylinder coordinates referred to the crack tip. The calculation is simplified by neglecting the escape of some hydrogen through the oxide-covered side faces of the specimen, and neglecting the influence of $\theta$. The concentration, $c_{H,crit}$, critical for the onset of crack propagation, as obtained by the computerized approximate solution of the system of equations, is also shown in Fig. 14.7. The experimental scatter of data is considerable, but the critical hydrogen concentration apparently is independent, or nearly independent of $K$. This confirms the supposition that an increase in $K_I$ leaves the peak value of $\sigma_{yy}$ in the plastic zone unchanged, changing only the size of the zone without much influence on crack propagation. The absolute value of he critical hydrogen concentration is very small, and certainly much too low for an overall lattice decohesion between crack faces. Instead, these results point to localized decohesion effects, caused by locally accumulated hydrogen atoms, triggering discontinuous overall crack growth by microcracks nucleation ahead of the main crack tip.

Essentially, these experiments determine critical values of the stress intensity factor, designated further below as $K_{ISCC}$, for the onset of subcritical crack growth for the special case of HISCC. For the atomistic interpretation of data, there apparently is a problem: If a change of $K_I$ leaves the peak value of $\sigma_{yy}$ in the plastic zone unchanged, changing only the radius of the plastic zone, $r_{plast}$, then it is difficult to see why crack extension should depend on $K_I$ at all. Three tentative explanations are available: a) The model describing the dependence of $r_{plast}$ on $K_I$ fails for small values of $K_I$. b) For small values of $r_{plast}$, it rather is $\sigma_{yy}$ outside the plastic zone which determines crack instability [17]. c) Crack instability is governed by a critical value of $r_{plast}$, such as the average distance between precipitates, or else between prior austenite grain boundaries [12]. At this point of theoretical considerations, the concepts of continuum mechanics and of atomistics obviously overlap, which makes conclusive reasoning difficult, but we omit the further discussion of this matter.

As noted above, and as more fully described in the Appendix, studies of subcritical crack growth rates, $v$, in the range $K_{SCC} \leq K_I \leq K_{IC}$ are typical for fracture mechanical research, and Fig. 17.5 in the Appendix schematically summarizes typical shapes of $\ln v$-$K_{ISCC}$ curves, with a steep increase in Stage I at the threshold value $K_{ISCC}$, leveling in Stage II either to $v = \text{const}$, independent of $K_I$, or to a weaker increase of $v$ with increasing $K_I$, and finally with rapid and completely unstable cracking in Rang III, beyond the fracture toughness in vacuum, $K_{IC}$. For the particular case of HISCC in dry hydrogen, Fig. 14.8 presents two experimental $\ln v$-$K_{ISCC}$ curves, although not for tensile Mode I but for torsional Mode II loading [1d,4]. The low-alloy steel shows a marked increase in rate of crack propagation in Range II, whereas for the maraging steel the plateau of the $\ln v$-$K_{ISCC}$ is well developed. Recalling the alleged independence of the peak value of $\sigma_{yy}$ of $K_I$, a plateau is, of course, what is expected from fracture mechanics.

As derived above, dissolved atomic hydrogen concentrates into the stressed and strained metal lattice in the plastic zone at the crack tip. This thermodynamic effect should be effective for running cracks as well. Therefore, hydrogen originally stored in the plastic zone, but left behind when the crack starts to propagate, will tend to follow the crack tip. If not irreversibly lost in deep traps,

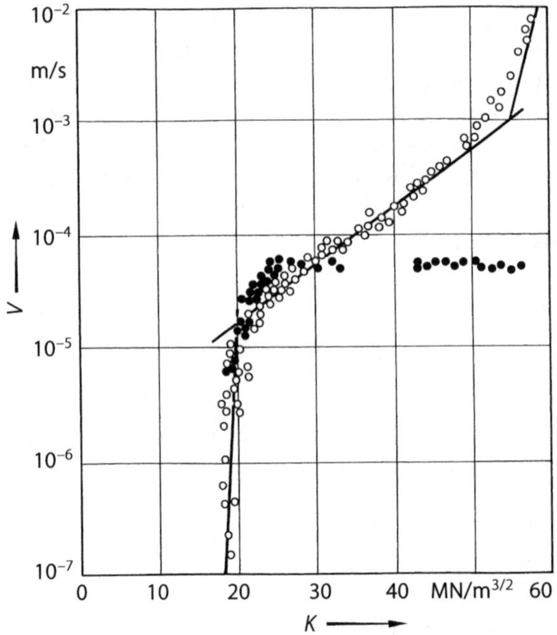


**Fig. 14.8.** Crack extension rates in notched steel specimens loaded in torsion in pure gaseous hydrogen at 24°C. *Open circles:* Low-alloy steel AISI-SAE 4130 with 1 Cr, 0.2 Mo, 0.3 Cu (wt.%); *closed circles:* Martensite-hardening austenitic stainless steel. (From Nelson and Williams, and from Hudak)

and for average crack extension rates not surpassing the diffusion rate, a small amount of embrittling hydrogen, once stored in the plastic zone, can move together with the crack, keeping embrittling effective without further hydrogen supply from the surrounding atmosphere. While this is an extreme case, the argument does show why, especially for high-strength steels, very little hydrogen is required for dangerous subcritical crack growth. Indeed, the total content in dissolved hydrogen in some high-strength steels undergoing cracking, including hydrogen irreversibly trapped and thus not available for accumulation in the plastic zone, was found to be 0.3 or 0.4 ppm, which is a very low value [10].

Above, the discussion always has implied that comparatively slow subcritical crack growth due to hydrogen uptake changes to unstable rapid crack extension whenever, for Mode I loading, $K_I$ is increased to beyond $K_{IC}$, which is fracture toughness as observed in vacuum. $K_{IC}$, of course, depends on materials properties and, thus, also on composition. In this context, it is interesting to compare the behaviour of material containing dissolved hydrogen in homogeneous distribution throughout the specimen, with initially hydrogen-free material taking up hydrogen by corrosion only at the crack tip. For this purpose, a DCB specimen of the low-alloy steels 90 MnV8, tempered to $R_p = 1530$ MPa was cathodically charged with low-activity hydrogen [6c,d]. The specimen was stored for sev-

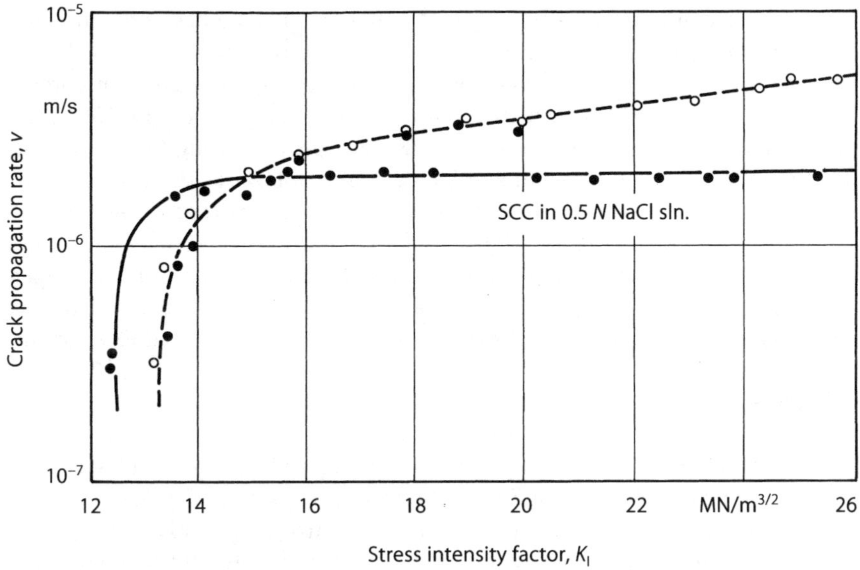


**Fig. 14.9.** Subcritical crack growth in DCB specimens of steel 90 MnV8, tempered to $R_p = 1530$ MPa. Broken curve observed in air for specimens pre-loaded with cathodic hydrogen; solid curve observed without hydrogen charging and testing in 0.5 N NaCl solution. (From Stellwag and Kaesche)

eral weeks, with hydrogen desorption prevented by a galvanized copper film, until a homogeneous hydrogen content was assured. Subsequent testing in air then produced the broken $\ln \upsilon$-$K_I$ curve in Fig. 14.9., clearly showing typical Range I and Range II crack extension behaviour, rather than unstable Range III high-rate rapid crack growth. The material, although not tested in gaseous hydrogen, or else in an aggressive electrolytic solution, thus shows subcritical crack growth, apparently owing to the mobility of interstitial hydrogen allowing for hydrogen accumulation in the plastic zone. For comparison, the solid curve is the $\ln \upsilon$-$K_I$ curve for the same, but hydrogen-free steel tested in 0.5 M NaCl solution, showing model subcritical crack growth behaviour owing to crack tip stress corrosion. Both cases are examples of HISCC, and in view of the similarity of behaviour a special subdivision into 'internal hydrogen embrittlement' (IHE)and 'hydrogen environment embrittlement' (HEE), as proposed in [18], is not necessary. At the same time, the two cases certainly differ in the amounts of dissolved hydrogen necessary to charge and embrittle the bulk volume of a specimen, or else only the plastic zone moving in front of the crack tip.

We repeat that in this section only an outline of basic hydrogen embrittlement effects has been presented. It doe not, however, apply to continuous tensile load testing, i.e. hydrogen-induced stress corrosion cracking only, but to hydrogen-enhanced failure by cyclic changing loads, i.e. to fatigue, as well. As an example, we quote investigations into fatigue of a low alloy steel in spheroidized condition with $R_p = 450$ MPa [6g–i]. TEM studies of the influence of hydrogen on the dislocation arrangement where performed in foils thinned from the interior of smooth cylindrical specimens, fracture tests with conventional precracked DCB specimens. Hydrogen pressures as low as $10^{-5}$ mbar were found to increase the rate of fatigue crack growth owing to an increased rate of microcrack nucleation at carbide/cementite interfaces in the plastic zone ahead of the main crack. Cyclic loading greatly favors hydrogen transport into the material by moving dislocations, which then strip off hydrogen at internal inhomogeneities, thus causing local decohesion. At the same time, hydrogen-induced changes in the dislocation arrangement enhance local plasticity and thus once more favor hydrogen ingress, as verified by permeation experiments.

## References

1. a) Subramanyan, P.K.: Electrochemical Aspects of Hydrogen in Metals. In: a) Comprehensive Treatise of Electrochemistry. (Bockris, J-O'M., Conway, B.E., Yeager, E., and White, R.E., eds.): Vol. IV, Electrochemical Materials Science. Plenum Press, NeW York, London, 1981; b) Embrittlement of Engineering Alloys. (Briant, C.L. and Banerji, S.K., eds.), in: Treatise on Materials Science and Technology (Herman, H., ed.). Vol. 25. Academic Press, New York, London, etc., (1983); c) Proc. Conf. Hydrogen Sulfide-Induced Environment Sensitive Fracture of Steels. Amsterdam 1986. Cor. sci. *27* (1987), Nr. 10/11; d) Proc.Int. Conf. Stress Corrosion Cracking and Hydrogen Embrittlement of Iron Base Alloys, Unieux-Firminy 1973. (Staehle, R.W., Hochmann, J., McCright, R.D., and Slater, J.E., eds.). NACE, Houston, 1977; e) Hydrogen Degradation of Ferrous Alloys. (Oriani, R.A., Hirsch, J.P., and Smialowski, M., eds.) Noyes Publications, Park Ridge, 1985; f) Proc. 4th Int. Conf. of the Effects of Hydrogen on the Behaviour of Materials , Jackson Lake Lodge, 1989 (Moody, N.R. and Thompson, A.W., eds.). The Minerals, Metals & Materials Society, Warrendale, 1990; g) Proc. 5th Int. Conf. of the Effects of Hydrogen on the Behaviour of Materials, Jackson Lake Lodge, 1994. (Thompson, A.W. and Moody, N.R., eds.) The Minerals, Metals & Materials Society, Warrendale, 1996; h) Proc. Int. Conf. Environment-Induced Cracking of Metals, Kohler, 1988. (Gangloff, R.P. and Ives, M.B., eds.). NACE, Houston, 1988; i) Wasserstoff u. Korosion. (Kuron, D., ed.), 2nd edn.. Verlag J. Kuron, Bonn, 2000
2. a) Hirth, J.P.: Metall. Trans. *11a* (1980) 861; b) Oriani, R.A.: Corr. NACE *43* (1987) 391; c) Birnbaum, H.K., in: loc. cit. 1f; d) Robertson, I.M.: Engng. Fracture Mechanics *86* (2001) 671; e) Delafoss, D. and Magnin, T.: ibid. p. 693; f) Taha, A. and Sofronis, P.: ibid. p. 801
3. Zapffe, C.: Trans. Amer. Soc. Metals *39* (1947) 191
4. a) Nelson, H.G.; Hydrogen Embrittlement. In: loc.cit. [1b], Vol. 25; b) Nelson, H.G. and Williams, D.P., in: loc. cit. [1d], p. 390; Hudak, S.J.: MS thesis, Lehigh University, 1972

5. a) Pöpperling, R. and Schwenk, W.: VDI-Berichte *365* (1980) 49; b) Haumann, W., Heller, W., Jungblut, H.-A., Pircher, H., Pöpperling, R., and Schwenk, W.: Stahl u. Eisen *107* (1987) 585; c) Pöpperling, R. and Schwenk, W.: Electrochim. acta *22* (1979) 121
6. a) Kaiser, H. and Kaesche, H., in: Proc. 9th Europ. Congr. on Corrosion, Utrecht, 1989. Stichting Nederlands Corrosie Centrum, Utrecht, 1989; b) Kaesche, H.: Werkstoffe u. Korr. *21* (1970) 185; c) Stellwag, B. and Kaesche, H.: Corr. NACE *35* (1979) 397; d) iid.: Werkstoffe u. Korr. *33* (1982) 274, 323; e) Maier, H.J. and Kaesche, H., in: Proc. 4th Int. Conf. Fatigue and Fatigue Thresholds, Honolulu 1990. Materials and Component Engng. Publ. Ltd., Birmingham, 1990, Vol. III, p. 1795; f) Maier, H., Popp, W., and Kaesche, H.: Acta met. *35* (1987) 875; g) iid., in: loc. cit. [1g], p. 343; h) Held, J. and Kaesche, H., in: Fatigue and Fracture Mechanics, 27th Vol., ASTM STP 1296. (Piascik, R.S., Newman, J.C., and Dowling, N.E., eds.) ASTM, West Conshohocken, 1997, p. 338
7. Naumann, F.K. and Carius, W.: Archiv Eisenhüttenwesen *30*
8. a) Jofa, Z.A. and Tomashova, G.: Zhur. Fiz. Khim. SSSR *34* (1960) 283; b) iid.: Corr. NACE *19* (1963) 13t
9. a) Devanathan, M.A.V. and Stachursky, Z.: Proc. Roy. Soc. *A270* (1962) 90; b) McBreen, J., Nannis, L., and Beck, W.: J. Electrochem. Soc. *113* (1966)
10. a) Pressouyre, C.G., Blondeau, R., Primom, G., and Cadiou, L., in: 1st. Int. Conf. Current Solutions to Hydrogen Problems in Steels. (Interrante, C.G.and Pressouyre, G.M., eds.). ASM, Washington, 1982; b) Jino, M., Nomura, N., Takezawa, H., and Gondom, M.: Rev. Métall. *76* (1976) 561C; c) Taira, T., Tsukada, T., Kobayashi, Y., Inagasi, H., and Watanabe, T.: Cor. NACE *37* (1981) 5; d) Nakai, Y., Kurahashi, H., Emi, T., and Haida, O.: Trans. ISIJ *19* (1979) 401
11. Coudreuse, L. and Charles, J., in: loc. cit. [1c], p. 1169
12. Rice, J.R., in: loc. cit. [1d] and earlier sources quoted therein
13. a) Oriani, R.A. and Josephic, P.H.: Acta met. *22*(1974) 1065; b) Oriani, R.A., in: loc.cit. [1d], p. 351; c) Li, J.C.M., Oriani, R.A., Darken, L.S.: Z. phys. Chemie N.F. *58* (1966) 18
14. a) Beecham, C.D.; Met. Trans. *3*(1972) 437; b) Lynch, S.P.: Scripta met. *13* (1979) 1051; *20* (1986) 1067; *21* (1987) 157; c) Birnbaum, H.K., in: loc. cit. [1h], p. 21
15. a) Bockris, J.O'M., Beck, W., Genshaw, M.A., Subramanyan, P.K., and Williams, F.S.: Acta met. *19* (1971) 1209; b) Marichev, V.A.: Protect. Metals *16* (1980) 427
16. a) Fromberg, R.P., Barnett, W.J., and Troiano, A.R.: Trans. ASM 47 (1955) 892; b) Blanchard, P., and Troiano, A.R.: Mémoires Scient.-Révue de Métall. 57 (1960) 409
17. Gerberich, W.W., Chen, Y.T., and John, C.S.: Metall. Transa. *6a* (1975) 1485
18. Symons, D.M.; Engng. Fracture Mechanics *68* (2001) 751

# 15 Stress Corrosion Cracking

## 15.1 General Aspects

### 15.1.1 Introduction

*Stress corrosion cracking* (SCC) is defined as crack nucleation and propagation in metals caused by the synergystic action of tensile stresses, either constant or slowly changing with time, together with crack-tip chemical reactions or other environment-induced crack-tip effects. In principle, dry hydrogen-induced cracking also is SCC, but we shall, in this chapter, assume that the environmental effect is electrolytic crack tip corrosion.

SCC cracks proceed with very little plastic plastic deformation, restricted to the plastic zone preceding the crack tip. Under sustained load a structure subject to SCC must eventually fail by overload of the remaining load-bearing section. This final fracture is no longer SCC, as it would also occur after critical reduction of the load bearing area by some corrosion process other than SCC, such as uniform thinning, or else pitting. In this context, intercrystalline corrosion is a special case, since grain boundary trenches may act as stress raising notches, which lower elongation to fracture and, thus, appear to point to brittle crack propagation [3]. True intercrystalline SCC, requiring tensile stresses in order to proceed, certainly does exist and will be described below, but tensile tests without check for intercrystalline stress-free corrosion have sometimes been misleading, when the mechanical stress actually was unimportant for the growth of intercrystalline trenches.

Some typical metallographic sections of stress corrosion cracks, are shown in Figs. 15.1–15.3 [4]. With respect to kinetics, it has become rather evident that a variety of SCC mechanisms are possible, depending on many conditions. The often quoted general rule that SCC is an effect of tensile stresses acting on a susceptible material in an aggressive environment, is a trivial re-formulation of the definition of SCC only, leaving all interesting questions unanswered.

Figure 15.4 shows a typical experimental setup for testing the stress corrosion susceptibility of comparatively ductile materials such as austenitic chromium nickel steels, or mild steels [4b]. Typically, smooth cylindrical specimens loaded in tension are used as the working electrode in the respective aggressive electrolytic solution, in connection with a counter electrode and a reference electrode. As also typical for reasonably advanced testing, the tensile specimen is kept at constant electrode potential by a potentiostat, and the polarizing current is recorded as a function of time. As indicated in the figure, temperature control is provided by a thermostat bath surrounding the electrolyte

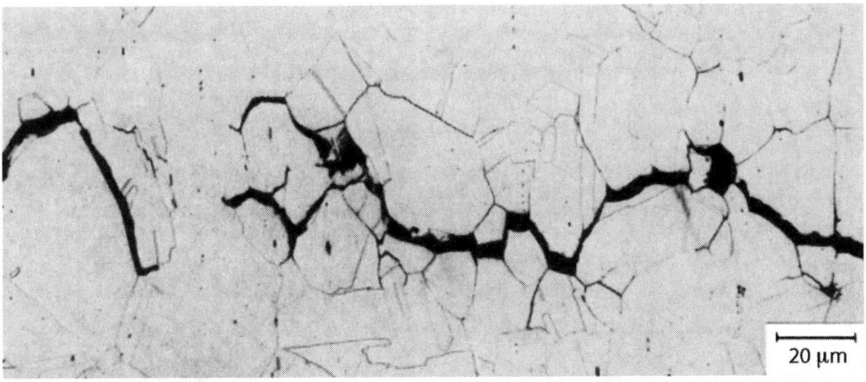


**Fig. 15.1.** intercrystalline stress corrosion crack in austenitic CrNi steel. Metallographic section. (From Herbsleb)

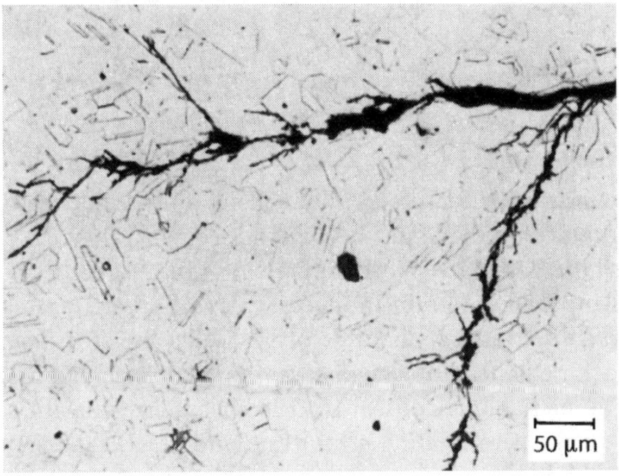


**Fig. 15.2.** Transcrystalline stress corrosion crack in austenitic CrNi steel. Metallographic section.(From Herbsleb)

vessel, and the solution is flushed with a gas of a desired composition. While this type of almost complete control of experimental conditions is highly desirable, it is often relaxed, aiming at lower costs especially for simultaneous routine checking of large numbers of specimens. In every case, SCC tests are classified according to the type of tensile loading: Comparatively cheap testing is available with specimens at constant deformation, such as U-bend, four-point-bend, or other, although this usually involves a complicated state of stress and strain. In the test described in Fig. 15.4, the tensile load nominally is uniaxial. The most frequently used variant then is *constant load testing*, CTL. In simple versions, the recorded property is the time to failure, $\tau_{\mathrm{fl}}$, as a function of load, $P$, and electrode potential, $\varepsilon$; in more elaborate versions, the absolute, or alternatively the relative elongation rates, $dl/dt$ or $\dot{\epsilon} = (1/l_{\circ})dl/dt$, are recorded during

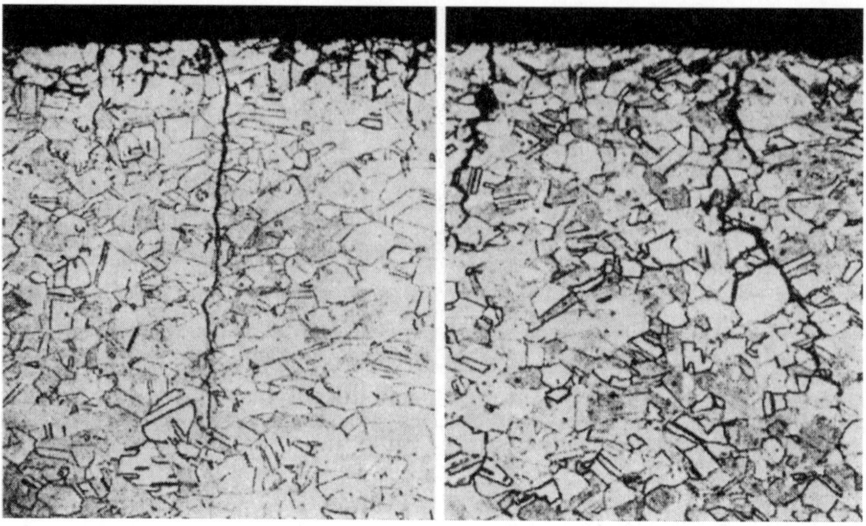

**Fig. 15.3.** Transcrystaline and intercrystalline stress corrosion cracks in $\alpha$-brass. Metallographic sections. (From Mattson)

the test. For ductile metals under tension, creep may considerably influence $\dot{\epsilon}$, especially at elevated temperatures, and should be determined separately.

A still more sophisticated method, particularly interesting with respect to the controversial issue of stress-induced vs. strain-induced SCC, *constant extension rate testing*, CERT. Here, recorded extension rates often really are crosshead speeds of the tensile test machine instead of specimen elongation rates, but this also usually is just as well as meticulous control of true specimen elongation, at any rate as long as present theories relating strain rate and crack extension rate are rather imperfect. It probably is more important to achieve, especially at very low crosshead speeds, really continuous elongation instead of stick-slip of the machine.

For brittle materials, fracture mechanical types of testing fatigue-precracked specimens have become important, as described in the Appendix, using *double cantilever beam* (DCB) specimens in general, and standardized CT *compact tension* (CT) specimen in particular. In the following, some typical results obtained by using a) constant load testing, b) constant strain rate testing, and c) fracture mechanical testing are presented, and this description is accompanied by the introduction of some of the more general aspects of the matter. After this, we shall turn to a preliminary overview of models describing SCC mechanisms. The description will then proceed to selected typical alloy/environment systems, but it should be noted that concepts and theories described in these sections will often apply, mutatis mutandis, to other systems as well, making the sectioning of the text somewhat artificial.

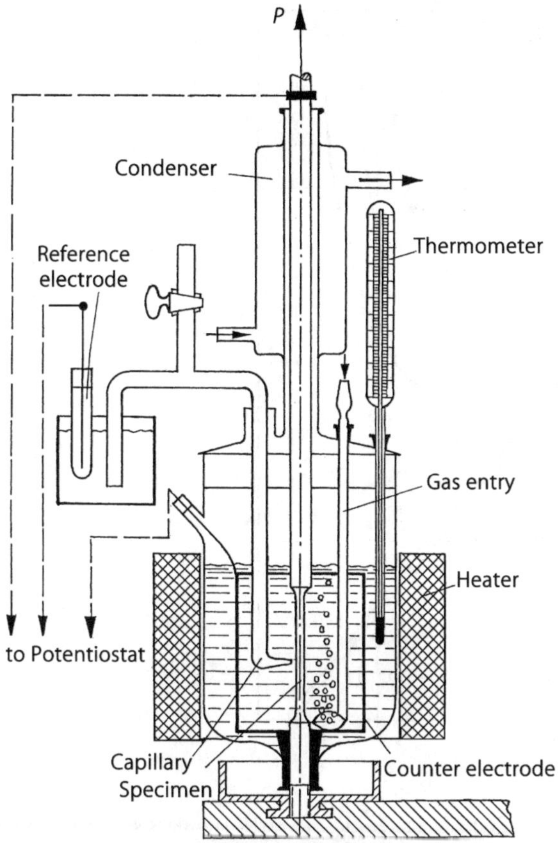


**Fig. 15.4.** Typical experimental setup for controlled-potential stress corrosion testing under constant tensile load in heated electrolytic solutions. (From Münster and Gräfen)

### 15.1.2
### Typical Results of SCC Testing

To discuss details of SCC behaviour as revealed by constant load testing, experiments with austenitic stainless steels are an appropriate example. These are known to be susceptible to transcrystalline (or 'transgranular') stress corrosion cracking (TGSCC) in dilute, hot, chloride containing electrolytes, limiting the use of the otherwise excellent material. Testing this susceptibility, as influenced by changes in composition, or heat treatment, or by other factors, therefore has always been important. With the aim of aggravating test conditions without changing reaction mechanisms, concentrated $MgCl_2$ solution, boiling at 154°C is a much used test medium, since it allows working at a comparatively high temperature but still at ambient pressure. The result of an early investigation into the behaviour of smooth tensile specimens reported in [5] is shown in Fig. 15.5,

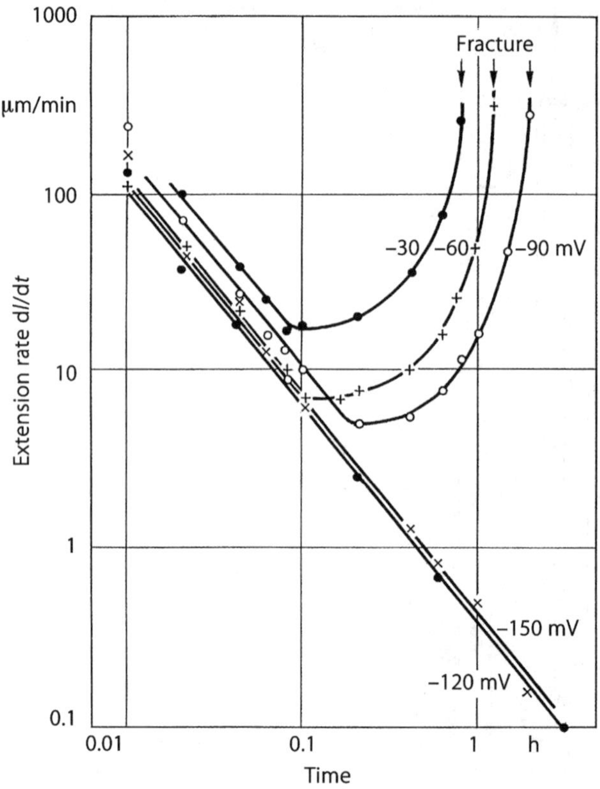


**Fig. 15.5.** The extension rate, dl/dt, of tensile austenitic 18 8 CrNi stainless steel specimens under constant load (200 MN $m^{-2}$) in deaerated boiling $MgCl_2$ solution, at different values of the electrode potential. Note pure creep at potentials below 100 mV, and onset of SCC beyond 90 mV. (From Ternes)

which unambiguously proves the importance of the electrode potential as a major factor: While at $\varepsilon \leq -0.15$ V the decay in elongation rate indicates decay of creep only, the onset of cracking by SCC at $-0.09$ V and beyond is clearly indicated, followed by a continuous increase of the strain rate, $\dot{\epsilon}$, due to increasing specific load on the decreasing load bearing area of the specimens.

Cracking is transcrystalline, with many cracks advancing simultaneously, and the ultimate plastic fracture is caused by the most advanced member of the family of cracks. We note, here, that neighboring cracks are mutually stress-relieving, which is a point to consider when, later, the dependence of crack propagation rates on local stress is discussed.

The curves also show that the time to failure, $\tau_{\text{fail}}$, can be interpreted as the sum of an incubation time and a crack extension time: $\tau_{\text{fail}} = \tau_{\text{inc}} + \tau_{\text{ext}}$. In the present case, where the two contributions are comparable, and the exact onset of cracking is difficult to determine, the separation is not very safe. Also, while the definition of 'crack propagation' is trivial, the events during 'crack incubation' are not obvious, and the same applies to 'crack nucleation' during this period, as the nature of crack nuclei is not well known. When, as in the present case, the specimens continuously creep, the constant load test also is a continuous extension test, albeit with a non-stationary extension rate, and it may well be that crack nucleation is caused or at least influenced by local plastic deformation processes.

In simplified versions, CLT is used with the simple aim of counting failure and no-failure cases, screening influences of small compositional or other changes in the materials tested. A disadvantage is that for the no-failure statement some convention is needed to end the test after, e.g., 1000 h. This difficulty is avoided in CERT, since it ends with fracture in every case. Apart from this, CLT is a more realistic type of testing, representing usual service conditions.

For constant strain rate testing smooth tensile specimen again are normally used, extended at a small constant rate until failure by fracture, and SCC is understood to have occurred during the test if the elongation at fracture $A_{fr}$, or the reduction of area at fracture , $Z_{fr}$, or the work performed until fracture, $W_{fr}$, as obtained from integrating the load/strain curves, are smaller than for fracture in an inert surrounding. In contrast to CLT, CERT thus is a type of testing under conditions not normally encountered in the engineering practice. The test can be aggravated by using precracked or precreviced specimens, but we shall ignore this variant in the following, except where otherwise noted.

The underlying concept for the interpretation of CERT data is that the the near-brittle propagation of stress corrosion cracks should show up through changes in the fracture properties, indicating an increased overall brittleness, if the macroscopic strain rate, $\dot{\epsilon}$, and the the microscopic crack growth rate, $v$, are comparable. Since SCC is comparatively slow, experimental $\dot{\epsilon}$ values also are low, and *slow strain rate testing* (SSRT) is another appropriate name for the procedure. Typically, experiments start with recording $A_{fr}$, or $Z_{fr}$, or $W_{fr}$ in air, and then proceed to testing in the desired environment in subsequent runs with changing values of $\dot{\epsilon}$. An early example of results thus obtained is presented in Fig. 15.6 for the case of the titanium alloy Ti-5Al-2Sn tested in air, in an aqueous NaCl solution, and in an alcoholic HCl solution [6a]. In air, fracture properties are scarcely influenced by the strain rate. In alcoholic acid solution, the influence of what is identified as SCC, as revealed by the decrease in elongation at fracture, $A_{fr}$, increases with decreasing $\dot{\epsilon}$. On first sight this appears intuitive: As the macroscopic strain rate decreases, the time for SCC to cause cracks to propagate increases as well, and so does the embrittling effect indicated by the lowering of $A_{fr}$. But a different effect is observed in aqueous salt solutions: While at intermediate values of $\dot{\epsilon}$ the influence on SCC again is clearly revealed, the embrittlement, as an indicator of SCC, once more disappears for very small values of $\dot{\epsilon}$. This is surprising, though not altogether non-intuitive. Rather, it corresponds to what is expected from a particular model, assuming crack initiation and crack propagation occurring at coarse slip steps, and thus requiring microscopic deformation processes. The consequence is that, apparently, SCC in the titanium alloy requires plastic deformation processes for SCC

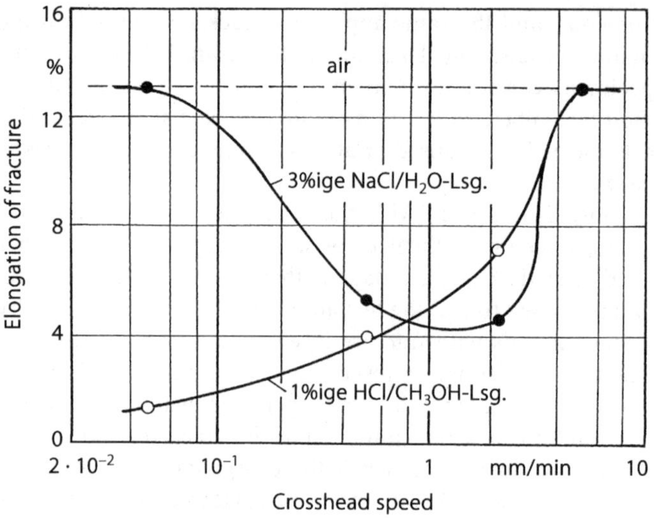


**Fig. 15.6.** Elongation at fracture of Ti-5Al-2Sn (wt.%)tensile test specimens observed in air, in aqueous NaCl solution and in alcoholic HCl solution as a function of the strain rate. (From Scully and Powell)

to start in aqueous chloride, but not in alcoholic acid solution. Observations of this type have led to distinguish *strain-induced stress corrosion* requiring, for cracks to grow, continuous, macroscopic plastic deformation processes, to extend, and *stress-induced stress corrosion* requiring stress only [1].

The distinction of the two classes does not exclude that stress-induced SCC still is triggered by microplastic deformation processes [7a,b,8]. Crack nucleation may involve such local deformation processes escaping detection except by closer microscopic observation, such as local deformation at precipitates emerging in the surface of the specimen. An example of nominally stress-induced SCC again is stainless steel, where SCC has been found to start while the load on tensile specimens was being reduced, i.e. at negative values of the strain rate [7a,b]. This effect is caused by micropitting occurring, under stress, at potentials below the pitting potential [8a]. While micropitting is abundant on slip bands, it is also, if only comparatively rarely, observed on non-slipped surface areas. A second case claimed to be stress- rather than strain-induced SCC is intercrystalline or intergranular stress corrosion cracking (IGSCC) of mild steel in carbonate solutions, which appears to be triggered by trenches along grain boundaries in the magnetite film covering the specimens [7f]. The point will be reconsidered further below.

For completeness, so-called *strain-induced crack corrosion* is added to the list of corrosion failures caused by cracks [7a]. Typically, this occurs in conventional power plants, damaging heater tubes internally protected against high-pressure high-temperature boiler feed water again by a thick magnetite layer. When these layers crack during shut-down periods by thermal contraction, corrosion starts attacking the bottom of crack trenches [7h]. In the following, we shall not explicitly return to this type of damage, although it is a limiting case of SCC, and although it is an important cause of service failure.

The classification of SCC types according to their dependence on stress or strain is somewhat artificial. For more basic aspects of the matter, it should be recognized that the variety in SCC mechanisms presents a spectrum of mixed influences of mechanical and non-mechanical contributions to cracking, ranging from crack extension predominantly by anodic crack tip metal dissolution assumed in ductile materials, as described below, to predominantly mechanical cleavage processes in brittle materials as already described for cracking in gaseous hydrogen [8c].

An important general aspect certainly is the crack extension rate. We note, in the first place, that for service failure cases the reconstruction of failure history based on the assumption of continuous crack growth during very long service times, sometimes of years, is unsafe. During accelerated laboratory testing, the situation is different, and continuous macroscopic crack growth easily available. Again, we expect to observe a wide range of values depending on conditions. As an example of a direct observation in a ductile material, Fig. 15.7 shows intercrystalline crack extension in a thin iron plate under practically constant load, notched to induce crack initiation at an appropriate place rather than for reasons of fracture mechanics. Cracking without incubation time was induced by boiling $Ca(NO_3)_2$ solution [9]. The crack extended at a rate increasing with time, apparently owing to the increase in load on the remaining bearing area, from ca. $10^{-6}$ to ca. $10^{-5}$ m s$^{-1}$. The case is probably one of crack extension by anodic crack-tip metal dissolution, and the crack-tip current density, $I_{tip}$, can be calculated from the extension rate. Using Faraday's law we find that, for iron dissolution to $Fe^{2+}$, crack propagation rates $\upsilon = 4 \times 10^{-7}, 4 \times 10^{-6}, 4 \times 10^{-5}$ m s$^{-1}$, correspond to dissolution current densities $I_{tip} = 1, 10, 100$ A cm$^{-2}$, respectively. These are very high values, but we have seen in Chap. 12 that pit initiation processes can reach 100 A cm$^{-2}$, and pitting-type attack at crack tips will, in fact, become an interesting issue further below. On the other hand, values far beyond 100 A cm$^{-2}$ can hardly be expected from true electrode processes; therefore, extension rates far beyond $10^{-5}$ m s$^{-1}$ are not entirely due to anodic metal dissolution, but require, instead, the cooperation of mechanical decohesion processes. These, then, may in the limiting case proceed by microspcopic steps, each at a rate close to the speed of sound.

Recording crack extension rates is also typical for fracture mechanical testing of inherently brittle materials. Figure 15.8 presents data for the case of a precipitation-hardening AlZnMgCu alloy, stressed in potassium iodide solutions at various levels of the salt concentration, including pure water, as a function of the stress level [10a]. The shape of the $\ln \upsilon$-$K_I$ curves is characteristic, showing the onset of subcritical crack growth beyond a threshold value, $K_{ISCC}$, far below the fracture toughness, $K_{IC}$, rapidly increasing through what is termed *Region I* of the curve with increasing $K_I$, but leveling in *Region II* to a constant value, which has already been described in Chap. 14 as one of the typical varieties in

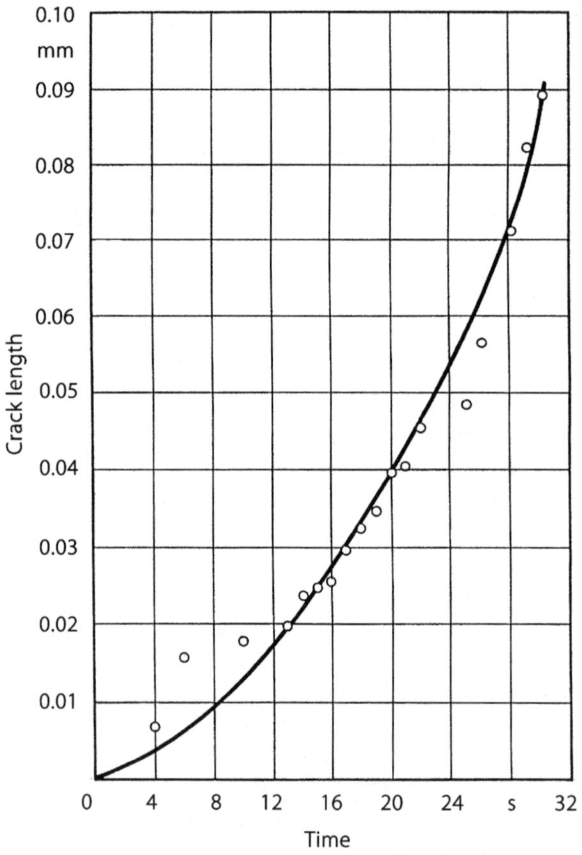


**Fig. 15.7.** Crack growth by intercrystalline stress corrosion cracking in a thin sheet of iron in boiling $Ca(NO_3)_2$ solution, as directly recorded by microscopic observation of the specimen surface. (From Engell, Bohnenkamp, and Bäumel)

subcritical fracture behaviour, for the particular case of HIC in gaseous hydrogen. In Fig. 15.8, the final change in a *Region III* to fast unstable fracture for $K_I$ approaching $K_{IC}$ is omitted.

We recall at this point that in Region II the $\ln v$-$K_I$ curves often show an increase of $v$ with $K_I$ rather than a plateau, and such cases will be discussed further below. Also 'brittleness' of materials as a prerequisite for straightforward fracture mechanical testing needs comment: The decisive property is the size of the plastic zone in front of the crack tip in relation to the dimensions of the test specimen. Therefore, fracture mechanical testing of, e.g., ductile austenitic stainless can be and is being performed, but with very large notched DCB specimens, and such tests produce $\ln v$-$K_I$ curves distinctly revealing Regions I and II [10b,11]. At the same time, fracture mechanical test records tend to smooth extension vs. time curves, thus masking discontinuous crack growth. A typical is, in fact, intercrystalline SCC of hardened aluminum alloys, as illustrated in

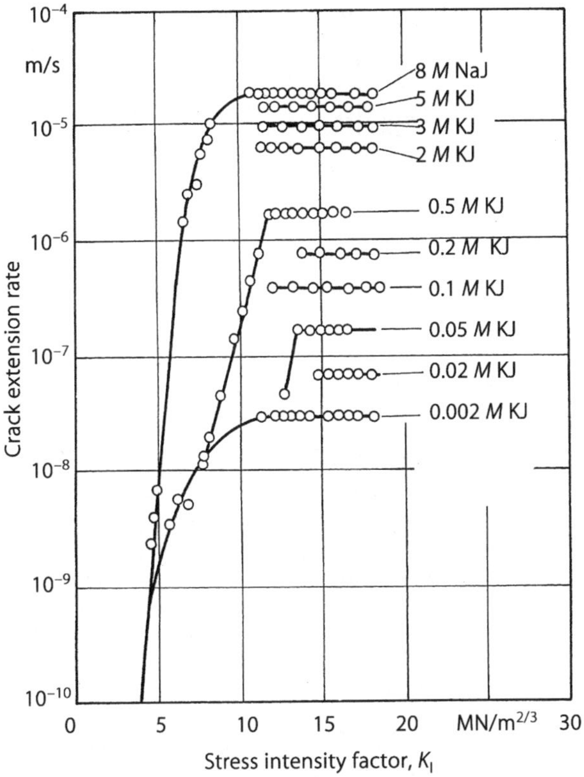


**Fig. 15.8.** Extension rates of intercrystalline stress corrosion cracks in double cantilever beam specimens of AlZnMgC, type 7079-T651, at 23° C, loaded in Mode I, in sodium iodide solutions at constant potential, $\varepsilon = -0.45$ V, and in distilled water, pH $\simeq$ 6. (From Speidel)

Fig. 15.9 for hardened AlZnMg3 in sodium chloride solution [6b]. We add that recording of acoustic emission produced by discontinuous cracking can be considerably refined by piezoelectric techniques, which have been widely used for this purpose [12].

### 15.1.3 Modeling Stress Corrosion Mechanisms

With respect to essential principles, three mechanistic models for stress corrosion mechanisms predominate: The first is continuous crack extension by anodic crack-tip metal dissolution alone; the second is discontinuous cracking owing to local hydrogen crack-tip embrittlement; and the third is discontinuous crack initiation in brittle surface films followed by stepwise crack extension into ductile zones. A fourth model, explaining crack extension by surface vacancy diffusion to the tip of cracks, is more controversial, as explained further below.

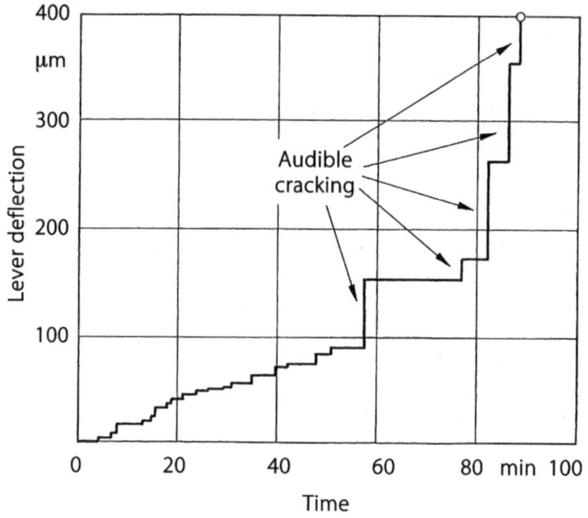


**Fig. 15.9.** Proving discontinuous crack growth in a hardened AlMgZn3 alloy tested in 3% NaCl solution. Testing was performed with notched specimens loaded by a lever in plane strain to 80% of the yield strength. Crack extension events are marked by discontinuous changes in the lever deflection. The last high extension crack events were audible. (From Watkinson and Scully)

Generally with respect to modeling, we repeat that we must not expect a universal, singular mechanism. Also, mechanisms may be mixed in many cases, possibly even with contributions from two or more mechanistic contributions varying with time.

A typical example of probably mixed control is SCC in hardened AlZnMg3, and Fig. 15.10 presents the supposed mechanism [10c]. Essentially, this is an application of the mechanism of pitting of aluminum in chloride solutions, as described in Chap. 12, to stress corrosion cracking, including the assumption of a layer of hydrated $AlCl_3$ at the pit bottom. On the whole, the crack is seen as a distorted pit. Hydrogen evolution in the crack is, in fact, directly observed, proving acidification of the crack electrolyte by hydrolysis of anodically dissolved aluminum ions. A part of the hydrogen atoms produced as an intermediate of the hydrogen deposition reaction is taken up in the metal, causing embrittlement of the crack tip region, and this is supposed to be the predominant effect for crack propagation probably in bursts. Since cathodic hydrogen evolution at the crack tip is the result of rapid anodic metal dissolution producing, by hydrolysis, the acid pit electrolyte, the effect can be described as *anodic hydrogen embrittlement*. Within this concept, discontinuous crack extension is readily explained as caused by delayed diffusion of absorbed atomic hydrogen into the zone ahead of the crack tip, periodically causing sufficient embrittlement to trigger crack extension. If, instead, crack extension is continuous, diffusion of hydrogen must be able to continuously accompany the moving crack tip.

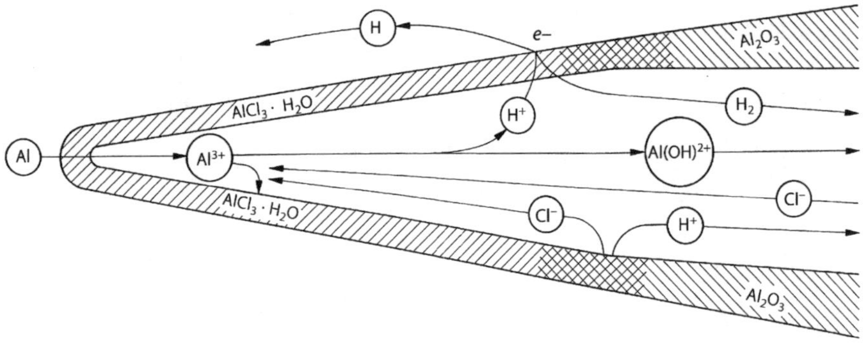


**Fig. 15.10.** The model for stress corrosion crack growth in hardened AlZnMg3. The crack tip is assumed active, while crack sides are repassivated by an oxide film. Note the minor role of anodic crack tip metal dissolution and the major role of accompanying uptake of embrittling hydrogen for crack extension presumably in bursts. Note also cathodic crack tip hydrogen evolution as the byproduct of anodic aluminum dissolution and crack tip electrolyte acidification by hydrolysis, as in pitting. (From Speidel and Hyatt)

If the mechanism of cracking in aluminum alloys of type AlZnMg or AlZnMgCu is close to that of pitting of aluminum, an apparently obvious conclusion is that cracking should no occur at electrode potentials below the pitting potential of the respective alloy. This, however, is a false supposition: Cracking, on the contrary, starts and proceed at distinctly more negative potentials [8d,e]. To explain this, the concepts developed for the mechanism of pitting must be supplemented by allowing for strain-induced or stress-induced micropitting events at crack tips.

To test SCC susceptibility, hydrogen uptake is often produced by overall cathodic charging of steel specimens loaded, afterwards or simultaneously, in the desired mode. To aggravate conditions, hydrogen uptake can be enhanced by adding 'electrode poisons' to the solution, as described in Chap. 14. We report, as an example, the behaviour of a high-strength martensitic steel which cracks in sodium chloride solutions without such additions, and does so as an effect of both cathodic and anodic external polarization [13]. Figure 15.11 shows the total current-potential curve, as determined by galvanostatic polarization experiments with a separate specimen for each experimental point, as numbered in the figure, and also shows times to failure, $\tau_{fl}$, observed during simple lever load tests with the same specimens. The decrease of $\tau$ with increasing cathodic polarization is easily explained as an overall increase in overall specimen brittleness by hydrogen uptake. On anodic polarization, a very steep increase in the rate of metal dissolution probably indicates chloride-induced pitting, and $\tau$ again abruptly decreases. Probably, this is not due to overall thinning of the specimens, but again caused by hydrogen uptake, now accompanying hydrogen evolution in pits by the hydrolysis/acidifying mechanism, i.e. 'anodic' hydrogen embrittlement.

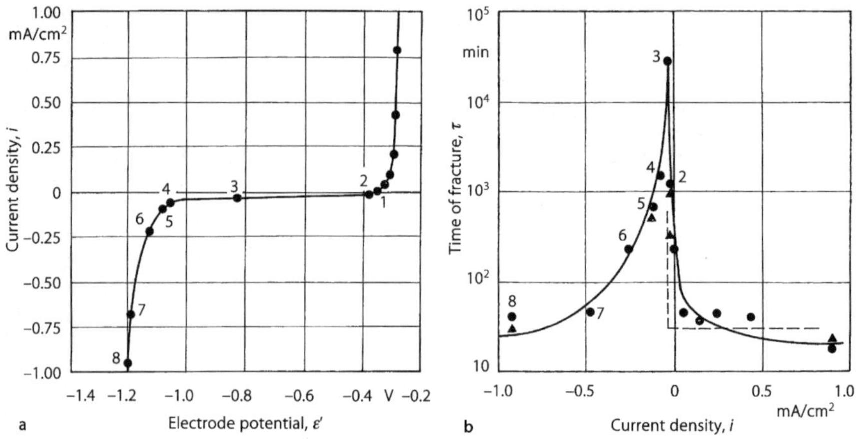


**Fig. 15.11.** **a** Current-potential curve for a martensitic chromium steel in deaerated NaCl solution at 25°C. Steel composition was 12.36 Cr, 1.02 Mo (wt.%), ultimate strength was 1735 MPa. The curve indicates exclusive cathodic hydrogen evolution at potentials below −0.8 V, and anodic iron dissolution. **b** Times to failure during lever tests with loads at 75% of the yield strength, plotted as a function of the current density recorded during external polarization. Potential referred to sat.cal.electrode. Specimen numbering is identical in **a** and**b**. (●● From Bhatt and Phelps ▲▲ from Wilde)

A general overview reveals that, except for hydrogen- induced cases, stress corrosion damages typically occur in passive metal and alloys. This observation has given rise to the crack propagation model described in Fig. 15.12 The mechanism applies to systems where crack-wall dissolution rates are by orders of magnitude slower than crack tip dissolution rates, and the obvious practical case is passive metals, with crack walls repassivated behind active crack tips. A second class is non-passive systems such as AuCu with protection at external surfaces and crack walls owing to layers enriched in the noble component by selective deallyoing, as described in Chap. 8. In every cases, the mechanism is self-perpetuating by continuous or discontinuous crack tip activation by plastic slip. Hence, the designation *slip-step dissolution model*. Truly continuous crack extension is then difficult to understand, as it requires a repassivation rate exactly equal to the crack propagation rate, adapting, for some unknown reason, to changes in the latter quantity with changes of the crack opening stress component. In principle, a mechanism can be devised where, as in pitting, a small electrolyte volume of critical concentration always continuously keeps the crack tip active, and continuously allows crack-wall repassivation beyond some small distance from the tip [8f]. We shall not, however, refer to measurements of repassivation rate, as they previously have been performed for titanium alloys activated by scratching experiments [14].

The frequency of discontinuous, intermittent crack extension events can be high enough for quasi-continuous crack growth. When, as frequently in ductile materials, stress corrosion produces groups of neighboring cracks, then the

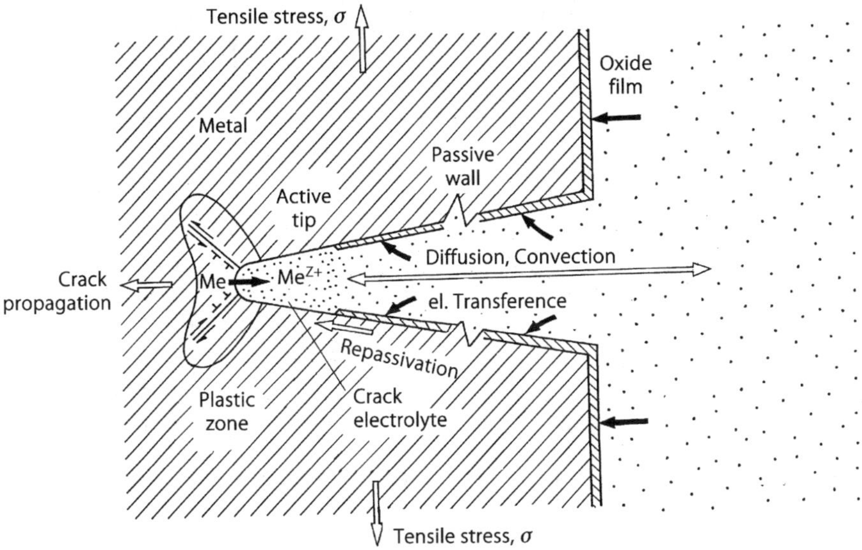


**Fig. 15.12.** Modeling stress corrosion crack extension in a passive metal by by quasi-continuous, intermittent crack tip slip activation followed by deformation-enhanced crack-tip metal dissolution. Note role of plastic deformation zone at crack tip, and protection of crack walls by repassivation following crack extension

overall specimen extension will be found to grow almost truly continuously, because growth steps of neighboring cracks overlap in time.

To explain observed high rates of crack propagation by crack tip anodic metal dissolution alone, correspondingly high values of $I_{tip}$ are required. These cannot be provided by superposition of anodic anodic and cathodic electrode reactions at the active crack tip alone, because, while $I_{tip}$ is large, the crack-tip current of oxygen reduction or hydrogen evolution is small. Under these circumstances, SCC proceeds either by anodic external polarization as during most of the controlled-potential experiments, or else by galvanic cell action with the active crack tip as anode, and crack walls and part of the surrounding metal surface as cathode providing, on the whole, the total cathodic current equivalent to the product of $I_{tip}$ and the active crack tip surface. This is the situation expected in practical service, and it was assumed in Fig. 15.12.

For crack advance by brittle decohesion steps due to local hydrogen embrittlement, or due to surface embrittlement by vacancy injection, this reasoning is not applicable. Even for crack advance by crack-tip metal dissolution only, the crack growth rate may exceed the equivalent of galvanic currents flowing to crack walls and the external surface, or to an external counter-electrode, and for the following reason: Consider a crack which at some moment has widened by an event of coarse glide depassivating the crack tip. The activated surface area will repassivate, and it may, in fact, do so by local, direct, non-electrolytic reaction with water: $Me + xH_2O \rightarrow MeO_x + xH_2$. Referred to the state before

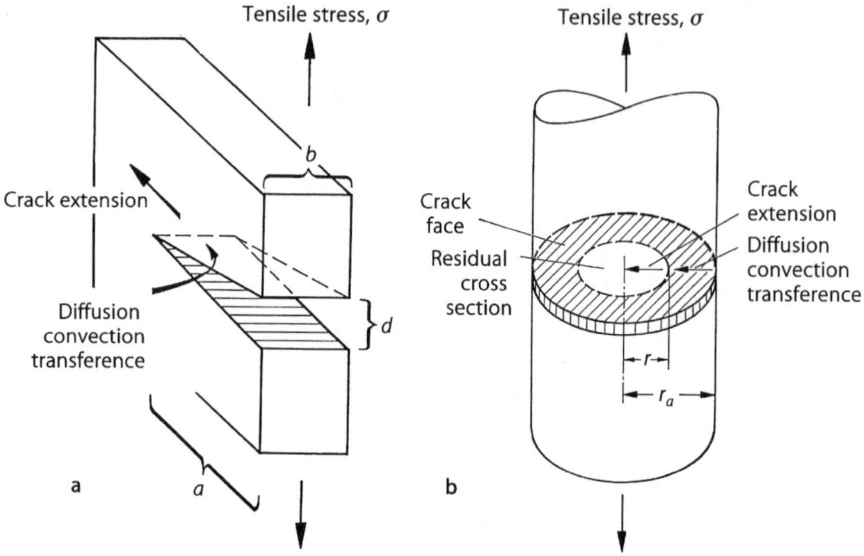


**Fig. 15.13a,b.** Considering diffusion and mass transport paths for externally polarized tensile specimens in electrolytic solutions. **a** Flat notched specimen, e.g. DCB specimen, where crack extension leaves the distance from the crack tip through the side crack electrolyte to the side walls constant with time. **b** Cylindrical specimen with a circular crack growing from the surface into the interior causing current and mass transport paths to increase with the same rate as the crack extension rate

coarse glide, the metal/oxide interface has now moved into the specimen interior by a distance $\Delta x$ equal to the thickness of the passivating layer, extending crack length by this same distance, without, however, involving a galvanic cell current to crack walls beyond the tip, or to the counter electrode of an external polarizing circuit. The importance of this point is quite controversial, as discussed further below.

The galvanic cell mechanism requires sufficiently rapid cathodic processes at crack walls, and for passive metals this means requiring sufficient semiconduction of the oxide films, or else, as observed for aluminum alloys with heterogeneous precipitates of intermetallic compounds (see further below), some other situation allowing for rapid cathodic reactions. For galvanic cell action as well as for SCC controlled by external polarization, a further common aspect is mass transfer and electric current flow in the electrolytic solution along the narrow crack interior, especially considering crack length increasing with time.

Concerning mass transport processes, crack geometry is an important factor which, however, is very variable for different types of test specimens: For flat, notched specimens used in fracture mechanical testing, the usual assumption is that the crack front, although curved, extends into both side walls. In this case, as seen in Fig. 15.13a, the length of mass macroscopic transport paths into the crack remain constant. When, instead, cracks grow into a cylindrical specimen,

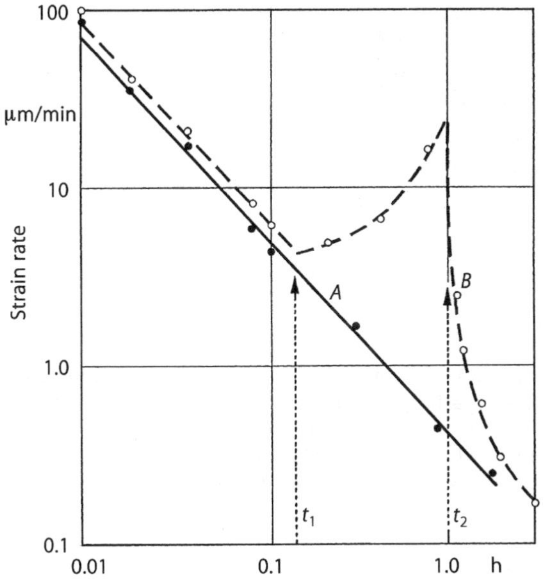


**Fig. 15.14.** Proving direct potential control of crack growth in cylindrical tensile test specimens of Fe-Cr19-Ni8 stainless steel loaded with 200 MPa in boiling $MgCl_2$ solution. Solid curve: Pure creep observed in a specimen cathodically protected at $\varepsilon = -130$ mV. Broken curve: Potential initially kept at $-130$ mV up to time $t_1$, switched to $-90$ mV at $t_1$, causing stress corrosion cracks to extend, and switched back to $-130$ mV at time $t_2$. Note interruption of crack growth by resetting the potential to cathodic protection, proving the change in potential to reach crack tips. (From Ternes)

as in conventional tensile testing, transport paths grow with time, and this type of conditions will usually be encountered in service.

When anodic rapid crack tip dissolution is decisive for crack extension in a system where crack walls are passive, the problem of calculating electric current flow and mass transport essentially is the same as for pitting, already described in Chap. 12. Sometimes, however, both ohmic voltages and concentration gradients along the flow path from the crack tip to the mouth are unimportant. This holds when crack yawning is considerable, i.e. when the crack mouth opening (CMOD) is large, and the the *crack tip opening displacement* (CTOD) is not negligible. In this case, which applies to stainless steels in boiling $MgCl_2$ solution, changes in potential, imposed to the crack mouth by a potentiostat, directly reach through to the crack tip. This is convincingly confirmed by constant load tests under potentiostatic control, changing from cathodic protection to anodic polarization and back to cathodic protection, presented in Fig. 15.14 [5]. In the first place, these observation also very convincingly prove an electrolytic mechanism of SCC in this system. They also prove that any ohmic potential drop along the crack interior must be small compared to 40 mV, and this, then, is due to crack yawning. A more quantitative approach to ohmic voltages along cracks is reported in [9b,c], and illustrated in Fig. 15.15. As typical values, CTOD =

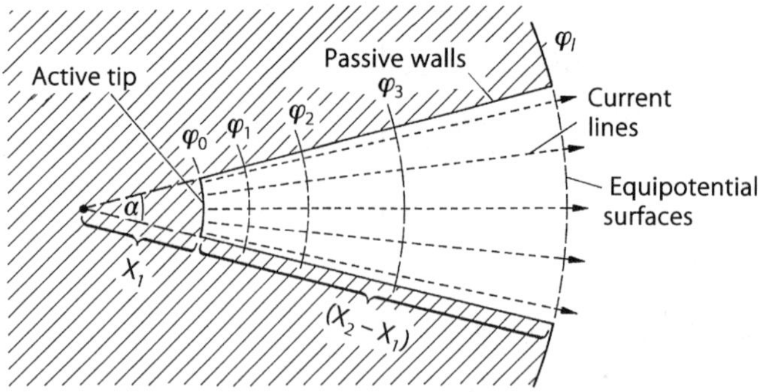


**Fig. 15.15.** Estimating Ohmic voltages, $\Delta\phi_{\text{Ohm}}$, in a wedge-shaped crack of opening angle $\alpha$. The crack length is $l = x_2 - x_1$, the crack tip opening displacement is $h$. Current flow is due to anodic metal dissolution at the crack tip only, and ionic transport along the crack interior to an external cathode, with no participation of repassivated crack walls through either passive metal dissolution or cathodic hydrogen deposition and/or oxygen reduction. (From Engell)

$10^{-6}$cm and a crack opening angle $\alpha = 5° = 0.087$ rad where assumed. Let $\kappa$ and $I_{\text{tip}}$ be solution conductivity and true crack tip metal dissolution current density, respectively. Since $x_1 \simeq (h/2) \times sin(\alpha/2)$, and $l \simeq x_2$, the ohmic voltage drop is

$$\Delta\phi_{\text{Ohm}} = (I_{\text{tip}}h/\alpha\kappa)\ \ln(x_2/x_1)\ . \tag{15.1}$$

For values $I_{\text{tip}} = 2.5$ A cm$^{-2}$, $l = 0.1$ cm, and, as in 5% NaCl solution, $\kappa = 6 \times 10^{-2}(\Omega\ \text{cm})^{-1}$, the calculated ohmic voltage is 4 mV, and thus small indeed. Owing to the assumed trapezoidal cross section of the crack, an increase in crack length does not much change this value. It is only for much sharper cracks, as in high-strength steels, that considerable ohmic voltages would be expected, but then we do not, for these materials, expect crack extension by anodic metal dissolution only. A different case to be considered further below is SCC in high-temperature very dilute solutions of low conductivity.

To calculate transport effects, the Nernst-Planck equations describing ionic fluxes in dilute electrolytic solutions by independent superposition of Ohm's law for electric currents and Fick's 1st law for diffusion transport are usually accepted as, hopefully, sufficiently accurate. For concentrated electrolytic solutions, these 'dilute solution' calculations are approximations at best, as already described in Chap. 12 on pitting.

A particularly simple model of a crack is a rectangular slot, and some essentials of stress corrosion crack properties can rather easily be demonstrated considering this simple geometry. The slot, with an active bottom and parallel passive walls, incidentally also is an alternative model for a corrosion pit, and we can, in fact, use calculations originally carried out for pitting rather than SCC [15]. Figure 15.16a gives the cross section of the slot, indicating the direction of the $x$-axis, pointing from the origin in the slot mouth to the slot

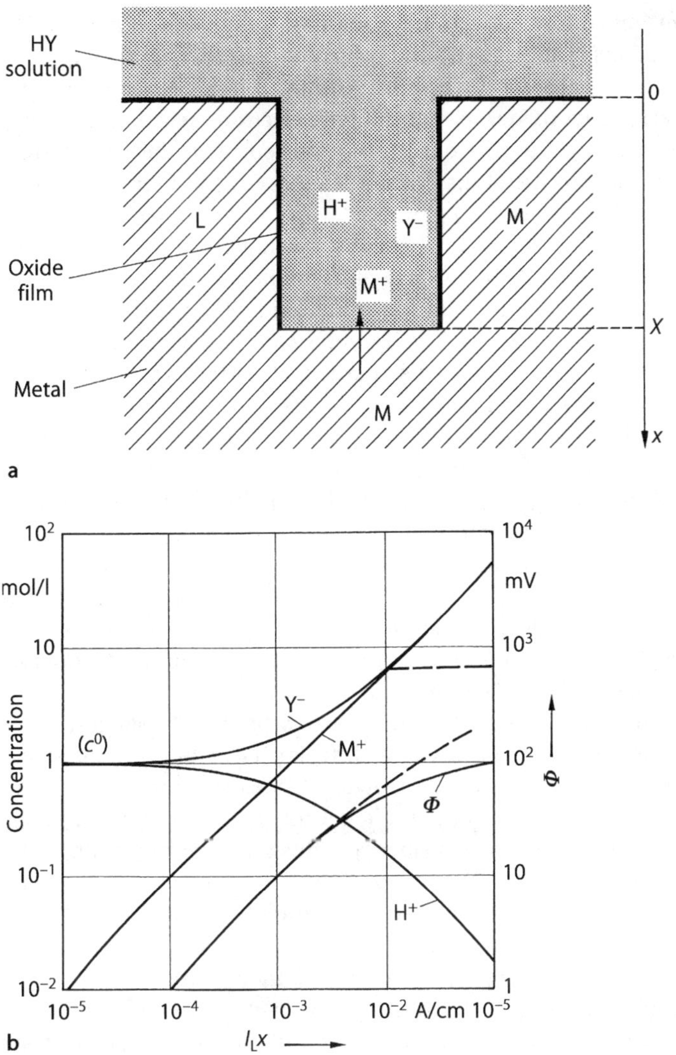


**Fig. 15.16. a** Cross section through a rectangular slot with inert walls and active bottom. **b** Distribution of concentrations, $c_i$, and ohmic voltage drop, $\Delta\phi$, in the slot as calculated for steady-state, assuming a non-receding bottom. Broken curves indicate concentrations and ohmic drop in the case of a solution saturated with respect to the salt MeY. (From Pickering and Frankenthal)

bottom, with the origin in the slot mouth. Assume a rectangular active slot bottom area, $f_{\text{tip}} = h \times b$ (width $\times$ breadth), which remains constant regardless of slot growth. On open circuit, without external polarization, the cathodic partial reactions would have to be considered as occurring at the outer surface of the specimen, but also at the crack walls, and a further point of consideration would be the metal dissolution from the passive surfaces. To simplify matters, and in fact without much loss concerning essentials, we introduce, instead, the following simplifications: The passive anodic current density, $i_{pass}$, is negligible. We assume inert crack walls, as for instance owing to lack of electronic conductivity of the passivating oxide, so that the only electrode reaction occurring is active anodic metal dissolution from the slot bottom with the true current density, $I_{tip}$. Anodic slot-tip metal dissolution is enforced by external galvanostatic polarization with the current $j_{an} = I_{tip} f_{tip}$. Any cathodic partial reaction at the slot tip is negligible, and the total current flowing through the slot from the tip and out of the mouth is $j_{an}$. Consider a simple anodic crack tip reaction: $Me \rightarrow Me^{z+} + ze^-$, into the solution of an acid, HY, which completely dissociates to cations $H^+$ and anions $Y^-$. Hydrolysis of dissolved metal cations and hydroxide or oxide deposition are disregarded. Nevertheless, repassivation is thought to exactly follow slot growth.

We wish to calculate the concentrations, $c_{Me^{z+}}$, $c_{H^+}$, and $c_{Y^-}$, and also the Galvani potential, $\phi$, along the interior of the slot. We note that measurable differences of $\phi$ are ohmic voltage drops. Outside the slot mouth, in the well conducting bulk solution, both concentration gradients and ohmic drops are zero; also we have $c_{H^+} = c_{Y^-} = c_\circ$, and $c_{Me^{z+}} = 0$. For convenience, we temporarily refer Galvani potentials to the slot mouth, setting $\phi_{x=0} = 0$. At $x = 0$, the ohmic drop is zero. The problem is one-dimensional, and we need to consider gradients in concentrations of substances labeled $i$, and gradients in potential as differentials, $dc_i/dx$ and $d\phi/dx$ only.

So far, we might continue handling the case as previously suggested for hemispherical pits in Chap. 13, assuming a diffusion layer of constant thickness in the slot mouth, with a slot electrolyte of constant composition underneath. This is not, however, a suitable model for a stress corrosion crack. Instead, it is more appropriate to consider a diffusion layer extending from the slot mouth to the slot tip. This then produces a major complication as compared to the previous model: The slot grows with the current density $I_{tip}$ and therefore, by Faraday's law, with the bottom displacement rate $v = dx_{bottom}/dt = I_{tip} M_{Me}/z_{Me} F \varrho_{Me}$, where $M_{Me}$ and $\varrho_{Me}$ are atomic weight and specific density of the metal Me, respectively. If now the diffusion layer grows accordingly, all gradients in concentrations and potential in the solution along the slot change already by this geometrical effect alone, and we have a receding-boundary diffusion and migration problem. By virtue of this, we have, in principle, to consider, for every species $i$, Fick's 2nd law in the form

$$\frac{\partial c_i}{\partial t} = D_i \frac{\partial}{\partial x}\left(\frac{\partial c_i}{\partial x} + c_i \frac{z_i F}{RT}\frac{\partial \phi}{\partial x}\right) . \tag{15.2}$$

At the same time, the flux density, $J_i$, of any species $i$ obeys the Nernst-Planck equations.

$$J_i = -D_i \left(\frac{\partial c_i}{\partial x} + c_i \frac{z_i F}{RT}\frac{\partial \phi}{\partial x}\right) . \tag{15.3}$$

Again, as in the case of pit growth in aluminum, an additional term of convective mass transport is considered negligible. For a rectangular slot, and in the absence of inside-slot electrode reactions, except anodic slot-tip metal dissolution, the condition of continuity of mass fluxes requires that for every $i$, $dI_i/dx = 0$. Without further simplification, the double system of partial differential equations is difficult to handle, except by computerized iteration. In [15], the problem was dismissed by neglecting the recession of the slot bottom, assuming $dx_{bottom}/dt = 0$, and thus also $\partial c_i/\partial t = 0$. We are then left with the Nernst-Planck equations only and, as for the case of pitting in aluminum, $J_{Me^{z+}} = I_{tip}/z_{Me^{z+}}F$, whereas $J_{H^+} = J_{Y^-} = 0$. For the analytical solution of the problem, conveniently expressed in terms of the variable $I_{tip} \times x$, see [15]. Figure 15.16 presents the result for $z_{Me^{z+}} = 1$; $D_{Me^{z+}} = 1 \times 10^{-5}$ cm$^2$ s$^{-1}$, and $c_\circ = 1$ mol L$^{-1}$. Assuming, for instance, $I_{tip} = 1$ A cm$^{-2}$, the ohmic drop down the slot is found to be less than 100 mV even for slot lengths of several 100 µm.

For completeness, we recall here an abnormal type of ohmic resistance in slots or cracks, already referred to previously. It is that hydrogen gas bubbles, when evolved by cathodic hydrogen deposition in the crack, may be sticking in the slot, interrupting mass transport and thus causing an apparent increase in resistance [15]. We shall not, in the following, further refer to this special case. Instead, we note that the calculated magnitude of the 'regular' ohmic drop, described above, is due to the assumption of a rectangular, i.e. parallel-wall slot. A certainly more realistic model for a corrosion crack is a tapering slot with wall spacing widening from tip to mouth. Then, the slot cross section increases from $f_{tip}$ to $f_{mouth}$, via intermediate values $f_{slot}(x)$. In steady state, and when the only electrode reaction still is slot-tip metal dissolution, the continuity condition now applies to fluxes $J_i \times f_{slot}$, rather than flux densities, i.e. $d(J_i \times f_{slot})/dx = 0$. We do not inspect the mathematics, which are not difficult for a regular trapezoidal slot. Qualitatively, it is intuitive that along the widening slot both the ohmic drop and the concentration gradients shrink. Therefore, $\Delta\phi_{Ohm}$ will be found smaller than estimated by the previous calculation, and the increase in concentration in both $Me^{z+}$ and $Y^-$ will concentrate towards the slot tip. For a quantitative treatment adapted to cracks in stainless steels, see e.g. [80]. More on trapezoidal pits in aluminum was already reported in Chap. 10.

Returning to rectangular slots, we see that the calculation does show the main effects of the interplay of electrode reactions, diffusion and migration on slot or crack electrolyte chemistry. It does not, however, aim at explaining observed slot or crack growth rates, or else at predictions of such rates, as they would be interesting especially for safety considerations. Therefore, we still are far from what would be most interesting, but we can outline what would be necessary to improve the situation: For any given set of conditions, the potential, either under external and by preference potentiostatic control at $\varepsilon_{pot}$, or else established as the open-circuit free corrosion potential, $\varepsilon_{corr}$, is the starting point. Considering the low current flow out of the crack mouth into the bulk solution to either a counter electrode, or else to the surrounding passive metal surface, we see that in well conducting, i.e. not very dilute solutions, the slot mouth or crack mouth potential, $\varepsilon_{mouth}$, coincides with either $\varepsilon_{pot}$, or $\varepsilon_{corr}$. The slot tip or crack tip potential, $\varepsilon_{corr}$, more negative than $\varepsilon_{mouth}$ by a difference $\Delta\phi$

as described above, co-determines $I_{tip}$ according to electrode kinetics as they apply to the crack-tip anodic metal dissolution. Other factors, including slot-bottom or crack-tip solution composition, and rate of local plastic deformation, enter these kinetics as well, and reliable suppositions concerning $I_{tip}$ appear difficult indeed. The usual assumption is Tafel kinetics, with a relation of the form $I_{tip} = const \times \exp(const' \, \varepsilon_{tip})$. Here, we must recall from Chap. 6 considerable inaccuracies encountered when attempting to predict corrosion rates from summarily determined polarization data. Even when the assumption of Tafel kinetics is better than arbitrary, but when the choice of values for exchange current densities and Tafel slopes is a matter of more or less intelligent guessing rather than experimental evidence, caution with respect to the application of such data for serious safety consideration appears very much suggested, indeed.

If, as normally in practical cases, the external electrode potential is left to adapt freely to conditions, the system is that of a large cathodic passive area in contact with a small crack tip anode. The outside potential is established by the superposition a) of the currents of anodic passive metal dissolution and of reduction of e.g. dissolved oxygen, and b) of the short-circuit current passing to the crack tip. The crack tip potential is governed by the kinetics of anodic active metal dissolution only. Crack walls are passive similar to the outside surrounding, but possibly at potentials shifting to the crack tip potential along the crack interior according to the ohmic voltage drop. Then, considerations apply as described in Chap. 11 for contact corrosion in the case of a very large cathode/anode area ratios.

Still for this case of non-control of electrode potentials by external polarization, an interesting point is the interdependence of crack tip, crack-mouth and external, outside potential. In the first place it can be expected that the polarizability, $R_\pi$, of crack tip anodes is much smaller than that of passive areas. If the crack mouth is surrounded by large areas of uncracked material, the electrode potential, referred to a reference electrode outside the crack equipped with a capillary tip adjacent to the metal surface, nevertheless will normally be the corrosion potential, $\varepsilon_{corr}$, as observed in the absence of cracks, owing to the large cathode/anode area ratio. This holds except for a small area surrounding the crack mouth, of a size depending on anode and cathode polarizabilities and on solution conductivity. So far, the situation is similar to that of an external potential fixed by a potentiostat, and the variable property is the crack tip potential only. If, instead, we are dealing with families of neighboring cracks, the combined tip anode area may become large enough to polarize the neighboring external cathode surface to a potential below the free corrosion potential observed in the absence of cracks.

Also, if the cathodic reaction accompanying corrosion is oxygen reduction, or reduction of some other dissolved species transported into the crack interior by diffusion only, it can be neglected inside the crack. The situation is different when one of the cathodic reactions is hydrogen deposition, restricted to the crack tip area again through the pitting mechanism, but then it is known, as described in Chap. 10, that this side effect, although decisive for crack propagation in the case of HISCC, does not contribute more than ca. 10% to the local active metal dissolution rate. In any analysis of crack extension mechanisms, 10%-effects usually are quite below the accuracy limits of either experimental evidence or theoretical calculations. Another matter is, however, local hydrolytic acidifying entering the kinetics of anodic metal dissolution, since crack tip acidity greatly decelerates anodic iron dissolution through the catalytic mechanisms described in Chap. 5.

The boundary condition of inert crack walls can be relaxed by allowing for noticeable passive current densities from crack walls and, possibly, crack wall hydrogen deposition. The passive current density is certainly small compared to the crack tip dissolution current density; still, passive metal dissolution from the comparatively large crack walls may noticeably contribute to overall metal dissolution into the crack. When, as observed for passive iron, $i_{pass}$ is constant, independent of potential, an ohmic voltage down the crack does not matter in this respect. When, instead, $i_{pass}$ depends on the potential, and when at the same time $\Delta\phi_{Ohm}$ is important, the current-potential function $i_{pass}(\varepsilon)$ may need attention. Originally, this situation was considered for rectangular cracks of width $h$ and breadth $b$, and for this case the problem then still is one-dimensional [16]. We continue using an $x$-axis pointing from the crack mouth at $x = 0$ to the crack tip, and we once more disregard gradients in solution conductivity by substituting a constant average value, $\kappa$. Let $j(x)$ be the electric current flowing in the crack electrolyte. The differential change in ohmic voltage, equal to the change of the Galvani potential and of the electrode potential along the crack wall, is $d\phi = d\varepsilon = j(x)dx/hb\kappa$. As the current passes along the crack walls, passive metal dissolution from both walls adds to it, causing a differential change $dj(x) = 2bi_{pass}dx$. This leads to a differential equation for the electrode potential distribution between crack tip and mouth: $d^2\varepsilon/dx^2 = 2i_{pass}/h\kappa$, which we do not further follow to its solution. We do note that, originally, this type of calculation was applied to SCC without recognizing crack wall passivity. Instead, crack walls and tips were both considered to be active and to obey Tafel kinetics. Crack growth was thought to be the result of a change in the values of the pre-exponential factors in the current-potential functions, assuming deformation-induced acceleration of crack tip corrosion. If now, for passive crack walls, the passive current-potential curve is known, the line of calculations essentially remains the same, as seen further below.

For brittle materials, the crack tip opening may become very small, tending to zero in the limiting case of ideal brittleness. In this case electrode kinetics may greatly change, when the spacing between crack walls close to the tip is less than the thickness of the electrical double layer. For dilute solutions, where the diffuse part of the double layer is large, and the plate capacitor model introduced in Chap. 5 fails, the effect may be important rather more often than considered at present. The situation, which is inspected in [17], is quite intriguing, but we shall leave it without further discussion owing to lack of further experimental evidence.

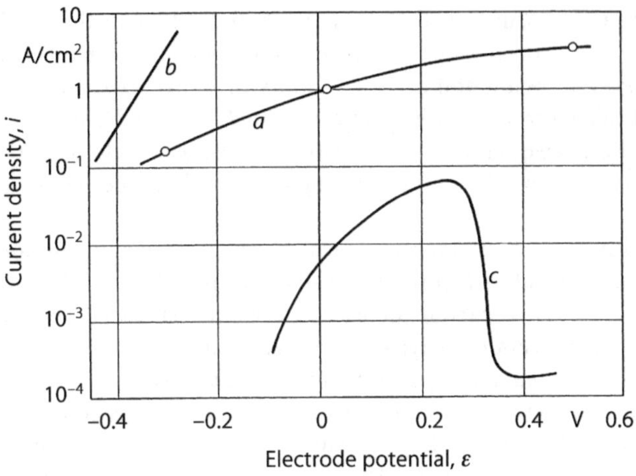


**Fig. 15.17.** An attempt to determine the kinetics of anodic titanium dissolution, $Ti \rightarrow Ti^{3+}$, in 12 M HCl solution from a bare metal surface produced by brittle fracture. *a* Peak current densities observed within 1 ms after fracture, including high Ohmic voltages. *b* Estimated bare-metal current-potential curve corrected for Ohmic voltages. *c* Steady-state current-potential curve for passive titanium in the same solution. (From Beck)

The major issue in SCC is crack-tip electrode kinetics, but direct evidence is not easily available. Assuming Tafel kinetics, i.e. exponential dependence of $I_{tip}$ on crack tip potential, high values of this current densities are expected, because the external electrode potential of a passive metal usually is far beyond the $Me/Me^{z+}$ equilibrium potential, and overpotentials $\eta_{Me} = \varepsilon_{tip} - E_{Me/Me^{z+}}$ tend to be high. The effect should be especially pronounced when, as for aluminum or titanium, the standard potential, $E^{\circ}_{Me/Me^{z+}}$, is very low. An interesting method to imitate crack tip conditions on larger surface areas is to activate passive electrodes by rapid and severe strain, and to record peak values of the repassivation current densities. Or else rapid straining is replaced by brittle fracture of electrodes, exposing fresh and initially active fracture faces to the solution. The result of an experiment of this type is shown in Fig. 15.17 for pure titanium fractured in concentrated hydrochloric acid [18]. In essence, the measurement gives a current-potential curve recorded for active metal dissolution from bare titanium. Since the metal dissolution currents are very high, this curve is heavily distorted by an ohmic voltage between the metal surface and the tip of the reference electrode. This, incidentally, also gives an idea of the magnitude of ohmic voltages expected down stress corrosion cracks. The indicated approximately true $i(\varepsilon)$ curve, which should be similar to the crack-tip current-potential curve, $I_{tip}(\varepsilon)$, may not be very accurate, but it does show values of the metal dissolution current density compatible with quite rapid crack extension. It should especially be noted that these peak current densities are by orders of magnitude higher than the critical passivating current density, $i_{crit}$, of titanium in the same solution, showing that, at least in this particular case, crack tip conditions probably are far from conditions of the normal active/passive transition.

For more ductile materials, most considerations concerning crack-tip conditions use the concepts of ductile deformation processes accompanying crack propagation in the plastic zone ahead of the crack tip. These slip processes will expose crystallographic slip steps to the aggressive solution, and on these provide steps half-crystal sites for easy metal ion transfer from the surface into the solution [6c].

This is sometimes described as dissolution 'in statu deformandi', or as 'mechano-chemical effect'. From this effect, an increase in the exchange current density, $i^{\circ}_{Me}$, is expected, and originally this was thought to be the origin of enhanced crack tip dissolution as compared to crack wall dissolution. Although this was an erroneous presupposition, the nevertheless quite useful suggestion was to try to rapidly strain metal electrodes at an overall rate compatible to what is expected at crack tips. Straining stainless steel wires with $\dot{\epsilon} = 108\%\ \text{min}^{-1}$ in concentrated, boiling, and well stirred $MgCl_2$ solutions already produce anodic currents up to 160 mA $cm^{-2}$ [6d]. Considering realistic, higher crack tip strain rates, anodic metal dissolution at crack tips up to ca. 1 A $cm^{-2}$ could be explained, close to what is required for modeling crack extension by electrolytic metal dissolution without further effects such as brittle fracture steps.

Anodic current increase caused by fast straining of wires was also observed for Al-7Mg in NaCl and in $Na_2SO_4$ solutions at pH 5.5 and pH 2 [6e]. As seen in Fig. 15.18 unstrained electrodes in sulfate solutions exhibit pure passive behavior, while in chloride solution a breakthrough potential is observed which in this case is the potential threshold of onset of intercrystalline corrosion. Straining leaves this threshold unchanged, but it lifts current densities by several order of magnitude regardless of potential. Qualitatively, this is what is needed to explain cracking in this alloy. The increase of the passive current density below the breakthrough potential indicates an increase in oxide dissolution rate, accompanied by film repair. The effect would merit reconfirmation and close inspection, but fast strain rate experiments apparently have fallen in disuse. Concerning SCC, the meaning of these observations remains obscure because, as already decribed previously, and as more fully discussed further below, intercrystalline SCC of aluminum alloys proceeds by discontinuous crack extension events.

Equally discontinuous transcrystalline stress corrosion crack extension has been observed in single crystals of $\alpha$-brass slowly bent in ammoniacal solution [19b]. This work merits particular consideration, as it emphasizes local embrittlement of an intrinsically ductile material by local vacancy injection accompanying dealloying. Ignoring details, the sequence of events, as revealed by direct microscopic observation of the walls of a cracking specimen is as schematically indicated in Fig. 15.19: a) The typical situation is that a wide crack has stopped in a band of heavy glide. Slip further in front of the crack tip is thought to be prevented by short-range order in the adjacent CuZn lattice. b) The crack tip widens, apparently by local corrosion, which presumably

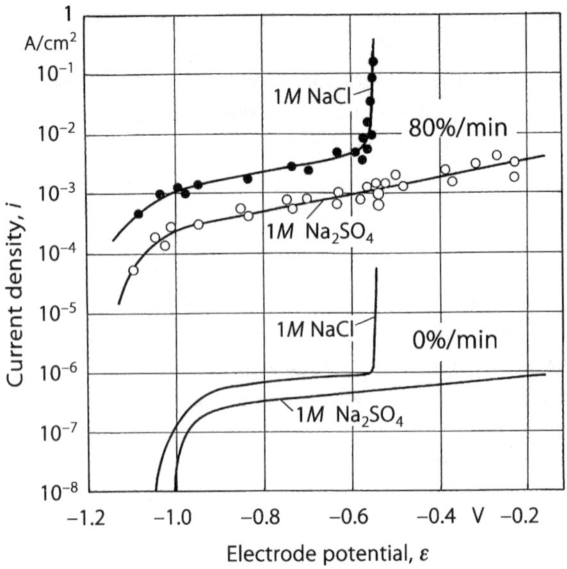


**Fig. 15.18.** Fast electrode straining experiments with Al-7 wt.% Mg wires, previously heat treated 4h at 360°C and water-cooled. Quasi-stationary current-potential curves were recorded in 1 M NaCl, resp. 1 M $Na_2SO_4$ solutions at pH 5.5. Upper curves were recorded with a strain rate of 80% min$^{-1}$, lower curves (experimental points omitted) without strain. (From Ford and Hoar)

is selective zinc dissolution injecting vacancies into the adjacent alloy region. c) Into the embrittled crack tip environment, a ' ghost crack' suddenly proceeds. This is a sudden brittle decohesion producing a hair-like crack without visible CTOD, and it stops in a band of previous weak glide which locally has destroyed the short range order. d) The hair crack widens by slip producing heavy glide, bringing the situation back to stage a), etc.

When crack extension is intermittent, the total extension time is the sum of arrest periods and growth periods:

$$\tau_{tot} = \Sigma\Delta\tau_{arr} + \Sigma\Delta\tau_{ext} \,. \tag{15.4}$$

When, furthermore, cracks grow by intermittent brittle decohesion steps, we expect $\Sigma\Delta\tau_{arr} \gg \Sigma\Delta\tau_{ext}$. In the work described above, single arrest periods, $\Delta\tau_{arr}$, of up to 10 s were observed, and thus were indeed dominant. In other cases, arrest periods may be much smaller, and crack extension quasi-continuous. Overall crack growth rates may, then, by far exceed possible rates of crack growth by anodic metal dissolution alone, but still much slower than overall true brittle cracking.

For SCC of polycrystalline brass, ammoniacal 'tarnishing solutions' producing visible dark oxide films, are distinguished from 'non-tarnishing solutions' where oxide films either are absent, or else transparent [20,21]. Considering the inherent brittleness of thick oxide films, intermittent crack growth is then

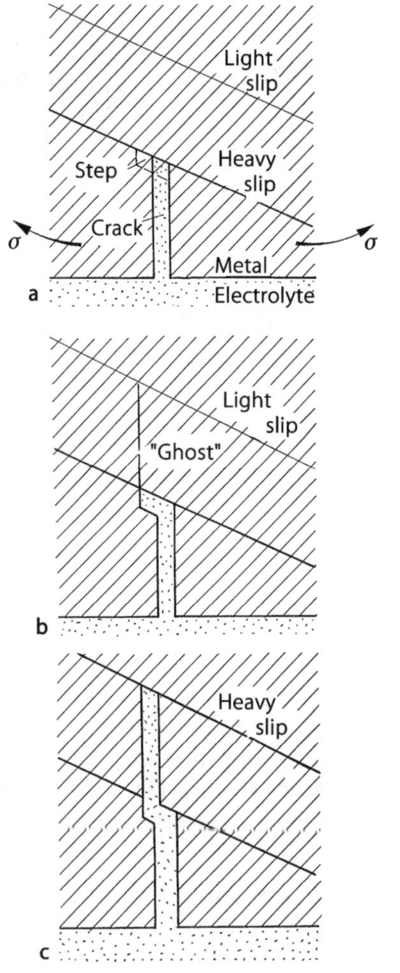


**Fig. 15.19.** Schematic drawings of successive stages of discontinuous transcrystalline crack growth in $\alpha$-brass single crystals loaded by bending in non-tarnishing ammoniacal solutions, showing triggering of crack growth from metal locally embrittled by dealloying corrosion. (From Edeleanu, but simplified)

modeled as in Fig. 15.20: In this case, SCC is intercrystalline, and it is thought to start by intercrystalline oxidative attack at grain boundaries by what mechanistically is low-temperature oxidation. Tensile stress will rupture the oxidized grain boundary, at the same time opening the crack by plastic slip, again exposing an emerging grain boundary. The mechanism thus again is repetitive.

For a long time, the favorite system for the application of the slip-step dissolution model was transcrystalline SCC in homogeneous austenitic stainless steel, although the records of both acoustic emission signals as well as of strain rates under constant load point to continuous crack extension [22a]. Of late, the preferred candidate for this mechanism rather is intercrystalline SCC in sensitized stainless steels. As applied to transcrystalline SCC in austenites, the main features of the model, illustrated in Fig. 15.21, were quite convincing: Slip occurring in a tensile specimen at the crack tip, owing to the peak in local

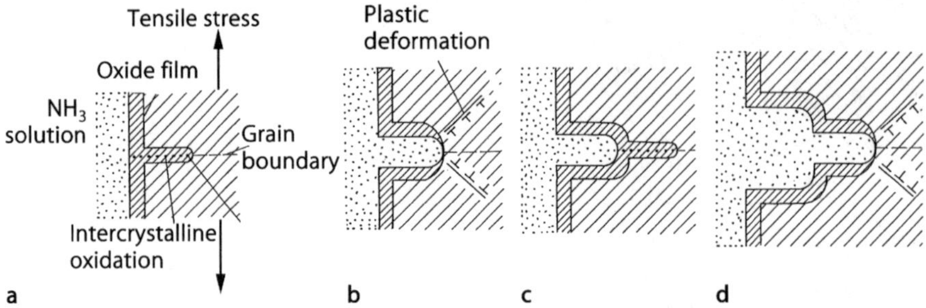


**Fig. 15.20.** Modeling discontinuous intercrystalline crack growth in polycrystalline $\alpha$-brass tensile specimens in tarnishing ammoniacal solutions. (From McEvily and Bond)

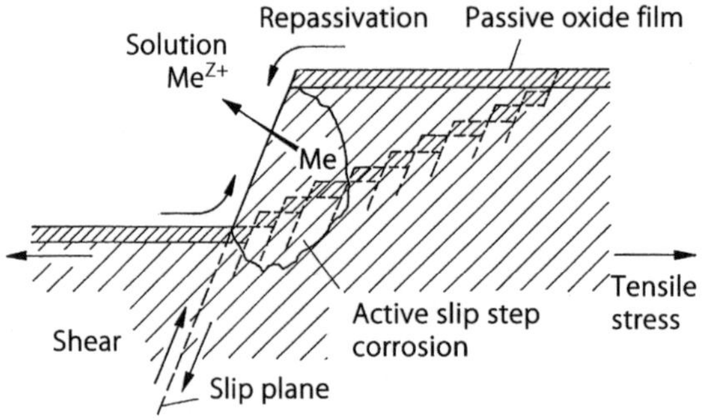


**Fig. 15.21.** The slip-step dissolution model for discontinuous crack extension: Important transient local metal dissolution from coarse slip steps before repassivation as compared to negligible effect of fine slip. (From Pickering and Swann)

stress in the plastic zone in front of the crack tip, ruptures the passive oxide film, and transient local anodic metal dissolution causes transient crack growth until repassivation stops the process. Obviously, this requires coarse slip steps of a height considerably surpassing the thickness of the passive oxide, which itself is unable to participate in slip. According to this concept, fine slip, as also indicated in the figure, does not support crack propagation. For quasi-continuous crack growth, systems of planes of coarse slip are supposed to be active [22b] and Fig. 15.22 shows a number of possible group slip events.

Coarse slip as a prerequisite for SCC susceptibility has been widely discussed in the literature, including the discussion of slip coarsening effects such as grain size, short range and distant range ordering and, especially stacking faults.

The slip-step dissolution model, while certainly attractive, does have its shortcomings, as local metal dissolution accompanying transient activation is not necessarily proceeding into the metal lattice in the direction required for crack extension. Also, local crack tip corrosion events such as microtunneling [6b,c,23] and micropitting [8a] are not easily incorporated in the model. Microtunnels aligned along stress corrosion cracks in stainless steel are shown in Fig. 15.23.

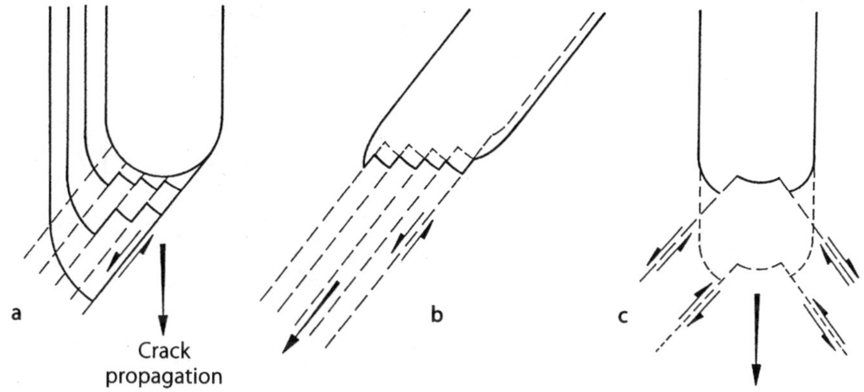


**Fig. 15.22a–c.** Quasi-contin-uous crack extension by transient metal dissolution from neighboring sites of coarse slip. Crack propagation **a** sloping relative to single slip system, **b** following the direction of single slip system, **c** following two simultaneously active slip systems. (From West)

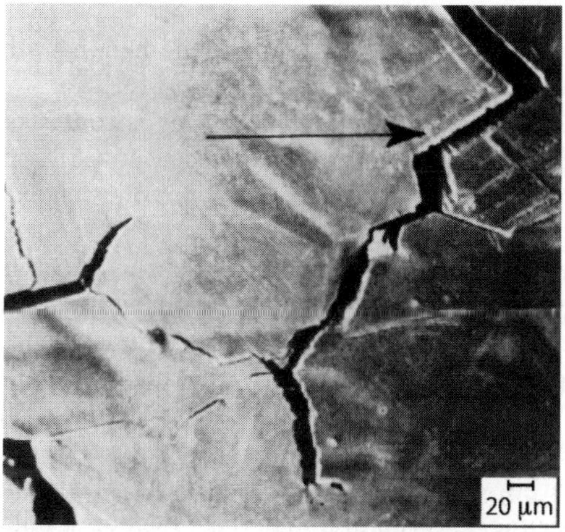


**Fig. 15.23.** Aligned corrosion tunnel mouths (→) accompanying trans-crystalline cracking of 18 Cr (Ni steel in $H_2SO_4$/NaCl solution. (From Scully)

Microtunneling was described in Chap. 8 as also typical for dealloying of homogeneous alloys such as CuPd, CuAu, and NiAu at critical potentials. In stainless steel, microtunnels aligned in a plane perpendicular to the crack opening stress component can easily supposed to participate in crack extension, as schematically shown in Fig. 15.24 [23]. The model has been refined using information obtained from TEM. Cracks were found to nucleate in the vicinity of nonmetallic inclusions on coarse slip steps by tunneling corrosion, creating a spongy structure, followed underneath by the formation of microslots approximately parallel to (110) faces, coalescing to crack nuclei of very small width,

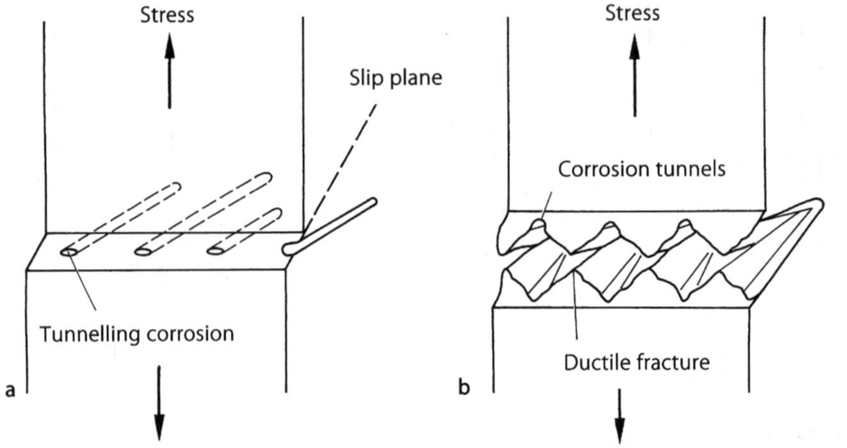


**Fig. 15.24.** The model of crack nucleation by rows of corrosion microtunnels in slip steps followed by ductile tearing of the remaining ligamens. (From Swann and Embury)

<10 nm. Tunneling again was supposed to be a dealloying phenomenon, with tunnel walls depleted in Ni and Fe and, thus, enriched in Cr and Mo.

The opinion that SCC susceptibility is a property of alloys only is contested by the observation of transcrystalline SCC in pure copper subjected to CERT in nitrite solutions [21c,24c,26]. Another interesting case is SCC in zone-refined, almost pure iron containing impurities at a level of only ca. 40 ppm [27]. Specimens were tempered in the austenite temperature field and quenched. In this state, impurities should have segregated to prior austenite grain boundaries, and ferrite grain boundaries should be pure iron. From CERT some evidence of SCC was obtained. By metallographic inspection, some cracking was identified as occurring, indeed, along prior austenite, but also along ferrite grain boundaries. In view of such observations it would seem rational to consider SCC susceptibility as not disappearing, but only tending to very low levels in pure metals [8f]. The point is academic, as structural materials, including mild steel, are, of course, all more or less highly alloyed.

Concerning homogeneous binary alloys, an extensive series of experiments screening a broad range of compositions has been described in [28]. As a typical example, Fig. 15.25 compiles data for CuAu solid-solution alloys observed under constant deflection in a variety of electrolytic solutions, all causing intergranular failure. As noted in Chap. 13, for these alloys a distinction must be made between intercrystalline corrosion and intercrystalline stress corrosion cracking, owing to the superposition of a 'static' and a 'dynamic' solid-solution alloy effect. There is, however, no doubt that the decrease in time to failure with increasing Au content to values as low as 1 s is due to intercrystalline SCC. The two solid-solution effects are both attributed to intergranular misfit dislocations: without tensile strain, preexisting misfit dislocations already cause intercrystalline susceptibility, and plastic strain adds the effect of continuously

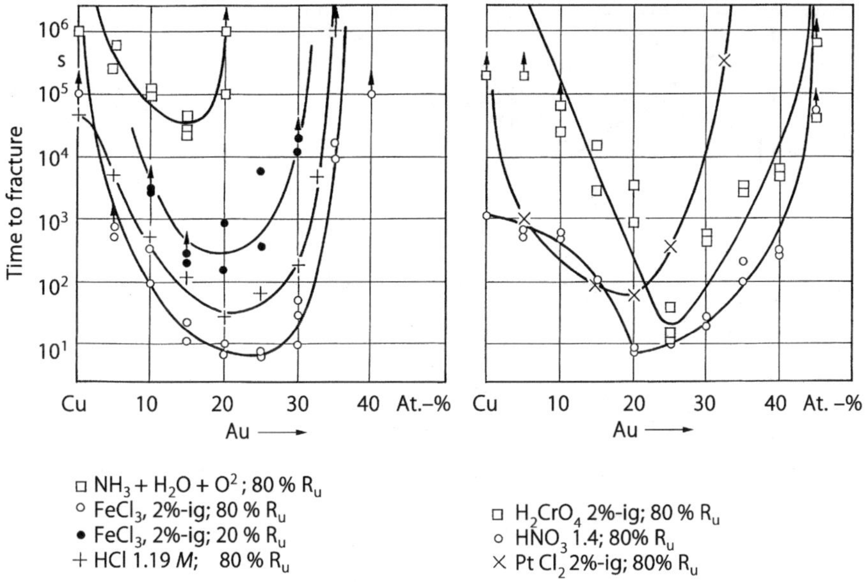


**Fig. 15.25.** Times to failure, $\tau_{fl}$, of tensile specimens prepared form CuAu solid- solution alloys in various test solutions. (From Graf and Buckel)

generating more dislocations [28c]. A conclusion from this supposed mechanism is that $\tau_{fail}$ curves should have a minimum at 50 at.% Au. Instead, the minimum always appears at much lower gold contents. The tentative and arbitrary assumption is that intercrystalline corrosion trenches at grain boundaries are blocked by too much gold remaining as a remnant of selective dealloying.

The existence of the two solid solution effects is well documented by comparing the influence of intercrystalline pre-corrosion without tensile stress on ultimate strength with the combined influence of pre-corrosion and SCC during tensile testing [28d]. As seen in Fig. 15.26, ultimate strength values as observed in 'dry' tensile tests decrease much less with increasing pre-corrosion time, than values observed during 'wet' tensile tests with specimens immersed in the aggressive electrolytic solution.

As the ductility of notched specimens decreases, so does CTOD, shrinking, in the limiting case of ideal brittleness, to one interatomic lattice spacing at the crack tip. Crack extension then only requires successive steps separating adjacent atoms in the crack tip. It has been early considered that adsorption of anions from the crack electrolyte onto pairs of metal atoms at the crack tip can then lower the interatomic bonding energy and thus increase the frequency of crack-propagating decohesion events [29]. It was also pointed out that anion adsorption increases with increasing potential, and that, therefore, observed influences of polarization on crack extension rates do not directly prove mechanisms requiring anodic metal dissolution. The model has remained hypothetical, as a direct proof of adsorption-enhanced crack extension is not available.

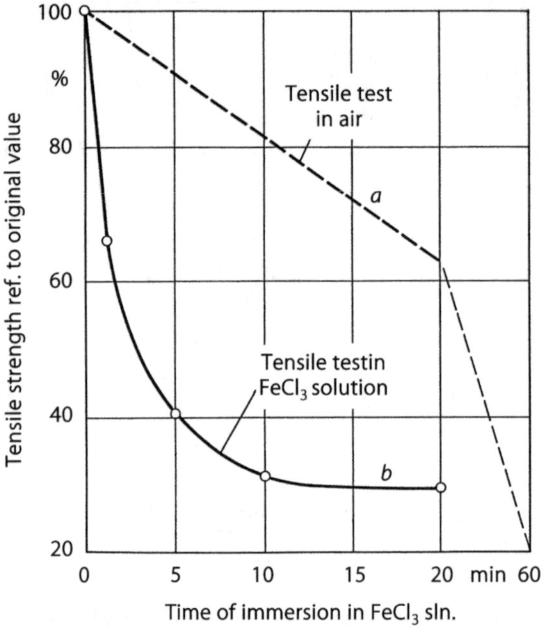


**Fig. 15.26.** The ultimate strength of Cu-10 wt.% Au tensile specimens pre-damaged by intercrystalline corrosion in 2% $FeCl_3$ solution during periods of length $t$. **a** Tensile tests in air with specimens rinsed and dried before loading, showing damage by pre-corrosion alone. **b** Strength of pre-corroded specimens, loaded in $FeCl_3$ solution, showing the added influence of intercrystalline SCC. (From Graf and Klatte)

Also in the direction of crack extension without concurrent anodic metal dissolution, the so-called 'surface-mobility stress corrosion cracking mechanism' postulates that crack extension occurs by migration of vacancies on the two crack walls into the crack tip [30]. According to this model, only elastic stress is relevant for crack extension. Surface self-diffusion governing vacancy flux into the crack tip is supposed to be influenced by interaction with the adjacent electrolytic solution. Using these assumptions, the authors relate crack extension rate, $v$, surface self-diffusion coefficient, $D_s$, average diffusion distance of vacancies, $L$, elastic stress at crack tip, $\sigma$, atom size, $a$, Boltzmann constant, $k$, and temperature, $T$, as follows:

$$v = (D_s/L)\left[\exp\left(\sigma a^3/kT\right) - 1\right] . \qquad (15.5)$$

To arrive at this expression, Fick's 1. law is applied by setting the rate of surface diffusion proportional to the gradient, $dc_\star/dx$, of the surface vacancy concentration in the direction of the crack tip. The gradient is approximated as the ratio, $(c_\star - c_\circ)/L$, of a characteristic difference in concentrations between strained and unstrained surface areas close to the crack tip, and a characteristic diffusion length. So far, the argument concerns kinetics only, but thermodynamics are now invoked by assuming an equilibrium relation $c_\star = c_\circ \exp(\sigma a^3/kT)$.

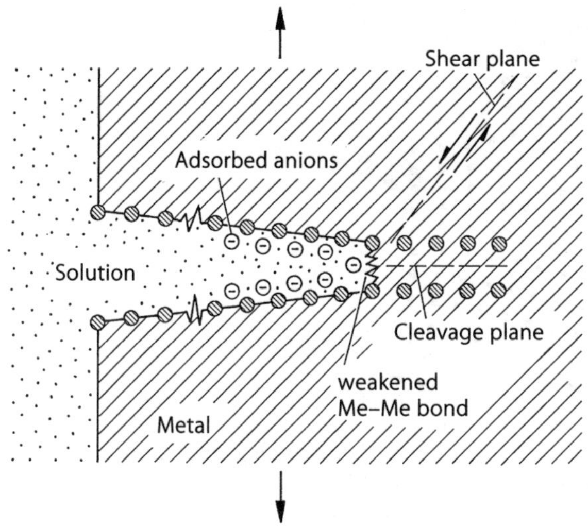


**Fig. 15.27.** Modeling brittle fracture eased by anion adsorption lowering metal-metal bond energy at crack tip

Here, $\sigma a^3$ is the reversible work to move, by a stress $\sigma$, a surface atom of area $a^2$ through a distance $a$. But this describes an equilibrium surface state, where the chemical potentials of vacancies in strained and unstrained areas are equal, and where the driving force for diffusion therefore is zero. The concept, while essentially interesting, thus appears to need reconsideration. For a review, see [46]. The model has also been critisized in [25b] as requiring a vacancy concentration higher than what can be expected from metal physics by a factor $10^{14}$. In a rebuttal it was justly pointed out that metal surfaces in contact with electrolytic solutions might well have a greatly increased vacancy concentration [30d]. In its present version, the model is applied to SCC of alloys subject to dealloying such as CuZn, AuAg, and AgCd, where selective dissolution of the less noble component certainly will leave a highly disordered surface. In this context, an important note is the well known detrimental influence of dissolved copper ions on transgranular SCC susceptibility of brass [21d,29c,30g,f]. This noble-metal cation effect has been confirmed for AgCd as well [30f]. Adding e.g. $Ag^+$ to a solution in contact with an electrode, whose surface is practically pure Ag, owing to dealloying, then fixes the open-circuit electrode potential at the equilibrium potential, $E_{Ag/Ag^+}$. Similar conditions, if somewhat complicated by $Cu^+/Cu^{2+}$ valence changes, hold for brass, but we continue discussing AgCd: Increasing the concentration of dissolved silver cations in a solution in contact with a dealloyed AgCd electrode also changes the electrode potential. An increase in SCC velocity can then be expected and is in fact observed during both slow and fast strain rate experiments, producing transcrystalline SCC. This accelerating effect is, however,

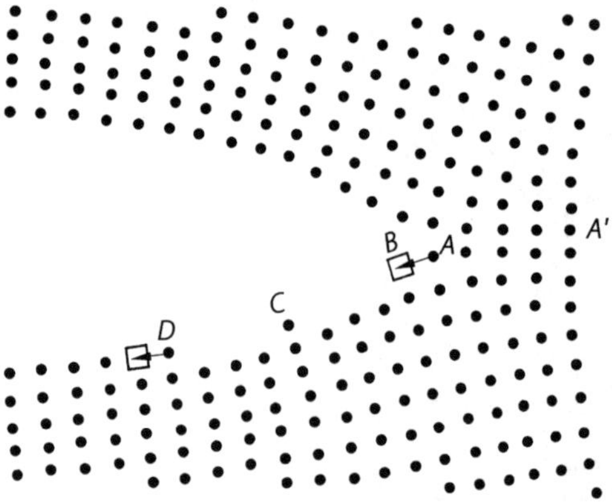


**Fig. 15.28.** The surface diffusion crack extension model, assuming crack growth by surface migration of vacancies to the crack tip. (From Galvele)

attributed to an increase of the equilibrium exchange current density, $i^{\circ}_{Me}$, of the relevant $Me/Me^{z+}$ electrode reaction. To gain quantitative insight, specimens fractured by CERT after a time interval $\tau$ where inspected for the depth, $d$, of brittle crack extension due to SCC, and the ratio $\tau/d$, i.e. an average crack extension rate, $\bar{v}$, was substituted in Eq. (15.4) instead of momentary values, $v$.

We insert here that this calculation of an average crack growth rate as the ratio of depth of ingress of SCC into a CERT specimen and time-to-fracture, while convenient to obtain and therefore widespread, is generally questionable, as will become more obvious as a point of major concern further below.

In the surface-mobility concept, $\sigma$ is simply set equal to overall tension. Average rack extension rates, as defined above, are observed to increase with increasing Zn content in CuZn and with increasing Cd content in AgCd alloys. Furthermore, $\ln \bar{v}$ is directly proportional to the logarithm, $\ln c_{Me^{z+}}$, for dissolved cations of the noble alloy component in the solution, and the slope of this relation compares well with the corresponding dependence of $\ln i^{\circ}_{Me}$ on $\ln c_{Me^{z+}}$. This correlation is seen as having direct physical meaning by a direct proportionality of $D_s$ to the product, $i^{\circ}_{Me}(1 - \gamma_{Me})$, where $\gamma_{Me}$ is the noble-component atomic fraction in the alloy.

## 15.2 Iron and Steels

### 15.2.1 Hydrogen-Induced SCC in High-Strength Steels

Hydrogen induced cracking, HIC, and hydrogen induced stress corrosion cracking, HISCC, have been introduced in Chap. 14 as two limiting cases of essentially

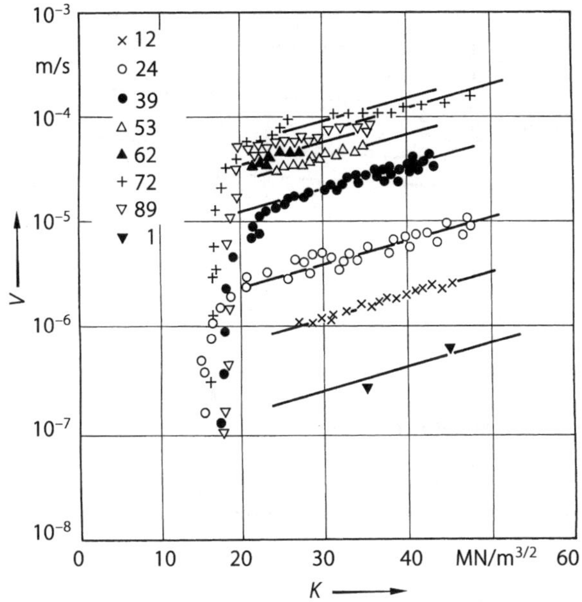


**Fig. 15.29.** The influence of solution temperature on the $\lg v$-$K_I$ curve of a low-alloy steel (AISI 4130, $R_p = 1330$ MPa) in distilled water. (From Nelson and Williams)

similar failure modes. We continue disregarding cases of mixed HIC and HISCC, except for the case of cracking of low-alloy steels in $H_2S$ solutions, relevant for oil field conditions, as discussed in [7i,j]. Here, two groups of steels are distinctly different: Low-alloy structural steels with $400 \leq R_p \leq 600$ MPa, and temper-hardened steels with $500 \leq R_p \leq 1100$ MPa. The minimum load causing indications of SCC was checked by tensile testing in sulfide solutions. It was observed to increase, when $R_p$ increased from 400 to ca. 600 MPa, indicating an increase in resistance to HIC. With further increased strength, this minimum load decreased rapidly, now probably indicating the onset of HISCC. Checking was continued covering steels with Rockwell hardness values between HRc 350 and 550, which embrittled in salt solutions even in the absence of sulfide as a promoter of hydrogen uptake, and with HRc >550, which failed already by exposure to pure water.

Results of fracture mechanical testing of DCB specimens of a high-strength steel in distilled water are shown in Fig. 15.29 [31]. The $\lg v$-$K_I$ curves clearly show the lower threshold, $K_{ISCC}$, of onset of subcritical crack extension of prenotched and precracked specimens and the steep increase in Region I of the curves, followed by what in this case is not a plateau, but a marked further increase in $\log v$ as a function of $K_I$ in Region II. In this region, the dependence of the crack extension rate on temperature is high, pointing to a correpondingly high activation energy of the rate-determining reaction step. Fig. 15.30 shows crack extension rates in DCB specimens of another high-strength steel

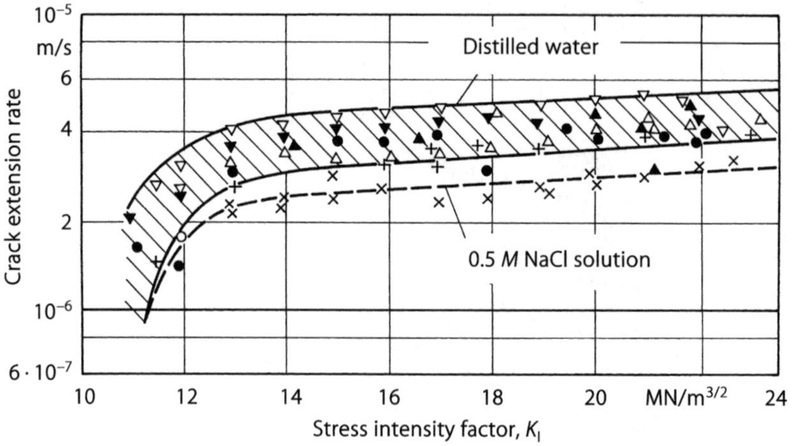


**Fig. 15.30.** Crack extension in a low-alloy steel (90 MnV8, $R_p$ 0 1530 MPa) in distilled water and in 0.5 M NaCl solution at ambient temperature. (From Stellwag and Kaesche)

in distilled water and in NaCl solution [8g], once more illustrating the frequent occurrence of a plateau in the lg $v$-$K_I$ curves in Region II. Here, crack extension depends only little on the salt content of the test solution, which is understandable, since low-alloy steels in neutral solutions are not passive and, therefore, do not require aggressive anions for local oxide film attack. At the same time, as explained in Chap. 14, only little uptake of hydrogen is necessary for local crack tip embrittlement, and this is provided already by little interaction with water.

As seen in Fig. 15.31, potassium chromate added to the test solution slows crack extension without much changing $K_{ISCC}$ or, for that matter, $K_{IC}$ [32]. Chromate and, with equal effect, benzoate, are typical passivating inhibitors, and the data shows that passivation inhibited crack extension in chloride solution. Added to distilled water, chromate was ineffective. We note these observations as hints concerning the largely unknown role of details of crack-tip electrode kinetics, but we omit a further discussion of this point.

Attempts to analyze the typical shape of ln $v$-$K_I$ curves usually concentrate on fracture-mechanical aspects. In this context it should once more be noted that plotting the logarithm of a kinetic property, $v$, as a function of $K_I$, which is a property describing the equilibrium state of elastic tension, is not directly obvious. Concerning Region II, we recall that an increase in $K_I$ probably changes the size of the plastic zone in front of the crack tip rather than the peak value of the crack opening component of stress. From this, a crack extension rate independent of $K_I$ beyond $K_{ISCC}$ would appear almost trivial. For high-strength materials, where the plastic zone diameter becomes very small, a dependence of the peak $\sigma_{yy}$-value on $K_I$ may however reappear, and this may explain slopes, $d \ln v / dK_I$, changing from approximately zero to considerably positive values. This has been more closely investigated for the steel 90 MnV8 tempered to

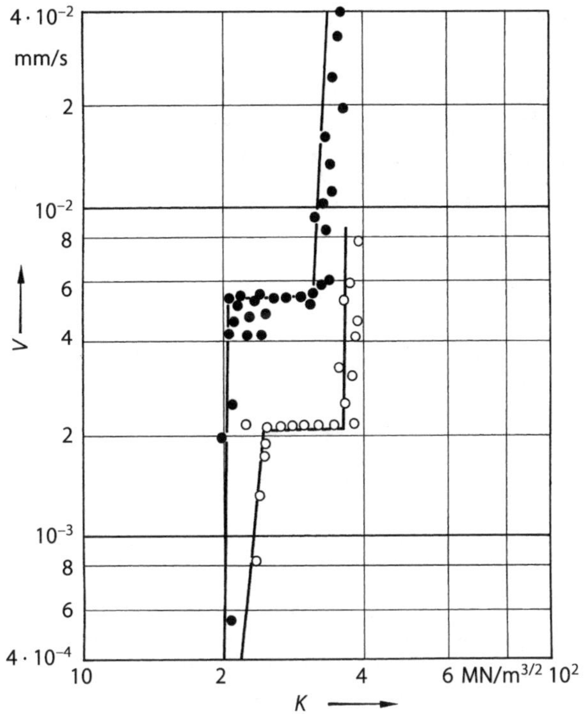


**Fig. 15.31.** Crack extension in a low-alloy martensitic steel (AISI 4340, ultimate strength 1900 MPa) in 025% NaCl solution, with (∘∘∘) and without (•••) added 0.01 M $K_2CrO_4$. Note complete record of threshold Region I (and $K_{SCC}$), plateau Region II, and unstable crack extension Region III ($K_{IC}$). (From Jaberi)

various degrees of hardness of a bainitic/martensitic structure. Cracking was intercrystalline, i.e. following former austenite grain boundaries. As shown in Fig. 15.32 [8g]. DCB specimens of this material readily fail by what presumably is HISCC. With increasing hardness, both $K_{ISCC}$ and $K_{IC}$ tend to lower values. The plateau in the curves observed for relatively low hardness values was tentatively explained as resulting from a rate-determining influence of lattice decohesion by stress-independent hydrogen uptake. When, instead, the hardness is high, the effect of local stress, now supposed to increase with $K_I$, takes over in determining the frequency of decohesion events at the crack tip, and $v$ tends to increase with $K_I$, at the same time surpassing rates observed for lower-hardness material, and running into Region III, approaching $K_{IC}$, at a lower level of $K_I$. It is obvious that, here, the viewpoint of continuum mechanics in fracture theory once more overlaps with the viewpoint of atomistics, and that this is difficult matter. At the same time, the slope of tw $\ln v$-$K_I$ curves has an important impact on crack morphology: When it tends to zero, cracks tend to branch [8h,10b].

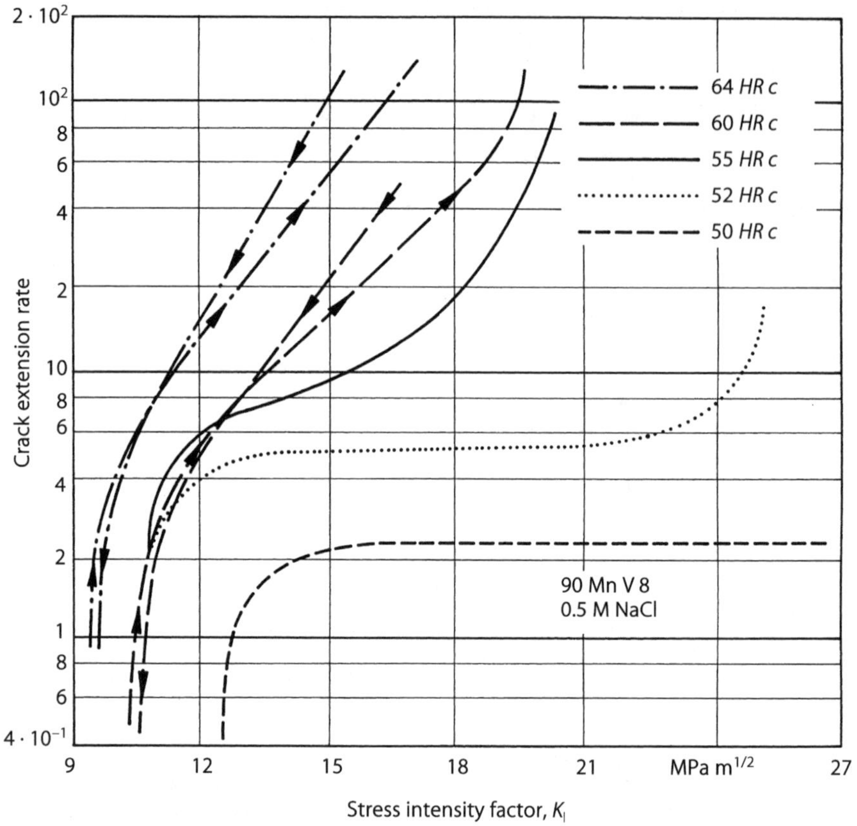


**Fig. 15.32.** Influence of strength on crack extension in a low-alloy steel (90 MnV8) in 0.5 M NaCl solution. The steel was heat treated to values of HRc varying from 50 to 65, corresponding to values of $R_p$ increasing from 1500 to 2500 MPa, and values of $K_{IC}$ decreasing from 27 to 19 MPa$\sqrt{m}$. (From Stellwag and Kaesche)

A notorious example of HISCC is stress corrosion cracking in high-strength steel tendons in prestressed concrete structures [33a]. Service failures are usually caused by improper techniques, leaving stressed tendons with insufficient concrete covering, or in badly injected ducts [8i,j,k,9c,33,34]. Out of contact with alkaline moisture, because without coat of concrete or mortar, steel surfaces will reactivate and start rusting accompanied by some slight hydrogen uptake sufficient to trigger cracking. A special case then is the protection of steel tendons by coating with zinc [8i]. When the zinc layer, or else the layered coat of FeZn alloys, has flaws exposing iron to the moisture present in concrete, the galvanic short-circuit between anodic iron and cathodic zinc, which cathodically protects iron against rusting, can cause cathodic hydrogen deposition on the steel surface and local hydrogen uptake. It is true that, according to permeation experiments (see Chap. 14), only low levels of interior hydrogen activity are then expected

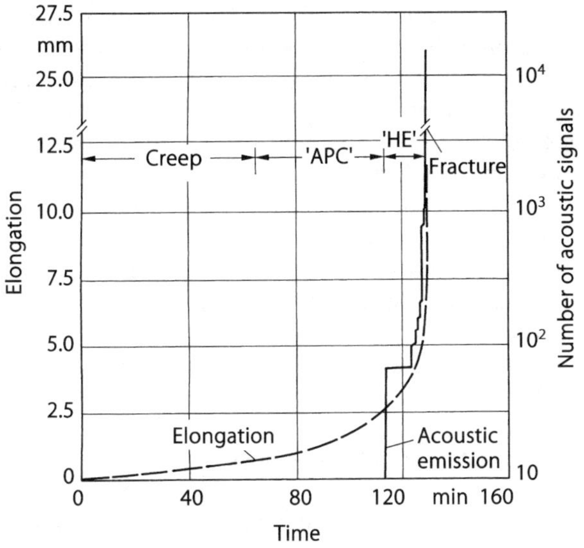


**Fig. 15.33.** Sound emis-sion accompanying change in length of a tensile specimen of a low-alloy steel (0.3 C, 1.3 Cr, 0.5 Mo, wt.%, ultimate strength 1620 MPa) during SCC in 60% $Ca(NO_3)_2$/3%$NH_4NO_3$ solution at 110°C. (From Yukawa and Tamura)

[33b,e]. Hence, it is only notched test specimens which are found susceptible to HISCC. Failures of prestressing steel tendons caused by cracking in contact with zinc are described in [34d]. A further point pertinent in this context is that, in contact with alkaline pore moisture at pH $\geq$ 12.7, contamination of the concrete by chloride (or sulfate) is not dangerous with respect to SCC, although chlorides always induce pitting [33f,g]. If moisture pH drops to 12.6 resp. 12.4, cracking of prefatigued specimens is, however induced by as little as 0.1%, rsp. 0.03% NaCl in the test solution. In this work, test conditions were very severe, using tendons with longitudinal precracks. We complete this picture by referring to experiments using pure $Ca(OH)_2$/$NaHCO_3$ solutions at 12.6 $\geq$ pH $\geq$ 7, modeling pore moisture conditions for concrete after prolonged exposition to air. SCC susceptibility was observed in quenched and tempered tendons, and in near-neutral solutions only [8j,k]. The rule that by applying proper techniques securing safe passivity prestressed tendons are well protected against HISCC is confirmed rather than contradicted by frequent cracks observed, after some 30 years of service, in an obsolete type of quenched and tempered prestressing steel of strength class 1420/1570 MPa. In this case, cracking probably already became incipient during exposure of prestressed tendons to atmospheric moisture before proper duct injection [34a,b,c].

Crack extension during HISCC proceeds exclusively by brittle decohesion. Anodic metal dissolution does not much contribute to crack lengthening, but only provides conditions of local hydrogen deposition. In the following section, caustic cracking in mild steel will be described as proceeding, instead, by anodic metal dissolution only, or 'anodic path SCC'. An intermediate case is high-strength steel under conditions described as causing 'caustic-cracking' in the following section. conditions. Figure 15.33 presents records of both length increase and acoustic emission as observed during SCC of a low-alloy high-strength steel in hot nitrate solution [35]. Straining of the tensile specimen initially causes creep, followed by 'silent' extension probably due to crack extension by anodic metal dissolution, and, finally, by audible discontinuous crack extension events probably due to hydrogen-induced brittle cleavage steps. Fractografic inspection of crack walls afterwards revealed that anodic crack extension at first was 'intercrystalline' along former austenite grain boundaries, turning to transcrystalline extension in the last cracking period. Confirming expectations, the change to HISCC occurs earlier for test specimens with increased hardness.

## 15.2.2 SCC of Mild Steel in Alkaline, Carbonate, and Nitrate Solutions

Together with 'season cracking' of brass in humid tropical climates, cracking of mild boiler steels was the SCC topic of most practical concern until the advent of stainless steels and of precipitation hardened aluminum alloys. In mild steel, cracking was observed in strongly alkaline, hot solutions, hence the description as 'caustic cracking', but hot concentrated fertilizer solutions proved to be dangerous as well. Hence the usual choice of hot concentrated alkali, or else boiling $Ca(NO_3)_2$ or $NH_4NO_3$ as test solutions. Testing was done using various types of constant load or constant deformation devices. Cracking in mild steel is intercrystalline, following former austenite grain boundaries. As in stainless steel, failure of specimens tested in in overall tension occurs by simultaneous growth of many cracks, in contrast to the typical failure mode of notched and prefatigued high-strength steels.

There is no doubt that, other than for HISCC in high-strength steels, we are now dealing with cracking in a passive metal. This is confirmed by the current-potential curve in Fig. 15.34, observed for pure iron monocrystals in boiling concentrated nitrate solution, and showing the free corrosion potential in the range of the anodic current plateau at $i_{\text{pass}} \simeq 1\ \mu A\ cm^{-2}$ [9]. Probably rather summarily, the overall corrosion reaction is supposed to be oxidation of iron to ferric oxide, accompanied by reduction of nitrate to nitrogen

$$10\text{Fe} + 6\text{NO}_3^- + 3\text{H}_2\text{O} \rightarrow 5\text{F}_2\text{O}_3 + 6\text{OH}^- + 3\text{N}_2\ . \qquad (15.6)$$

Detailed investigations into the electrode kinetics of this reaction are lacking. An additional intermediate formation of hydrogen, though probable, is difficult to prove.

Figure 15.35 shows that during potentiostatic constant load tests the time to failure of smooth tensile specimens decreases not only with increasing load, which, of course, is natural, but also with increasing electrode potential, and this is strong indication of a mechanism governed by anodic crack-tip iron dissolution. On closer inspection, the dependence of the time to failure, $\tau_{\text{fail}}$,

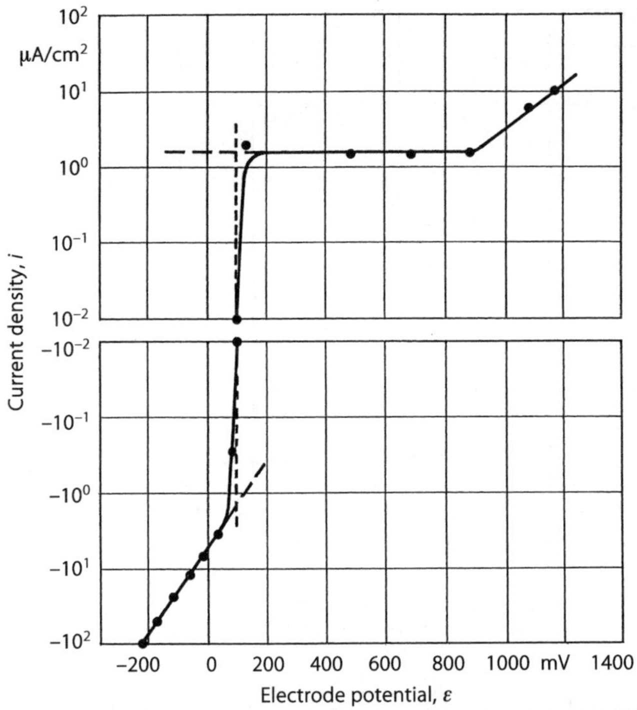


**Fig. 15.34.** Stationary current-potential curve for monocrystalline pure iron in deaerated 55% $Ca(NO_3)_2$ solution boiling at 115°C. Note position of the free corrosion potential at +0.09 V, in the range of stable passivity. (From Engell and Bäumel)

on the potential is not simply what would be expected: From a potential of complete cathodic protection against SCC at ca. −0.05 V, $\tau_{\text{fail}}$ drops to very low values over a range in potential of ca. 0.2 V, but a considerable further increase in potential (and, thus, anodic overpotential) has only little effect.

Tensile smooth specimens of mild steel usually are fully plastified and thus outside the scope of linear fracture-mechanical calculations. The elasto-plastic variant of fracture mechanics would be applicable, but was not yet considered at the time. Instead, the interpretation of observations was quantified by empirically setting the crack extension rate, $v$, proportional to the difference of the load on the bearing cross section remaining at time $t$, i.e. $\sigma(t)$, and a lower threshold load: $v = k(\sigma(t) - \sigma_\circ)$. To calculate $\tau_{\text{fail}}$, a cylindrical specimen of radius $r_\circ$ under longitudinal stress is considered. $v$ is set equal to the rate, $-dr/dt$, of decrease in radius of the load bearing area, neglecting the simultaneous growth of many cracks. The time, $t$, can then explicitly be expressed as related to $k, \sigma_\circ, r_\circ$, and to the ratio of actual stress and the true ultimate strength of the notched specimen, $\sigma/\sigma^*_m$. Owing to local stress increase beyond notches, $\sigma^*_m/R_m > 1$, and this ratio is a separate experimental quantity. The time to failure is $\tau_{\text{fail}} = (r_\circ)/k\sigma_\circ) f(\sigma/\sigma^*_m)$. In this relation, $f(\sigma/\sigma^*_m)$ is an an-

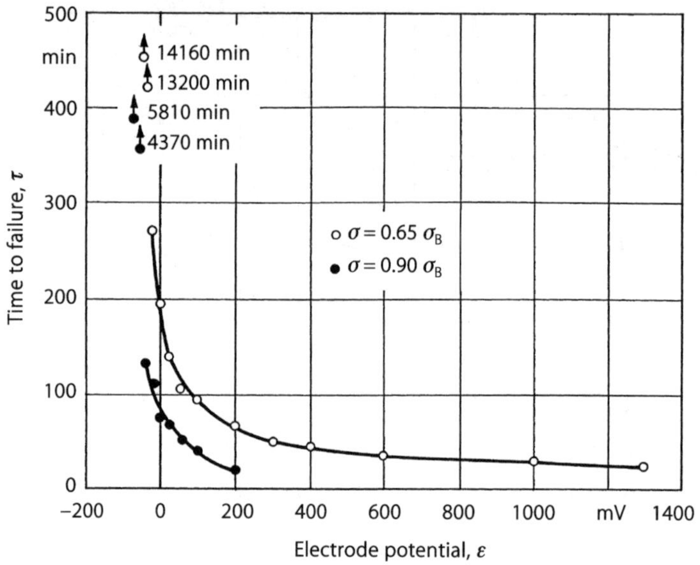


**Fig. 15.35.** Times to failure of mild steel at two values of constant load in boiling 55%$Ca(NO_3)_2$ solution a a function of the electrode potential. (From Bäumel and Engell)

alytical expression approximated by exp($-const \times \sigma$). The expression derived for $\tau_{fl}$ represents experimental results very well. Of the two fit parameters, $k$ is a weak, and $\sigma_\circ$ a strong function of the electrode potential, and this shows the importance of potentiostatic control for reproducible results.

More interesting clues as to the mechanism in SCC of mild steel exposed to hot salt solutions were revealed by current-potential measurements as shown in Fig. 15.36 [36a,b]. For unstressed specimens, the passive current density increases by an order of magnitude as the potential passes through a threshold, $\varepsilon_{thrs} \simeq 0.95$ V. Metallographic inspection reveals that this increase in polarizing current is due to the onset of intercrystalline corrosion (IC). When now a load of 0.2 $R_m$ is applied, this threshold potential for onset of IC shifts to ca. 0.4 V. When specimens are left unpolarized, this load is below $\sigma_\circ$, because the open-circuit corrosion potential, $\varepsilon_{corr} < \varepsilon_{thrs}$, and at open circuit the specimen remains immune with respect to SCC. When, in this case, the potential is shifted to beyond $\varepsilon_{thrs}$ by external polarization, specimens suffer intercrystalline corrosion and eventually fail. They probably would have failed at every potential exceeding the shifted threshold, but this was not checked in this work. When, however, the load is increased to 0.4 $R_m$, the threshold $\varepsilon_{thrs}$ shifts to $\varepsilon_{corr}$, and intercrystalline cracking is observed on open circuit, i.e. under usual test conditions. Thus, SCC in this case is IC caused by a critical stress-induced shift of the threshold potential critical for the onset of IC [9f,36a,b].

Both confirmation and some refinement of this view was furnished by work described in [81], proving that stress-induced IC and intercrystalline SCC are

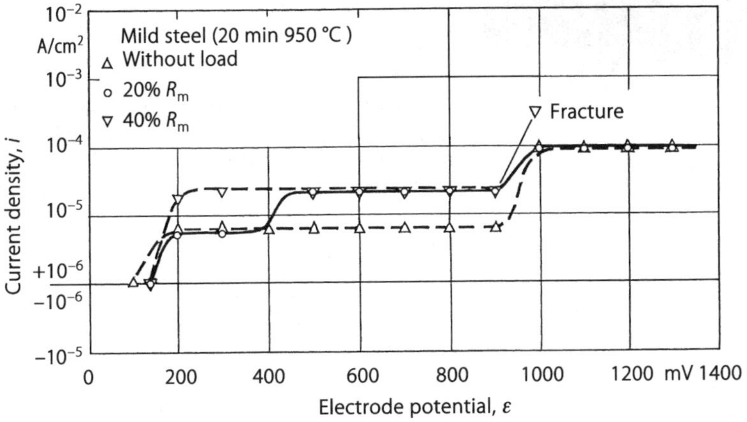


**Fig. 15.36.** The stationary current-potential curve of mild steel in boiling 55% $Ca(NO_3)_2$ solution as observed at zero stress, at 20%, and at 40% of ultimate strength, $R_m$. (From Münster and Gräfen)

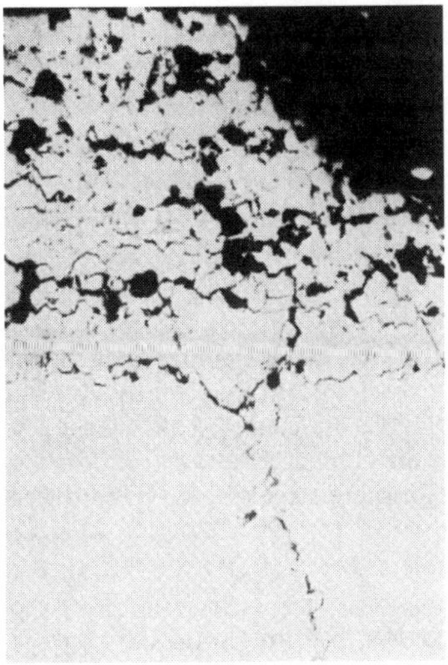

**Fig. 15.37.** Metallographic section through a mild steel specimen (0.08C, 0.004N, wt.%, normalized) stressed in boiling 55, $Ca(NO_3)_2$ solution, showing intercrystalline corrosion and intercrystalline stress corrosion cracking developed within 168 h. The load was 0.2 $R_m$, the electrode potential −0.85 V. (From Schneider, Kessler, and Kaesche)

superimposed processes. At 0.2 $R_m$, the threshold potential was 0.7 V, but specimens polarized to 0.85 V still showed IC only. Polarization to beyond 0.85 V caused additional intercrystalline SCC. As seen in Fig. 15.37, the two effect can very well be distinguished. An increase in load caused intercrystalline SCC to overtake IC, until at 0.7 $R_m$ rapid intercrystalline cracking with no indication of general IC occurred. Below $\varepsilon_{thrs}$, internal oxidation was observed. As documented in Fig. 15.38, this corrosive attack slowly proceeds into the crystallites, advancing more rapidly along grain boundaries.

Judging from extensive experience, susceptibility to SCC in mild steels is mainly dominated by the degree of carbon and nitrogen segregation to grain

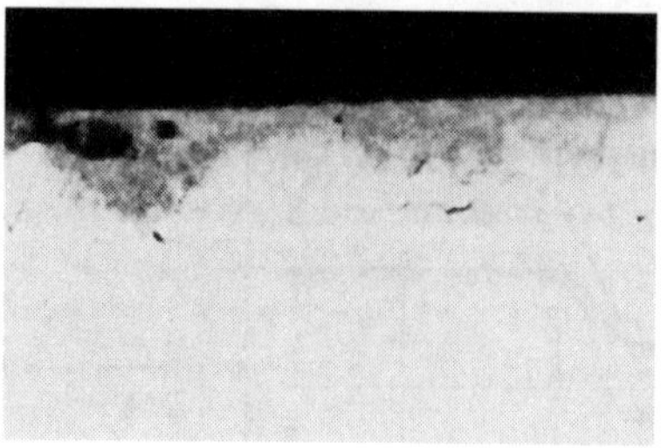

**Fig. 15.38.** Metallographic section through the surface of an unstressed mild steel specimen after anodic polarization at −0.6 V during 24 h in boiling 55% + $Ca(NO_3)_2$ solution, showing internal oxidation. (From Schneider, Kessler, and Kaesche)

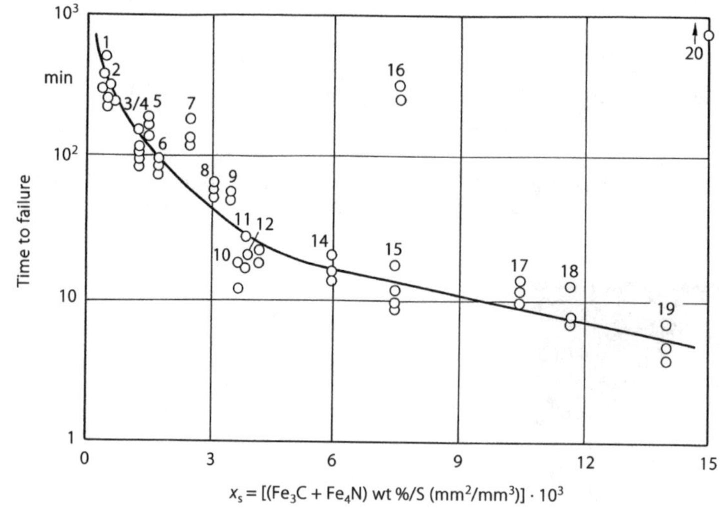


**Fig. 15.39.** Times to failure of mild steels tensile specimens with varying contents of carbon and nitrogen, and with varying average grain size. The abscissa is grain surface coverage with C and N. Specimens had been normalizing at 920°C, air cooled, cold worked to 85%, recrystallized at various temperatures and again air cooled. The test solution was 55% $Ca(NO_3)_2$ at 110°C, and the load was 70% of the ultimate strength, $R_m$. (From Graf and Becker)

boundaries [37a–c]. When the overall content is below 5 ppm C and 0.3 ppm N, susceptibility is very low or negligible, increasing to a maximum beyond these content levels, but once more disappearing for C > 0.3% apparently owing to binding of carbon in the perlite structure. Both heat treatment and cold work strongly influence susceptibility. Apparently, most or many of these effects can be rationalized under the heading of carbon and nitrogen availability at grain boundaries. This has been closely investigated in [28e], using steel specimens with $0.002 \leq \%C \leq 0.33$ and $0.001 \leq \%N \leq 0.014$, recrystallized to grain diameters between 6.5 and 243 μm. From the equilibrium concentrations, $\gamma_{Fe_3C} + \gamma_{Fe_4N}$, the grain boundary content in C and N was estimated, dividing $\Sigma\gamma_i$ by the specific grain boundary area. The plot of times to failure data as a function of the grain boundary coverages thus calculated coverage is shown in Fig. 15.39, and the correlation between the two properties is obvious.

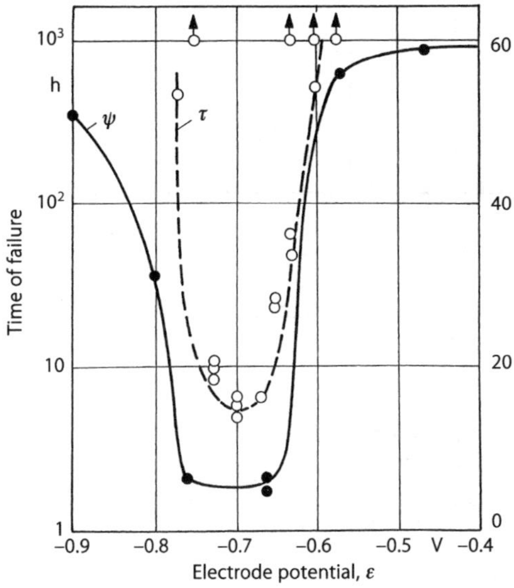


**Fig. 15.40.** Open circles: Times to failure for an unalloyed steel (0.02C and 0.006N, wt.%, tempered at 930°C and air-cooled, loaded in tension with 300 MPa in boiling 33% NaOH solution. (Data from Bohnenkamp). *Solid circles:* Reduction of area at fracture for an unalloyed steel (0.08C, 0.002N, tempered 650 h at 1080, and 17 h at 650°C) strained with constant rate $4.6 \times 10^{-6}$ $m^{-1}$ in boiling 35% NaOH solution. (Data from Humphries and Parkins)

Regardless of stress or composition, intercrystalline SCC in unalloyed steels, as observed in boiling calcium nitrate, solution disappears in the range of electrode potentials beyond ca.1.3 V. The effect may be due to the onset of transpassivity but is not well understood at present. Also, it is of little practical importance, because the free corrosion potential will not normally drift to such positive values. The situation is quite different in hot alkaline solutions. Here, susceptibility is restricted to a surprisingly narrow range of potentials, and the effect has practical importance, indeed [9g,36d,37]. To illustrate this, Fig. 15.40 combines data times to failure data observed during CLT and of reduction in area at failure during CERT for unalloyed steels in boiling concentrated NaOH solutions [9g,37d]. Intercrystalline SCC is observed in the range $-0.8 \lesssim \varepsilon \lesssim -0.6$ V only.

Concerning kinetics, a current-potential curve for iron dissolution recorded in hot alkaline solution, shown in Fig. 15.41, suggests that the restricted SCC susceptibility range, as described above, coincides with the active/passive transition range indicated by this curve [38]. Apparently, SCC requires the presence of a passivating oxide film of insufficient protective quality. Also for comparison, Fig. 15.42 completes the potential-pH diagram for iron from Chap. 3, where the high-pH side had been ignored [39]. Here, intermediate dissolution of iron to $HFeO_2^-$, or $Fe(OH)_3^-$ anions is considered, followed by deposition

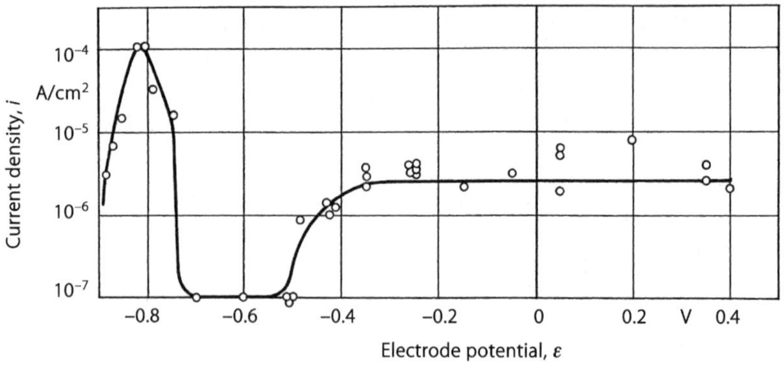


**Fig. 15.41.** Partial current potential curve, $i_{Fe}(\varepsilon)$, for anodic iorn dissolution in ca. 4 M KOH solution at 90°C. (from Simon and Schwarz)

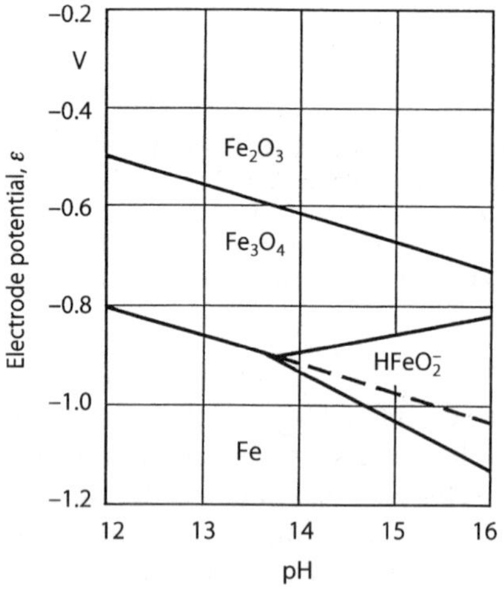


**Fig. 15.42.** Section of the potential-pH diagram for iron in alkaline solutions. The concentration of dissolved iron is assumed $10^{-6}$ mol $kg^{-1}$,; and calculations are for 25°C.(From Pourbaix)

of magnetite, $Fe_3O_4$, and conversion of magnetite to, as assumed in this case, ferric oxide $\alpha$-$Fe_2O_3$. We pass a discussion of the accuracy of these data, and we also pass a recalculation for elevated temperatures (see Chap. 3). Suffice it to observe that the active/passive transition potential range in Fig. 15.41 roughly coincides with the magnetite/ferric oxide transition range in the potential-pH diagram. This observation is qualitatively confirmed by experiments combining CERT and polarization measurements in [9g]. A working hypothesis to explain intercrystalline SCC of mild steel in hot alkaline solutions is that, in the active/passive transition range, grain boundary passivation lags behind grain face passivation. This, however, is reasoning not particularly well supported by independent evidence as to local properties of passive films. Also, the hypothesis obviously fails as an explanation of SCC of mild steel in hot nitrate solutions far beyond the active/passive transition potential. We also have so far ignored that in concentrated alkaline solutions intercrystalline SCC reappears when the potential is raised beyond ca. +0.6 V [40]. Cracking in this region still is intercrystalline, although insufficient grain boundary passivity is not obvious, and this type of SCC events probably corresponds to SCC in hot nitrate solutions.

Figure 15.41 presents the stationary current-potential curve for iron dissolution in an alkaline solution. It should be recognized that in such solutions steady states of passivity of iron are established only slowly. Under these conditions, non-stationary curves recorded by potentiokinetic current-potential scans potential very much depend on the scan rate, and these effects may be particularly important in the active/passive transition range. Since, supposedly, this also is the SCC range,just this effect of time-dependent current-potential curves may be used to identify SCC susceptibility, and this has been proposed as convenient test method [37d]. Another mean of screening for SCC susceptibility is rapid straining of working electrodes during current-potential measurements.

When conditions for SCC depend on the electrode potential, the position of the free corrosion potential, as governed by the superposition of partial current voltage curves, is decisive for the fate of inherently susceptible structural materials under tension. In the case of unalloyed steel in hot alkaline solutions the curious observation is that SCC susceptibility disappears not not only in complete absence, but also in abundant presence of dissolved oxygen SCC. It is only at intermediate levels of dissolved oxygen, that $\varepsilon_{corr}$ happens to fall into the range of susceptibility [36]. Therefore, both cathodic and anodic protection by polarization with external currents can safely protect steel in alkaline solutions indefinitely, and this has been shown for an alkali electrolysis plant [36], as well as for a caustic solution evaporator [41].

In comparatively weakly alkaline carbonate/bicarbonate solutions, the susceptibility of mild steel is similar to that observed in concentrated alkaline solutions. Again, intercrystalline SCC is observed in a comparatively narrow active/passive transition rage in potential only [7g,37e–h]. This has been the cause of unexpected damage to underground pipelines. Routinely, these lines, if carrying oil or gas, are cathodically protected in an effort to counteract external corrosion at accidental defects in the line coatings. This proved fatal, when protection brought the electrode potential down to the desired negative level where general corrosion is negligible, but also into the range of SCC susceptibility. For details, including accompanying 'strain induced cracking', as different from SCC, see the literature [37e-g]. One particular observation merits, however, special attention: It is the evidence of continuing plastic deformation as prerequisite for the onset of SCC, and Fig. 15.43 presents data obtained for the behaviour of a line-pipe steel again in hot carbonate/bicarbonate solution, in this case during constant load tests with records of the deflection of the loading beam caused by both creep and SCC [37e]. Owing to the narrow range of potentials for SCC susceptibility, specimens can easily be protected cathodically against the onset of SCC, and if protection continues until virtually complete

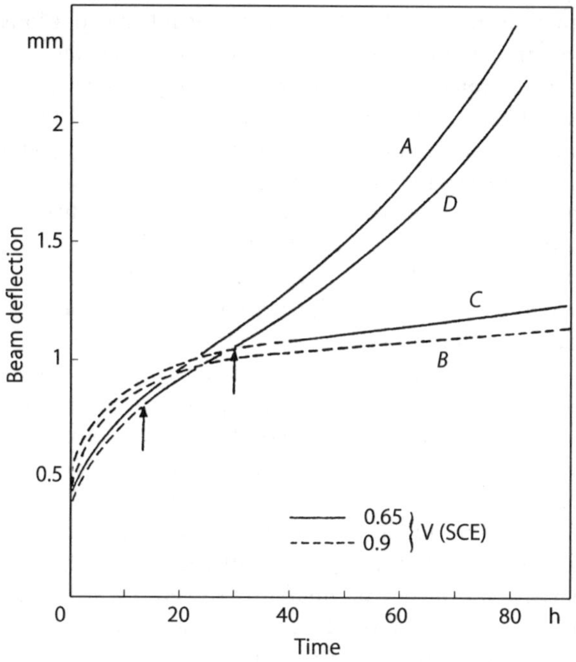


**Fig. 15.43.** Beam deflection-time curves for constant-load tests with notched and prefatigued cantilever beam specimens of a pipeline steel in 1Ð N $Na_2CO_3$/1 N $NaHCO_3$ solution at 75°C. Potentials are referred to sat.cal. el. *A* Decaying creep with superposed SCC at −0.65 V, in the range of potentials of possible intercrystalline SCC. *B* Decaying creep without SCC in a specimen protected cathodically against SCC at −0.9°C. *C* Creep in a specimen cathodically protected during 30 h followed (indicated by arrow) by polarization to a potential of SCC susceptibility: No SCC, owing to virtually complete decay of creep. *D* as *C*, but with polarization into the range of SCC susceptibility before decay of creep, causing onset of SCC. (From Parkins and Fessler)

decay of creep, a following polarization into the range of potentials of SCC susceptibility does not cause onset of SCC. If polarized directly from the start, or else before virtually complete decay of creep, into the range of SCC susceptibility, SCC is seen to start readily. We thus have here an example of crack extension requiring plastic crack-tip deformation. The specimens, incidentally, were prenotched and prefatigued to create a sharp precrack.

So far, the quantitative role of anodic crack-tip metal dissolution in determining crack extension has been a matter of speculation only. This problem has, however, been attacked using a very suitable model system, which is SCC in FeSi bicrystals, i.e. in specimens consisting of two monocrystalline halves, possessing exactly one suitably oriented susceptible grain boundary [42]. Figure 15.44 gives both the crystallographic orientations involved, and a photograph of a specimen showing momentary crack arrest in the grain boundary and also, on the back, ductile glide. From this observation, it is concluded that intercrystalline crack propagation proceeds by steps of ductile coarse glide through small distances,

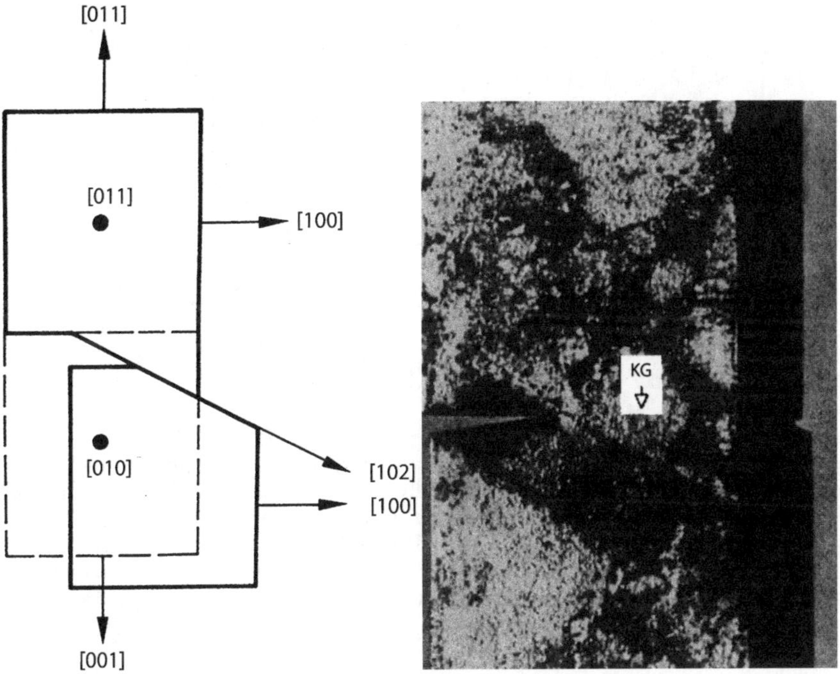


a Slip on (211) plane b

**Fig. 15.44a,b.** Stress corrosion crack advance in a FeSi bicrystal exposed to hot $(NH_4)_2CO_3$ solution. **a** Schematic drawing showing crystallographic planes and directions. **b** Direct photograph of the crack after tip advance along the asymmetrical 45°-< 100 > tilt grain boundary. Note accompanying plastic deformation as indicated by the relative displacement of the crystal halves visible at the back of the specimen. (From Stenzel, Vehoff, and Neumann)

$\Delta a_d$, each followed by a step of crack growth by corrosion, $\Delta a_c$. Corrosion steps are calculated from Faraday's law applied to the total electric charge transported by the anodic current during transient metal dissolution. Since SCC is intercrystalline, anodic metal dissolution occurs along the grain boundary only, and lateral dissolution is negligible. Anodic crack extension steps appear brittle, if the width of the susceptible seam along grain boundaries is very narrow. This width is a point of speculation which at present is left without further discussion. By pairs of subsequent ductile and brittle steps, the crack extends by $\Delta_a = \Delta a_d + \Delta a_c$, thus sharpening the tip as compared to ductile tearing alone.

For the following, compare Fig. 15.45. Let $\alpha_d$ be the crack tip angle observed when crack extension is by ductile glide alone. Together with the change in crack tip opening displacement (CTOD), indicated as $\Delta\delta$ we have $\Delta a_d = 2 \cot(\alpha_d/2)\,\Delta\delta$. When brittle extension steps are added, the crack tip angle decreases to $\alpha$, where

$$\cot(\alpha/2) = 2(\Delta a_d + \Delta a_c)/\Delta\delta) = \cot(\alpha_d/2) + 2\Delta a_c/\Delta\delta\ . \qquad (15.7)$$

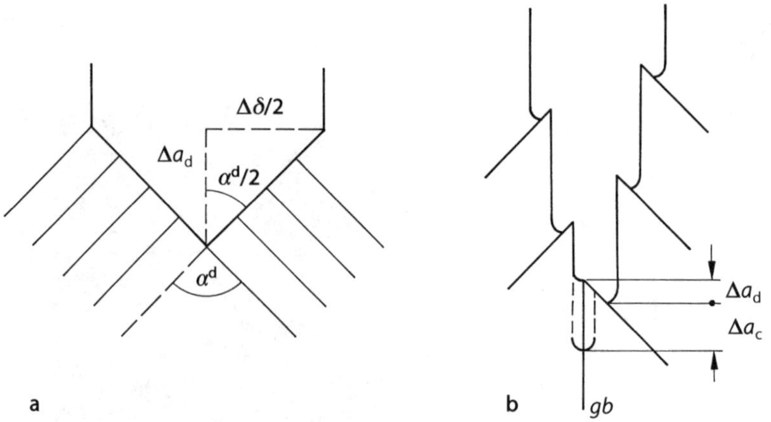


**Fig. 15.45a,b.** Explaining the mechanism of intercrystalline stress corrosion crack extension in mild steel: **a** Crack growth in ideally ductile material in inert surroundings, without corrosion, by alternating coarse glide events. The crack tip opening angle, $\alpha^d$, is equal to the angle between the two planes of the active glide system. CTOD is $\Delta\delta$. **b** Intercrystalline SCC as caused by transient slip step anodic iron dissolution extending the crack by $\Delta a_c$, along a grain boundary, $gb$, assuming negligible lateral iron dissolution. Note sharpening of the crack. (from Stenzel, Vehoff, and Neumann)

At this point the important question is, whether in the average time interval, $t_s$, between two coarse glide events repassivation reaches the crack tip. Introducing an average time interval, $t_p$, characterizing the repassivation rate, the authors arrive at an expression describing the increments in crack length caused by transient active metal dissolution as

$$\Delta a_c(t) = \Delta a_\circ(1 - \exp(-t_s/t_p). \tag{15.8}$$

As seen further below, similar relations, albeit with exponential decay functions replaced by negative power-law functions, are typical for other attempts to describe crack extension as caused by successive slip steps or film rupture steps. They always are based on the model of crack extension by successive decaying steps of anodic crack tip metal dissolution, every single one essentially calculated from Faraday's law. The succession of other, and especially of brittle non-Faradaic decohesion steps, will, however probably lead to numerically similar expressions. It will then be up to fractography to really discern the origin of crack growth. In this respect, the present case of intercrystalline cracking of bicrystals is unusually straightforward and convincing, clearly showing decisive anodic slip step dissolution.

Independent of metal dissolution and repassivation, coarse glide events occur after an average characteristic time interval, $t_s$. Let $\delta_s$ be the average change in CTOD observed during $t_s$. Introducing the average rate of change of CTOD as $\Delta\delta_s/t_s$, the authors arrive at

$$\bar{v} = \frac{\Delta a}{t_p} = \left(\frac{1}{2}\cot\frac{\alpha_d}{2} + \frac{\Delta a_\circ}{\delta_p}\left[(1 - \exp\left(-\frac{t_s}{t_p}\right)\right]\right)\frac{\delta_s}{t_s}\,. \tag{15.9}$$

From the evaluation of data obtained by these experiments, one of several conclusions is that, typically, repassivation reaches the crack tip between two successive glide events, making, as expected, crack extension discontinuous.

A different, but essentially equivalent approach to calculate $\bar{v}$ as the result of successive coarse slip steps causing transient metal dissolution during oxide film repair as discussed in a series of papers [43]. Consider a crack tip momentarily widened by a coarse slip event. Assume, arbitrarily, instead of an exponential, a negative power law describing the transient decay of anodic metal dissolution: $I_{\text{tip}} = I_{\text{tip}}^{\circ} \times (t/t_{\circ})^{-n}$. For the meaning of the exponent $n$, see further below. Let $t_f$ be the time for a single dissolution/repassivation cycle. The average density of the metal dissolution current during a cycle then is:

$$\bar{I}_{\text{tip}} = \frac{\int_0^{t_f} I_{\text{tip}}\,dt}{t_f} = \frac{I_{\text{tip}}^{\circ}\, t_{\circ}^{n}}{1-n}\, t_f^{-n} \,. \tag{15.10}$$

Let $\epsilon_{\text{crit}}$ and $\dot{\epsilon}_{\text{tip}}$ denote critical strain causing rupture of the brittle oxide film at the crack tip, and crack tip strain rate, respectively, so that $t_f = \epsilon_{\text{crit}}/\dot{\epsilon}_{\text{tip}}$. Considering Faraday's law to convert current densities to crack advance rates for a metal with atomic weight $M$ and density $\rho$, dissolving to $z$-valent cations, we arrive at

$$\bar{v} = \frac{M\, I_{\text{tip}}^{\circ}\, t_{\circ}^{n}}{zF\rho\,(1-n)\,\epsilon_{\text{crit}}}\,(\dot{\epsilon}_{\text{tip}})^{n} \,. \tag{15.11}$$

Further below, this relation will be seen to be important for attempts to predict crack extension rates in power plant steels. In the context of the present section, it should be equivalent to Eq. (15.9), if corrected for crack extension components by ductile glide alone. As an obvious natural consequence of the slip dissolution model, Eq. (15.11) relates crack tip strain rate and crack extension rate. The necessary next step then is to relate $\dot{\epsilon}_{\text{tip}}$ to the overall strain rate, $\dot{\epsilon}$, but this is difficult, especially because $\dot{\epsilon}_{\text{tip}}$ may depend on crack length. For constant load conditions, $(\dot{\epsilon}_{\text{tip}})_{\text{CTL}}$ thus is a function of both the rate of creep, $\dot{\epsilon}_{\text{creep}}$, and of the rate of crack extension, $\bar{v}$. For low strain-rate conditions the applied strain rate, $\dot{\epsilon}_{\text{appl}}$, enters the expression for $(\dot{\epsilon}_{\text{tip}})_{\text{SSRT}}$. We ignore details, referring to discussions reviewed in [46a,43c]. The approximate solutions of the problem are

$$\text{a)}\quad (\dot{\epsilon}_{\text{tip}})_{\text{CTL}} = A_1\,\dot{\epsilon}_{\text{creep}} + A_2\,\bar{v}/A_3 \,; \tag{15.12}$$

$$\text{b)}\quad (\dot{\epsilon}_{\text{tip}})_{\text{SSRT}} = B_1\,\dot{\epsilon}_{\text{appl}} + B_2\,\bar{v}/A_3 \,.$$

In most practical cases of SCC of mild steels in hot salt solutions, it is typical that cracks tend to appear in colonies, distributing stress and strain at crack tips among members. For such a colony of $N$ cracks along the gauge length of the tensile specimen, and for slow strain rate conditions, an expression

$$\text{a)}\quad (\dot{\epsilon}_{\text{tip}})_{\text{SSRT}} = (C_1/N)\,\dot{\epsilon}_{\text{appl}} + C_2\,\bar{v}\,\ln(C_3/N) \tag{15.13}$$

has been proposed [37i,j]. It shows that at high applied strain rates crack extension by SCC has little effect on crack tip strain rate and, thus, on SCC kinetics, while at low strain rates the influence of crack extension by SCC on SCC kinetics predominates. If this is sound reasoning, it explains why lower threshold values of $\dot{\epsilon}_{appl}$ limiting the onset of SCC are not, or else not always observed. Still more complications rendering realistic crack extension rate calculations risky arise from crack coalescence, causing a slow decrease of the number of colony members, increasing the stress concentration at the remaining fewer crack tips.

### 15.2.3 Stainless Steels

A number of observations concerning SCC of austenitic stainless steels have already been reported in the introduction to this chapter as more generally typical, or else as marking typical steps in the development of stress corrosion science. We continue the list by noting attempts to determine the anodic current density, $I_{tip}$, of crack tip metal dissolution in austenitic stainless steels by recording peak currents observed during repassivation of metal surfaces fractured in electrolytic solutions of supposedly typical crack tip composition [44,45], similar to what has been described for titanium previously. On closer inspection, the exact interpretation of results of such experiments is, however, difficult: In the ideal case, fracturing and wetting of the newly created surfaces would be momentary events; and, also ideally, a potentiostat would fix the potential directly from the start. Creating the initially non-existent double layer within zero time then would require a current pulse of infinite height, which is impossible. For less stringent conditions, the situation is somewhat relaxed, but recording of peaks of charging currents for momentarily fractured surfaces is intricate matter in any case, mainly owing to capacitive effects adding to the anodic current of metal dissolution. The point is reviewed in [46a].

Crack initiation events, rather more complicated than predicted by the slip dissolution/repassivation model, are illustrated in Fig. 15.46. These are micropits, in every case grown below the steady-state pitting potential of the material in this solution, but very much different in frequency, depending on the load on the tensile specimens below or beyond the yield strength [8a]. Using EDX, it was further observed that pit interiors are depleted in Ni and Fe and enriched in Cr, indicating the precipitation of chromium oxide. For similar and refined results obtained with TEM see [6f,23d].

SCC of stainless steel is usually tested with small, smooth specimens loaded to beyond the yield strength, but testing of DCB specimens large enough to allow the application of linear fracture mechanics theory have revealed interesting results as well. As seen in Fig. 15.47, $\log v$-$K_I$ curves for sufficiently large specimens clearly show a very well defined threshold in Region I, and model independence of $v$ of $K_I$ in the plateau Region II [10d]. More information from fracture-mechanical testing is presented in Fig. 15.48 revealing a distinct dependence of $K_{ISCC}$ on the electrode potential, in this case referred to the open-circuit corrosion potential [11]. With increasing polarization, $\pi$, i.e. increasing potential, $\varepsilon$, the overpotential, $\eta_{Me}$ also increases, and the observed shift in $K_{ISCC}$

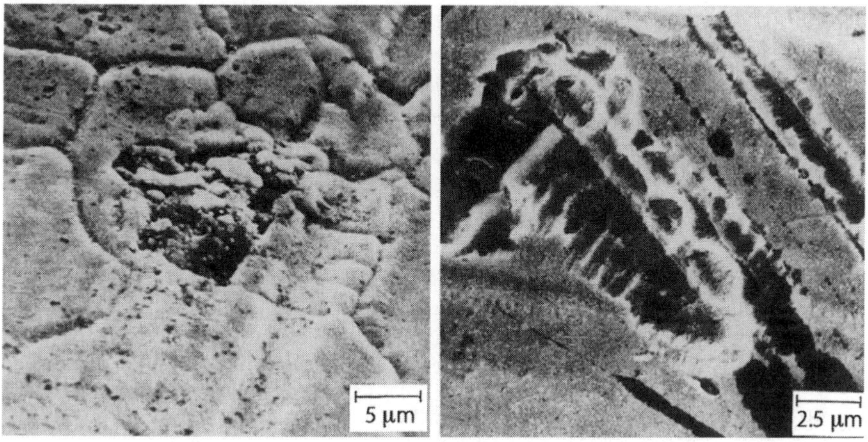


a b

**Fig. 15.46a,b.** Concerning precursors of crack nuclei in tensile specimens of 18 9 CrNi steel in hot $MgCl_2$ solution. Specimens were under potentiostatic control below the steady-state pitting potential. Note **a** isolated micropits observed when the load was below the yield strength, and **b** accumulated micropits on glide steps observed beyond the yield strength. (From Kessler and Kaesche)

certainly points to an effect of electrode kinetics. Considering micropitting and microtunneling as typical for onset of SCC, this behaviour may well reflect the role of electrochemical mechanisms of nucleation processes, similar to what is observed for pitting as well as for selective dealloying corrosion. It should be noted at this point that microtunneling and micropitting are not directly identical with crack extension itself; instead, they are triggering events. Ductile tearing of ligaments between tunnels has already been noted above as one possibility of crack extension. Considering the observation of faceted crack walls apparently indicating brittle steps in crack extension, attention to local embrittlement by dealloying and/or hydrogen uptake is once more suggested. This, then, is quite different from what the slip step dissolution model implies.

In the case of intercrystalline SCC, crack advance by anodic metal dissolution without important lateral crack broadening by metal dissolution can be understood to be following a preexisting path of narrow width, as governed by the sensitizing mechanism. In the case of sensitized austenites, this would be the width of grain boundary zones depleted in chromium. For transcrystalline SCC, the situation is different, since narrow pre-existing zones directing crack growth are not obvious. To explain crack sharpness in terms of the simple slip step dissolution model, an 'asymmetry parameter' of anodic metal dissolution must be arbitrarily invoked, tacitly or explicitly, to explain observations [42d], leaving the situation rather unsatisfactory.

Regardless of reservations noted with respect to the quantification of observations by the slip-step dissolution model, anodic crack tip metal dissolution certainly is important for transcrystalline SCC of stainless steels. This is easily

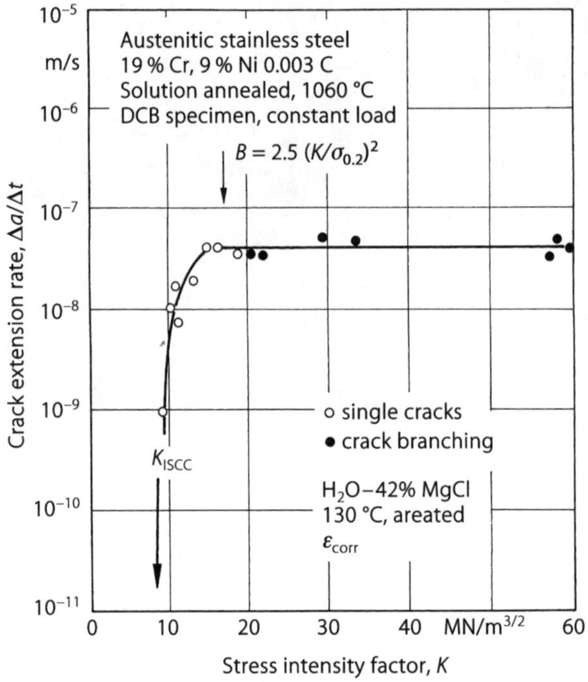


**Fig. 15.47.** The extension rate of transcrystalline cracks in hot $MgCl_2$ solutions as a function of the stress intensity factor,$K_I$. in large DCB specimens. (From Speidel)

verified by the direct control of onset and progress of SCC by external polarization. In the ductile material overall crack widening is considerable, making ohmic voltage drops along the crack small. Complementing data in Fig. 15.5, Fig. 15.49 presents data of time to failure values, $\tau_{fail}$, for an 18 9 CrNi steel in boiling magnesium chloride solution as a function of the electrode potential, for different levels of constant tensile loads on smooth cylindrical specimens [5]. Apparently, a threshold potential exists, increasing with decreasing load, below which SCC is no longer observed. Beyond this limit, $\tau_{fail}$, rapidly drops. As far as can be told from this figure, $\tau_{fail}$ reaches a fairly constant low value already at potentials only ca. 0.05 to 0.1 V more positive than the threshold value. As to electrode kinetics, this is surprising, but from simple measurements of this type finer details of SCC mechanisms are not really available.

Considering that SCC requires only low values of tensile load, a question of considerable practical importance is whether there really is a lower limit of load, below which SCC of stainless steels is no longer observed. This has been checked using specimens of 18 10 CrNi steel (Nr. 1.4301) carefully controlled for absence of internal stresses. The result was that SCC susceptibility disappears for loads below 50 MPa, which, of course, is a dangerously low value [7k–m]. Temperature is also important: in magnesium chloride solution, SCC is not observed below 40-45°C. This temperature limit was not, however, valid in a catastrophic case of

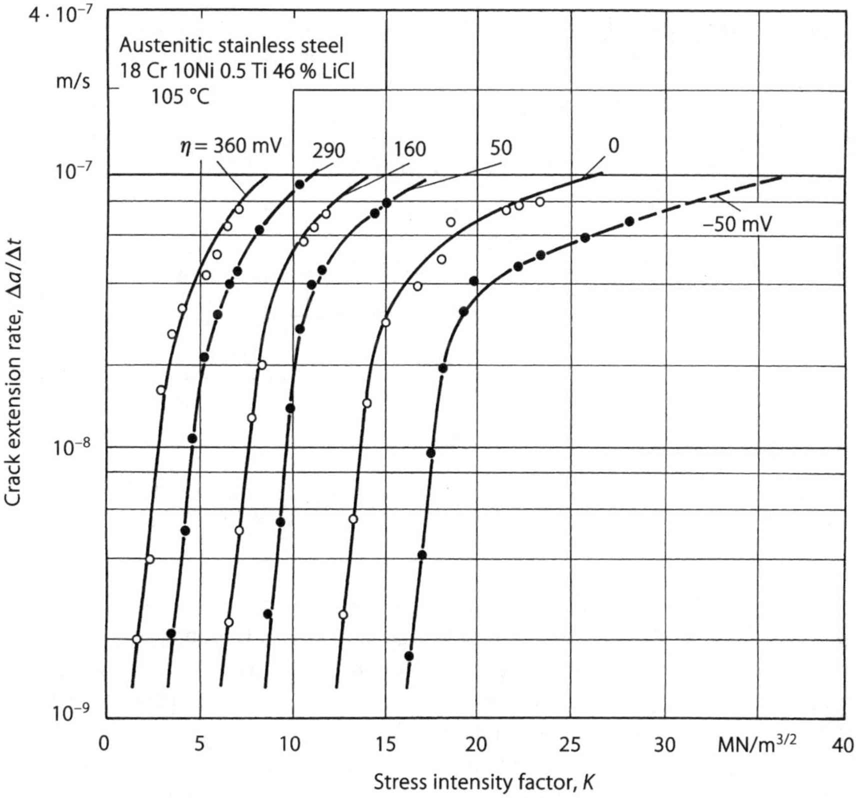


**Fig. 15.48.** The potentiostatic extension rate of transcrystalline cracks in hot LiCl solutions as a function of the stress intensity factor, $K_I$, in large 18 8 CrNi steel DCB specimens. $\pi$ is the polarization, $\varepsilon - \varepsilon_{corr}$, referred to the open-circuit corrosion potential. (From Eremiáš and Maričev)

SCC of stainless steel wires bearing a ceiling suspended over a swimming pool filled with chlorinated water. In this case, chloride and hypochlorite bearing moisture at ambient temperature sufficed to crack the wires [47].

The stationary current-potential curve for 18 9 CrNi steel in 35% $MgCl_2$ solution at 120°C is shown in Fig. 15.50 [8a]. On first sight, this suggests the behaviour of an active metal, but in the range of potentials below −0.1 V it is, instead, the current-potential curve of a passive metal, where the passive current density, $i_{pass}$, is negligible compared to the current density of cathodic hydrogen deposition. The steep increase in anodic metal dissolution rate at −0.08 V marks the pitting potential, $\varepsilon_{pit}$, of the steel in this solution. For the definition of this steady-state pitting potential, see Chap. 12. The typical current peak usually accompanying the active/passive transition is not visible, because the critical passivating current density, $i_{crit}$, also is negligible compared to the cathodic current of hydrogen deposition. Transient higher currents indicating

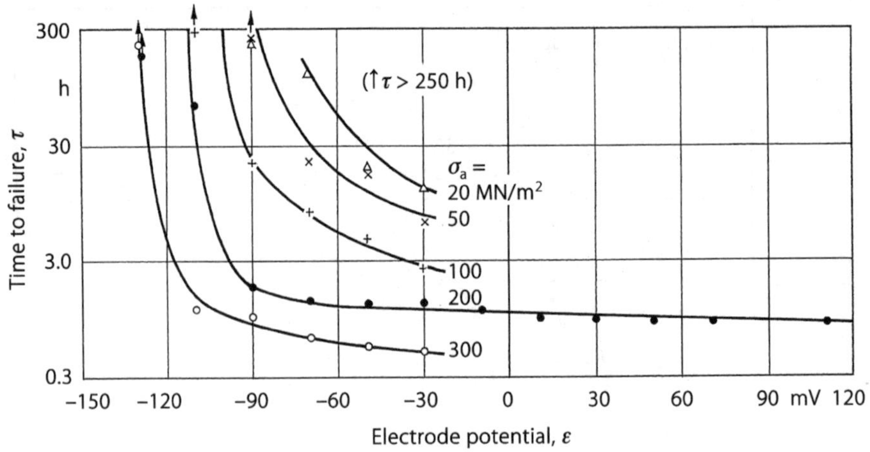


**Fig. 15.49.** The time to failure of smooth cylindrical tensile specimens of 18 9 CrNi steel in hot $MgCl_2$ solutions as a function of tensile load and electrode potential. (From Ternes)

the position of the passivation potential in this solution $\varepsilon_p \simeq -0.1$ V, can be observed by rapidly scanning the potential range. For the absolute magnitude of $i_{\text{pass}}$, see [5]. As seen in the figure, the open-circuit corrosion potential, $\varepsilon_{\text{corr}}$, coincides with the pitting potential. SCC without massive pitting, but apparently still triggered by micropitting events, occurs in the range $\varepsilon_p \leq \varepsilon \leq \varepsilon_{\text{pit}}$. At and beyond $\varepsilon_{\text{pit}}$, general roughening of the surface is superposed to SCC. Below −0.12 V, SCC is not observed [5,6h,i,28f,48], and this indicates that SCC of stainless steels again is a phenomenon characteristic for the passive state only.

As deviating from usual conditions for SCC, we note observations of transcrystalline SCC occurring in 0.005 N $H_2SO_4$/NaCl solutions at ambient temperature [7k,n]. As reported by the authors, SCC is, in this case, limited to a narrow range of electrode potentials directly below the active/passive transition, in fact pointing to SCC in the active state. In this range of potentials, the oxide film is supposed to be replaced by a layer of, perhaps, $NiCl_2$. and this appears to leave the alloy surface in a state of quasi-passivity with a high anodic current density of metal dissolution. In boiling magnesium chloride solution, a similar behaviour is reported for potentials again directly below the active/passive transition. According to [49] intercrystalline SCC is observed in $MgCl_2$ solutions only below 135 °C, and according to both [49] and [50a] the SCC path depends on strain rate. The $MgCl_2$ test solution is tricky as well, because it tends to change in pH on aging. As reported in [30j] both intercrystalline and transcrystalline SCC are observed in freshly prepared solutions at 100°C, and SCC susceptibility does not require an electrode potential exceeding a limiting threshold value. The threshold reappears and intercrystalline SCC disappears when the solution is aged. More complications are reported for HCl/NaCl solutions at high acidity: Here, the free corrosion potential, $\varepsilon_{\text{corr}}$, is below the

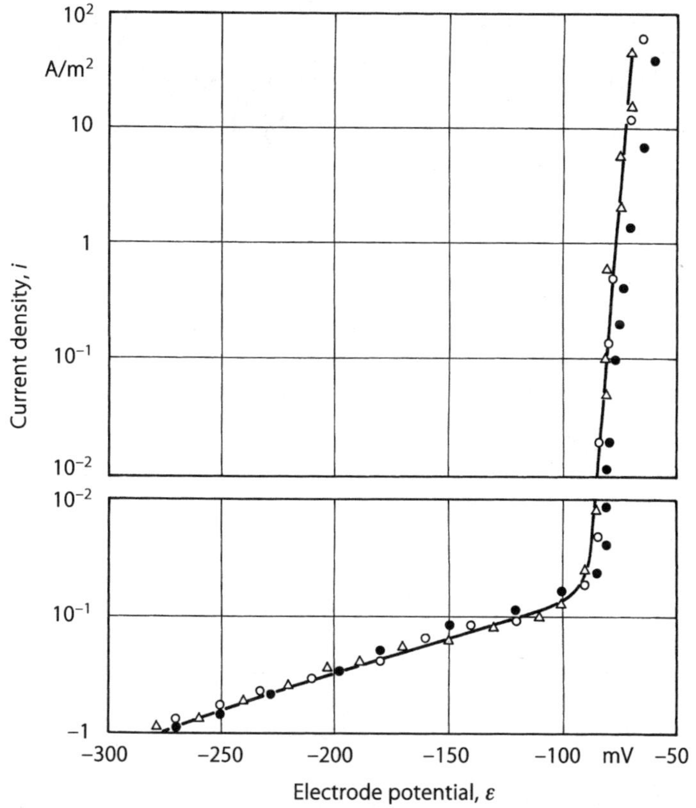


**Fig. 15.50.** The stationary current-potential curve for 18 9 CrNi steel in 35% $MgCl_2$ solution at 120°C. Electrodes had been cold worked 10% (*solid squares*), 20% (*open squares*) and 40% (*triangles*), respectively. (From Kessler and Kaesche)

pitting potential, $\varepsilon_{pit}$, and SCC is observed close to both potentials, but not in the intermediate range of potentials [30k,l].

To explain crack growth rates observed in stainless steels by anodic crack-tip metal dissolution alone, average current densities, $I_{tip}$, of the order of 1 A $cm^{-2}$ are required These are large but still acceptable values. Hydrogen bubbles are observed escaping from cracks, proving cathodic hydrogen deposition inside, probably in the crack tip region [49].

Since hydrogen deposition is observed to accompany crack propagation, the possible role of hydrogen embrittlement of the plastic zone in front of the corrosion crack tip merits attention [8a,m,n,51b,52]. In austenitic steels, hydrogen solubility is relatively high, but hydrogen mobility is relatively low. This suggests that for hydrogen embrittlement, if effective at all, dislocation transport into the metal is important. The model, as devised on this assumption, is described in Fig. 15.51 [8a,m,n]:

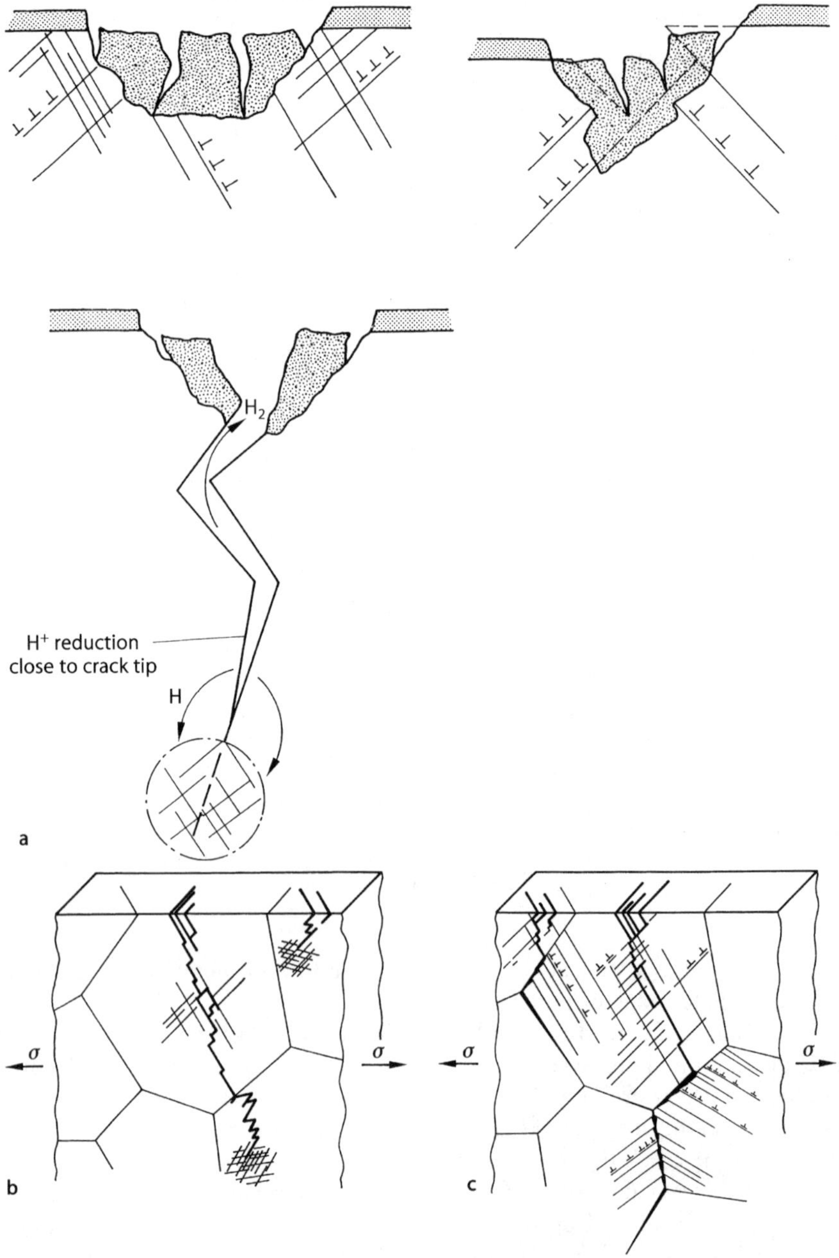


**Fig. 15.51a–c.** Modeling stress corrosion cracking in austenitic stainless steel in hot chloride solution using the 'anodic hydrogen embrittlement' concept. Crack nucleation by micropitting on, e.g., glide steps, including acidifying of the pit and crack tip electrolyte. **b** Transcrystalline crack extension in weakly deformed material. **c** Change to intercrystalline cracking in heavily deformed material. (From Kaesche and Kessler)

Stress corrosion is supposed to start by the formation of micropits on steps of coarse glide. In these pits, chromium oxide is assumed to accumulate. The next step, which is microtunneling starting from strings of micropits and ending in the formation of microslots, is not shown. Eventually, an advancing crack is nucleated. The crack-tip electrolyte is acidified by hydrolysis of dissolved metal salts produced, as during pitting, by anodic metal dissolution. An electric current flows to the crack mouth and to either, in the case of potentiostatic external polarization, a counter electrode, or else, in the case of open circuit corrosion, to the outside surface of the steel, where some cathodic reaction then is part of the overall corrosion reaction.

At the crack tip, cathodic hydrogen deposition proceeds at a rate presumably low in comparison to the rate of anodic metal dissolution. Nevertheless, the crack extension rate is thought to exceed the Faraday-equivalent anodic crack-tip metal dissolution rate, and crack extension proceeds mostly by intermittent steps of brittle fracture. Matters may be further complicated by intermittent repassivation down to the crack tip and intermittent activation of glide steps.

The model implies a possible change from transcrystalline to intercrystalline SCC in non-sensitized stainless steel, i.e. without the influence of grain boundary zones depleted in passivating chromium by chromium carbide precipitation. The suggestion is that hydrogen transported by dislocations is stripped from dislocations and trapped in grain boundaries, causing grain boundary embrittlement.

High-voltage TEM has temporarily modified this picture [51a]: Comparing cracks caused by SCC and by deliberate hydrogen charging, respectively, the conclusion now is that at low levels of tension, SCC proceeds predominantly by anodic slip step dissolution alone, while at high levels hydrogen-induced fracturing of martensite plates contributes to crack extension. This concept has been contested in [51b], and fractographic REM and TEM evidence now is again re-interpreted as pointing to crack extension by successive steps of hydrogen-induced microcracks, and this explicitly applies to test temperatures up to at least 250 °C.

Judging from the variety of observations, the best conjecture probably is that the mechanism of transcrystalline SCC in stainless, non-sensitized steels is variabel. A coherent model, then, has to imply that cracking process by varying contributions of microsteps $\Delta a_{\mathrm{dct}}$ of ductile deformation or creep, $\Delta a_{\mathrm{an}}$ of anodic slip step dissolution, and $\Delta a_{\mathrm{brt}}$ of brittle decohesion [8a,m,,n]. Decohesion steps probably are induced by hydrogen embrittlement, but vacancy injection during dealloying may also needs consideration [45b]. Contributions to decohesion by vacancy surface diffusion into the crack tip would further add to possibilities.

For most of the extensive studies and experiences of metallurgical and metal-physical influences on SCC susceptibility of stainless steels, the reader is referred to comprehensive reviews. For details concerning nitrogen in stainless steels, and concerning properties of austenitic/feritic 'duplex', and of martensitic steels see e.g. [2y]. Here, we are particularly interested in martensite plates occurring in nominally austenitic alloys. In this context we refer, in the first place, to the well known increase in SCC resistance gained by an increase in the nickel content beyond 8 wt.%, and especially by changing from stainless steels to nickel-base alloys. On first sight, this appears paradox, since low-nickel ferritic CrNi steels are not susceptible to transcrystalline SCC. Nickel, as well as carbon, stabilizes the austenitic structure. But nickel also increases the stacking fault energy, and this, then introduces a property which has been much discussed in connection with stress corrosion cracking. In the face-centered cubic lattice of austenites, stacking faults are stacks of lattice planes in hexagonal sequence, and the cause of their appearance is dissociation of dislocations. Dissociated dislocations cannot easily change between glide systems, and this restriction of movement favors coarse glide. This, then also favors coarse slip, which, for the slip-step dissolution model, is the very cause of SCC susceptibility. Hence the interest in observing, by TEM, plain stacks of dislocations indicating coarse glide, piled up in the glide plane against local obstacles, and thus creating local stress [53]. An increase in stacking fault energy, for instance by an increase in the nickel content, then disfavors SCC.

The stacking fault effect has more consequences [9h,24d,54c]. The same structural conditions which favor coarse glide and stacking faults, also favor the diffusionless, maretensitic changes of crystal structure from austenite, via hexagonal $\epsilon$-martensite, to body centered $\epsilon'$-martensite. Martensitic transformation is further favored by both plastic deformation and hydrogen uptake at crack tips. $\epsilon$-martensite has indeed been identified on crack walls by electron diffraction [54b].

Brittle fracture of martensitic plates in front of the crack tip certainly is a possible mechanism of crack extension. At the same time, anodic dissolution of martensitic phases also is more rapid that of the parent austenite [54c]. A further possibility is decohesion at martensite/austenite grain boundaries.

For a given set of alloy properties, stress state, temperature and composition of the electrolytic solution, the existence of a threshold electrode potential, $\varepsilon_{SCC}$, for the onset of SCC is a well established observation. When the level of tensile stress is high, this threshold appears to coincide with the active/pasive transition potential, $\varepsilon_{pass}$. As tension decreases, the threshold moves to more positive values closer to the pitting potential. During tests of SCC susceptibility without external polarization, the position of $\varepsilon_{corr}$, as determined by the kinetics of electrolytic corrosion, i.e. by the superposition of partial voltage curves, relativ to $\varepsilon_{SCC}$, then decides the fate of tensile specimens. This is confirmed by data reported in Table 15.1 presenting results of simple SCC tests in boiling $MgCl_2$ solutions without and with additions of corrosion inhibitors [29e]. The table also shows that $\varepsilon_{SCC}$ itself depends on solution composition. This was interpreted as fitting into the concept of crack extension by crack tip ion adsorption, but displacement of activating chloride ions in the adsorption layer by predominant inhibitor adsorption would seem to explain observations just as well.

In this case, the influence of solution electrochemistry on SCC susceptibility, though not well understood, is obvious, because the alloy properties where con-

**Table 15.1.** Threshold electrode potentials, $\varepsilon_{SCC}$, critical for the onset of transcrystalline stress corrosion cracking in 18 8 CrNi steel (AISI 304) in $MgCl_2$ solution boiling at 130°C, containing inhibiting acetate and nitrate additions, compared to open-circuit corrosion potentials, $\varepsilon_{corr}$, observed after several hours for specimens bent to constant deformation in the same solution. (From Uhlig and Cook)

| Solution | $\varepsilon_{SCC}$ V | $\varepsilon_{corr}$, V | cracking |
|---|---|---|---|
| $MgCl_2$ | - 0.145 | - 0.11 | yes |
| +0.1%$CH_3COONa$ | - 0.132 | - 0.12 | yes |
| +2%$CH_3COONa$ | - 0.116 | - 0.12 | no |
| +2%$NaNO_3$ | - 0.090 | - 0.06 | yes |
| +5%$NaNO_3$ | - 0.070 | - 0.06 | no |

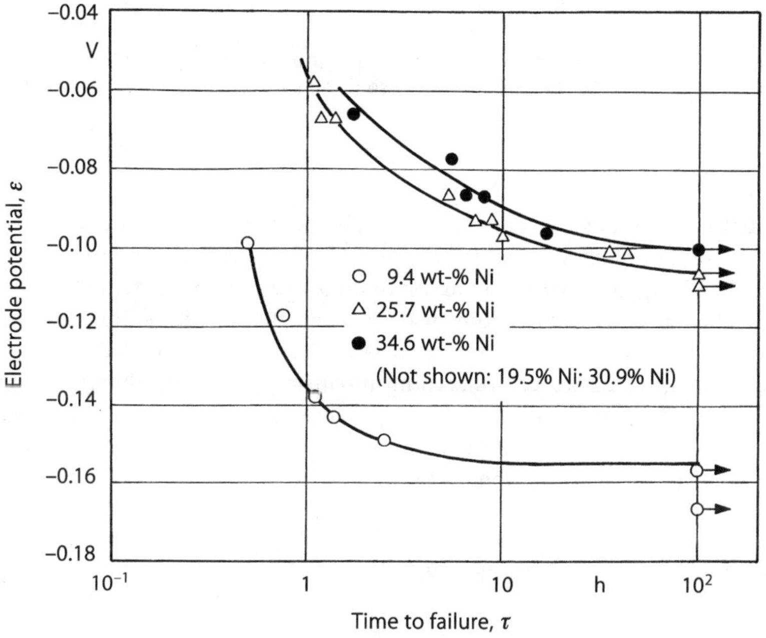


**Fig. 15.52.** Testing U-bend 20% Cr stainless steel specimens at constant deformation in $MgCl_2$ solution boiling at 130°C. Time to failure values are plotted as function of the electrode potential. (from Lee and Uhlig)

stant. When, instead, the solution composition is constant, but the alloy composition is varied, attributing changes in alloy composition to metal-physical parameters is tempting and frequently attempted, but often erroneous. To illustrate this, Fig. 15.52 shows time to failure data for specimens of steels with increasing nickel content plotted as a function of the electrode potential, and it is seen that even for high nickel levels U-bend specimens fail by SCC, if only

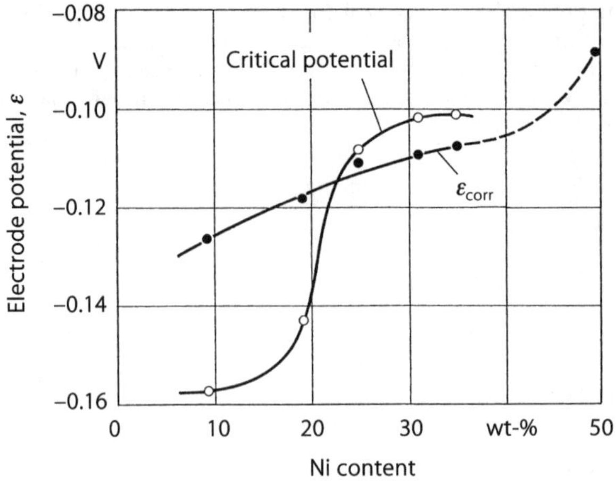


**Fig. 15.53.** Critical potentials for the onset of SCC, $\varepsilon_{SCC}$, and open-circuit corrosion potentials $\varepsilon_{corr}$, of 20% Cr stainless steel in hot $MgCl_2$ solution as a function of the Ni content. (From Lee and Uhlig)

the potential is made positive enough [29e]. The data confirm the existence of a critical threshold potential, $\varepsilon_{SCC}$, for the onset of SCC. Such values are replotted in Fig. 15.53 together with the corresponding open-circuit potentials, $\varepsilon_{corr}$, observed in the same solution. From this it follows that in the boiling magnesium chloride solutions, tensile specimens would be observed to fail only for nickel contents up to 22 wt.%. Concerning intrinsic SCC susceptibility, this is an incidental result, and metal-physical conclusions would be misleading, since $\varepsilon_{corr}$ can be shifted to values positive to $\varepsilon_{SCC}$ by adding a sufficiently strong oxidizing agent to the test solution.

### 15.2.4 SCC in High-Temperature, High-Pressure Water

SCC in steels and nickel-base materials, exposed to high-temperature, nearly pure water or extremely dilute electrolytic solutions as power plant coolants, has been a matter of much concern, dispute, and research, mainly concerning the safe operation of nuclear power plants. The topic is the object of a vast amount of expert literature, not always very easy to review, and certainly far beyond the scope of a review in a textbook. In the following, we therefore present only a selection of such experimental data and approaches to theoretical interpretation, as seem appropriate for an attempt to put the topic into its place within the more general field of corrosion science. Doing so, quite a number of second-order details will remain ignored, although they may be important for safety considerations and safety regulations.

**Table 15.2.** Slow strain rate data for a ferritic steel, A533B, quenched and tempered, in high temperature deionized water at 288°C. Ultimate strength at room temperature: 610 MN $m^{-2}$; composition: 0.185C, 1.395Mn, 0.006S, 0.655Ni, 0.130Cr, 0.485Mo (wt.%). Reduction in area at fracture, RA, in air: ca. 71%. Strain rate: $2 \times 10^{-6}$ $s^{-1}$. 'Crack velocities' are calculated as ratio of deepest SCC crack length and time to failure. 'ppb' = $\mu g\ kg^{-1}$. (From Congleton, Shoji, and Parkins)

| Average $O_2$ (ppb) | $\varepsilon_{corr}$ (V, SHE) | RA% | Max. crack length(($\mu$m)) | No. of cracks | crack velocity (mm $s^{-1}$) |
|---|---|---|---|---|---|
| 0 | −0.850 | 69.8 | 0 | 0 | 0 |
| 15 | −0.25 | 76.0 | 0 | 0 | 0 |
| 40 | −0.18 | 70.8 | 0 | 0 | $2.7 \times 10^{-7}$ |
| 100 | −0.155 | 69.0 | 20 | 1 | $1.75 \times 10^{-7}$ |
| 200 | −0.135 | 65.2 | 30 | 1 | $2.77 \times 10^{-7}$ |
| 600 | −0.11 | 15.0 | 500 | 3 | $6.98 \times 10^{-6}$ |
| 2650 | −0.007 | 17.2 | 560 | 3 | $1.29 \times 10^{-5}$ |

In contrast to electrolytic solutions normally encountered in industrial services, very dilute boiler feed water, and nuclear reactor coolants as well, have only poor electric conductivity. For this reason, there is a conspicuous lack in data on electrode kinetics, because current-potential curves usually are heavily distorted by ohmic voltages between working and reference electrodes. Frequently this difficulty is evaded by adding conductivity-enhancing but, hopefully, otherwise inert electrolytes, notably borate buffers, to the test solutions, expecting electrode kinetics to remain unchanged as compared to, e.g., pure water. The procedure is somewhat doubtful at least if it comes to considerations of galvanic coupling such as that of active crack tips and external specimen surfaces. Often, the situation may be such that from measurements of electrode potentials in low-conductivity solutions alone, without investigations into polarization characteristics, some useful information is available.

A further distinct difference in electrode kinetics of corrosion of steel, and also of nickel base alloys, in high-temperature water as compared to ambient-temperature solutions, is the change of electrode surface states from ultra-thin passive oxide layers to much thicker, although still oxidic layers protecting these materials in power plant water loops, as described in Chap. 10. The respond of these thicker layers to changes of the electrode potential will usually be sluggish.

For an important example, we turn to *low alloy steels* in high-temperature, high-pressure water in general, and to slow strain rate experiments with the nuclear pressure vessel steel A533B, containing Mn, Ni, Cr and Mo in small amounts, in particular. From more extensive data reported in [37i], Table 15.2 lists data collected at 288°C during slow strain rate experiments with smooth tensile specimens in a small refreshing autoclave containing water of constantly controlled oxygen content. As seen from the table, the oxygen content determines the extent of SCC incidents, revealed by an important decrease in area at fracture. As established by fractography, cracking was transcrystalline and decidedly brittle. The oxygen content also determines the corrosion potential:

At zero and at low oxygen content, cracking is exclusively ductile, and $\varepsilon_{corr}$ is $\leq -0.18$ V; at higher oxygen content, $\varepsilon_{corr}$ shifts to considerably more positive values, and SCC becomes intense. The authors interpret the data as indicating severe SCC only at oxygen contents $\gtrsim$400 ppb, but this is not our concern presently. Also, crack extension rates are calculated as ratios of depth of brittle crack extension over total time to failure, i.e. over test duration, which also is debatable, but again this is ignored for the moment. It is more important that a) $\varepsilon_{corr}$ is essentially controlled by the oxygen content, i.e. by the superposition of the cathodic current-potential curve for oxygen reduction with the anodic curve for metal dissolution, and that b) stress corrosion crack initiation or/and crack growth depend on the potential, indicating initiation or/and growth kinetics governed or co-governed by anodic crack-tip metal dissolution. The point has been checked by external anodic polarization in low-oxygen, and cathodic polarization in high-oxygen solutions. Owing to high ohmic voltage drops in the scarcely conductive water, such measurements do not give very useful values of the polarization. Still the main effect was unambiguous: SCC can be initiated by anodic polarization and prohibited by cathodic protection, confirming the role of the potential [37i].

We insert that this influence of oxygen content on transcrystalline SCC in low-alloy steels exposed to high-temperature water has been the subject of much controversy directly bearing on safety considerations. The claim has been that, statistically, the correlation between oxygen content and SCC incidences is unsafe [10e,f,g]. For a review of the controversy, see [55]. It will be seen below that SCC can indeed be observed at very low oxygen levels, if test conditions are extremely severe, but this, then raises the question of legitimate, viz. reasonable, limits for test procedures.

Figure 15.54 presents data on onset of SCC in slowly strained specimens of steel A533B exposed to high-temperature water as a function of the corrosion potential [37i]. Data from previous work, reported in [56] were used for this diagram , but this is incidental. Apparently, there is, at any rate, a threshold potential for the onset of SCC, which, if at all, depends only weakly on temperature. Because beyond the threshold potential the oxide layer is observed to contain increasing amounts of ferric oxide, the threshold has tentatively been interpreted as coinciding with the magnetite/ferric oxide transition potential [37i].

Of course, slow strain rate testing is a particularly severe testing procedure, if periodically occurring oxide film breaking is required for crack growth. Also, it is irrealistic as compared to service conditions, except for the cases of tensile periods during very slow frequency fatigue, as described further below. Still, SSRT is quite revealing concerning SCC mechanisms, and we continue by describing test results reported in [8p,q]. In this work, the pressure steel SA508Cl2 containing Mn, Cr, Mo and Ni was studied. The water was slightly alkaline by addition of some ammonia to pH 8.5. In contrast to similar work reported in [57a], and to most other work in the field, the test solution was circulated in a closed loop at high values of the Reynolds number, corresponding to service conditions. As reviewed in [55], water circulation increases stress corrosion re-

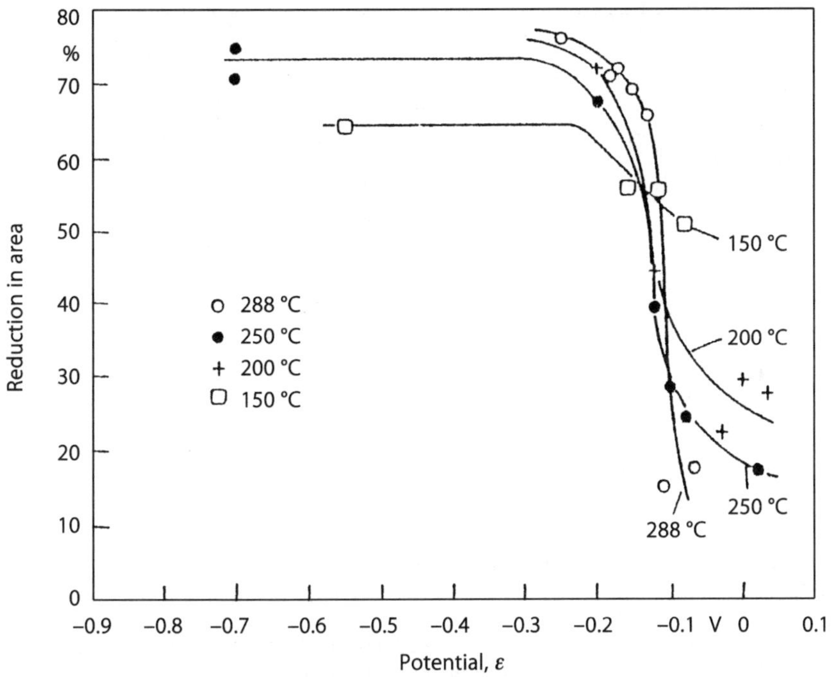


**Fig. 15.54.** The reduction in area at fracture observed during slow strain rate testing of steel A533B in water at different temperatures plotted as a function of the corrosion potential as inferred from the oxygen content. (From Congleton, Shoji, and Parkins, using data from Indig, Weber, and Weinstein)

sistance. This, incidentally, merits a note: If, at given oxygen concentration, SCC is expected to set on only when a threshold electrode potential is passed, then water circulation would be expected to be detrimental, as decreasing the concentration overpotential of oxygen reduction, shifting the corrosion potential to more positive, and thus more dangerous values. On the other hand, the same effect would enhance the growth of the passive oxide film, counteracting oxide film breakdown. Furthermore, water circulation should diminish the danger of differential aeration cells, and this may be another effect to be considered. On the whole, the point appears to merit future closer experimental inspection.

Three melts with different levels of the sulfur content were inspected, because manganese sulfide is known to be a decisive factor in stress corrosion susceptibility. As revealed by metallography, the test material was bainitic with elongated MnS inclusions mostly oriented perpendicular to the tension and thus acting as stress raisers. From this work, Fig. 15.55 gives examples of the typical influence of the oxygen content in the test solution on the corrosion potential, once more showing the shift by more than 0.5 V from nominally oxygen-free to oxygen-rich water. We note here the steep increase of the potential between

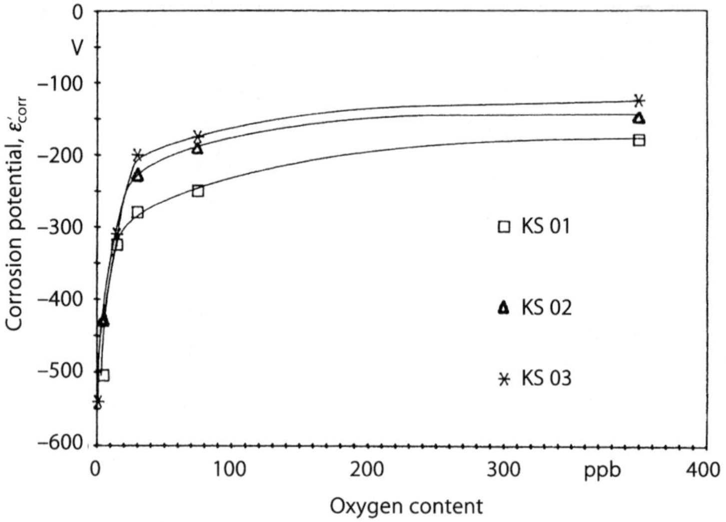


**Fig. 15.55.** The corrosion potential of three melts of steel SA508Cl2 (0.21–0.25C, 0.69–0.99Mn, 0.41–0.48Cr, 0.62–0.75Mo, 0.71–1.21Ni, wt.%., yield strength 430–500, and ultimate strength 586–625 MPa) as function of the oxygen content in slightly alkaline water at 220°C. Sulfur content in KS01, 02, and 03: 0.022, 005, and 0.011%, respectively. Reynolds number of water flow $7\times10^4$. Water conductivity at working temperature $4\mu S\ cm^{-1}$. Note initial very steep increase of $\varepsilon_{corr}$. (From Rippstein and Kaesche)

0 and 50 ppb of oxygen. At 100 and at 160°C, a potential threshold for the onset at ca. −0.2 V for the two melts high in sulfur, and at ca. −0.1 V for the melt low in sulfur, was clearly marked by a distinct decrease in reduction of area at fracture. At 220°C the threshold was less conspicuous but should exist between −0.4 to −0.3 V. SCC was transcrystalline and brittle, producing fan or thumb-nail crack faces originating from MnS inclusions and sometimes following strings of such inclusions.

In extensive series of experiments reported in [8p], the strain rate was varied between $\dot{\varepsilon} = 5 \times 10^{-6}$ and $2 \times 10^{-5}\ s^{-1}$. Straining was interrupted after various periods of time, $\tau$, before finale rupture at time to failure, $\tau_{fail}$. The specimens were then rapidly strained to purely mechanical fracture, and the fracture surface was inspected for the average depth, $\bar{x}$, of the stress corrosion cracks at time $\tau$. For tests in oxygen containing and, thus, SCC promoting solutions, the surprising result was that cracking never started before a critical elongation, $A_{crit}$, of 6.5% was reached, which is 95% of the elongation at fracture in air and indeed a very high deformation, unlikely ever to occur in service. A typical test result is shown in Fig. 15.56. This means that in every case a crack incubation period, $\tau_{inc}$, preceded the onset of SCC, and that it can be expressed as $\tau_{inc} = A_{crit}/\dot{\varepsilon}$. From this it follows that the true average crack growth rate, $\bar{v}_{true}$, determined from the stress corrosion crack depth, $\bar{x}$, observed at $\tau$, is

$$\bar{v}_{true} = \bar{x}/(\tau - \tau_{inc}) = \bar{x}/(\tau - A_{crit}/\dot{\varepsilon})\ . \qquad (15.14)$$

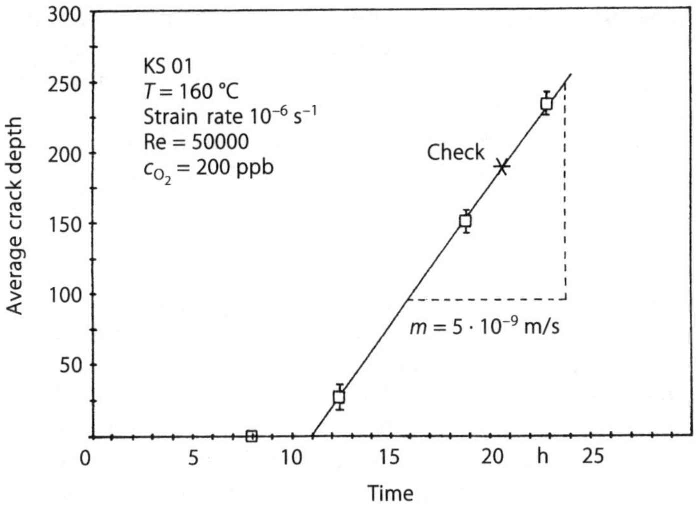


**Fig. 15.56.** Average depth of brittle SCC cracks observed during slow strain rate testing of steel melt KS01 in water at 160°C. □: Values obtained by continuous straining at constant rate. $*$: Value obtained by initial rapid straining to 6.5% elongation, followed by slow straining for 10 min. All values obtained after interrupting slow straining before final failure of specimens. (From Rippstein and Kaesche)

For $\tau = \tau_{\text{fail}}$ at the end of slow strain rate tests, this gives values for $\bar{v}_{\text{true}}(\tau_{\text{fail}})$ which are considerably higher, and certainly closer to reality than conventional values $\bar{v} = \bar{x}(\tau_{\text{fail}})/\tau_{\text{fail}}$.

The factor $A_{\text{crit}}$ was found to be independent of sulfur content, temperature, and strain rate. It thus marks a critical degree of specimen deformation, necessary for the onset of SCC. The independence of this critical deformation of the strain rate was further confirmed by rapid prestraining specimens in the test solution to $A_{\text{crit}}$, and then switching to slow straining. Counting $(\tau - \tau_{\text{inc}})$ from this moment, this correctly produced the point marked '$*$' in the figure.

For all $\dot{\varepsilon}$ values, the same true average crack growth rate, $\bar{v}_{\text{true}}$, was observed, while conventional $\bar{v}$ values increased with increasing strain rate. Considering Eq. (15.12b), or else (15.13), this appears to indicate a situation where the influence of the applied strain on the true crack tip strain rate is negligible, compared to the influence of crack growth itself. If or when it is true that $\bar{v}$ or, more precisely, $\bar{v}_{\text{true,}}$ dominates $\dot{\varepsilon}_{\text{tip}}$, equations of the type of Eq. (15.11), or generally any other relation of the form $\bar{v}_{\text{true,}} = f(\dot{\varepsilon}_{\text{tip}})$, are not predictive concerning SCC rates. More specifically, measured values of crack growth rates then are unsuitable to prove, or else disprove the validity of Eq. (15.11) and the underlying slip step or oxide film rupture model. Instead, fractographic inspection of crack walls suggests that crack-tip hydrogen embrittlement, similar to HISCC in high strength steels, is rate-determining rather than transient anodic metal dissolution.

For more $\bar{v}_{\mathrm{true}}$ values, see [8p,q]. From this work we further note that specimens tested at $\dot{\varepsilon} = 2 \times 10^{-5}\ \mathrm{s}^{-1}$ at 100°C never showed SCC, but did so at $\dot{\varepsilon} = 2 \times 10^{-6}\ \mathrm{s}^{-1}$. As inferred from the change of the critical oxygen concentration as a function of temperature, the threshold potential from the onset of SCC increases with increasing temperature to more positive values. A particular result needs attention: At a test temperature of 220°C and a strain rate of $1 \times 10^{-6}\ \mathrm{s}^{-1}$, the critical oxygen content of the test, while not exactly determined, was found to possibly be as low as 10 ppb, probably due to the steep increase of the corrosion potential shown in Fig. 15.55.

The importance of MnS inclusions was confirmed by testing a specimen which by severe machining happened to be free of inclusions in or close to the surface, and which failed by completely ductile fracture without any indication of SCC. Concerning the role of these inclusions, the usual assumption is that in addition to the mechanical effect already described there also is a detrimental influence by chemical dissolution of the sulfides into the crack-tip electrolyte [43]. Within the concept of HISCC, this effect would seem quite natural, since sulfide solutions, as described in Chap. 14, promote hydrogen uptake of steels. If, instead, transient metal dissolution is more important, catalysis of the anodic iron charge transfer reaction by $HS^-$, similar to catalysis by $OH^-$, would conveniently enter the argumentation. Published suggestions are rather more complicated, assuming inhibition of repassivation by adsorbed sulfide ions, or by a local decrease in pH. Definite proof of either mechanism apparently is not available at present. In [8p], a variant of assumptions is offered, which is crevice corrosion at inclusion/matrix interfaces creating an acid sulfide crevice electrolyte promoting hydrogen uptake and onset of SCC at crevice tips.

For thick-wall pressure vessels and the like, fracture-mechanical, i.e. constant-load testing of notched and prefatigued DCB specimens, by preference large enough to allow the application of linear-elastic fracture mechanics, certainly is a more realistic procedure than slow strain rate testing, except for low-cycle fatigue (LCF) or other dynamic loading similar to LCF. For the latter cases, see Chap. 16. Concerning low-alloy steels of the pressure vessel type, extensive, careful tests with CT 50 specimens in quasi-stagnant, but otherwise well controlled water at 240°Care described in [58]. In these tests, specimens were slowly heated to the working temperature, then exposed without load to the high-temperature water for an accommodation period of 7 days to allow for the formation of the protective magnetite film, and only then loaded stepwise to the desired level. After 30, resp. 230, resp. 100 h the load was removed and the setup was left to cool for two days. The specimens were then fractured and sectioned for fractographic inspection of the depth of the SCC crack between the fatigue precrack and the final mechanical fracture. Since SCC crack fronts were irregular, the largest observed depths was used to calculate the maximum average crack growth rates. As shown in Fig. 15.57, these rates invariably decreased with increasing test time from ca. $2 \times 10^{-7}\ \mathrm{mm\ s}^{-1}$ after 30 h to ca. $1.5 \times 10^{-8}\ \mathrm{mm\ s}^{-1}$ after 1000 h, indicating a systematic decay of crack growth rates following initial loading. Neither the sulfur content of the steels nor the

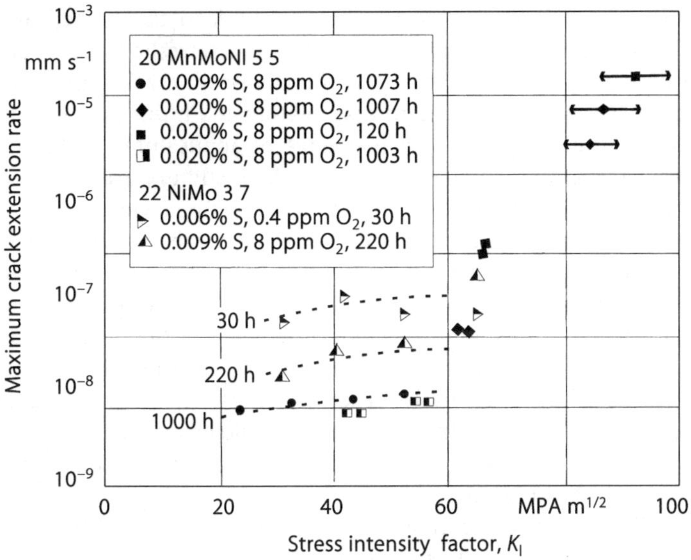


**Fig. 15.57.** Maximum average growth rates in CT 50 specimens of two pressure vessel steels (German 20 MnMoNi 5 5 and 22 NiMoCr 3 7; sulfur content 0.009 resp. 0.020%; yield strength 455 resp. 527 MPa, ultimate strength 596 resp. 670 MPa) in oxygen-containing deionized water at 240°C. (Redrawn from Läpple, Blind, and Deimel)

oxygen content of the water had an influence on this result. Also, there is little doubt that on further prolonging the test the decay would have continued, showing SCC to be, under these conditions, a transient phenomenon, irrelevant with respect to vessel-wall safety. All this applies over the range of $K_{\mathrm{I}}$ values up to 60MPa$\sqrt{m}$, which coincides with the range of applicability of linear-elastic fracture mechanics. Crack growth rates increase to unacceptable values only at equally unacceptable values of $K_{\mathrm{I}}$ close to $K_{\mathrm{IJ}}$.

None of the observations so far reported for SCC of low-alloy steel in high-temperature water support the slip-step dissolution model on which Eq. (15.11) is based. For practical applications, this same relation is, however, simplified to $\bar{v} = const \times \dot{\epsilon}_{\mathrm{tip}}^{n}$ [43]. If both const and $n$ are taken as empirical functions of system parameters such as temperature, steel composition, etc., this is a versatile expression suitable to adapt to most any cracking mechanism, provided that strain rather than stress is the predominant parameter. Another suggestion is to relate, for fracture mechanical testing, the crack tip strain rate and the stress intensity factor: $\dot{\epsilon}_{\mathrm{tip}} = const' \times K_{\mathrm{I}}^{4}$ [43], As observed in [55], this expression neglects the possibility of exhaustion of creep between crack extension steps, and thus the possibility of decay of crack extension rates under constant load conditions. Regardless of this difficulty, relations of this type are still proposed, if only as worst-case estimates, with *const'* depending on the sulfur content; hence the distinction between 'low sulfur' and 'high sulfur' conditions, extensively discussed in the literature.

The observation of crack growth decay under static load in notched and prefatigued specimens in high-temperature high-pressure refreshed and circulating water has been fully confirmed in [59a]. In this work, crack growth, as observed after initial loading, was found to cease within 100 h during constant load tests with $K_I \leq 60$ MPa $\sqrt{m}$, i.e. in the range of applicability of linear-elastic fracture mechanics. Matters changed only in distinctly impure water, or at loads increased into the range of elasto-plastic specimen behaviour.

If crack extension steps involve brittle lattice decohesion, inherent brittleness of the material will normally enhance SCC susceptibility. In this connection, an interesting point is dynamic strain aging (DSA) [59b,61]. DSA is the result of immobilization of dislocations, and thus embrittlement, by slow interaction with interstitials, such as atomic nitrogen and carbon, but possibly also atomic hydrogen. It our context, this possibly is an important effect, when the aging rate is comparable to the strain rate. In the case of low-alloy steels, DSA is most marked at ca. 250°C, which, in fact, also is the temperature range considered most dangerous for SCC in these materials.

If SCC in low-alloy pressure vessel steels depends on continued straining, we are again close to conditions effective during tensile periods of low-cycle fatigue, and the topic will therefore reappear in Chap. 16. In this context, a new variant of slow strain rate testing merits notion already presently. It is *constant-rate rising displacement testing* (RDT) of notched and prefatigued CT specimens in the range of elasto-plastic fracture mechanical behaviour. Apparently, SCC thresholds determined by this methods coincide with those obtained from constant load testing, while crack extension rates are much higher [62]. A logical consequence of these observations is that cracks, started to run by RDT, should, though perhaps with some delay, stop to run after switching to constant load. Investigations into this behaviour are being pursued presently [63], and apparently they have the expected result. If so, it confirms that SCC of low-alloy steels in high-temperature, high-pressure water is definitely strain-induced rather than stress-induced, settling a previously open discussion [8r].

A second prominent topic concerning SCC in high-temperature, high- pressure water particularly as water coolant for boiling water nuclear reactors is intergranular stress corrosion cracking (IGSCC) in *stainless steel.* We disregard stainless steel cladding of pressure vessels and stainless steel core components, and consider welded pipes of the primary coolant circle only. Again, the following is a brief outline only, using selected information from the wealth of published contributions in [60] and many other sources, with the aim to find general lines of present arguments concerning mechanisms, rather than applications.

From previous sections in this chapter we recall that susceptibility to transcrystalline stress corrosion cracking (TGSCC) in high-temperature chloride-rich solutions is a characteristic deficiency of stainless steel. Also, it has already been described that plastic strain enhances crack initiation, but that strings of micropits can induce SCC without strain as well, if only at a much lower rate.

Furthermore, examples of transition from TGSCC to IGSCC in stainless steels were noted.

In the presence of dissolved chlorides, transcrystalline cracking is predominant. As examples, we quote TGSCC in titanium-stabilized 18Cr10Ni steel (German x6 CrNiTi 18 10, Nr. 1.4541) in 0.01 M NaCl solutions (170 ppm) at temperatures up to 250°C. The material was steam generator pipes, which had cracked under the influence of residual internal stresses [52a,c,d]. In this work, crack extension rates were not determined. Instead, the test autoclave was equipped for hydrogen permeation measurements as described in Chap. 14, using a stainless steel test tube to record the permeation current across the tube wall. In a typical case, hydrogen permeation sets in after a delay of 31 days, increasing within 2 h to a peak hydrogen flux of $30 \times 10^{-12}$ mol $cm^{-2}$ $s^{-1}$. This was considered to point to crack-tip hydrogen embrittlement as the cause of cracking. Careful TEM inspection proved that hydrogen-induced martensite also was involved.

In a further series of these experiments, samples of the steels 17Cr10Ni (AISI 321), 21Crr33Ni ('Incoloy 800'), and samples of the nickel-base alloys 59Ni 30Cr8Fe ('Incoloy 690') and 73Ni16Cr9Fe ('Incoloy 600') were tested for both overall hydrogen embrittlement and for SCC susceptibility. For the first point, samples were cathodically charged with hydrogen from solutions containing promoters of hydrogen uptake. Embrittlement was characterized by recording the decrease in strain to ultimate failure, and was found to increase with increasing nickel content of the samples. In contrast to this, SCC susceptibility decreases in the same direction, confirming general experience concerning nickel in steels and in nickel-base alloys. The authors nevertheless continue to argue that cracking is by HISCC, and they do so on the grounds of TEM micrographs of crack-tip areas, showing brittle decohesion cracks in stainless steels, but not in nickel-base alloys. Crack initiation in steels usually was associated with corrosion pits in the surface. A very useful further result was that, as determined by hot extraction of hydrogen, the crack-zone hydrogen content in stress corroded AISI 321 was 380 ppm, compared to 7.6 ppm in the base material. For Incoloy 800, the quite similar numbers were 240 nd 8.1 ppm, respectively. Clearly, this is substantial support for a HISCC mechanism.

From permeation data obtained at comparatively low temperatures (25 to 85°C), the diffusivity of interstitial hydrogen in the austenitic fcc lattice was extrapolated, applying the Arrhenius equation, to $D_H(288°C) = 2.15 \times 10^{-8}$ $cm^2$ $s^{-1}$ for AISI 321, and $8.74 \times 10^{-8}$ $cm^2$ $s^{-1}$ for Incoloy 800. This compares well with a value $1.96 \times 10^{-8}$ $cm^2$ $s^{-1}$ reported for AISI 321 at 250°C in [64]. These data probably show that, owing to the higher diffusivity, hydrogen taken up into the crack tip zone in Incoloy 800 spreads into a larger volume as compared to AISI 321, thus lowering the crack tip hydrogen available for brittle decohesion processes. This predominance of hydrogen diffusivity compared to overall hydrogen uptake then removes the conflict between data on susceptibility observed during corrosion, and data on strain-to-failure of cathodically charged samples.

For more on SCC of stainless steel in high-temperature, high-pressure water containing 5 ppm chloride ($3 \times 10^{-4}$ mol $L^{-1}$), rather close to accidental impurities in nominally pure water, at 100 to 300°C, see [37k–n]. In this work, the steel was AISI 316SS, which is non-stabilized 17Cr10Ni2Mo, sensitized by heating to 700°C during 30 h. This treatment causes susceptibility with respect to intergranular corrosion (IGC) due to grain boundary chromium depletion by chromium carbide precipitation (see Chap. 13). IGC susceptibility was verified by the EPR technique also described previously, and SCC was tested by slow straining smooth specimens at controlled potentials with strain rates varying between $1.5 \times 10^{-6} \leq \dot{\epsilon} \leq 4.2 \times 10^{-6}$ $s^{-1}$. Flowing and stagnant test solutions were used alternatively. Average crack growth rates, once more measured as ratios of depth of SCC cracks over time to failure increased with increasing potential, from $10^{-7}$ mm $s^{-1}$ at ca. −0.65 V to $10^{-5}$ mm $s^{-1}$ at ca. −0.4 V. Ignoring possible influences of varying initiation periods, i.e. ignoring true growth rates, this trend is believed to point to a metal-dissolution controlled crack extension mechanism.

Figure 15.58 compiles data on threshold potentials for the onset of transgranular SCC as a function temperature, and also for the transition from transgranular to intergranular, or to mixed trans- and intergranular SCC, especially showing that IGSCC requires more positive potentials. From previous experience described in [37l], it is known that strains required for initiation of IGSCC in sensitized steel are relatively small and insensitive to temperature, compared to strains required for TGSCC in non-sensitized material. In Fig. 15.58, data for pure TGSCC in non-sensitized steel, as described in [37k], are included, surprisingly showing a gap in the potential/temperature field, where cracking was not observed. The authors report that in the low-potential region cracks remained short, and that they may be artefacts of the straining procedure rather than true SCC.

As for low-alloy steels, the dependence of either TGSCC or IGSCC on the electrode potential is well established. It indicates that in practical service, where the potential is the open-circuit corrosion potential, the oxygen content, driving $\varepsilon_{corr}$ to increasingly positive values, will be the decisive detrimental factor. Also it should be recalled from Chap. 14 that high-temperature passivating, or else quasi-passivating oxide films are much thicker than films at ambient temperature. Therefore, if SCC initiation requires mechanical film rupture, film brittleness will be an important parameter. In [37n], results of Auger spectroscopy of films formed at high temperature are reported. In this work, the films were slowly thinned by argon sputtering, in order to obtain information about both film thickness and composition. Films were found to vary in thickness between 100 and 780 nm. One supposition is that brittleness of these films increases with increasing thickness, and that crack initiation by film rupture would become easier as films grow thicker. But SCC susceptibility also correlated with an enhanced nickel content of oxide films, which so far leaves the issue undecided.

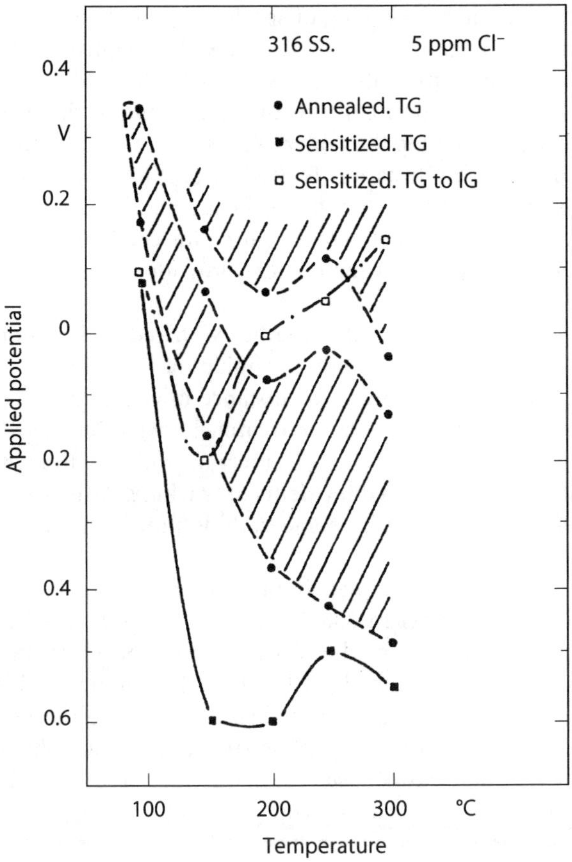


**Fig. 15.58.** Threshold potentials for slow-strain-rate SCC of sensitized 17Cr10Ni2Mo stainless steel in 5ppm NaCl solution as function of temperature. *Solid squares:* Minimum potential for onset of TGSCC; Open squares: Minimum potential for change to IGSCC, or to mixed TGSCC and IGSCC. (From Congleton and Young). *Open squares:* Domaines of TGSCC in non-sensitized steel of same composition, showing an immunity gap for intermediate potentials and temperatures. (From Congleton, Zheng, and Hua)

SCC of stainless steels in nuclear reactor coolants always is intercrystalline. Considering what has been described for dilute chloride solutions above, this may be seen as being the limiting case of SCC in high-temperature, high-pressure waters, observed when the water content in dissolved electrolytes tends to zero. Originally, the occurrence of cracks in coolant loops was explained as being due to accidental sensitizing of the non-stabilized austenitic steel, then widely used [43k]. This, then, was explained as stress- or strain-assisted intercrystalline corrosion owing to grain boundary chromium depletion. If so, the mechanism would involve activation of narrow grain boundary zones in contact with large grain matrix areas, but apparently this mechanism was not clearly established, and some other unspecified weakening of sensitized grain boundaries with respect to corrosion may be operative. Sensitation is well explained by the observation that intergranular cracks invariably are associated with weld seams, indicating chromium carbide precipitation close to the fusion line in the heat affected zone (HAZ), where during cooling the material passes through a critical temperature range. As a consequence, tests for TGSCC susceptibility used the complete stock of methods as described in Chap. 13, including Strauss, Huey and especially EPR tests. Residual internal stresses required for crack extension are left in the material from weld strains, which, added to code stresses, may locally exceed the yield strength.

As an obvious counter-remedy stainless steels either with reduced carbon content, or else stabilized with respect to carbide precipitation by titanium or niobium additions became the preferred choice for instance in Germany [65a]. Also, welding techniques and post-welding surface treatments were improved. This certainly reduced the number of incidents of cracking, without, though, completely removing the danger of occasional incidents of cracking. As revealed by EPR tests, slight sensitation apparently by some carbide precipitation was indeed still involved [65d].

But of late, there is evidence of occasional, if rare occurrence of IGSCC of definitely non-sensitized stainless steel in 288°C water of nearly theoretical purity, and even in deaerated water, i.e. in the absence of dissolved oxygen [43g–i,59b,65]. As before, cracking is confined to the HAZ. The origin of this intercrystalline susceptibility is not quite clear, but then it should be recalled that in other cases, as for instance TGSCC in stainless steels, the reasons for a particular, in this case transcrystalline path chosen for continuous crack growth, are not directly obvious either. In [43i], the synergistic action of residual strains from weld shrinkage in zones close to the fusion line, and high corrosion potentials are seen as major factors, but grain boundary chromium depletion, in this case not by carbide precipitation but by thermal segregation, may also be important. The possibility of local hydrogen embrittlement is excluded in [43j] on the grounds of hydrogen permeation measurements. In view of what has been explained above concerning data on hydrogen uptake from permeation experiments as compared to crack tip zone hydrogen contents obtained from hot extraction, the argument may not be quite conclusive.

Reproducing IGSCC in unsensitized stainless steel in laboratory tests is no easy task. As described in [43g], the difficulty is that prefatiguing notched double cantilever beam probes produces a transcrystalline precrack which cannot easily turn into an intercrystalline corrosion crack. A suitable procedure is prefatiguing a specimen initially at 1 Hz with a load amplitude $\Delta K$ with a maximum, $K_{max}$, somewhat below the test value of e.g. 27.5 MPa $\sqrt{m}$, then slowing down fatiguing to 0.001 Hz, eventually holding, for some brief time, the load during every cycle, and finally switching to full static load. Some cracks thus initiated proved to be able to continuous growth during long testing periods. As shown in Fig. 15.59, a typical growth rate value in oxygen containing water was $1.5 \times 10^{-7}$ mm $s^{-1}$. Reducing the oxygen level from 2000 to 95 ppb decelerated the crack to 2.8 ×

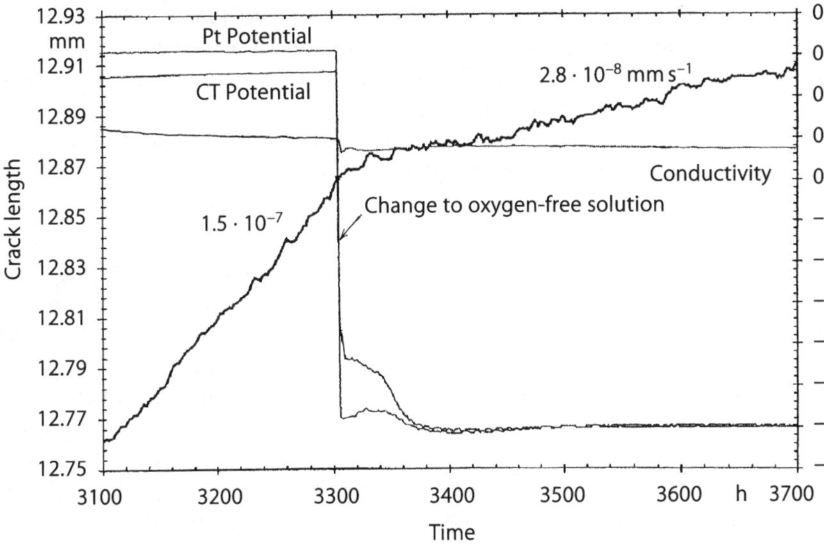


**Fig. 15.59.** IGSCC crack extension in notched and prefatigued CT specimens of unsensitized 18Cr9Ni stainless steel at 288° C in pure water initially containing 2000 ppb (2 ppm) $O_2$ h, changing to 95 ppb $O_2$ for the following 400 h. $K_I = 25$ ksi $\text{In}^{1/2}$. Crack growth recorded by the potential drop method. Also shown: Open-circuit corrosion potential ('CT potential') of the steel, redox potential of the solution ('Pt Potential'), and solution conductivity. (From Andresen, Angeliu, Catlin, Young, and Horn)

$10^{-8}$ mm $s^{-1}$, but crack growth remained continuous and even constant in rate during hundreds of hours. As also recorded in the figure, a change to a low-level oxygen content caused the corrosion potential to drop by almost 1 V. Incidentally, $\varepsilon_{corr}$ was quite close to the so-called 'redox potential' in the solution. This is a curiosity in power plant chemistry, namely the potential of an inert platinum electrode in the same solution. In hypothetical equilibrium, this should be either the oxygen, or else the hydrogen equilibrium potential in the solution; in the usual non-equilibrium it is an accidental potential in between, which apparently often qualitatively reflects the content in oxygen or other oxidizing species.

Whether sustained crack growth under true service conditions really must be expected, is at least open to question, if not directly improbable. Should the danger be real, a quantitative prediction of crack growth rates and, thus, life time of precracked structures would, of course, be a point of major interest. This, then, leads back to calculating crack growth rates from principles of electrode kinetics and mass transfer already introduced further above. The problem, as applied especially to crack growth in sensitized stainless steel, has been and is being dealt with by several groups [43l,m,46b,c,66]. An invariable attribute of all contributions is the presupposition of cracks extending by the slip step or oxide film rupture mechanism, with growth controlled by transient anodic metal dissolution current, as calculated by Eq. (15.10). As described previously, this disregards any possibility of brittle and plastic contributions to crack growth, which weakens the deductions, which are, however, claimed to be fully supported by experimental evidence.

A characteristic property of the simple slot or crack growth model presented previously is that anodic metal dissolution is confined to the slot or crack tip, and that crack walls are practically inert. The cathodic partial reaction sustaining anodic crack-tip dissolution is either occurring at an external counter electrode or, the metal surface surrounding the slot or crack opening, and this then is open-circuit corrosion. In the latter case the system is a galvanic cell and, since the external cathodic reaction usually will be reduction of dissolved oxygen, a differential aeration cell. The model readily explains the detrimental effect of oxygen contaminating the water. We also have already indicated the possible contribution of crack tip activation/repasivation cycles to crack growth in addition to galvanic cell action. The local overall corrosion reaction would be of the type $Me+nH_2O \rightarrow MeO_n+(n/2)H_2$. According to one source [43l,m], this is, in fact, the predominant contribution to crack growth. In this context, the proposal that during IGSCC of Ni-16Cr-8Fe ('Incoloy 600') in reactor coolants the chemical local reaction proceeds by intergranular internal oxidation is quite interesting [55b]. The mechanism is reminiscent of what has been observed for IGSCC of mild steel (see Fig. 15.38), but it has been refused in a recent review [57b].

The opposite opinion, which is predominance of galvanic coupling of the crack tip as anode, and the outer surface of the metal as cathode, is advocated in [66a]. Of late, the conflict of opinions appears relaxed by a compromise: Galvanic differential aeration cell action still is considered predominant, controlled by the oxygen concentration in the bulk solution. At the same time, hydrogen deposition by reduction of hydrogen ions and/or water is now assumed to occur at internal crack walls as well, increasing in rate towards the crack tip, according to the increase in ohmic drop depressing the local electrode potential in this direction [46b,c,66b-e]. Also, slow anodic metal dissolution from passive crack walls is recognized as possibly non-negligible, if only at current densities between $10^{-7}$ and $10^{-6}$ A $cm^{-2}$, assumed to be independent of potential. These inside-crack conditions require a change in the system of Eqs. (15.2) and (15.3): The current flowing in the crack electrolyte is no longer simply equal to the anodic crack-tip dissolution current, $j_{tip}$, i.e. $I_{tip} \times f_{tip}$. Instead, currents $j_{H^+}$, $j_{H_2O}$, and $j_{pass}$ of crack-wall electrode reactions contribute to the overall current, changing in magnitude as function of the position in the crack. Accordingly, sources and sinks, $S_{H^+}$,$S_{H_2O}$, and $S_{Me^{z+}}$ (mol $cm^{-3}$ $s^{-1}$), must be added to Eqs. (15.2).

We are, at this point, talking in terms of current strengths, $j$, rather than current densities, $i$, because in both [46b] and [66d] trapezoidal cracks are considered, further complicating matters concerning mass conservation. When linear elastic fracture mechanics are applicable, tapering cracks can be described in terms of crack mouth opening displacement, CMOD, and crack tip opening displacement, CTOD. We have, in the first place, CTOD $= 0.6(1 - \nu^2)K_I/E\,\sigma_y$, where E, $\sigma_y$ and $\nu$ are Young's modulus, yield strength, and Poisson's ratio, respectively. From CTOD, CMOD is calculated, and the cross section at position $x$, $f_{crack}(x) = b \times h(x)$, of a crack of constant breadth $b$, length $a$ and half-width $h(x)$ is found using the tangent of the crack opening angle:

$$f_{crack}(x) = 2b\,(\text{CTOD} + (x/a)(\text{CMOD} - \text{CTOD}))\ . \tag{15.15}$$

All electrode reactions, except passive metal dissolution, are considered to obey Tafel kinetics, i.e. governed by charge-transfer kinetics. Thus rate equations of the form

$$\text{a)}\quad I_{tip} = A_1 \exp(B_1\,\varepsilon); \qquad \text{b)}\quad i_{H^+} = A_2\,c_{H^+} \exp(-B_2\,\varepsilon) \tag{15.16}$$

$$\text{c)}\quad i_{H_2O} = A_3 \exp(-B_3\,\varepsilon); \qquad \text{d)}\quad i_{O_2} = A_4\,c_{O_2} \exp(-B_4\,\varepsilon)\ ,$$

are consistently used. Even if equations of this type are valid throughout, which is not a safe assumption, empirical, or semi-empirical, or arbitrary values must be substituted for the many rate constants and Tafel slopes. Concerning crack tip metal dissolution, we recall that, if a succession of transient metal dissolution processes is assumed, $I_{tip}$ in the first place enters Eq. (15.10), determining an average current density $\bar{I}_{tip}$. Adapted to fracture mechanics, this equation is used in the form

$$\bar{I}_{tip} = \text{const}\ I_{tip}(t_\circ K_I^4)^n/(1 - n)\,, \tag{15.17}$$

adding still more adjustable parameters to the calculation [66a]. While the danger of errors concerning rate-determining reaction steps apparently is never considered, the broad margin of inaccuracy in numerical values of kinetic parameters is recognized in [46b]. Therefore, while the principles of the calculations are thought to be fundamental, numerical results are, so far, presented only to illustrate the procedure. Such model calculations assume 288°C, and water slightly contaminated by traces of NaCl. The oxygen content of the bulk solution, i.e. the factor determining the galvanic cell current, is left unspecified. Inside-crack conditions are discussed mostly in terms of the crack mouth potential, $\varepsilon_{mouth}$. This is a somewhat difficult point because, owing to the low conductivity of the water, galvanic currents flowing from the crack mouth to the surrounding cathodic areas depress the crack mouth potential to values negativ with respect to the corrosion potential of positions remote to the mouth by a considerable ohmic drop. The discussion of this point makes specialist's reading in [46c] and [66e] and will not be pursued presently. In any case, the calculations as yet stop short of the principal aim of predicting, for given metallurgical properties – steel composition, degree of accidental sensitation in non-stabilized, intergranular dislocation pile-up in stabilized steels, magnitude of residual stresses in weld seames, and the like – the rate of extension of preexisting cracks as function of service conditions.

We recall that these calculations apply to cracks growing, during long periods of time, continuously, or quasi-continuously. Whether this modeling is applicable to long-term corrosion crack growth in power plant components is questionable, and, as noted previously, improbable. Also, crack initiation, the prerequisite of crack advance, but principally disregarded in fracture mechanical testing, also may be avoided by suitable post-weld surface and heat treatment [65e]. Therefore, even if occasional detrimental water contamination is considered possible, component integrity between inspections should be safe in well-controlled and well-treated plants.

## 15.3 Titanium and Titanium Alloys

As already described in Chap. 15, *titanium* is a valve metal with excellent corrosion resistance owing to passivity by an oxide film of low electronic conductivity. The metal is used as structural material in the chemical industry, behaving very well except for occasional crevice corrosion. It also is a candidate for protecting radioactive waste disposal. In the sixties, when commercial supersonic aircraft development was en vogue, much attention and labour was being paid to titanium alloys for body and fuselage construction, and extensive research into SCC of e.g Ti-8Al-1Mo-1V (wt.%) ('Ti-8-1-1'), was reported, decaying with the decay of expectations [67].

Concerning electrode kinetics of titanium alloy corrosion, not very much information is available. Figure 15.60 is an example of rapidly scanned current-potential curves [68]. Tentatively, the two maxima can be attributed to the successive formation of a double-layer oxide film, consisting of $Ti_2O_3/TiO_2$, and this would make the passive film similar to what is assumed for pure titanium. An interesting point is the pitting potential as defined in Chap. 13. While in 3%NaCl solution pure titanium pits at potentials beyond ca.10 V only, pitting in Ti-8-1-1 is observed already beyond ca.2 V, but pitting still is a very unlikely event. The situation changes in methanol/hydrochloric acid solutions, because pitting of, for instance, Ti-6Al-4V in $CH_3OH$/0.1 N HCl is now observed already beyond ca. −0.13 V [69]. As the water content is increased, the onset of pitting shifts to more positive potentials. Since water promotes passivation by oxide films, the shift may be due to an increase in film thickness with the water content of the alcohol.

SCC in slowly strained Ti-5Al-2Sn in aqueous and in methanolic chloride solution has already been described further above (see Fig. 15.6). Turning to methanolic solutions is, in this book, a digression, but in principle $CH_3OH$ is a solvent similar to $H_2O$. For practical applications, the behaviour of commercially pure titanium in methanol contaminated with both water and hydrochloric acid is an interesting case [70]. As shown in Fig. 15.61, stressed thin titanium foils crack already in the presence of $10^{-5}$ M HCl, provided that the alcohol is contaminated with water, while more water inhibits SCC. Similar behaviour is reported for methanol/bromide and methanol/iodide solutions [71a,b], and for halide-contaminated higher alcohols [71c]. For otherwise constant conditions, time to failure values of strained foils increases with increasing alcohol chain length. This may indicate a correlation of crack growth rates and solu-

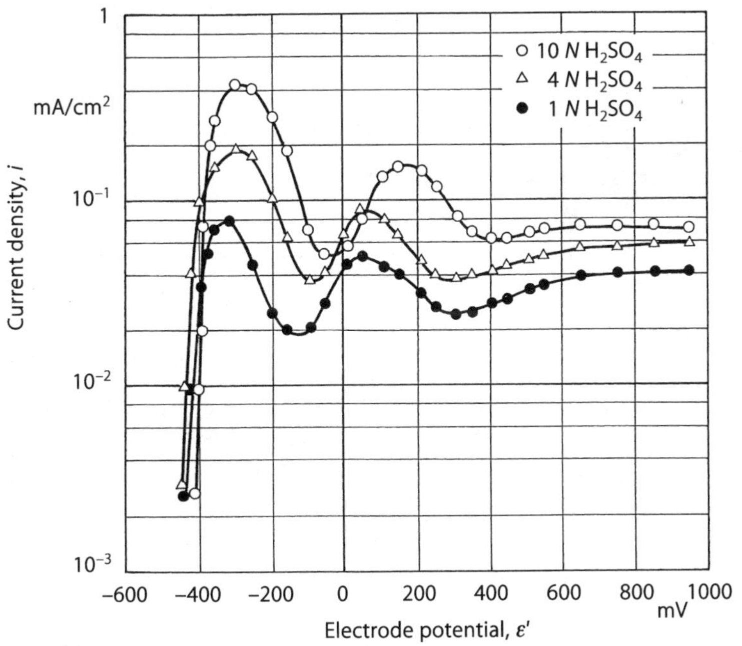


**Fig. 15.60.** Current-potential curves, obtained by rapid potentiodynamic scanning, for Ti-8Al-1Mo-1V in sulfuric acid solutions. $\varepsilon'$ referred to sat.cal.el. (from Chen, Beck, and Fontana)

tion viscosity, perhaps pointing to rate-determining diffusion of some reaction participant to or from the crack tip. The minimum in time to failure values for water contents between 0.1 and 1% should also make an interesting clue as to mechanisms, but further evidence is not as yet available.

In nominally pure titanium, cracking is intercrystalline. Similar to what was observed for mild steel, intercrystalline corrosion (IC) from metallographic sections of strained specimens in Fig. 15.62, where water and chloride contamination decreases in the sequence of pictures from left to right, susceptibility with respect to general IC decreases with increasing purity of methanol. leaving only IGSCC susceptibility [72]. The role of water is also evident from the current-potential curves in Fig. 15.63: In pure $CH_3OH$/NaCl solutions with water contents up to 4%, electrode polarizability is very low, and currents increase with time regardless of the potential, probably indicating an early onset of pitting. At 10% $H_2O$, passivity is stable, and currents decrease with time at every potential up to what presumably is the pitting potential, ca. 0.17 V. At 50% $H_2O$, stable passivity is established at least up to 2 V and thus under all normal service conditions. All this suggests that intergranular corrosion of titanium is caused by local, perhaps pitting attack of chloride at grain boundaries, inhibited by protective passivating film growing in the presence of sufficient amounts of dissolved water in the alcohol. Figure 15.64 shows, in the first place,

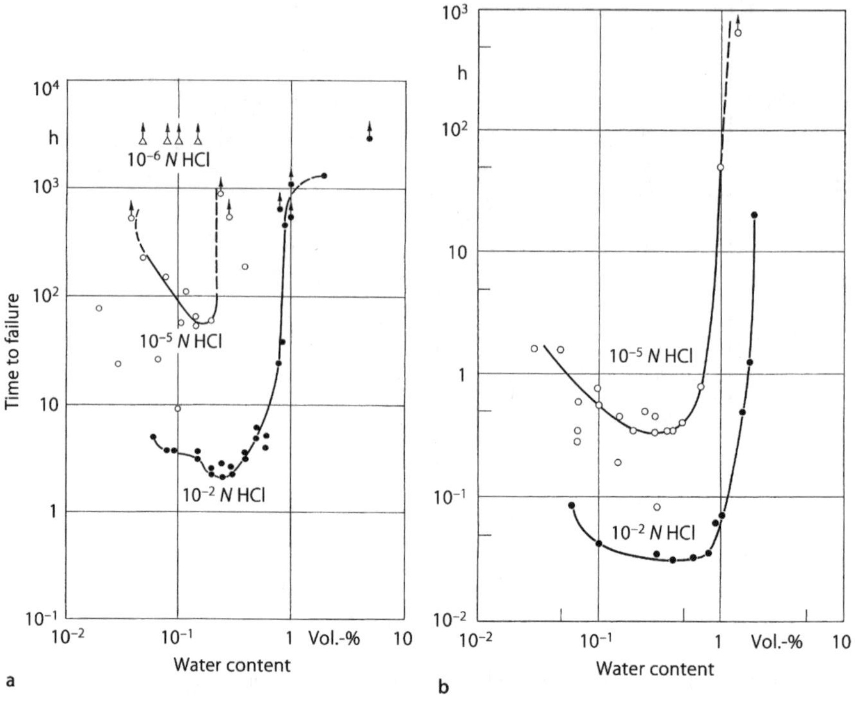


**Fig. 15.61a,b.** The time to failure of thin foils, loaded in tension with 75% of the yield strength at 30°C in methanol containing hydrochloric acid, as function of the water content, at different contents of hydrochloric acid. **a** Commercially pure Ti (0.07Fe, 0.02C, 0.01N, 0.04H, wt.%). **b** Ti-13V-11Cr-3Al. (From Haney and Wearmouth)

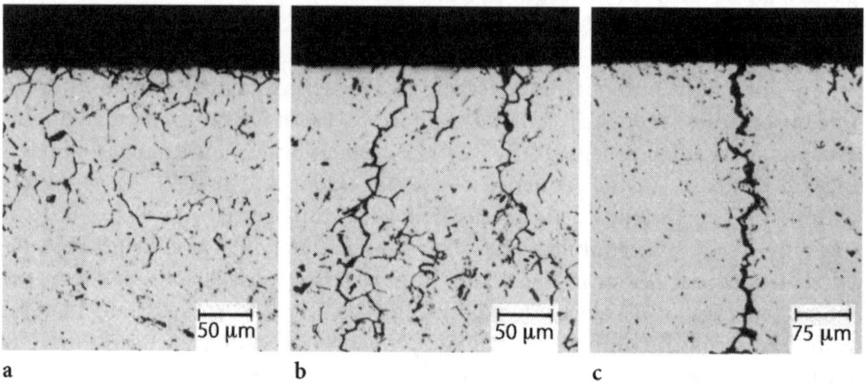


**Fig. 15.62a–c.** Metallographic sections of titanium samples (0.1Fe, 0.0008N, 0.006H, wt.%) corroded in methanol contaminated with water and sodium chloride. **a** 0.1% $H_2O$/0.1 M NaCl; **b** 0.1% $H_2O/10^{-5}$ M NaCl; **c** 10 ppm $H_2O$, no NaCl. (From Mazza)

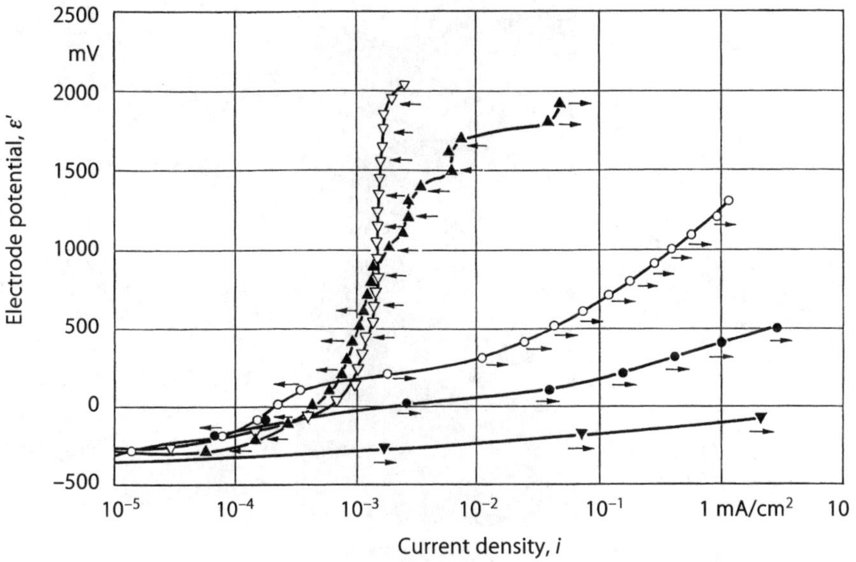


**Fig. 15.63.** Anodic current-potential curves for commercially pure titanium in methanol containing 0.01 mol $L^{-1}$ NaCl. Water content: 0.1% (▼), 2% (●), 4% (○), 10% (▲), 50% (▽). Arrows indicate currents increasing (→), resp. decreasing (←) with time at constant potential. $\varepsilon'$ referred to sat.cal.el. (From Mazza)

the fracture surface of a U-bend specimen of titanium containing 0.024% Fe and 40 ppm H damaged by pure IGSCC in acidic methanol, and subsequently brought to complete fracture in air [73a]. Tested under the same conditions, a sample of oxygen-reach titanium (0.015% Fe, 2900 ppm O, 65 ppm H) already failed during exposure to acidic methanol, and the crack surface clearly reveals transcrystalline cleavage faces, i.e. TGSCC accompanying IGSCC. We note, without showing fractographs, that some TGSCC is also observed in low-oxygen but cold worked titanium. A further interesting point evident from the figure is that transcrystalline cleavage appears to be enhanced by titanium hydride precipitates produced by a 700°C temper.

So far, the topic was mostly SCC of pure, or commercially pure, or oxygen rich $\alpha$-phase Ti. For $\beta$-phase alloys, SCC of smooth specimens is again observed in acidic methanol solutions. As also shown in Fig. 15.61, the dependence of the times to failure during constant-load tensile testing of thin foils is much the same as in commercially pure material.

Crack morphology in $\alpha$-phase alloy Ti-5Al-2.5Sn is much the same in either methanolic or aqueous solutions [73b]. This also holds for the $\alpha/\beta$-phase Ti-8Al-1Mo-1V. In this alloy, cleavage planes are vicinal to basal hexagonal (0001) faces of $\alpha$-crystallites, and crack propagation appears to be impeded by both $\beta$-crystallites and martensitic $\beta$-phase particles. On the whole, crack propagation, once initiated, appears to proceed by similar mechanisms in both alcoholic and aqueous solution.

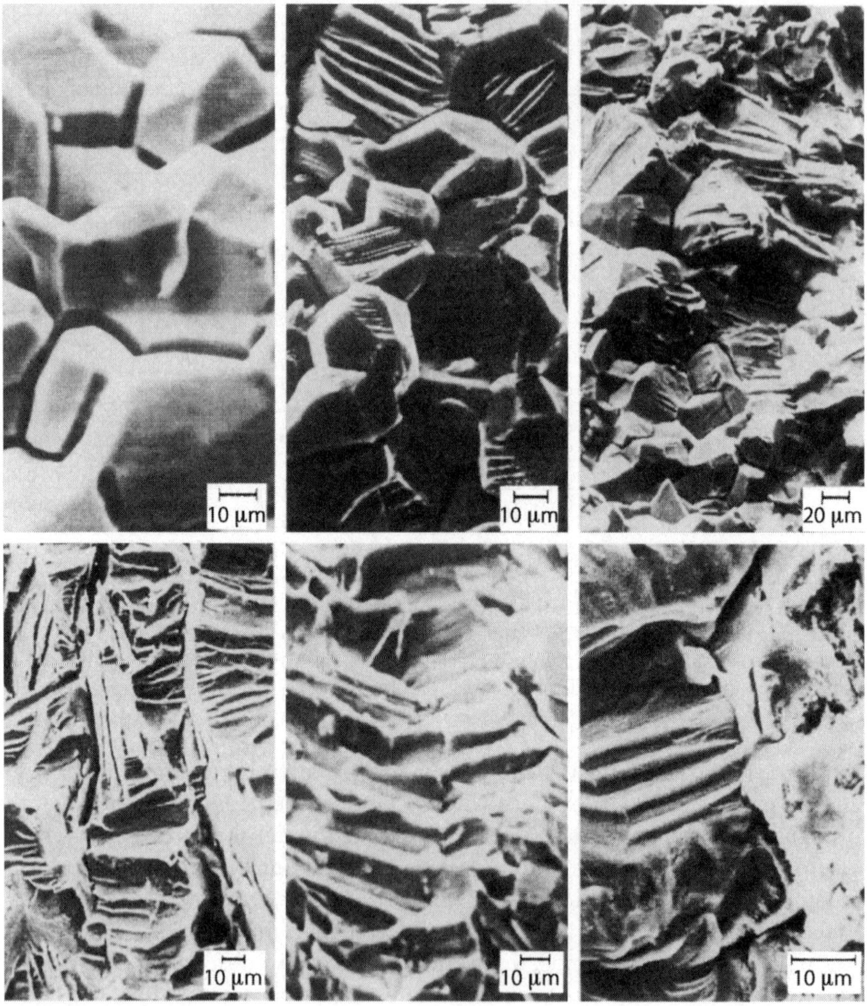


**Fig. 15.64.** Typical slow strain-rate stress-corrosion fracture surfaces in titanium and TiAlSn. **Top row:** Titanium in methanol/hydrochloric acid solutions. *Left:* IGSCC of commercially pure Ti; *center:* IGSCC and TGCC of oxygen-rich titanium, *right:* IGSCC and TGSCC of oxygen-rich Ti heat-treated to contain $TiH_2$ precipitates. **Bottom row:** *left:* Ti-5Al-2.5Sn fractured in aqueous 3% NaCl solution; *center:* oxygen-rich Ti containing $TiH_2$ precipitates, fractured in the same solution; *right:* Ti-5Al-2.5Sn, precorroded without applied stress in methanol/hydrocloric acid solution, fractured in air. (Sections of SEM fractographs from Powell and Scully)

We insert here, that SCC of titanium materials has also been observed in carbon tetrachloride [73b,c,74a], in fluoro-chloro-hydrocarbons [74b], and in nitrogen tetroxide [74c], but we pass these experiences as too far outside the scope of this text.

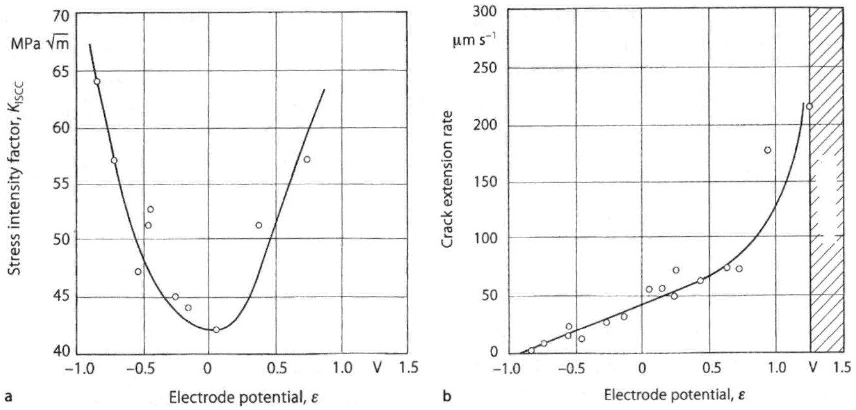


**Fig. 15.65.** Potentiostatic fracture-mechanical testing of Ti-6Al-4V DCB specimens in 3.5% NaCl solution: *Left:* Threshold stress intensity factor $K_{ISCC}$ for onset of crack growth in Region I of the $\ln \upsilon$-$K_{K_I}$ curve. *Right:* Crack extension rate in the plateau of Region II. (From Vogelhuber)

A more generally interesting item is the discrepancy of behaviour in methanolic and in aqueous solutions observed, for instance, during slow strain rate testing of smooth TiAlSn specimens (see Fig. 15.6). These, and similar observations for other alloys [75], imply that true constant load testing would reveal SCC susceptibility of these alloys in methanolic, but not in aqueous solutions. Again, as for low-alloy steels in high-temperature water, the reason for this behaviour probably is strain-induced crack initiation. It has, in fact, been reported for Ti-7Al-2Nb-1Ta that the loading mode is decisive: Testing notched and prefatigued DCB specimens, SCC was readily observed at low values of the stress intensity, but only when applying the load to specimens already exposed to the test solution [76].

Fracture mechanical testing of titanium alloys usually reveals the typical $\ln \upsilon$-$K_I$ curves, passing through a steep increase at $K_{ISCC}$ in Region I to a plateau value in Region II, before reaching the fracture toughness, $K_{IC}$ [75b,c]. For simplicity, we pass the occasional observation of an additional step in the plateau Region II. For Ti-8Al-1Mo-1V, curves of similar shape but with widely different Region II-values of crack propagation rates in Region II are observed in environments ranging from pure methanol ($\upsilon \simeq 10^{-6}$ m s$^{-1}$), carbon tetrachloride, chloride aqueous salt solutions, hydrochloric acid, and salt melts to mercury ($\upsilon \simeq 10^{-1}$ m s$^{-1}$). For 1 M KCl solution, a typical value is $\upsilon \simeq 10^{-4}$ m s$^{-1}$), and this certainly leaves doubts as to the possibility to explain crack propagation by anodic crack tip metal dissolution only. Results of potentiostatic fracture mechanical testing of Ti-6Al-4V in NaCl solution are shown in Fig. 15.65 [8t]. While the increase of crack propagation rate values parallel to an increase in electrode potential up to 0.7 V would be considered normal, the following steep increase, and especially the drop to negligible SCC susceptibility at potentials beyond 1.2 V is surprising. Also, the minimum in $K_{ISCC}$-values at 0 V indicates a change in mechanism of crack initiation. We may recall that titanium chemistry is complicated, because the metal and its alloys tend to form oxides as well as hydrides. Changes in surface chemistry reflecting this complexity may well influence crack initiation.

Further above, attempts to determine true current densities of anodic dissolution, Ti → $Ti^{3+}$ from 'naked' cleavage planes were shown in Fig. 15.17. These results suggest that crack-tip dissolution current densities, $I_{tip,}$ are sufficient to cause saturation of the crack-tip electrolyte with respect to dissolved titanium chloride [18b,c]. From this, a model of crack propagation very similar to that described in Fig. 15.10 for aluminum alloys, was proposed. The Nernst-Planck equations were considered applicable for mass transport from crack tip to crack mouth. The crack interior was supposed to have two regions: Close to the tip, the crack electrolyte is mainly a concentrated $TiCl_3$ solution, and titanium-ion mass transfer is by diffusion and migration of $Ti^{3+}$. Beyond a small distance of e.g. 0.1 cm from the tip, the crack electrolyte composition is similar to that of the bulk electrolyte, and transport of $Ti^{3+}$ is by diffusion only, because electrolytic migration becomes negligible in the presence of excess bulk solution components. The model is able to explain the order of magnitude of experimental crack growth rates.

Considering the chemical affinity of titanium to hydrogen, the possibility of hydrogen embrittlement causing or contributing to crack extension has long been recognized especially for transgranular cleavage processes. For intergranular cracking, an anodic metal dissolution mechanism was the preferred supposition [37o,73e]. From more recent evidence it now appears that HISCC may well be another or even the sole cause not only of transgranular but also of intergranular crack extension [73e–g]. This was inferred from experiments with low-oxygen, commercially pure titanium anodically polarized for 25 or more hours in acid methanol. When specimens were strain-tested in air immediately after removal from the solution, intergranular brittle cracking occurred at rates $10^{-3} \lesssim \dot{\varepsilon} \lesssim 10^{-1}\ s^{-1}$. Aging for 20 h at 150°C previous to testing suppressed the embrittling effect, apparently owing to desorption of hydrogen. Slow strain rate testing in acidic methanol with crosshead speeds between 1.67 nm $s^{-1}$ and 1667 μm $s^{-1}$ revealed predominantly intergranular cracking with some transgranular cleavages, and average crack extension rates up to 1.2 mm $h^{-1}$. Adding a selenium salt to the solution enhanced, and adding a platinum salt retarded SCC. Since selenium is an electrode poison, enhancing hydrogen uptake at metal surfaces, while platinum, on the contrary, catalyses recombination and thus release of adsorbed atomic hydrogen, these observations point to HISCC as at least contributing to IGSCC. We note that in this context strain-enhanced embrittlement caused by hydrogen transport into the metal lattice by moving dislocations is also discussed as promoting SCC. The origin of atomic hydrogen would be cathodic crack-tip hydrogen deposition superposed to anodic metal dissolution, owing to a probably very negative crack-tip electrode potential. Embrittlement should be due to interstitial hydrogen, as described in Chap. 14.

Hydride precipitation ahead of the crack tip would be a different matter, but apparently there is no evidence pointing in this direction.

Hydriding titanium structural materials by overall cathodic electrolysis causing well defined heterogeneous hydride precipitates acting as local stress raisers is well documented [73]. As briefly noted in Chap. 14, loss in strength due to internal hydride precipitation in titanium has been the object of some concern for the electrolytic recovery of uranium and plutonium from nitric acid solutions, but was shown to be tolerable. For details, see [8u–x].

## 15.4 Precipitation Hardening Aluminum Alloys

Alloys of *aluminum* with zinc, magnesium and copper heat-treated to high strength, and especially high strength/weight ratios are structural materials of considerable importance. Strength increase is due to precipitation hardening. Unfortunately, increase in strength is usually accompanied by increase in corrosion suceptibility, and especially in susceptibility with respect to intercrystalline or intergranular stress corrosion cracking, IGSCC. For reviews, see [78,79,80a]

A typical case is Al-5Zn-3Mg (wt.%) [80b–k]: Tempered to maximum strength, SCC susceptibility is excessive, and stepped temper programs must instead be used to attain an acceptable compromise between strength and corrosion resistance. Generally, the preference therefore is to use AlZn5Mg2 and AlZn5Mg1, which are more easily handled.

AlZn5Mg3 is, in fact, one of two prototypes of aluminum alloys susceptible to SCC. As described in more detail further below, IGSCC of this alloy is observed at electrode potentials distinctly below the potential range for pitting or intergranular corrosion of unstressed material, as reviewed in Chaps. 12 and 13. For comparison, Fig. 15.66 shows that the breakthrough potential for intergranular pitting attack depends on previous temper [79c]. It also shows the broad range in potentials between anodic attack and cathodic hydrogen evolution, where current densities are virtually zero, indicating passivity with negligible passive current density, as well as a negligible current density of oxygen reduction at the insulating oxide film. This, then, is the very range in potential where IGSCC is observed. We shall see presently that in this range SCC proceeds by hydrogen embrittlement.

The second prototype is Al-4 wt.% Cu. In this case, IGSCC is stress enhanced intercrystalline corrosion, which already occurs in unstressed material, and the threshold potential for intercrystalline corrosion also is the threshold potential for IGSCC [80h,81]. The mechanism therefore essentially is the same for IGSCC and IC, i.e. probably crack propagation by anodic crack tip metal dissolution.

Concerning SCC susceptibility, the commercial AlCu5Mg2 alloy used in the aircraft industry is similar to AlCu4 [81]. At the same time, the corrosion resistance of this alloy is unsatisfactory already without danger of SCC. The data presented in Fig. 15.67 readily explain the situation [80j]: Apart from the usual dependence of the breakthrough potential on the chloride content of the so-

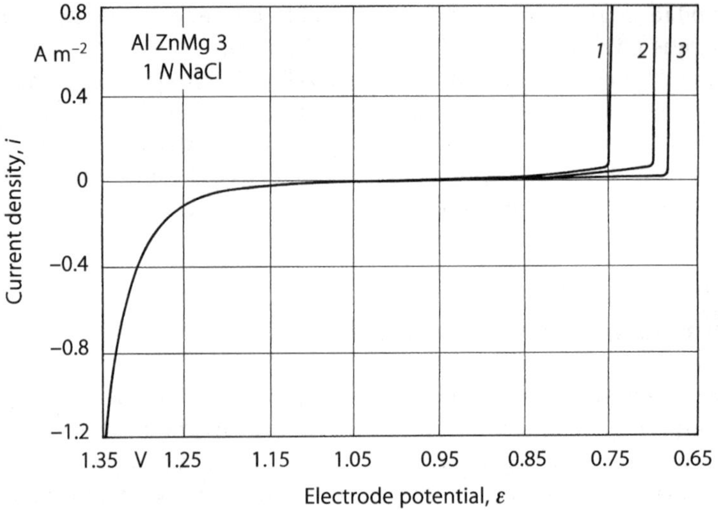


**Fig. 15.66.** Steady-state current-poten-tial curves for pure AlZn5Mg3 in aerated 1N NaCl solutions at ambient temperature. *Curve (1):* Specimens solution-tempered at 753K; *(2):* 100 h at 373K; *(3):* 15 min 353 K + 100 h 453+K. (From Richter and Kaesche)

lutions, identical for aerated and deaerated solutions, the curves clearly show a cathodic oxygen reduction current, in fact a diffusion-limited plateau of this current, at what certainly is passive AlCuMg. The reason is that intermetallic precipitates, easily discerned under a microscope, pierce the passive oxide film, and act as local cathodes for oxygen reduction. In comparison to pure aluminum, the material therefore is rather ill-behaved concerning corrosion resistance, and it is for this reason that aircraft structural material of this type usually is plated with aluminum.

Apparently, the SCC susceptibility of commercial AlZnMgCu alloys is similar to that of AlZnMg alloys. Results of fracture-mechanical testing of the commercial alloy AlZnMgCu have already been presented in Fig. 15.8. A point deserving particular attention is that, while the chloride concentration of the test solution does play a major role, crack growth, though very slow, is observed already in distilled water. The data confirm previous experience for constant load testing of hardened commercial AlZnMg3 shown in Fig. 15.68 [82a,b]. In these experiments, smooth specimens were tested. The overall decrease of time to failure values with increasing load agrees with expectations; the observation of an absolute lower threshold load required for the onset of SCC is interesting, but then the possible influence of the electrode potential remains to be investigated. This also applies to fracture-mechanical experiments with commercial Al-5Zn-2Mg-2Cu (Alloy 7022) shown in Fig. 15.69 [83]. The decrease in $K_{ISCC}$, indicating growing SCC susceptibility, caused by increasing temperatures of solution heat treatment reflects the increase in strength of the material, and so does the over-

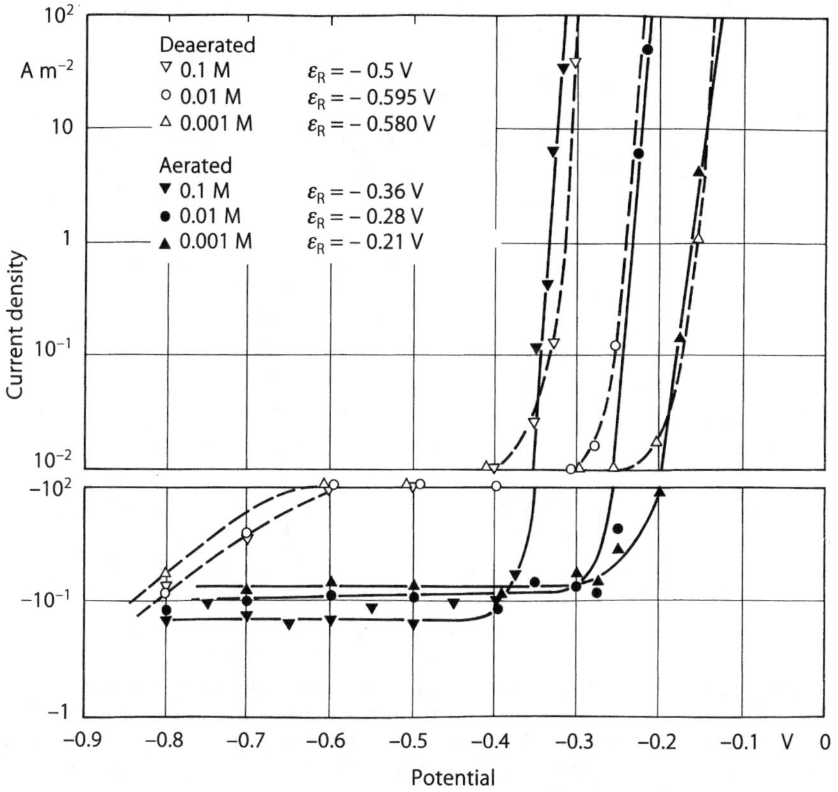


**Fig. 15.67.** Steady-state current-potential curves for commercial Al 4.5wt.% Cu–2wt.% Mg (temper T 351) in NaCl solutions at 30°C. Open symbols: Solutions deaerated; *solid symbols:* solutions aerated. (From Nikol and Kaesche)

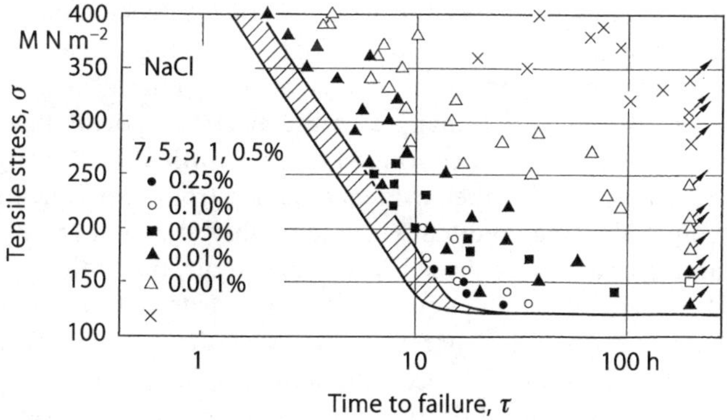


**Fig. 15.68.** The dependence of the time to failure of hardened AlZnMg3 specimens on the constant tensile load in aerated NaCl solutions, and in distilled water. (From Gruhl)

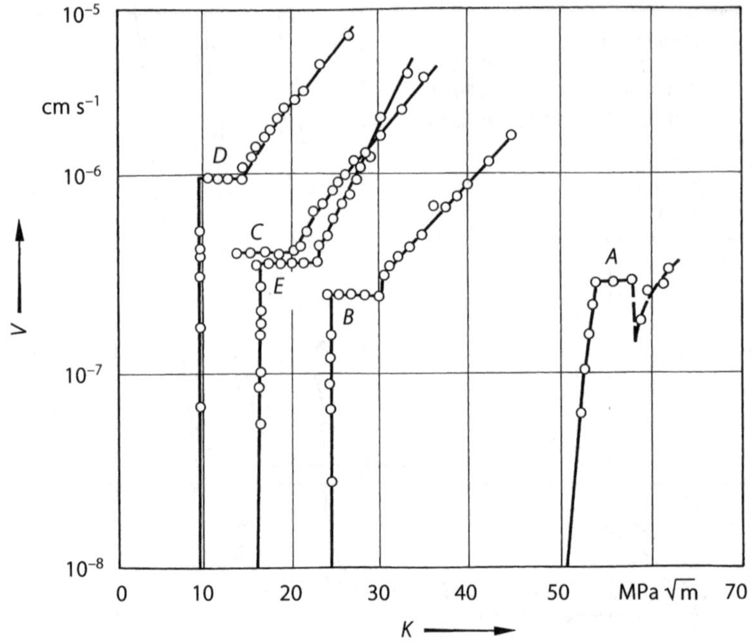


**Fig. 15.69.** Crack growth rates in bolted double cantilever beam specimens of Alloy 7075 in 3.5% NaCl solution in air at ambient temperature. The crack opened in the short transverse direction of the rolled material. Solution annealing for 2 h at A: 666 K, B: 711 K, C: 755 K, D: 800 K, and E: 800 K + 600 K was always followed by 24 h at 394 K. (From Shastry, Levy, and Joshi)

all trend to higher crack growth rates at the onset of the plateau Region II of the curves as well as the final increase to $K_{IC}$ in Region III.

For both AlZnMg and AlZnMgCu alloys, the accepted crack extension mechanism is crack advance by brittle steps through hydrogen-embrittled zones beyond the crack tip. This was concluded from distinct memory effects, such as delayed relaxation of growth rates after changes of the electrode potential, as well as from quite simple observations of continued crack growth after blowing the solution out of cracks [80b,j]. Further evidence comes from experiments with AlZnMg pipes, tempered to maximum SCC susceptibility (120 h at 363 K), with an external circumferential notch, loaded in tension in air, but with NaCl solution inside [82c,d]. To secure unambiguous observations, external corrosion was prevented by galvanizing with zinc. Intercrystalline corrosion was shown to set in, after several hours, at the root of the external notch, i.e. in air, obviously triggered by atomic hydrogen produced during internal corrosion of the pipe, diffusing through the wall to the outside.

Crack growth by this mechanism may still be quasi-continuous, as described previously. However, as already demonstrated in Fig. 15.9, crack growth by discontinuous microscopic steps at considerable intervals is readily observed

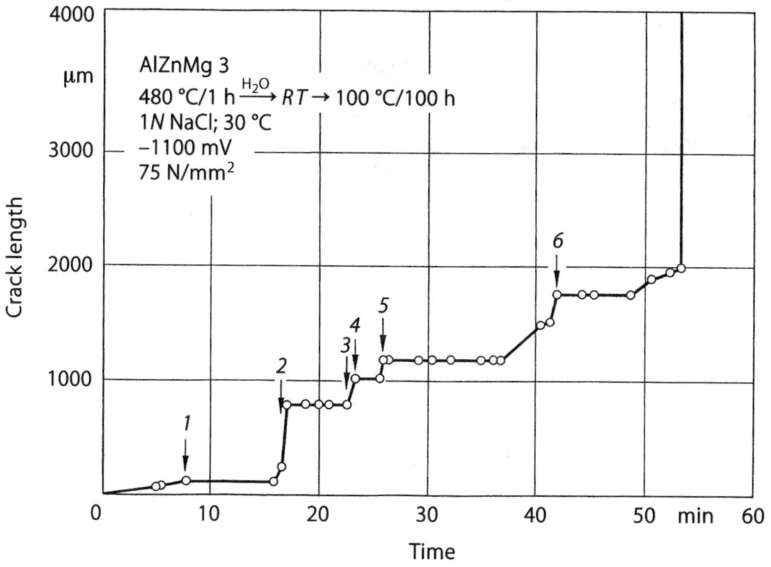


**Fig. 15.70.** Intermittent intercrystalline crack growth in a thin specimen of aged pure AlZnMg3, notched and loaded in tension in 1 M NaCl solution at the controlled potential −1.1 V. (From Richter and Kaesche)

[6b,80b,82]. In thin foils, these comparatively wide steps can be directly recorded by microphotography, as documented in Fig. 15.70 [80c]. During this experiment the specimen failed after 53 min, but the true crack running time was extremely short. Under such conditions, average values of crack growth rates, calculated for some time, $\tau$, and corresponding crack length, $a(\tau)$, as fraction $\bar{v} = a(\tau)/\tau$, while useful for estimating times to failure, do not throw light on rate-determining steps during true crack growth periods. Sometimes, $\ln \bar{v}$-values are evaluated from smoothed, and even rather rigorously smoothed fracture-mechanical crack extension data in Region II of $ln\bar{v}$-$K_I$ diagrams, and plotted over (1/T). The procedure has occasionally been used to obtain, from the slope, $-A_{act}/R$, of this Arrhenius plot, $A_{act}$ as an apparent activation energy. This, then, certainly is not the activation energy of the rate determining step during true crack growth periods; instead it may, for instance, characterize processes determining arrest periods of crack growth instead. Still, such data have been considered as indicating that crack-rates are determined by hydrogen diffusion at temperatures below, and by ad-atom surface diffusion beyond 50°C [78c]. Concerning the supposed strength of the surface diffusion model, referred to in this work, see further above.

True momentary crack extension rates have been described to be as high as 0.2 m $s^{-1}$ [82e,]; elsewhere, lower rates of ca. 10 cm $s^{-1}$ have been reported [84].

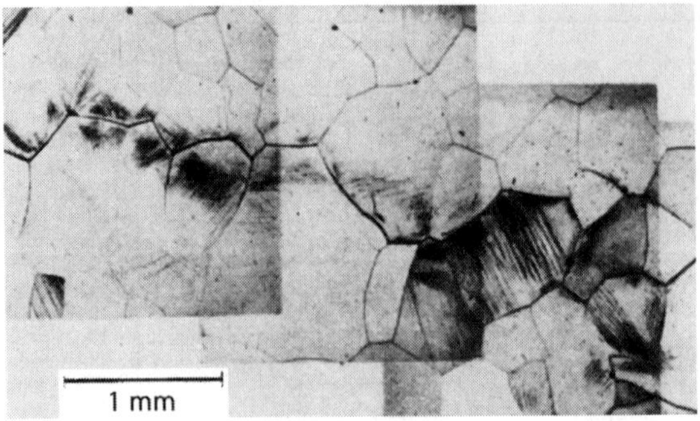


**Fig. 15.71.** Metallographic evidence of coarse glide accompanying crack growth in pure AlZnMg3 tensile SCC specimens. Etched cross sections, with glide bands decorated by galvanizing. (From Engell and Speidel)

The sequence of short steps of rapid of crack growth followed by long arrest periods can be understood from the concept of hydrogen embrittlement assuming that cracks momentarily overshoot brittle areas and then wait for hydrogen uptake to catch up during slow crack tip local corrosion. Some interstitial hydrogen taken up previously will also migrate to the stressed crack tip area [80k]. Careful metallographic inspection has confirmed that in every case crack arrest occurs at grain-boundary triple points where the crack-opening stress splits into two weaker components [80c].

From metallographic inspection of fractured constant-load specimens, it was also concluded that the area of the brittle SCC part of the fracture is proportional to the square root of the time to failure, and this was considered as evidence of rate-determining hydrogen diffusion [82c].

Generally, crack arrest during most of the crack growth fits into the picture of slow hydrogen uptake up to a critical concentration able to trigger brittle extension steps. Similar to the case of HISCC in high-strength steels, the details of hydrogen action are then still open to discussion. One possibility is hydrogen-enhanced local plasticity (HELP), as noted further above, caused by dislocation injection into the crack tip region, and thus microscopic plastic deformation of the crack tip region, rather than decohesion of metal/metal bonds by interstitial hydrogen, determines crack extension [85].

As documented in Fig. 15.71 crack growth is accompanied by coarse glide [9b]. The same phenomenon is readily visible in situ when cracks grow in thin foils [80f–g], and it gives cause to consider the contribution of slip step, or else oxide film rupture processes to the mechanism of cracking: Also considering the influence of the halide concentration of the elctrolytic solution on SCC propagation, documented in Fig. 15.8, the suggestion is that crack initiation, needing repetition after each arrest period, is caused by local corrosion processes similar to pitting, although occurring, under the influence of strain, below the pitting potential. Local corrosion then acidifies the crack tip electrolytic solution as postulated previously, followed by local hydrogen deposition via atomic hydrogen,

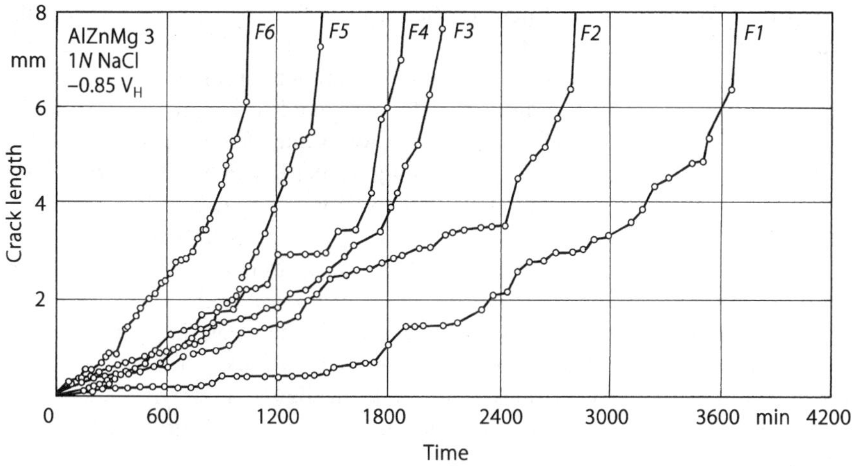


**Fig. 15.72.** Potentiostatic crack propagation in pure AlZnMg3 notched specimens under constant load, exposed to 1 M NaCl solution at 30°C, $\varepsilon = -0.85$ V. Specimens heat treated for 15 min at 753 K and quenched in water, followed by step-aging for $x$ min at 373 K below, and further 60 min at 456 K beyond the Guinier-Preston solvus temperature. $F_1 \ldots F_6$: $x = 15,25,50,100,150,500$ min. (From Richter and Kaesche)

and hydrogen uptake into the metal. This then is a more detailed description of slow events during arrest periods of crack extension.

For any given chloride concentration and temperature of an aggressive electrolytic solution, overall crack extension rates in AlZnMg alloys very much depend on metallurgical factors such as composition, heat treatment, cold work, and direction of stress relativ to grain texture. We pass most of the systematic investigations into this matter, including the influence of additions of Si, Fe, Cu, Mn, Cr, Zn, Ti, and Zr, described in [80f–g], also including AlLi alloys such as AlLiCuMgCr (AA 8090C) [80l,m]. One interesting observation is that pure AlZnMg1 specimens readily crack, while specimens of the corresponding commercial alloy are next to immune with respect to SCC . Also interesting is the influence of step-aging on crack propagation in pure AlZnMg3 specimens shown in Fig. 15.72 [80c]. After this heat treatment, the alloy has narrow, relatively soft precipitation-free grain boundary zones (PFZ). With increasing duration of the first aging step, the width of these zones decreases, and so does their capability to accommodate dislocations piling up at grain boundaries. This then increases the strain acting on grain boundaries, and also increases SCC susceptibility. As revealed by TEM, the effect is caused by an increase in the width of brittle rapid extension steps during cracking.

We note here that, in contrast to AlCu alloys, the susceptibility of AlZnMg alloys to stress-free intercrystalline corrosion does not parallel SCC susceptibility. Instead, step-aging decreases the concentration of both Zn and Mg in the PFZ. This shifts the grain-boundary breakthrough potential to more noble values, decreasing the range of corrosion potentials were intercrystalline attack is expected. We recall from Chap. 12 that such shifts very accurately reflect the grain boundary zone composition.

An interesting case is the behaviour of commercial alloys such as AlZn-MgCu (AA 7075), and AlCuMg (AA 2024) tempered to low SCC susceptibility by temper T7351, which is solution treating, water quenching, stress relieving by mechanical 3% deformation, and step-aging at 380 K and 441 K. In this state, crack propagation rates in notched and prefatigued specimens in Region II of the $\ln \bar{v}$-$K_I$ curves are very low for both types of alloys, between $10^{-9}$ and $10^{-10}$ m $s^{-1}$ [80a]. Further above, pure AlCu alloys were described as immune only below the threshold potential for onset of intercrystalline corrosion. For the alloys now discussed, the effect is different: When polarized to potentials in the range of IC, preexisting cracks are blunted by local crack-tip corrosion, further diminishing growth rates which were negligible anyway. The same incidentally applies to specimens polarized into the potential range of 'cathodic corrosion'.

## 15.5
## General Aspects Continued

Typically, stress corrosion is dangerous for structural metallic materials alloyed to achieve high strength. Dealing with alloys, particular questions concerning the influence of selective metal dissolution arise. In this section, this aspect of the matter will be more explicitly discussed, mostly using experimental evidence concerning brasses and CoNi alloys. Also, we take occasion to more explicitly discuss the initiation of brittle crack steps from surfaces embrittled by vacancy injection rather than by hydrogen uptake.

A much investigated case is SCC of $\alpha$-brasses [86]. Usually, the electrolytic solution is ammoniacal, corresponding to the practical case of 'season cracking', and this will remain the main topic presently, although SCC of brasses in a number of other electrolytic solutions has also been reported. Typically, testing is carried out in Mattson's ammoniacal copper salt solution, and results have already been presented in Figs. 15.3 and 15.19. Concerning crack paths, the usual assumption, already noted above, was that there exist two types of test solutions, one 'tarnishing', causing the precipitation of thick, visible dark $Cu_2O$ surface films, the other 'non-tarnishing', leaving the surface bright. Tarnishing solutions where thought to cause intergranular, non-tarnishing, transgranular cracking. The distinction is, however, arbitrary, since TGSCC is also observed in tarnishing solutions, while passivating, thin oxide films are present on the alloy surface in non-tarnishing solutions [87].

As to the causes for cracks to choose an inter- or else a transgranular propagation path, also influenced by tin or arsenic as minor alloy components, straightforward evidence is not available. For intercrystalline cracking, the preferred opinion is that kinetics are governed by anodic crack-tip metal dissolution. Other authors point out that overall crack extension may begin with intercrystalline microcracking caused by dislocation pile-up at grain bound-

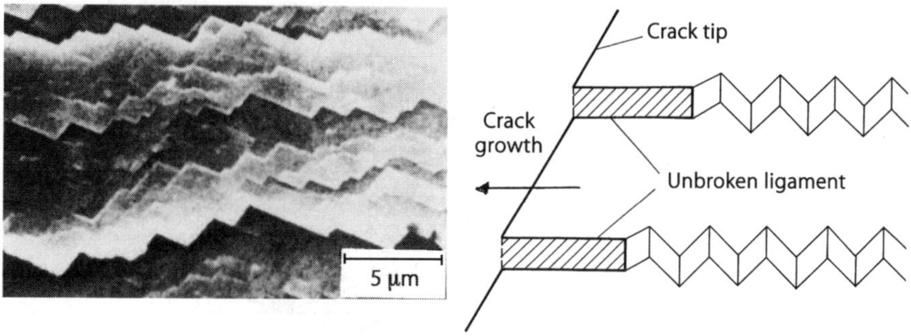


**Fig. 15.73.** Transcrystalline stress corrosion crack surface in admiralty brass cracked in ammoniacal solution. *Left:* REM picture of a crack surface area, looking onto (011) facets. *Right:* Schematic drawing showing sawtooth steps separating cleavage faces and still unbroken ligament areas at the crack front. (From Bursle and Pugh)

aries, followed by crack coalescence [85c]. Transcrystalline cracking is usually explained as a sequence of brittle extension steps followed by intermittent local crack tip corrosion, as also already described further above. This view is corroborated by fractographic evidence presented in Fig. 15.73, showing brittle crack faces without any indication of anodic metal dissolution. The width of steps is ca. 10 µm, that of time intervals between steps ca. 100 s [21b]. Brittle steps probably proceed close to the speed of sound, but the overall average crack extension rate is only of the order 0.1 µm $s^{-1}$. More details of the crack propagation mechanism are schematically indicated in Fig. 15.74, explaining the typical appearance of fan-shape or thumb-nail fracture surfaces. Similar observations have been reported for CrNi austenites cracking in hot $MgCl_2$ solutions, were the width of steps is ca. 0.5 µm, and step time intervals are ca. 15s [21e]. Judging from more observations of brittle facets on crack surfaces in zirconium, magnesium, MgAl-, AlZnMg- and Ti-alloys, the predominance of brittle crack extension steps for overall crack propagation apparently is quite generally typical at least for transcrystalline SCC [26]. This then adds more evidence against the slip-step dissolution model, where anodic metal dissolution is claimed to predominate .

From fractographs alone, it would be difficult to distinguish crack-tip embrittlement by hydrogen uptake or else by vacancy injection. However, for Cu-30 wt.%Zn in 1 M $NH_3$/0.8 g $L^{-1}$ $Cu^{2+}$ solutions, pH 12, the free corrosion potential is ca. −0.15 V. At this potential, the hydrogen equilibrium pressure is ca. $10^{-19}$ bar, and this makes hydrogen uptake very unlikely, especially since in ammoniacal solution an important crack-tip electrolyte acidification cannot be expected. Therefore, hydrogen embrittlement is ruled out in this case [21]. In such solution, copper and zinc dissolution proceed by the complexing reactions

$$\mathrm{Cu} + 2\mathrm{NH_3} \rightarrow \mathrm{Cu(NH_3)_2^+} + \mathrm{e^-}; \quad \mathrm{Zn} + 4\mathrm{NH_3} \rightarrow \mathrm{Zn(NH_3)_4^{2+}} + 2\mathrm{e^-}. \quad (15.18)$$

Subsequently, dissolved oxygen present in aerated solutions oxidizes the cupro-complex to the cupri-complex in a bulk-solution homogeneous reaction:

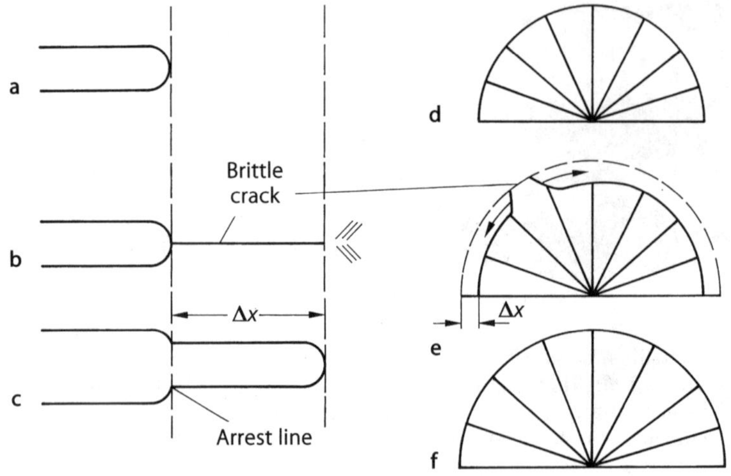


**Fig. 15.74.** Stepped crack extension in brass, as inferred from REM inspection of fracture surfaces after SCC in ammoniacal solutions. (a,b,c) Crack at rest, followed by brittle rapid extension, and slow crack widening; the crack is seen in cross section. (d,e,f) thumbnail crack facet as seen on the crack surface at rest, after brittle crack extension and growth to new rest position. (From Beggs, Hahn, and Pugh)

$$2Cu(NH_3)_2^+ + 1/2O_2 + H_2O + 4NH_3 \rightarrow 2Cu(NH_3)_4^{2+} + 2OH^- . \qquad (15.19)$$

In the test solutions, the cupri-complex has an equilibrium potential of ca. 0.05 V for the cathodic partial reaction

$$Cu(NH_3)_4^{2+} + e^- \rightarrow Cu(NH_3)_2^+ + 2NH_3 , \qquad (15.20)$$

and the superposition of partial reactions according to Eqs. (15.18) and (15.20) fixes the free corrosion potential in the range of simultaneous copper and zinc dissolution [21f]. At the same time, the important difference of the standard potentials, $E^\circ_{Zn/Zn^{2+}}$ and $E^\circ_{Cu/Cu^{2+}}$, should cause predominant zinc dissolution, leaving a surface depleted with respect to zinc and thus enriched in embrittling vacancies.

More details of the surface embrittlement mechanism are reported in [45c–f], in particular with respect to transcrystalline SCC of $\alpha$-CuZn and $\alpha$-CuAl single crystals in cupro-ammoniacal solutions. Here, the free corrosion potential is approximately equal to the equilibrium potential of copper dissolution and redeposition. For this type of alloys we recall from Chap. 8 the existence of a threshold electrode potential for the onset of rapid de-zincification or de-aluminisation, increasing with increasing copper content. In the present case, this threshold was identified by scratching the surface and recording current-time transients at different potentials.

As shown in Fig. 15.75, a threshold electrode potential exists for the onset of transcrystalline SCC as well, and in fact coincides with the threshold value for rapid dealloying [45f]. The explanation is that dealloying embrittles the

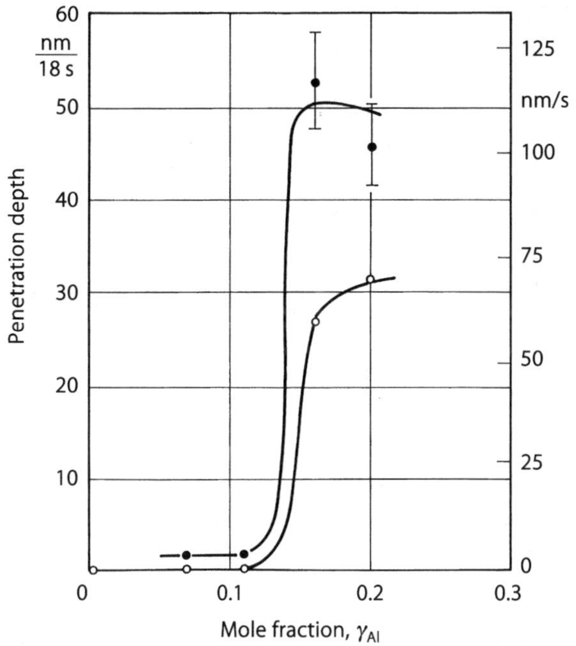


**Fig. 15.75.** Comparing thresholds for de-aluminisation and for transcrystalline stress corrosion cracking of homogeneous CuAl alloys in deaerated ammoniacal $Cu^+$ solutions. Solid circles: Dealloying composition limit as revealed by scratching experiments and considering the depth of selective attack. Open circles: Average crack growth rates observed for notched and slowly strained specimens. (From Sieradzki, Kim, Cole, and Newman)

surface down to a depth of ca. 20 to 50 nm, that internal stresses in the surface layer trigger brittle crack extension steps which, once running, continue to grow into the alloy to a depth of up to 2 µm. The repetition of the process drives further overall crack growth. Clearly, this is an elaboration of the model of cracking in brasses presented further above. It is versatile, since by the same mechanism cracks may be initiated in other brittle surface films as well, as for instance in oxide films on copper, thus possibly explaining, for instance, occasional observations of cracks in pure copper. With regard to dealloying of surface zones, the percolation theory, as also briefly described in Chap. 8, again is interesting.

For homogeneous CuAu alloys, the relation between dealloying and SCC susceptibility is different [88]. Figure 15.76 shows data on times to failure and elongation at fracture for Al-25Cu observed at constant potentials in deaerated NaCl solutions. The threshold potential for rapid onset of copper dissolution is ca. 0.7 V, but SCC is observed at potentials far below this value. During straining, intermittent crack extension was observed, and confirmed afterwards by inspection of the fracture surfaces. The width of single extension steps was

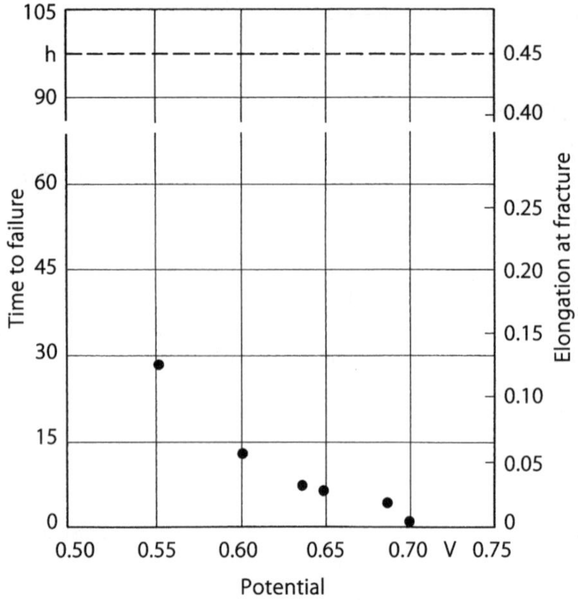


**Fig. 15.76.** Times to failure and elongation at fracture of smooth monocrystalline tensile specimens of homogeneous Cu-25 at.%Au in NaCl solution as function of the electrode potential. Slow strain rate $1.3 \times 10^{-6}$ $s^{-1}$; stress direction vicinal to <110>; Threshold electrode potential for rapid copper dissolution ca. 0.7 V. (From Flanagan, Lee, Zhu, and Lichter)

between 75 and 150 μm and thus quite large; time intervals between steps changed between 5 and 180 s; the average overall crack extension rate was ca. 1 μm $s^{-1}$. Again this is much more than expected from anodic crack-tip metal dissolution alone. Since hydrogen embrittlement can be ruled out, a sequence of brittle fracture steps triggered in dealloyed surface films is once more suggested.

Homogeneous NiPd alloys (see Chap. 8) are susceptible to both stress-free intercrystalline or intergranular corrosion (IGC) and intercrystalline or intergranular stress corrosion cracking (IGSCC) [80n,o]. For any Pd content of the alloy, and for any chloride concentration of the electrolytic solutions, both types of corrosion have a distinct threshold electrode potential. As shown in Fig. 15.77, the penetration rates of both stress-free IGC corrosion and IGSCC increase with the potential beyond the respective threshold. The threshold itself, however, is distinctly more negative for IGSCC than for IGC. In the potential range of simultaneous IGSCC and IGC, the former is faster by an order of magnitude. Therefore, if IGC is expected to be equivalent in rate to local anodic metal dissolution, IGSC is very likely to proceed by a different mechanism. Should this again be discontinuous crack propagation, involving brittle extension steps followed by crack-tip selective corrosion, previous views on general properties of intercrystalline versus transcrystalline cracking mechanisms would need reconsideration.

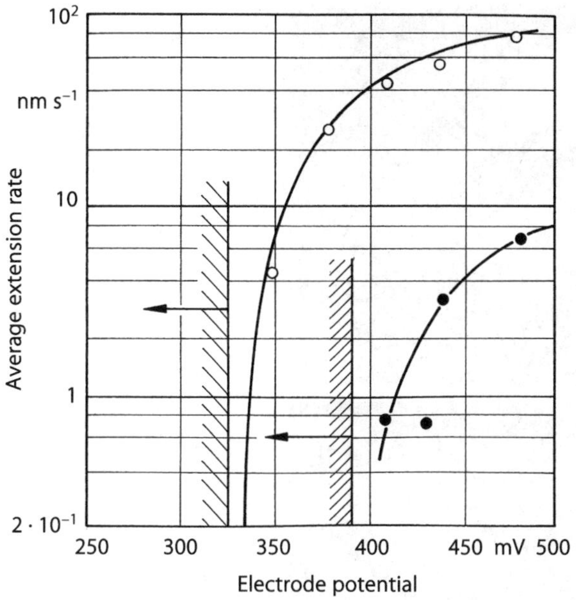


**Fig. 15.77.** Comparing average extension rates of intercrystalline stress corrosion cracks (open circles), and average penetration rates of stress-free intercrystalline corrosion (solid circles) in Ni-10 at.%Pd. Experiments at controlled potential in weakly acid M/10 NaCl solutions with notched, flat tensile specimens loaded to 90% of the yield strength. (From Kaiser, Lenz, and Kaesche)

In NiCoPd alloys with low Co and low Pd content both IGC and IGSCC are observed, with again slow IGC and rapid IGSCC, as before [80n,o]. At higher Co contents the same holds, provided that the Pd content is not lower than 8at.%. The picture changes when the Pd content is down to 2%. In this case, and if, at the same time, the Co content is also low, SCC still is intercrystalline. If, however, the Co content is increased, crack paths turn transcrystalline. Then, cracks pass through layers left spongy by extensive tunneling corrosion, or through coarse glide bands, and/or through martensite plates. Two examples are shown in Fig. 15.78. As explained in Chap. 13, the tendency to coarse glide and to martensitic transformation is enhanced by the decrease in stacking fault energy accompanying increasing contents in Co. Here, the more interesting observation is SCC cracks propagate with the same rate as stress-free tunneling corrosion, or stress-free selective corrosion of martensite plates. Therefore, what has the appearance of SCC, now really is the trivial cracking of brittle phases, with no special synergystic effect of stress or strain accelerating corrosion.

**Fig. 15.78a,b.** Transcrystalline cracking observed in tensile specimens of NiCoPd tested for 30 resp. 34 h at 0.085 V in 0.1 M NaCl solution. **a** Cracks in Ni-59Co-2Pd, spongy by extensive micro-tunneling corrosion. **b** Cracks in deformation-induced martensite plates in Ni-69Co-2Pd. (From Kaiser, Lenz, and Kaesche)

## References

1. Mittlg. AGK Arbeitskreise 'Wasserstoff-induzierte Werkstoffschäden' und 'Spannungs- u. Schwingungsrißkorrosion': Werkstoffe u. Korr. *37* (1985) 45
2. a) Symp. Spannungs- und Dehnungs-induzierte Korr., Baden-Baden 1978: Werkstoffe u. Korr. *29* (1978) Nr. 11; b) Proc. Symp. Electrochemical Corrosion Testing, San Francisco 1970. (Mansfeld, F. and Bertocci, U., eds.). ASTM, Philadelphia, 1981; c) Proc. 8th Int. Conf. Metallic Corrosion, Mainz 1981. DECHEMA, Frankfurt, 1981; d) Symp. Korrosion u. Bruch, Frankfurt 1988. DVM, Berlin, 1988; e) Tagung Spannungsrißkorrosion in Wasser, Zürich 1988. SVMT, Zürich 1988; f) Proc. Conf. Theory of Stress Corrosion Cracking, Ericeira 1971. (Scully, J., ed.). NATO Scientifique Affairs Div., Brussels, 1971; g) Proc. 4th Int. Conf. Metallic Corrosion, Amsterdam 1969. NACE, Houston, 1969; h) Proc. Int. Conf. Stress Corrosion Cracking and Hydrogen Embrittlement in Iron Base Alloys, Unieux-Firminy

1973. (Staehle, R.W., Hochmann, J., McCright, R.D., and Slater, J.C., eds.): NACE, Houston, 1979; i) Advances in Corrosion Science and Technology. (Fontana,M.G. and Staehle, R.W., eds.). Vol. 2. Plenum Press, New York, 1972; j) Korrosion 22. DGM Tagung Spannungsrißkorrosion, Frankfurt 1968. (Kaesche, H., ed.). Verlag Chemie, Weinheim, 1969; k) U.R. Evans Conf. Localized Corrosion, Williamsburgh 1971. NACE, Houston, 1974; l) Proc. Int. Conf. Environment-Sensitive Cracking of Materials, Guildford 1977. (Swann, P.R., Ford, F.P., and Westwood, A.R.C., eds.). The Metals Society, London 1977; m) Proc. Conf. Corrosion Chemistry within Pits, Cevices and Cracks, Teddington 1984. (Turnbull, A. ed.). HMSO, London, 1987; n) Proc. Conf. Fundamental Aspects of Stress Corrosion Cracking, Columbus 1967. (Staehle, R.W., Forty, A.J., and Van Rooyen, D., eds.). NACE, Houston, 1969; o) Comprehensive Treatise on Electrochemistry. (Bockris, J-O'M., Conway, B.E., Yeager, E., White, R.E., eds.), Vol. IV: Electrochemical Materials Science. Plenum Press, New York, London, 1981; p) Recent Advances in Stress Corrosion. (Bresle, A., ed.). Esslete AB, Stockholm, 1961; q) Proc. Symp. Environment-Sensitive Fracture of Engineering Materials, Chicago 1977. (Fouroulis, Z.A., ed.). The Metallurgical Soc. of AIME, Warrendale, 1977; r) High Strength Materials. ( Zackay, V.F., ed.). J. Wiley, New York, 1965; s) Electrodeposition and Corrosion Processes. Van Nostrand, New York 1965; t) Physical Metallurgy of Stress Corrosion Failure. (Rhodin, T.N., ed.). Interscience Publs., New York, London 1959; u) Modern Aspects of Electrochemistry (Bockris, J.O'M., Conway, B.E., and White, R., eds.). Vol.27. Plenum Press, New York 1995; v) Proc. Meeting Modeling Environmental Effects on Crack Growth Processes, Toronto 1985. (Jones, R.H. and Gerberich, W.W., eds.). TMS-AIME, Warrendale, 1986; w) Hydrogen Degradation of Ferrous Alloys. (Oriani, R.A., Hirth, J.P., and Smialowski, M., eds.). Noyes Publications, Park Ridge, 1985; x) Moderne Stähle. (Uggowitzer, P.J., ed.). Schweizer Akademie der Wissenschaften Zürich 1987; y) Spannungsrißkorrosion von Stahl in Wasser. (Speidel, M., Magdowski, R.M., and Uggowitzer, P.J., eds.). Schweizer Akademie der Wissenschaften Zürich 1987; z) Proc. Int. Conf. Environment-Induced Cracking of Metals, Kohler, 1988. (Gangloff, R.P. and Ives, M.B., eds.). NACE, Houston 1990; za) Proc. Int. Conf. Life Prediction of Corrodible Structures, Cambridge 1994. (Parkins, R.N., ed.). NACE, Houston, 1994; zb) Proc. Int. Conf. Corrosion-Deformation Interactions, Fontainebleau 1992.(Magnin, F. and Gras, J.M., eds.). Les Éditions de Physique, Les Ulis, 1993; zc) Proc. Int. Conf. Hydrogen Transport and Cracking in Metals, Teddington 1994
3. v. Meysenbug, C.M. and Litzkendorf, M.: Werkstoffe u. Korr. *19* (1970) 901
4. a) Herbsleb, G.: private communication 1989; b) Münster, Gräfen, H.: Archiv Eisenhüttenwesen em 36(1965) 277; c) Mattson, E.: Electrochim. acta *3* (1961) 279
5. Brauns, E. and Ternes, W.: Werkstoffe u. Korr. *19* (1968) 1; Ternes, W.: Doctoral thesis, University Clausthal, 1963

6. a) Scully, J.C. and Powell, D.T.: Corr. NACE *25* (1969) 483; b) Watkinson, F.E. and Scully, J.C., in: loc. cit. [5f], p. 140; c) Hoar, T.P. in: loc. cit. [2n], p.98; d) Scully, J.C. and Hoar, T.P., in: Proc. 2nd Int. Congress Metallic Corrosion, New York 1963. NACE, Houston,1966, p. 66; e) Hoar, T.P. and Ford, P.J.: J. Electrochem. Soc. *120* (1973) 1013; f) Scully, J.C., in: loc. cit. [2e], 127; g) id., in: loc. cit. [2h], p. 496; h) Hines, J.G. and Hoar, T.P.: J. Appl. Chem. (1958) 764; i) Hoar, T.P. and West, I.M.: Proc. Ro. Soc. *A268* (1962) 304
7. a) Herbsleb, G. and Pöpperling, R., in: loc. cit. [2a], p. 740; b) Herbsleb, G. and Schwenk, W. Corr. NACE *41* (1985) 431; c) Schwenk, W., in: loc. cit. [2a], p. 740; d) iid.: Werkstoffe u. Korr. *31* (1984) 171; e) iid. in: loc .cit. [2c], p. 536; f) Wendler-Kalsch, E. in: loc. cit. [2a], p. 703; g) id.: Werkstoffe u. Korr. *31* (1980) 534; h) Krägeloh, E. and Navab-Motlagh, M.M.: Z. Werkstofftechnik *9* (1979) 400; i) Haumann, W., Heller, H., Jungblut, H.-A., Pircher, H., Pöpperling, R., and Schwennk, W.: Stahl u.Eisen *107* (1987) 585; j) Pöpperling, R., Schwenk. W., and Venketesvarlu, J.: Werkstoffe u. Korr. *36* (1985) 389; k) Herbsleb, G., Pfeiffer, G., and Ternes, H.: Werkstoffe u. Korr. *30* (1979) 322; l) Herbsleb, G.and Pfeiffer, B.: ibid. *35* (1984) 254 -m) Herbsleb,G. and Schwenk, W.: ibid. *18* (1967) 521; n) Frölich, Th., Schwenk, W., and Dahl, W.: Materials and Corrosion *48* (1997) 207
8. a) Kessler, K.J. and Kaesche, H., in: loc.cit. [2b]; b) iid.: Werkstoffe u. Korr. *35* (1985) 171; c) iid., in: loc. cit. [2c], p.536; d) Berggreen, J., Engell, H.J., and Kaesche, H.: Werkstoffe u. Korr. *26* (1975) 599; e) Richter, J. and Kaesche, H.: ibid. *28* (1977) 602; f) Kaesche, H.: Z. Metallkunde *67* (1976) 439; g) Stellwag, B. and Kaesche, H.: Werkstoffe u. Korr. *33* (1982) 274, 323; h) iid.: ibid. *35* (1979) 397; i) Kaesche, H.: Archiv Eisenhüttenwesen *36* (1965) 911; j) Stoll, F. and Kaesche, H., in: loc. cit. [2c], p.530; k) iid., in: Proc. Asian Pacific CorrosionControl Conf., Taipei 1983. APMACA, Taipeh, 1983, p.19; l) Schneider, M., Kessler, K.J., and Kaesche, H.: Archiv Eisenhüttenwesen *50* (1979) 379; m) Kessler, K.J. and Kaesche, H.: Werkstoffe u. Korr. *31* (1980) 907; n) iid., in: Wasserstoff in Metallen. Ergebnisse eines Schwerpunktprogramms. Deutsche Forschungsgemeinschaft. Bad Godesberg 1985, p. 427; o) Grötsch, G., Wendler-Kalsch, E., and Kaesche, H., in: loc. cit. [1l] p. 301; p) Rippstein, K. and Kaesche, H.: Corr. sci. *29* (1989) 517; q) Gladen, H. and Kaesche, H.: Nuclear Engng. and Design *151* (1994) 463; r) Kaesche, H., in: loc. cit. [2z]; s) id.: Z. Metallkunde *64* (1973) 593; t) Vogelhuber, G.: Diploma thesis, University Erlangen, 1981; u) Kaiser, H. and Kaesche, H., in: Proc. 9th Eur. Congr. Corrosion, Utrecht 1989. NCC Stichting Nederlands Corrosie centrum. Vol.2, PL 118; v) Fiebiger, C., Kilian, R., Degelmann, E., Kaiser, H., and Kaesche, H.: Werkstoffe u. Korr. *40* (1989) 695; w) Hack, T., Popp, W., and Kaesche, H.: ibid. *42* (1991) 396; x) Habermann, M.E., Doctoral Thesis, University Erlangen-Nürnberg 1995
9. a) Engell, H.J., Bohnenkamp,K., and Bäumel,A.: Archiv Eisenhütten-wesen *33* (1962) 285; b) Engell, H.J. and Speidel, M.O., in: loc. cit.[2j]; c) Engell, H.J., in: loc. cit. [1f], p.68 - d) id.: Stahl u. Eisen *98* (1978) 637; e) Engell,

H.-J. and Bäumel, a., in: loc. cit. [2s], p. 341 f) Bäumel, A. and Engell, H.-J.: Archiv Eisenhüttenwesen *32* (1961) 379; g) Bohnenkamp, K.: ibid. *39* (1968) 361; h) Schreiber, F. and Engell, H.-J.: Werkstoffe u. Korr. *23* (1972) 175
10. a) Speidel, M.O., in: loc.cit. [2f], p.140; b) id.: Corr. NACE *33* (1977) 199; c) Speidel, M.O. and Hyatt, M.V., in: loc. cit. [2h], p. 115; d) Speidel, M.O.: Corr. NACE *33* (1977) 199; e) Speidel, M.O. and Magdowski, R.M.: Corrosion 86, Houston 1986, Paper No. 329; f) iid.: Corrosion 88, St.Louis 1988, Paper No. 283; g) Magdowski, R.M. and Speidel, M.O., in: loc. cit. [1y], p. 117
11. Eremiáš, B. and Maričev, V.V.: Corr. sci. *20* (1980) 307
12. See, e.g., Tatro, C.A., Liptal, R.G., and Moon, D.W. in: loc. cit. [2h], p. 509
13. a) Bhatt, H.J. and Phelps, E.H.: Corr. NACE *17* (1961) 430t; b) Wilde, B.E.: ibid. *27* (1961) 602
14. a) Buhl, H.: Corr. sci. *13* (1973) 639; b) Rätzer-Scheibe, H.: Werkstoffe u. Korr. *26* (1975) 2
15. Pickering, H.W. and Frankenthal, R.P.: J. Electrochem. Soc. *119* (1972) 1297
16. a) Shuck, R.R. and Swerdlow, J.L. in: loc. cit. [2k], p.190, 208; b) Doig, P. and Flewitt, P.F., in: loc. cit. [2l], p. 113; c) iid.: Corr. NACE *37* (1981) 378; d) Melville, P.H.: British Corr. J. *14* (1979) 15
17. Cherepanov, G.P.: Stress Corrosion Cracking. In loc. cit. [1o]
18. Beck, T.R., in: loc. cit. [2f], p.64; b)id.: Corr. NACE *30* (1974)408; c) id., in: Proc. Conf. Predictive Capabilities in Environmental Assisted Cracking. (Rungta, R., ed.). ASME 1985, p. 177
19. a) Edeleanu, C.: J. Inst. Metals 80 (1952) 187; b) id., in: loc. cit. [2p], p. 79, see also Forty, A.J., in this volume.
20. McEvily, A.J. and Bond, A.P.: J. Electrochem. Soc. *112* (1965) 131
21. a) Pugh, E.N., in: loc. cit.[1f], p. 418; b) Bursle, A., Pugh, E.N., in: loc. cit. [2f], p. 18; c) Pugh, E.N.: Corr. NACE *41* (1968) 517; d) Pugh, E.N. and Westwood, A.R.C.: Phil. Mag. *13* (1966) 167 ; e) Beggs, D.V., Hahn, M.T., and Pugh, E.N., in: Hydrogen Embrittlement and Stress Corrosion Cracking Metals Park. ASTM, 1984, p. 181; f) Bertocci, K., and Pugh, E.N., in: loc. cit. [2m], p. 144
22. a) West, J.M. and Farman, L.: British Corr. J. *1* (1966) 67; b) West, J.M., in: loc. cit. [2q]
23. a) Swann, P.R.: Corr. NACE *19* (1963) 102; b) Pickering, H. W. and Swann, P.R.: ibid. p. 373; c) Swann, P.R. and Embury, in: loc.cit. [2q], p. 327; d) Swann, P.R., in: loc. cit. [2e], p.113; e) Sicock, J.M. and Swann, P.R., in: loc. cit. [2p], p.133
24. a) Staehle, R.W. and Smith, T.J.: Corr. NACE *23*(1967) 117; b) Staehle, R.W., in: loc. cit. [2h], p. 180; c) Pednakar, S.P., Agrawal, A.K., Chang, H.E., and Staehle, R.W.: J. Electrochem. Soc. *126* (1979) 701; d) Latanision, R.M. and Staehle, R.W., in: loc. cit. [2n], p. 214
25. a) Sieradzki, K., Sabatini, R.L., and Newman, R.C.: Met. Trans. *15A* (1984) 1941; b) Sieradzki, K. and Friedersdorf, F.J.: Corr. sci. *36* (1994) 669
26. Meletis, E.I. and Hochman, R.F.: Cor. sci. *24* (1984) 843
27. Flis, J.: Brit. Corr. J. *3* (1968) 183

28. a) See e.g. Graf, L., in: loc. cit. [2f], p. 399; b) Graf, L. and Budke, J.: Z. Metallkunde *46* (1955) 378; c) Frank, W. and Graf, L.: ibid. *66* (1975) 555; d) Graf, L. and Klatte, H.: ibid. *46* (1955) 673; e) Graf, L. and Becker, H.: Z. f. Metallkunde *62* (1971) 685; f) Graf, L. and Springe, G.: Werkstoffe u. Korr. *20* (1969) 323
29. a) Uhlig, H.H., in: loc. cit. [2t], p. 1; b) id., in: loc. cit. [2n], p. 86; c) Uhlig, H.H., Gupta, K., and Lang, W.: J. Electrochem. Soc. *122* (1975) 343; d) Uhlig, H.H. and Cook, E.W.: ibid. *116* (1969) 173; e) Lee, H.H. and Uhlig, H.H.: ibid. *117* (1969) 18
30. a) Galvele, J.R.: J. Electrochem. Soc. *133* (1986) 953; b) id.: Corr. sci. *35* (1993) 419; c) id., in: loc. cit. [[2ut], p. 233; d) id.: Corr. sci. *36* (1994) 901; e) Montoto, M.L., Duffò. G.S., and Galvele,J.R.: ibid. p. 1805; f) Galvele, J.R. and Duffó, G.S.: ibid. *39* (1997) 605; g) Giordano, C.M., Duffó, G.S., and Galvele, J.R.: ibid. p. 1915; h) Galvele, J.R.: ibid. *40* (1998) 3537; i) Alvarez, M.G., Fernandez,S.A., and Galvele, J.R.: ibid. *42* (2000) 739; j) Manfredi,C., Maier, I.A., and Galvele, J.R.: Corr. sci. *22* (1987) 887; k) Maier, I.A., Manfredi,C., and Galvele, J.R.: ibid.*25* (1985) 15; l) Maier, I.A., López-Pérez, E., and Galvele, J.R.: ibid. *22* (1982) 537
31. Nelson, H.G. and Williams, D.P., in: loc. cit. [2h], p. 390
32. Jaberi, J.: Brit. Corr. J. *20* (1985) 133
33. a) Riecke, E.:: Archiv Eisenhüttenwesen *44* (1973) 647; b) id.: Werkstoffe u. Korr. *30* (1979) 610; c) Richards, W.: Zement, Kalk, Gips *22* (1969) 447; d) Nürnberger, U.: Forschung Straßenbau u. Verkehrstechnik Nr. 308 (1980); e) Riecke, E. and Johnen, B.: Werkstoffe u. Korr. *37* (1986) 310; f) McGuinn, K.and Griffith, J.R.: Brit. Corr. J. *12* (1977) 152; g) Eickemeyer, J.: Corr. sci. *18* (1973) 397; h) Nürnberger, U.: Korrosion und Korrosionsschutz im Bauwesen, Band I. Bauverlag, Wiesbaden, Berlin, 1995
34. a) Mietz, J., Fischer, J., and Isecke, B.: Materials and Corr. *50* (1999) 535; b) Mietz, J.: ibid. *51* (2000 80); c) Nürnberger, U.: ibid. *48* (1997)602; d) Menzel, K. and Onuseit, G.: ibid. *47* (1996) 42
35. Okada, H., Yukawa, K.-T., and Tamura, H.: Corr. NACE *32* (1976)201
36. a) Münster, R. and Gräfen, H.: Archiv Eisenhüttenwesen *36* (1965) 277; b) Gräfen, H.: Corr. sci. *7* (1967) 177; c) Gräfen, H. and Kuron, D.: Archiv Eisenhüttenwesen *36* (1965) 285; d) Rädecker, W. and Gräfen, H.: Stahl u. Eisen *76* (1958) 1616
37. a) Parkins, R.N., in: loc. cit. [2n], p. 361; b) id., in: loc. cit. [2f], p. 167; c) id., in: loc. cit. [2h], p. 601; d) Humphries, M.J. and Parkins, R.N., in: loc. cit. [2m], p. 384; e) Parkins, R.N. and Fessler, R.R.: Engng. Appl. *1* (1978) 80; f) Fessler, R.R., Markworth, A.J., and Parkins, R.N.: Corr. NACE *39* (1983) 20; g) Parkins, R.N., O'Dell, C.S., and Fessler, R.R.: Corr. sci. *24* (1984) 343; h) Parkins, R.N. and Zhou, S.: ibid. *39* (1997) 159; i) Congleton, J., Shoji, T. and Parkins, R.N.: ibid. *25* 1986) 633; j) Parkins, R.N.: ibid. *29* (1989) 1019; k) Congleton, J., Zheng, W., and Hua, H.: ibid. *30* (1990) 555, 651; l) Yang,W., Zhao,M., and Congleton, J.: Corr. sci. *33* (1992) 89; m) Yang, W., Congleton, J., Kohnen-Chari,O., and Sajdl, P.: ibid. p. 375; n) Congleton,J.

and Yang, W.: ibid. *37* (1995) 429; o) Ebtehaj,K. Hardie, D., and Parkins, R.N.: ibid. *25* (1985) 415
38. Simon, W. and Schwarz, W.: Ber. Bunsenges. phys. Chemie *67* (1963)108
39. Pourbaix, M.: Atlas d'Équilibres Électrochimiques. Gauthier-Villars, Paris, 1963
40. Venczel, J. and Wranglen, G.: Corr. sci. *4* (1964) 137
41. Paulekat, F.: Comment on Herbsleb, G. and Schwenk, W.: Stahl u. Eisen *90* (1970) 903
42. a) Stenzel, H., Vehoff, H., and Neumann, P., in: loc. cit. [2v], p.225; b) Vehoff, H. and Neumann, P., in: loc. cit. [2w], p. 686; c) Vehoff, H., Stenzel, H., and Neumann, P.: Z. f. Metallkunde *78* (1987) 550; d) Gröschel, F.: Doctoral thesis, RWTH Aachen, 1981
43. a) Andresen, P.L.: Corr. NACE *44*(1988) 450; b) Andresen, P.L. and Ford, F.P.: Mat. Sci. Engng. *A103* (1988) 167; c) Ford, F.P.., in: loc. cit. [2y], p. 139; d) Ford, F. P. and Andresen, P.L., in: Corrosion 89, New Orleans 1989. NACE, Houston 1989, Paper No. 498; e) Ford, F.P.: Corr. NACE *52* (1996) 375; f) Ford, F.P. and Hickling, J., in: loc. cit. [2z], p. 855; g) Andresen, P.L., Angeliu, Th.M., Catlin, W.R., Young, L.M., and Horn,. R.M., in: Corrosion 2000. NACE, Houston, 2000, Paper No. 203; h) Andresen, P.L., Angeliu, Th.M., and Young, L., in: Corrosion 2001, NACE, Houston, 1991,paper 1228; i) Angeliu, T.M., in: Corrosion 2001, NACE, Houston 2001, Paper 1121; j) Andresen, P.L., Angeliu, Th.M., and Young, L.M.: ibid. Paper 1228; k) Andresen, P.L., in: EPRI International Workshop on Low Temperature Sensitation, Palo Alto 1982, Paper No. 1; l) Ford, F.P. and Andresen, P.L., in: loc. cit. [60e], p.789; m) Andresen, P.L., in: Corrosion 1996. NACE, Houston 1996
44. a) Wei, R.P., Gao, M., and Xu, P.Y.: J. Electrochem. Soc. *136* (1989) 1835; b) Wei, R.P, and Dao, M.: ibid. *138* (1991)2601; c) Burstein, G.T. and Gao, G.: ibid. p. 2627
45. a) Kelly, R.G. and Newman, R.C.: J. Electrochem. Soc. *137* (1990) 357; b) Newman, R.C., Corderman, R.R., and Sieradzki, K.: Brit. Corr. J. *24* (1989) 143; c) Sieradzki, K. and Newman, R.C.: Phil. Mag. *A 51* (1985) 95; d) Newman, R.C.and Sieradzki, K., in: loc. cit. [2n], p. 199; e) iid.: Proc. NATO Advanced Research Workshop Chemistry and Physics of Fracture. (Latanision, E.M. and Jones, R.H., eds.). ASI Series E. No. 130. Martinus Nijhoff, Dordrecht, Dordrecht, 1987; f) Sieradzki, K., Kim. J.S., Cole, A.T., and Newman, R.C.: J. Electrochem. Soc. *134* (1987) 1635
46. a) Turnbull, A.: Corr. sci. *34* (1993) 921; b) id.: ibid. *39* (1997) 189; c) id.: ibid. *43* (2001) 193
47. a) EMPA; Schweizer Ingenieur u. Architektur *21* (1985) 509; b) Herbsleb, G. and Theiler, F.: Werkstoffe u. Korr. *40* (1989) 467; c) Arnold, N., Gümpel, P., and Heitz, T.W.: Materials and Corr. *49* (1998); *50* (1999) 140
48. Barnartt, S. and Van Rooyen, D.: J. Electrochem.Soc. *108* (1961) 222
49. Smith, J.A., Peterson, M.H., and Brown, B.F.: Corr. NACE *26* (1970) 540
50. Stalder, F. and Duquette, D.J.: Corr. NACE *33* (1976) 67

51. a) Takano, M., in: Proc. 5th Int. Congr. Metallic Corrosion, Tokio 1972. NACE, Houston, 1974; b) Nakayama, T.and Takano, M.: Corr. NACE *36* (1980) 530
52. a) Engelmann, H.-J. and Mummert, K.: Werkstoffe u. Korr. *43* (1992) 161; b)iid.: ibid. *44* (1993) 454; c) Engelmann, H.-J., in: loc. cit. [2zb], p. 237; d) Engelmann, H.-J. and Mummert, K., Schwarz, S., and Uhlemann, M., in: loc. cit. [2zc]; e) Uhlemann, M., Engelmann, H.-J., and Mummert, K.: Materials and Corr. *46* (1995) 63
53. See e.g. a) Latanision, R.M. and Westwood, A.R.C., in: Advances in Corrosion Science and Technology. (Fontana, M.G. and Staehle, R.W., eds.). Vol. 1. Plenum Press, New York, 1970; b) Staehle, R.W., in: loc. cit. [2f], p. 223
54. a) Vaughn, D.A., Phalen, D.I., Peteron, C.L., and Boyd, W.K.: Corr. NACE *19* (1963) 315; b) Birley, S.S. and Tromans, D.: ibid. *27* (1971) 63; c) Honkasolo, A.: ibid. *29* (1973) 237
55. a) Scott, P.M. and Tice, D.R.: Nuclear Engng. and Design *119* (1990) 399; b) Scott, P.M. and LeCalvar, M., in: loc. cit. [60f], p. 657
56. Indig, M.E., Weber, J.E., and Weinstein, D., in: Reviews on Coatings and Corrosion. (Parkins, R.N., ed.), Vol. 5. Freund. Publ. House, Israel 1982, p. 173
57. a) Choi, H., Beck, F.H., Szklarska-Smialowska,S., and Macdonald, D.D.: Corr. NACE *38* (1982) 136; b) Rebak, R.B. and Szklarska-Smialowska, Z.: Corr. sci. *38* (1996) 971
58. a) Läpple, V., Blind, D., and Deimel, P.: VGB Kraftwerkstechnik *74* (1994) 797; b) Läpple, V. and Deimel, P.: Werkstoffe u. Korr. *45* (1994) 583; c) Kussmaul, K., Blind, D., and Läpple, V.: Nuclear Engng. and Design *168* (1997) 53
59. a) Heldt, J. and Seifert, H-P.: Nucl. Engng. and Design *206* (2001) 57; b) Hänninen, H., Yagodzinskyy, Y., Tarasenko, O., Seifert, H-P., Ehrnstén, U., Aaltonen, P., in: loc. cit. [60d]
60. Proc. Int. Symposia on Environmental Degradation of Materials in Nuclear Power Systems - Water reactors: a) 7th Symp., Breckenridge 1995. (Andresen, G. and Brown, P., eds.). NACE, Houston 1995; b) 8th Symp., Amelia Island 1997. NACE Houston 1997; c) 9th Symp., Newport Beach 1999. (Ford, P., Bruemmer, S.M., and Was, G.S., eds.). TMS Warrendale, 2000; d) 10th Symp., Lake Tahoe 2001. (Ford, F.P., Was, G., Nelson, J., and Parsons, P., eds.). (in press); e) 5th Symp. 1988 (Theus, J.L., Weeks, J.R., eds.). TMS, 1988; f) 6th Symp., San Diego 1993. TMMMS Society, Warrendale 1993
61. Atkinson, J.D. and Yu, J.: Fatigue Fract. Engng. Mater. Struct. *20* (1997) 1
62. See e.g.Ruščák, A., B. and Dietzel, W., in: loc. cit. [60c], p. 885
63. Ehrnsten, U., Foehl, J., Gomez-Briceno, D., Weissenberg, T., Roth, A., Ritter, S., Seifert, H.P., and Zamboch, M., in: Proc. FISA Conf. EU Research in Reactor Safety, Luxemburg 2001
64. Xiukui,S., Jian, X., and Yiyi, L.: Mats. Sci. and Engng. *114* (1989) 179

65. a) Kilian,R., Brümmer, G., Hoffmann, H., Ilg, U., Maier, V., Wachter, O., and Widera, M., in: 26th MPA Seminar, Stuttgart 2000; b) Kilian,R., Eberle, U., Brümmer, G., Hoffmann, H., Ilg, U., and Widera, M., in: loc.cit. [60d]; c) Kilian, R,, Brümmer, G., Ilg,U., Meier, V., Teichmann,H., and Wachter, O., in: loc. cit. [60a], p. 529; d) Kilian, R., Eberle, U., Brümmer, G., Hoffmann, H., Ilg, U., Maier, V., and Widera, M., in: loc.cit. [60c]; e) Kilian, R.: private communication 2001
66. a) Macdonald, D.D. and Urquidi-Macdonald, M.: Corr. sci. *32* (1991) 51; b) Manahan, M-P., Sr., Macdonald, D.D., and Peterson, A.J.: ibid. *37* (1995) 189; c) Macdonald, D.D.: ibid.*38* (1996) 1003; d) Engelhardt, G.R., Macdonald, D.D., and Urquidi-Macdonald, M.: ibid. *41* (1999) 2267; e) iid.: ibid.*43* (2001) 195
67. See e.g. a) Kaesche, H.: loc. cit. [8s]; b) Wanhill, R.J.H.: British Corr. J. *10* (1975) 69
68. Chen, C., Beck, F.H., and Fontana, M.G.: Corr. NACE: *26* (1970) 135
69. Zhang, X.G., Verecken, J.: Corr. NACE *46* (1990) 136
70. Haney, E.G.. and Wearmouth, W.R.: Corr. NACE *25* (1969) 87
71. a) Sedriks, A.J., Slattery, P.W., and Pugh, E.N., in: loc.cit. [2m], p.673; b) Sedriks, A.J., Green, J.A., and Slattery, P.W.: Corr.NACE *24* (1968) 172; c) Sedriks, A.J.: ibid. *25* (1969) 207
72. Mazza, F.: Werkstoffe u. Korr. *20* (1969) 199
73. a) Powell, D.T. and Scully, J.C.: Corr. NACE *25* (1969) 483; b) Sanderson, G., Powell, D., and Scully, J.C., in: loc.cit.[2n], p.638 c) Scully, J.C. and Powell, D.T.: Corr.sci. *10* (1970) 719; d) Scully, J.C., in: loc. cit.[2f], p.127; e) Hollis, A.C. and Scully, J.C.: ibid. *34* (1993) 837; f) Simbi, D.J. and Scully, J.C.: Corr. sci. *34* (1993) 1743; g) iid.: Corr NACE *53* (1997); Kermani, M. and Scully, J.C.: Corr. sci. *18* (1978) 883
74. a) Leith, J.R., Hightower, J.W., and Harkins, C.G.: Corr. NACE *26* (1970) 377; b) Raymond, L. and Russel, R.J.: Corr. NACE *25* (1969) 250; c) Boyd, W.K., in: loc. cit. [2m], p. 593
75. a) Blackburn, M.J. and Williams, J.C., in: loc.cit. [2n], p. 620; b) Feeney, J.A. and Blackburn, M.J., in: loc. cit. [2f], p.355; c) Feeney, J.A., Blackburn, M.J., and Beck, T.R., in: Advances in Corrosion Science and Technology. (Fontana, M.G. and Staehle, R.W., eds.). Vol. 3. Plenum Press, New York 1973, p. 67
76. Leckie, H.P.: Corr. NACE: *23* (1967) 187
77. See e.g. Tomashov, N.D., Modestova, S.T., Galuznov, Borisova, E.A., and Zoztov, V.L., in: Corrosion of Metals and Alloys. Coll.1. N.L.L.S.T., Boston Spa 1964
78. a) Speidel, M.O. and Hyatt, M.V., in: loc. cit. [2h], p.115; b) Speidel, M.O., in: loc. cit. [2f]; c) Vogt, H. and Speidel, M.O.: Corr. sci. *40* (1998) 251
79. Holroyd, N.J.H., in: loc. cit. [2z], p.311
80. a) Friedrich, H., Kilian, H., and Kaesche, H., in: Proc. NATO Adv. Res. Workshop Modelling Aqueous Corrosion, Plymouth 1993. (Threthewey, K.R. and Roberge, P.R., eds.). NATO ASI Series E, Vol. 266. Kluwer Acad. Publ.,

Dordrecht, etc., 1994, p. 239; b) Berggreen, J., Engell, H.-J., and Kaesche, H.: Werkstoffe u. Korr. *26* (1975) 599; c) Richter, J. and Kaesche, H.: ibid. *28* (1977) 602; d) Stoll, F., Hornig, W., Richter, J., and Kaesche, H.: ibid. *29* (1974) 585; e) Richter, J., Hornig, W., and Kaesche, H.: ibid. *35* (1984) 23; f) Rippstein, K., Baumgärtner, M., Konys, J., and Kaesche, H.: Z. Metallkunde *75* (1984) 291; g) Baumgärtner, M. and Kaesche, H.: Corr. NACE *44* (1988) 231; h) Wenzel, G., Knörnschild, G., and Kaesche, H.: Werkstoffe u. Korr. *42* (1991) 449; i) iid.: Extended Abstract N3 129, Spring Meeting Electrochem Soc., Honolulu 1993; j) Nikol, Th. and Kaesche, H.: Preprints Conf. Eurocorr, Karlsruhe 1987. DECHEMA, Frankfurt, 1987, p. 505; k) David, W., Stellwag, B., and Kaesche, H. *59* (1983) E 147; l) Roth, A. and Kaesche, H., in: Proc. 5th AlLi Conf., Williamsburg 1989, Vol. III; m) iid., in: Proc. 6th AlLi Conf., Garmisch-Partenkirchen 1991; n) Kaiser, H., Lenz. E., and Kaesche, H., in: loc. cit. [2k]; o) Lenz, E.: Doctoral Thesis, University Erlangen-Nürnberg 1978

81. Sugimoto, K., Hoshino, K, Kageyama, M., Kageyama, S., and Sawada, Y.: Corr. sci. *15* (1975) 709
82. a) Gruhl, W.: Metall *17* (1963) 197; b) id.: Z. Metallkunde *53* (1963) 670; c) Ratke, L. and Gruhl, W.: Werkstoffe u. Korr. *31* (1980) 768; d) iid.: Z. Metallkunde *71* (1980) 568; e) Brungs, D. and Gruhl, W.: Metall *24* (1970) 217
83. Shastry, C.R., Levy, M., and Joshi, A.: Corr. sci. *21* (1981) 673
84. a) Haynie, F.H. and Boyd, W.K., in: loc. cit. [2n], p. 580; b) Gest, R.J. and Trioano, A.R.: Corr. NACE *30* (1974) 274
85. Lynch, S.P.: ibid. *24* (1984) 375
86. a) Pugh, E.N.: Corr. NACE *41* (1985) 517; b) Bertocci, U. and Pugh, E.N., in: loc. cit. [2m], p. 273; c) Bertocci, U., Pugh, E.N., and Ricker, R.E., in: loc. cit. [2z], p. 273; d) Beggs, D.V., Hahn, M.T., and Pugh, E.N., in: Proc. Conf. Hydrogen Embrittlement and Stress Corrosion Cracking, 1984. ASTM Metals Park 1984; e) Hahn, M:T. and Pugh, E.N.: Corr. NACE *36* (1980) 380
87. a) Wendler-Kalsch, E., in: loc. cit. [2l), p. 102; Pinchback, T.R., Clough, S.P., and Heldt, L.A.: Metal Trans. *6A* (1975) 1479; c) Takano, M., Teramoto, K., and Nakayama, T.: Corr. sci. *21* (1981) 459
88. a) Cassagne T.B., Flanagan, W.F., and Lichter, B.D.: Metal Trans. *17A* (1986) 703; b) Lichter, G.B., Flanagan, W.F., Lee, J.B., and Zhu, M., in: loc. cit. [2z], p. 251

# 16 Corrosion Fatigue

## 16.1 General Aspects

*Fatigue*, in general, is the loss in strength of structural materials caused by cyclic loads [1]. A possible dangerous outcome is *fatigue cracking*, i.e. the initiation and propagation of quasi-brittle cracks under the influence of loads cycled with an amplitude, $\Delta\sigma = \sigma_{max} - \sigma_{min}$, where $\sigma_{max}$ is lower than the static ultimate strength, $R_m$, of smooth samples, or else with an amplitude, $\Delta K = K_{max} - K_{min}$, of the stress intensity factor, where $K_{max}$ is lower than the fracture toughness, $K_C$, of notched samples. Fatigue cracking is a widespread source of failure of structural materials. In contrast to stress corrosion cracking, it is a variant of 'subcritical' crack growth which does not primarily require simultaneous rack-tip corrosion processes. As long as it is not enhanced by an environment, it is not *corrosion fatigue* and, therefore not a problem of corrosion science but of metal physics. Also, corresponding to the prime importance of the subject, the expert literature collected in monographs, in meeting volumes, and in journals is huge and far beyond the scope of the present text.

*Corrosion fatigue* (CF), then, is the combined effect of corrosion and fatigue damage, if corrosion enhances fatigue cracking, and this narrower field is the topic of this chapter. Again, although on a smaller scale, the expert literature is voluminous, and we shall concentrate in the following on basic aspects rather than applications. For more general reviews, see e.g. [3].

In practice, and, in fact, in contrast to the size of the present chapter, damage by CF is much more frequent than damage by stress corrosion cracking. According to experience in a large chemical plant, SCC is a problem essentially restricted to stainless steel equipment, while CF is of concern to practically all structural materials [4]. For danger of failure by CF in the aircraft and in the power plant industry see [5].

One traditional method to characterize CF susceptibility is to count the number of load cycles, $N_{fail}$, sustained until failure by fraction. In standard fatigue testing procedures, smooth or notched specimens are usually subjected to symmetrical tension/compression cycles with zero mean load; but finite positive mean loading is also used, especially when the object of testing is corrosion fatigue. Other useful loading modes are cyclic bending with and without rotation. In any case, $N_{fail}$ is determined as function of the load amplitude, $\Delta\sigma$, till fracture. By this procedure, fatigue crack initiation and fatigue crack propagation cannot be separated, though crack initiation will occupy most of the cycles to failure. The characteristic plot, $N_{fail}(\Delta\sigma)$, is the characteristic *Wöhler curve*, and Fig. 16.1 shows a typical example for a chromium steel with 0.2C and 13.7Cr (wt.%) [5]. The increase of $N_{fail}$ with decreasing $\Delta\sigma$ is typical and, of course,

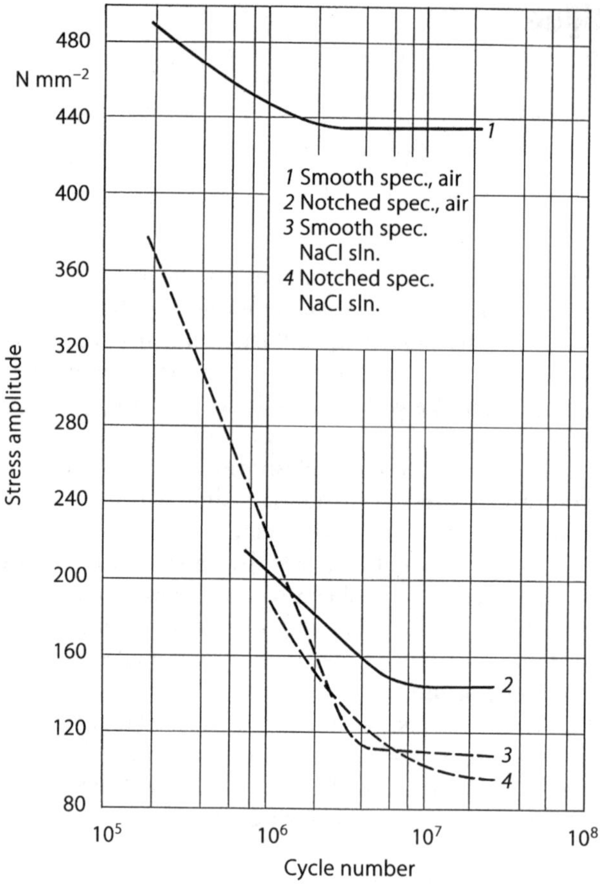


**Fig. 16.1.** Wöhler curves for a chromium steel obtained by alternate-bend testing in air and in NaCl solution, of smooth and of notched specimens. (From Lipp)

expected. If there is, as for this material observed at least in air, a minimum limit of the load amplitude, $\Delta\sigma = R_{\text{perm}}$, below which the number of load cycles sustained tends to infinite, we have a very useful behaviour, characterized by a distinct *fatigue endurance limit* or, more briefly, *fatigue limit.* It is typical for carbon steels fatigued in inert environments, and thus for a very large group of structural materials and for many environments, but it is not the general case. Instead, the number $N_{\text{fail}}$ frequently remains finite down to very low values of amplitudes $\Delta\sigma = R_{\text{time}}$. In such cases, safety considerations must be based on what is designated as *fatigue strength,* defined as load amplitude sustained for, e.g., $10^8$ cycles.

In the figure, the Wöhler curve for smooth specimens in NaCl solution shows a dramatic detrimental influence of an aggressive environment on fatigue strength, and this clearly illustrates the effect of corrosion fatigue. At the same time, notched specimens show poor fatigue strength already in inert environment without corrosion, and corrosion does not very much further affect this behaviour, except for an important detail: Clearly, the safe 'fatigue limit' disappears, leaving, as safety criterion, only the more restricted 'fatigue strength'. For the present case, the explanation of these effects probably is simple: In a neutral aerated chloride solution, a Cr-rich steel is passive, but subject to pitting, as described in Chap. 12. Pits in an otherwise smooth surface probably have an effect similar to that of notches. We shall see that similarly simple explanations apply to more, but by far not to all types of CF. We also see that surface roughness is a decisive factor for nominally smooth specimens, rendering results subject to surface finish and thus to somewhat accidental properties of test specimens.

A generally made distinction is made between *low-cycle fatigue*, (LCF), and *high-cycle fatigue* (HCF). This distinguishes cases where the load amplitude is high, causing $N_{fail}$ to drop to ca. $10^3$ or $10^4$, from others where $N_{fail}$ is higher by orders of magnitude. The distinction is important not only in practice, but for the essentials of fatigue mechanics as well: For large values of $N_{fail}$, the load amplitude $\Delta\sigma$ is low, and this is the normal range of cyclic loading occurring in practice, with $\sigma_{max}$ distinctly below the yield strength. $R_p$ must not, however, be confused with the true elasticity limit, because cyclic pure elastic deformation is endured indefinitely in any case. It is the cyclic accumulation of microplastic deformation processes reaching, for high numbers of cycles, very high values, which governs fatigue. It is for this reason that much attention is paid to the *cyclic stress-strain curve*, which for a quasi-stationary state of fatigue at intermediate cycle numbers relates plastic strain, $\Delta\epsilon_{plast}$ and load amplitude, $\Delta\sigma$. This also leads to the *Coffin-Manson* equation relating $\Delta\epsilon_{plast}$ and $N_{fail}$. Writing $\Delta\epsilon_{plast} = const \times N_{fail}^{-m}$, two factors, *const* and $m$, denoted as 'fatigue ductility factor' and 'fatigue ductility exponent', are introduced. In the case of LCF, i.e. when the experimental elongation cycles, $\Delta\epsilon$, are close to $\Delta\epsilon_{plast}$, *const* is close to the yield strength, and $m$ usually between 0.5 and 0.7. For HCF, the application of the equation is less obvious and in fact often replaced by an empirical relation $\Delta\sigma = const' \times (2N_{fail})^{-n}$. The 'fatigue strength coefficient', *const'*, is close to the tensile strength of the material, the 'fatigue strength exponent', $n$, close to 0.1.

Both fatigue crack propagation and fatigue crack initiation involve the localized accumulation of irreversible microplastic events [7,8]. Reversible dislocation movement has no such effect, and this explains, in principle, the existence of threshold values of the load amplitude for the onset of fatigue. The nature of ductile microevents varies from the appearance of so-called persistent glide bands to growth of pores at inclusions beyond the crack tip and to many other effects. Therefore, the possible influences of crack-tip corrosion processes are quite variable as well, and some of these are easily predictable. As an example, Fig. 16.2 shows the number of cycles to fracture for a 13% Cr steel in dilute sulfuric acid under potentiostatic control. A steel with this chromium content is activate below ca. 0.0 V, passive between ca. 0.0 and 1.1 V, and transpassive beyond this potential threshold. At −0.1 V, which also is the open-circuit corro-

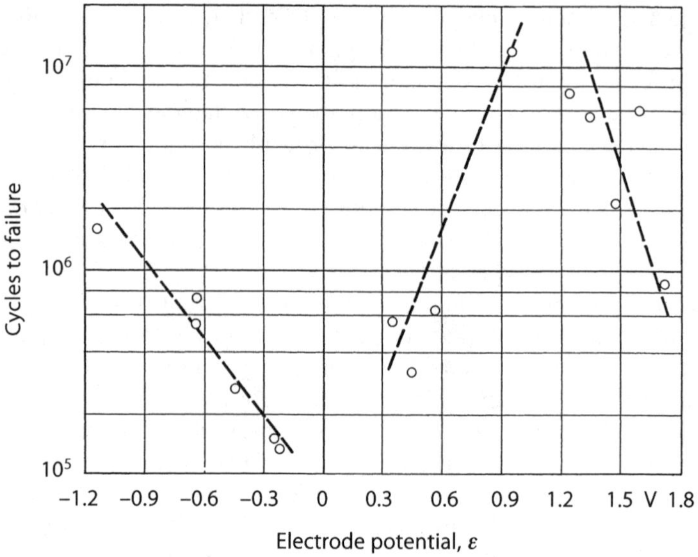


**Fig. 16.2.** Number of cycles to fracture observed for a 13% chromium steel in deaerated dilute sulfuric acid, pH 2, 20°C, at 90% of the fatigue limit in air, a function of the electrode potential. (From Spähn)

sion potential, active acid corrosion is rapid, and fatigue fracture occurs already after $N_{\text{fail}} \simeq 10^5$ cycles. Cathodic polarization and corresponding corrosion protection decreases fatigue susceptibility, raising $N_{\text{fail}}$. A very marked increase is also observed when the steel is passive and the rate of anodic metal dissolution, therefore, negligible. When transpassive rapid metal dissolution sets in, $N_{\text{fail}}$ again drops to low values.

For comparison, Fig. 16.3 shows the dependence of the fatigue limit of an unalloyed steel in NaCl solutions at solution pH values between 1 and 12 [9a]. As evident by comparing the data in Fig. 6.12 for corrosion rates of mild steel in aerated solutions, these curves again directly reflect the change in corrosion rate form rapid acid to oxygen corrosion and, eventually, spontaneous passivation. Interesting, but equally predictable is the change of cycles to fracture as a function of the electrode potential for the same steel in neutral salt solution, shown in Fig. 16.4 [9b] for three levels of the stress amplitude. At $\Delta\sigma = 270\,\text{MN}\,\text{m}^{-2}$ the amplitude is below the fatigue limit in air. Therefore, fatigue occurs by corrosion only, and is completely suppressed by a shift of the electrode potential to sufficiently negative values assuring safe cathodic protection. At 318 and 360 MN $\text{m}^{-2}$ the load amplitude exceeds the fatigue limit in air. Therefore, while the acceleration of fatigue by corrosion can be suppressed as before, fatigue still proceeds as in an inert environment.

Finally we remember that in neutral solutions corrosion of unalloyed steel is oxygen corrosion only and, therefore, negligible in the absence of dissolved oxygen. As documented in Fig. 16.5, fatigue of unalloyed steel in deaerated

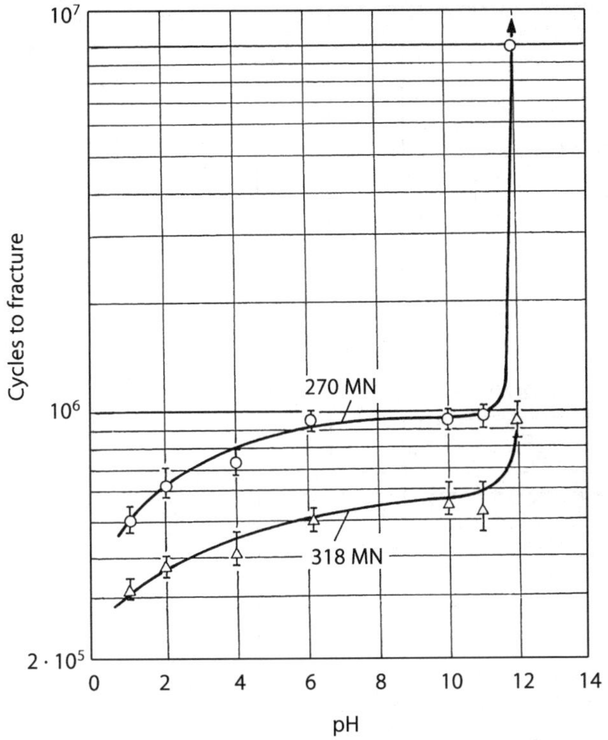


**Fig. 16.3.** Number of cycles to fracture of a low-carbon unalloyed steel in NaCl solutions with alkali resp. acid additions, as function of solution pH. (From Duquette and Uhlig)

chloride solutions then is practically the same as in air, while in solutions open to air oxygen corrosion causes a severe worsening of the situation, especially destroying the safe fatigue limit [9b]. The observation that in air at nominally 93% relative humidity no corrosive influence was observed proves that no moisture condensation occurred, or else that any moisture film was clean enough to prevent oxide-film breakdown on the specimens.

## 16.2 The Fracture Mechanical Approach

In this section, the topic is fatigue of notched and prefatigued specimens. We assume that the dimensions of the plastic zone in the material beyond the crack tip are small, compared to the size of the specimens tested, and that linear-elastic fracture mechanics are applicable. Typically, this applies to comparatively high-strength materials such as steels, aluminum alloys and titanium alloys [10]. For the fracture mechanical approach, it is standard practice to plot extensions of the crack length, $a$, observed per load cycle, that is the property $da/dN$, as function

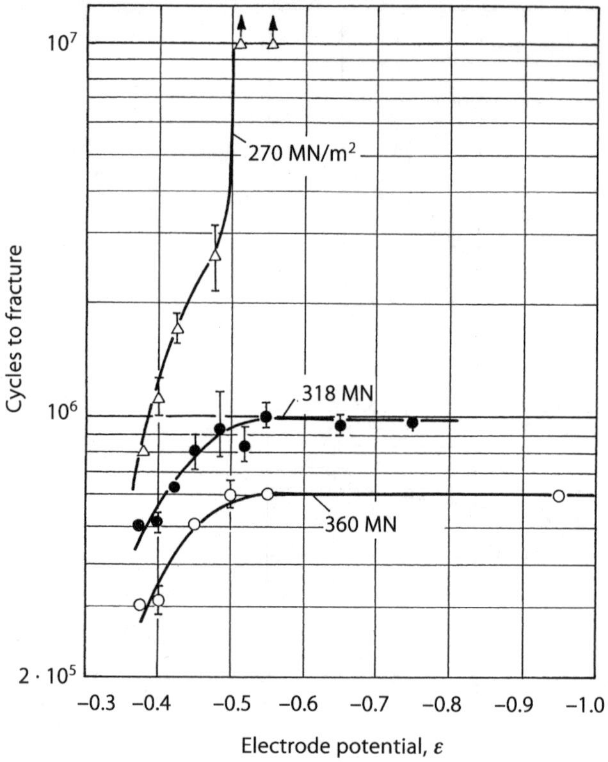


**Fig. 16.4.** The number of cycles to fracture of a low-carbon unalloyed steel in NaCl solutions with alkali resp. acid additions, as function of solution pH. (From Duquette and Uhlig)

of the amplitude, $\Delta K_I$, of the stress intensity factor. Figure 16.6 schematically shows the three main types of corrosion fatigue usually encountered [11a–c,12]:

a) By *true corrosion fatigue*, the crack extension per cycle is increased, as compared to fatigue in an inert environment, for all values of $\Delta K_I$, indicating an acceleration of crack extension under all conditions.
b) *Stress corrosion fatigue* is a process where purely mechanical fatigue alternates with true stress during tensile periods of load cycles, requiring that $(K_I)_{min}$ reaches or exceeds $K_{ISCC}$.
c) Finally, these two types may overlap to *mixed true and stress corrosion fatigue.*

When true fatigue and stress corrosion fatigue occur simultaneously, the cycle frequency and the average tensile load during each cycle are important quantities. For tension-controlled experiments, the ratio $R = \sigma_{min}/\sigma_{max}$ usually is used to characterize the load cycles. At low frequency and at the same time high average static stress, with $R \to 1$, the cycling experiment comes close to constant load testing, and SCC is expected to dominate the behaviour. At

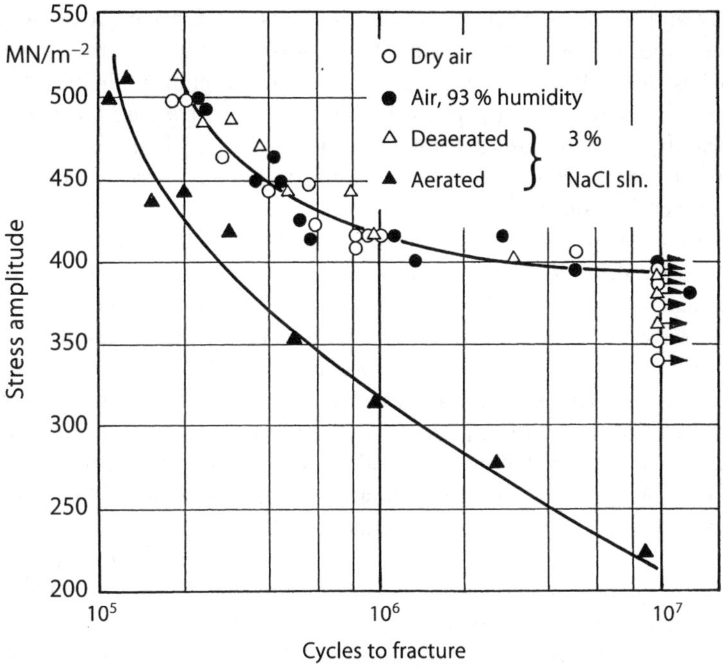


**Fig. 16.5.** Wöhler curves for unalloyed steel AISI 4140 observed in air, in aerated and in deaerated NaCl solutions, respectively. (From Lee and Uhlig)

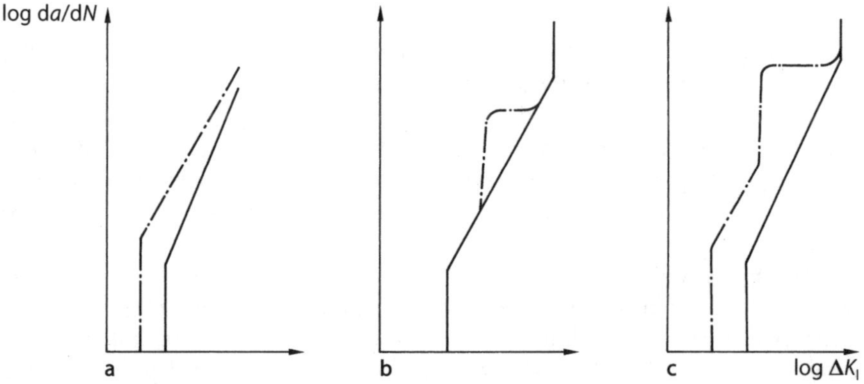


**Fig. 16.6a–c.** Fracture mechanical plots of cyclic crack extension rates, $da/dN$, as function of the stress intensity factor amplitude, $\Delta K_I$, showing the three essential types of corrosion fatigue: **a** True corrosion fatigue, **b** stress corrosion fatigue, and **c** superposition of true and stress corrosion fatigue. *Solid curves:* Fatigue in an inert environment

high frequency and at the same time low average stress component, $R \to -1$), true fatigue is expected to be predominant. For less pronounced conditions, the fatigue behaviour will tend to appear complicated. To gain guidelines, the classification of concurrent damage processes by the *process competition model* [10c] or, alternatively, by the *process superposition model* [13] is useful.

In the case of process competition, true fatigue and stress corrosion fatigue compete as independent mechanisms without mutual influence, and crack propagation is determined by the faster of the two. In process superposition, both mechanisms operate simultaneously and interfere, resulting in an acceleration of the crack propagation rate exceeding the simple sum. Apparently, competition and superposition are both encountered in practice [10,13]. As an example of relevant experiments, Fig. 16.7 combines crack extension data and fractographic evidence of fatigue experiments with a manganese-vanadium steel tempered to high strength, in chloride solution [11a,b]. As described in Chap. 15, this steel cracks in tension by intercrystalline HISCC along prior austenite grain boundaries. In this case, $K_{ISCC}$ was 12.5 MPa $\sqrt{m}$, and the plateau value of the crack extension rate in Region II of the $\ln v$-$K_I$ curve was ca. 2 µm $s^{-1}$. If, at sufficiently high values of the load ratio, $R$, stress corrosion does replace true fatigue, we can relate $v$, frequency, $f$, and cyclic crack extension rate by $da/dN = v \times f$, and we should be able to quantitatively predict the position and the height of the shoulder in the $\ln da/dN$-$\ln \Delta K_I$ curve. Except for some side effects blurring the sharp step, which we pass presently, the data confirm expectations proving, for this case, process competition. Also, the effect is readily explained by the inserts presenting fractographic evidence: SCC, replacing fatigue in the intermediate range of $\Delta K$ values, is intercrystalline, while true fatigue outside this range is transcrystalline, and these two fracture modes are, of course, unable to mix. In each case it then is the faster of the two mechanisms which takes over crack propagation.

As shown in Fig. 16.8 fracture mechanical fatigue testing of a low-alloy pressure vessel steel in water at 288°C produces similar $\ln da/dN$-$\ln \Delta K_I$ relations [14]. The curves drawn through the experimental points have again been calculated using the process competition model. The conclusion that corrosion fatigue of pressure vessel steels is generally stress corrosion fatigue and, therefore, as described in Chap. 15, probably hydrogen-induced, thus appears justified.

Analogous and therefore probably for the same reasons as stress corrosion cracking, hydrogen-induced corrosion fatigue of high-strength steels is enhanced already by cycling in dry hydrogen. This is demonstrated for the steel 90 MnV8 in Fig. 16.9, [11e,f] also showing that water vapour and even dry oxygen are to some degree effective as well. For hydrogen the supposition certainly is that, similar to the case of static-stress corrosion cracking, but enhanced by the cyclic repetition of deformation, hydrogen atoms are swept from the adsorption layer into the metal by crack-tip dislocation transport. Crack extension then occurs by decohesion, probably at matrix/precipitate interfaces, where atomic hydrogen is stripped from dislocations and concentrated to val-

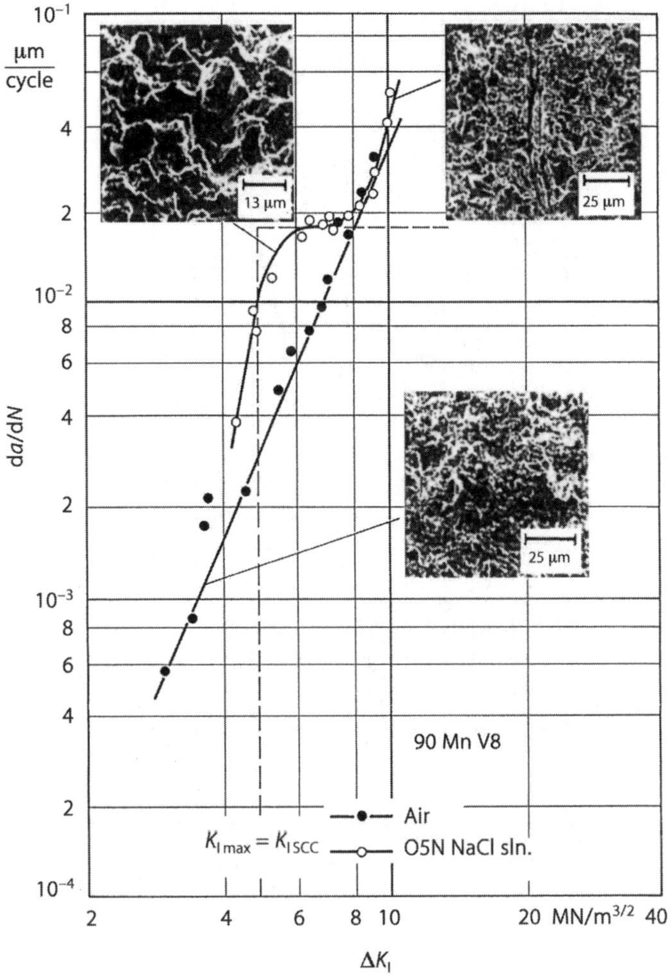


**Fig. 16.7.** Fatigue crack propagation in DCB specimens of steel 90 MnV8 ($R_p = 1530$ MPa, $K_{IC} = 27 MPa \sqrt{m}$) cycled with 109 Hz and R = 0.6 in 0.5 M NaCl solution. *Dashed lines:* vertical: $\Delta K_I$ for $K_{max} = K_{SCC}$; horizontal: $da/dN$ as calculated from the plateau Region II of the SCC crack propagation rate. *Inserts:* Fractographic evidence. (From Deimling, Stellwag, and Kaesche)

ues beyond saturation. The effect of water vapor essentially is the same, with adsorbed water molecules reacting to atomic hydrogen and iron hydroxide. The effect of oxygen, less obvious because oxygen is not expected to enter the metal lattice, is probably due to an influence of crack-tip surface oxydation on local microplasticity. A similar, but much more thoroughly investigated effect is reported in [12] for intercrystalline fatigue cracking of nickel bicrystals with controlled desorientation of the single grain boundary: In pure material, crack initiation was observed in persistent glide bands. When, instead, the grain

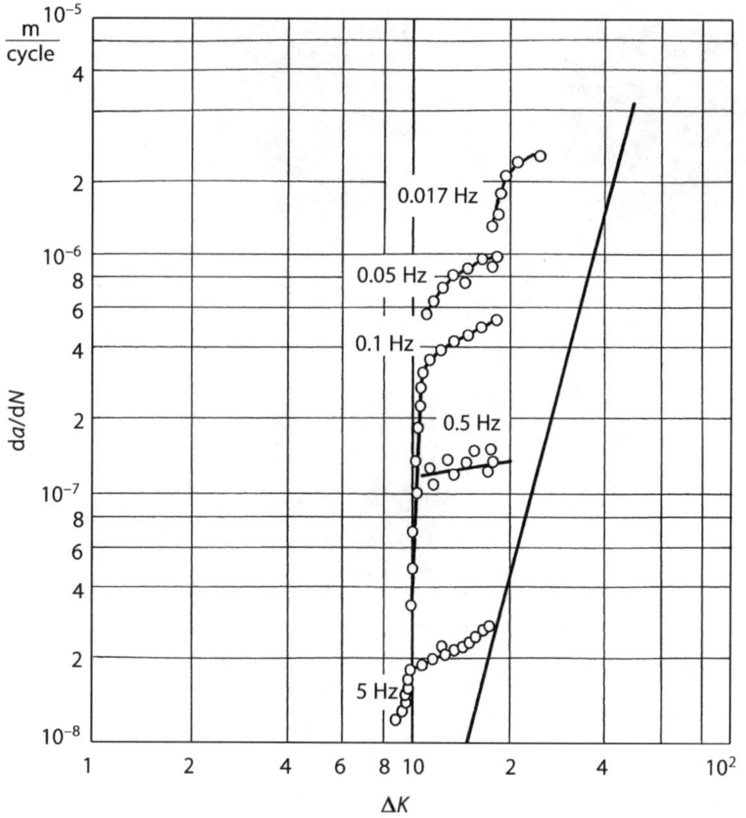


**Fig. 16.8.** Influence of cycling frequency on fatigue crack propagation in a reactor pressure vessel steel in water at 288°C. (Curves calculated by Speidel using data from Atkinson, Tice, and Scott)

boundary was contaminated with sulfur, cracking rapidly set in at much lower hydrogen pressure.

Low cycle fatigue of low-alloy pressure vessel steels in high-temperature, high-pressure reactor coolants is a topic much discussed along similar lines as stress corrosion cracking in the same system [14a, 15]. For a recent study of static SCC, slow rising dynamic SCC or very low-frequency fatigue, and both low-cycle and high-cycle fatigue of 20 MnMoNi 5 5 and two similar steels in refreshed water at temperatures between 150 and 288°C see [16]. Generally, crack extension is discussed as the result of, in principle, three contributions per load cycle: the purely mechanical contribution, $(da/dN)_{\text{inert}}$, as it would be observed in inert environments, the true corrosion fatigue contribution, $(da/dN)_{\text{corr}}$, simply enhancing mechanical fatigue, and the stress corrosion contribution, $(da/dN)_{\text{corr}}$. Depending on the relative role assigned to the contributions, models reflect process competition or process superposition. A widely used model is

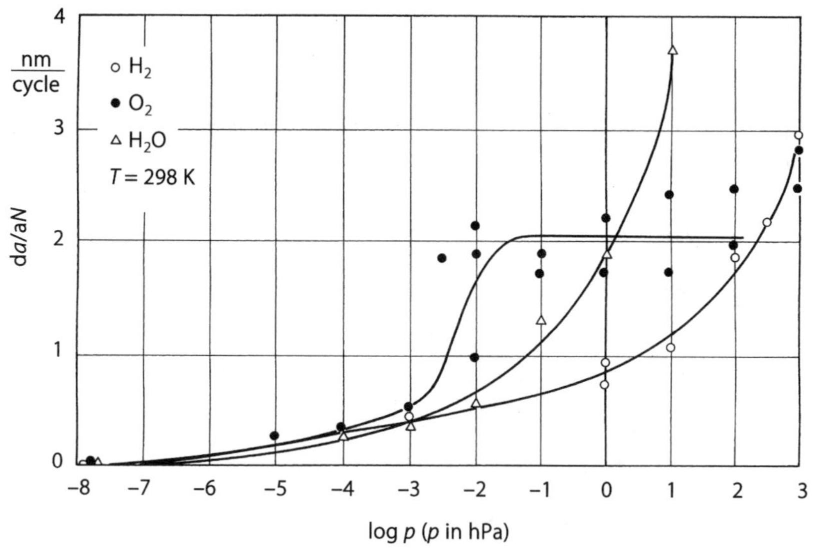


**Fig. 16.9.** The influence of gaseous hydrogen, oxygen, and water vapor on fatigue crack propagation rates in DCB specimens of steel 90 MnV8, tempered to yield strength 650MPa, as function of pressure. Experimental points recorded at constant $\Delta K_I$ in Region I of the $\ln da/dN$-$\ln \Delta K_I$ curve. (From Popp and Kaesche)

process competition with stress corrosion fatigue assumed predominant during tension periods of low-frequency cycles. As described in Chap. 15 for the case of SCC under static or slowly increasing load, the slip-step dissolution model is invoked regardless of relevant doubts, and for transient events of anodic metal dissolution during tension periods again a negative-power law, $I_{tip} \sim t^{-n}$, is applied.

As an example of competition and superposition of true corrosion fatigue and stress corrosion fatigue, crack propagation in a pipeline steel cycled in carbonate/bicarbonate solutions has been thoroughly investigated [17]. As described in Chap. 15, SCC is, in this case, strain-induced. Therefore, stress corrosion fatigue is expected to add to true corrosion fatigue during tensile periods of low-frequency cycling. Since SCC is intergranular, and fatigue transgranular, the two contributions can be distinguished from fractographic inspection. Figure 16.10 summarizes the results, plotting the average stress intensity factor, $\bar{K}_I = (K_{max} + K_{Imin})/2$, on the abscissa, and the amplitude, $\Delta K_I$, on the ordinate. $\Delta K_I = 0$ indicates straightforward testing for SCC, with $K_{ISCC} = 21$ MPa $\sqrt{m}$. When, at small values of the load amplitude, the static load component remains high, cracking remains intergranular along prior austenite grain boundaries, i.e. cracking still is SCC. As the amplitude increases, crack propagation becomes mixed, indicating both SCC and true corrosion fatigue, until, at high load amplitudes, transgranular pure corrosion fatigue predominates. Most of these observations correspond to what is expected, still we note that the persistence of

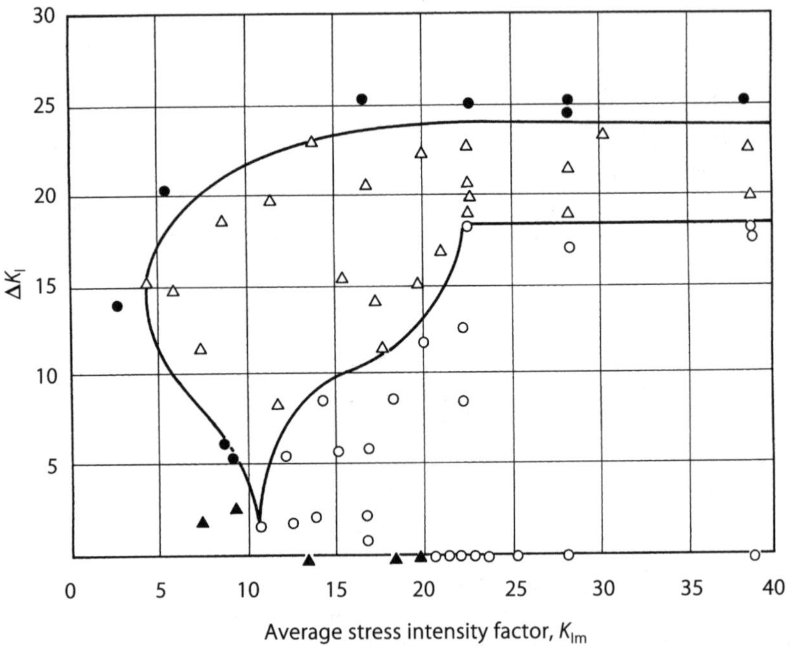


**Fig. 16.10.** Modes of fatigue cracking of a low-alloy pipeline steel in 0.5 Na2$CO_3$/$NaHCO_3$ solution at 75°C. Electrode potential −0.4 V; cycling frequency 0.19 Hz. ▲: no cracks; ○: intergranular crack; ●: transgranular cracks; △: mixed mode. (From Parkins and Greenwell)

the SCC component at very low values of $\bar{K}_I$, for $K_{max} < K_{ISCC}$, is not expected. Here and elsewhere for low-frequency experiments, the complication probably is that during every cycle the true momentary rate of deformation continuously cycles around the nominal value $\dot{\epsilon}$,. This, then, suggests the use of other than sinusoidal load profiles. As an example we quote investigations into fatigue of stainless steel (AISI 304) in water containing 8 ppm dissolved oxygen, at 288°C [18]. For fracture mechanical testing, DCB specimens were used either solution heat treated, or sensitized with respect to intercrystalline corrosion by a 650°C temper. Load profiles were a) positive sawtooths (positive ramps of load followed by rapid removal of load, b) negative sawtooths (rapid loads followed by negative ramps), c) trapezoidal with constant load periods interrupted by rapid loading resp. unloading, and pulsed without constant load. Air fatigue was quite normal, following a *Paris* equation of the type $\ln(da/dN) = \text{const} + m \times \ln \Delta K_I$ in every case. Down to frequencies $f \geq 10^{-2}$ Hz the influence of corrosion remained low. It became important at still lower frequencies, increasing $da/dN$ Hz by the factor 15 at $10^{-4}$ Hz. The influence was found to depend on the load profile, being most pronounced for positive sawtooth loading.

In these experiments, crack propagation in solution treated specimens remained transgranular as in air, and the effect is true corrosion fatigue. For any ramp increase of the load during a time interval $\Delta\tau$, the crack propagation rates $(da/dN)_{corr}$ in hot water and $(da/dN)_{air}$ were correlated as

$$(da/dN)_{corr} \sim (\Delta\tau)^n (da/dN)_{air} . \tag{16.1}$$

In sensibilized specimens, intercrystalline crack propagation became increasingly important with decreasing cycling frequency, and the $\ln(da/dN)$-$ln(\Delta K_I)$ curves developed shoulders typical for the superposition or competition of SCC. Formally, this behaviour is described by

$$da/dt = (da/dt)_{SCC} + f \times (da/dN)_{corr} . \tag{16.2}$$

For trapezoidal load cycles, where the load is constant with brief interruptions by negative and positive ramps, $(da/dt)_{SCC}$ could be available from static SCC tests. For the case of positive sawtooth load ramps, this method failed: Now, $(da/dt)_{SCC}$ as taken from static tests was much too low. Apparently, a differentiation between static and dynamic stress corrosion behaviour is required, distinguishing rates $(da/dt)_{statSCC}$ and $(da/dt)_{dynSCC}$. This leads back to the more general aspect of the influence of crack-tip strain on crack extension and will not further be discussed presently. In [18], a relation $(da/dt)_{dynSCC} \sim f^n (da/dt)_{statSCC}$ is proposed as a useful correlation. As usual, intercrystalline crack extension is supposed to proceed by transient anodic metal dissolution from activated glide steps.

In [19], fatigue tests are reported for a martensitic steel (HT 80, containing Cu, Ni, Cr, Mo; $R_p = 784$ MPa), for a ferritic-perlitic plain steel ($R_p = 372$ MPa), and for an austenitic CrNi steel. Test frequencies were 0.3 and 50 Hz, R = 0.1, and the test solution was 3% NaCl at ambient temperature. In the first place, this work revealed that it is important to consider crack closure effects, and to correct nominal $\Delta K_I$ amplitudes accordingly. Also, an interpretation of true corrosion fatigue is offered as follows: For cyclic periods of tensile strain, transient anodic metal dissolution is invoked as usual. The same metal dissolution process then is thought to impede back-movement of dislocations. Since the steels are not passive in neutral salt solutions, the decay of metal dissolution following slip steps cannot be explained as repassivation. Instead, hydroxylation according to: $Fe + H_2O \rightarrow Fe(OH)_{ads} + H^+ + e^-$, of glide steps is assumed as a precursor to passivity. This combined consideration of possible kinetic details, closure effects, and inhibition of reverse glide may, on the whole, be open to discussion, but it is less summary than the usual much simpler descriptions of either SCC or fatigue crack-tip electrode kinetics.

As pointed oupt in Chap. 15, precipitation hardening aluminum alloys differ in SCC behaviour, depending on composition: SCC as typical for AlCu alloys sets in beyond the threshold potential for intercrystalline corrosion observed without stress as well, while SCC of the type represented by AlZnMg alloys sets in far below this threshold potential. The same difference in behaviour once more shows up in fatigue. Figure 16.11 illustrates the influence of a corrosive environment on fatigue crack propagation in hardened AlCuMg2 [11g]. In the first place, these curves indicate that, at low amplitudes of $\Delta K_I$, a limiting value, $\Delta K_{ICF}$, exists for the onset of fatigue. This effect, apparently typical for other

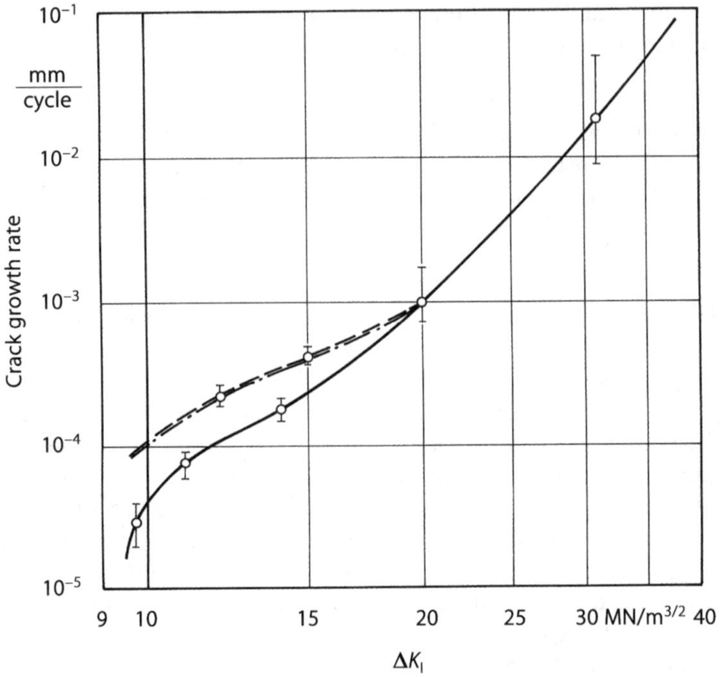


**Fig. 16.11.** Fatigue crack extension in precipitation hardened DCB specimens of Al-4.3Cu-1.5Mg (wt.%., AA 2024, temper T351) at ambient temperature. f = 0.1 Hz; R = 0.05. Curves continually recorded, variation observed in different runs as indicated by error limits. (—-): vacuum ($10^{-9}$ hPa), — · — air, 60% rel. humidity; (— — —): 0.1 M NaCl solution (flushed with air). (From Nikol and Kaesche)

aluminum alloys as well, is, however, observed already for purely mechanical fatigue, and not discussed presently.

As compared to fatigue in vacuo, an environmental influence is noticed at low stress amplitudes, $\Delta K_I$, only, but already in relatively dry air, and it is not further enhanced by changing to NaCl solution. Cracking is transcrystalline throughout, and the case clearly is one of true corrosion fatigue. The same apparently applies to fatigue of Al-1Cu-1Mg (AA 6013 T6), which has been tested for fatigue in probably slightly moist air and in NaCl solution [20]. In this case, fatigue behaviour is complicated by the appearance of secondary cracks. At the same time, hydrogen uptake was checked and confirmed by post-fatigue hot extraction, pointing to hydrogen embrittlement contributing to corrosion fatigue. The hydrogen content of the crack-tip zone was found to depend on cyclic frequency, with a maximum at ca. 1 Hz.

As expected, fatigue behaviour of hardened AlZnMg3 is distinctly different. Figure 16.12 shows the now very important effect of changing from air to water and to electrolytic halide solutions [10]. The supposition, suggested by these curves, that stress corrosion fatigue at least contributes to fatigue crack propa-

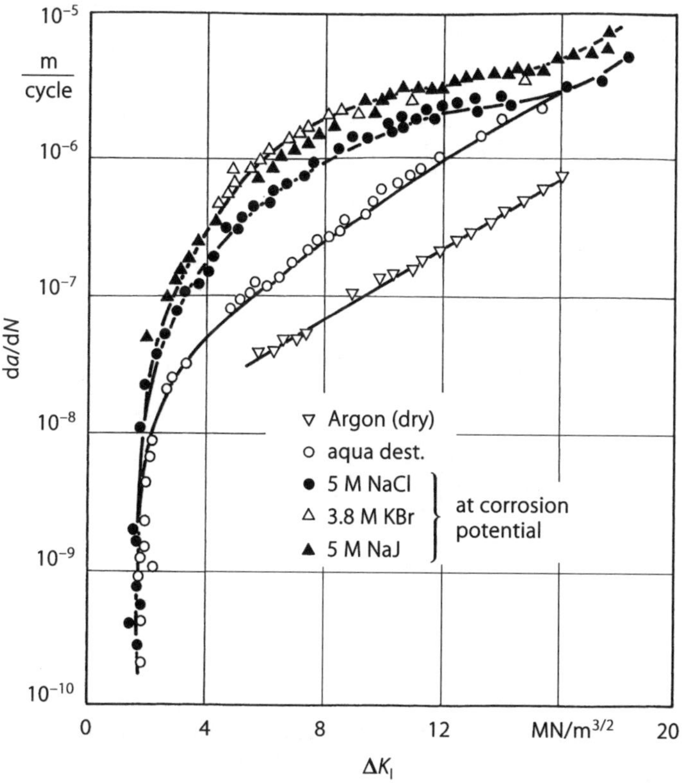


**Fig. 16.12.** Fatigue crack propagation in DCB specimens of hardened AlZnMg3 (AA 7079, temper T+651) in different aggressive media at ambient temperature. (From Speidel, Blackburn, Beck, and Feeney)

gation has been thoroughly substantiated by fractographic inspection of crack faces observed in fatigued, hardened AlZnMg3 (AA 7017 T 651) [21]. In the f-$\Delta K_I$ diagram in Fig. 16.13 the domaines for three predominant crack morphologies are outlinded: Ductile transcrystalline (Type A), brittle transcrystalline (B), and brittle intercrystalline (C). The explanation offered by the authors is that Type A is mechanical fatigue without environmental influence, whereas both Type B and Type C indicate hydrogen-induced fatigue damage. Here, Type B is true corrosion fatigue, i.e. local crack-tip hydrogen uptake, while embrittling the plastic zone ahead of the crack, does not change the crack path, which remains transcrystalline, and this, then, is not stress corrosion fatigue in the strict definition of the term. On further decreasing $\Delta K_I$, intercrystalline, again hydrogen-induced SCC eventually does overtake hydrogen-induced true fatigue cracking, turning extension to intercrystalline paths. Considering crack extension and required diffusion rates, a diffusion coefficient $D_H \simeq 3 \times 10^{-13}$ m$^2$ s$^{-1}$ is estimated, which reasonably well compares with literature values.

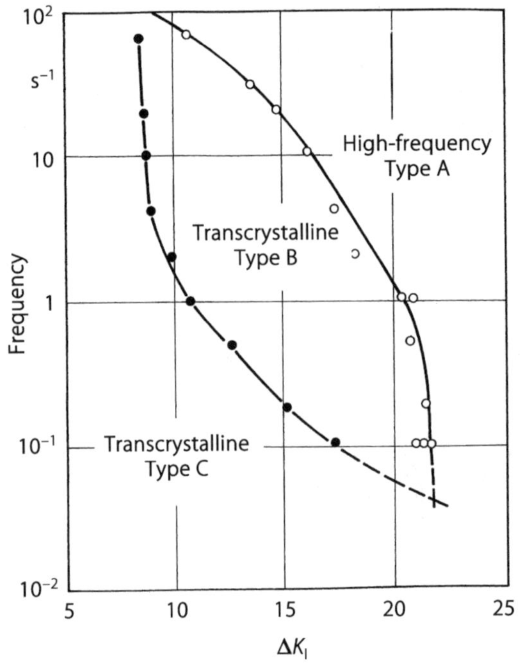


**Fig. 16.13.** Fatigue of AlZnMg3 (7017-T651): Metallographic crack morphology observed in DCB specimens fatigued in seawater. Type A: Predominantly ductile, transcrystalline; Type B: Predominantly brittle, transcrystalline; Type C: Predominantly intercrystalline crack faces. (From Holroyd and Davie)

Fatigue of pure Al-5Zn-1Mg in water-vapor/air mixtures has been investigated by fracture mechanical testing with regard to the influence of hardening precipitations on the morphology of fatigue cracking [22a,b]. In the presence of coherent $\eta'$-precipitates, the shear component of stress causes transcrystalline crack extension in persistent glide bands. Instead, non-crystallografic cleavage caused by the crack-opening component, $\sigma_{yy}$, of the stress is observed when overaging has lead to coarse $MgZn_2$ precipitates. Data on the influence of water vapor in air on fatigue of the same alloy tempered to peak hardness are shown in Fig. 16.14 [22]. For water partial pressures exceeding 8 hPa, fatigue crack extension is considerably enhanced, and cracking changes from trans- to intercrystalline propagation, indicating a change from true corrosion fatigue to stress corrosion fatigue. The decrease in $\Delta K_{ICF}$ accompanying the acceleration of fatigue is correlated with the water partial pressure, $p$, by the relation $\Delta K_{ICF} = a \times \sqrt{(b/p)}$, with empirical constants $a$ and $b$. Memory effects, as revealed by changing from moist to dry atmosphere, once more suggest that stress corrosion fatigue is definitely hydrogen-induced. Furthermore, the observations again point to dynamic stress corrosion fatigue, with $(da/dt)_{dynSCC}$ by far exceeding $(da/dt)_{statSCC}$. Comparing Figs. 16.12 and 16.14, it is interesting to note that $\Delta K_{ICF}$, while shifted to lower values by an increase in water vapor

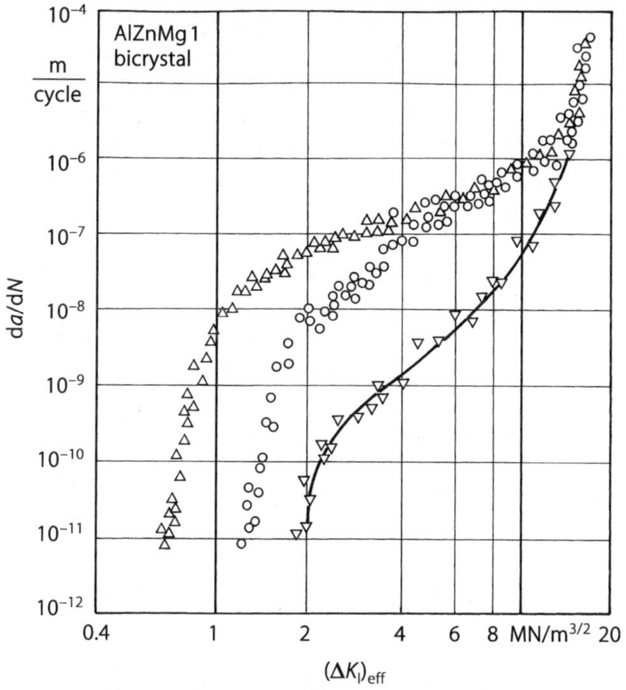


**Fig. 16.14.** Fatigue crack propagation in pure bicrystals of Al-4.5Zn-1.25Mg (wt.%) tempered to peak hardness for 100 h at 135°C. DCB specimens tested at 100 Hz, R $= -1$, $\Delta K_{\mathrm{eff}} = K_{\max}$. Water partial pressure: 31 (△), 12 (□), and $\leq 8$ hPa (▽). (From Niegel, Gutladt, and Gerold)

pressure of a gaseous atmosphere, remains unaffected by a change from pure condensed water to concentrated halide concentrations. As an item of practical importance, more knowledge of this detail would seem desirable.

From Chap. 15 we recall, finally, that for SCC in AlZnMg alloys the properties of precipitation-free zones (PFZ) at grain boundaries are important. For AlZnMg1 alloys it is observed that overaging minimizes the susceptibility with respect to stress corrosion fatigue [22c]. Probably, this points to broadening of PPFZs reducing stress by piled-up dislocations, and depletion of PFZs with respect to dissolved zinc

## References

1. See e.g. a) Konf. Ermüdungsverhalten metallischer Werkstoffe (Munz, D., ed.). DGM Informationsgesellschaft Verlag, Oberursel, 1985; b) Hertzberg, R.W.: Deformation and Fracture Mechanics of Engineering Materials. J. Wiley & Sons, New York etc., 1983
2. a) Proc. Conf. Corrosion Fatigue: Chemistry, Mechanics and Microstructure, Storrs 1971. (Devereux, O., McEvily, A.J., and Staehle, R.W., eds.). NACE,

Houston 1972; b) Subcritical Crack Growth Due to Fatigue, Stress Corrosion and Creep. (Larsson, L., ed.). Elsevier Appl. Science Publ., 1984; c) Proc. Int. Conf. Stress Corrosion Cracking and Hydrogen Embrittlement of Iron Base Alloys, Unieux-Firminy 1973. (Staehle, R.W., Hochmann, J., McCright, R.D., and Slater, J.C., eds.). NACE, Houston, 1979; d) Diskusionstagung Angewandte Bruchuntersuchung u. Schadensaufklärung, Ismaning 1975. Allianz Zentrum f. Technik, Ismaning, 1975; e) Proc. 2nd Int. Conf. Low Cycle Fatigue and Elasto-Plastic Behaviour of Materials, München 1987. (Rie, K.-T., ed.). Elsevier Appl. Science Publ., 1987; f) Environmental-Induced Cracking of Metals, Kohler 1988. (Gangloff, R.P. and Ives, M.B., eds.). NACE, Houston, 1990; g) Proc. Conf. Mechanism of Environment-Sensitive Cracking of Materials, Guildford, 1977. (Swann, P.R., Ford, F.P., Westwood, A.R.C., eds.). The Metals Soc., London, 1977

3. a) Duquette, D.J., in: loc. cit. [2f], p. 45; b) Gangloff, R.P., ibid. p. 55
4. Spähn, H., in: loc. cit. [1a], p. 40; loc. cit. [2b], p. 275; loc. cit. [2d], p. 59
5. a) see eg. a) Cohen, B., in: loc. cit. [2a], p. 65; b) Brown, B.F., in: loc. cit. [2a], p. 25; c) Mogford, I.L. and Jones, D.G., in: loc. cit. [2a], p. 30
6. Lipp, H.-J.; Werkstoffe u. Korr. *26* (1975) 825
7. See e.g. a) Mughrabi, H., in: loc. cit. [1a], p. 7; b) Laird, C., Duquette, D.J., in: loc. cit. [2a], p. 88
8. Stanzl, S.E., in: loc. cit. [1a], p. 107
9. a) Duquette, D.J. and Uhlig, H.H.: Trans. ASM *62* (1969) 839; b) Lee, H.H. and Uhlig, H.H.: Metal Trans. *3* (1969) 2949
10. a) Speidel, M.O., Blackburn, M.J., Beck,T.R., and Feeney, J.A., in: loc. cit. [2a], p. 324; b) McEvily, A.J., Wei, R.P., in: loc. cit. [2a], p. 381; c) Austen, J.M. and Walker, E.F., in: loc. cit. [2g], p. 334
11. a) Deimling, H.-J., Stellwag, B., and Kaesche, H.: Z. Werkstofftechnik *12* (1981) 51; b) Stellwag, B. and Kaesche, H.: Corr. NACE *35* (1979) 397; c) iid.; Werkstoffe u. Korr. *33* (1982) 323; d) Maier, H.J., Popp, W., and Kaesche, H.: Acta met. *35* (1987) 875; e) Popp, W. and Kaesche, H., in: loc. cit. [2e], p. 383; f) iid.; Steel Research *61* (1990) 490; g) Nikol, Th. and Kaesche, H., in: Proc. Eurocorr, Karlsruhe 1987, DECHEMA, Frankfurt, 1987, p. 505
12. Vehoff, H., Laird, C., and Duquette, D.J.; Acta met. *35* (1987) 2877
13. a) Wei, R.P. and Landes, J.D.; Mat. Res. Stand. *9* (1969) 25; b) Wei, R.P. and Gallagher, J.P., in: loc. cit. [2a], p. 409
14. a) Atkinson, J.D., Tice, T.R., and Scott, P.M., in: Proc. 2nd IAEA Specialists' Meeting on Subcritical Crack Growth, Sendai 1986. Report NUREG/CP-0067. Vol. 1. National Technical Information Service, Springfield, 1986, p. 251; b) Speidel, M.O.; J. Mater. Engng. *9* (1987) 157
15. a) Scott, P.M., Truswell, A.E., and Druce, S.G.: Corr. NACE *40* (1984) 351; b) Ford, F.P.: Mechanism of Environmental Cracking in Systems Peculiar to the Power Generation Industry, Final Report EPRI Contract RP 1332-1; Epri Report NP-2589 (1982)

16. Seifert, H.P., Ritter, S., and Heldt, J., in: Proc. 10th Int. Conf. Environmental Degradation of Materials in Nuclear Power Systems-Water Reactors, Lake Tahoe, 2001. (Ford, F.P., Was, G.S.,Nelson, L.J., and Parsons, P., organizers). NACE Houston (in press)
17. Parkins, R.N. and Greenwell, B.S.: Metal sci. *11* (1977) 405
18. Kawakubo, T., Hishida, M., Amano, K, and Katsuda, M.: Corr. NACE *36* (1980) 638
19. Matsuda, H., Matsuoka, S., Nishijima, S., and Shimodaria, M.; Corr. sci. *28* (1988) 433
20. Nocke, K., Bergner, F., Bersch, H., Haase, I., Worch, H., Tempus, G., and Loechelt, E.: Materials and Corr. *51* (2000) 628
21. Holroyd, N.J.H. and Hardie, D.; Corr. sci. *23* (1983) 527
22. a) Ling, P., Gutladt, H.-J., and Gerold, V., in: Proc. ECFG Fracture Control of Engineering Structures, Amsterdam 1986. (Van Elst, H.C., and Bakker, A., eds.). Chameleon Press, London 1986, Vol. III, p. 1493; b) iid., in: loc. cit. [2e], p. 687; c) Niegel, A., Gutladt, H.-J., and Gerold, V., in: Proc. 3rd Conf. Fatigue and Thresholds, Charlottesville, 1987. (Ritchie, R.O. and Starke, E.A., eds.). EMAS Advisory Service Ltd., Warley, 1987, Vol. III, p. 1229

# 17 Appendix

## 17.1
## Anodic and Cathodic Protection

*Anodic protection* is the technique of minimizing corrosion by external anodic polarization of metallic structures into the potential range were the metal or the alloy in question is passive. The method, therefore, implies the change from active e.g. acid corrosion to passivity, and it is useful only if the passive corrosion rate, as indicated by the passive current density, $i_{pass}$ defined in Chap. 10, is negligible. Cases of obvious practical relevance are iron and steel, especially high-alloy chromium-nickel steels in acid solutions. In the range of passivity, and in the absence of oxidizing agents in the solution other than $H^+$, the total electrolytic current density is equal to $i_{pass}$,. The anodic protective external current, $j_{prot}$, then is equal to the product $A \times i_{pass}$ of the surface, $A$, of the structure and the passive current density, and often will be very low, rendering the technique extremely useful. Recalling anodic oxygen evolution at passive iron and transpassivity of chromium steels and stainless steels, it is obvious that a potentiostatic polarizing circuit normally will be required in order to avoid driving the potential to undesirable anodic electrode reactions other than passive metal dissolution. Also, when passivity implies danger of pitting, or intercrystalline corrosion, or stress corrosion cracking, unskilled application of anodic protection may be a serious error. When for the same, potentially fatal corrosion processes threshold potentials exist, as especially for pitting, potentiostatic sub-threshold protection may still be a safe possibility. At any rate it is easily seen, that anodic protection does require knowledge and careful consideration of electrode kinetics. When applicable, it is, of course, very useful and economic. As an early example, work reported in [1] may be quoted showing that CrNi stainless steel, which does not spontaneously passivate in 50% $H_2SO_4$ and, instead, fails by rapid acid corrosion, can be fully protected by anodic polarization into the passive potential range. Once passivated, the protective current density is 2.5 $\mu$A $cm^{-2}$. This application in the chemical industry is typical. For other than simple geometries, the uniform or near-uniform distribution of the protective current supplied by the potentiostat to the circuit consisting of auxiliary counter electrodes as cathodes and the walls of solution containing vessels as anodes may need particular attention. These are problems of electrical engineering whose details we pass. Suffice it to note that in this direction the high polarization resistance, $R_\pi$, of passive metal/solution interfaces is very useful, as it levels inhomogeneities of the current distribution between auxiliary cathodes and vessel walls to be protected. Again an early experience was that relatively narrow pipes between different parts of a complete sulfonating plant were safely passivated by anodic protection of the vessels [2]. Attention must then be paid to accidental activation inside pipes and the like, where current paths to the auxiliary cathode are long: The polarization resistance of non-passivated interfaces where acid corrosion sets in is low, and the local current distribution may become dominated by the Ohmic resistance along the current path in the solution, preventing sufficient repassivating current supply [3].

Anodic protection has found an unusual application for the protection of unalloyed steel in hot alkaline solutions against stress corrosion cracking [4]. As described in Chap. 15, this is a case of SCC where crack initiation and propagation is confined to an intermediate range in electrode potentials close to the active/passive transition potential. SCC damage of this type became a problem in the operation of a large chlorine-alkali electrolysis plant, because in certain parts, depending on the position in the chain of electrolysis cells, the wall electrode potential happened to be in the critical range. The problem was solved by suitable wiring, securing a shift of the potential to values positive to the critical range, and thus in effect by anodic protection.

While anodic protection is applicable for very specific metal/solution systems, where the metal is passivatable and, if passivated, not in danger of pitting, or intercrystalline attack, or stress corrosion cracking, *cathodic protection*, opposite in polarity, has much less restricted limits of applications. External cathodic polarization by currents either impressed from an external voltage source, or else generated in a short-circuit galvanic couple including the structure to be protected as the cathode and sacrificial zinc or magnesium anodes, is a widespread method, protecting huge numbers of buried vessels and immense lengths of buried pipes or electric ducts against corrosion by soil moisture, or of ship hulls, ballast tanks, offshore oil drilling rigs etc.. As a rule, the damage to be prevented is oxygen corrosion, with oxygen dissolved in soil moisture or in seawater from the atmosphere. In the vast majority of cases, except for buried electric ducts, the metal to be considered is unalloyed steel.

For the following, we mainly consider buried pipes, and especially oil pipelines. These are a matter of considerable public concern because contamination of ground water by oil leaking from pipes perforated by corrosion cannot be permitted. By this argument, safe corrosion protection, assuming extrem conditions, is the important issue. In this context we recall from Chap. 6 that for iron and low-alloy steel in neutral or approximately neutral electrolytic solutions the corrosion rate, i.e. the rate of rusting, of otherwise unprotected surfaces is limited by the limits of oxygen diffusion from the bulk of the solution to the metal surface. This directly points to the answer of the question how to dimension the protective current density: From what has been explained in Chap. 6, polarization with the cathodic diffusion-limited current density of oxygen reduction, $i_{O_2,D}$, will drive the electrode potential into a range where anodic active iron dissolution becomes negligibly slow.

Since the moisture in soil pores is a stagnant electrolyte, $i_{D,O_2}$ will be very low, because the thickness of the effective diffusion layer will be of the order of the distance between the buried structure and the soil surface. This may seem to ease matters considerably, but even for very small current densities required for safe protection, the total current strength to protect, for instance, thousands of miles of buried pipes is uneconomic, and in fact cathodic protection never is applied to bare pipes, but to pipes already protected by thick organic coatings instead. Under these conditions, cathodic protection is the mean to satisfy the legal requirement of near-to-ideal safety by adding, to an already practically safe coating, cathodic protection taking care of accidental flaws in this coating. By this procedure, corrosion protection is reasonably redundant. Under these conditions, the protective current is no longer estimated from the total metal surface and some specific current density. Also, we do not really need to consider complicating factors such as large-dimension galvanic cells between pipe sections in different types of soil, etc. Instead, cathodic protection of buried structures is generally installed applying either expert practical experience, or applying a protective-potential criterion stating that, again by expert experience, protection is safe if the electrode potential of the buried metal is equal or below −0.85 V, referred to a $Cu/CuSO_4$ electrode in a saturated $CuSO_4$ solution.

Referred to the standard hydrogen electrode, this protective potential is −0.53 V. For a consideration of the implications of this value described in [5], assume, in the first place, that the damaging anodic electrode reaction is $Me \rightarrow Me^{z+} + ze^-$, and also assume a definite bulk concentration, $c_{Me^{z+}}$, of dissolved metal cations, and usual ambient temperatures. Disregarding, as appropriate for a safety consideration, details such as activity coefficients differing from unity, the equilibrium potential for metal dissolution and redeposition is $\varepsilon = E_{Me/Me^{z+}} = E^{\circ}_{Me/Me^{z+}} + (0.06/z)\lg c_{Me^{z+}}$. At this, or below this potential, metal dissolution and thus corrosion is impossible. In practice, however, $c_{Me^{z+}}$ is ill defined or even virtually zero, and the problem is what reasonably to substitute as its value. At this point, consider that in steady state, since anodic electrolytic migration does not occur, $Me^{z+}$ cations once transferred across the electrical double layer, move into the bulk of the solution by diffusion only. Applying, again as appropriate for order-of-magnitude effects, the Nernst model of a diffusion layer of finite thickness, $\delta_{Nernst}$, the rate of steady-state diffusion is proportional to the difference, $(c_{Me^{z+}})_* - (c_{Me^{z+}})_\circ$, in $Me^{z+}$ concentration across the diffusion layer. Since $(c_{Me^{z+}})_\circ \simeq 0$, the cations diffuse at the rate $v_{diff} = D_{Me^{z+}}(c_{Me^{z+}})_*/\delta_{Nernst}$, and in steady state this also is the rate of corrosion. A reasonable value to be substituted is $D_{Me^{z+}} \simeq 10^{-5}\ cm^2\ s^{-1}$. For soil corrosion, the length $\delta_{Nernst}$ may be anything up to meters, so to chose the value 10 mm is very conservative, indeed. Concerning $(c_{Me^{z+}})_*$ we now require, to guarantee safety, that the rate of corrosion of the metal must not exceed a very low value, e.g. 10 µm/year or $4.4 \times 10^{-12}$ mol $cm^{-2}$ $s^{-1}$, corresponding to a corrosion current density of ca. 1 $\mu A\ cm^{-2}$. From this it follows that the average rate of corrosion, $v_{corr}$, does not exceed 10 µm/year, if $(c_{Me^{z+}})_* \leq 4.4 \times 10^{-6}$ mol $L^{-1}$. To make safety even more convincing, we lower this number to $1 \times 10^{-6}$ mol $L^{-1}$, thus finally motivating what has been described as standard, if tacitly introduced practice in the use of potential-pH diagrams in Chap. 3. The next step to consider is that electrode potential, equilibrium potential and overpotential of anodic metal dissolution are related as $\varepsilon = E^{\circ}_{Me/Me^{z+}} + (0.06/z)\log(c_{Me^{z+}})_* + \eta_{Me}$.

In this equation, concentration overpotential already is included by substituting $(c_{Me^{z+}})_*$ for $c_{Me^{z+}}$, and other overpotential contributions are negligible at the extremely low current densities tolerated for safe protection. Therefore, the protective potential which must not be exceeded in order to satisfy requirements is $E^\circ_{Me/Me^{z+}} + (0.06/z)\log 10^{-6}$. For iron dissolving with $z = 2$, this is $\varepsilon \lesssim -0.62$ V. Numerically, this limit is similar to that used in practice based on extended experience rather than theory, showing that, although crude, the derivation makes sense. For other metal dissolution reactions, the protective potential may change. This certainly applies to cases of formation of strongly complexed ions, where the standard potential shifts to much more negative values or, respectively, where the thermodynamic activity of dissolved free $Me^{z+}$ ions becomes exceedingly small. Also, when the reaction product is a solid compound, e.g. an oxide, the electrode potential, including the protective potential, may be calculated from the metal/oxide equilibrium, or else again from the true concentration of dissolved metal in contact, at the metal surface, with the oxide [6]. In this case, solution pH at the surface enters the calculation, and this, then, adds a complication to circumstances because the cathodic protective impressed current will increase pH locally, which also is a source of danger to organic coatings [8].

Sometimes, the application of cathodic protection appears odd, when, as in the case of unalloyed steel in moist concrete, the structural metal is passive owing to the alkalinity of the aerated moisture. In this context, an interesting point is the carbonisation of concrete, i.e. the reaction of free calcium hydroxide in concrete with atmospheric carbon dioxide, $Ca(OH)_2 + CO_2 \rightarrow CaCO_3 + H_2O$, tending to neutralize alkalinity and to activate steel rebar. A cathodic current locally producing alkalinity may counteract this effect. Also, a cathodic current will cause anions in the moisture in concrete to move, in the course of cathodic current flow to rebar surfaces, away from these surfaces, and this may be a useful effect when it comes to damage of pitting corrosion by chloride from antifreeze additions to the concrete. Here, we recall from Chap. 12 that for pitting of low-alloy steel in alkaline solutions, a threshold pitting potential is well defined, and by potentiostatic cathodic protection pitting can be suppressed without shifting the potential to below the active/passive transition [6b,7b]. This also avoids another obvious danger: The cathodic protective current density is practically equal to the cathodic diffusion-limited current density of reduction of dissolved oxygen. On further increasing the current beyond this limit, which in practice may not easily be recognized from current-potential measurements, the electrode potential shifts into the range of cathodic hydrogen evolution by reduction of $H_2O$. When, as for tendons in prestressed concrete structures, the steel is susceptible to hydrogen embrittlement, this effect is, of course, highly undesirable.

Cathodic protection is comprehensively reviewed in [9], and for both more details and general aspects of the matter, reference is made to this source. Concerning corrosion protection of steel in concrete, a special new topic, still in an experimental stadium but interesting, is electrolytic rehabilitation of concrete layers defective owing either to carbonization or to chloride content. In the pores of moistened concrete, cathodic hydrogen-depositing currents flowing from suitable counter-electrode nets on the concrete surface to the steel reinforcements will tend to cause a) an increase of pH at the reinforcements surfaces and b) electrolytic migration of anions away from these surfaces. The first effect counteracts the loss in alkalinity owing to carbonization by atmospheric carbon dioxide, the second counteracts local pitting attack caused by chloride in the pore moisture. If, by subsequent application of a suitable coating on the concrete surface, re-carbonization and/or re-entry of anti-freeze chlorides is prevented, the method should, in principle, be considerably extend the service life of concrete structures. For a state-of-the-art report see [10].

### References

1. Tomashov, N.D.: Z. Elektrochemie *62* (1958) 717
2. Riggs, O.L., Hutchison, M., and Conger, N.L.: Corr. NACE *16* (1960) 102
3. Edeleanu, C.: Chem. and Ind. *1961* 301
4. Gräfen, H. and Kuron, D.: Archiv Eisenhüttenwesen *36* (1965) 285
5. Wagner, C.: J. Electrochem.Soc. *99* (1952) 1
6. a) Pourbaix, M.: Corr. NACE *25* (1969) 267; b) Revue Roumaine de Chemie *17* (1972) 239
7. a) Korrosion 11. Kathodischer Korosionsschutz. (Wiederholt, W. and Kaesche, H., eds.) Verlag Chemie, Weinheim, 1959; b) Kaesche, H., in: Proc. 4th Int. Conf. Metallic Corosion, Amsterdam 1969. (Hamnner, N.E., eed.). Nace, Houston, 1971, p. 15
8. a) Winkelmann, D., in: loc. cit. [7a], p. 58; b) Engell, H.-J. and Forchhammer, P.: Corr. sci. *5* (1965) 479
9. Handbuch des kathodischen Korrosionsschutzes. (v. Baeckmann, W. and Schwenk, W., eds.). 4th edn., Wiley VCh, Weinheim 1989
10. Electrochemical Rehabilitation Methods for Reinforced Concrete Structures. (Mietz, J., ed.). European Federation of Corrosion Publ. Nr. 24. IOM Communications Ltd., London, 1998

## 17.2 Mass Transport by Diffusion

In Chap. 5, the rate of diffusion of dissolved species, notably $H^+$ ions and $O_2$, to electrode surfaces was calculated using the Nernst model of a diffusion layer of thickness $\delta_{\mathrm{Nernst}}$ or, more briefly, '$\delta$', between the metal surface and a well stirred bulk of the solution. In this layer, concentrations of reacting species are thought to change, in steady state, linearly from bulk values $(c_i)_\circ$, for distances $x \geq \delta$ from the surface, to values $(c_i)_*$ for $x = 0$ on the surface. This diffusion layer must not be confused with the electric double layer still underneath. For electrolytic solutions, where the simple capacitor model for the latter layer is applicable, this distinction is unambiguous, since the diffusion layer thickness is larger than the double layer thickness by orders of magnitude. For very dilute solutions, the situation is different due to the large extent of the diffuse part of the double layer, but this will be disregarded presently. In the diffusion layer, transport of ions is by electrolytic migration and by diffusion only, since the Nernst concept implies the absence of any convection inside the diffusion layer.

For convenience only, we now ignore ions and consider transport of uncharged species only, thus concentrating on diffusion only. To be more specific, we mainly refer to diffusion of oxygen, and subsequent reduction by the electrode reaction $O_2 + 2H_2O + 4e^- \rightarrow 4OH^-$. Using Eqs. (5.37) and (5.36), but omitting '$O_2$' subscripts, we then have, for the current strength, $j$, for oxygen reduction in general, and $j_D$ for the diffusion-limited current strength in particular, at an electrode of surface $A$, the expressions $j = -zFAD(c_\circ - c_*)/\delta$, and $j_D = -zFA(c_\circ/\delta)$, respectively.

Considering that $\delta$, though defined as a diffusion layer thickness, really is an empirical fit parameter, it is appropriate to define a *mass transfer coefficient*, $\beta = D/\delta$, and to write $j = -zFA\beta(c_\circ - c_*)$; $j_D = -zFA\beta c_\circ$. In real diffusion layers, the change of the oxygen concentration, $c$, will be a steady function of the the distance, $x$, from the metal surface. The correct formulation then is that at $x = 0$ the true local current density, $I$, is proportional to the local gradient $\partial c/\partial \vec{n}$ of the concentration along the normal, $\vec{n}$, on the surface: $I = -zF(\partial c/\partial \vec{n})_*$. In the general case of nonuniform current densitiesy, the current strength then is obtained by integration $\int I(A)dA$. From this it follows that

$$\delta = \frac{A(c_\circ - c_*)}{\int_A (\partial c/\partial \vec{n})_* dA}\,; \quad \beta = \frac{D\int_A (\partial c/\partial \vec{n})_* dA}{A(c_\circ - c_*)}\,. \tag{17.1}$$

While analytical solutions usually are difficult to obtain, similarity theory often helps to find appropriate relations. As described in the textbooks on heat and/or mass transfer, the coefficient $\beta$ is proportional to the ratio $D/L$, of the diffusion coefficient and some geometric length, $L$, characteristic for the problem at hand [1]. The influence of the bulk flow rate, $v$ m s$^{-1}$, is expressed as an influence of the characteristic dimensionless *Peclet number of mass transfer* defined as $Pe' = vL/D$, the influence of the kinematic viscosity, $\nu$ m$^2$ s$^{-1}$, of the solution as an influence of the equally dimensionless *Reynolds number*, $Re = vL/\nu$. The transfer coefficient then is proportional to D/L and some function of both $Pe'$ and $Re$:

$$\beta = (D/L)f\{Pe',Re\}\,. \tag{17.2}$$

Introducing the *Sherwood number*, $Sh = \beta L/D$, this expression reduces to $Sh = f\{Pe',Re\}$.

Previously, similarity theory has found extensive applications in electrode kinetics [1b,2]. Direct applications to corrosion problems are, however, rare [3], and we comment, instead, on some general derivations concerning electrode kinetics including corrosion processes.

Consider, then, a large flat platinum electrode in a quiescent elctrolytic solution of infinite depth, containing dissolved oxygen at the concentration $c_\circ$.

Assume that oxygen reduction is the only possible electrode reaction, setting in on polarizing at time $t = 0$. Assume also that on polarizing the electrode to a potential in the range of diffusion-limited cathodic oxygen reduction, the current immediately is $j = j_D$, because also immediately $c_* = 0$.

The problem is one-dimensional with an $x$-axis pointing in the direction of $\vec{n}$, and Fick's 2nd law reads $\partial c/\partial t = D(\partial^2 c/\partial x^2)$. For the system with infinite depth of the liquid phase, the mathematical solution is relatively simple: For any time, $t$, and any distance, $x$, from the metal surface, we have

$$c/c_\circ = \mathrm{erf}\{\lambda\}\ ; \quad \text{where} \quad \lambda = \frac{\mathrm{x}}{2\sqrt{\mathrm{Dt}}}\ , \quad \mathrm{erf}\lambda = \frac{2}{\sqrt{\pi}} \int_0^\lambda \mathrm{e}^{-\mathrm{w}^2} \mathrm{dw}\ . \tag{17.3}$$

Values of the *error function*, erf$\{\lambda\}$, are tabulated. Briefly, we note that for $\lambda = 0$, erf$\{\lambda\} = 0$; for $\lambda \gtrsim 3, erf\{\lambda\} = 1$. We note that the behaviour of the system is fully governed by the dimensionless characteristic number $x/2\sqrt{Dt}$.

In our present case, we find by differentiation that $(\partial c/\partial \vec{n})_* = (\partial c/\partial x)_* = c_\circ/\sqrt{\pi Dt}$. From this, the true current density, $I_D$, the total current strength, $j_D$ for a cathode of area $A$, and the transfer coefficient, $\beta$, of diffusion-limited oxygen reduction are calculated as

$$I_D = -zFc_\circ\sqrt{D/\pi t}\ ; \quad j_D = AI_D\ ; \quad \beta = \sqrt{D/\pi t}\ . \tag{17.4}$$

For $t \to 0$, the current tends to zero because the concentration profile flattens indefinitely. The situation is typically different for an electrode in the shape of a small sphere of radius $r_\circ$. In spherical coordinates the diffusion equation now is $\partial c/\partial r = D(\partial^2 c/\partial r^2 + 2\partial c/r\partial r)$. Using the *error function complement*, $erfc\{\lambda\} \equiv 1 - erf\{\lambda\}$, the solution to the problem now is

$$c/c_\circ = 1 - (r/r_\circ)erfc\{\lambda\}\ ; \quad \lambda = (r - r_\circ)/2\sqrt{Dt}\ , \tag{17.5}$$

and

$$j_D = -zAFc_\circ\left(\sqrt{D/\pi t} + D/r_\circ\right);. \tag{17.6}$$

For short times the result is the same as before; but for $t \to \infty$ a residual current remains. Qualitatively, the same applies, as an effect essentially of concave edges, to other fully immersed small electrodes, such as small flags, often used for experiments in electrode kinetics. The effect is, however, small compared to the influence of some residual convection usually encountered in any nominally stagnant solution.

A more practical problem is the case of diffusion-limited cathodic electrode reactions from a solution flowing past a metal surface at the vectorial rate $\vec{v}$. Consider, as an example,oxygen reduction from a solution flowing past a rectangular flat platinum plate. As in Fig. 17.1, the plate has an infinite length along the $y$ axis, but a straight, 'leading' edge at $y = 0$. Let $\pm b/2$ be the plate width, and $l$ a particular plate length. The depth of the solution is infinite, and solution flow is parallel to the plate in the $y$ direction, with $\vec{v} = v_y$ constant, regardless of the distance from the plate $x$, including $x = 0$. It is important to note that this implies the assumption of a solution with no internal friction.

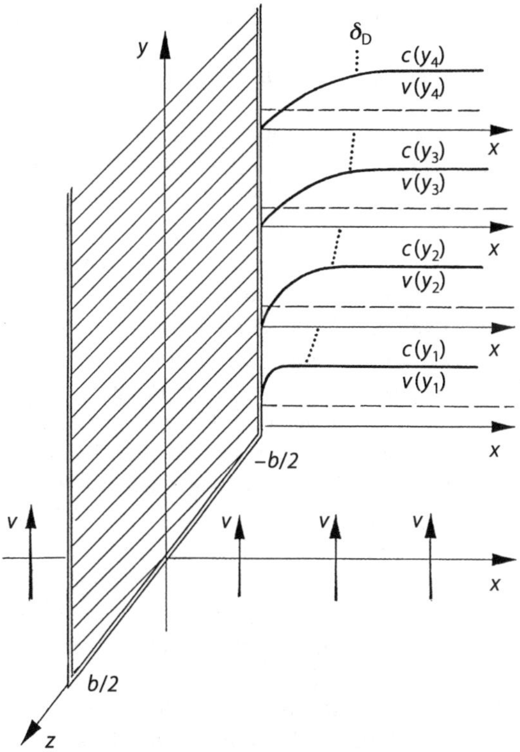


**Fig. 17.1.** Schematically illustrating mass transport from a flow of frictionless liquid phase to a flat electrode plane, assuming diffusion-limited cathodic reaction at the metal surface. $v(y_1)...v(y_4)$ : Constant flow velocity distribution, $c(y_1)...c(y_4)$ : concentration profile developing along the metal surface, indicating development of a diffusion layer of increasing thickness

Consider now a layer of the liquid phase of height $dy$ between two planes parallel to the $xz$ plane, moving past the surface at the rate $v_y$. Before arriving from negative values of $y$ at $y = 0$, the concentration of dissolved oxygen is $c_\circ$ regardless of $x$, including $x = 0$. But the plate is polarized to an electrode potential in the range of rate-determining limiting diffusion; and in the moving solution layer this reduction to $c_* = 0$ occurs immediately on arrival of the liquid layer at $y = 0$. Let the time be counted from this moment as $t = 0$. It is then obvious that along the metal surface a diffusion profile of dissolved oxygen develops in what is a diffusion layer of increasing depth. At a distance $y = v \times t$ the concentration profile is the same as everywhere on a flat surface in quiescent solution at time $t$ [4]. This derivation neglects a small contribution to mass transport along $y$ owing to the gradient $\partial c/\partial y$ developing in the $y$ direction, but we ignore this effect.

Substituting, therefore, in Eq. (17.5) $y/v_y$ for $t$, we arrive at

$$I_D = -zFc_\circ\sqrt{v_y D/\pi y}\ ; \quad j_D = -2zFbl\sqrt{v_y D/\pi l} \quad \beta = 2\sqrt{v_y D/\pi l}\ . \tag{17.7}$$

Defining, for this case, $Sh \equiv \beta l/D$, the result is summarized as $Sh = 1.13\sqrt{Pe'}$.

In reality, friction in the solution, while negligible in the bulk of the solution, cannot be neglected close to the metal surface. Here, and already without mass transfer by an electrode reaction, a hydrodynamic boundary layer, the *Prandtl layer* develops, and $v_y$ drops from the bulk value to zero at $x = 0$. When, now, an electrode reaction is superposed, the diffusion boundary layer develops in the hydrodynamic layer, and the local value of $I_D$ depends on the local gradient in $v_y$. Conditions then very much depend on whether liquid flow is laminar, or else turbulent. Laminar flow is expected for $Re_l = vl/\nu \lesssim 10^5$. Assume, for the case of the flat plate of length $l$ introduced above, $v = 1$ cm s$^{-1}$ and, as for water at ambient temperature, $\nu = 10^{-6}$ m$^2$ s$^{-1}$. Solution flow then is laminar for $l \lesssim 0.1$ m, which is safe for experiments with small electrodes, but not for flow in large technical equipment.

For laminar flow, conditions still are comparatively simple, if along the metal surface the flow velocity profile is constant. This applies to flow through a tube of diameter $d$, at distances $l$ from the opening exceeding an initial 'inlet length', beyond which the Prandtl layer is 'fully developed', and completely fills the tube cross section. Tube flow is laminar for values $Re \lesssim 2.3 \times 10^3$ and mass transport to the tube surface then still is by diffusion only, related to $Pe'_d = vd/D$ and to $d/l$. As long as, at the same time, the diffusion boundary layer does not yet also completely fill the tube cross section, $c = c_\circ$ in the center axis of the tube. In this case, $Sh_d = 1.27(d/l)\sqrt[3]{Pe'_d}$.

For flow along a plate in an infinite solution volume, at any distance $l$ from the leading edge, as well as for flow in a tube in the 'inlet length', the Prandtl layer grows along the metal surface, and the flow velocity has a component in the $x$-direction. In this case, $\beta$ depends on both $Re$ and $Pe'$ by some function $Sh = f\{Re, Pe'\}$.

At this point, it is convenient to define the *Schmidt number*, $Sc = Pe'/Re = \nu/D$. For dilute aqueous solutions, $D \simeq 10^{-9}\, m^2\, s^{-1}$ is a typical value for species other then $H^+$, and $Sc \simeq 10^3$. For $H^+$, where $D_{H^+}$ is of the order $10^{-8}$m$^2$ s$^{-1}$,Sc $\simeq 10^2$.

For $Re > 10$, but laminar flow, the analytical solution to the problem of diffusion-limited cathodic electrode reaction at the flat plate with an edge in the flow is

$$Sh = 0.68\sqrt[3]{Sc}\sqrt{Re} = 0.68\sqrt{Pe'}/\sqrt[6]{Sc}\ . \tag{17.8}$$

Compared to the solution for frictionless flow, $Sh$ is halved when $Sc = 10^2$, and reduced to one fifth for $Sc = 10^3$.

In Chap. 6, a circular disk with radius $r$ rotating in the surface of an electrolytic solution, with a flow pattern created by this rotation in an otherwise unstirred solution, was described as a valuable tool for investigations into electrode kinetics. Defining, as appropriate to this system, $Sh_r = \beta r/D$, and $Re_r = \omega r^2/\nu$, the solution is

$$Sh_r = 0.62\sqrt[3]{Sc}\sqrt{Re_r}\ . \tag{17.9}$$

From this follows that $\beta = 0.62\, D^{2/3}\omega^{1/2}\nu^{-1/6}$, as used in Chap. 5.

We pass commenting on conditions of turbulent flow with the brief note that, generally, an expression of the form $Sh = const \times (Sc)^m(Re^n)$ is found useful to summarize empirical relations. We also pass flow as induced by gravitational forces. The point is interesting, when massive electrode reactions, such as large-scale galvanizing reactions, change the density of the solutions at an electrode surface. For the definition of the now suitable dimensionless *Grashoff number*, relating density changes, viscosity and gravitation, see [5]; applications are described in [6].

While the role of the error function, $erf\{x/\sqrt{Dt}\}$, for solutions of one-dimensional diffusion problems is generally typical, analytical solutions for other than semi-infinite systems are no longer comparatively simple. A typical case is hydrogen permeation through a membrane, very convenient for the study of hydrogen diffusion in metals, especially in steels. Results of this type of experiments have been described in Chap. 14. Conditions can be chosen such that in the membrane with thickness $L$, at the front inside, $x = 0$, the concentration of dissolved atomic hydrogen, $c_{\mathrm{H}}^{(0)}$, is equal to the saturation concentration, $(c_{\mathrm{H}})_{\mathrm{sat}}$, determined by the cathodic charging current density, $(i_{\mathrm{H}})^{(0)}$. Furthermore, conditions can be chosen such that at the back inside, at $x = L$, every hydrogen atom arriving by diffusion through the membrane is immediately ionized, so that $c_{\mathrm{H}}^{(L)} = 0$. The time-dependent concentration profile of dissolved hydrogen in the membrane, $c_{\mathrm{H}}(x,t)$ is obtained by Fourier transformation [7a]; or by Laplace transformation [7b] providing the solution:

$$\frac{c_{\mathrm{H(x,t)}}}{c_{\mathrm{H}}^{(\circ)}} = \sum_{n=0}^{\infty}(-1)^n \mathrm{erfc}\left(\frac{\mathrm{x + 2nh}}{2\sqrt{\mathrm{Dt}}}\right) - \sum_{\mathrm{n}=0}^{\infty}(-1)^{\mathrm{n}}\mathrm{erfc}\left(\frac{\mathrm{2L(n+1) - x}}{2\sqrt{\mathrm{Dt}}}\right). \tag{17.10}$$

The measured quantity is the potentiostatic anodic current density of hydrogen ionisation at the membrane backside, $i_{\mathrm{H}} = -D_{\mathrm{H}}F(dc_{\mathrm{H}}/dx)_{\mathrm{L}}$. In steady state, the concentration profile in the membrane is linear, and

$$(i_{\mathrm{H}})_{t\to\infty} = D_{\mathrm{H}}Fc_{\mathrm{H}}^{(\circ)}/L\ . \tag{17.11}$$

The ratio $(i_{\mathrm{H}})_t/(i_{\mathrm{H}})_{t\to\infty}$ can be expanded into an infinite series, whose first term,

$$\frac{i_{\mathrm{H}})_t}{(i_{\mathrm{H}})_{t\to\infty}} = \frac{2L}{\sqrt{\pi D_{\mathrm{H}}t}}\exp\left(-\frac{L^2}{4D_{\mathrm{H}}t}\right), \tag{17.12}$$

is a sufficient approximation up to $(i_{\mathrm{H}})_t/(i_{\mathrm{H}})_{t\to\infty} \lesssim 0.965$. At time $t_{0.5}$ where $(i_{\mathrm{H}})_t/(i_{\mathrm{H}})_{t\to\infty} = 1/2$, the diffusion coefficient is obtained from $D_{\mathrm{H}} = 0.138L^2/t_{0.5}$. Therefore, the membrane diffusion experiment provides, separately, both the diffusion coefficient, $D_{\mathrm{H}}$, and the saturation concentration, $c_{\mathrm{H}}^{(\circ)}$, and thus is very convenient.

## References


1. See e.g. a) Eckert, A.: Einführung in den Wärme- und Stoffaustausch. Springer Verlag, Heidelberg etc., 1959; b) Levich, V.: Physicochemical Hydrodynamics. Prentice Hall, Englewood Cliffs, 1962
2. a) Vielstich, W.: Z. Elektrochemie *57* (1953) 646; b) Ibl, N.: Electrochim. acta *1* (1959) 117; c) Hsueh, L. and Newman, J.: ibid. *12* (1967) 417, 429; d) For gas-evolving electrode reactions, see Ibl, N., Adam, E., Venczel, J., and Schalch, E.: Chem. Ing. Technik *43* (1971) 202
3. See e.g. a) Makrides, A.C. and Hackerman, N.J.: J. Electrochem. Soc. 105 (1958) 156; b) Moshtev, R.V., Budevsky, E.B., and Christova, N.J.: Corr. sci. *3* (1963) 125; c) Venczel, J., Knutson, L., Wranglen, G.: ibid. *4* (1964) 1; d) Heitz, E.: Werkstoffe u. Korr. *15* (1964) 63; e) id.: Electrochim. acta *10* (1965) 49; f) Mahato, B.K., Steward, F.R., Shemilt, L.W.: Corr. sci. *8* (1968) 173, 737; g) Bohnenkamp, K.: Archiv Eisenhüttenwesen *47* (1976) 253
4. Krischer, O. and Kast,W.: Die wissesnschaftlichen Grundlagen der Trocknungstechnik. 4. Auflage. Springer Verlag, Berlin etc., 1978
5. a) Wagner, C.: J. Phys. Coll. Chem. *53* (1949) 161; b) id.: J. Electrochem. Soc. *95* (1948) 161
6. a) Böhm, U., Ibl, N., and Frei, A.M.: Electrochim. acta *11* (1966) 421; b) Böhm, U. and Ibl, N.: ibid. *13* (1968) 891; c) Marchiano, S.B. and Arvia, A.J.: ibid. *13* (1968) 1657, *14* (1969) 741, *15* (1970); d) Wragg, A.A. and Ross, T.K.: ibid. *12* (1967) 1421; e) Wragg, A.A.: ibid.*13* (1968) 2159; f) Taylor, J.L. and Manratty, T.J.: ibid.*19* (1974) 529
7. a) Devanathan, M.A.V. and Stachursky, Z.: loc. cit.[14.9a]; b) McBreen, J., Nannis, and Beck, W.: loc. cit. [Chap. 14,9b]


## 17.3 Applications of Fracture Mechanics

*Fracture mechanics* is the continuum-mechanical treatment of crack growth processes in stressed structural materials, and the important point is the stability or instability of a pre-existing crack with respect to propagation. For safety considerations, the existence of such precracks of some characteristic length, $a_o$, is presupposed. Instability with respect to propagation is indicated, if by growth of the crack length, $a$, the free enthalpy or Gibbs energy, $G$, of the system is decreased, that is, if $dG/da < 0$. The energy change involved in crack extension has two positive components: plastic deformation work of the ligament in front of the crack tip, and increase in surface energy owing to increase in crack surfaces. The sum of these contributions is the *crack resistance.* Without corrosion, the only negative contribution is the decrease in elastic energy of the system. The rate of growth of an unstable crack is determined by kinetics, and cannot, without non-thermodynamic assumptions, be predicted from the energy release rate. This also applies to *subcritical crack growth* of structures loaded in tension in environments favoring crack extension by corrosion, i.e. by chemical reactions increasing the negative component of the energy output.

In an ideal brittle body the tip of a crack is sharp, and any contribution of plastic deformation to crack growth energy disappears. For the range of linear-elastic response of the material in question to stress, we then have the simple basis of *linear-elastic fracture mechanics* (LEFM). The main and, of course, unrealistic consequence is an infinite stress concentration at the ideally sharp crack edge, causing zero strength of the precracked material regardless of the crack length, $a_\circ$. In reality, a plastically deformed small zone always precedes the crack tip in the ligament. If the diameter of this zone is sufficiently small, LEFM is still applicable for the surrounding material. Conditions where, instead, the characteristic diameter of the plastic zone is comparable to the other dimensions of the precracked body, are considered by *elasto-plastic fracture mechanics* (EPFM) which we shall leave without further discussion except for considerations concerning stresses in the plastic zone.

As always, the aim of corrosion science is to establish the chemical kinetics of corrosion processes. In the case of *stress corrosion cracking*, (SCC), and *corrosion fatigue*, (CF), the interplay of atomistic, non-continuous chemical events, dominating crack growth, are crack-tip events in the plastic zone, and thus outside the range of applicability of LEFM. The, hopefully, connecting link is the *stress intensity factor*, $K$, describing the equilibrium distribution of the stress components in the elastically strained body outside the plastic zone as function of the distance from the stress-raising crack crack edge or tip.

Consider in Fig. 17.2 a *double cantilever beam* (DCB) tensile specimen, side-notched and prefatigued to sharpen the notch. The load, in Mode I (as compared to shear load in Mode II and torsional load in Mode III,) is applied through bores on both sides of the notch. The notch together with the fatigue precrack form the 'crack' in the sense of fracture mechanics, whose stability or instability with respect to propagation is investigated.

For this geometry, and for small values of $r$, the tensile components and the shear component of the stress tensor are approximated as

$$\text{a)}\quad \sigma_{yy} = \frac{K_\mathrm{I}}{\sqrt{2\pi r}} \cos\frac{\theta}{2}\left(1 + \sin\frac{\theta}{2}\sin\frac{3\theta}{2}\right)\;;$$

$$\text{b)}\quad \sigma_{xx} = \frac{K_\mathrm{I}}{\sqrt{2\pi r}} \cos\frac{\theta}{2}\left(1 - \sin\frac{\theta}{2}\sin\frac{3\theta}{2}\right)\;; \tag{17.13}$$

$$\text{c)}\quad \tau_{xy} = \frac{K_\mathrm{I}}{\sqrt{2\pi r}} \cos\frac{\theta}{2}\sin\frac{\theta}{2}\cos\frac{3\theta}{2}\;.$$

In the $xy$ plane, where $\theta = 0$, $\sigma_{yy} = K_\mathrm{I}/\sqrt{2\pi r}$. This still predicts a crack opening stress component, $\sigma_{yy}$, tending to infinity for $r$, which is the distance from the crack edge, tending to zero. In reality the stress peak is relieved by the formation of the small-diameter plastic zone beyond the crack edge. Still, the steep increase of stress close to the edge is real and tends to strain the specimen accordingly in all directions. But in the depth of the specimen, contraction in the

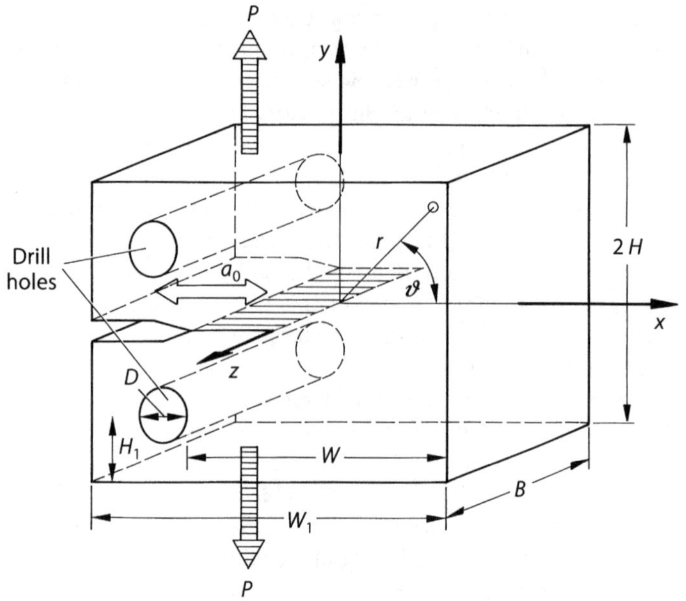


**Fig. 17.2.** DCB (*double cantilever beam* probe) for fracture mechanical testing under the tensile load, $P$, applied to bores adjacent to the surfaces of a mechanical notch including a fatigue precrack. Cartesian coordinates $x,y,z$ and cylinder coordinates $r,\theta,z$. Probe height, $2H$; width, $B$; overall length, $W_1$; effective length referred to points of load attack, $W$; 'crack' length, $a_\circ$

$z$-direction direction is impossible. This 'plastic constraint' leaves only strain components in the $xy$-plane. Therefore, while stress in the interior is triaxial, strain is only biaxial; hence the designation of this state as *plane strain.* In the side faces of the specimen, deformation in the $z$-direction is possible, relieving stress in this direction. Local stress, therefore is biaxial, and the state is *plane stress.* Plane strain is the state of maximum local stresses and, thus, maximum susceptibility to cracking. Therefore, and also adapted to conditions for thick walls of pressure vessels, DCB probes are used as *compact tension probes* with dimensions standardized to secure predominant plane strain. The condition for applicability of LEFM is that $K_\mathrm{I} \gtrsim 2.5(K_{\mathrm{I}/R_\mathrm{P}})^2$.

In Eqs. (17.13), the *stress intensity factor,* $K_\mathrm{I}(\mathrm{Mpa}\sqrt{\mathrm{m}})$, determines the steep increase of the crack-opening stress component on approach to the crack edge. For a center crack of length $2a$ in a sheet of otherwise infinite extension, $K_\mathrm{I} = \sigma\sqrt{\pi a}$. For DCB probes, this expression changes to $K_\mathrm{I} = \beta\sigma\sqrt{\pi a}$. $\beta$ is a factor depending on the ratio $a/W$. For $W$, see the figure. Values of $\beta$ are tabulated for different specimen geometries; we briefly note that for $a \ll W, \beta = 1.12$. Figure 17.3 schematically illustrates the calculated and the real change in $\sigma_{yy}$ on approach to the crack edge, and also the typical 'dog's bone' shape of the plastic zone. In the plastic zone, $\sigma_{yy}$ passes through a peak value, because directly at the crack edge the state again is one of biaxial, i.e. plane stress, and

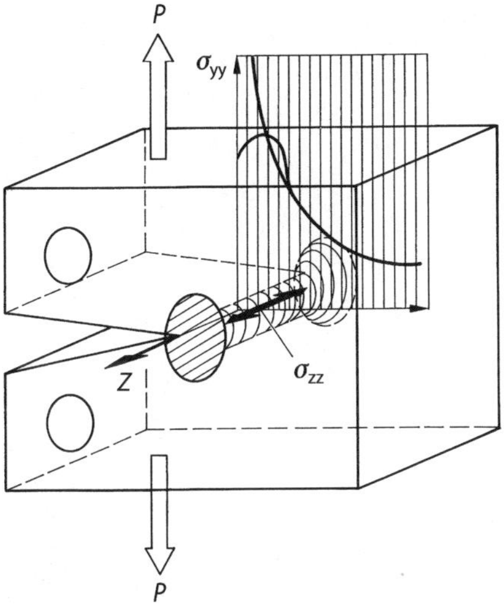


**Fig. 17.3.** The crack-opening stress component, $\sigma_{yy}$, in the $xy$-plane at $\theta = 0$ as function of the distance $x = r$ from the crack edge, considering microplastic deformation in the ductile zone beyond the edge, Note 'dog's bone' shape of this zone caused by plastic constraint in the interior with predominant plane strain, and plane stress close to the surface

the plastic constraint disappears. For the diameter of the zone consider, in the first place, a state of plane stress without plastic constraint. Plastic flow will set in whenever $\sigma_{yy}$ increases to the conventional yield strength, $R_p$. In this case, the radius of the plastic zone is $(r_{\text{pl}})_{\text{pl.stress}} \simeq (1/2\pi)(K_{\text{I}}/R_p)^2$. In a state of plain strain, with full plastic constraint, the local stress exceeds the nominal yield strength by a factor of about 3. If there is work hardening, the factor increases to about 3.5. In the region of plane strain the radius of the plastic zone is $(r_{\text{pl}})_{\text{pl.strain}} \simeq (1/3)(r_{\text{pl}})_{\text{pl.stress}}$.

On closer inspection, the plastic zone is double-club shaped instead of cylindrical, as schematically depicted in Fig. 17.4 according to [2]. The figure also indicates the *crack tip opening displacement*, CTOD, caused at the tip by the microplastic deformation. The magnitude is CTOD $\simeq \text{K}_{\text{I}}^2/\text{E}\sigma$, where $E$ is Young's modulus. Stress, $\sigma_{yy}$, is counted in units of $R_p$; distance, $x$, is normalized in units of $(K_{\text{I}}/R_p)^2$ on the assumption that stresses are related to this ratio rather than to $K_{\text{I}}$. This then implies that according to this concept the peak value of $\sigma_{yy}$ is independent of $K_{\text{I}}$, and that the true effect of an increase of $K_{\text{I}}$ is an increase in the diameter of the plastic zone only. Concerning the two curves drawn in the figure we note that the origin of the strengthening coefficient, $n$, is a relation $\sigma = K(\epsilon)^n_{\text{true}}$ between stress and true strain.

This model may not be the only possible one. In any case it is useful to note that $K_{\text{I}}$ is a factor entering the calculation of local stress intensity, and is not stress intensity itself. The importance of the factor is based on the similarity principle that, for otherwise similar conditions, fracture mechanical properties,

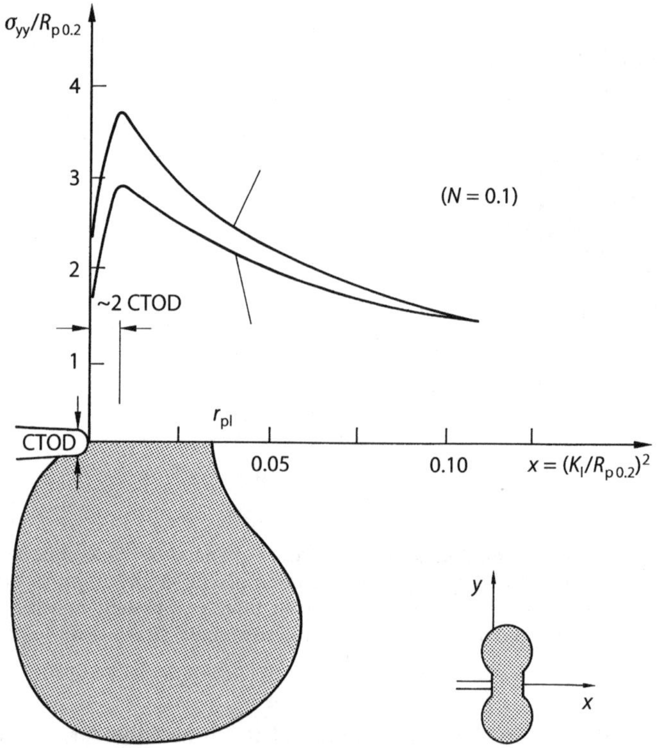


**Fig. 17.4.** Approximate crack-opening stress component, $\sigma_{yy}$ in the $xz$-plane at $\theta = 0$ as function of the distance from the crack-tip, for fully ductile, and for work hardening material. The abscissa is normalized to values $x = (K_I/R_p)^2$. Note distance, $\simeq 2 \times CTOD$, of the stress peak from the crack tip. Lower half: Contour of the plastic zone on the same scale. *Insert:* Smaller full section through the plastic zone. (From Rice)

including values of the ultimate *fracture toughness*, $K_{IC}$, are the same for equal values of $\sigma\sqrt{a}$. By this analogy, test results are claimed to be directly applicable to service conditions.

Without corrosion, a crack in a strained body is stable up to a critical value of the stress intensity factor, $K_{IC}$, designated as *fracture toughness*. Theoretically, it is related to a critical energy release rate: $G_{IC} = (1-\nu^2)K_{IC}^2/E$. The detrimental effect of stress corrosion then is a shift of the critical stress intensity factor to a value $K_{ISCC}$ below $K_{IC}$. At the same time, crack extension rates observed in the range of this stress corrosion cracking are low and well measurable. This is the reason for the widespread use of plotting $\ln v$ -$K_I$ or $\ln v$-$\ln K_I$ curves. Figure 17.5 schematically shows the typical range of such curves observed in an aggressive environment: With increasing $K_I$, crack propagation sets in at $K_{ISCC}$, rapidly increasing in *Region I* to a sometimes flat and sometimes inclined shoulder in *Region II*, until in *Region III* unstable rapid crack growth sets in at $K_{IC}$, which is the same as in an inert environment.

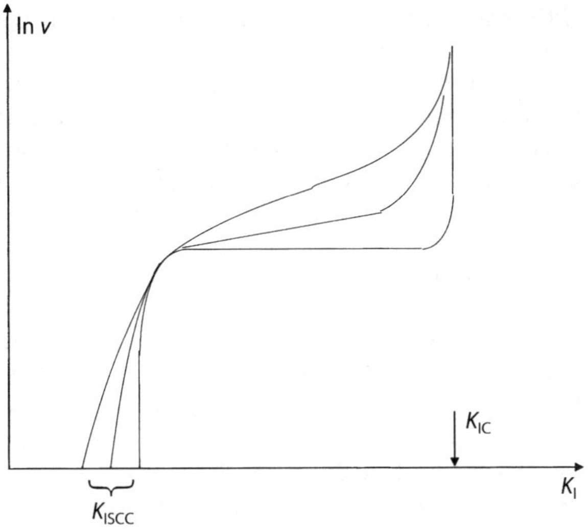


**Fig. 17.5.** Typical dependence of the logarithm of the crack extension rate, $v$, in notched and prefatigued DCB specimens, exposed to aggressive environments, on the stress intensity factor, $K_I$

The practical importance of these measurements is obvious. It also is useful that fracture mechanical testing provides data on crack propagation rates, although with a tendency to overlook discontinuous extension steps as they occur in many cases described in preceding chapters. Sometimes, crack growth in DCB probes can be directly observed in the specimen's side faces. More frequently, the *compliance*, relating, by calibration, *crack mouth opening displacement*, CMOD, and crack length is recorded. Or else, crack length is determined with a 'potential probe' by monitoring the ohmic resistance for electric current flow from specimen top to specimen bottom across the load-bearing section, which decreases with increasing crack length. A further convenient method to determine $K_{ISCC}$ is to use specimens bolted to constant CMOD, and to observe the crack extension. As the crack grows, the stress intensity factor now decreases, until growth comes to standstill, when $K_I = K_{ISCC}$. This method has been found useful fro observing stress corrosion cracking in aluminum alloys [3].

As yet, the quantitative interpretation of $\ln v$-$K_I$ or $\ln v$-$\ln K_I$ curves with respect to mechanisms of stress corrosion cracking is open to discussion. Since the crack growth rate is a kinetic property, whereas the stress intensity factor enters the equilibrium distribution of stresses and displacements, this situation is not really surprising.

Turning to *fatigue*, we continue to limit considerations to cases where LEFM are applicable. Also we suppose that the cyclic load cycles are sinusoidal, with a frequency, $f$, a load amplitude $\Delta\sigma = \sigma_{max} - \sigma_{min}$, and a constant-load component, $\sigma_\circ$, either zero or positive:

$$\sigma(t) = \Delta\sigma/2) \sin(2\pi f t) \ . \tag{17.14}$$

Cracking is considered in DCB specimens, i.e. for Mode I cracking with the crack opening stress component $\sigma_{yy}$. A complication is *crack closure*, occurring during each cycle for some stress level below the constant-load component. $\sigma_{clr} < \sigma_{\circ}$. The crack is closed for some value $\sigma_{clr} \leq \sigma_{\circ}$, and the fatigue crack is propagated by an amplitude $\Delta\sigma_{eff} = \sigma_{max} - \sigma_{clr}$ only, which is smaller than $\Delta\sigma$.

Typical quantities to be considered are crack length, $a$, amplitude of the stress intensity factor, $\Delta K_I = K_{Imax} - K_{Imin}$ and effective amplitude $(\Delta K_I)_{eff} = K_{Imax} - K_{Iclr}$. The number of cycles is $N$; the ratio $K_{Imin}/K_{Imax}$ is designated as R. Since $da/dN$ is crack extension per cycle, the crack length is $a = a_{\circ} + \int (da/dN)dN$.

Without corrosion, fatigue crack extension normally is independent of the cycling frequency, $f$, very much to the advantage of testing procedures, which can be brought to typical high values of $N = 10^6$ or $10^7$ within short experimental times. In contrast, corrosive influences on fatigue usually increase with decreasing cycling frequency, as described in Chap. 16. For 'stress corrosion fatigue' the reason is particularly obvious. For 'true corrosion fatigue' the effect is qualitatively similar, and this points to a rather simple influence of slow crack-tip corrosion, possible only during and for the duration of open-crack periods.

Usually, the records of fracture-mechanical fatigue testing are plotted in $\ln N$-$\Delta K_I$ or $\ln N$-$\ln \Delta K_I$ diagrams. Figure 17.6 schematically shows the typical shape of such curves as observed already in inert environments, divided into *Region I*, where fatigue crack propagation sets in at a threshold, $\Delta K_{IF}$, *Region II*, where a *Paris equation*, $v \sim (da/dN)^m$, is obeyed and Region III, where unstable crack propagation causes rapid failure. Often observed numerical values are $2 \lesssim \Delta K_{IF} \lesssim 12$ MPa$\sqrt{\text{m}}$, and $m \lesssim 4$. Region II usually begins at rates, $da/dN$, between $10^{-6}$ and $10^{-7}$ m/cycle. $da/dN$ tends to increase with with increasing Young's modulus, $E$. Temperature is not very important, but microstructure is, especially in the transition Ranges I/II and, although to a lesser extent, II/III as well. An important property particularly with respect to crack closure effects is the load ratio, $R$. Frequently, the experimental choice is $R = -1$, indicating symmetrical tension/pressure cycling. In this case, crack opening tensile periods of stress are only half of every cycle, and the effective load amplitude is $(\Delta K_I)_{eff} = 0.5\Delta K_I$ For experiments with $R > -1$, the effective load amplitude increases as $(\Delta K_I)_{eff}/\Delta K_I \simeq 0.5 + 0.4\ R$ [4].

A more precise calculation of crack closure effects is difficult, owing to influences of factors such as plastic deformation, crack surface roughness, and crack surface oxidation. For large positive values of $R$ the diameter of the plastic zone in front of the crack tip is roughly the same as for static load, whereas for small values it is smaller, with $(r_{pl})_{cyclic} \simeq (1/4)(r_{pl}))_{static}$.

In practice, the presence of macroscopic notches in the surface of fatigued structures is as unlikely as for structures under static load. Instead, surface irregularities caused by localized corrosion may act as stress raisers, or else this action is due to brittle precipitates in or close to the surface. The size of such surface irregularities will often be too small for regular fracture mechanics to

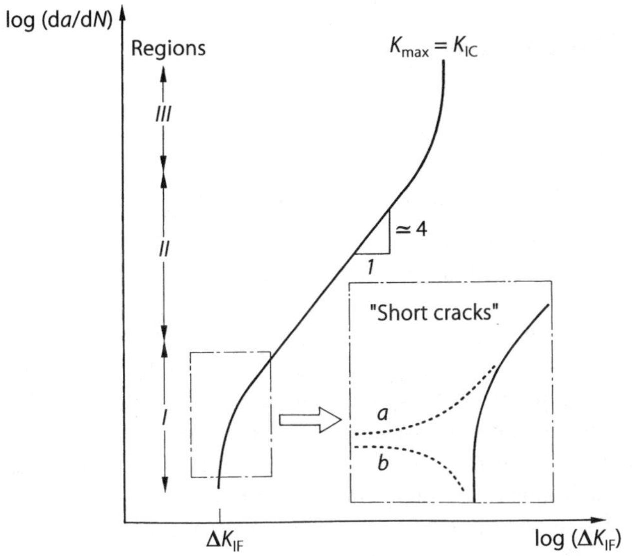


**Fig. 17.6.** Typical dependence of the logarithm of fatigue crack extension step width, $da/dN$, in DCB specimens on the amplitude of the stress intensity factor, $\Delta K_I$, as observed in inert environments, practically independent of cycling frequency. Note threshold value $\Delta K_{IF}$ in Regions I; linear dependence in Region II ('Paris region'), and rapid unstable crack propagation in Region III on approaching $K_{Imax} \to K_{IC}$. *Insert:* 'Short crack' growing *a* too fast as compared with long cracks, *b* showing transient growth only

be applicable. As a result, irregular *short crack behavior* then is recorded, as indicated in the insert of Fig. 17.6. Short cracks may grow faster than predicted from growth of large cracks, but others may show only transient growth.

## References

1. a) See e.g. a) Broeck, D.: Elementary Engineering Fracture Mechanics. Martinus Nijhoff Publ., Dordrecht, 1986; b) Schwalbe, K.H.: Bruchmechanik metallischer Werkstoffe. Hanser Verlag, München, 1980; c) Spähn, H. and Lenz, H.W.: J. Mat. Technology *4* (1973) 16, 351
2. Rice, R.C., in: Proc. Int. Conf. Stress Corosion Cracking and Hydrogen Embrittlement of Iron Base Alloys, Unieux-Firminy 1973. (Staehle, R.W., Hochmann, J., McCright, and Slater, J.E., eds.). NACE, Houston, 1977, p. 11, and other sources quoted therein.
3. Speidel, M.O. and Hyatt, M.V.: Stress Corrosion Cracking of High-Strength Aluminum Alloys. Advances in Corrosion Science and Technology. (Fontana, G.M. and Staehle, R.W., eds.). Vol. 2. Plenum Press, New York, London, 1972, p. 115
4. Stanzl, S.E., in: Symp. Ermüdungsverhalten metallischer Werkstoffe, 1984. (Munz, D., ed.). DGM Informationsgesellschaft 1985, p. 107

## 17.4 Electrode Impedance Spectroscopy

### 17.4.1 Introduction

*Impedance spectroscopy* is a powerful tool to study electrode reaction kinetics. The main point is the dependence of complex polarization resistances on the frequency of low-amplitude sinusoidal perturbations. currents. The usual procedure is to perturb a direct-current (d.c.) polarization current, or a constant electrode potential, by a superposed sinusoidal potential signal applied by a potentiostat, causing an alternating-current (a.c.) oscillation of the polarizing current, and to record the response of the system. The signal frequency is varied between sometimes very wide, and not always safe limits, e.g. between $10^{-3}$ and $\geq 10^4$Hz. As the the stationary current-potential curve may be slowly scanned, while also scanning, at each quasi-stationary potential, the response to the perturbation over the full range of frequencies, this type of experiments produces huge amounts of data waiting for interpretation.

For small-amplitude sinusoidal perturbations of the potential, the response of the current is sinusoidal as well, as for an electric circuits containing resistors, $R$, capacitors, $C$, and inductors, $L$, only. The system response thus corresponds to the complex a.c. resistance or the *impedance*, $\vec{Z}$, of the system. Often, though not always, the response of the metal/solution interface to the perturbation can be then modeled by constructing what is designated as an 'equivalent circuit' of passive electric components, $R$,$C$, and $L$.

Impedance spectroscopy is only one of many possibilities to inspect electrode kinetics by recording system response to polarizing signals of variable shape. Owing to the properties of capacitors and inductors, which both store information, i.e. preceding history, all are subjects of the general *dynamic systems theory* [1a]. The particular topic of impedance spectroscopy of electrodes is fully described in a wealth of literature [2–5], and the following is a brief outline only. We especially note, that of the prerequisites for well behaved systems, which mainly are causality and invariance with respect to time, the second often holds for corroded electrodes only approximately, and this may sometimes cause doubts with respect to very-low frequency data. At the same time, very-high frequency data may sometimes suffer from other spurious side effects. Also, current-potential functions encountered in electrode kinetics usually are non-linear and, therefore, approximately linear, as required for system theory to apply, only for perturbation amplitudes not exceeding some millivolts.

From Chap. 5 we recall that, except for very dilute solutions which we exclude from further consideration, the electric double layer at a meta/solution interface can be modeled as a leaking parallel-plate capacitor. This makes the *double layer capacity*, $C_{dl}$, a well defined and frequency-independent property, which depends on the electrode potential, with a minimum at the point of zero charge (pzc). For investigations into the structure of the double layer, $C_{dl}$ is the important quantitity; for studies of electrode kinetics, including corrosion kinetics, it is not interesting, but always part of the measured impedance, $\vec{Z}_{\text{total}}$, and thus complicating matters. The interesting part of the impedance, associated with the electrode reactions across the double layer, is the *Faraday impedance*, or 'faradaic' impedance, $\vec{Z}_{\text{farad}}$.

We recall, at this point, the definition of the *polarization resistance*, $R_\pi\,(\Omega\text{m}^2)$ of electrodes, introduced in Chap. 5 as a measure of the tangent to stationary current-potential curves, equivalent to a direct current resistance. In simple cases, the impedance $\vec{Z}_{\text{farad}}$ has no capacitive and/or inductive components, and is simply equal to $R_\pi$, which we now re-index to $R_{\text{trf}}$, indicating the property of a resistance caused by an electron transfer reaction.

A second, still comparatively simple case is that the impedance has only one ohmic component, corresponding to the charge-transfer reaction, $R_{\text{trf}}(\Omega\text{m}^2)$, and one capacitive component, $C(\text{Farad}\,\text{m}^{-2})$, equal to $C_{\text{dl}}$. The frequency response of this system then is the same as that of a circuit containing a capacitor with $C_{\text{dl}}$ in parallel with a resistor, $R_{\text{trf}}$.

## 17.4.2 The Basic Equivalent Circuit

The electrolytic cell used for impedance spectroscopy is the same three-electrode arrangement as always used for current-potential measurements. Therefore, the measured electrode potential will contain a small ohmic contribution caused by an ohmic resistance, $R_{\text{sl}}$, in the solution between the tip of the Haber-Luggin capillary of the reference electrode and the metal surface. The impedance is than equivalent to the circuit schematicaly shown in Fig. 17.7.

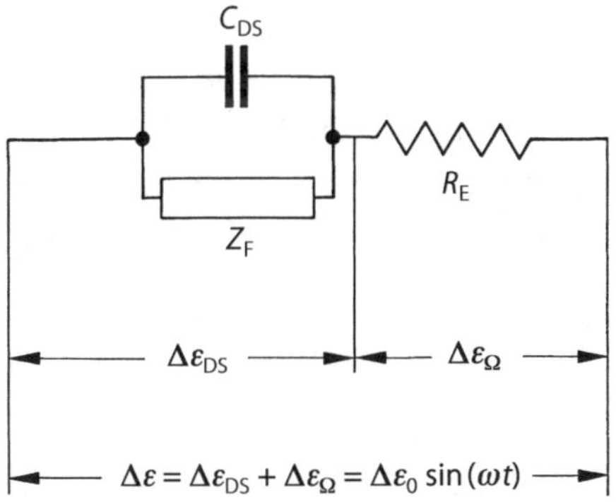


**Fig. 17.7.** An electric equivalent circuit modeling a metal/solution impedance consisting of the double layer capacitance, $C_{\text{dl}}$, the electrolytic resistance between reference electrode tip and double layer, $R_{\text{sln}}$, and a complex faradaic impedance, $\vec{Z}_{\text{elr}}$ caused by the transfer electrode reaction. Note that $\vec{Z}_{\text{elr}}$ is the only interesting quantity. In a simple case, $\vec{Z}_{\text{elr}}$ is a constant charge transfer resistance, $R_{\text{trf}}$

Generally, but subject to the suppositions listed above, a sinusoidal perturbation of the electrode potential, $\Delta\varepsilon$, with the circular frequency $\omega = 2\pi f$ causes a phase shift, $\Phi$, of the current response, varying with $\omega$.

$$\text{a)} \quad \Delta\varepsilon = \Delta\varepsilon_{\circ} \sin(\omega t) \quad \rightarrow \quad \text{b)} \quad \Delta i = \Delta I_{\circ} \sin(\omega t - \Phi(\omega)) \,, \tag{17.15}$$

When the electric circuit contains a resistor only, the current response, $\Delta i_R$, is in phase with the potential perturbation, i.e. $\Phi = 0$.

$$\Delta i = \Delta i_R = \frac{\Delta\varepsilon_{\circ}}{R} \sin(\omega t) \,. \tag{17.16}$$

The resistance, of course, is $\Delta\varepsilon / \Delta i = \Delta\varepsilon_{\circ} / \Delta i_{\circ} = R$. For a capacitor, $C$, alone,

$$\Delta i = \Delta i_C = C \frac{d\varepsilon}{dt} = \omega C \Delta\varepsilon_0 \sin(\omega t + \frac{\pi}{2}) \,; \tag{17.17}$$

the phase angle is $\Phi = -\pi/2$; the ratio $\Delta\varepsilon_{\circ} / \Delta i_{\circ} = 1/\omega C$. For an inductor, $L$, alone, the corresponding expression is

$$\Delta i = \Delta i_L = \frac{1}{L} \int \Delta\varepsilon dt = \frac{\Delta\varepsilon_{\circ}}{\omega L} \sin(\omega t - \frac{\pi}{2}) \,; \tag{17.18}$$

the phase angle is $+\ \pi/2$, the ratio $\Delta\varepsilon_{\circ} / \Delta i_{\circ} = \omega L$. For more complicated circuits, the phase angle always is between $-\pi/2$ and $+\ \pi/2$; signal and response always have the same frequency; and at constant frequency both the ratio $\Delta\varepsilon_{\circ} / \Delta i_{\circ}$ the phase angle do not change with time.

The aim of impedance spectroscopy is to deduce, from the behaviour of the system under investigation, a combination of impedance components whose frequency response to the signal coincides with the experimental data. As shown further below, the combination does not necessarily contain only $R$,$C$, and $L$ components, but for the moment it is convenient to postpone this circumstance.

Any impedance is fully described in polar coordinates by an absolute value and a phase angle. This is equivalent to describing $\vec{Z}$ as a two-component vector with an absolute length and a direction. Even more convenient is the notation of $\vec{Z}$ as a complex number, with $\mathrm{j} = \sqrt{-1}$, either in the form of a pointer, $\vec{Z} = Z \exp(\mathrm{j}\,\Phi)$, or in the complex plane, plotting the 'real' part, $Re(\vec{Z})$, and the 'imaginary' part, $Im(\vec{Z})$, on rectangular axes, as $\vec{Z} = Re(\vec{Z}) + \mathrm{j}\, Im(\vec{Z})$. This also implies rewriting Eq. (17.15) into

$$\text{a)} \quad \Delta\vec{\varepsilon} = \Delta\varepsilon_{\circ} \exp(\mathrm{j}\,\omega t) \,, \quad \rightarrow \quad \text{b)} \quad \Delta\vec{i} = \Delta i_{\circ} \exp(\mathrm{j}\,(\omega t - \Phi)) \,. \tag{17.19}$$

We note, for use further below, that $\Delta\vec{i}$ can be expressed as

$$\Delta\vec{i} = \Delta i \exp(-\mathrm{j}\,\Phi) \exp(\mathrm{j}\,\omega t) = f_i(\omega) \Delta\varepsilon \,, \tag{17.20}$$

defining a function $f_i(\omega)$ relating signal and response. The advantage of this notation is that for two impedances, $\vec{Z}_1$ and $\vec{Z}_2$, in series we now have $\vec{Z} = \vec{Z}_1 + \vec{Z}_2$, and for parallel impedances , $1/\vec{Z} = 1/\vec{Z}_1 + 1/\vec{Z}_2$. For single resistors, capacitors and inductors we now write

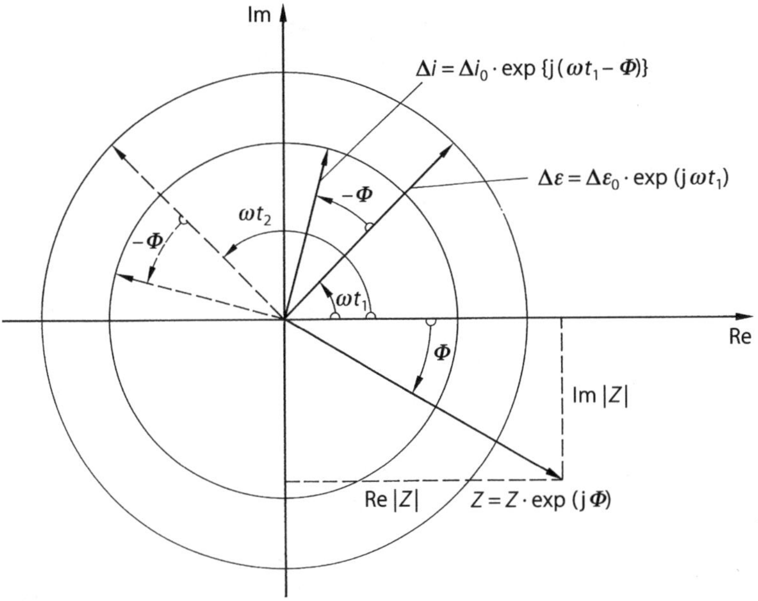


**Fig. 17.8.** The complex pointers $\Delta\vec{\varepsilon}$ and $\Delta\vec{i}$, rotating with equal frequency, $\omega$, in the complex plane. For constant $\omega$, the complex impedance, $\vec{Z}$ is constant

$$\text{a)}\quad \vec{Z}_R = R \qquad \text{b)}\quad \vec{Z}_C = (1/\omega C)\exp(-j\pi/2) = 1/j\,\omega C\,; \tag{17.21}$$

$$\text{c)}\quad \vec{Z}_L = \omega L \exp(j\,\pi/2) = j\,\omega L\,.$$

The quantity $1/\vec{Z}$ is often designated as 'admittance', $\vec{Y}$.

Concerning the magnitude of measured effects, consider an arrangement, where the solution resistance between reference electrode tip and metal surface is ca.$1\Omega\,m^2$. Assume a faradaic resistance approximately equal to a polarization resistance of e.g. 0.1 $\Omega$ $m^2$, and a double layer capacity, $C_{dl,}$ of 0.1 $\mu$F $m^{-2}$. The apparent capacitance for values $\omega = 1$, resp. $10^3$, resp. $10^6$ $s^{-1}$ then is $10^{-1}$, resp. $10^{-4}$, resp. $10^{-7}$ $\Omega$ $m^2$. Thus, the capacitance of the double layer rapidly tends to zero with increasing frequency, shunting the parallel transfer resistance, and leaving the solution resistance, $R_{sln}$, as the only, unfortunately unimportant measured quantity.

For an RC circuit, with a resistance $R_1$ parallel to a capacitance $C$, the impedance is calculated by solving the relation $1/\vec{Z}_{RC} = 1/R_1 + j\omega C$ for $\vec{Z}_1$ :

$$\vec{Z}_{RC} = R_1/(1 + j\,\omega R_1 C)\,. \tag{17.22}$$

The impedance of a circuit with a resistance, $R_2$, in series with the $R_1$ C circuit is $\vec{Z}_{RC} + R_2$. Separating real and imaginary parts leads to

$$\text{a)}\; Re(\vec{Z}) = R_2 + \frac{R_1}{1 + (\omega R_1 C)^2}\,, \quad \text{b)}\; Im(\vec{Z}) = -\frac{\omega R_1^2 C}{1 + (\omega R_1 C)^2}\,, \tag{17.23}$$

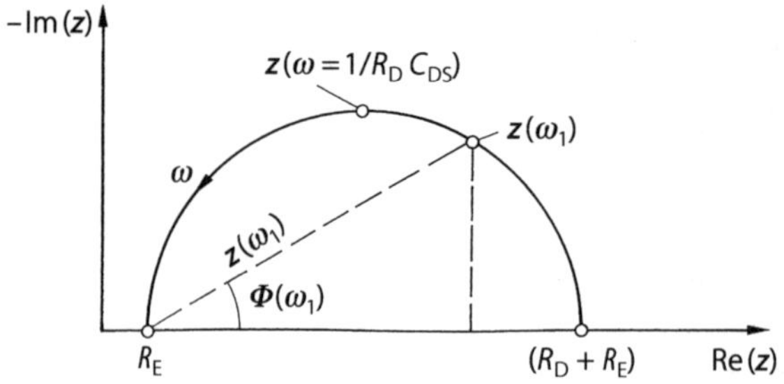


**Fig. 17.9.** Nyquist plot of the complex impedance of a circuit with parallel capacitance $C_{dl}$ and resistance $R_{trf}$, and series resistance $R_{sln}$

and this rearranges into

$$\left(Re(\vec{Z}) - \frac{2R_1 + R_2}{2}\right)^2 + \left(Im(\vec{Z})\right)^2 = \left(\frac{R_1}{2}\right)^2 . \tag{17.24}$$

This is the equation of a semicircle in rectangular $Re(\vec{Z})$-$Im(\vec{Z})$ coordinates, with radius $r = R_1/2$ and the center at distance $(R_1/2 + R_2)$ from the origin on the $Re(\vec{Z})$ axis. The usual plot is the *Nyquist diagram*, with abscissa $Re(\vec{Z})$ and ordinate $-Im(\vec{Z})$. Figure 17.9 is this graph for $R_1 = R_{trf}$, $R_2 = R_{sln}$, and $C = C_{dl}$. The imaginary component is zero for both $\omega = 0$ and $\omega = \infty$. For values of $\omega$ decreasing from $\infty$ to 0, points on the curve move from right to left. Phase angle and absolute value of this impedance are

$$\text{a)}\ \Phi = -artan(Im(\vec{Z}))/Re(\vec{Z}) , \quad \text{b)}\ Z = \left(Im^2(\vec{Z}) + Re^2(\vec{Z})\right)^{1/2} , \tag{17.25}$$

respectively. Using this, *Bode diagrams,* which are double plots of $\ln Z$ and $\Phi$ on the ordinate over $\ln \omega$ on the abscissa, are often preferred to the Nyquist diagrams. We pass writing the analytical expressions for $\ln Z(\omega)$ and $\Phi(\omega)$, and directly show a Bode plot in Fig. 17.10.

### 17.4.3 Real Electrode Impedances

As a typical example of the experimental frequency response of a corroded electrode, Fig. 17.11 shows the Nyquist diagram for active acid corrosion of iron in the absence and in the presence of an adsorption inhibitor (see Chap. 7) [6a]. The main effect of the inhibitor is an important increase of the ohmic component of the transfer resistance caused by inhibitor adsorption, corresponding to the typical increase in polarizability accompanying the decrease in corrosion rate. In fact, the impedance, extrapolated to $\omega \to 0$, together with the experimental Tafel slopes of the partial current-potential curves, give the same value for the corrosion rate as the direct current polarizability, and the colorimetric analysis of the iron content of the solution confirms this result.

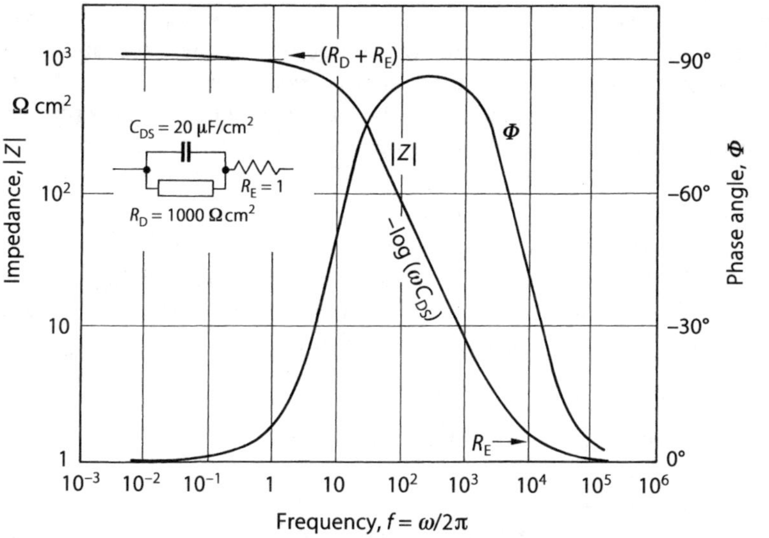


**Fig. 17.10.** Bode plot of the complex impedance of a circuit with parallel capacitance $C_{dl}$ and resistance $R_{trf}$, and series resistance $R_{sln}$. Values as indicated in the figure

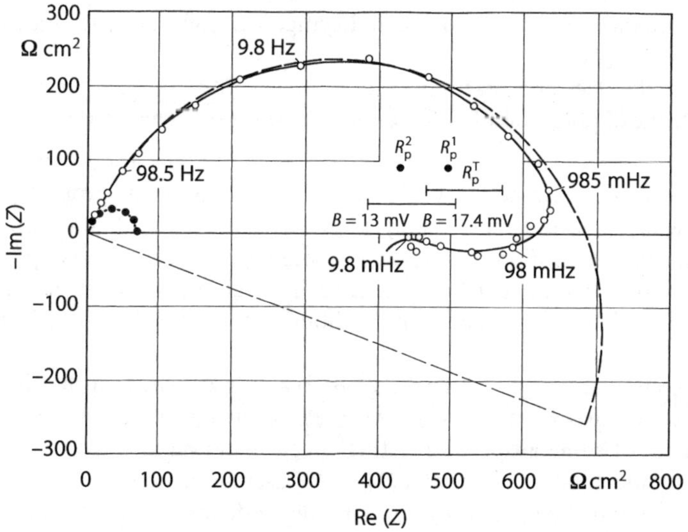


**Fig. 17.11.** Experimental Nyquist plot of the complex impedance of an iron disc rotating at 60 rvm in aerated 0.5 M $H_2SO_4$ solution at 298 K. *Solid circles:* uninhibited solution; *open circles:* Effect of adding 10 mM tri-phenyl-benzyl-phosphonium-chloride. (From Lorenz and Mansfeld)

A second effect is the change of the phase angle at low frequencies to negative values, which indicates an inductive component of the impedance. This behaviour is generally characteristic for acid corrosion of iron [6b,7]. We do not, already at this point, comment on the origin of inductances, except by briefly noting that the effect probably is due to a decrease in inhibitor surface coverage, as described in Chap. 7 for phenylthiourea or carbon monoxide.

The figure also sows that the semicircle reflecting the parallel RC behaviour of an RC circuit at high frequencies has a center below, instead of on the $Re(\vec{Z})$ axis. This 'depressed circle behaviour' is often observed [5,6,8]. It is not well understood, but probably caused by the roughness of solid metal surfaces. Formally, the effect can be accounted for by changing Eq. (17.22) to

$$\vec{Z} = R_1/(1 + (j\omega R_1 C)^m) \,. \tag{17.26}$$

Concerning the origin of the correcting coefficient, $m$, see [4,6d]. A typical value is 0.8.

The two curves presented in Fig. 17.11 are relatively simple examples of of impedance spectra. More complicated behaviour caused by both transient and stationary electrode processes, reflected in more and varying capacitive and inductive components of the faradaic transfer impedance is often encountered, for instance during studies of frequency response to a perturbation in the range of the active/passive transition of iron electrodes. For extensive work in this direction see [6,7]. We pass these details and describe, instead, the essential properties of impedance elements involved, following, with some simplifications, competent reviews in [4].

Assume a simple electrode reaction such as $Me \rightarrow Me^{z+} + ze^-$, far from equilibrium, with negligible rate of reverse metal deposition. The current density is $i = \vec{i}_{Me}$, and we assume it to be constant on the electrode surface. Transfer coefficient, $\alpha$, and reaction valency, $n_e$, are also constant properties. All of the following concerns the Faraday impedance only; the double layer capacitance, $C_{dl}$, and the solution resistance, $R_{sln}$, are both disregarded, supposing that they have been eliminated from the experimental impedance including both. Similarly, the electrode potential, $\varepsilon$, is thought corrected for the trivial ohmic component $i \times R_{sln}$.

Typically, the current density of the electrode reaction depends on the electrode potential, $\varepsilon$, on the concentration, $c$, of some intermediate reacting species, on the coverage, $\theta$, of some adsorbed reacting species, and on a rate constant, $k$, according to some function $i = kf(c,\theta,\varepsilon)$. We assume a rate-determining charge-transfer reaction, causing an exponential dependence of the current on the electrode potential.

Also typically, the concentration of a rate determining species and the electrode coverage with an adsorbed species will depend on the current density and on time: $c = c(i,t), \theta = \theta(i,t)$, whereas the rate constant is expected to depend on the potential and, possibly, also on time: $k = k(\varepsilon,t)$. For small-amplitude perturbations we expect the quantities $c,\theta$, and k to respond to the $\Delta\varepsilon$ signal with the frequency of the signal, $\omega$, but with finite and different phase shifts,

$\Phi_c$,$\Phi_\theta$, and $\Phi_k$. In analogy to Eq. (17.20) we also define functions $f_c(\omega)$,$f_\theta(\omega)$, and $f_k(\omega)$, such that

$$\text{a) } \Delta\vec{c} = \Delta c \exp(-\Phi_c) \exp(j\omega t) = -f_c(\omega)\,, \tag{17.27}$$

$$\text{b) } \Delta\vec{\theta} = \Delta\theta \exp(-\Phi_c) \exp(j\omega t) = -f_\theta(\omega)\,,$$

$$\text{c) } \Delta\vec{k} = \Delta k \exp(-\Phi_c) \exp(j\omega t) = -f_k(\omega)\,.$$

The total differential of the current density $i$ is

$$di = \frac{i}{k}dk + \frac{\partial i}{\partial \varepsilon}d\varepsilon + \frac{\partial i}{\partial \theta}d\theta + \frac{\partial i}{\partial c}dc\,. \tag{17.28}$$

Combining Eqs. (17.20), (17.27), and (17.28) we find that

$$\vec{Z}_{\text{farad}} = \left(1 + \frac{\partial i}{\partial c}f_c + \frac{\partial i}{\partial \theta}f_\theta\right)\left(\frac{\partial i}{\partial \varepsilon} + \frac{i}{k}f_k\right)^{-1}\,. \tag{17.29}$$

The first term in the denominator of this expression is the reciprocal polarization resistance, $1/R_{\text{trf}}$, the second the reciprocal *relaxation impedance*, $1/\vec{Z}_k$, more closely considered further below. When (see also below) $\partial k/\partial\varepsilon = 0$, the expression reduces to

$$\vec{Z}_{\text{farad}} = R_{\text{trf}} + \vec{Z}_c + \vec{Z}_\theta\,,\vec{Z}_k \to \infty)\,, \quad \textit{for } \vec{Z}_k \to \infty\,, \tag{17.30}$$

with elements *concentration impedance*, $\vec{Z}_c$, and *adsorption impedance*, $\vec{Z}_\theta$, defined as

$$\text{a) } \vec{Z}_c = R_{\text{trf}}(\partial i/\partial c)\,f_c \quad \text{b) } \vec{Z}_\theta = R_{\text{trf}}(\partial i/\partial\theta\, f_\theta\,. \tag{17.31}$$

Another likely case is

$$1/\vec{Z}_{\text{farad}} = 1/R_{\text{trf}} + 1/\vec{Z}_k \quad \textit{for } \vec{Z}_c = \vec{Z}_\theta = 0\,. \tag{17.32}$$

We recall that the expressions presented above have been deduced for a simple electrode reaction. When, as for instance always in corrosion, more than one electrode reaction occurs, often also involving filmed or coated electrodes, generalized expressions will obviously tend to become difficult to handle, and we shall not continue this description presently.

### 17.4.4 Laplace and Fourier Transforms

An outstanding early contribution to electrode kinetics was the deduction of the *Warburg impedance*, dating to 1901. We attend to this impedance element mainly to illustrate the role of *Laplace* and *Fourier transforms* as convenient tools for handling more complicated impedance problems.

Consider once more a simple $Me/Me^{z+}$ electrode not, however, far from, but at equilibrium and, therefore, at the equilibrium potential $\varepsilon = E_{Me/Me^{z+}}$. At rest, the $Me^{z+}$ concentration is $c_\circ$, and the current density of metal dissolution and redeposition, $i = 0$. To disturb this state, a sinusoidal small-amplitude current signal $\Delta\vec{i} = \Delta i \exp(\mathrm{j}\,\omega t)$ is applied, causing the potential to respond by $\Delta\vec{\varepsilon} = \Delta\varepsilon \exp(\mathrm{j}\,\omega t + \Phi)$. This is an oscillation of the overpotential, $\eta$, for the electrode reaction.

Assume, now, rate determining transfer of $Me^{z+}$ cations in the solution to and from the metal surface, and choose, by assuming an excess content of an inert electrolyte in the solution, conditions such that this transfer is by diffusion only. The electrode is large and flat, so that the flux is one-dimensional in the direction of the normal on the surface, and $i = -zF(dc/dx)_{x=0}$. Therefore, the concentration at the surface, $c_\star$, oscillates with frequency $\omega$, amplitude $(\Delta c)_{x=0}$, and phase angle $\Phi_c$. When the solution is at rest, this oscillation of the concentration will extend as a damped wave into the solution, with an amplitude tending to $\Delta c \to 0$, for $x \to \infty$.

When, instead, the solution is stirred, the concentration wave disappears beyond the hydrodynamic double layer. This case, which leads to the calculation of the *Nernst* instead of the *Warburg* impedance, will be ignored in the following.

The concentration, $c$, changes continuously with time everywhere. Therefore, Fick's 2nd law, $\partial c/\partial t = D(\partial^2 c/\partial x^2)$, must be solved with $c(x,t=0) = c_\circ$, and also $c(x \to \infty,t) = c_\circ$.

If $Me^{z+}$ diffusion is rate determining, the overpotential is concentration polarization (see Chap. 5), and $\eta = (RT/zF)\ln(c_\star/c_\circ)$. For small values of $(c_\star - c_\circ)$, the approximation $\ln(c_\star/c_\circ) = (c_\star - c_\circ)/c_\circ$ holds, and

$$\eta = (RT/zFc_\circ)(c_\star - c_\circ)\,. \tag{17.33}$$

The elegant method to find the solution for Ficks 2nd law is to apply *Laplace transformation* [1c]. The principle is to transform a function, $f(t)$, of time into a function $\vec{L}\{f(t)\} = F(s)$ of an arbitrary complex Variable $s = a + \mathrm{j}\,b$, defined by the relation

$$\vec{L}\{f(t)\} \equiv \int_0^\infty f(t)\exp(-st)dt = F(s)\,. \tag{17.34}$$

Some useful properties of the transformation are the following:

$$\text{a)}\quad \vec{L}\{f(x)f(t)\} = f(x)\vec{L}\{f(t)\}\,; \tag{17.35}$$

$$\text{b)}\quad \vec{L}\{f(t) + g(t)\} = \vec{L}\{f(t) + \vec{L}\{g(t)\}\,;$$

$$\text{c)}\quad \vec{L}\left(\frac{\partial f(x,t)}{\partial x}\right) = \frac{\partial}{\partial x}\left(\vec{L}\{f(x,t)\}\right)\,;$$

$$\text{d)}\quad \vec{L}\left\{\frac{\partial f(t)}{\partial t}\right\} = s\vec{L}\{f(t)\} - f(t=+0)\,.$$

The '+0' sign indicates that $t = 0$ is approached from positive values of $t$.

Laplace correspondences are extensively tabulated. Typical examples are $\exp(kt) \to 1/(s-k)$, $\sin(\omega t) \to \omega/(s^2+\omega^2)$, and $1/\sqrt{\pi t} \to 1/\sqrt{s}$. By the transformation, differentiation change to multiplication, directly incorporating initial conditions. Fick's 2nd law, a partial differential equation, changes to an ordinary differential equation for the Laplace transform, $\vec{L}\{c(x,t)\}$ of the concentration function, which we briefly note as $C(s)$ :

$$sC(s) - c_\circ = D\frac{d^2C(s)}{dx^2} \tag{17.36}$$

A particular integral of this equation is $C(s) = \vec{L}\{c_\circ\} = c_\circ/s$. The solution appropriate to the present problem is obtained by adding the general solution, which is

$$C(s) = A\exp\left(x\sqrt{s/D}\right) + B\exp\left(-x\sqrt{s/D}\right) + c_\circ/s\ . \tag{17.37}$$

In this equation, the constant $A = 0$, since otherwise $C(s) \to \infty$ for $x \to \infty$, which does not make sense. The transform, $\vec{L}\{i\}$, or briefly $I(s)$, of the electrode reaction current is

$$I(s) = -zFD\left(\frac{\partial C(s)}{\partial x}\right)_{x=0}\ . \tag{17.38}$$

This makes $B = I(s)/(zf\sqrt{sD})$, and also, for $x = 0$:

$$(C(s))_{x=0} - c_\circ/s = I(s)/(zf\sqrt{sD})\ . \tag{17.39}$$

The overpotential, $\eta$, expressed by Eq. (17.33), transforms to $\vec{L}\{\eta\} = H(s)$, where

$$H(s) = \frac{RT}{zFc_\circ}\left((C(s))_{x=0} - \frac{c_\circ}{s}\right) = \frac{RT}{z^2F^2\sqrt{sD}c_\circ}I(s)\ . \tag{17.40}$$

Thus the Laplace transform of the current density is related to the Laplace transform of the overpotential or, more generally, of the electrode potential, by a *transfer function*, $G(s)$, which in the present case is $RT/(z^2F^2)\sqrt{sD}c_\circ$.

Surprisingly, this already is the desired result, because, by a theorem of system theory, the transfer function relating the Laplace transforms of current and potential, with $s = \mathrm{j}\,\omega$, is identical with the impedance, $\vec{Z}_{\text{farad}}$. Considering that $\sqrt{\mathrm{j}} = \exp(\mathrm{j}\,\pi/4)$, and contracting $RT/(z^2F^2)\sqrt{D}c_\circ = W$, the Warburg impedance thus is

$$\vec{Z}_{\text{farad}} = \frac{W}{\sqrt{\omega}}\exp(-\mathrm{j}\,\pi/4)\ . \tag{17.41}$$

This circuit element, proportional to $\sim 1/\sqrt{\omega}$, with a $-\pi/4$ phase shift independent of frequency, has no analogy in electrical engineering. In a Nyquist plot, it is a straight line at an angle of 45°. Combined with other impedance elements, the Warburg impedance will show up at low frequencies only, and disappear in well stirred solutions as well as in solutions concentrated with respect to $Me^{z+}$ ions.

In the same way, the often tedious re-transformation from the $s$-domain to the time-domain can generally be dispense with. We insert, at this point a further theorem valid for systems in steady-state response. It states that in steady state the system behaviour, which so far was considered perturbed since $t = 0$, is the same as perturbed since $t = -\infty$. This leads to the two-sided Laplace transform by expanding the integral in Eq. (17.33) to $-\infty \leq t \leq +\infty$. Limiting, furthermore, the values of the complex variable, $s$, to $j\omega t$, we then are using *Fourier transformation*, with transforms defined as

$$\vec{F}\{f(t)\} \equiv \int_{-\infty}^{+\infty} f(t)\exp(-j\omega t)dt\ . \tag{17.42}$$

The next fundamental theorem is that for any change, $U(t)$, of the voltage applied to a circuit causing a response, $j(t)$, of the current, the impedance, explicitly in terms of the variable $\omega$, i.e. in the *frequency domain*, is $\vec{Z}(\omega) = \vec{F}\{U(t)\}(\vec{F}\{j(t)\})^{-1}$.

Instead of using convenient theorems, the overpotential, $\eta$, can, if desired, be explicitly obtained as function of time, $\eta(t)$, by re-transformation designated as $\vec{L}^{-1}\{H(s)\}$. We pass this procedure and only note that in our case the result is

$$\Delta\varepsilon = \frac{RT}{z^2F^2\sqrt{\omega D}c_\circ}\left(\frac{1}{\sqrt{2}}\sin\omega t - \frac{1}{\sqrt{2}}\cos\omega t\right)\Delta i\ , \tag{17.43}$$

equivalent, of course, to Eq. (17.41).

### 17.4.5 Adsorption and Relaxation Impedance

Concerning the *adsorption impedance*, assume an electrode reaction involving the adsorption of an intermediate reaction product on the electrode surface. A typical example is hydrogen evolution with intermediate adsorption of atomic hydrogen, $H_{ad}$. Instead of the coverage, $\theta$, we now prefer to consider the surface concentration, which we denote as $\Gamma$. Following a discussion in [4], we also assume that the rate of change of $\Gamma$, i.e. the quantity $d\Gamma/dt$ or, briefly, $\dot{\Gamma}$, depends on the electrode potential, $\varepsilon$, on the concentration, $c$, of some reaction partner such as, in our example, hydrogen ions $H^+$, and also on $\Gamma$. As before, $\omega$ is the frequency of a low-amplitude sinusoidal polarizing signal. The total differential then is

$$\Delta\dot{\Gamma}(\omega) = \frac{\partial\dot{\Gamma}}{\partial\varepsilon}\Delta\varepsilon(\omega) + \frac{\partial\dot{\Gamma}}{\partial\Gamma}\Delta\Gamma(\omega) + \frac{\partial\dot{\Gamma}}{\partial c}\Delta c(\omega)\ . \tag{17.44}$$

During an impedance measurement, $\dot{\Gamma}$ is related to the (small) amplitude of the $\Gamma$-wave: $\dot{\Gamma} = j\omega t\Delta\Gamma$. Also, because $\dot{\Gamma}\,(\mathrm{mol\,m^{-2}\,s^{-1}})$ changes by an electrode reaction with valence $n_e$, a transient faradaic current is observed: $i_\Gamma = n_eF\dot{\Gamma}(\mathrm{A\,m^{-2}})$.

For simplicity, we disregard the third term on the right side of the equation, which would give rise to a Warburg impedance. We then have

$$\Delta i_\Gamma = n_eF\frac{\partial\dot{\Gamma}}{\partial\Gamma}\Delta\varepsilon + \frac{\partial\dot{\Gamma}}{\partial\Gamma}\frac{\Delta i_\Gamma}{j\omega}\ . \tag{17.45}$$

This can be rearranged to show that the adsorption impedance, $\vec{Z}_\Gamma$, is represented by a circuit with an adsorption resistance, $R_\Gamma = 1/n_e F(\partial\Gamma/\dot{\partial}\varepsilon$, in series with an adsorption capacitance $C_\Gamma = -n_e F(\partial\dot{\Gamma}/\partial\varepsilon)/(\partial\dot{\Gamma}/\partial\Gamma)$. We ignore the further evaluation, as in every practical case it requires the knowledge of the adsorption isotherm and of other details of reaction kinetics.

Concerning the *relaxation impedance*, consider the case of a metal undergoing anodic dissolution with the current density $i$ as a function, $f(\varepsilon)$, of the electrode potential. The function itself may be, though not necessarily, an exponential equation. The complete equation for $i$ then usually is the product of some factor, $K$,, for instance pre-exponential, and $f(\varepsilon)$. Usually, $K$ itself is the product of a rate constant, $k$, as for instance given by an Arrhenius equation, and powers of concentrations of reaction participants. In the preceding chapters, constants $k$ were considered to be independent of potential, but there may be cases where this does not apply. There may be, instead, a dependence for instance on the morphology of the metal surface, which may change with the electrode potential. Again, these changes will usually lag behind the changes in potential, and since $k$ enters the current-potential relation, this then causes a contribution to the Faraday impedance.

An arbitrary but plausible case is that the rate, $\dot{k}$, of change of the rate constant caused by a perturbation of the potential, is proportional to a difference between momentary and steady-state values

$$-\dot{k} = (1/\tau)(k_{\mathrm{t}} - k_{\mathrm{stat}}) , \tag{17.46}$$

where $\tau$ is a characteristic relaxation time. The quantities $k_{\mathrm{t}}$, and $k_{\mathrm{stat}}$ will oscillate with amplitudes $\Delta k_{\mathrm{t}}$ and $\Delta k_{\mathrm{stat}}$. $\Delta k_{\mathrm{stat}}$ should be in phase with $\Delta\varepsilon$, so that $k_{\mathrm{stat}} = (dk_{\mathrm{stat}}/d)\varepsilon)\Delta\varepsilon$. $\Delta k_{\mathrm{t}}$ oscillates at the same frequency, but out of phase. Since in any case $\dot{k} = \mathrm{j}\omega\Delta k$,

$$-\frac{1}{\tau}\left(\Delta k - \frac{dk_{\mathrm{stat}}}{d\varepsilon}\Delta\varepsilon\right) = \mathrm{j}\omega t . \tag{17.47}$$

According to Eqs. (17.27c) and (17.29)

$$\text{a)}\quad f_k = \frac{1}{1+\mathrm{j}\,\omega t}\frac{dk_{\mathrm{stat}}}{d\varepsilon} , \qquad \text{b)}\quad \vec{Z}_k = \frac{1+\mathrm{j}\,\omega t}{i(d\ln k_{\mathrm{stat}}/d\varepsilon} . \tag{17.48}$$

This is the impedance of a resistance $R_k$ in series with an inductance, $\vec{X}_k$, where

$$\text{a)}\quad R_k = 1/i(d\ln k_{\mathrm{stat}}/d\varepsilon) , \qquad \text{b)}\quad \vec{X}_k = \mathrm{j}\,\omega\tau R_k = \mathrm{j}\,\omega L_k . \tag{17.49}$$

The inductance is important only for frequencies below $1/2\pi\tau$, and this is one reason for the appearance of inductive impedance components predominantly in the low-frequency range. Numerically, both $R_k$ and $\vec{L}_k$, which can be positive or negative quantities, are determined by the reaction mechanism.

### 17.4.6
### Filmed and Coated Electrodes

A broad field of applied impedance spectroscopy is the inspection of oxide films and coatings on metal surfaces. In every case, reactions through the metal/film or metal/coating interface, within the film or coating, and through the film/solution or coating/solution interface may then contribute to the impedance, and the separation may not always succeed. Still, as in the following examples, rather simple equivalent circuits often suffice to model the impedance.

Many such studies are concerned with the quality of films on anodically oxidized aluminum. Fig. 17.12 schematically shows a cross section through the oxide film with a porous layer on top of a non-porous primary layer, together with equivalent circuits assumed for a) anodizing without, and b) anodizing with subsequent healing of pores [9].

The anodized film is assumed to be a non-conducting, inert dielectric with capacitance, $C_o$, small owing to the large film thickness. The air-formed primary film is thought to allow some electrode transfer reactions characterized by a resistance $R_{trf} = R_1$ parallel to a constant capacitance, $C_2$. Pores are filled with electrolytic solution, contributing an ohmic resistance, $R_{sln} = R_2$, to the impedance. Healing of pores, which in the experiments was achieved by soaking with nickel acetate solution, causes pores to be narrowed or blocked by hydrated aluminum oxide. This effect is modeled by introducing additional impedance elements, $R_{1,1}$, $R_{1,2}$, $C_{1,1}$, $C_{1,2}$.

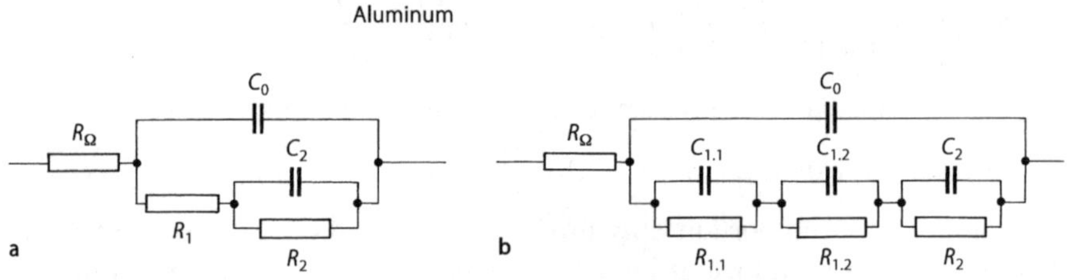


**Fig. 17.12a,b.** *Upper part:* Section through an oxide film, as obtained by anodizing aluminum in acid solution. A porous thick secondary film has grown on the non-porous primary air-formed barrier film. *Lower part:* Equivalent circuits modeling the impedance **a** without additional healing of pores, **b** with pores narrowed and/or blocked by healing in nickel acetate solution. (From Hoar and Wood)

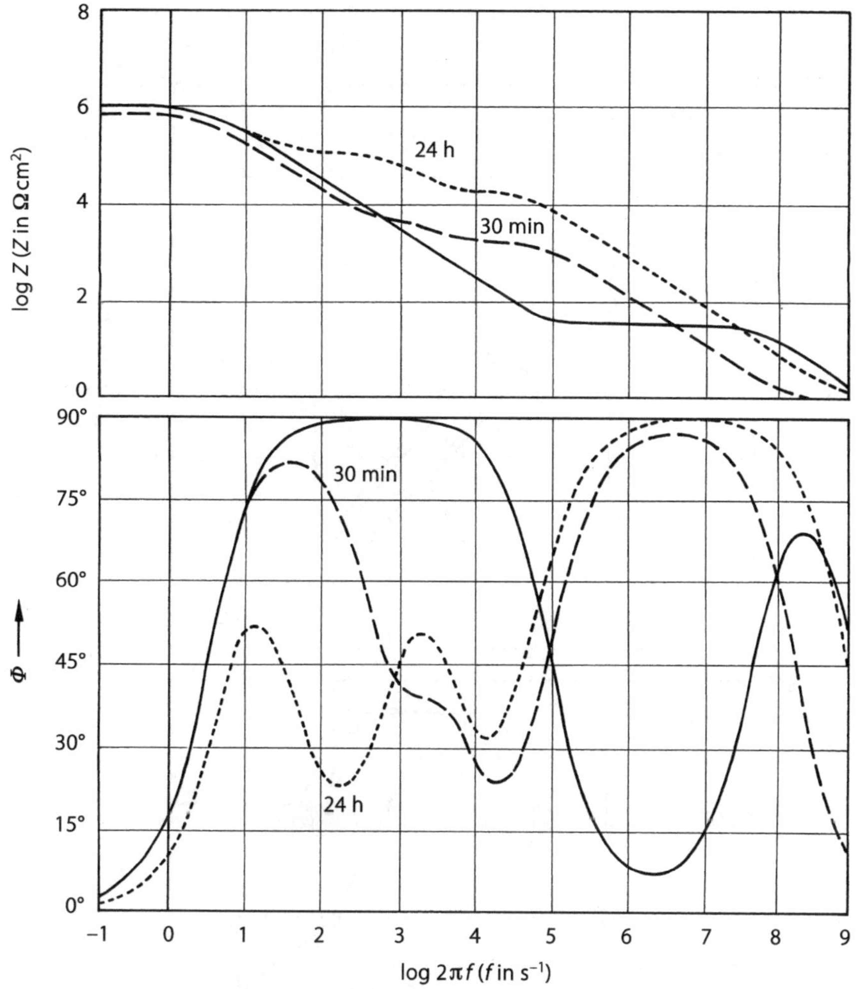


**Fig. 17.13.** Bode plot of the impedance of anodized aluminum as a function of the frequency of an applied a.c. signal. *Upper part:* Logarithm of the value $|\vec{Z}|$; *lower part:* phase angle, Φ. Porous anodized film (——); film healed for 30 min (- - - -), resp. 24 h (- - - - - ). (From Mansfeld and Kendig)

Using data as reported in [9] for frequencies up to $10^5$Hz, the complete Bode plot of the frequency response of the anodized specimens was calculated in [6c], and Fig. 17.13 presents the result extrapolated to frequencies beyond the range of experiments. For comparison, Fig. 17.14 is the quite similar Bode plot as evaluated by experimental impedance spectroscopy with the aluminum alloy 2024 in NaCl solution, again showing the effect of pore healing, in this case by soaking in bichromate solution [6g].

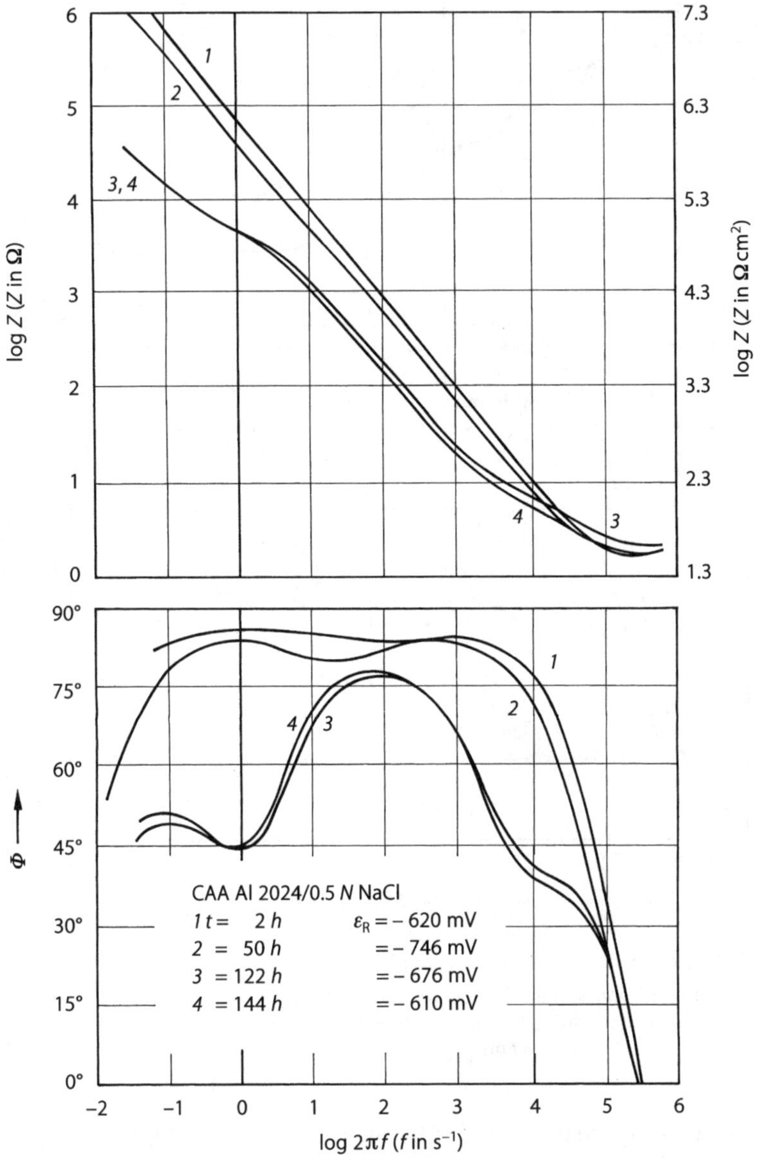


**Fig. 17.14.** Bode plot of of the impedance of anodized alloy 2024, healed in bichromate solution as function of the frequency of an applied a.c. signal, as observed after increasing times of immersion in aerated 0.5 M NaCl solution. *Upper part:* Logarithm of the value $|\vec{Z}|$; *lower part:* Phase angle, Φ. (From Mansfeld and Kendig)

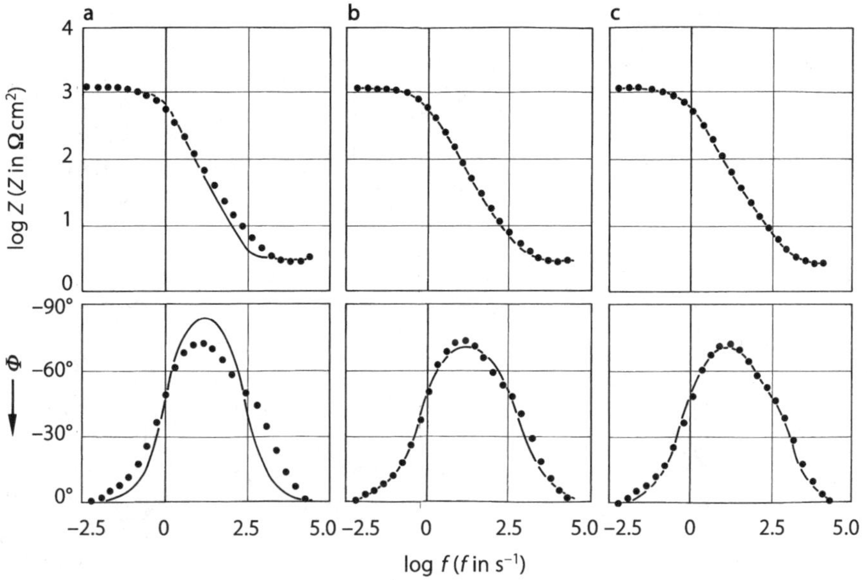


**Fig. 17.15.** Bode plots of lg $|\vec{Z}|$ and Φ for rotating disks of carbon steel in deaerated, weakly acid chloride solution, pH 3.5, at ambient temperature, as function of the frequency of an AC signal. *Solid circles:* Experimental points. *Solid curves calculated from equivalent circuits:* **a** Eqs. (17.23)–(17.25): Transfer resistance parallel to double layer capacitance. **b** Eq. (17.26): Transfer resistance parallel to double layer capacitance as before, added suppressed circle correction. **c** Assuming coverage, Θ, with an inert film, and the transfer reaction restricted to the pore area, (1-Θ), also considering an ohmic pore resistance. The parallel film and pore impedances are in series with an ohmic resistance between reference electrode tip and working electrode surface. (From Mitzlaff, Hoffmann, Jüttner, and Lorenz)

As a further example, Fig. 17.15 is the experimental impedance spectrum of a rotating steel disc rusting in a deaerated, weakly acid solution, together with theoretical curves calculated with increasing refinement [6d]. The first calculation applies the simple parallel RC circuit. The second improves the calculation by introducing a suppressed circle correction, and this, then already very closely models the experimental results without, in fact, any particular assumptions specific to the particular system. With respect to kinetics, the authors suppose, in a third step, the presence of a porous, otherwise inert oxide layer, adding a parallel film capacitance to the faradaic electrode reaction impedance now restricted to the pore bottoms. Current flow through the pores adds an ohmic component in series to the faradaic impedance. Also considering the ohmic resistance between the reference electrode tip and the working electrode surface, the calculation now results in curves, which nearly perfectly models experimental data. Therefore, further refining of the theory is not warranted by the impedance response of the system. A further refinement, more closely inspecting conditions for current flow through and out of pores, produces a still

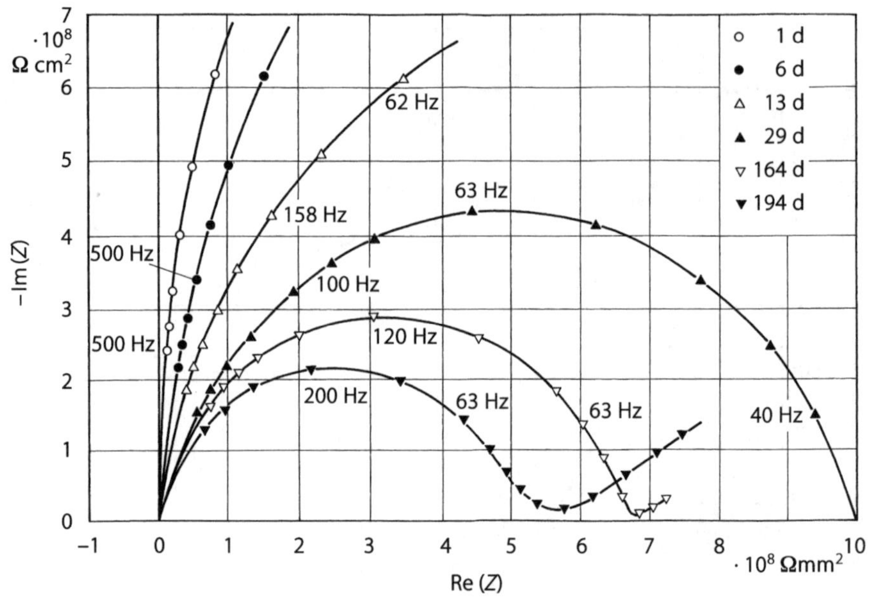


**Fig. 17.16.** Nyquist plot of the impedance of a coal tar epoxy coating, thickness ca. 60 μm, on mild steel in aerated % NaCl solution observed for exposure times varying between 1 and 194 days. (From Scantelbury, Ho, and Eden)

more elaborate mathematical apparatus offering already twelve fit parameters. All this does not yet take account for possible and probable contributions to the impedance by valence changes in the rust layer, as described in Chap. 9.

A typical practical application of impedance measurements is monitoring the behaviour of protective paint coatings on, mostly, mild steel. As an example, Fig. 17.16 shows the slow change of the impedance spectrum of a coating on steel exposed to NaCl solution [10]. The main observation is a slow shift to smaller values of the ohmic resistance, apparently due to water uptake in the coating. An interesting detail is the appearance of a Warburg impedance at low frequencies, probably reflecting the kinetics of ion migration through the wetted coating.

Impedance data for wet ion-exchanging coatings on metals have been reported in [11]. The aim was to separate influences of electrode reactions at the metal/coating interface, of ionic migration through the coating, and of coating/solution interface exchange reactions. Figure 17.17 schematically illustrates the spectrum expected for a simple redox reaction at the metal/coating interface. It should be noted that, in the case of external polarization with an electrolysing current, the distribution of the Galvani potential across the system now has three separate components,: the potential differences, $\Phi_{\mathrm{metal/coat}}$, at the metal/coating interface, $\Phi_{\mathrm{coat,coat}}$, in the coating, i.e. an ohmic voltage, and a step of the potential, $\Phi_{\mathrm{coat/sln}}$, which is governed by the ion exchange equilibrium between the coating and the solution, and designated as a *Donnan* potential. The fig-

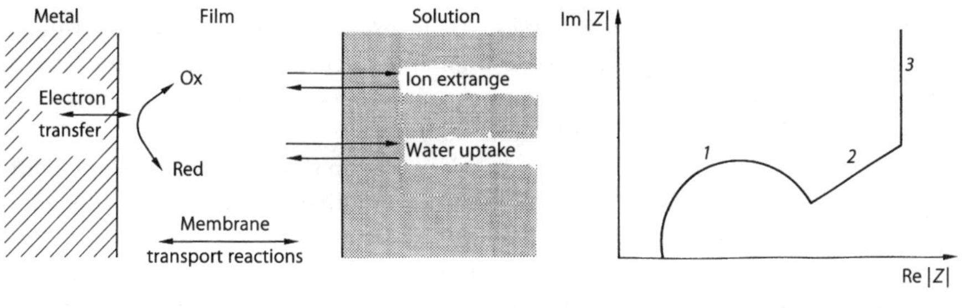


**Fig. 17.17.** Phase diagram of a membrane-type coating on an inert but electron transferring metal electrode in in an electrolytic solution, showing redox reaction at the metal surface, ion transport through the membrane and through the membrane/solution interface, and moisture take up of the membrane. (From Dobelhofer)

ure also shows the expected frequency dependence of the impedance: At high frequencies, the metal/coating electrode reaction should govern the behaviour, and the typical semicircle should appear. At lower frequencies, migration in the coating should cause a Warburg term to take over, due to a concentration wave extending into the wet coating. When, at extremely low frequencies, the depth of this wave exceeds the thickness of the coating, the ion exchanging film behaves as a dielectric medium of constant capacitance. This holds for the simple case that the metal/coating charge-transfer reaction does not contribute complex low-frequency relaxation components.

## 17.4.7
## Some Recent Contributions

Of late, the study of electrode kinetics by recording a.c impedance spectra is gaining increasing recognition and publicity. For more recent reviews and proceeding volumes, see [12]. Concerning essentials, the application of the *Kramers-Kronig* theorem for complex impedances merits notion: The theorem states that the real part of any impedance can be computed from a complete set of data for the imaginary part covering, in principle, the whole range in frequencies from 0 to $\infty$, and vice versa. Since for $\omega \to 0$ the real part is the direct current polarization resistance, $R_\pi$, which, in turn, is directly related to the corrosion current density, $i_{\mathrm{corr}}$, the method offers a further possibility to monitor the corrosion rate from impedance data. Concerning useful algorithms, see [13]. As described in Chap. 6, the evaluation of $i_{\mathrm{corr}}$ from $R_\pi$ requires the knowledge of the equations for the partial electrode reaction currents. For more details, see [6f-h,12a,b,13].

More mundane are applications of impedance spectroscopy to monitoring the performance of protective coatings. As explained in the preceding subsection, contributions to the impedance from the coating film, from pores in the film, from metal/film corrosion reactions and from diffusion of reacting species through the film can be studied separately if the relaxation time constants have sufficiently different numerical values. As an example, we quote investigations into the possibilities to replace toxic pigments such as red lead or hexavalent chromates in coatings on steel by non-toxic substances such as, in this case, metallic aluminum flakes or zinc phosphate [14].

The difficult aim of replacing hexavalent chromates as protective pigments especially for conversion coatings on aluminum alloys is also pursued in [6h,i] As indicated by impedance records, as well as by direct observation, the pitting resistance of a variety of aluminum alloys anodized in boric acid, sealed in hot cerium nitrate solution, and post-treated by anodic polarization in sodium molybdate solution, was found equivalent to that obtained by conventional chromate conversion coatings. Similar results where obtained with passive layers incorporating cerium and molybdate compounds on stainless steels [6j].

## References

1. a) Unbehauen, R.: Systemtheorie. Eine Darstellung für Ingenieure. 5. Auflage. Oldenburg Verlag, Wien, 1990; b) See textbooks on electrical engineering; c) Föllinger, O.: Laplace- und Fourier-Transformation. 3. Auflage. AEG Telefunken, Berlin, Frankfurt, 1982
2. Vetter, K.J.: Elektrochemische Kinetik. Springer, Berlin, 1961. Translation: Electrochemical Kinetics. Academic Press, New York, 1967; b) Bard, A.J., Faulkner, L.R.: Electrochemical Methods. Fundamental Applications. J. Wiley & Sons; New York etc., 1980; c) Greef, R., Peat, R., Peter, L.M., Pletcher, D., and Robinson, J.: Instrumental Methods in Electrochemistry. E. Horwood Ltd. (Div. of J. Wiley); Chichester, 1985; d) Proc. Symp. Electrochemical Corrosion Testing, San Francisco, 1979. (Mansfeld, F., and Bertocci, U., eds.). ASTM, Philadelphia, 1981
3. a) Epelboin, I., Keddam, M.: J. Electrochem. Soc. *117* (1970) 547; b) Deslouis, C., Epelboin, I., Keddam, M., and Lestrade, J.C.; J. Electroanal. Chem. *28* (1970) 57; c) Epelboin, I., Keddam, M., and Takenouti, H.J.: J. Appl. Electrochem. *2* (1972) 71
4. a) Göhr, H. and Meissner, W.: Z. phys. Chemie N.F. *93* (1974) 217; b) Göhr, H.: Ber. Bunsenges. Phys. Chemie *85* (191981) 274; c) id.; DECHEMA-Monographien Bd. 90. Verlag Chemie, Weinheim, 1981, p. 1; d) Göhr, H. and Schiller, C.-A.: Z. phys. Chemie N. F. *148* (1986) 105
5. a) Macdonald, D.D. and McKubre, M.C.: Impedance Measurements in Electrochemical Systems. In: Modern Aspects of Electrochemistry. (Bockris, J.O'M., Conway, B.E., andWhite, R.E., eds.). Plenum Press, New York, 1982; b) Gabrielli, C.: Identification of Electrochemical Processes by Frequency Response Analysis. Issue 2, 1984. Solartron Instruments, Farnborough, 1984; c) Jüttner, K., Lorenz. W,J., and Paatsch, W., Kendig, M., and Mansfeld, F.: Werkstoffe u. Korr. *36* (1985) 120; d) Impedance Specroscopy. (Mcdonald, D.D., ed.). J. Wiley & Sons, New York etc., 1988
6. a) Lorenz, W.J., Mansfeld, F.: Corr. sci. *21* (1981) 647; b) Schweickert, H., Lorenz, W.J. and Friedburg, H.J.: J. Electrochem. Soc. *127* (1980) 1693; c)

Mansfeld, F., Kendig, M.W.; Werkstoffe u.Korr. *36* (1985) 473; d) Mizlaff, M., Hofmann, H.N., Jüttner, K., and Lorenz, W.J.: Ber. Bunsenges. Phys. Chemie *92* (1988) 1234; e) Schweickert, H., Lorenz, W.J., and Friedburg, H.: J. Electrochem. Soc. *128* (1981) 1 1295; f) Kendig, M.W. and Mansfeld, F.: Corr. NACE *39* (1983) 466; g) Mansfeld, F. and Kendig, M.W.: Werkstoffe u. Korr. *34* (1983) 397; h) iid.: Mats. and Corr. *49* (1999) 475; i) Mansfeld, F. and Wang, Y.: Mats. sci. and Engng. *A198* (1995) 51; j) Mansfeld, F., Zhang, G. and Chen, C.: Plating and Surface Finishing *84* (1997) 72; k) Mansfeld, F., Breslin, C.B., Pardo, A., and Pérez, F.J.: Surface Coatings and Technology *90* (1997) 224

7. a) Epelboin, I., Gabrielli, C., Keddam, M., and Takenouti,H., in: loc. cit. [2d] – b) Keddam, M., Mattos,O.r., and Takenouti, H.; J.Electrochem. Soc. *128* (1981) 257,266,1294 – c) iid.; Electrochim. acta *31* (1986) 1147,1159 – e) iid.: J.Electrochem. Soc. *128* (1981),1294
8. Mulders, W.H. and Sluyters, J.H.: Electrochim. acta *33* (1988) 308
9. Hoar, T.P. and Wood, G.C.: Electrochim.acta *7* (1963) 333
10. Scantlebury, J.D., Ho, K.N., and Eden, D.A., in: loc. cit. [2d], p. 187
11. Dobelhofer, K. and Armstrong, R.D.: Electrochim. acta *33* (1988)
12. a) Mansfeld, F. and Lorenz, W.J.: Electrochemical Impedance Spectroscopy(EIS): Application in Corrosion Science and Technology. In: Techniques for Characterization of Electrodes and Electrochemical Processes. J. Wiley, 1991, p. 581; b) Kendig, M. and S. Jeanjaquet: Simplified and rapid Analysis of Paint Impedance. In: Advances in Corrosion Protection by Organic coatings (III). The Electrochemical Society Pro. Vol. 97-41, 1997, p. 101; c) EIS' 98 Proc. Impedance Spectroscopy. (Mattos, O.R., ed.). Electrochim. acta *44* Nr. 24 (1999)
13. Rochini, G.; Mats. and Corr. a) 49 (1998) 764; b) *50* (1999) 475
14. van Westing, F.P.M., Ferrari, G.M., and de Witt, J.H.W.: Corr. sci. *36* (1994) 1323, and previous contributions quoted therein

## 17.5 Electrode Noise Spectroscopy

In Chap. 12 on pitting of passive metals, early work recording statistical fluctuations of the electrode potential of aluminum during galvanostatic anodic polarization in chloride solutions has already been described [1]. In modern language, this was recording low-frequency, galvanostatic electrode noise. The early example was later readily explained by the role of the galvanostatic pitting potential, $\varepsilon_{pit}$, as the steady-state potential where existing pits grow with the true local current density $I_{pit}(\varepsilon_{pit})$, and where the pit nucleation rate is zero, $\upsilon_{pit} = 0$. The apparent galvanostatic current density is the product of total active pit area and local dissolution current density, $A_{pit} \times I_{pit}(\varepsilon_{pit})$, and the complication is that, owing to growth of existing pits, $A_{pit}$ always tends to exceed this product. Therefore, some pits will repassivate, momentarily leaving the product at a value $< A_{pit} \times I_{pit}(\varepsilon_{pit})$ and , therefore, c= 0. Now, some new pits are nucleated, causing the potential to return to $\varepsilon_{pit}$, etc.. When the solution is diluted with respect to dissolved chloride, and when $\upsilon_{pit}$ is small, the potential-time record shows discrete spikes, indicating single pit nucleation events. It is true tat, usually, the term 'noise', although not clearly defined, rather better applies to random ensembles of smaller and more frequent effects stochastically overlapping in time. But this would be the situation expected for similar experiments in solutions rich in dissolved chloride causing, beyond the pitting potential, pit nucleation at high values of $\upsilon_{pit}$.

A second topic in Chap. 12 was pitting of passive iron in neutral and acid chloride solutions. We add that noise preceding and accompanying the onset of rapid pitting has been the object of a number of publications [2,3]. As already explained, the mechanism of pit incubation appears well understood [3]: Fluctuating island adsorption causes current-less local film thinning; this in turn causes an acceleration of ionic migration through the film and, thus, a transient increase of the faradaic anodic film-growing current. Eventually, at points where the film occasionally is thinned to zero thickness, and where, at the same time, a screw dislocation emerges in the metal surface, rapid active metal dissolution sets in. The situation has been closely inspected with respect to potentiostatic noise in neutral acetate/acetic acid buffer solutions containing 0.5 M NaCl, where at sufficiently positive values of the potential the pit initiation rate is indeed high, although acetate appears to inhibit pitting to some degree. [3c]. Noise measurements, notoriously subject to spurious effects, where closely controlled by careful experimentation, especially by using very small working electrodes only. In chloride-free solution, the root mean squares of the potentiostatic steady state passive current density, resp. of the accompanying current noise were of the order of several nA $cm^{-2}$, and several pA $cm^{-2}$, respectively. We note that the noise signals observed in this work were smaller than those reported for the same system previously [3a,b]. The noise signal was quasi-continuous and, thus, suitable for spectral analysis. To change from the time domain to the frequency domain, a fast Fourier transformation (FFT) was applied to the records of the current as a function of time. In this context, it has become standard practice to plot the signal power density, $S = A^2/f$, i.e. the square of, in this case, the current noise amplitude, divided by the frequency, $f$, as function of $f$. Figure 17.18 shows the development of the *current power density spectrum* (PDS) with time, observed after a potential step from active metal dissolution to passivation. This spectrum can be semi-quantitatively explained considering, as an approximation, a simple metal/solution equivalent circuit with an ohmic transfer resistance, $R$, parallel to a double layer capacitance, $C$, in series with a small ohmic solution resistance, $R_\Omega$, between working electrode surface and reference electrode capillary tip, as in Fig. 17.7 in the preceding section.

In the next step, consider that the current noise is caused by a potential noise of power density $S_\varepsilon/V^2\,Hz^{-1}$, added to the potential imposed by the external potentiostat, as an electric circuit element apparently to be placed between the RC circuit and $R_\Omega$. We now introduce the frequency-dependent *noise resistance* defined as

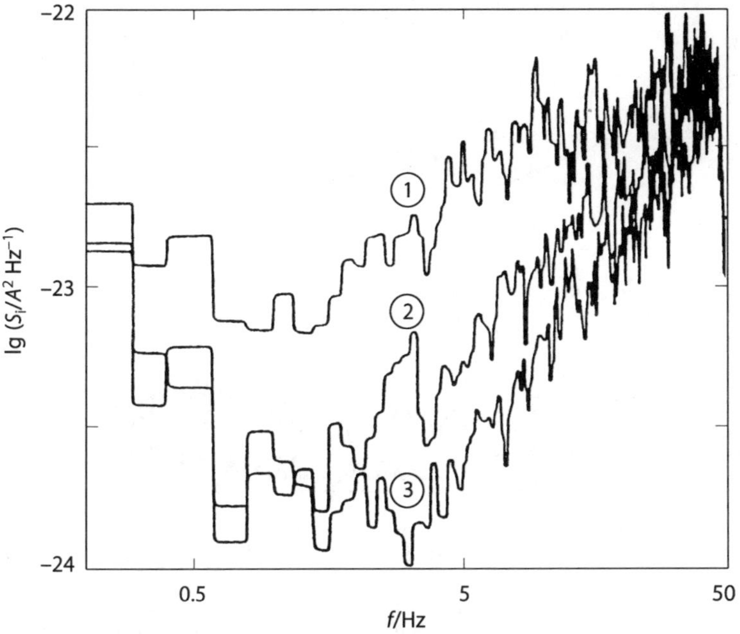


**Fig. 17.18.** Spectral densities, $S_j/\mathrm{A}^2\ \mathrm{Hz}^{-1}$, of the current noise at passive iron electrodes with area 1.33 mm$^2$ in 0.3 sodium acetate/acetic acid buffer solution, pH 7,2, at 298.15 K, for various times $t$ after a step from the activation potential $-0.15$ V to 1.068 V. (1) $t = 3$ min, $j_{\text{pass}} = 25$ nA; (2) 65 min, 3 nA; (3) 300 min, 1 nA. (From Nachstedt and Heusler)

$$|\vec{Z}|^2(f) = S_\varepsilon / S_j\ . \tag{17.50}$$

The nature of the source of random small voltage fluctuations remains unknown. Tentatively, we may assume that this is 'white' potential noise, independent of frequency: $S_\varepsilon(f) = S_\varepsilon$. From Eq. (17.23) in the preceding section, $|\vec{Z}|^2$ is readily evaluated. From this expression follows that

$$S_j(f) = S_\varepsilon / [R_\Omega^2 + R(R + 2R_\Omega)(1 + \omega^2 R^2 C^2)^{-1}]\ , \tag{17.51}$$

where $\omega = 2\pi f$.

Passing the quantitative evaluation, we note three typical ranges of frequency, disregarding the very-low frequency $1/f$-decay seen in Fig. 17.18. In the range $0.5 \le f \le 2$Hz, the large capacitance $1/\omega C$ is shunted by $R$, and $S_j \sim 1/(R + R_\Omega)^2$, independent of frequency. Beyond ca. 20 Hz, $R$ is shunted by the low capacitance $1/\omega C$, and $S_j \sim 1/R_\Omega^2$, again independent of $f$. In an intermediate region, $S_j \sim \omega^2 C^2$, proportional to $f^2$. Directly after rapidly changing to a solution containing dissolved chloride, the root mean square of the current increased to ca. 45 pA, leveling to ca. 10 pA afterwards, still before the onset of rapid pitting. During this period, the current noise immediately rose by an order of magnitude. Figure 17.19, showing three stage in the development of $S_j$

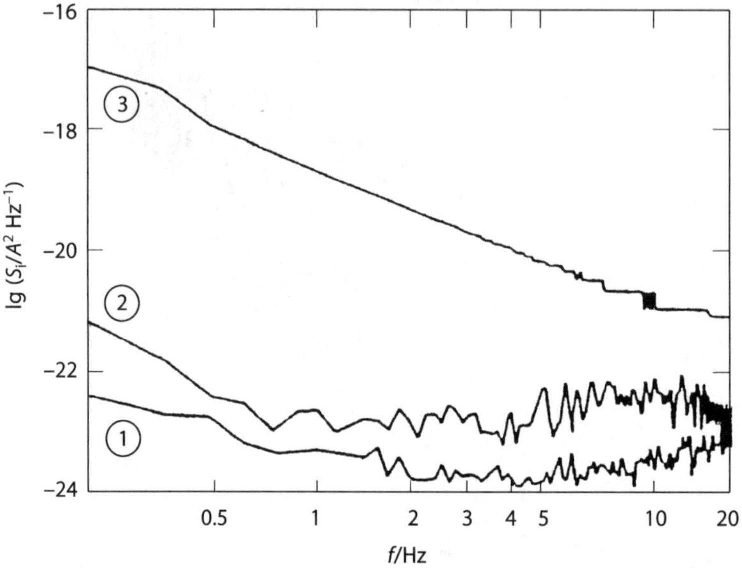


**Fig. 17.19.** Spectral density, $S_j$, of the current noise at passive iron electrode with area 0.79 $mm^2$ in 0.3 sodium acetate/acetic acid buffer, pH 7.2 at 298.15 K, at 0.28 V. (1) In steady passive state, $j = 0.3$ nA; (2) during the incubation period shortly after exchange of the solution for one containing 0.1 M NaCl; (3) during stable pit growth ca. 1 min after onset of pitting. (From Nachstedt and Heusler)

before and after chloride addition, includes the PDS observed after the onset of pitting, but this $1/f$-behaviour is not the interesting point. As also shown, the PDS in the incubation time, while grown to distinctly higher values than before, remains essentially similar to the pre-incubation time spectrum. This agrees with the concept that incubation implies random chloride island adsorption locally thinning the passive oxide film without otherwise changing the film properties. For more quantitative details, see [3c].

In the work so far described, the instrumentation used to record noise was the same working/reference/counter three-electrode arrangement as for other current-potential measurements. For recording the noise resistance, other arrangements have come into use. These are either three-electrode systems with two identical working electrodes and one reference electrode, or four identical working electrodes [4a,b,5,6]. In every case, the aim is to simultaneously acquire $S_j$ and $S_\varepsilon$, measuring current and potential fluctuations at pairs of working electrodes and reference electrodes, or pairs of identical working electrodes, using a zero resistance ammeter and a high-input impedance voltmeter, respectively.

Noise analysis is seen as a potential method to continuously monitor corrosion processes, including, somewhat surprisingly, rates of distinctly local events such as the onset of stress corrosion cracking. As an example, we quote investigations into electrode noise at SCC specimens in slightly oxygenated high-temperature, high-pressure water at 288°C, i.e. for power-plant conditions [7].

As described in Chap. 15, SCC of precracked and loaded specimens is, in principle, possible at sufficiently positive values of the electrode potential, as possibly established in actual service by an irregularly high oxygen content of the solution. For both stainless steel and low-alloy steel, the onset of crack extension apparently was, in fact, marked by an increase in noise level. Applications of noise spectroscopy to non-localized surface corrosion effects probably are less intricate. An example is checking the overall degradation of protective coatings immersed in electrolytic solutions. For a comparison of the relative merits of impedance and noise spectroscopy, see [4c,d].

## References


1. a) Ergang, R. and Masing, G.: Nachr. Akad. Wiss. Göttingnen *1* (1946) 62; b) Ergang, R. and Möhling, M.; Z. Elektrochemie em 56 (1952) 9
2. a) Williams, D.E., Westcott, C., and Fleischmann, M.: J. Eelectrochm. Soc. *132* (1985) 1796; b) iid.: J. Electrocanal. Chem. *180* (1984) 549; c) Bertocci, U. and Ye. Y.X.: J. Electrochem. Soc. *131* (1984) 1011; d) Okamoto, G., Tachibana, K., Nishiyama, S., and Sugita, T., in: Proc. USA-Japan Seminar Passivity and Its Breakdown On Iron and Iron Base Alloys, Honolulu. (Staehle, R.N. and Okada, N., eds.). NACE, Houston, 1976, p. 106; d) Holliger, R. and Böhni, H., in: Proc. Conf. Computer Aided Acquisition and Analysis of Corrosion Data. (Kendig, M., Bertocci, U., and Strut, J.E., eds.).The Electochem. Soc. PV 85-3, Pennington, 1985, p. 200
3. a) Doelling, R. and Heusler, K.E.: Z. phys. Chemie *189* (1984) 39; b) K.E. Heusler and L. Fischer; Werkstoffe u. Korr. *27* (1976) 551, 697; c) Nachstedt, K. and Heusler, K.E.: Corr. sci. *33* (1988) 311
4. a) Mansfeld, F. and Xiao, H.J.: J. Electrochem. Soc. *140* (1993) 2205; b) iid.: Werkstoffe u. Korr. *46* (1995) 3; c) Mansfeld, F. and Xiao, X.: J. Electrochem. Soc. *141* (1994) 2332; d) Mansfeld, F., Han, L.T., Lee, C.C., and Zhang, G.: Electrochim. acta *43* (1998) 2933
5. Bertocci, U., Gabrielli, C., Huet, F., and Keddam, M.: J. Electrochem. Soc. *144* (1997) 31, 37, 2786
6. a) Göllner, J. and Burkert, A.: Mats. and Corr. *49* (1998) 614; b) Göllner, J., Burkert, A., Deimel, P., and Klenke, K.: ibid. p. 671
7. a) Stewart, J., Wells, D.B., Scott, P.M., and Williams, D.E.: Cor. sci. *33* (1991) 73; b) Hickling, J., Taylor, D.F., and Andresen, P.L.: Mats. and Corr. *49* (1998) 651; c) Dorsch, T., Kilian, R., and Wendler-Kalsch, E.: ibid. p. 659

# Index